RECONSTRUCTION

THE FRANCIS PARKMAN PRIZE

The Society of American Historians was founded in 1939 by Allan Nevins and others to stimulate literary distinction in historical writing. In its early years, the Society sought to create a magazine for popular history; this effort culminated in the establishment of *American Heritage* in the early 1950s. The Society then resolved to advance its mission by offering an annual prize for the best-written book in American history. To name the prize after Francis Parkman seemed nearly a foregone conclusion.

Parkman was born in 1823, the grandson of one of Boston's wealthiest merchants. Throughout his life, Parkman was plagued with nervous disorders and problems with his eyes. Yet he loved the outdoors and exulted in sojourns into the wild. "I was haunted with wilderness images day and night," he recalled, one reason why his evocations of frontier life were so powerful.

But he also was drawn to history. As a sophomore at Harvard he resolved on his life's work—a history of the struggle between France and Britain for mastery of North America. But first he trekked to the West and wrote *The Oregon Trail* (1849), a youthful work that exhibited the masterful writing that became his hallmark. The first volume of *France and England in North America* was published in 1865, the ninth and final volume in 1892.

Some of Parkman's judgments have been superseded by modern scholarship, and, in matters of style, he overused the concept of opposition (the most significant being the contrast between authoritarian and Catholic France, and democratic and Protestant Britain). Henry Adams was a more sophisticated thinker, and perhaps a more profound writer, but Parkman was the great American historian of the nineteenth century.

The first Parkman Prize was awarded in 1958 to Arthur M. Schlesinger, Jr., for his *Crisis of the Old Order*, published the previous year. And now the Society of American Historians is pleased to be working together with History Book Club to develop their exclusive new editions of Parkman Prize winners from the past—allowing readers to rediscover what many consider to be some of the best writing on the American story.

Mark C. Carnes
Executive Secretary
The Society of American Historians

RECONSTRUCTION

AMERICA'S UNFINISHED REVOLUTION: 1863-1877

Eric Foner

With a New Introduction by the Author

FRANCIS PARKMAN PRIZE EDITION
HISTORY BOOK CLUB
NEW YORK

This edition was especially created in 2005 for History Book Club by arrangement with HarperCollins Publishers Inc.

Designer: Sidney Feinberg
Copy editor: Ann Finlayson
Indexer: Judith Hancock
Maps by George Colbert

The map that appears on page 126 is adapted from Sam Hilliard, *The Atlas of Antebellum Southern Agriculture*, by permission of Louisiana State University Press. Copyright © 1984 by Louisiana State University Press.

Printed in the United States of America

For Lynn

Contents

CONTENTS

Illustrations

ILLUSTRATIONS

MAPS

Introduction to the
Francis Parkman Prize Edition

T HE reissue of a book is always a source of personal gratification for the author, as well as a less welcome reminder of the passage of time. It also offers the occasion for some reflections on how the book was originally written, and why it remains relevant today.

It was the late Richard Morris, a distinguished scholar of early American history, who in 1975 asked me to write the volume on Reconstruction for the New American Nation series, of which he and Henry Steele Commager were editors. The invitation was totally unexpected. True, my first book dealt with the pre-Civil War Republican party, many of whose leaders went on to play pivotal roles in Reconstruction. But in 1975, I was nearing completion of a book on Tom Paine and was planning to embark on a history of American radicalism. I had written nothing on Reconstruction except for an essay on Thaddeus Stevens, the leader of the era's Radical Republicans. But I had long had an interest in this, the most controversial and misunderstood period in all of American history.

My first effort to express my opinions about Reconstruction came two decades earlier, in my ninth grade American history class at Long Beach High School, in the suburb of New York City where I grew up. Our teacher was Mrs. Bertha Berryman, affectionately known among the students as Big Bertha (after a piece of World War I artillery). Following the then-dominant view of the era, Mrs. Berryman described the Reconstruction Act of 1867, which gave the right to vote to black men in the South, as the worst law in all of American history. I raised my hand and proposed that the Alien and Sedition Acts were "worse." Whereupon Mrs. Berryman replied, "If you don't like the way I'm teaching, Eric, why don't you come in tomorrow and give your own lesson on Reconstruction?"

This I proceeded to do, in a presentation based largely on W.E.B. Du Bois's monumental history, *Black Reconstruction in America*. Although largely ignored by historians when it first appeared in 1935, *Black Reconstruction* was a brilliant forerunner of modern interpretations of the Civil War era. Du Bois insisted that Reconstruction must be understood as an episode in the struggle for genuine democracy—political and economic—in the United States. This is how I presented it. At the end of the class, Mrs. Berryman announced: "Class, you have heard me, and you have heard Eric. Now let us vote to see who was right about Reconstruction." Only one student voted for me, my intrepid friend Neil Kleinman.

It therefore seemed almost preordained when Morris offered me the chance to get even with Mrs. Berryman. I soon discovered that I had agreed to take on a project with a checkered past. In 1948, Howard K. Beale agreed to do the book; he died eleven years later without having written a word. He was succeeded by David Donald, who in 1974 abandoned the project to devote himself to a more manageable one, a biography of Thomas Wolfe.

Most books in the New American Nation series summarize the current state of historical scholarship. I assumed I could do a year or two of reading and complete the book soon afterwards. In fact, it took about ten years to research and write. The turning point in my conceptualization of the project came in 1978, when I was invited to teach for a semester at the University of South Carolina, in Columbia. There, in the State Archives, I encountered 121 thickly packed boxes of correspondence received by the state's Reconstruction governors. The letters contained an incredibly rich record, almost entirely untapped by scholars, of utopian hopes and shattered dreams, of the grievances and aspirations of black and white Carolinians attempting to rebuild their lives after the Civil War and the abolition of slavery, of struggles for human dignity and ignoble violence by the Ku Klux Klan.

When I began working on the book, scholars inspired by the civil rights revolution of the 1960s had begun the process of overturning the traditional view of Reconstruction. Every year, as I was doing my own research, important volumes appeared on one or another aspect of the era's history. My own book draws heavily on this generation of scholarship. But after my experience in South Carolina, I realized that to tell the story of Reconstruction I could not rely on available works of history, impressive as so many of them were, but would have to delve into the archives to recover the texture of life at the local level. My research took me to libraries across the country. Like Du Bois decades earlier, I became convinced that the freed people were the central actors in the drama of Reconstruction. Rather than simply victims of manipulation or passive recipients of the actions of others, they were agents of change, whose demand for individual and community autonomy helped to establish the agenda of Reconstruction politics.

In 1982, I returned to Columbia University, where I had received my doctorate, to teach, and here over the next few years the book was written. There is a certain irony in the fact that a Columbia historian produced this new history of Reconstruction, exemplified by the fact that my research expenses were partly covered by the department's Dunning Fund and much of my reading took place in Burgess Library. For it was at Columbia at the turn of the century that William A. Dunning and John W. Burgess had established the traditional school of Reconstruction scholarship, teaching that blacks were "children" incapable of appreciating the freedom that had been thrust upon them, and that the North did a "monstrous thing" in granting them the right to vote. The views of the Dunning School helped freeze the white South for generations in unalterable opposition to any change in race relations, and justified decades of Northern indifference to Southern nullification of the 14th and 15th Amendments.

Reconstruction: America's Unfinished Revolution aspires to be a work of historical synthesis. It draws on both the scholarship that overturned the Dunning School, and my own extensive research. At a time when the study of history seems in danger of fragmenting into numerous subfields, it seeks to address the period after the Civil War as a whole. Slavery was simultaneously a system of labor, politics, and race relations. Thus, the book tries to weave into a coherent narrative the political struggle over Reconstruction, the transition from slavery to free labor in the South, and the evolution of a new system of race relations, and to delineate how they interacted with each other. It deals with Reconstruction at all levels of society, from debates in Congress to struggles on individual plantations. And while the focus of Reconstruction lay in the South, the North and West also experienced dramatic changes in this era, and I try to address events in those regions as well.

With the publication of *Reconstruction*, I assumed I would turn my scholarly attention to other areas. But things did not turn out that way. In the course of my research, I had gathered an immense file of biographical information about black political leaders in the postwar South— justices of the peace, sheriffs, and state legislators, as well as Congressmen and U.S. Senators—most of them unknown even to scholars. In 1993, I brought this information together in *Freedom's Lawmakers*, a directory containing capsule biographies of some 1,500 individuals. Generations of historians had ignored or denigrated these black officeholders, citing their alleged incompetence to justify the violent over-throw of Reconstruction and the South's long history of disenfranchising black voters. My hope was to put these men, as it were, on the map of history, to make available the basic data concerning their lives and to bury irrevocably such misconceptions as that Reconstruction's leaders were illiterate, propertyless, and incompetent.

In addition, with Olivia Mahoney of the Chicago Historical Society, I served as co-curator of a museum exhibition, "America's Reconstruction,"—the first ever to be devoted exclusively to the period—that opened in 1995 at the Virginia Historical Society in Richmond. It subsequently traveled to venues in New York City, Columbia, Raleigh, Tallahassee, and Chicago, and has recently been digitalized and can be viewed on the Internet. In 2004, I was an adviser for the first PBS television documentary series on Reconstruction.

In the seventeen years since the book appeared, new scholarship has been published on numerous aspects of the Reconstruction era. Studies of the experience of individual states and localities—such as Richard Zuczek's history of Reconstruction in South Carolina and Michael Fitzgerald's account of the era in Mobile, Alabama—have added significantly to our knowledge of the period. Work on the development of new labor systems in the tobacco and sugar districts of the South have complemented previous studies of the rise of sharecropping in the cotton areas. Literature on the legal and constitutional changes of Reconstruction has continued to flourish, spurred by debates over the interpretation of the 14th Amendment in our own time. Probably the most exciting new direction in Reconstruction scholarship has been the literature on the experience of women, white and black, in the postwar South. In ways that my book only briefly suggested, Reconstruction profoundly affected gender relations and family structures. No one, however, has as yet produced an account of Reconstruction that brings together this new work in a coherent new synthetic interpretation.

Today, the greatest obstacle of an understanding of Reconstruction is not accumulated myths and misconceptions, as in the past, but widespread ignorance. In 1990, the Department of Education conducted a survey of 16,000 graduating seniors in American high schools, who were asked to identify various terms or issues in American history. Reconstruction received the lowest score on the entire test—only one student in five could correctly identify it. Ignorance of Reconstruction is unfortunate because, whether we realize it or not, Reconstruction remains a part of our lives, nearly a century and a half after the Civil War ended.

The historian C. Vann Woodward first called the civil rights movement the Second Reconstruction. Although history never really repeats itself, the parallels between the period after the Civil War and the 1950s and 1960s are very dramatic, as are the retreats from the Reconstruction ideal of racial justice and social equality in the latter decades of the nineteenth century and again in our own time. Today, Congress and the Supreme Court still debate questions arising from Reconstruction laws and constitutional amendments. The rights of American citizens, the proper roles of the state and federal governments, the possibility of interracial political coalitions, affirmative action, reparations for slavery, the proper ways for the government to protect citizens against terrorist violence, the relationship between political and economic democracy— these and other issues of our own time are Reconstruction issues. Or, at least, they cannot be properly understood without knowledge of how they were debated during the Reconstruction era. So long as questions placed on the national agenda during Reconstruction remain unresolved, the era will remain relevant to modern-day America.

—Eric Foner
November 2004

Abbreviations Used in Footnotes

AC—Annual Cyclopedia
AgH—Agricultural History
AHR—American Historical Review
A1HQ—Alabama Historical Quarterly
AMA—American Missionary Association
ArkHQ—Arkansas Historical Quarterly
ASDAH—Alabama State Department of Archives and History
CG—Congressional Globe
CR—Congressional Record
CWH—Civil War History
DU—Duke University
ETHSP—East Tennessee Historical Society Publications
F1HQ—Florida Historical Quarterly
FSSP—Freedmen and Southern Society Project, University of
        Maryland (with document identification number)
GaHQ—Georgia Historical Quarterly
GDAH—Georgia Department of Archives and History
HL—Huntington Library
HSPa—Historical Society of Pennsylvania
HU—Houghton Library, Harvard University
IndMH—Indiana Magazine of History
JAH—Journal of American History
JEcH—Journal of Economic History
JISHS—Journal of the Illinois State Historical Society
JMH—Journal of Mississippi History
JNH—Journal of Negro History
JSH—Journal of Southern History
JSocH—Journal of Social History
LaH—Louisiana History
LaHQ—Louisiana Historical Quarterly
LC—Library of Congress

LML—Lawson McGhee Library
LSU—Louisiana State University
MDAH—Mississippi Department of Archives and History
MHS—Massachusetts Historical Society
MoHR—Missouri Historical Review
MVHR—Mississippi Valley Historical Review
NA—National Archives
NCHR—North Carolina Historical Review
NCDAH—North Carolina Division of Archives and History
NYPL—New York Public Library
OHQ—Ohio Historical Quarterly
PaH—Pennsylvania History
PaMHB—Pennsylvania Magazine of History and Biography
PMHS—Publications of the Mississippi Historical Society
RG 105—Record Group 105: Records of the Bureau of Refugees,
    Freedmen, and Abandoned Lands
RG 393—Record Group 393: Records of the United States Army
    Continental Commands
SAQ—South Atlantic Quarterly
SC—Sophia Smith Collection, Smith College
SCDA—South Carolina Department of Archives
SCHM—South Carolina Historical Magazine
SCHS—South Carolina Historical Society
SHSW—State Historical Society of Wisconsin
SS—Southern Studies
SWHQ—Southwestern Historical Quarterly
THQ—Tennessee Historical Quarterly
TSLA—Tennessee State Library and Archives
UGa—University of Georgia
UNC—Southern Historical Collection, University of North
    Carolina
USC—South Caroliniana Library, University of South Carolina
UTx—Eugene C. Barker Texas History Center, University of Texas
VaMHB—Virginia Magazine of History and Biography
WMH—Wisconsin Magazine of History
WVaH—West Virginia History

Editors' Introduction

P ROBABLY no other chapter of American history has been the subject, one might say the victim, of such varied and conflicting interpretations as what attempts to give unity and coherence to the era we call
Reconstruction. Even the chronology is chaotic. Did the process begin
with the bizarre creation of West Virginia in 1861—or should that be
dated 1863? Did it conclude with the Compromise of 1877 or was its true
conclusion Brown v. Topeka in 1954? Was its central theme political—
the reconstruction of the old Union, or was it legal and constitutional—
the revolutionary Fourteenth Amendment that still functions as an instrument of revolution? Was its central theme social and moral—the end of
slavery, or did the realities of slavery persist for another half century or
more? Was its significance fundamentally in what has been called the
Emergence of Modern America—into the *Triumphant Democracy* that Andrew Carnegie celebrated, or was it rather the emergence of America to
world power—or certainly to Pacific power? Or might it all be interpreted
in philosophical terms—the Age of Darwin and Spencer, of Lester Ward
and William Jones, who contributed so much to reconstructing American
thought?

"Reconstruction" embraced, of course, all these chapters of our history—a conclusion illustrated by successive generations of historians
from James Ford Rhodes, Ellis Oberholtzer, John Burggs, and Vernon
Parrington to the schools of William Dunning, W.D.B. Du Bois, Walter
Fleming, and Allan Nevins.

It is to this distinguished lineage of Reconstruction scholars that Professor Foner belongs, and in nothing is he more distinguished than in his
independence and originality. The most striking feature of that independence is his insistence that the Negro was the central figure and the most
effective in Reconstruction: in this he was, to be sure, anticipated by the
great Negro leader, Du Bois. To the support of his thesis, Mr. Foner has
brought a prodigious body of evidence, organized it not only skillfully but
also, we may almost say, with stylistic genius, and produced what is a

scholarly convincing Reconstruction of what is indubitably the most controversial chapter in our history.

HENRY STEELE COMMAGER
RICHARD B. MORRIS

Preface

REVISING interpretations of the past is intrinsic to the study of history. But no part of the American experience has, in the last twenty-five years, seen a broadly accepted point of view so completely overturned as Reconstruction—the violent, dramatic, and still controversial era that followed the Civil War. Since the early 1960s, a profound alteration of the place of blacks within American society, newly uncovered evidence, and changing definitions of history itself have combined to transform our understanding of race relations, politics, and economic change during Reconstruction. Yet despite this change in consciousness, so to speak, historians have yet to produce a coherent new portrait of the era.

The scholarly study of Reconstruction began early in this century with the work of William Dunning, John W. Burgess, and their students. The interpretation elaborated by the Dunning School may be briefly summarized as follows. When the Civil War ended, the white South genuinely accepted the reality of military defeat, stood ready to do justice to the emancipated slaves, and desired above all a quick reintegration into the fabric of national life. Before his death, Abraham Lincoln had embarked on a course of sectional reconciliation, and during Presidential Reconstruction (1865–67) his successor, Andrew Johnson, attempted to carry out Lincoln's magnanimous policies. Johnson's efforts were opposed and eventually thwarted by the Radical Republicans in Congress. Motivated by an irrational hatred of Southern "rebels" and the desire to consolidate their party's national ascendancy, the Radicals in 1867 swept aside the Southern governments Johnson had established and fastened black suffrage upon the defeated South. There followed the sordid period of Congressional or Radical Reconstruction (1867–77), an era of corruption presided over by unscrupulous "carpetbaggers" from the North, unprincipled Southern white "scalawags," and ignorant freedmen. After much needless suffering, the South's white community banded together to overthrow these governments and restore "home rule" (a euphemism for white supremacy). All told, Reconstruction was the darkest page in the saga of American history.[1]

The fundamental underpinning of this interpretation was the conviction, to quote one member of the Dunning School, of "negro incapacity." The childlike blacks, these scholars insisted, were unprepared for freedom and incapable of properly exercising the political rights Northerners had thrust upon them. The fact that blacks took part in government, wrote E. Merton Coulter in the last full-scale history of Reconstruction written entirely within the Dunning tradition, was a "diabolical" development, "to be remembered, shuddered at, and execrated." Yet while these works abounded in horrified references to "negro rule" and "negro government," blacks in fact played little role in the narratives. Their aspirations, if mentioned at all, were ridiculed, and their role in shaping the course of events during Reconstruction ignored. When these writers spoke of "the South" or "the people," they meant whites. Blacks appeared either as passive victims of white manipulation or as an unthinking people whose "animal natures" threatened the stability of civilized society.[2]

During the 1920s and 1930s, new studies of Johnson's career and new investigations of the economic wellsprings of Republican policy reinforced the prevailing disdain for Reconstruction. Despite their critique of Republican rule in the South, Dunning and Burgess had placed much of the blame for the postwar political impasse on Johnson, who, they charged, had failed to recognize that Congress had a perfect right to insist on legal and constitutional changes that would "reap the just fruits of their triumph over secession and slavery." Johnson's new biographers, however, portrayed him as a courageous defender of constitutional liberty, whose actions stood above reproach. Simultaneously, historians of the Progressive School, who viewed political ideologies as little more than masks for crass economic ends, further undermined the Radicals' reputation by portraying them as agents of Northern capitalism, who cynically used the issue of black rights to fasten economic subordination upon the defeated South.[3]

1. General accounts of Reconstruction in the tradition of the Dunning School include William A. Dunning, *Reconstruction, Political and Economic 1865–1877* (New York, 1907), Walter L. Fleming, *The Sequel of Appomattox* (New Haven, 1919), Claude G. Bowers, *The Tragic Era* (Cambridge, Mass., 1929), E. Merton Coulter, *The South During Reconstruction 1865–1877* (Baton Rouge, 1947). See also Philip R. Muller, "Look Back Without Anger: A Reappraisal of William A. Dunning," *JAH*, 61 (January 1974), 325–38.

2. Charles W. Ramsdell, *Reconstruction in Texas* (New York, 1910), 48; Coulter, *South During Reconstruction*, 60, 141; Walter L. Fleming, *Civil War and Reconstruction in Alabama* (New York, 1905); E. Merton Coulter, *William G. Brownlow, Fighting Parson of the Southern Highlands* (Chapel Hill, 1937), 352.

3. John W. Burgess, *Reconstruction and the Constitution 1866–1876* (New York, 1902), 54; Robert W. Winston, *Andrew Johnson: Plebeian and Patriot* (New York, 1926); George F. Milton, *The Age of Hate: Andrew Johnson and the Radicals* (New York, 1930); Howard K. Beale, *The Critical Year: A Study of Andrew Johnson and Reconstruction* (New York, 1930).

From the first appearance of the Dunning School, dissenting voices had been raised, initially by a handful of survivors of the Reconstruction era and the small fraternity of black historians. In 1935, the black activist and scholar, W. E. B. Du Bois, published *Black Reconstruction in America*, a monumental study that portrayed Reconstruction as an idealistic effort to construct a democratic, interracial political order from the ashes of slavery, as well as a phase in a prolonged struggle between capital and labor for control of the South's economic resources. His book closed with an indictment of a profession whose writings had ignored the testimony of the principal actor in the drama of Reconstruction—the emancipated slave—and sacrificed scholarly objectivity on the altar of racial bias. "One fact and one alone," Du Bois wrote, "explains the attitude of most recent writers toward Reconstruction; they cannot conceive of Negroes as men." In many ways, *Black Reconstruction* anticipated the findings of modern scholarship. At the time, however, it was largely ignored.[4]

Despite its remarkable longevity and powerful hold on the popular imagination, the demise of the traditional interpretation was inevitable. Once objective scholarship and modern experience rendered its racist assumptions untenable, familiar evidence read very differently, new questions suddenly came into prominence, and the entire edifice of the Dunning School had to fall. Indeed, only a few years after the publication of *Black Reconstruction*, Howard K. Beale, who had earlier helped to discredit Radical Republicans' motives, called for a sweeping reassessment of Southern Reconstruction. Historians, he insisted, must rethink the prevailing idea that the South owed "a debt of gratitude" to the restorers of white supremacy; even more profoundly, they must free themselves from the conviction that "their race must bar Negroes from social and economic equality." During the 1940s and 1950s, a growing number of historians, taking up the revisionist challenge announced by Du Bois and Beale, offered sympathetic accounts of the once-despised freedmen, Southern white Republicans, and Northern policymakers.[5]

It required, however, not simply the evolution of scholarship but a profound change in the nation's politics and racial attitudes to deal the final blow to the Dunning School. If the traditional interpretation reflected, and helped to legitimize, the racial order of a society in which blacks were disenfranchised and subjected to discrimination in every

4. W. E. B. Du Bois, *Black Reconstruction in America* (New York, 1935). See also A. A. Taylor, "Historians of the Reconstruction," *JNH*, 23 (January 1938), 16–34.

5. Howard K. Beale, "On Rewriting Reconstruction History," *AHR*, 45 (July 1940), 807–27; David H. Donald, "The Scalawag in Mississippi Reconstruction," *JSH*, 10 (November 1944), 447–60; Vernon L. Wharton, *The Negro in Mississippi 1865–1890* (Chapel Hill, 1947); John H. Cox and LaWanda Cox, "General O. O. Howard and the 'Misrepresented Bureau'," *JSH*, 19 (November 1953), 427–56; Stanley Coben, "Northeastern Business and Radical Reconstruction: A Re-Examination," *MVHR*, 46 (June 1959), 67–90. For the state of the field in 1959 see Bernard A. Weisberger, "The Dark and Bloody Ground of Reconstruction Historiography," *JSH*, 25 (November 1959), 427–47.

aspect of their lives, Reconstruction revisionism bore the mark of the modern civil rights movement. In the 1960s the revisionist wave broke over the field, destroying, in rapid succession, every assumption of the traditional viewpoint. First, scholars presented a drastically revised account of national politics. New works portrayed Andrew Johnson as a stubborn, racist politician incapable of responding to the unprecedented situation that confronted him as President, and acquitted the Radicals—reborn as idealistic reformers genuinely committed to black rights—of vindictive motives and the charge of being the stalking-horses of Northern capitalism. Moreover, Reconstruction legislation was shown to be not simply the product of a Radical cabal, but a program that enjoyed broad support both in Congress and the North at large.[6]

Even more startling was the revised portrait of Republican rule in the South. So ingrained was the old racist version of Reconstruction that it took an entire decade of scholarship to prove the essentially negative contentions that "Negro rule" was a myth and that Reconstruction represented more than "the blackout of honest government." The establishment of public school systems, the granting of equal citizenship to blacks, the effort to revitalize the devastated Southern economy—these were commendable achievements, which refuted the traditional description of the period as a "tragic era" of rampant misgovernment. Revisionists pointed out as well that corruption in the Reconstruction South paled before that of the Tweed Ring, Crédit Mobilier scandal, and Whiskey Rings in the post–Civil War North. By the end of the 1960s, the old interpretation had been completely reversed. Radical Republicans and Southern freedmen were now the heroes, white supremacist Redeemers the villains, and Reconstruction was a time of extraordinary social and political progress for blacks. If the era was "tragic," revisionists insisted, it was because change did not go far enough, especially in the area of Southern land reform.[7]

Even when revisionism was at its height, however, its more optimistic findings were challenged. Shocked by the resistance to racial progress in the 1960s and the deep-seated economic problems the Second Reconstruction failed to solve, influential historians portrayed change in the post–Civil War years as fundamentally "superficial." Persistent racism,

6. Eric L. McKitrick, *Andrew Johnson and Reconstruction* (Chicago, 1960); W. R. Brock, *An American Crisis* (London, 1963); LaWanda Cox and John H. Cox, *Politics, Principle, and Prejudice 1865–1866* (New York, 1963); James M. McPherson, *The Struggle for Equality: Abolitionists and the Negro in the Civil War and Reconstruction* (Princeton, 1964); Hans L. Trefousse, *The Radical Republicans: Lincoln's Vanguard for Racial Justice* (New York, 1969).

7. Willie Lee Rose, *Rehearsal for Reconstruction: The Port Royal Experiment* (Indianapolis, 1964); Joel Williamson, *After Slavery: The Negro in South Carolina During Reconstruction, 1861–1877* (Chapel Hill, 1965); Otto H. Olsen, *Carpetbagger's Crusade: The Life of Albion Winegar Tourgée* (Baltimore, 1965); Robert Cruden, *The Negro in Reconstruction* (Englewood Cliffs, N.J., 1969). Kenneth M. Stampp, *The Era of Reconstruction 1865–1877* (New York, 1965) was the most influential summary of the revisionist outlook.

these postrevisionist scholars argued, had negated efforts to extend justice to blacks, and the failure to distribute land prevented the freedmen
from achieving true autonomy and made their civil and political rights all
but meaningless. In the 1970s and 1980s, a new generation of scholars,
black and white, extended this skeptical view to virtually every aspect of
the period. Recent studies of Reconstruction politics and ideology have
stressed the "conservatism" of Republican policymakers, even at the
height of Radical influence, and the continued hold of racism and federalism despite the extension of citizenship rights to blacks and the enhanced
scope of national authority. Studies of federal policy in the South portrayed the army and Freedmen's Bureau as working hand in glove with
former slaveowners to thwart the freedmen's aspirations and force them
to return to plantation labor. At the same time, studies of Southern social
history emphasized the survival of the old planter class and the continui-
ties between the Old South and the New. The postrevisionist interpretation represented a striking departure from nearly all previous accounts
of the period, for whatever their differences, traditional and revisionist
historians at least agreed that Reconstruction was a time of radical
change. Recent work, on the other hand, questioned whether anything of
enduring importance occurred at all. Summing up a decade of writing,
C. Vann Woodward observed in 1979 that historians now understood
"how essentially nonrevolutionary and conservative Reconstruction really was."[8]

   In emphasizing that Reconstruction was part of the ongoing evolution
of Southern society rather than a passing phenomenon, the postrevision-
ists made a salutary contribution to the study of the period. The description of Reconstruction as "conservative," however, did not seem altogether persuasive when one reflected that it took the nation fully a
century to implement its most basic demands, while others are yet to be
fulfilled. Nor did the theme of continuity yield a fully convincing portrait
of an era that contemporaries all agreed was both turbulent and wrenching in its social and political change. Over a half-century ago, Charles and
Mary Beard coined the term "The Second American Revolution" to

   8. August Meier, "Negroes in the First and Second Reconstructions of the South," *CWH*,
13 (June 1967), 114; C. Vann Woodward, "Seeds of Failure in Radical Race Policy," in
Harold M. Hyman, ed., *New Frontiers of the American Reconstruction* (Urbana, Ill., 1966),
125–47; William S. McFeely, *Yankee Stepfather: General O. O. Howard and the Freedmen* (New
Haven, 1968); Michael L. Benedict, "Preserving the Constitution: The Conservative Basis
of Radical Reconstruction," *JAH*, 61 (June 1974), 65–90; Harold M. Hyman, *A More Perfect
Union: The Impact of the Civil War and Reconstruction on the Constitution* (New York, 1973);
William Gillette, *Retreat from Reconstruction 1869–1879* (Baton Rouge, 1979); Leon F. Lit-
wack, *Been in the Storm So Long: The Aftermath of Slavery* (New York, 1979); Jonathan M.
Wiener, *Social Origins of the New South: Alabama 1860–1885* (Baton Rouge, 1978); C. Vann
Woodward, review of *The Confederate Nation, New Republic*, March 17, 1979, 26. See also Eric
Foner, "Reconstruction Revisited," *Reviews in American History*, 10 (December 1982), 82–
100.

describe a transfer in power, wrought by the Civil War, from the South's "planting aristocracy" to "Northern capitalists and free farmers." And in the latest shift in interpretive premises, attention to changes in the relative power of social classes has again become a central concern of historical writing. Indeed, the term "revolution" has reappeared in the most recent literature as a way of describing the Civil War and Reconstruction. Unlike the Beards, however, who all but ignored the black experience, modern scholars tend to view emancipation itself as among the most revolutionary aspects of the period.[9]

If anything is clear from this brief account, it is that despite the remarkable burst of creativity that discredited the Dunning School interpretation, historians have yet to produce a coherent account of Reconstruction to take its place. In part, this "failure" arises from the very vitality of the field, for the abundance, and continuing publication, of new literature vastly complicates the task of synthesis. But it also reflects a problem that has marked the recent study of American history as a whole. The past generation has witnessed an unprecedented expansion and redefinition of historical study, as an older preoccupation with institutions, politics, and ideas has given way to a host of new "social" concerns. One result of the new attention to the experience of blacks, women, and labor, and to subjects like family structure, social mobility, and popular culture, has been to enrich immeasurably our understanding of the nation's history. Another, however, has been a fragmentation of historical scholarship and a retreat from the idea that a coherent vision of the past is even possible. As with other periods, the literature on Reconstruction seems divided between broad studies of national politics that all but ignore the social ferment within the South, and detailed accounts of individual communities or states, or of a single aspect of Southern life, isolated from the era's political history and from a broader national or regional context. For all their faults, it is ironic that the best Dunning studies did, at least, attempt to synthesize the social, political, and economic aspects of the period.

In a sense, this book aims to combine the Dunning School's aspiration to a broad interpretive framework with the findings and concerns of recent scholarship—to provide, that is, a coherent, comprehensive modern account of Reconstruction. This effort necessarily touches on a multitude of issues, but certain broad themes unify the narrative. The first is the centrality of the black experience. Rather than passive victims of the actions of others or simply a "problem" confronting white society, blacks were active agents in the making of Reconstruction. During the Civil War,

9. Charles A. Beard and Mary R. Beard, *The Rise of American Civilization* (New York, 1933 ed.), Chapter 28; Ira Berlin et al., eds., *Freedom: A Documentary History of Emancipation 1861–1867* (New York, 1982–); Eric Foner, *Nothing But Freedom: Emancipation and Its Legacy* (Baton Rouge, 1983); Peyton McCrary, "The Party of Revolution: Republican Ideas about Politics and Social Change," *CWH*, 30 (December 1984), 330–50; Barbara J. Fields, *Slavery and Freedom on the Middle Ground: Maryland During the Nineteenth Century* (New Haven, 1985).

their actions helped force the nation down the road to emancipation, and in the aftermath of that conflict, their quest for individual and community autonomy did much to establish Reconstruction's political and economic agenda. Although thwarted in their bid for land, blacks seized the opportunity created by the end of slavery to establish as much independence as possible in their working lives, consolidate their families and communities, and stake a claim to equal citizenship. Black participation in Southern public life after 1867 was the most radical development of the Reconstruction years, a massive experiment in interracial democracy without precedent in the history of this or any other country that abolished slavery in the nineteenth century. And the pages that follow pay special attention both to the political mobilization of the black community and to the emergence and changing composition of a black political leadership that seized upon America's republican values as a weapon for attacking the nation's racial caste system.

The transformation of slaves into free laborers and equal citizens was the most dramatic example of the social and political changes unleashed by the Civil War and emancipation. A second purpose of this study is to trace the ways in which Southern society as a whole was remodeled, and to do so without neglecting the local variations in different parts of the South. Indeed, the black experience cannot be understood without considering how the status of white planters, merchants, and yeomen, and their relations with one another, changed over time. By the end of Reconstruction, a new Southern class structure and several new systems of organizing labor were well on their way to being consolidated. I have tried to show, however, that instead of proceeding in a linear, predetermined fashion, these developments arose from a complex series of interactions among blacks and whites, Northerners and Southerners, in which victories were often tentative and outcomes subject to challenge and revision. The ongoing process of social and economic change, moreover, was intimately related to the politics of Reconstruction, for various groups of blacks and whites sought to use state and local government to promote their own interests and define their place in the region's new social order.

The evolution of racial attitudes and patterns of race relations, and the complex interconnection of race and class in the postwar South, form a third theme of this book. Racism was pervasive in mid-nineteenth-century America and at both the regional and national levels constituted a powerful barrier to change. Yet despite racism, a significant number of Southern whites were willing to link their political fortunes with those of blacks, and Northern Republicans came, for a time, to associate the fate of the former slaves with their party's raison d'être and the meaning of Union victory in the Civil War. Moreover, in the critical, interrelated issues of land and labor and the persistent conflict between planters' desire to reexert control over their labor force and blacks' quest for economic independence, race and class were inextricably linked. As a Washington newspaper noted in 1868, "It is impossible to separate the question of color from the question of labor, for the reason that the majority of the laborers . . . throughout the Southern States are colored people, and nearly all the colored people are at present laborers."[10] Thus, instead of viewing racism as a deus ex machina that independently explains the course of events and Reconstruction's demise, I view it as an intrinsic part of the process of historical development, which affected and was affected by changes in the social and political order.

The chapters that follow also seek to place the Southern story within a national context. The book's fourth theme is the emergence during the Civil War and Reconstruction of a national state possessing vastly expanded authority and a new set of purposes, including an unprecedented commitment to the ideal of a national citizenship whose equal rights belonged to all Americans regardless of race. Originating in wartime exigencies, the activist state (paralleled at the local level both North and South) came to embody the reforming impulse deeply rooted in postwar politics. And Reconstruction produced enduring changes in the laws and Constitution that fundamentally altered federal–state relations and redefined the meaning of American citizenship. Yet because it threatened traditions of local autonomy, produced political corruption and rising taxes, and was so closely associated with the new rights of blacks, the rise of the state inspired powerful opposition, which, in turn, weakened support for Reconstruction.

Finally, this study examines how changes in the North's economy and class structure affected Reconstruction. Many of the processes and issues central to Southern Reconstruction—the consolidation of a new class structure, changes in the position of blacks, conflicts over access to the region's economic resources—were also present, in different forms, in the North. These developments brought new political issues to the fore and undermined the free labor ideology that had inspired efforts to remake Southern society. That the Reconstruction of the North receives less attention than its Southern counterpart reflects, in part, the absence of a detailed historical literature on either the region's social and political structure in these years, or the relationship between changes there and events in the South. It also recognizes that events in the South remain the heart of the Reconstruction drama. Nonetheless, Reconstruction cannot be fully understood without attention to its distinctively Northern and national dimensions.

Beyond the desire to provide a new account of Reconstruction, this study has an additional purpose—to demonstrate the possibility, and

10. Washington *Weekly Chronicle*, August 15, 1868.

value, of transcending the present compartmentalization of historical study into "social" and "political" components, and of historical writing into "narrative" and "analytical" modes. Some practitioners of the "new" history have expressed fear that the very notion of "synthesis" suggests a return to the excessively broad generalizations and narrow political focus of an earlier era.[11] This is not my intention. Rather, my aim is to view the period as a whole, integrating the social, political, and economic aspects of Reconstruction into a coherent, analytical narrative.

Like nearly every aspect of the period, the chronological definition of Reconstruction remains open to dispute. I have chosen the conventional date of 1877 to close this account (with a brief look at the New South that followed), because although the process of social change did not abruptly end in that year, the fall of the South's last Republican governments and the removal of federal troops from a role in regional politics marked a definitive turning point in American history. My opening is more unusual, for the book begins not in 1865 (or, as others would insist, in 1861), but with the Emancipation Proclamation of 1863. I do this to emphasize the Proclamation's importance in uniting two major themes of this study— grass-roots black activity and the newly empowered national state—and to indicate that Reconstruction was not merely a specific time period, but the beginning of an extended historical process: the adjustment of American society to the end of slavery. The destruction of the central institution of antebellum Southern life permanently transformed the war's character, and produced far-reaching conflicts and debates over the role former slaves and their descendants would play in American life and the meaning of the freedom they had acquired. These were the questions on which Reconstruction persistently turned.

Over a century ago, prodded by the demands of four million men and women just emerging from slavery, Americans made their first attempt to live up to the noble professions of their political creed—something few societies have ever done. The effort produced a sweeping redefinition of the nation's public life and a violent reaction that ultimately destroyed much, but by no means all, of what had been accomplished. From the enforcement of the rights of citizens to the stubborn problems of economic and racial justice, the issues central to Reconstruction are as old as the American republic, and as contemporary as the inequalities that still afflict our society.

11. Eric H. Monkkonen, "The Dangers of Synthesis," *AHR*, 91 (December 1986), 1146–57.

RECONSTRUCTION

CHAPTER 1

The World the War Made

The Coming of Emancipation

ON January 1, 1863, after a winter storm swept up the east coast of the United States, the sun rose in a cloudless sky over Washington, D.C. At the White House, Abraham Lincoln spent most of the day welcoming guests to the traditional New Year's reception. Finally, in the late afternoon, as he had pledged to do 100 days before, the President retired to his office to sign the Emancipation Proclamation. Excluded from its purview were the 450,000 slaves in Delaware, Kentucky, Maryland, and Missouri (border slave states that remained within the Union), 275,000 in Union-occupied Tennessee, and tens of thousands more in portions of Louisiana and Virginia under the control of federal armies. But, the Proclamation decreed, the remainder of the nation's slave population, well over 3 million men, women, and children, "are and henceforth shall be free."[1]

Throughout the North and the Union-occupied South, January 1 was a day of celebration. An immense gathering, including black and white abolitionist leaders, stood vigil at Boston's Tremont Temple, awaiting word that the Proclamation had been signed. It was nearly midnight when the news arrived; wild cheering followed, and a black preacher led the throng in singing "Sound the loud timbrel o'er Egypt's dark sea, Jehovah hath triumphed, his people are free." At a camp for fugitive slaves in the nation's capital, a black man "testified" about the sale, years before, of his daughter, exclaiming, "Now, no more dat! . . . Dey can't sell my wife and child any more, bless de Lord!" Farther south, at Beaufort, an enclave of federal control off the South Carolina coast, there were prayers and speeches and the freedmen sang "My Country 'Tis of Thee." To Charlotte Forten, a young black woman who had journeyed from her

1. New York *Tribune*, January 2, 1863; Washington *Daily National Intelligencer*, January 2, 1863; James D. Richardson, ed., *A Compilation of the Messages and Papers of the Presidents 1789–1897* (Washington, D.C., 1896–99), 5:157–59.

1

native Philadelphia to teach the former slaves, "it all seemed . . . like a brilliant dream." Even in areas exempted from the Proclamation, blacks celebrated, realizing that if slavery perished in Mississippi and South Carolina, it could hardly survive in Kentucky, Tennessee, and a few parishes of Louisiana.[2]

Nearly two and a half centuries had passed since twenty black men and women were landed in Virginia from a Dutch ship. From this tiny seed had grown the poisoned fruit of plantation slavery, which, in profound and contradictory ways, shaped the course of American development. Even as slavery mocked the ideals of a nation supposedly dedicated to liberty and equality, slave labor played an indispensable part in its rapid growth, expanding westward with the young republic, producing the cotton that fueled the early industrial revolution. In the South, slavery spawned a distinctive regional ruling class (an "aristocracy without nobility" one Southern-born writer called it) and powerfully shaped the economy, race relations, politics, religion, and the law. Its influence was pervasive: "Nothing escaped, nothing and no one."[3] In the North, where slavery had been abolished during and after the American Revolution, emerged abolition, the greatest protest movement of the age. The slavery question divided the nation's churches, sundered political ties between the sections, and finally shattered the bonds of Union. On the principle of opposing the further expansion of slavery, a new political party rose to power in the 1850s, placing in the White House a son of the slave state Kentucky, who had grown to manhood on the free Illinois prairies and believed the United States could not endure forever half slave and half free. In the crisis that followed Lincoln's election, eleven slave states seceded from the Union, precipitating in 1861 the bloodiest war the Western Hemisphere has ever known.

To those who had led the movement for abolition, and to slaves throughout the South, the Emancipation Proclamation not only culminated decades of struggle but evoked Christian visions of resurrection and redemption, of an era of unbounded progress for a nation purged at last of the sin of slavery. Even the staid editors of the *New York Times* believed it marked a watershed in American life, "an era in the history . . . of this country and the world." For emancipation meant more than the end of a labor system, more even than the uncompensated liquidation of the nation's largest concentration of private property ("the most stupendous act of sequestration in the history of Anglo-Saxon jurispru-

2. [Frederick Douglass], *Life and Times of Frederick Douglass* (New York, 1962 ed.), 352–53; John Hope Franklin, *The Emancipation Proclamation* (New York, 1963), 89–128; W. E. B. Du Bois, *Black Reconstruction in America* (New York, 1935), 301–302; Ray Billington, ed., *The Journal of Charlotte Forten* (New York, 1953), 171–72.
3. David H. Strother Diary, August 9, 1869, West Virginia University; Frank Tannenbaum, *Slave and Citizen* (New York, 1946), 117. Tannenbaum refers here to Brazil, but his observation is even more applicable to the American South.

dence," as Charles and Mary Beard described it).[4] The demise of slavery inevitably threw open the most basic questions of the polity, economy, and society. Begun to preserve the Union, the Civil War now portended a far-reaching transformation in Southern life and a redefinition of the place of blacks in American society and of the very meaning of freedom in the American republic.

In one sense, however, the Proclamation only confirmed what was already happening on farms and plantations throughout the South. War, it has been said, is the midwife of revolution, and well before 1863 the disintegration of slavery had begun. Whatever politicians and military commanders might decree, slaves saw the war as heralding the long-awaited end of bondage. Three years into the conflict, Gen. William T. Sherman encountered a black Georgian who summed up the slaves' understanding of the war from its outset: "He said . . . he had been looking for the 'angel of the Lord' ever since he was knee-high, and, though we professed to be fighting for the Union, he supposed that slavery was the cause, and that our success was to be his freedom."[5] Based on this conviction, the slaves took actions that propelled a reluctant white America down the road to abolition.

As the Union Army occupied territory on the periphery of the Confederacy, first in Virginia, then in Tennessee, Louisiana, and elsewhere, slaves by the thousands headed for the Union lines. Union enclaves like Fortress Monroe, Beaufort, and New Orleans became havens for runaway slaves and bases for expeditions into the interior that further disrupted the plantation regime. Even in the heart of the Confederacy, far from Union lines, the conflict undermined the South's "peculiar institution." Their "grapevine telegraph" kept many slaves remarkably well informed about the war's progress. In one part of Mississippi, slaves even organized Lincoln's Legal Loyal League to spread word of the Emancipation Proclamation. Southern armies impressed tens of thousands of slaves into service as laborers, taking them far from their home plantations, offering opportunities for escape, and widening the horizons of those who returned home. The drain of white men into military service left plantations under the control of planters' wives and elderly and infirm men, whose authority slaves increasingly felt able to challenge. Reports of "demoralized" and "insubordinate" behavior multiplied throughout the South. Six months after the war began, slaves in one Kentucky town marched through the streets at night, shouting hurrahs for Lincoln.[6]

4. *New York Times*, January 3, 1863; Charles A. Beard and Mary R. Beard, *The Rise of American Civilization* (New York, 1933 ed.), 100.

5. Vincent Harding, *There Is a River: The Black Struggle for Freedom in America* (New York, 1981), 221–26; [William T. Sherman] *Memoirs of General W. T. Sherman, Written by Himself* (New York, 1891 ed.), 2:180–81.

6. Ira Berlin et al., eds., *Freedom: A Documentary History of Emancipation* (New York, 1982–), Ser. 1, 1:1–30, 59–66; Leon F. Litwack, *Been in the Storm So Long: The Aftermath of Slavery* (New

But generally, it was the arrival of federal soldiers that spelled havoc for the slave regime, for blacks quickly grasped that the presence of occupying troops destroyed the coercive power of both the individual master and the slaveholding community. A Virginia coachman, informed by soldiers in 1862 that he was free, "went straight to his master's chamber, dressed himself in his best clothes, put on his best watch and chain . . . and insolently informed him that he might for the future drive his own coach." On Magnolia plantation in Louisiana, the arrival of the Union Army in 1862 sparked a work stoppage and worse: "We have a terrible state of affairs here negroes refusing to work. . . . The negroes have erected a gallows in the quarters and give as an excuse for it that they are told they must drive their master . . . off the plantation hang their master etc. and that then they will be free." Here in the sugar country, where large gangs of slaves labored in some of the South's most wretched conditions, blacks sacked planters' homes and, months before the Emancipation Proclamation, refused to work unless paid wages. Slavery, wrote a Northern reporter in November 1862, "is forever destroyed and worthless, no matter what Mr. Lincoln or anyone else may say on the subject."[7]

"Meanwhile," in the words of W. E. B. Du Bois, "with perplexed and laggard steps, the United States government followed in the footsteps of the black slave." The slaves' determination to seize the opportunity presented by the war initially proved an embarrassment to the Lincoln administration and a burden to the army. Lincoln fully appreciated, as he would observe in his second inaugural address, that slavery was "somehow" the cause of the war.[8] But he also understood the vital importance of keeping the border slave states in the Union, generating support among the broadest constituency in the North, and weakening the Confederacy by holding out to irresolute Southerners the possibility that they could return to the Union with their property, including slaves, intact. In 1861 the restoration of the Union, not emancipation, was the cause that generated the widest support for the war effort.

Thus, in the early days of the war, the administration insisted that slavery had little to do with the conflict. When Congress assembled in special session in July 1861, one of its first acts was to pass, nearly

York, 1979), 22–27; George P. Rawick, ed., *The American Slave: A Composite Autobiography* (Westport, Conn., 1972–79), Supplement 1, 6:8–12; James L. Roark, *Masters Without Slaves: Southern Planters in the Civil War and Reconstruction* (New York, 1977), 89–90; E. Merton Coulter, *The Civil War and Readjustment in Kentucky* (Chapel Hill, 1926), 247.

7. Bell I. Wiley, *Southern Negroes 1861–1865* (New Haven, 1938), 73–74; J. Carlyle Sitterson, *Sugar Country* (Lexington, Ky., 1953), 207–11; William F. Messner, *Freedmen and the Ideology of Free Labor: Louisiana 1862–1865* (Lafayette, La., 1978), 35. Charles P. Roland, *Louisiana Sugar Plantations During the American Civil War* (Leiden, Neth., 1957), 92–98, who also describes these events, states that slaves flocked to federal lines out of "childlike curiosity."

8. Du Bois, *Black Reconstruction*, 81–82; Richardson, ed., *Messages and Papers*, 6:276.

unanimously, the Crittenden Resolution, affirming that the "established institutions" of the seceding states were not to be a military target. Throughout 1861, army commanders ordered their camps closed to fugitive slaves and some actually returned them to their owners, a policy that caused Gov. John A. Andrew to protest: "Massachusetts does not send her citizens forth to become the hunters of men." Yet as the Confederacy set slaves to work as military laborers, and the presence of Union soldiers precipitated large-scale desertion of the plantations, the early policy quickly unraveled. Increasingly, military authorities adopted the plan, inaugurated in Virginia by Gen. Benjamin F. Butler, of designating fugitive slaves as "contraband of war." Instead of being either emancipated or returned to their owners, they would be employed as laborers for the Union armies.[9]

Then, too, an influential segment of the Northern public—abolitionists and Radical Republicans—recognized that secession offered a golden opportunity to strike a fatal blow at slavery. "We have entered upon a struggle," wrote a Massachusetts abolitionist four days after the firing on Fort Sumter, "which ought not to be allowed to end until the Slave Power is completely subjugated, and *emancipation made certain.*" Black abolitionist Frederick Douglass repeatedly called for the liberation and arming of the slaves, insisting from the outset, "The Negro is the key of the situation— the pivot upon which the whole rebellion turns." Carl Schurz, who had fled his native Germany after the abortive revolution of 1848 and emerged as a leading antislavery lecturer in the 1850s, later remarked that emancipation would have come "even if there had not been a single abolitionist in America before the war." But the pressure of antislavery men and women had its impact. With traditional policies unable to produce victory, abolitionists and Radicals offered a coherent analysis of the conflict and a plausible means of weakening the rebellion. Most of all, they kept at the forefront of Northern politics the question of the struggle's ultimate purpose.[10]

The steps by which Congress and the President moved toward abolition have often been chronicled. As the danger of secession by the border states receded, the collapse of slavery accelerated, and the manpower needs of the Union armies increased, pressure mounted for emancipation. In March 1862, Congress enacted an article of war expressly prohib-

9. Herman Belz, *Reconstructing the Union: Theory and Practice During the Civil War* (Ithaca, N.Y., 1969), 19–28; Berlin et al., eds., *Freedom*, Ser. 1, 1:15–16, 353–54; Louis S. Gerteis, *From Contraband to Freedman: Federal Policy Toward Southern Blacks, 1861–1865* (Westport, Conn., 1973), 11–13.

10. Samuel Gridley Howe to Charles Sumner, April 16, 1861, Samuel Gridley Howe Papers, HU; Philip S. Foner, ed., *The Life and Writings of Frederick Douglass* (New York, 1950–55), 3:13, 94; Frederic Bancroft, ed., *Speeches, Correspondence and Political Papers of Carl Schurz* (New York, 1913), 1:232; James M. McPherson, *The Struggle for Equality: Abolitionists and the Negro in the Civil War and Reconstruction* (Princeton, 1964), 59–82. A more extensive discussion of Radical Republicanism will be found in Chapter 6.

iting the army from returning fugitives to their masters. Then came abolition in the District of Columbia (with compensation for loyal owners) and the territories, followed by the Second Confiscation Act, liberating slaves who resided in Union-occupied territory or escaped to Union lines, if their masters were disloyal.

Seeking to hold the political middle ground, even as that ground shifted to the left, Lincoln searched for a formula that would initiate the emancipation process but not alienate conservatives and Southern Unionists. First, he urged the border states to adopt measures for gradual, compensated emancipation, promising generous financial aid from the federal government. But he found no takers, even in tiny Delaware with fewer than 2,000 slaves. In a widely publicized conference, Lincoln urged Northern black leaders to support the colonization of freedmen in Central America or the Caribbean, insisting "there is an unwillingness on the part of our people, harsh as it may be for you colored people to remain with us." Whether his embrace of colonization stemmed from genuine conviction, uncharacteristic naïveté, or political calculation (an attempt to neutralize fears that emancipation would produce an influx of blacks into the free states), Lincoln's plans came to naught. But to the very end of 1862, he held out the possibility of compensation and colonization, raising both ideas in his December message to Congress, and adding a thinly veiled suggestion that Northern states possessed the authority to exclude freedmen from their territory. As late as December, the President signed an agreement with an entrepreneur of dubious character for the settlement of 5,000 blacks on an island off Haiti. (Four hundred hapless souls did in fact reach Île à Vache; those fortunate enough to survive returned to the United States in 1864.)[11]

It is tempting to interpret the evolution of Lincoln's policy as the vacillation of a man desperate to avoid the role history had thrust upon him. This, however, would be unfair, for Lincoln genuinely abhorred slavery. He shared, it is true, many of the racial prejudices of his time and accepted without dissent the racial discriminations so widespread in both sections. But Frederick Douglass, who had encountered racism even within abolitionist ranks, considered Lincoln a fundamentally decent individual. "He treated me as a man," Douglass remarked in 1864, "he did not let me feel for a moment that there was any difference in the color of our skins." It is probably most accurate to say that Lincoln, neither an egalitarian in a modern sense nor a man paralyzed, like so many of his contemporaries, by racial fears and prejudices, did not approach any policy, even emancipation, primarily in terms of its impact upon blacks;

11. V. Jacque Voegeli, *Free but Not Equal: The Midwest and the Negro during the Civil War* (Chicago, 1967), 39, 44–45, 66, 97; Richardson, ed., *Messages*, 6:54, 128, 136–37, 140–41. There is an excellent account of the coming of emancipation in Peter J. Parish, *The American Civil War* (New York, 1975), 226–61.

for him, winning the war always remained paramount. The Emancipation Proclamation itself, with its exemption of Union-held areas, reflected not only Lincoln's effort to make emancipation legally unassailable, but also his determination to retain the backing of the millions of Northerners who cared little about abolition but might support an act essential to military victory.[12]

Most important of all, however, Lincoln understood that the war had created a fluid situation that placed a premium upon flexibility and made far-reaching change inevitable. As Wisconsin Sen. Timothy O. Howe explained in December 1861, change had become the order of the day: "Don't anchor yourself to any policy. Don't tie up to any platform. The very foundations of the Government are cracking. . . . No mere policy or platform can outlast this storm." The Proclamation represented a turning point in national policy as well as in the character of the war. For the first time tying Union success to abolition—a commitment from which Lincoln never retreated—it ignored entirely both compensation and colonization, and for the first time authorized the large-scale enlistment of black soldiers. In effect, it transformed a war of armies into a conflict of societies, ensuring that Union victory would produce a social revolution within the South. In such a struggle, compromise was impossible; the war must now continue until the unconditional surrender of one side or the other. Even in areas exempted from the Proclamation, the Union Army henceforth acted as a liberating force. Indeed, a federal army officer in Tennessee flatly declared in 1863: "Slavery is dead; that is the first thing. That is what we all begin with here, who know the state of affairs." In December 1861 Lincoln had admonished Congress that the Civil War must not degenerate into "a violent and remorseless revolutionary struggle." The Emancipation Proclamation announced that this was precisely what it must become.[13]

Of the Proclamation's provisions, few were more radical in their implications or more essential to breathing life into the promise of emancipation than the massive enrollment of blacks into military service. Preliminary steps had been taken in 1862, since as the army moved into the South, it required a seemingly endless stream of laborers to construct fortifications and additional soldiers to guard its ever-lengthening supply lines. The reservoir of black manpower could not be ignored, but it was

12. John Eaton, *Grant, Lincoln and the Freedmen* (New York, 1907), 175; Roy P. Basler, ed., *The Collected Works of Abraham Lincoln* (New Brunswick, 1953), 6:428–29. For a sympathetic portrait of Lincoln's progress toward emancipation, see LaWanda Cox, *Lincoln and Black Freedom: A Study in Presidential Leadership* (Columbia, S.C., 1981), 3–7, 22–36; for a less favorable view, George M. Fredrickson, "A Man but not a Brother: Abraham Lincoln and Racial Equality," *JSH*, 41 (February 1975), 39–58.

13. Timothy O. Howe to James H. Howe, December 31, 1861, Timothy O. Howe Papers, SHSW; Stephen V. Ash, *Middle Tennessee Society Transformed, 1860–1870: War and Peace in the Upper South* (Baton Rouge, 1987), Chapter 6; Richardson, ed., *Messages and Papers*, 6:54.

only with the Emancipation Proclamation that the enlistment of blacks began in earnest. Massachusetts Governor Andrew commissioned a group of prominent black abolitionists to tour the North for recruits, and other Northern governors quickly followed suit. In the South, especially in the Mississippi Valley under the direction of Gen. Lorenzo Thomas, former slaves by the thousands were enlisted. By the war's end, some 180,000 blacks had served in the Union Army—over one fifth of the nation's adult male black population under age forty-five. The highest percentage originated in the border states, where enlistment was, for most of the war, the only route to freedom. Nearly 60 percent of eligible Kentucky blacks served in the armed forces. Here, military service pushed the Union's commitment to abolition beyond the terms of the Proclamation to embrace, first, black soldiers, and, shortly before the war's end, their families as well. Well before its legal demise, slavery in the border states had been fatally undermined by the enlistment of black men in the army.[14]

Within the army, black soldiers were anything but equal to white. Organized into segregated regiments, they often found themselves subjected to abuse from white officers. Initially, black enlistment was intended to free whites for combat; accordingly, black recruits received less pay than white and were assigned largely to fatigue duty, construction work, and menial labor, with few opportunities to demonstrate their martial talents. Even after proving themselves in battle, blacks could not advance into the ranks of commissioned officers until 1865. In the end, only about 100 (including chaplains and surgeons) obtained commissions.[15]

Nonetheless, black soldiers played a crucial role not only in winning the Civil War, but in defining the war's consequences. Their service helped transform the nation's treatment of blacks and blacks' conception of themselves. The "logical result" of their military service, one Senator observed in 1864, was that "the black man is henceforth to assume a new *status* among us." For the first time in American history, large numbers of blacks were treated as equals before the law—if only military law. In army courts blacks could testify against whites (something unheard of throughout the South and in many Northern states), and former slaves for the first time saw the impersonal sovereignty of the law supersede the personal authority of a master. The galling issue of unequal pay sparked a movement that familiarized former slaves with the process of petition and protest, and resulted in a signal victory when Congress in 1864

14. Berlin et al., eds., *Freedom*, Ser. 2, 1–15, 116–26, 185, 191–97; Mary F. Berry, *Military Necessity and Civil Rights Policy: Black Citizenship and the Constitution, 1861–1868* (Port Washington, N.Y., 1977), 41–57. A series of military orders and acts of Congress in 1864 and early 1865 extended freedom to all blacks who enrolled in the army from the border states, and in March 1865 Congress emancipated their families.

15. Berlin et al., eds., *Freedom*, Ser. 2, 1, 40, 303–12, 483–87.

enacted a measure for equality in pay, bounties, and other compensation. It was in the army that large numbers of former slaves first learned to read and write, either from teachers employed by Northern aid societies or in classrooms and literary clubs established and funded by the soldiers themselves. "A large portion of the regiment have been going to school during the winter months," wrote a black sergeant from Virginia in March 1865. "Surely this is a mighty and progressive age in which we live."[16]

From Oliver Cromwell's New Model Army to the militias raised during the American Revolution to guerrilla armies of our own day, military service has often been a politicizing and radicalizing experience. Lincoln's "thinking bayonets" (his term for Union soldiers) debated among themselves the issues of war and emancipation. As the army penetrated the heart of the Deep South and encountered the full reality of plantation slavery, soldiers became imbued with abolition sentiment. "Since I am here," one Democratic colonel wrote from Louisiana, "I have learned and seen . . . what the horrors of slavery was. . . . Never hereafter will I either speak or vote in favor of slavery." For black troops, particularly the vast majority just emerging from slavery, the army's impact was especially profound. "No negro who has ever been a soldier," wrote a Northern official in 1865, "can again be imposed upon; they have learnt what it is to be free and they will infuse their feelings into others." Black troops flaunted their contempt for symbols of bondage, and relished the opportunity to exert authority over Southern whites. One soldier celebrated his ability to walk "fearlessly and boldly through the streets [of New Orleans] . . . without being required to take off his cap at every step." Another, recognizing his former master among a group of military prisoners, exclaimed: "Hello massa; bottom rail top dis time!"[17]

For black soldiers, military service meant more than the opportunity to help save the Union, more even than their own freedom and the destruction of slavery as an institution. For men of talent and ambition, the army flung open a door to advancement and respectability. From the army would come many of the black political leaders of Reconstruction, including at least forty-one delegates to state constitutional conventions, sixty-four legislators, three lieutenant governors, and four Congressmen. One group of discharged black soldiers formed "the Council" after the war to

16. Herman Belz, *A New Birth of Freedom: The Republican Party and Freedmen's Rights 1861–1866* (Westport, Conn., 1976), 24; Berlin et al., eds., *Freedom,* Ser. 2, 28–29, 362–68, 433–42, 611–13; Berry, *Military Necessity,* 62–74; John W. Blassingame, "The Union Army as an Educational Institution for Negroes," *Journal of Negro Education,* 34 (Summer 1965), 152–59; Cam Walker, "Corinth: The Story of a Contraband Camp," *CWH,* 20 (March 1974), 15; *Christian Recorder,* March 18, 1865.
17. Harold M. Hyman and William M. Wiecek, *Equal Justice Under Law: Constitutional Development 1835–1875* (New York, 1982), 244; Frank Byrne and Jean P. Soman, eds., *Your True Marcus: The Civil War Letters of a Jewish Colonel* (Kent, Ohio, 1985), 315–16; George D. Reynolds to Stuart Eldridge, October 5, 1865, Registered Letters Received, Ser. 2054, Ms. Asst. Comr., RG 105, NA [FSSP A-9074]; Litwack, *Been in the Storm,* 96–102.

collect information on the condition of Louisiana's freedmen, "look after their contracts" with white employers, and explain their legal rights. In time, the black contribution to the Union war effort would fade from the nation's collective memory, but it remained a vital part of the black community's sense of its own history. "They say," an Alabama planter reported in 1867, "the Yankees never could have whipped the South without the aid of the negroes." Here was a crucial justification for blacks' self-confident claim to equal citizenship during Reconstruction, a claim anticipated in the soldiers' long battle for equal pay during the war. At the Arkansas constitutional convention of 1868, former slave William Murphey held his silence for weeks, in deference to more accomplished white delegates (who, he pointed out, "have obtained the means of education by the black man's sweat"). But when some of these delegates questioned blacks' right to the suffrage, Murphey felt compelled to protest: "Has not the man who conquers upon the field of battle, gained any rights? Have we gained none by the sacrifice of our brethren?"[18]

The Emancipation Proclamation and the presence of black troops ensured that, in the last two years of the war, Union soldiers acted as an army of liberation. As the Civil War drew to a close, the disintegration of slavery accelerated, even as masters clung tenaciously to the institution. Early in 1865, slaves were still being bought and sold in areas as yet unoccupied by Northern troops. By then, however, more than 1 million blacks were within Union lines inside the Confederacy, and another 700,-000 lived in states of the border and Upper South where slavery was dead or dying. Even in Confederate territory, planters were negotiating wage and share agreements to induce their increasingly recalcitrant laborers to return to the fields. Seemingly minor incidents told of slavery's death throes, for example, the Mississippi black who responded early in 1865 to a planter's salutation "Howdy, Uncle" with an angry "Call me Mister." By 1865, no matter who won the Civil War, slavery was doomed.[19]

For upholders of the South's "peculiar institution," the Civil War was a terrible moment of truth. The most perceptive among them suddenly realized they had never really known their slaves at all. "I believed that these people were content, happy, and attached to their masters," South Carolina rice planter A. L. Taveau confessed two months after the war's close. But if this were the case, why did the slaves desert their masters

18. 46th Congress, 2d Session, Senate Report 693, 75–76; John H. Parrish to Henry Watson, Jr., June 20, 1867, Henry Watson, Jr. Papers, DU; *Debates and Proceedings of the Convention which Assembled at Little Rock, January 7, 1868* . . . (Little Rock, 1868), 629. The black Congressmen who served in the Union armed forces were Charles E. Nash, Robert Smalls, Josiah T. Walls, and John R. Lynch, the latter an army cook. Sen. Hiram Revels helped recruit black soldiers.

19. W. McKee Evans, *Ballots and Fence Rails: Reconstruction on the Lower Cape Fear* (Chapel Hill, 1967), 20–21; Litwack, *Been in the Storm*, 36; Gerteis, *From Contraband to Freedman*, 193; P. L. Rainwater, ed., "Letters of James Lusk Alcorn," *JSH*, 3 (May 1937), 207.

"in [their] moment of need and flock to an enemy, whom they did not know?" Blacks, Taveau now understood, had, for generations, been "looking for the Man of Universal Freedom."[20]

The Inner Civil War

Like a massive earthquake, the Civil War and the destruction of slavery permanently altered the landscape of Southern life, exposing and widening fault lines that had lain barely visible, just beneath the surface. White society was transformed no less fully than black, as traditional animosities grew more acute, long-standing conflicts acquired altered meanings, and new groups emerged into political consciousness.

From the earliest days of settlement, there had never been a single white South, and in the nineteenth century the region as a whole, and each state within it, was divided into areas with sharply differing political economies. The plantation belt, which encompassed the South's most fertile lands, supported a flourishing agriculture integrated into the world market for cotton, rice, sugar, and tobacco. It contained the majority of slaves, as well as the planters who dominated Southern society and politics and commanded most of the region's wealth and economic resources. A larger number of white Southerners lived in the upcountry, an area of small farmers and herdsmen who owned few or no slaves and engaged largely in mixed and subsistence agriculture. The upcountry itself encompassed the Piedmont, where slavery was a significant presence, and the mountains and hill country, where small communities of white families lived in frontier conditions, isolated from the rest of the South. Although tenancy was by no means unknown in the upcountry and the high price of cotton in the 1850s drew some Piedmont counties into staple production, market relations only partly dominated the upcountry economy. Self-sufficiency remained the primary goal of the area's farm families, a large majority of whom owned their land and worked it with their own labor, without resort to slaves or wageworkers. Most families owned a few head of cattle, sheep, and hogs, which, under the South's open-range system, roamed freely on both public and private land and supplied a major part of food requirements. Little currency circulated, barter was common, and upcountry families dressed in "home-spun cloth, the product of the spinning wheel and hand loom." This economic order gave rise to a distinctive subculture that celebrated mutuality, egalitarianism (for whites), and proud independence.[21]

20. Eugene D. Genovese, *Roll, Jordan, Roll: The World the Slaves Made* (New York, 1974), 112; Joel Williamson, *After Slavery: The Negro in South Carolina During Reconstruction, 1861–1877* (Chapel Hill, 1965), 70. Taveau wrote several versions of this letter, one of which was published in the New York *Tribune*, June 10, 1865.

21. Gavin Wright, *The Political Economy of the Cotton South* (New York, 1978), 34; Steven Hahn, *The Roots of Southern Populism: Yeoman Farmers and the Transformation of the Georgia*

Within the South, state borders did not coincide with lines of economic specialization. The Appalachian South, a vast mountain region of extraordinary beauty, with towering peaks and lush verdant valleys, stretched from western Virginia through parts of Kentucky, Tennessee, North Carolina, Georgia, and Alabama. The cotton kingdom, dominated by slave plantations, reached from the Carolinas southwest into Louisiana and eastern Texas. Virginia, with the largest slave population of any state, also contained a mountainous panhandle tied economically to Pennsylvania, Ohio, and Maryland. Tennessee had a western region with numerous cotton plantations, a middle section with prosperous medium-sized farms growing corn and livestock for the market, and a large mountainous area to the east, with small subsistence-oriented farms and few slaves. In Alabama, one plantation region, in the Tennessee River valley, was separated by mountains from another, the "black belt" (so named originally for its dark, fertile soil), whose ten counties contained half the state's slave population. To the south of the black belt lay the wiregrass and piney woods, peopled mainly by poor whites. There was little economic intercourse between northern and southern Alabama, for no railroad pierced the mountains; the planters and farmers of northern Alabama traded with Nashville, Louisville, and Cincinnati, while the black belt lay within the economic orbit of Mobile and New Orleans.[22]

Southern politics had long reflected these intrastate divisions. There was persistent debate about the apportionment of state legislatures, with yeoman regions like western North Carolina demanding a "white basis" while planters insisted upon counting at least some slaves. But these conflicts rarely challenged planters' domination of state politics and almost never called into question slavery itself. Many small farmers (nearly half in both Mississippi and South Carolina) owned a slave or two, and even in the mountains, slavery was "firmly entrenched" among a small but influential local elite: the few large-scale farmers, professional men, merchants, and small-town entrepreneurs. Outside the plantation belt, however, the majority of yeomen had little economic stake in the institution and "were not dazzled and blinded by the overpowering influence" of the planter class, as an East Tennessee writer observed after the war. Yet slavery affected society everywhere in the South, and even mountaineers shared many attitudes with the planters, beginning with a commitment to white supremacy. The decentralized structure of Southern

Upcountry, 1850–1890 (New York, 1983), 15–85; Henry W. Warren, *Reminiscences of a Mississippi Carpetbagger* (Holden, Mass., 1914), 29; Samuel H. Lockett, *Louisiana As It Is*, edited by Lauren C. Post (Baton Rouge, 1969), 46–47; John S. Otto, "Southern 'Plain Folk' Agriculture," *Plantation Society in the Americas*, 2 (April 1983), 30–34.

22. Allison G. Freehling, *Drift Toward Dissolution: The Virginia Slavery Debate of 1831–1832* (Baton Rouge, 1982), 34, 242; Blanche H. Clark, *The Tennessee Yeoman 1840–1860* (Nashville, 1942), 8; Peter Kolchin, *First Freedom: The Responses of Alabama's Blacks to Emancipation and Reconstruction* (Westport, Conn., 1972), 12–14.

politics accorded upcountry yeomen authority over their local affairs, and the low rate of antebellum Southern taxation meant that planter-dominated state governments did not appear as a financial burden. So long as slavery and planter rule did not interfere with the yeomanry's self-sufficient agriculture and local independence, the latent class conflict among whites failed to find coherent expression.[23]

It was in the secession crisis and Civil War that large numbers of upcountry yeomen discovered themselves as a political class. The elections for delegates to secession conventions in the winter of 1860–61 produced massive repudiations of disunion in yeoman areas. Once the war had begun, most of the upcountry rallied to the Confederate cause. But from the outset, disloyalty was rife in the Southern mountains. Its western counties seceded from Virginia in 1861 and two years later reentered the Union as a separate state. In East Tennessee, long conscious of its remoteness from the rest of the state, supporters of the Confederacy formed a small minority, except among "the wealthier and more cultured class." A convention of Unionists in June 1861 called for the region's secession from the state (an idea dating back to the proposed state of Franklin in the 1780s), but since East Tennessee, unlike West Virginia, was not occupied by Union armies until well into the war, it never achieved political autonomy.[24]

East Tennessee would remain the most conspicuous example of discontent within the Confederacy. From this area of bridge burning and other acts of armed resistance, thousands of men made their way through the mountains to enlist in the Union army. But other mountain counties also rejected secession from the outset. One citizen of Winston County in the northern Alabama hill country believed yeomen had no business fighting for a planter-dominated Confederacy: "All tha want is to git you . . . to fight for their infurnal negroes and after you do their fightin' you may kiss their hine parts for o tha care." On July 4, 1861, a convention of 3,000 residents voted to take Winston out of the Confederacy. Mountainous Rabun County, Georgia, was "almost a unit against secession," and secret Union societies flourished in the Ozark mountains of northern Arkansas, from which 8,000 men eventually joined the federal army.[25]

23. Armstead L. Robinson, "Beyond the Realm of Social Consensus: New Meanings of Reconstruction for American History," *JAH*, 68 (September 1981), 280–81, 287; John C. Inscoe, "Mountain Masters: Slaveholding in Western North Carolina," *NCHR*, 61 (April 1984), 143–73; John M. Price, "Slavery in Winn Parish," *LaH*, 8 (Spring 1967), 137–49; Oliver P. Temple, *East Tennessee and the Civil War* (Cincinnati, 1899), 554; Hahn, *Roots of Southern Populism*, 89, 107–15; Gordon B. McKinney, *Southern Mountain Republicans: 1865–1900* (Chapel Hill, 1978), 12–14.

24. Paul Escott, "Southern Yeomen and the Confederacy," *SAQ*, 77 (Spring 1978), 147–49; J. G. de Roulhac Hamilton, ed., *The Papers of Randolph Abbott Shotwell* (Raleigh, 1926–36), 2:265; Charles F. Bryan, Jr., "A Gathering of Tories: The East Tennessee Convention of 1861," *THQ*, 39 (Spring 1980), 27–48.

25. E. Merton Coulter, *William G. Brownlow: Fighting Parson of the Southern Highlands* (Chapel Hill, 1937), 166–75; Virginia Hamilton, *Alabama* (New York, 1977), 24–25; Wesley S. Thompson, *The Free State of Winston: A History of Winston County, Alabama* (Winfield, Ala.,

Discontent developed more slowly outside the mountains, with their cohesive communities of intense local loyalties, where slaves comprised only a tiny fraction of the population. It was not simply devotion to the Union, but the impact of the war and the consequences of Confederate policies, that awakened peace sentiment and social conflict. In any society, war demands sacrifice, and public support often rests on the conviction that sacrifice is equitably shared. But the Confederate government increasingly molded its policies in the interest of the planter class. Slavery, Confederate Vice President Alexander H. Stephens proudly affirmed, was the cornerstone of the Confederacy. Accordingly, slavery's disintegration compelled the Confederate government to take steps to preserve the institution, and these policies, in turn, sundered white society.

The impression that planters were not bearing their fair share of the war's burdens spread quickly in the upcountry. Committed to Southern independence, most planters were also devoted to the survival of plantation slavery, and when these goals clashed, the latter often took precedence. After a burst of Confederate patriotism in 1861, increasing numbers of planters resisted calls for a shift from cotton to food production, even as the course of the war and the drain of manpower undermined the subsistence economy of the upcountry, threatening soldiers' families with destitution. When Union forces occupied New Orleans in 1862 and extended their control of the Mississippi Valley in 1863, large numbers of planters, merchants, and factors salvaged their fortunes by engaging in cotton traffic with the Yankee occupiers. In few was self-interest as unalloyed as James L. Alcorn (Mississippi's future Republican governor), who, after a brief stint in the Southern army, retired to his plantation, smuggled contraband cotton into Northern hands, and invested the profits in land and Union currency. But it was widely resented that, as a Richmond newspaper put it, many "rampant cotton and sugar planters, who were so early and furiously in the field for secession," quickly took oaths of allegiance during the war and resumed raising cotton "in partnership with their Yankee *protectors.*" Other planters resisted the impressment of their slaves to build military fortifications and, to the end, opposed calls for the enlistment of blacks in the Confederate army, afraid, an Alabama newspaper later explained, "to risk the loss of their property."[26]

1968), 3–4; G. M. Netherland to James Johnson, July 22, 1865, Georgia Governor's Papers, UGa; Ted R. Worley, "The Arkansas Peace Society of 1861: A Study in Mountain Unionism," *JSH,* 24 (November 1958), 445–55; Carl N. Degler, *The Other South: Southern Dissenters in the Nineteenth Century* (New York, 1974), 169–74.

26. Roark, *Masters Without Slaves,* 42–45; Armstead Robinson, *Bitter Fruits of Bondage: The Demise of Slavery and the Collapse of the Confederacy* (forthcoming), Chapters 2, 4; Harold D. Woodman, *King Cotton and His Retainers: Financing and Marketing the Cotton Crop of the South, 1800–1925* (Lexington, Ky., 1968), 200, 213–35; Wiley, *Southern Negroes,* 121; Mobile *Register,* January 26, 1871.

Even more devastating for upcountry morale, however, were policies of the Confederate government. The upcountry became convinced that it bore an unfair share of taxation; it particularly resented the tax-in-kind and the policy of impressment that authorized military officers to appropriate farm goods to feed the army. Planters, to be sure, now paid a higher proportion of their own income in taxes than before the war, but they suffered far less severely from such seizures, which undermined the yeomanry's subsistence agriculture. During the war, poverty descended upon thousands of upcountry families, especially those with men in the army. Food riots broke out in Virginia and North Carolina, and in Randolph County, Alabama, crowds of women seized government stores of corn "to prevent starvation of themselves and families." But, above all, it was the organization of conscription that convinced many yeomen that the struggle for Southern independence had become "a rich man's war and a poor man's fight." Beginning in 1862, the Confederacy enacted the first conscription laws in American history, including provisions that a draftee could avoid service by producing a substitute, and that one able-bodied white male would be exempted for every twenty slaves. This "class legislation" was deeply resented in the upcountry, for the cost of a substitute quickly rose far beyond the means of most white families, while the "twenty Negro" provision, a response to the decline of discipline on the plantations, allowed many overseers and planters' sons to escape military service. Even though this provision was subsequently repealed, conscription still bore more heavily on the yeomanry, which depended on the labor of the entire family for subsistence, than on planter families supported by the labor of slaves.[27]

In large areas of the Southern upcountry, initial enthusiasm for the war was succeeded by disillusionment, draft evasion, and eventually outright resistance to Confederate authority—a civil war within the Civil War. Beginning in 1863, desertion became a "crying evil" for the Confederate Army. By war's end, more than 100,000 men had deserted, almost entirely, one officer observed, from among "the poorest class of non slaveholders whose labor is indispensable to the daily support of their families." In the hill counties and piney woods of Mississippi, bands of deserters hid from Confederate authorities, and organizations like Choctaw County's Loyal League worked to "break up the war by advising desertion, robbing the families of those who remained in the army, and keeping the Federal authorities advised" of Confederate military movements. Northern Alabama, generally enthusiastic about the Confederacy

27. Berlin et al., eds., *Freedom*, Ser. 1, 1:663–82, 757; Paul D. Escott, *After Secession: Jefferson Davis and the Failure of Confederate Nationalism* (Baton Rouge, 1978), 94–168; Peter Wallenstein, *From Slave South to New South: Public Policy in Nineteenth-Century Georgia* (Chapel Hill, 1987), 117–18, 125–27; Robinson, *Bitter Fruits of Bondage*, Chapters 4–5.

in 1861, was the scene two years later of widespread opposition to conscription and the war. And in western and central North Carolina, whose white inhabitants also supported the Confederacy at the outset, Unionists organized the Heroes of America, numbering perhaps 10,000 men, who established an "underground railroad" to enable Unionists to escape to federal lines. Alexander H. Jones, a Hendersonville newspaper editor and leader of the Heroes of America, secretly printed an address that expressed pointedly the class resentment rising to the surface of southern life:

> This great national strife originated with men and measures that were . . . opposed to a democratic form of government. . . . The fact is, these *bombastic, highfalutin* aristocratic fools have been in the habit of driving negroes and poor helpless white people until they think . . . that they themselves are superior; [and] hate, deride and suspicion the poor.[28]

More than ever before, the Southern upcountry was divided against itself between 1861 and 1865. Yeomen supplied both the bulk of Confederate soldiers and the majority of deserters and draft resisters. Lying at the war's strategic crossroads, portions of upcountry Tennessee, Alabama, and Mississippi were laid waste by the march of opposing armies. In other areas, marauding bands of deserters plundered the farms and workshops of Confederate sympathizers, driving off livestock and destroying crops, while Confederate troops and vigilantes routed Union families from their homes. In this internal civil war, atrocities were committed on both sides, but since the bulk of the upcountry remained within Confederate lines for most of the war, Unionists suffered more intensely. In East Tennessee, hundreds were imprisoned by military tribunals and their property seized "to the total impoverishment of the sufferers." In a remote valley in Appalachian North Carolina, Confederate soldiers murdered thirteen Unionist prisoners in cold blood. Throughout the upcountry, Unionists abandoned their homes to hide from conscription officers and Confederate sheriffs, who hunted them, as they had once hunted runaway slaves, with bloodhounds; some found refuge in the very mountain caves that had once sheltered fugitives from bondage.[29]

28. Richard Reid, "A Test Case of the 'Crying Evil': Desertion Among North Carolina Troops During the Civil War," *NCHR*, 58 (Summer 1981), 234–62; William T. Blain, "Banner Unionism in Mississippi, Choctaw County 1861–1869," *Mississippi Quarterly*, 29 (September 1976), 209–13; Bessie Martin, *Desertion of Alabama Troops from the Confederate Army* (New York, 1932), 43–51, 107–16, 150–51; W. J. Brantley to Benjamin G. Humphreys, August 22, 1866, Mississippi Governor's Papers, MDAH; Richard A. McLemore, ed., *A History of Mississippi* (Hattiesburg, 1973), 1:519–25; Durwang Long, "Unanimity and Disloyalty in Secessionist Alabama," *CWH*, 11 (September 1965), 268; William T. Auman and David D. Scarboro, "The Heroes of America in Civil War North Carolina," *NCHR*, 58 (Autumn 1981), 327–63; Alexander H. Jones, *Knocking at the Door* (Washington, 1866), 7–21.
29. Emory M. Thomas, *The Confederate Nation 1861–1865* (New York, 1979), 226; Martin, *Alabama Troops*, 153–90; McKinney, *Southern Mountain Republicans*, 19–24; William G. Brownlow to Andrew Johnson, August 7, 1865, Tennessee Governor's Papers, TSLA; Philip S.

For Southerners loyal to the Union, "the war had been a hell that they would not forget." Long after the end of fighting, bitter memories of persecution would remain, and tales would be told and retold of the fortitude and suffering of Union families. "We could fill a book with facts of wrongs done to our people. . . ." an Alabama Unionist told a Congressional committee in 1866. "You have no idea of the strength of principle and devotion these people exhibited towards the national government." A Tennessean recounted the same history: "They were driven from their homes . . . persecuted like wild beasts by the rebel authorities, and hunted down in the mountains; they were hanged on the gallows, shot down and robbed. . . . Perhaps no people on the face of the earth were ever more persecuted than were the loyal people of East Tennessee."[30]

Thus the war permanently redrew the economic and political map of the white South. Military devastation and the Confederacy's economic policies plunged much of the upcountry into poverty, thereby threatening the yeomanry's economic independence and opening the door to the postwar spread of cotton cultivation and tenancy. Yeoman disaffection shattered the political hegemony of the planter class, separating "the lower and uneducated class," according to one Georgia planter, "from the more wealthy and more enlightened portion of our population." The war ended the upcountry's isolation, weakened its localism, and awakened its political self-consciousness. Out of the Union opposition would come many of the most prominent white Republican leaders of Reconstruction. The party's Southern governors would include Edmund J. Davis, who during the war raised the First Texas Cavalry for the Union Army, William W. Holden, whose unsuccessful 1864 peace campaign for governor of North Carolina became the backbone of white Republicanism in the state, and William G. Brownlow, a circuit-riding Methodist preacher and Knoxville editor, who personified the bitter hatred for rebels so pervasive in East Tennessee. (He would, Brownlow wrote in 1864, arm "every wolf, panther, catamount, and bear in the mountains of America . . . every rattlesnake and crocodile . . . every devil in Hell, and turn them loose upon the Confederacy" in order to win the war.)[31]

The South's inner civil war bequeathed to Reconstruction explosive political issues, unresolved questions, and broad opportunities for

Paludan, *Victims: A Civil War Story* (Knoxville, 1981), 84–98; David Dodge, "The Cave Dwellers of the Confederacy," *Atlantic Monthly*, 68 (October 1891), 514–21; Warren, *Reminiscences*, 40.

30. Carl H. Moneyhon, *Republicanism in Reconstruction Texas* (Austin, 1980), 18; 39th Congress, 1st Session, House Report 30, pt. 3:14, pt. 1:115.

31. William H. Stiles to Elizabeth Stiles, July 1, 1865, Mackey-Stiles Papers, UNC, James A. Baggett, "Origins of Upper South Scalawag Leadership," *CWH*, 29 (March 1983), 65–69; James A. Baggett, "Origins of Early Texas Republican Party Leadership," *JSH*, 40 (August 1974), 447–49; Otto H. Olsen, "North Carolina: An Incongruous Presence," in Otto H. Olsen, ed., *Reconstruction and Redemption in the South* (Baton Rouge, 1980), 157–59; James W. Patton, *Unionism and Reconstruction in Tennessee 1860–1869* (Chapel Hill, 1934), 56–57.

change. Regions like East Tennessee and western North Carolina and individual counties in the hill country of other states would embrace the Republican party after the Civil War and remain strongholds well into the twentieth century. Their loyalty first to the Union and then to Republicanism did not, however, imply abolitionist sentiment during the war (although they were perfectly willing to see slavery sacrificed to preserve the Union) or a commitment to the rights of blacks thereafter. Upcountry Unionism was essentially defensive, a response to the undermining of local autonomy and economic self-sufficiency rather than a coherent program for the social reconstruction of the South. Its basis, related Northern reporter Sidney Andrews in the fall of 1865, was "hatred of those who went into the Rebellion" and of "a certain ruling class" that had brought upon the region the devastating impact of the Civil War.[32]

The North's Transformation

For the Union as well as the Confederacy, the Civil War was a time of change. The Northern states did not experience a revolution as far-reaching as emancipation, but no aspect of life emerged untouched from the conflict. Reaffirming and strengthening some existing tendencies in Northern life, the war transformed others and laid the groundwork for further change. The policies of a national government whose powers were magnified each year the war continued offered unparalleled economic opportunities to some Northerners while spurring determined opposition among others. As in the South, how Northerners reacted to the war and its consequences reflected prior divisions of class, race, and politics, even as these were themselves reshaped by the conflict.

If economic devastation stalked the South, for the North the Civil War was a time of unprecedented prosperity. Nourished by wartime inflation, the profits of industry boomed, as did income from speculative ventures in the stock and gold markets. Since the outbreak of the war, Massachusetts Sen. Henry Wilson observed in 1867, "the loyal States have accumulated more capital, have added more to their wealth, than during any previous seven years in the history of the country." While all branches of industry prospered, those most closely tied to the war effort expanded most rapidly. Railroads thrived on carrying troops and supplies, and profited from the closing of the Mississippi River. The war's first year, a railroad journal reported, was "one of the most prosperous that has ever been known," and thereafter earnings doubled and redoubled. On a tide of demand from the army, the meat-packing industry boomed; Chicago, the city of railroad and slaughterhouse, experienced

32. Sidney Andrews, *The South Since the War* (Boston, 1866), 111–12.

unprecedented growth in population, construction, banking, and manufacturing. By 1865 it stood unchallenged as the Midwest's preeminent commercial center. The woolen mills of New England and the mid-Atlantic states worked day and night to meet the military's demand for blankets and uniforms, reaping enormous profits. (So did a Philadelphia textile firm noted for "the superiority of its mourning goods.")[33]

Agriculture also flourished, for even as farm boys by the thousands were drawn into the army, the frontier of cultivation pushed westward, with machinery and immigration replacing lost labor. During the war, Iowa furnished 75,000 men to the Union Army and Wisconsin 90,000, yet in both states population, improved acreage, grain production, and farm income continued to grow. Many farmers, as agricultural machinery magnate Cyrus McCormick complained, took advantage of inflation to liquidate mortgages and other debts; they "pursued [their creditors] in triumph and paid them without mercy." McCormick, however, also knew how to take advantage of the war, borrowing large sums in order to hoard raw materials, and buying up farmland and urban real estate with as small a down payment as possible. By 1865 he was Chicago's largest landlord.[34]

"In April, 1861," James A. Garfield told the House of Representatives three years after the end of the war, "there began in this country an industrial revolution . . . as far-reaching in its consequences as the political and military revolution through which we have passed." Garfield exaggerated the war's immediate economic impact, for even before the outbreak of fighting, the North was well on its way toward becoming an industrial region with a widespread factory system, an integrated railroad and telegraph network, and a permanent laboring class. During the war, mechanization and the concentration of capital proceeded apace in some sectors—notably boot and shoe production, meat-packing, and agriculture—and hundreds of new factories were constructed, but most manufacturing remained organized on a small-scale basis. The Northern economy stood poised between the ages of iron and steel, between an older competitive capitalism and the world of large-scale concentrated industry looming on the horizon. Indeed, the war witnessed "the culmination of

33. CG, 40th Congress, 2d Session, 246; Victor S. Clark, History of Manufactures in the United States (New York, 1929 ed.), 3:9–37; Henry C. Hubbart, The Older Middle West (New York, 1936), 218–22; Emerson D. Fite, Social and Industrial Conditions in the North During the Civil War (New York, 1910), 42–44, 151; John H. Keiser, Building for the Centuries: Illinois, 1865 to 1898 (Urbana, Ill., 1977), 152, 182; Wyatt W. Belcher, The Economic Rivalry Between St. Louis and Chicago 1850–1880 (New York, 1947), 139–53; Philip Scranton, Proprietary Capitalism: The Textile Manufacture of Philadelphia (New York, 1983), 279–84.

34. Wayne D. Rasmussen, "The Civil War: A Catalyst of Agricultural Revolution," AgH, 39 (October 1965), 187–95; Clyde O. Ruggles, "The Economic Basis of the Greenback Movement in Iowa and Wisconsin," Proceedings of the Mississippi Valley Historical Association, 6 (1912–13), 142–50; William T. Hutchinson, Cyrus Hall McCormick (New York, 1930–35), 2:112–22; Eugene Lerner, "Investment Uncertainty During the Civil War—a Note on the McCormick Brothers," JEcH, 16 (March 1956), 34–40.

the great age of free and unrestricted enterprise," as thousands of small firms sprang into existence to ride the crest of wartime profits, and even traditional craft production experienced a renewed, albeit brief, lease on life.[35]

Nonetheless, Garfield was not wrong to sense the war's profound economic consequences. Beneath dry statistics of commodity output lie deeper structural changes wrought by the Civil War. The very size of wartime profits permanently affected the scale and financing of industrial enterprise, channeling income into the hands of men prepared to invest in economic expansion. In many industries, the "balance of financial power . . . shifted from the merchant's counting house to the factory office," as the profit boom enabled manufacturers to liquidate long-standing debts, thereby establishing their independence of merchant capital and achieving full control of their own operations. The vast scale of profits, production, marketing, and financing, and the experience of coordinating local enterprises in a common national purpose, taught "daily lessons in the value of organization." The government itself promoted consolidation in the telegraph industry (since instantaneous communication was hampered by the large number of competing firms), and the establishment of the nation's first major monopoly by Western Union in 1866 flowed directly from the war experience. Indeed, throughout the business world, as one contemporary observed, the war "created a new order of ideas. . . . The sudden rise of great fortunes, the . . . concentration of great capitals, private and public, the elevation of speculators and adventurers of every sort to the command of millions of money . . . all these brought new men and new dangers to the front." Men in their twenties and early thirties acquired the skills and the fortunes that would shortly enable them to revolutionize American industry. Philip D. Armour earned millions by supplying beef to the Union armies and helped make Chicago the nation's meat-packing center. Thomas A. Scott perfected modern techniques of railroad management and finance, and John D. Rockefeller, in 1860 a twenty-one-year-old commission merchant, also profited handsomely from government contracts and began uniting refineries into a petroleum empire.[36]

35. Burke A. Hinsdale, ed., *The Works of James Abram Garfield* (Boston, 1882–83), 1:287–88; Thomas C. Cochran, "Did the Civil War Retard Industrialization?" *MVHR*, 48 (September 1961), 197–210; David T. Gilchrist and W. David Lewis, eds., *Economic Change in the Civil War Era* (Greenville, Del., 1965), 1–22, 30, 148–51; Saul Engelbourg, "The Impact of the Civil War on Manufacturing Enterprise," *Business History*, 21 (July 1979), 150–54; Glenn Porter and Harold C. Livesay, *Merchants and Manufacturers* (Baltimore, 1971), 133; Fite, *Social and Industrial Conditions*, 155–60.

36. Porter and Livesay, *Merchants and Manufacturers*, 116–27; Morton Keller, *Affairs of State: Public Life in Late Nineteenth Century America* (Cambridge, Mass., 1977), 11; Simeon E. Baldwin, "Recent Changes in Our State Constitutions," *Journal of Social Science*, 10 (1879), 136; Thomas C. Cochran and William Miller, *The Age of Enterprise: A Social History of Industrial America* (New York, 1942), 115–16; Matthew Josephson, *The Robber Barons* (New

Accelerating the emergence of an American industrial bourgeoisie, the war tied the fortunes of this class to the Republican party and the national state. More was involved than profits wrung from government contracts, for faced with the war's unprecedented financial demands, Congress adopted economic policies that promoted further industrial expansion and permanently altered the conditions of capital accumulation. To mobilize the financial resources of the Union, the government created a national paper currency, an enormous national debt, and a national banking system. To raise funds it increased the tariff and imposed new taxes on nearly every branch of production and consumption. To help compensate for the drain of men into the army, a federal bureau was established to encourage immigration under labor contracts. To promote agricultural development, the Homestead Act offered free lands to settlers on the public domain, and the Land Grant College Act assisted the states in establishing "agricultural and mechanical colleges." And to further consolidate the Union, Congress lavished enormous grants of public land and government bonds upon internal improvements, most notably the transcontinental railroad, which, when completed in 1869, expanded the national market, facilitated the penetration of capital into the West, and heralded the final doom of the Plains Indians.

Not all these measures, of course, proved effective. Contract labor never supplied more than a tiny fraction of the needs of Northern industry, and the economic impact of the tariff and homestead remain a matter of dispute. But taken together, the policies of the Union embodied a spirit of national economic activism unprecedented in the antebellum years. As with emancipation, however, economic policy originated in the crucible of war; instead of reflecting a comprehensive plan or an intention to chart the course of future growth, it aimed first and foremost to mobilize the nation's resources in order to finance the conflict. A few of these measures, notably the Homestead Act and land grants for a transcontinental railroad, had long enjoyed broad support in the North and might well have been enacted even in peacetime, but the majority were inconceivable apart from the war (and, of course, the absence of Southern representatives). Nonetheless, once in place, each had consequences that reverberated throughout Northern society, shifting the balance of economic power, sowing the seeds of future conflict, and bequeathing to the postwar world, as economist David A. Wells noted, "never-ending public discussion."[37]

York, 1934), 32–65, 102; David Hawke, *John D.: The Founding Father* (New York, 1980), 30–32.

37. Leonard P. Curry, *Blueprint for Modern America: Non-Military Legislation of the First Civil War Congress* (Nashville, 1968); Charlotte Erickson, *American Industry and the European Immigrant 1860–1885* (Cambridge, Mass., 1957), 3–11; Edward C. Kirkland, *Industry Comes of Age: Business, Labor and Public Policy, 1860–1897* (New York, 1961), 19; David A. Wells, "The Reform of Local Taxation," *North American Review*, 122 (April 1876), 357–403.

As in the Confederacy, the principle that wartime sacrifices should be equitably shared clashed with the need to accommodate the demands of Northerners who commanded the economic resources indispensable to the war effort. Those who could mobilize these resources benefited most fully from economic legislation, their gains imbued with the aura of patriotism rather than appearing as mere selfishness. Whatever their effect upon individual businesses, the high tariffs, at first regarded as temporary but later formalized in a system of "extreme protection," worked to the benefit of the industrial sector and to the detriment of agriculture. Taxation was highly regressive, with receipts from excise levies on items of consumption like liquor and tobacco far exceeding those generated by the new income tax. The issuance of over $400 million in paper money, legal tender for virtually all public and private debts (an unprecedented exercise of federal authority in a country that had never had a national currency), sparked a rapid increase in prices that inflated profits and reduced real wages, thus redistributing income upward in the social scale. Moreover, to facilitate the sale of its bonds, the government declared the interest tax-exempt and payable in gold. Thanks to a national advertising campaign organized by Jay Cooke, who recruited a small army of agents to market government securities and invoked God, country, and manifest destiny to make their purchase seem a patriotic duty, perhaps 1 million Northerners ended up owning shares in a national debt that by war's end amounted to over $2 billion. But most bonds were held by wealthy individuals and financial institutions, who reaped the windfall from interest paid in gold at a time when depreciating paper money was employed for all other transactions (except customs duties).[38]

In order to create a guaranteed market for these bonds, the federal government established a national banking system that further solidified the existing imbalance of financial power. Negotiated, in effect, between the Lincoln administration and leading Eastern bankers, a series of laws provided for the granting of federal charters, including the right to issue currency, to banks holding specified amounts of bonds. A tax of 10 cents on each dollar effectively ended the printing of money by state-chartered banks. The minimum capital requirement of $50,000 and a proviso barring national banks from holding mortgages on land restricted these institutions to large cities. The system both promoted the consolidation of a national capital market essential to future investment in industry and commerce and placed its control firmly in the hands of Wall Street. As a result, the West and, after the war, the South, found themselves starved

38. F. W. Taussig, *The Tariff History of the United States* (New York, 1931 ed.), 155; Fite, *Social and Industrial Conditions*, 119; Harry N. Scheiber, "Economic Change in the Civil War Era: An Analysis of Recent Studies," *CWH*, 11 (December 1965), 407–409; Irwin Unger, *The Greenback Era: A Social and Political History of American Finance 1865–1897* (Princeton, 1964), 16–17; Kirkland, *Industry Comes of Age*, 20–25.

of capital. Like all the wartime economic policies, the banking and currency measures created a set of unresolved problems that would bedevil national politics for years to come—how to equalize bank note circulation in different regions of the country, how to reduce the disparity between the greenback dollar and the gold dollar and eventually resume specie payments, and how to pay off the massive national debt.[39]

In their unprecedented expansion of federal power and their effort to impose organization upon a decentralized economy and fragmented polity, these measures reflected what might be called the birth of the modern American state. On the eve of the Civil War, the federal government was "in a state of impotence," its conception of its duties little changed since the days of Washington and Jefferson. Most functions of government were handled at the state and local level; one could live out one's life without ever encountering an official representative of national authority. But the exigencies of war created, as Sen. George S. Boutwell later put it, a "new government," with a greatly expanded income, bureaucracy, and set of responsibilities.[40]

Emancipation, ending the personal sovereignty of master over slave, made all Americans equally subject to the authority of the national state. A government whose regular army in 1860 numbered only 16,000 introduced conscription (with the opportunity to avoid service by paying $300 or, as in the South, providing a substitute) and found itself training, equipping, and coordinating the activities of millions of men. The federal budget, amounting to $63 million in 1860, rose to well over $1 billion by 1865. At war's end the federal bureaucracy, with 53,000 employees including new Custom House officials, internal revenue agents, clerks, and inspectors, was the largest employer in the nation. And the generous system of pensions for Union veterans and their dependents inaugurated in 1865 (and amounting by the turn of the century to one third of the government's entire budget) created both a vast new patronage machine for the Republican party and a broad constituency committed to maintaining the integrity of the national state (at least as far as its ability to raise revenue was concerned).[41]

39. Bray Hammond, *Sovereignty and an Empty Purse: Banks and Politics in the Civil War* (Princeton, 1970), 140–42; John A. James, *Money and Capital Markets in Postbellum America* (Princeton, 1978), 27–28; David M. Gische, "The New York City Banks and the Development of the National Banking System," *American Journal of Legal History*, 23 (January 1979), 21–67; George L. Anderson, "Western Attitudes Toward National Banks, 1873–1874," *MVHR*, 33 (September 1936), 205–16; Richard Sylla, "Federal Policy, Banking Market Structure, and Capital Mobilization in the United States, 1863–1913," *JEcH*, 29 (December 1969), 657–81.

40. Hammond, *Sovereignty*, 57; Harold M. Hyman, "Reconstruction and Political-Constitutional Institutions: The Popular Expression," in Harold M. Hyman, ed., *New Frontiers of the American Reconstruction* (Urbana, Ill., 1966), 11; *CR* 43rd Congress, 1st Session, 4116.

41. Peter D. Hall, *The Organization of American Culture, 1700–1900* (New York, 1982), 227; Paul B. Trescott, "Federal Government Receipts and Expenditures, 1861–1875," *JEcH*, 26 (June 1966), 207; U.S. Bureau of the Census, *Historical Statistics of the United States, Colonial*

The Civil War, observed Orestes Brownson, a New England social critic, forced the United States for the first time to confront the implications of its own nationality. "Among nations," Brownson wrote in 1865, "no one has more need of full knowledge of itself, than the United States, and no one has hitherto had less. It has hardly had a distinct consciousness of its own national existence." The war changed all this. Especially after the Emancipation Proclamation clothed national authority with an indisputably moral purpose, Republicans exalted the state's continuing maturation as one of the most salutary of the war's consequences. "The policy of this country," declared Sen. John Sherman in defending the new currency and banking systems, "ought to be to make everything national as far as possible; to nationalize our country so that we shall love our country."[42]

Within Congress, Sherman's broad nationalism was embraced, above all, by the Radical Republicans, whose advocacy of strong federal action against slavery carried over into support for measures—like the tax on notes issued by state banks—that expanded federal authority in other realms. The world, declared the Radical Chicago *Tribune*, commenting on the draft, needed to be shown that the American government "can confidently command the support of its citizens, and make the duty of the individual to the state a debt to be collected." Such views were forcefully echoed among Northern reformers and intellectuals. Temporarily submerged in support for the war effort and the newly empowered national state were the principled individualism of antebellum abolitionists and their alienation from a federal government dominated by the Slave Power. For reformers, the nation had become the "custodian of freedom," and some questioned whether the states deserved continued existence at all. The war vindicated their conviction, itself a product of the slavery controversy, that freedom stood in greater danger of abridgment from local than national authority (a startling reversal of the founding fathers' belief, enshrined in the Bill of Rights, that centralized power posed the major threat to individual liberties). In the full flush of nationalist exuberance, a magazine founded in 1865 by antislavery crusaders and entitled, significantly, *The Nation*, announced in its second number:

> The issue of the war marks an epoch by the consolidation of nationality under democratic forms. . . . This territorial, political, and historical oneness of the nation is now ratified by the blood of thousands of her sons. . . . The

Times to 1970 (Washington, D.C., 1975), 1104; Ari Hoogenboom, *Outlawing the Spoils: A History of the Civil Service Reform Movement 1865–1883* (Urbana, Ill., 1961), 1–5; Richard E. Bensel, *Sectionalism and American Political Development 1880–1980* (Madison, Wis., 1984), 61–67.

42. Orestes Brownson, *The American Republic* (Boston, 1865), 2; Hammond, *Sovereignty*, 333.

prime issue of the war was between nationality one and indivisible, and the loose and changeable federation of independent States.[43]

As never before, the war mobilized the energies of Northern reformers, imbuing their lives with a renewed sense of purpose. Especially among women, it inspired an outpouring of voluntary undertakings. Much of their activity centered on the United States Sanitary Commission, which organized medical relief and other services for soldiers, and the freedmen's aid movement, an attempt to help the government cope with the daunting problem of the destitute blacks liberated from slavery. Some 200,000 women were mobilized by local societies throughout the North, raising money and gathering supplies for soldiers, and sending books, clothing, and food to the freedmen. Although men controlled these organizations at the highest levels, the war years inculcated among these women a heightened interest in public events and a sense of independence and accomplishment, while also offering training in organization. By the end of the war, the freedmen's aid movement had sent hundreds of thousands of dollars, and more than 1,000 teachers, most of them women, had taught over 200,000 Southern black pupils. The experience left a complex legacy for postwar women's reform. It produced a generation of leaders—like Annie Wittenmyer of Iowa, subsequently the first president of the Women's Christian Temperance Union, and Josephine Shaw Lowell, later prominent in various New York charities—who believed women should direct their energy toward social welfare activities. At the same time, even though the prewar agitation for the suffrage came to a virtual standstill, the war enlarged the ranks of women who resented their legal and political subordination to men and believed themselves entitled to the vote in recognition of their contribution to Union victory and the end of slavery.[44]

If the war opened doors of opportunity for women, it held out hope for an even more radical transformation in the condition of the tiny, despised black population of the free states. Numbering fewer than a quarter million in 1860, blacks comprised less than 2 percent of the North's population, yet they found themselves subjected to discrimina-

43. Allan G. Bogue, *The Earnest Men: Republicans of the Civil War Senate* (Ithaca, N.Y., 1981), 313–25; Chicago *Tribune*, August 5, 1862; George M. Fredrickson, *The Inner Civil War: Northern Intellectuals and the Crisis of the Union* (New York, 1965); Harold M. Hyman, *A More Perfect Union: The Impact of the Civil War and Reconstruction on the Constitution* (New York, 1973), 382; *Nation*, July 13, 1865.

44. Robert H. Bremner, *The Public Good: Philanthropy and Welfare in the Civil War Era* (New York, 1980), 37–54, 94–96; Ronald E. Butchart, *Northern Schools, Southern Blacks, and Reconstruction: Freedmen's Education, 1862–1875* (Westport, Conn., 1980), 3–6; McPherson, *Struggle for Equality*, 171–72; Rejean Attie, "'A Swindling Concern': The United States Sanitary Commission and the Northern Female Public, 1861–1865" (unpub. diss., Columbia University, 1987); Steven M. Buechler, *The Transformation of the Woman Suffrage Movement: The Case of Illinois, 1850–1920* (New Brunswick, 1986), 4–5, 65; Ruth A. Gallagher, "Annie Tyler Wittenmyer," *Iowa Journal of History and Politics*, 29 (October 1931), 518–69.

tion in every aspect of their lives. Barred in most states from the suffrage, schools, and public accommodations, confined by and large to menial occupations, living in the poorest, unhealthiest quarters of cities like New York, Philadelphia, and Cincinnati, reminded daily of the racial prejudice that seemed as pervasive in the free states as in the slave, many Northern blacks had by the 1850s all but despaired of ever finding a secure and equal place within American life. Indeed, the political conflict between free and slave societies seemed to deepen racial anxieties within the North. The rise of political antislavery in the 1840s and 1850s was accompanied by the emergence of white supremacy as a central tenet of the Northern Democratic party, and by decisions by Iowa, Illinois, Indiana, and Oregon to close their borders entirely to blacks, reflecting the fear that, if slavery weakened, the North might face an influx of black migrants.[45]

The small black political leadership of ministers, professionals, and members of abolitionist societies, had long searched for a means of improving the condition of Northern blacks while at the same time striking a blow against slavery. Most embraced what one historian calls the Great Tradition—an affirmation of Americanism that insisted that blacks formed an integral part of the nation and were entitled to the same rights and opportunities white citizens enjoyed. Free blacks were advised to forsake menial occupations, educate themselves and their children, and live unimpeachably moral lives, thus "elevating" the race, disproving the idea of black inferiority, and demonstrating themselves worthy of citizenship. In the 1850s, however, a growing number of black leaders came to espouse emigration to Africa or the Caribbean, reflecting both an incipient racial nationalism and pessimism about black prospects in this country. Rejecting entirely the Great Tradition, H. Ford Douglass pointedly reminded one black convention that far from being an aberration, a "foreign element" in American life, slavery had received the sanction of the founding fathers and was "completely interwoven into the passions and prejudices of the American people."[46]

The Civil War produced an abrupt shift from the pessimism of the 1850s to a renewed spirit of patriotism and a restored faith in the larger society. Even before the Emancipation Proclamation, a black Californian foresaw the dawning of a new day for his people:

Everything among us indicates a change in our condition, and [we must] prepare to act in a different sphere from that in which we have heretofore acted. . . . Our relation to this government is changing daily. . . . Old things

45. Leon F. Litwack, *North of Slavery: The Negro in the Free States, 1790–1860* (Chicago, 1961).
46. Harding, *There Is a River,* 117–39, 172–94; George A. Levesque, "Boston's Black Brahmin: Dr. John S. Rock," *CWH,* 26 (December 1980), 337–42; Frederick Cooper, "Elevating the Race: The Social Thought of Black Leaders 1827–1850," *American Quarterly,* 24 (December 1972), 604–26.

are passing away, and eventually old prejudices must follow. The revolution has begun, and time alone must decide where it is to end.[47]

Emancipation further transformed the black response to American nationality, dealing the death blow, at least for this generation, to ideas of emigration. Symbolic, perhaps, was the fact that Martin R. Delany, the "father of black nationalism" and an advocate during the 1850s of emigration, recruited blacks for the Union Army and then joined himself. "I am proud to be an American citizen," declared black abolitionist Robert Purvis in May 1863, recalling how, when the federal government was "a slaveholding oligarchy," he had denounced it as "the basest despotism" on earth. Frederick Douglass, throughout his life the most insistent advocate of the Great Tradition, emerged as black America's premier spokesman. Welcomed at the White House, his speeches widely reprinted in the Northern press, Douglass believed his life exemplified how America might move beyond racism to a society founded upon universal human rights. Throughout the war Douglass insisted that the logical and essential corollaries of emancipation were the end of all color discrimination, the establishment of equality before the law, and the enfranchisement of the black population—the "full and complete adoption" of blacks "into the great national family of America."[48]

Meeting in October 1864, a national black convention reflected the optimism rekindled by the Civil War. By then the demise of slavery was far advanced, and black troops had proved themselves on the battlefield. The 145 men who gathered at Syracuse included Richard H. Cain, Francis L. Cardozo, and Jonathan J. Wright, later prominent leaders in Reconstruction South Carolina, and Jonathan C. Gibbs, subsequently Florida's secretary of state. The convention's spirit was very much that of the Great Tradition. Henry Highland Garnet reaffirmed his belief in "Negro nationality," but his was a lonely voice, drowned in a sea of support for "the fundamental principles of this Government" and "acknowledged American ideas." Written by Douglass, the convention's address reflected his priorities—abolition, equality before the law, and suffrage—and the·convention established a National Equal Rights League, with branches in every state, to press for these goals. What little was said on economic matters bore the mark of the antebellum self-help tradition, the convention urging upon the freedmen frugality, "the accumulation of property," and religious and moral self-improvement. One delegate lamented that the freedmen's material interests "had not been sufficiently considered by the Convention."[49]

47. Harding, *There Is a River*, 233–35; David M. Katzman, *Before the Ghetto: Black Detroit in the Nineteenth Century* (Urbana, Ill., 1973), 47; James M. McPherson, ed., *The Negro's Civil War* (New York, 1967), 251–52.

48. Philip S. Foner, ed., *The Voice of Black America* (New York, 1975 ed.), 1:293–94, 318; Foner, ed., *Douglass*, 3:292, 348–52, 396.

49. *Proceedings of National Convention of Colored Men Held in the City of Syracuse, N. Y.* (Boston, 1864), 4–6, 15, 19, 26, 44–62; Harding, *There Is a River*, 246–49.

Nonetheless, the national convention galvanized a black assault upon the Northern color line that, in the war's final months, won some modest but impressive victories. In February 1865, John S. Rock of Boston became the first black lawyer admitted to the bar of the Supreme Court. (Only eight years earlier, in the Dred Scott case, the Court had denied that any black person could be a citizen of the United States.) Slowly, the North's racial barriers began to fall. In 1863, California for the first time permitted blacks to testify in criminal cases; early in 1865 Illinois repealed its laws barring blacks from entering the state, serving on juries, or testifying in court, while Ohio eliminated the last of its discriminatory "black laws." And in May 1865, Massachusetts passed the first comprehensive public accommodations law in American history. In January 1865 the issue of segregated transport became a national cause célèbre when Robert Smalls, a black war hero, was ejected from a Philadelphia streetcar and forced to walk several miles to the navy yard where the *Planter*, the ship he had spirited from Charleston harbor nearly three years earlier, was undergoing repairs. Despite concerted pressure by the city's blacks and white allies, including banker Jay Cooke, integration did not come to Philadelphia transport until 1867, but New York City, San Francisco, Cincinnati, and Cleveland all desegregated their streetcars during the war.[50]

Under the pressure of emancipation, black military service, and Northern black protest, racial prejudice bent, but did not break. No Northern state was added to the five in New England that had allowed blacks to vote on equal terms with whites before the war. Yet for reformers, the initial progress was an augury of the reborn republic they believed was emerging from the war. "Now that God has smitten slavery unto death," pronounced Congregational clergyman Edward Beecher, "He has opened the way for the redemption and sanctification of our whole social system." Essential to this transformation would be some as yet undefined measure of equality before the law for blacks, and the "regeneration" of the South, transplanting there, as Frederick Douglass put it, "the higher civilization of the North."[51]

Republicans had brought into the war an ideology grounded in a conviction of the superiority of free to slave labor, which saw the distinctive

50. Philip S. Foner and George E. Walker, eds., *Proceedings of the Black State Conventions, 1840–1865* (Philadelphia, 1979), 1:164–66, 349, 2:14; Levesque, "Rock," 336; Robert J. Chandler, "Friends in Time of Need: Republicans and Black Civil Rights in California During the Civil War Era," *Arizona and the West*, 11 (Winter 1982), 329; Roger D. Bridges, "Equality Deferred: Civil Rights for Illinois Blacks, 1865–1885," *JISHS*, 74 (Spring 1981), 82–87; Milton R. Konvitz, *A Century of Civil Rights* (New York, 1961), 155–56; Okon E. Uya, *From Slavery to Public Service: Robert Smalls 1839–1915* (New York, 1971), 26–27; Philip S. Foner, "The Battle to End Discrimination Against Negroes on Philadelphia Streetcars," *PaH*, 60 (July 1973), 261–90, (October 1973), 355–79.

51. Reinhold Niebuhr, *The Kingdom of God in America*, (New York, 1959), 157; Foner, ed., *Voice of Black America*, 1:321–24.

quality of Northern society as the opportunity it offered the wage laborer to rise to the status of independent farmer or craftsman. At the war's outset, Lincoln placed the struggle firmly within the familiar context of the free labor ideology. Slavery, he insisted, embodied the idea that the condition of the worker should remain forever the same; in the North, by contrast, there was "no such . . . thing as a free man being fixed for life in the condition of a hired laborer. . . . Men, with their families . . . work for themselves on their farms, in their houses, and in their shops, taking the whole product to themselves, and asking no favors of capital on the one hand nor of hired laborers or slaves on the other." Here was a social vision already being rendered obsolete by the industrial revolution and the appearance of a class of permanent wage laborers. Indeed, even as the war vindicated the free labor ideology, it strengthened tendencies that inexorably transformed the society of small producers from which that ideology had sprung, and undermined the social assumptions on which it rested. "The youngest of us," abolitionist Wendell Phillips wrote in 1864, "are never again to see the republic in which we were born."[52]

Yet the world that spawned the free labor ideology remained close enough in time, and its assumptions authentic enough in the experience of men like Lincoln and the millions of small-town and rural men and women who still made up the majority of the North's population, for the ideology to retain a broad plausibility. Sanctified by the North's triumph, the free labor ideology would emerge from the war further strengthened as a definition of the good society, an underpinning of Republican party policy, and a starting point for discussions of the postwar South. For decades, the antislavery movement had hammered home the free labor indictment of slavery: Freedom meant prosperity and progress; bondage produced only stagnation. Now the prospect opened of an affluent, democratic, free labor South, with small farms replacing the great plantations and Northern capital and migrants energizing the society. "Let us have such a peace," the Chicago *Tribune* exclaimed as the war neared its conclusion, "as shall open up the South forever for free labor." And presiding over the transformation would stand a benevolent and powerful national state. With "the whole continent opened to free labor and Northern enterprise," an abolitionist journal exulted just two days after the Emancipation Proclamation, "the imagination can hardly exaggerate the glory and power of the American republic. Its greatness will overshadow the world."[53]

Thus the Civil War consolidated the national state while identifying

52. Eric Foner, *Free Soil, Free Labor, Free Men: The Ideology of the Republican Party Before the Civil War* (New York, 1970), 11–72; Richardson, ed., *Messages and Papers*, 5:122; *Liberator*, May 20, 1864.
53. Mark M. Krug, *Lyman Trumbull: Conservative Radical* (New York, 1965), 257; *National Anti-Slavery Standard*, January 3, 1863.

that state, via emancipation, with the interests of humanity in general, and, more prosaically, with a coalition of diverse groups and classes. An emerging industrial bourgeoisie, adherents of the Republican party, men and women of the reform milieu, a Northern black community demanding a new status in American life—all these found reason to embrace the changes brought on by the war. But these very developments also galvanized a wartime opposition that reverberated in the postwar world. The enrichment of industrialists and bondholders appeared unfair to workers who saw their real income devastated by inflation. The process of national state formation clashed with cherished traditions of local autonomy and cultural diversity. And even small improvements in the status of Northern blacks, not to mention the vast changes implied by emancipation, stirred ugly counterattacks from advocates of white supremacy.

For large numbers of Northern workers, the war was an economic disaster, as a flood of paper money and the regressive tax system combined to produce a massive decline in real income. Wages did rise, but they did not keep pace with inflation. Skilled laborers, especially in industries tied to government contracts, where high demand produced a temporary labor shortage, made out better than most. As iron magnate Abram Hewitt explained, "The capitalist yielded, and if the government was his customer, he made the government suffer." But for the unskilled and the independent craftsman, real income plummeted. From this situation came the rebirth of a labor movement devastated by the depression of 1857. Multiethnic citywide labor organizations appeared in New York, Chicago, St. Louis, San Francisco, and other cities, and strikes became common even in states like Wisconsin where they had been virtually unknown before the war. In St. Louis, the labor-owned *Daily Press*, founded in 1864 as a result of a printers' strike, enjoyed a circulation of 20,000 and helped organize a local labor party that won over one third of the mayoral vote in 1865.[54]

Generally, however, the Democratic party, the preeminent conservative institution of the era, reaped the political harvest of opposition to the changes wrought by the war. Tainted with disloyalty in Republican eyes, unable to develop a coherent alternative to the policies of the Lincoln administration (it remained throughout the period, as one historian puts it, a "party of negations"), the Democracy's very survival is a matter of

54. David Montgomery, *Beyond Equality: Labor and the Radical Republicans 1862–1872* (New York, 1967), 91–92; Philip R. P. Coehlo and James F. Shepherd, "Regional Differences in Real Wages: The United States, 1851–1880," *Explorations in Economic History*, 13 (April 1976), 212–13; Reuben A. Kessel and Armen A. Alachian, "Real Wages in the North during the Civil War: Mitchell's Data Reinterpreted," *Journal of Law and Economics*, 2 (October 1959), 95–113; *U. K. Parliamentary Session Papers*, 1867, 32, c. 3857; Frederick Merk, *Economic History of Wisconsin During the Civil War Decade* (Madison, Wis., 1916), 162–63; David Roediger, "Racism, Reconstruction, and the Labor Press: The Rise and Fall of the St. Louis *Daily Press*, 1864–1866," *Science and Society*, 42 (Summer 1978), 156–64.

some surprise, testifying to the resilience of the deeply rooted traditions the war threatened to undermine. The party emerged from the war remarkably intact despite the defection of a number of prominent War Democrats and an internal division between those who supported the war effort, while criticizing specific administration policies, and others who advocated immediate peace. Its greatest strength lay in areas like the "butternut" farming regions of the Ohio Valley, closely tied to the South and bypassed by wartime economic expansion, and among urban Catholic immigrants and other voters hostile to the perfectionist reform tradition, with its impulse toward cultural homogeneity.[55]

To unite these groups, the Democracy built upon an ideological appeal developed in the 1850s, which identified the Republican party as an agent of economic privilege and political centralization, and a threat to individual liberty and the tradition of limited government. The Lincoln administration's economic policies, Democrats charged, enriched Northeastern capitalists at the expense of farmers and laborers, and spawned an enormous group of parasitic nonproducers—notably bondholders and stock market speculators—to the detriment of "the industrious poor." During the war, Democrats perfected economic appeals based upon the inequity of protective tariffs, high railroad freight rates, and state and federal aid to private corporations, that would provide the staples of agrarian protest in years to come. And the party benefited from widespread resentment over the use of troops to suppress strikes, a dramatic illustration of what many workers perceived as the federal government's partiality toward capital. The army was sent to Cold Springs, New York, to quell a strike at an arms factory, and prohibited the organization of workers in St. Louis war-production industries. Under the guise of putting down resistance to the draft, troops occupied Schuylkill County, Pennsylvania, in the nation's most productive anthracite coal region, and suppressed strikes by immigrant workers protesting wage cuts by Republican mine owners.[56]

The potent cry of white supremacy provided the final ideological glue in the Democratic coalition. Sometimes the appeal to race was oblique. The Democratic slogan, "The Union as It Is, the Constitution as It Was," had as its unstated corollary, blacks as they were—that is, as slaves. Often, it was remarkably direct. "Slavery is dead," the Cincinnati *Enquirer* announced at the end of the war, "the negro is not, there is the misfortune."

55. William R. Brock, "Reconstruction and the American Party System," in George M. Fredrickson, ed., *A Nation Divided* (Minneapolis, 1975), 84; Montgomery, *Beyond Equality*, 45–58; Joel Silbey, *A Respectable Minority: The Democratic Party in the Civil War Era* (New York, 1977), 56–59, 91–112, 149–54, 166–72.

56. Silbey, *Respectable Minority*, 25–27, 70–82; New York *World*, March 19, 1867; *CG* 39th Congress, 1st Session, 154; Frank L. Klement, "Economic Aspects of Middle Western Copperheadism," *Historian*, 14 (Autumn 1951), 27–44; Montgomery, *Beyond Equality*, 98–101; Grace Palladino, "The Poor Man's Fight: Draft Resistance and Labor Organization in Schuylkill County, Pennsylvania, 1860–1865" (unpub. diss., University of Pittsburgh, 1983).

The Emancipation Proclamation provoked lurid Democratic descriptions of an impending black inundation of the Midwest. In Indiana, one group of Democratic women paraded before an election with banners emblazoned: "Fathers, save us from nigger husbands." As Georges Clemenceau, reporting on Reconstruction for a French newspaper, observed after the war, "Any Democrat who did not manage to hint that the negro is a degenerate gorilla would be considered lacking in enthusiasm."[57]

All the elements of opposition to the war and its consequences came together for a few terrifying days in July 1863. The New York City draft riot, the largest civil insurrection in American history apart from the South's rebellion itself, originated in resentment over conscription—the quintessential example of the aggrandizement of federal power and, with its $300 commutation clause, of "class legislation." But it reflected as well resentment toward the emerging industrial bourgeoisie and the Republican party that appeared as its handmaiden, and violent hostility to emancipation, abolitionists, and blacks.

Having risen to dominance through command of the Southern cotton trade and ties to the West via the Erie Canal, New York in 1863 was the nation's commercial entrepôt and preeminent manufacturing center. A city sharply divided along lines of class, ethnicity, and politics, New York possessed a tightly knit Protestant mercantile and manufacturing elite identified with the war effort through the Union League Club, a large population of skilled artisans and journeymen, and a vast, impoverished immigrant working class aligned with the Democratic party. New York harbored every kind of antiwar sentiment, from the stance of opposition to both the Lincoln Administration and the Confederacy of wealthy Democrats like banker August Belmont, to the inflammatory pro-Southern rhetoric of former Mayor Fernando Wood and the extreme racism of Democratic journals like the New York *Caucasian*.

Beginning with an attack on a conscription office, the riot quickly developed into a wholesale assault upon all the symbols of the new order being created by the Republican party and the Civil War. Its targets included government officials, factories and docks (some, the scene of recent strikes), the opulent homes of the city's Republican elite, and such symbols of the reform spirit as the Colored Orphan Asylum, which was burned to the ground. Above all, the riot degenerated into a virtual racial pogrom, with uncounted numbers of blacks murdered on the streets or driven to take refuge in Central Park or across the river in New Jersey. Mattie Griffith, a white abolitionist, watched from her window as the immigrant working class—or, in her way of putting it, "the strange,

57. Forrest G. Wood, *Black Scare: The Racist Response to Emancipation and Reconstruction* (Berkeley, 1968), 18–34; J. A. Lemcke, *Reminiscences of an Indianian* (Indianapolis, 1905), 196; Georges Clemenceau, *American Reconstruction*, edited by Fernand Baldensperger, translated by Margaret MacVeagh (New York, 1928), 131.

wretched, abandoned creatures that flocked out from their dens and lairs"—took to the streets to commit acts of unimaginable cruelty upon the city's black population:

> A child of 3 years of age was thrown from a 4th story window and instantly killed. A woman one hour after her confinement was set upon and beaten with her tender babe in her arms. . . . Children were torn from their mother's embrace and their brains blown out in the very face of the afflicted mother. Men were burnt by slow fires.

Only the arrival of troops fresh from the Union victory at Gettysburg restored order to the city.[58]

Exposing the class and racial tensions lying just beneath the surface of the city's life and exacerbated by the war experience, the draft riot haunted New York's elite long after its suppression, serving as a reminder of the threat posed by a "dangerous class" whose existence could no longer be denied. It spurred efforts by the Union League and other elite organizations to reform city government and strengthen the forces of order, and to improve the conditions of New York's black population. A number of firms publicly announced their intention to replace Irish workmen with blacks, the Union League succeeded in integrating the city's streetcars, and in March 1864 a massive reception was organized for New York's black soldiers, an "astonishing" change, observed the *New York Times,* from the time eight months earlier when "the African race in this city were literally hunted down like wild beasts."[59] Yet these events did not erase the memory of the four July days when a mob took control of the nation's commercial capital. Nor did they wipe out the troubling questions raised by the New York riot: Could a society in which racial hatred ran so deep secure justice for the emancipated slaves? What would it mean to remake Southern society in the Northern image when the North was itself so bitterly divided by the changes brought on by the Civil War?

Like the inner civil war within the Confederacy, the draft riot underscored the fact that, for North and South alike, the war's legacy was fraught with ambiguity. "Happily," an abolitionist journal announced in 1863, "it is not in the power of man . . . to stop the course of events, after they have once been set out on a career of revolution." Yet the war's most conspicuous legacies—the preservation of the Union and the abolition of slavery—posed a host of unanswered questions. And their wartime corol-

58. Paul Migliore, "Business of Union: The New York Merchant Community and the Civil War" (unpub. diss., Columbia University, 1975), Chapters 1–3; Montgomery, *Beyond Equality,* 103–106; Adrian Cook, *The Armies of the Streets: The New York City Draft Riots of 1863* (Lexington, Ky., 1974); Mattie Griffith to Mary Estlin, July 27, 1863, Estlin Papers, Dr. Williams's Library, London.

59. Migliore, "Business of Union," Chapter 7; Henry W. Bellows, *Historical Sketch of the Union League Club of New York* (New York, 1879), 55–57; *New York Times,* March 7, 1864.

laries—a more powerful national state and a growing sense that blacks were entitled to some measure of civil equality—produced their own countervailing tendencies, as localism, laissez-faire, and racism, persistent forces in nineteenth-century American life, reasserted themselves.[60]

All Americans, nonetheless, shared a common sense of having lived through events that had transformed their world. "Southern newspaper articles of three or four years ago make me feel very old. . . ." New Yorker George Templeton Strong confided in his diary in 1865. "We have lived a century of common life since then." The Civil War would remain the central event of this generation's lives, creating and solidifying political loyalties, permeating the language with martial imagery (from the Salvation Army to "captains of industry"), defining the issues, from the nation's financial system to the rights of former slaves, that would shape political debate.[61]

Thus, two societies, each divided internally, entered the Reconstruction years to confront the myriad consequences of the Civil War. As Sidney Breese, an Illinois jurist and politician, observed, all Americans "must live in the world the War made."[62]

60. *National Anti-Slavery Standard,* January 10, 1863; Keller, *Affairs of State,* 39–42. Although the Civil War did not produce a "third war aim" (alongside Union and emancipation) of a national commitment to racial equality in any modern sense (an idea proposed and subsequently retracted by C. Vann Woodward), by the end of the conflict a commitment to civil equality for all citizens had become widespread in Republican ranks. C. Vann Woodward, *American Counterpoint* (Boston, 1976), 159–62; Belz, *New Birth of Freedom,* 25–30.
61. Allan Nevins and Milton H. Thomas, eds., *The Diary of George Templeton Strong* (New York, 1952), 3:601, 4:14.
62. Hyman, "Reconstruction and Political-Constitutional Institutions," 1.

CHAPTER 2

Rehearsals for Reconstruction

Dilemmas of Wartime Reconstruction

OF the Civil War's innumerable legacies, none proved so divisive as the series of questions that came to form the essence of Reconstruction. On what terms should the defeated Confederacy be reunited with the Union? Who should establish these terms, Congress or the President? What system of labor should replace plantation slavery? What should be the place of blacks in the political and social life of the South and of the nation at large? These and other issues were raised almost from the war's first days, and became increasingly urgent as slavery disintegrated and the prospect of Northern victory increased. One definitive conclusion emerged from the war: The reconstructed South would be a society without slavery. But even this raised as many questions as it answered.

The Emancipation Proclamation permanently transformed not only the character of the Civil War, but the problem of Reconstruction. For it suggested that even if, as Lincoln maintained, the rebelling Southern states remained theoretically a part of the Union, they could not resume their erstwhile position without acknowledging the destruction of slavery—a requirement that implied far-reaching changes in Southern society and politics. But nearly a year elapsed between emancipation and Lincoln's first announcement of a comprehensive program for Reconstruction. Issued on December 8, 1863, his Proclamation of Amnesty and Reconstruction offered full pardon and the restoration of all rights "except as to slaves" to persons who resumed their allegiance by taking an oath of future loyalty, and pledged to accept the abolition of slavery. A few groups, including high ranking civil and military officers of the Confederacy, were excluded. When in any state the number of loyal Southerners, thus defined, amounted to 10 percent of the votes cast in 1860, this minority could establish a new state government. Its constitution must abolish slavery, but it could adopt temporary measures regarding blacks "consistent . . . with their present condition as a laboring, landless, and homeless class." Such a government would then be entitled to representation at Washington, although Lincoln was careful to note that each House of Congress retained the authority to judge the qualifications of its own members.

A few abolitionists criticized this 10 Percent Plan for making no provision for suffrage or equality before the law, or defining any role whatever for blacks in the Reconstruction process. The proclamation, remarked Wendell Phillips, "frees the slave and ignores the negro." Clearly, Lincoln did not understand emancipation as a social revolution, or believe that Reconstruction entailed social and political changes beyond the abolition of slavery. He seems to have assumed that the South's former Whigs, many of whom, although large slaveholders, had been reluctant secessionists, would step forward to accept his lenient terms. Black suffrage would undoubtedly alienate such men, while the veiled invitation to return to the Union in order to oversee and regulate the transition from slave to free labor might well attract them. In actual operation, particularly in Louisiana, the plan quickly stirred Radical Republican opposition, but at its announcement, few demurred. For on the crucial question of 1863—whether emancipation must be a condition of Reconstruction—Lincoln and the Radicals agreed. For both, the definition of Southern loyalty now encompassed not merely a willingness to rejoin the Union, but acceptance of the slaves' freedom.[1]

It would be a mistake to see the 10 Percent Plan as a hard and fast policy from which Lincoln was determined never to deviate. Rather than as a design for a reconstructed South, it might better be viewed as a device to shorten the war and solidify white support for emancipation. As functioning governments, those established under the terms of Lincoln's proclamation would partake of the absurd. Inverted pyramids, the New York *World* called them, for a few thousand voters would control the destiny of entire states. Excluding blacks and the disloyal majority of Southern whites altogether, the plan could hardly promise stable government for a postwar South. But in strictly military terms, for 10 percent of the voters of 1860 to renounce their loyalty to the Confederacy would indeed be an achievement, augmenting the Union war effort, undermining the Confederacy's will to fight (and, as an added bonus, possibly adding a few electoral votes to the Republican column in 1864).[2] No one, least of all Lincoln, believed that the 1863 proclamation laid out a com-

1. James D. Richardson, ed., *A Compilation of the Messages and Papers of the Presidents 1789–1897* (Washington, D.C., 1896–99), 6:190, 213–15; Herman Belz, *Reconstructing the Union: Theory and Practice During the Civil War* (Ithaca, N.Y., 1969), 159–64, 188; David Donald, *Charles Sumner and the Rights of Man* (New York, 1970), 178–79; Kenneth M. Stampp, *The Era of Reconstruction* (New York, 1965), 32–42.

2. William B. Hesseltine, *Lincoln's Plan of Reconstruction* (Chicago, 1967 ed.), 96–97; Harold M. Hyman, *Era of the Oath: Northern Loyalty Tests During the Civil War and Reconstruction* (Philadelphia, 1954), 44–49.

prehensive blueprint for the postwar South. But the very process of establishing loyal governments had unanticipated consequences, producing serious divisions among Southern Unionists, creating political forums in which long-excluded groups stepped forward to claim a share of political power, and inspiring demands by blacks and their Radical allies for even more far-reaching changes in Southern life.

The four slave states and part of a fifth that had remained within the Union and thus were unaffected by the Emancipation Proclamation and 10 Percent Plan were the first to reveal the revolutionary implications of tying Reconstruction to abolition, as well as the bitter strength of resistance to change. The percentage of blacks in the border population was far lower than in the Deep South, and slavery less central to the economy. Indeed, in the prewar decades the peculiar institution had atrophied in the older border states, with the number of slaves actually declining in Delaware and Maryland. Further west, in Kentucky and Missouri, slavery had continued to expand, but all the border states contained large and rapidly growing regions organized on a free labor basis and susceptible to antislavery politics. Nonetheless, in all the loyal slave states, slaveholders had dominated antebellum politics. Throughout the war, Delaware and Kentucky clung to the decaying body of slavery. West Virginia, Maryland, and Missouri, by contrast, underwent internal reconstructions that brought to power new classes eager to overturn slavery and revolutionize state politics. The experience of the border states provided an early indication of the potential and the limitations of a Reconstruction that excluded the participation of blacks.

Resistance to change proved greatest in Kentucky, the border state with the largest number of slaveholding families—mostly small farmers engaged in mixed agriculture, rather than staple-producing planters. After a brief attempt at "neutrality" in 1861, Kentucky became firmly committed to the Union, but throughout the war remained under the control of a conservative Unionist coalition that steadfastly opposed all federal policies that threatened to undermine slavery. State officials denounced the Emancipation Proclamation as unconstitutional and refused to recognize the liberty of any black person "claiming or pretending to be free" under its terms. The enlistment of black troops, rather than any act of Kentucky herself, sealed the fate of slavery in the Bluegrass State, but at the war's end more than 65,000 blacks remained in bondage. In fact, slavery did not officially end until the ratification of the Thirteenth Amendment, which, to the last, Kentucky's legislature opposed.[3]

3. Ira Berlin et al., eds., *Freedom: A Documentary History of Emancipation 1861–1867* (New York, 1982–) Ser. 1, 1:493–518; Victor B. Howard, *Black Liberation in Kentucky: Emancipation and Freedom, 1862–1884* (Lexington, Ky., 1983); 36–61; E. Merton Coulter, *The Civil War and Readjustment in Kentucky* (Chapel Hill, 1926), 49–54, 161–81.

The Union party of Kentucky, remarked Gen. John M. Palmer, the federal military commander in the state, "is distinguished by its timidity." Slavery was more deeply entrenched here than elsewhere in the border region, and the elements that would transform Kentucky's neighbors were either weak or absent. Although loyal to the Union, the state's mountain region remained politically inactive; there was no major city (like Baltimore or St. Louis) with an antislavery cadre ready to take the lead in reconstructing the state; and the traditional political leadership retained sufficient unity to fend off challenges to its authority. Elsewhere in the border region, apart from tiny Delaware, deep internal tensions and the course of the war itself combined to open the door to emancipation and political revolution.[4]

The first Southern state to abolish slavery at least partially of its own volition was West Virginia, where blacks comprised only 5 percent of the population. Reconstruction, in a sense, began in 1861 when a convention of Unionists meeting in Wheeling repudiated Virginia's secession and chose Francis H. Pierpont, a railroad attorney and coal mine operator, as the state's legitimate governor. In 1863 West Virginia was admitted to the Union as a separate state, with the proviso that it abolish slavery. A popular referendum then approved a plan whereby all blacks born after July 4, 1863, would enjoy freedom. By the end of the war, complete emancipation had been enacted.[5]

The creation of West Virginia represented both the culmination of deep-rooted sectional divisions within Virginia and the overthrow of the western region's own antebellum elite, which had generally supported secession. Statehood brought new men to the fore, who enacted long-demanded democratic reforms, including free public education, the secret ballot, annual elections, and changes in the structure of local government. A coalition of groups hostile to planter domination created the state: on the one hand, mountaineers who cherished their communities' isolation and self-sufficiency; on the other, entrepreneurial Whigs attuned to the free labor ideology and modern notions of progress, who blamed Virginia planters for retarding the region's development and preventing its integration into the expanding capitalist economy of the North. Both groups were devoted to the Union and resentful of the antebellum elite, and neither was much concerned with the rights of the state's black minority. Indeed, only the insistence of Congress forced

4. Berlin et al., eds., *Freedom*, Ser. 1, 1:493–94; James B. Murphy, "Slavery and Freedom in Appalachia: Kentucky as a Demographic Case Study," *Register of the Kentucky Historical Society*, 80 (Spring 1982), 168; Ross A. Webb, "Kentucky: 'Pariah Among the Elect'," in Richard O. Curry, ed., *Radicalism, Racism, and Party Realignment: The Border States During Reconstruction* (Baltimore, 1969), 106.

5. Charles H. Ambler, *Francis H. Pierpont* (Chapel Hill, 1937), 40–41, 81, 99, 162–69, 179, 202; Forest Talbot, "Some Legislative and Legal Aspects of the Negro Question in West Virginia During the Civil War and Reconstruction, Part I," *WVaH*, 24 (April 1963), 8–19.

the deletion from the new constitution of a clause barring blacks from entering the state.[6]

Lincoln carried West Virginia in 1864. But in a state whose boundaries had been drawn so as to include not only the staunchly Unionist mountain area, but twenty-six counties farther south that had voted for secession, Republicans feared the political threat posed by returning Confederate soldiers. Instead of enfranchising black voters, who, in any case, represented only a tiny fraction of the population, Republicans, early in 1865, enacted proscriptive legislation that enabled them to retain power for the next five years. Voters were required to take oaths affirming their past and future loyalty to the Union, thus disqualifying thousands in the counties bordering on Virginia. (A similar set of policies was adopted by the "restored government" of Virginia, which Pierpont had established in Alexandria under the watchful eye of federal troops. In 1864 a "constitutional convention" of sixteen delegates abolished slavery, restricted suffrage to "loyal" whites, and provided for a system of public education apparently intended to be closed to blacks, since it would be financed by taxes on whites alone.)[7]

The amalgam of hostility to the antebellum regime, commitment to democratic change for Unionist whites, reluctance to push beyond emancipation as far as blacks were concerned, and reliance on wholesale proscription of former Confederates to retain political power characterized wartime Unionism throughout the border region. Nowhere was this more evident than in Maryland, a state as divided internally as any in the South. Its 87,000 slaves (nearly equaled in number by free blacks) were concentrated in the counties of southern Maryland, whose large tobacco plantations recalled the social order of the Deep South. The area was economically stagnant, but its political leaders dominated the state, thanks to an archaic system of legislative apportionment that reduced the influence of Baltimore and the rapidly growing white farming counties to its north and west.[8]

Occupied by federal troops from the outset of the war, Maryland experienced earlier than other states the disintegration of slavery from within, and the mobilization of free blacks against the institution. It also witnessed the rapid growth of emancipationist sentiment among the

6. Francis N. Thorpe, ed., *The Federal and State Constitutions* (Washington, D.C., 1909), 7:4018, 4031–32; John A. Williams, "The New Dominion and the Old: Ante-Bellum and Statehood Politics as the Background of West Virginia's 'Bourbon Democracy,' " *WVaH*, 33 (July 1972), 342–52.

7. Richard O. Curry, "Crisis Politics in West Virginia, 1861–1870," in Curry, ed., *Radicalism, Racism, and Realignment*, 83–90; Milton Gerofsky, "Reconstruction in West Virginia, Part I," *WVaH*, 6 (July 1945), 300–306; Ambler, *Pierpont*, 208–22; Thorpe, ed., *Federal and State Constitutions*, 7:3861.

8. Barbara J. Fields, *Slavery and Freedom on the Middle Ground: Maryland During the Nineteenth Century* (New Haven, 1985), 1–22; Charles L. Wagandt, ed., "The Civil War Journal of Dr. Samuel A. Harrison," *CWH*, 13 (July 1967), 134.

white population. The "great army in blue," remarked antislavery leader Hugh Lennox Bond, brought in its wake "a great army of ideas." These found a receptive audience among the small farmers of northwestern Maryland and the manufacturers and white laborers of Baltimore. "It seems to give great satisfaction to the laboring whites," one slaveholder noted as the flight of blacks to Union lines and their enlistment in the army undermined slavery, "that the non laboring slave owners are losing their slaves, and they too will be reduced to the necessity of going into the fields." Unlike Kentucky, Maryland produced a brilliant Radical leader, Henry Winter Davis, who echoed the free labor critique of slavery, promising that abolition would create a dynamic and prosperous society, freed from "the domination of . . . property over people, of aristocratic privilege over republican equality." By 1863, conservative Unionists had come to accept the inevitability of emancipation, but believed that owners should be reimbursed for their loss. Replied Davis: "Their compensation is the cleared lands of all Southern Maryland, where every thing that smiles and blossoms is the work of the negro that they tore from Africa."[9]

Bolstered by loyalty oaths administered to voters by army provost marshals, Unionists committed to immediate and uncompensated emancipation swept the Maryland elections of 1863 and called a constitutional convention to reconstruct the state. Abolition headed the agenda, but the convention also voiced long-standing sectional and class grievances. It established a free, tax-supported public school system, exempted property worth up to $500 from seizure for debt, and by basing legislative representation on the white population alone, drastically reduced the power of the plantation counties. Voting was confined to those who could take a strict loyalty oath, which included an avowal that one had never expressed a "desire" for the Confederacy's triumph. The delegates celebrated the dawn of a new era, a "free and regenerated" Maryland.[10]

Except among a small group of emancipationists, however, little concern was evinced for the fate of the former slaves. Many delegates felt compelled to deny that voting for abolition implied "any sympathy with *negro equality.*" The school system excluded blacks, the "white basis" of representation seemed to rule out a role for blacks in state politics, and the legislature did nothing to alter a prewar statute that authorized local courts to apprentice free black children, even over the objections of their parents. Within a month of November 1, 1864, the date of emancipation,

9. Richard P. Fuke, "Hugh Lennox Bond and Radical Republican Ideology," *JSH*, 45 (November 1979), 583–84; Wagandt, ed., "Civil War Journal," 136; Jean H. Baker, *The Politics of Continuity: Maryland Political Parties from 1858 to 1870* (Baltimore, 1973), 77–85; Charles L. Wagandt, *The Mighty Revolution: Negro Emancipation in Maryland, 1862–1864* (Baltimore, 1964), 143; Henry Winter Davis, *Speeches and Addresses* (New York, 1867), 392.

10. Wagandt, *Mighty Revolution*, 157, 184, 195, 222; *The Debates of the Constitutional Convention of the State of Maryland* (Annapolis, 1864), 1:592; Thorpe, ed., *Federal and State Constitutions*, 3:1746–47, 1752, 1757, 1773.

literally thousands of former slaves had been bound to white masters, an injustice that galvanized the black community to protest and bedeviled relations between Maryland and federal authorities for years to come. All in all, as one Maryland Unionist remarked after the constitution won narrow approval in a September 1864 referendum:

> [It] must be a source of mortification that emancipation has . . . not been from high principle, . . . but party spirit, vengeful feeling against disloyal slaveholders, and regard for material interest. There has been no expression, at least in this community, of regard for the negro—for human rights, but . . . many expressive of the great prosperity to result to the state by a change of the system of labor.[11]

Lincoln carried Maryland in 1864, but the new constitution passed by only a narrow margin, underscoring the emancipationists' precarious hold on political power. Early in 1865, the antislavery party moved to solidify its position by disenfranchising all who had served in the Southern armies or given support by "open deed or word" to the Confederacy. No thought was given to expanding the Unionist base by allowing the black 20 percent of Maryland's population to vote.[12]

The last border state to experience an internal reconstruction was Missouri, like Maryland a state divided between slave and free economies and possessing a great industrial city that harbored abolitionist sentiments. Indeed the presence of a large population of German immigrants, many of them exiles from the failed revolution of 1848 who identified the Slave Power with the landed aristocracy of Europe, gave the democratic revolution a significant base of white support. Rural Missouri, moreover, was ravaged by guerrilla fighting that embittered many white farmers and Ozark mountaineers against slaveholders and the Confederacy, further strengthening antislavery forces.[13]

Almost from the outset of the war, Missouri Unionists were divided by what Lincoln called a "pestilent factional quarrel." Conservatives hoped to preserve as much of the old order as possible, while Radicals demanded emancipation, the arming of blacks, and the disenfranchisement of rebels. Southern society must yield to that of the North, declared the *Missouri Democrat*'s Radical editor, B. Gratz Brown: "This is progress, this

11. *Maryland Convention Debates*, 1:552–53; Richard P. Fuke, "The Baltimore Association for the Moral and Educational Improvement of the Colored People, 1864–1870," *Maryland Historical Magazine*, 66 (Winter 1971), 369–71; *Maryland Laws 1865*, 285; Richard P. Fuke, "A Reform Mentality: Federal Policy Toward Black Marylanders, 1864–1868," *CWH*, 22 (September 1976), 222–25; Herbert G. Gutman, *The Black Family in Slavery and Freedom 1750–1925* (New York, 1976), 402–10; Wagandt, ed., "Civil War Journal," 285.

12. Wagandt, *Mighty Revolution*, 258; *AC*, 1865, 526.

13. Berlin et al., eds., *Freedom*, Ser. 1, 1:395–412; Steven Rowan, ed., *Germans for a Free Missouri: Translations from the St. Louis Radical Press, 1857–1862* (Columbia, Mo., 1983), vii–viii, 30–31, 104; Fred DeArmond, "Reconstruction in Missouri," *MoHR*, 41 (April 1967), 365–71.

is the Revolution." Despite the President's effort to remain impartial, the Emancipation Proclamation, and the collapse of slavery in the state as blacks fled the plantations and enrolled in the army, strengthened the Radicals' position. In 1864, aided by loyalty oaths required of prospective voters, Radical Thomas C. Fletcher was elected governor and a convention called to devise a plan of emancipation.[14]

The delegates who gathered in January 1865 represented a cross section of the diverse coalition of outsiders now reaching for political power: St. Louis Germans, poor Ozark farmers, and small-town merchants and professionals. Few had ever held political office. Like those of Maryland and West Virginia, the new constitution embodied democratic reforms, including a state-supported educational system and the abolition of imprisonment for debt. After abolishing slavery, the convention adjourned amid a chorus of "John Brown's Body" (a sign of the changes wrought since Brown had battled proslavery Missourians on the Kansas prairies a decade earlier). Thanks to Radical lawyer Charles Drake, the constitution took greater cognizance of the implications of emancipation than those of other states in the border, mandating racial equality in property rights and access to the courts, empowering the legislature to establish schools for blacks, and guaranteeing their right to testify in court. Drake himself favored black suffrage, but feared that such a provision would ensure the constitution's defeat.

In Missouri, as in the rest of the border states, the Radicals depended not on black votes but on a comprehensive system of registration and test oaths to bar "rebels" from voting and officeholding and "secure the control of the State to its loyal people." Indeed, the Missouri Radicals went further than their counterparts in other states in attempting to solidify their hold on power. They barred the "disloyal" from acting as teachers, lawyers, or ministers, declared hundreds of state and local offices vacant, to be filled by appointment of the governor, and prohibited the legislature from altering the disenfranchising provisions before 1871. Some delegates wanted to go even further, proposing to confiscate the property of secessionists in order to compensate Unionists for their wartime losses.

Here were men acting as if in the midst of a revolution. The nation's founding fathers had created a republic in which citizens were presumed to possess the quality of "virtue"—defined in the eighteenth century as the ability to subordinate individual self-interest to the public good. The Missouri Radicals—like Unionists throughout the border region—attempted, in effect, to forge a new polity with an updated test of virtue: past loyalty to the Union. The strongest support for excluding "rebels" from citizenship in the new Missouri came from the guerrilla-ravaged

14. Norma L. Peterson, *Freedom and Franchise: The Political Career of B. Gratz Brown* (Columbia, Mo., 1965), 106–16, 121–34; Thomas S. Barclay, *The Liberal Republican Movement in Missouri 1865–1871* (Columbia, Mo., 1926), 5–7.

areas of the north and southwest, where legal niceties had long since given way before the depredations of Confederate raiders. German delegates, preferring to enfranchise blacks rather than to bar Confederates, denounced the proscriptive clauses. In June 1865 the constitution was approved by a narrow margin, and Missouri emerged from the war a free state, but a profoundly divided society.[15]

Thus the end of the Civil War found the border states in the throes of change. Slavery was dead or dying, politics in turmoil. Outside Kentucky and Delaware, legal abolition ratified the destruction of slavery brought about by the actions of blacks, the policies of the Lincoln administration, and the course of the war, and provided the opportunity for groups that lacked influence under the slave regime to seize political power. Radical Unionists disenfranchised Confederate sympathizers to strengthen their precarious hold on office, acts that would taint their rule with illegitimacy. At the end of the Civil War, twenty-three Republicans represented border states in Congress (there had been only one in 1860, elected from St. Louis). It remained to be seen whether these men were harbingers of a political revolution destined to sweep the entire South, or temporary anomalies—the products of military control and disenfranchisement.

Unlike the border states, the Confederate Upper South, especially Tennessee, experienced wartime Reconstruction under the auspices of direct military rule. But in many ways, events followed a similar course. By an accident of war, Tennessee's Reconstruction began not in the staunchly Unionist eastern mountains, but in the middle and western parts of the state, where slavery was deeply entrenched and Confederate sentiment dominant. After the Union capture of Nashville early in February 1862, Lincoln appointed Andrew Johnson military governor. Johnson's decision to remain in the Senate after Tennessee seceded had made him a national symbol of what both he and the Republican North supposed to be a legion of courageous Southern Unionists. Having risen to prominence in the Democratic party, Johnson now found himself heading a Union movement composed largely of former Whigs. He quickly won their admiration and outraged secessionists by removing Nashville's mayor and city council for refusing to take an oath of allegiance. When, shortly thereafter, a supporter of the Confederacy won election as circuit judge, Johnson ordered his arrest and appointed his rival to the office. Johnson soon took to using the phrase that won him a national reputation for Radicalism: "Treason must be made odious and traitors punished."[16]

15. David D. March, "Charles D. Drake and the Constitutional Convention of 1865," *MoHR*, 44 (January 1954), 110–23; Barclay, *Liberal Republican Movement*, 11–33; William E. Parrish, *Missouri Under Radical Rule 1865–1870* (Columbia, Mo., 1965), 15–32; *CG*, 41st Congress, 3d Session, Appendix, 2; Thorpe, ed. *Federal and State Constitutions*, 4:2194–95, 2200, 2212.

16. Leroy P. Graf and Ralph W. Haskins, eds., *The Papers of Andrew Johnson* (Knoxville, 1967–), 5: xix–liii.

Time would reveal that Johnson's Radicalism was cut from a different cloth than that of Northerners who wore the same label, but this was not completely evident during the Civil War. Although Tennessee was exempted from its terms, the Emancipation Proclamation split the state's Unionists, driving some to embrace the Confederacy and others to demand a restoration of the antebellum status quo. "Unconditional Unionists," like East Tennessee's William G. Brownlow, endorsed the Proclamation, and Johnson followed suit. By the end of 1863, Johnson had declared for abolition in Tennessee. His conversion, however, was based less on concern for the slave than hatred of the Confederacy and of the slaveholders he believed had dragged poor whites unwillingly into rebellion. As he remarked to General Palmer: "Damn the Negroes, I am fighting those traitorous aristocrats, their masters." Yet, as middle Tennessee whites remained resolutely pro-Confederate, while Nashville's free black community mobilized to support his administration and put an end to slavery, Johnson's prejudices softened. By 1864 Johnson was speaking of elevating Tennesseans of both races. He now regarded "the wide world as my home, and every honest man, be he white or colored, as my brother." Addressing a black gathering in October, Johnson unilaterally decreed the end of slavery in Tennessee. "I will indeed be your Moses," he went on, "and lead you through the Red Sea of war and bondage to a fairer future of liberty and peace."[17]

In November 1864 Johnson was elected Vice President by the Republicans (temporarily rechristened the Union party). His presence on the ticket symbolized the party's determination to reward Southern Unionists and extend Republican organization into the South. But Johnson's own efforts to inaugurate Reconstruction under the 10 Percent Plan had achieved nothing, partly because he had added to Lincoln's lenient oath of future loyalty "a hard oath—a tight oath," whereby prospective voters were required to pledge that they had desired both the defeat of the Confederacy and the abolition of slavery in Tennessee. Before leaving for Washington, Johnson took direct action to reconstruct the state. Bypassing elections, he endorsed the assembling of a self-appointed convention of unconditional Unionists. This, in turn, adopted a constitutional amendment to abolish slavery that won the nearly unanimous approval of the 25,000 white Tennesseans permitted to vote in a February 1865 referendum. In March, Johnson assumed the Vice Presidency, and William G. Brownlow was elected the first governor of a free Tennessee. Its support confined al-

17. Verton M. Queener, "The Origins of the Republican Party in East Tennessee," *ETHSP*, 13 (1941), 77–78; Peter Maslowski, *Treason Must Be Made Odious: Military Occupation and Wartime Reconstruction in Nashville, Tennessee, 1862–1865* (Millwood, N. Y., 1978), 81–85; John Cimprich, "Military Governor Johnson and Tennessee Blacks, 1862–65," *THQ*, 39 (Winter 1980), 460–68; James W. Patton, *Unionism and Reconstruction in Tennessee 1860–1869* (Chapel Hill, 1934), 47–48.

most exclusively to the eastern part of the state, the Brownlow regime moved swiftly to consolidate its power by securing the ballot box, in the governor's words, against "the approach of treason." A franchise law limited the right to vote to white males "publicly known to have entertained unconditional Union sentiments" throughout the war. As for blacks, Brownlow urged Congress to set aside a territory for their settlement as a "nation of freedmen."[18]

Of all the states where wartime Reconstruction was attempted, only Louisiana lay in the heart of the Confederacy. Here Lincoln invested the greatest hopes, and here he suffered the greatest disappointments. In a way, it was a cruel trick of fate that decreed that the Reconstruction of the Deep South should be attempted in a state "long divided along economic, cultural and racial lines . . . its politics . . . faction-ridden, corrupt, and occasionally violent." Louisiana's white population included not only planters and upcountry yeomen, but a unique configuration of religious and ethnic groups, including Protestant hill farmers, urban Irish immigrants, French and Spanish Creoles, and Cajuns living in its southern bayous.[19]

In contrast to Tennessee, however, the initial federal occupation of Louisiana at least took place in a Unionist stronghold. When troops under the command of Gen. Benjamin F. Butler seized New Orleans in April 1862, the Union came into possession of the South's largest and most distinctive city. The nation's premier cotton port, New Orleans was embedded in a web of commercial relations with New York City and London and in 1860 had been carried by John Bell on a platform advocating the preservation of the Union. Its population of 144,000 was predominantly white, nearly half of whom were foreign-born. The city also included a sizable contingent of Northerners prominent in banking, mercantile activities, and the professions. Among its black population of 25,000 was a remarkable community of 11,000 free persons of color, many of them prosperous and well-educated. The Union Army also took control of the sugar parishes of southeastern Louisiana, whose planters, unlike the cotton aristocrats of the northern part of the state, depended upon national tariff protection and tended to be pro-Union Whigs. Many sugar planters had gone with the Confederacy in 1861, but now hastened

18. James E. Sefton, *Andrew Johnson and the Uses of Constitutional Power* (Boston, 1980), 93–100; Thomas B. Alexander, *Political Reconstruction in Tennessee* (Nashville, 1950), 18–31, 47, 73–74; *Knoxville Whig*, April 19, 1865. Arkansas also experienced a wartime Reconstruction. Its new constitution, drafted by mountain Unionists in 1864, abolished slavery but allowed the apprenticeship of black minors and sought to bar additional blacks from entering the state. The new state government passed laws disenfranchising most Confederates. Ruth C. Cowan, "Reorganization of Federal Arkansas, 1862–1865," *ArkHQ*, 18 (Summer 1959), 33–50.

19. George Rable, "Republican Albatross: The Louisiana Question, National Politics, and the Failure of Reconstruction," *LaH*, 23 (Spring 1982), 110; William I. Hair, *Bourbonism and Agrarian Protest: Louisiana Politics 1877–1900* (Baton Rouge, 1969), 74.

to reaffirm allegiance to the Union, partly in the hope of retaining possession of their slaves.[20]

Thus, the prospects for creating a Unionist movement in southern Louisiana seemed auspicious. Butler moved to broaden local support by distributing beef and sugar seized by his troops to the poor of New Orleans and organizing a massive project to clean the city's streets, thereby providing public employment for the immigrant working class and combating yellow fever. (It was the last time this "filthy pesthole" was cleaned until the turn of the century.) In the fall of 1862, a Union Association, later called the Free State Association, was organized in New Orleans, and in December two of its members, Michael Hahn and Benjamin Flanders, were sent to Congress in an election in which nearly 8,000 votes were cast, a number that represented over half the turnout in the city's last antebellum election.[21]

As in other states, however, Louisiana's Unionists were divided. Conservatives, notably sugar planters and wealthy merchants, hoped somehow to preserve slavery and, when that idea was overtaken by events, believed planters should receive compensation for their slaves and retain their traditional grip on political power. The Free State Association embodied the more radical view that a free Louisiana should be more than simply the old order without slavery. The two new Congressmen reflected its diverse membership: immigrants, artisans, small merchants, reform-oriented professionals and intellectuals, Northerners (or men educated in the free states and married to Northern women), and federal officeholders. Hahn was a native of Bavaria attuned to such classic nineteenth-century liberal ideas as free trade, universal public education, and religious rationalism; the New Hampshire-born Flanders, a teacher, newspaper editor, and railroad officer in New Orleans before the war, had been exiled by the Confederate government because of his vociferous opposition to secession. Such men accepted the free labor ideology and saw emancipation as the key to remolding the backward South in the image of the progressive North. For them and their associates, the Civil War was a genuine revolution that offered the opportunity to overthrow a reactionary and aristocratic ruling class.[22]

In August 1863 Lincoln endorsed the program of the Free State As-

20. Peyton McCrary, *Abraham Lincoln and Reconstruction: The Louisiana Experiment* (Princeton, 1978), 22–25, 160; William W. Chenault and Robert C. Reinders, "The Northern-Born Community of New Orleans in the 1850's," *JAH*, 51 (September 1964), 232–47; J. Carlyle Sitterson, *Sugar Country* (Lexington, Ky., 1953), 204.

21. McCrary, *Abraham Lincoln and Reconstruction*, 78, 100; Joe G. Taylor, *Louisiana Reconstructed, 1863–1877* (Baton Rouge, 1974), 410; Roger W. Shugg, *Origins of Class Struggle in Louisiana* (Baton Rouge, 1939), 185–88.

22. McCrary, *Abraham Lincoln and Reconstruction*, 96–97, 168; Amos E. Simpson and Vaughan B. Baker, "Michael Hahn: Steady Patriot," *LaH*, 13 (Summer 1972), 229–32; Ted Tunnell, *Crucible of Reconstruction: War, Radicalism and Race in Louisiana 1862–1877* (Baton Rouge, 1981), 8–25.

sociation, urging Gen. Nathaniel P. Banks, who had succeeded Butler, to organize a constitutional convention that would abolish slavery in Louisiana. But with much of the state under Confederate control and pro-Southern sentiment still widespread in Union-held areas, preparations advanced slowly. With its lenient terms, the 10 Percent Plan of December was motivated, in part, by Lincoln's desire to speed up Reconstruction in Louisiana. But the plan exacerbated an emerging split between Radicals and moderates in the Free State movement. Many factors contributed to this increasingly acrimonious division, including the clash of strong personalities, the question of civilian or military control of a free state government, and the Radicals' desire to put off a constitutional convention until emancipation enjoyed broader support. Presidential ambition further muddied the political waters. Secretary of the Treasury Salmon P. Chase employed his extensive patronage powers to build up a following in Louisiana, while Banks, who supported the President's bid for reelection, simultaneously imagined himself replacing a faltering Lincoln in the White House. But increasingly, the split came to focus on the rights, if any, to be enjoyed by blacks in the new Louisiana.[23]

In New Orleans lived the largest free black community of the Deep South. The wealth, social standing, education, and unique history of this community set it apart not only from the slaves, but from most other free persons of color. The majority were the light-skinned descendants of unions between French settlers and black women or of wealthy mulatto emigrants from Haiti, and identified more fully with European than American culture. Many spoke only French and educated their children at private academies in New Orleans, or in Paris. Although barred from the suffrage, they enjoyed far more rights than free blacks in other states, including the right to travel without restriction and testify in court against whites, and had a self-conscious military tradition dating back to their participation under Andrew Jackson in the Battle of New Orleans. On the eve of the Civil War, they owned some $2 million worth of property and dominated skilled crafts like bricklaying, cigarmaking, carpentry, and shoemaking. At the apex of this community stood men of extensive wealth, like Aristide Mary, who owned real estate valued at $30,000, and Antoine Dubuclet (later the Reconstruction state treasurer), a sugar planter with over 100 slaves. After conversing with one cultured and wealthy free black in 1865, Carl Schurz concluded: "There is no country of the world, save this, in which he would not be received as a gentleman of the upper class."[24]

23. LaWanda Cox, *Lincoln and Black Freedom: A Study in Presidential Leadership* (Columbia, S.C., 1981), 59–69; Tunnell, *Crucible of Reconstruction*, 26–50.

24. Laura Foner, "The Free People of Color in Louisiana and St. Domingue," *JSocH*, 3 (Summer 1970), 406–30; New Orleans *Louisianian*, February 20, 1875; David C. Rankin,

This self-conscious community, with a strong sense of its collective history and a network of privately supported schools, orphanages, and benevolent societies, was well-positioned to advance its interests under Union rule. When General Butler appeared, free blacks offered their military services to the Union, and after some hesitation, the general enrolled them, in separate units with their own officers, in the army. In February 1863, however, eighteen officers resigned, claiming they had been treated with "scorn and contempt" by white soldiers. Nine months later, a mass meeting of free blacks heard speakers, including P. B. S. Pinchback, a future governor of the state, who had resigned a captaincy in one of Butler's black units, call for political rights for the free community. At this point, the free blacks were speaking only for themselves, for, as one who knew them well later recounted:

> They tended to separate their struggle from that of the Negroes; some believed that they would achieve their cause more quickly if they abandoned the black to his fate. In their eyes, they were nearer to the white man; they were more advanced than the slave in all respects. . . . A strange error in a society in which prejudice weighed equally against all those who had African blood in their veins, no matter how small the amount.[25]

By January 1864 Lincoln appears to have privately endorsed the enrollment of freeborn blacks as voters in Louisiana. But to General Banks, black suffrage, however limited, was not only personally distasteful, but a threat to his efforts to win white support for reconstructing Louisiana under the 10 Percent Plan. Having been made "master" of the situation by an impatient Lincoln, Banks plied the President with optimistic reports about the growth of Unionist sentiment, while warning that "revolutions which are not controlled and held within reasonable limits, produce counter-revolutions." In February an election for state officials was held under the prewar Louisiana constitution (which recognized slavery), and Banks threw his full support to the moderate Free State group, now headed by Michael Hahn. The poor showing of Radical candidate Benjamin Flanders, who finished third behind a pro-slavery Unionist, led many of his supporters to repudiate the entire Louisiana experiment. But Banks

---

"The Impact of the Civil War on the Free Colored Community of New Orleans," *Perspectives in American History*, 11 (1977–78), 380–83; Charles Vincent, "Aspects of the Family and Public Life of Antoine Dubuclet: Louisiana's Black State Treasurer, 1868–1878," *JNH*, 66 (Spring 1981), 26–28; Joseph Schafer, ed., *Intimate Letters of Carl Schurz 1841–1869* (Madison, Wis., 1928), 351.

25. Manoj K. Joshi and Joseph P. Reidy, "'To Come Forward and Aid in Putting Down This Unholy Rebellion': The Officers of Louisiana's Free Black Native Guard During the Civil War Era," *SS*, 21 (Fall 1982), 328–36; Charles Vincent, *Black Legislators in Louisiana During Reconstruction* (Baton Rouge, 1976), 7–14, 19–20; Jean-Charles Houzeau, *My Passage at the New Orleans "Tribune": A Memoir of the Civil War Era*, edited by David C. Rankin, translated by Gerard F. Denault (Baton Rouge, 1984), 81.

pressed ahead with plans for a constitutional convention to deal the death blow to slavery in the state.[26]

Meanwhile, two representatives of the free black community, Arnold Bertonneau, a wealthy wine dealer, and Jean Baptiste Roudanez, a plantation engineer, arrived in Washington to present a petition for the suffrage. On March 13, 1864, the day after they met with Lincoln, the President wrote Governor Hahn concerning the coming convention: "I barely suggest for your private consideration, whether some of the colored people not be let in—as for instance, the very intelligent, and especially those who have fought gallantly in our ranks. . . . But this is only a suggestion, not to the public, but to you alone." Hardly a ringing endorsement of black suffrage, Lincoln's letter nonetheless illustrated the capacity for both growth and compromise that was the hallmark of his political leadership. This quality, unfortunately, was in short supply in Louisiana.[27]

The constitutional convention ratified the overthrow of Louisiana's old order. The assembly's composition would not have surprised anyone familiar with the leadership of radical movements in contemporary Europe, for the delegates included reform-minded professionals, small businessmen, artisans, civil servants, and a sprinkling of farmers and laborers. The planter class, which, as one delegate put it, had governed the state "for the sole and exclusive benefit of slaveholders," was conspicuous by its absence. Reflecting the urban orientation of the Unionist coalition, the constitution made New Orleans the state capital and sharply increased the city's power in the legislature by basing representation upon voting population rather than total number of inhabitants (thus reducing the power of the plantation counties). In addition, the new constitution established a minimum wage and nine-hour day on public works and adopted a progressive income tax and a system of free public education. And, of course, slavery was abolished. Delegate after delegate excoriated the planters and hailed emancipation as "the true liberation and emancipation of the poor white laboring classes of the South." To be sure, slavery had long since disintegrated in New Orleans and the surrounding parishes. As federal judge Edward H. Durrell, the convention's president, remarked: "If you think slavery exists, go out in the streets and see if you can get your slave to obey you." But he, too, spoke of legal abolition as "the commence-

26. Cox, *Lincoln*, 80–93; Fred H. Harrington, *Fighting Politician: Major General N. P. Banks* (Philadelphia, 1948), 143–46; McCrary, *Abraham Lincoln and Reconstruction*, 186, 224–26.

27. David C. Rankin, "The Origins of Black Leadership in New Orleans During Reconstruction," *JSH*, 40 (May 1974), 139; Houzeau, *My Passage*, 25n.; Roy F. Basler, ed., *The Collected Works of Abraham Lincoln* (New Brunswick, 1953), 7:243. The petition originally called for the vote for Louisiana blacks free before the Civil War, but Charles Sumner persuaded Bertonneau and Roudanez to add a call for suffrage for all Louisiana blacks, "whether born slave or free, especially those who have vindicated their right to vote by bearing arms." Ted Tunnell, "Free Negroes and the Freedmen: Black Politics in New Orleans During the Civil War" *SS* (Spring 1980), 16–17.

ment of a new era in civilization . . . [a] dividing line between the old and worn out past and the new and glorious future."[28]

When it came to the role of blacks in free Louisiana, however, the ideas of the "old and worn out past" displayed remarkable resiliency. "Prejudice against the colored people is exhibited continually," reported a correspondent of Secretary Chase, "prejudice bitter and vulgar." It was not surprising, perhaps, that the few conservatives at the convention defended slavery as "the most perfect, humane and satisfactory" system of labor "that has ever been devised." But Radicals were shocked when men who favored abolition demanded the expulsion of all blacks from the state, even though, as one delegate pointed out, black troops were at that very moment guarding the convention hall. The convention petitioned Congress to compensate loyal planters for their slaves, and while rejecting any disenfranchisement of whites (in the hope of attracting support from Confederates), it entirely ignored Lincoln's "suggestion" concerning limited black suffrage. Only determined pressure from Governor Hahn and General Banks produced clauses allowing the legislature to extend the suffrage and affording blacks access to state-supported education. One Radical delegate expressed scorn for "these half-way men who are afraid to . . . meet the exigencies of the times and lag behind in this hour of revolution," but the convention, like the February elections, revealed the weakness of the Radical faction. The result was to widen the breach in Unionist ranks, turn the Radicals ever more sharply against the Banks government, and propel them, within a few months, down the road to universal manhood suffrage.[29]

Land and Labor During the Civil War

Of the many questions raised by emancipation, none was more crucial to the future place of both blacks and whites in Southern society than how the region's economy would henceforth be organized. Slavery had been, first and foremost, a system of labor. And while all Republicans agreed that "free labor" must replace slave, few were certain how the transition should be accomplished. "If the [Emancipation] Proclamation makes the slaves actually free," declared the New York Times in January 1863, "there will come the further duty of making them work. . . ." "All this," the Times admitted, "opens a vast and most difficult subject."[30]

28. McCrary, Abraham Lincoln and Reconstruction, 245–53; Debates in the Convention for the Revision and Amendment of the Constitution of the State of Louisiana (New Orleans, 1864), 190, 546, 627; Shugg, Origins of Class Struggle, 203–205.
29. "Diary and Correspondence of Salmon P. Chase," Annual Report of the American Historical Association, 1902, 438; Louisiana Convention Debates, 155, 213–14, 394, 556; Cox, Lincoln, 97–99.
30. New York Times, January 3, 1863.

As the war progressed, the Union army found itself in control of territory ranging from coastal Virginia and South Carolina to the plantation belt along the Mississippi River. Legal title to land under army control was uncertain. Theoretically, the Second Confiscation Act of 1862 raised the prospect of the wholesale forfeiture of property owned by Confederates. This penalty, however, could be imposed only after court proceedings in individual cases, and a clause added at Lincoln's insistence provided that the loss of property would be limited to lifetime of the owner, and not affect his or her heirs. The President had no enthusiasm for large-scale confiscation that, he feared, would undermine efforts to win the support of loyal planters and other Southern whites, and the act remained largely unenforced. Far more land came into federal hands from seizures for nonpayment of taxes (in which case it could be sold at auction), or as abandoned property (which the Treasury Department would then administer). How to dispose of this land, coupled with the organization of its black labor, became points of conflict as former slaves, former slaveholders, military commanders, and Northern entrepreneurs and reformers sought, in their various ways, to influence the wartime transition to free labor.[31]

By 1865, hundreds of thousands of slaves in different parts of the South had become, under federal auspices, free workers. The most famous of these "rehearsals for Reconstruction" occurred on the South Carolina Sea Islands. When the U.S. Navy occupied Port Royal in November 1861, virtually all the white inhabitants (comprising less than one fifth of the area's population), fled to the mainland, leaving behind a community of some 10,000 slaves long accustomed to organizing their own labor. The system of labor employed on mainland rice and Sea Island cotton plantations, in which slaves were assigned daily tasks, completion of which left them free to cultivate their own crops, hunt, fish, or enjoy leisure time, gave these blacks a unique control over the pace and length of their workday. It also enabled slaves to acquire small amounts of property by selling to their masters or in nearby towns crops raised on their own time.[32]

Sea Island blacks, it quickly became clear, possessed their own definition of the meaning of freedom. When the planters fled, the slaves sacked the big houses and destroyed cotton gins; they then commenced planting corn and potatoes for their own subsistence, but evinced considerable resistance to growing the "slave crop," cotton, which "had enriched the masters, but had not fed them." But blacks were not to chart their own

31. Paul W. Gates, *Agriculture and the Civil War* (New York, 1965), 362–70; Charles Fairman, *Reconstruction and Reunion 1864–88: Part One* (New York, 1971), 796.
32. Willie Lee Rose, *Rehearsal for Reconstruction: The Port Royal Experiment* (Indianapolis, 1964), xv; Eric Foner, *Nothing But Freedom: Emancipation and Its Legacy* (Baton Rouge, 1983), 78–79; Philip D. Morgan, "Work and Culture: The Task System and the World of Lowcountry Blacks, 1770 to 1880," *William and Mary Quarterly*, 39 (October 1982), 587–93.

path to free labor, for in the navy's wake came a white host from the North—military officers, Treasury agents, Northern investors, and a squad of young teachers and missionaries known collectively as Gideon's Band, the men fresh from Harvard, Yale, or divinity school, the women from careers as teachers and work in the abolitionist movement. Each group had its own ideas about how the transition to freedom should take place and how to judge its success or failure. And the entire Sea Island experiment took place in a blaze of publicity, as the area became a mecca for newspapermen, government investigators, and others hoping to learn how the freedmen adjusted to the end of slavery.[33]

Perhaps the most dramatic part of the story was the encounter between idealistic young reformers and Sea Island freedmen. The Gideonites had had little previous contact with blacks, aside from the occasional abolitionist speaker or minister. Their expectations were shaped both by reading *Uncle Tom's Cabin* and other embodiments of midcentury "romantic racialism," and by the free labor ideology. The first taught that blacks, while superior to whites in religious devotion, lacked the "manly," aggressive traits of Anglo-Saxons. The latter suggested that the debilitating effects of slavery—the destruction of families and family "instincts," a penchant for lying and stealing, and a lack of self-reliance—could be overcome through education, and that blacks could be prepared to take their place in the competitive world of the marketplace. Clearly, the Gideonites carried with them paternalistic attitudes; yet most of these young men and women genuinely desired to assist the freedmen. Many supported blacks' desire to acquire land, and even sympathized with their reluctance to plant cotton. "The negro can see plainly enough," wrote teacher Laura M. Towne, "that the proceeds of the cotton will never get in black pockets."[34]

The most highly publicized Northerners on the islands, the Gideonites were also the least powerful. More influential were Treasury officials, army officers, and those, lured by the fabulously high price of cotton, who proposed to employ the ex-slaves as paid plantation laborers. The relative power of these groups became apparent in 1863 and 1864, when Treasury agents auctioned Sea Island land seized for nonpayment of taxes. Despite efforts by the Gideonites to secure some kind of preferential treatment for blacks, only a small portion of the land went to groups

33. Rose, *Rehearsal*, 40–48, 79, 237; Elizabeth W. Pearson, ed., *Letters From Port Royal Written at the Time of the Civil War* (Boston, 1906), 181; Edward L. Pierce, "The Freedmen at Port Royal," *Atlantic Monthly*, 12 (September 1863), 299.

34. Joseph A. Mills, "Motives and Behaviors of Northern Teachers in the South During Reconstruction," *Negro History Bulletin*, 42 (January 1979), 7; George M. Fredrickson, *The Black Image in the White Mind: The Debate on Afro-American Character and Destiny, 1817–1914* (New York, 1971), 97–129; William H. Pease and Jane H. Pease, *Black Utopia* (Madison, Wis., 1963), 131–33; Rose, *Rehearsal*, 90–92, 217–18; Rupert S. Holland, ed., *Letters and Diary of Laura M. Towne* (Cambridge, Mass., 1912), 20.

of freedmen, who pooled their meager resources to purchase it. Many plantations ended up in the hands of army officers, government officials, and Northern speculators and cotton companies. Eleven plantations were purchased by a consortium of Boston investors that included Edward Atkinson, agent for six Massachusetts textile firms, and Edward S. Philbrick, assistant superintendent of the Boston & Worcester Railroad.[35]

Motivating the likes of Atkinson and Philbrick was a typically American combination of reform spirit and desire for profit. One abolitionist journal called Philbrick "a second Wilberforce and Astor united into one." In the eyes of these antislavery entrepreneurs, Port Royal offered the perfect opportunity to demonstrate that "the abandonment of slavery did not imply the abandonment of cotton" and that blacks would work more efficiently and profitably as free laborers than as slaves. An early advocate of wartime emancipation, Atkinson had long viewed slavery as a violation of "sound principles of political economy," since it relied upon coercion to elicit labor, rather than the promise of reward and advancement. Although he considered the desire of blacks to cultivate food rather than cotton thoroughly misguided, he envisioned the freedmen eventually acquiring land through "the ordinary workings of our system of land tenure." In the meantime, by working for wages, they would internalize a market orientation, so that as landowners, they would become productive cotton farmers and a "large new market . . . [for] Northern manufacturers."[36]

Sent to the Sea Islands to oversee the experiment, Philbrick sought to create a model free-labor environment, with blacks neither exploited by their employers nor lapsing into dependency upon the government. In order to "multiply their simple wants" and stimulate their desire for cash wages, he established plantation stores, placing a variety of "knick-knacks and household comforts" from the North within the freedmen's reach. And he opposed efforts to allow blacks access to land at below the market price, insisting "no man . . . appreciates property who does not work for it." He failed to consider the possibility that the former slaves had worked for the land during their 250 years of bondage.[37]

Was the free labor experiment a success? One Gideonite, William C. Gannett, believed so, pointing to an improvement in black living conditions—wooden chimneys replaced by brick, better clothing, a more var-

35. Rose, *Rehearsal*, 64–68, 200–15, 272–96, 313; Pierce, "Freedmen at Port Royal," 310; Joel Williamson, *After Slavery: The Negro in South Carolina During Reconstruction, 1861–1877* (Chapel Hill, 1965), 54–58.

36. Pease and Pease, *Black Utopia*, 157; Rose, *Rehearsal*, 36–38, 50; Edward Atkinson, *Cheap Cotton by Free Labor: by a Cotton Manufacturer* (Boston, 1861), 3–5, 49; [Edward Atkinson] "The Future Supply of Cotton," *North American Review*, 98 (April 1864), 495–97.

37. Eric Foner, *Politics and Ideology in the Age of the Civil War* (New York, 1980), 108–109; Pearson, ed., *Letters from Port Royal*, 219–21, 245, 276–77; George W. Smith, "Some Northern Wartime Attitudes Toward the Post-Bellum South," *JSH*, 10 (August 1944), 262.

ied diet. Philbrick himself remained uncertain. Personally, it was lucrative enough, earning him $20,000 in 1863 alone. But Philbrick believed that "the amount of cotton planted will always be a pretty sure index to the state of industry of the people," and the freedmen continued to prefer growing provision crops to cotton. By 1865, concluding that blacks "will not produce as much cotton in this generation as they did five years ago," he divided his plantations into small parcels, sold them to the laborers, and returned to Massachusetts. Philbrick's employees may have viewed the outcome more favorably, since on other plantations acquired by Northerners, freedmen were required by the army to sign labor contracts or leave the premises. In the end, the experiment underscored both the ambiguities within the concept of "free labor" itself and the conflicting interests lurking beneath the common aspiration of reconstructing Southern society. As Northern investors understood the term, "free labor" meant working for wages on plantations; to blacks it meant farming their own land, and living largely independent of the marketplace.[38]

Despite the attention lavished upon it by both contemporaries and historians, the Sea Island experience was anything but typical.[39] In some sense, it took place in a social and political vacuum. It involved a relatively small number of freedmen and remained insulated from both the presence of the former slaveowners and, until late in the war, the disruptive impact of the massive flight of slaves to Union lines. Idealistic reformers played a greater role here than elsewhere in the South, and the issue of wooing local white support did not arise. Far larger in scope, more diverse in its cast of characters, and more indicative of the future course of Southern labor relations was the system that evolved in southern Louisiana. Here, as in the Sea Islands, slavery disintegrated as federal troops approached and blacks aspired to own the land. But occupied Louisiana also contained a large group of Unionist planters, who called upon the army to enforce plantation discipline. The federal military commander, Gen. Nathaniel P. Banks, was thoroughly committed to ending slavery. But he remained convinced that maintaining the plantation system would relieve the army of the burden of caring for black refugees, restore the vitality of the state's economy, and assist in creating a

38. [William Gannett] "The Freedmen at Port Royal," North American Review, 101 (July, 1865), 23; Pease and Pease, Black Utopia, 147–54; Pearson, ed., Letters from Port Royal, 275; Rose, Rehearsal, 226–28; Williamson, After Slavery, 68.

39. Another "rehearsal" similar in some respects to that in the Sea Islands took place in the Hampton region of tidewater Virginia, which was captured by Union troops even earlier than Port Royal. Here, too, Northern missionaries and teachers came to organize schools and assist the freedmen in various ways and, as in the Sea Islands, tensions developed between the missionaries and the freedmen. A small number of black families were settled on abandoned land by the army, but most were required to sign labor contracts or face military discipline. Robert F. Engs, Freedom's First Generation: Black Hampton, Virginia, 1861–1890 (Philadelphia, 1979), 30–78; Richard L. Morton, " 'Contrabands' and Quakers in the Virginia Peninsula, 1862–1869," VaMHB, 61 (October 1953), 419–29.

Free State movement with broad support among the white population.

The policy of transforming slaves into paid workers on Louisiana's sugar plantations, under conditions mandated by the army, was initiated in 1862 by Gen. Benjamin F. Butler. A response to the flight of slaves from the plantations and the insubordination of those who remained, Butler's policy required blacks to continue to labor on the estates of loyal masters, where they would receive wages according to a fixed schedule, as well as food, medical care, and provision for the aged and infirm. Corporal punishment was prohibited, but army provost marshals could discipline blacks for refusing to work. Abandoned plantations, meanwhile, would be leased to Northern investors. Legally, these black employees were still slaves, but Butler's plan inescapably suggested that the transition to free labor had begun.[40]

Butler's successor extended this labor system throughout occupied Louisiana. In January 1863, Banks met with a group of loyal planters and bluntly informed them that their previous experience in managing slave laborers was worthless, their "theories, prejudices and opinions based on the old system" out of date, and that they must adapt themselves to a new order resting on free labor. Yet when Banks issued labor regulations for the coming year, many critics charged that they bore a marked resemblance to slavery. The former slaves, he announced, must avoid vagrancy and idleness, and the army would "induce" them to enter into yearly contracts with planters, for which they would receive 5 percent of the proceeds of the year's crop or a wage of $3 per month, as well as food, shelter, and medical care. Once hired, the blacks were forbidden to leave the plantations without permission of their employers.[41]

Depending upon one's point of view, Banks's system was either, as he saw it, "the first step in the transition from slave to free labor," or a cynical device to win planters' support for Reconstruction by using the army to restore plantation discipline. The pass system and the fact that blacks had no choice but to sign contracts and little leeway in negotiating terms, led many critics to charge that "the relation of master and slave . . . is the same as heretofore." The zeal of provost marshals in rounding up "vagrant" blacks produced complaints that the army was acting more like a slave patrol than an agent of emancipation. Yet Chaplain George H. Hepworth, one of Banks's labor superintendents, insisted in 1864 that "the whole plan was devised and executed for the well-being of the negro alone." Banks later claimed to have sent a group of free blacks into the plantation belt "to ascertain what the negroes wanted." His emissaries discovered, Banks related, that the freedmen's conception of freedom

40. William F. Messner, *Freedmen and the Ideology of Free Labor: Louisiana 1862–1865* (Lafayette, La., 1978), 21–39; Sitterson, *Sugar Country*, 219–21.
41. Sitterson, *Sugar Country*, 219; Messner, *Freedmen and Free Labor*, 54; McCrary, *Abraham Lincoln and Reconstruction*, 115–21.

emphasized above all the sanctity of the family, education for their children, the end of corporal punishment, and payment of reasonable wages. All these, the general insisted, were provided for by his policy, especially after revised regulations in 1864 substantially increased wages and required planters to supply laborers with garden plots, permitted the freedmen to choose their employers, and allowed black children to attend schools financed by a property tax.[42]

A compulsory system of free labor was an anomaly born of the exigencies of war, ideology, and politics. Hepworth likened it to the apprenticeship established in 1834 to ease the transition from slavery to freedom in the British Caribbean—an unintentionally apt comparison since both arrangements were halfway houses that satisfied no one. Blacks, who resented having yearly contracts forced upon them and found the wages inadequate, labored irregularly, devoting much of their time to their garden plots and often refusing to obey their employers altogether. "They work less, have less respect, are less orderly than ever," one planter complained. Planters, for their part, believed the ban on corporal punishment rendered the entire system useless since blacks, they were convinced, would never labor efficiently without it. Banks's system, moreover, was always subordinate to military needs, for, when required, the army impressed black laborers into its ranks directly from the plantations. The system also became a political football in a conflict between the army and Treasury Department for control of abandoned property and in the factional fighting of Free State Louisiana. For all these reasons, the labor situation remained chaotic and the army never succeeded in reviving the state's agriculture. The rich sugar parishes, wrote one planter, were enveloped in "darkness and gloom . . . plantations abandoned, fences and buildings destroyed, . . . the negroes conscripted into the army or wandering about. . . . Such is war, civil war."[43]

At its peak, Banks's system involved some 50,000 laborers on nearly 1,500 estates, working either directly for the government or for individual planters under contracts supervised by the army. And after the fall of Vicksburg, when Union forces took control of the vast cotton belt along

42. Cecil D. Eby, Jr., ed., *A Virginia Yankee in the Civil War* (Chapel Hill, 1961), 148–50; "Diary and Correspondence of Chase," 378–79; C. Peter Ripley, *Slaves and Freedmen in Civil War Louisiana* (Baton Rouge, 1976), 90–95; George H. Hepworth, *The Whip, Hoe, and Sword* (Boston, 1864), 27–28; Nathaniel P. Banks, *Emancipated Labor in Louisiana* (n.p., 1864), 6–7; B. I. Wiley, "Vicissitudes of Early Reconstruction Farming in the Lower Mississippi Valley," *JSH*, 3 (November 1937), 443–45; McCrary, *Abraham Lincoln and Reconstruction*, 155–56.

43. McCrary, *Abraham Lincoln and Reconstruction*, 149–50; Foner, *Nothing But Freedom*, 16–18; Messner, *Freedmen and Free Labor*, 74; Sitterson, *Sugar Country*, 212–13, 224; J. Carlyle Sitterson, "The Transition from Slave to Free Economy on the William J. Minor Plantations," *AgH*, 17 (January 1943), 218–20; Louis S. Gerteis, *From Contraband to Freedman: Federal Policy Toward Southern Blacks, 1861–1865* (Westport, Conn., 1973), 88–92; Charles P. Roland, *Louisiana Sugar Plantations During the American Civil War* (Leiden, Neth., 1957), 127.

the Mississippi River with a black population of over 700,000, the system was extended to the entire Mississippi Valley. Here, the army's first concern was not the labor system per se, but the masses of slaves who fled to Union lines. In November 1862 General Grant had appointed John Eaton, a Dartmouth graduate and former superintendent of schools in Toledo, "to take charge of the contrabands." Already, black fugitives and the families of black military laborers were congregating in makeshift shanty towns ringing army posts in Tennessee and northern Mississippi. Eaton established a network of "contraband camps" and "home farms" where schools were set up, medical care provided, and temporary employment offered. After the fall of Vicksburg, the camps were overwhelmed by the black exodus from Mississippi plantations; the cabins became overcrowded, disease was rife, the death rate soared.[44]

In the spring of 1863, Gen. Lorenzo Thomas devised a plan to lease plantations along the Mississippi River to loyal men from the North, who would hire black laborers on terms prescribed by the army. The system expanded greatly after the fall of Vicksburg. Thomas hoped to consolidate Union control of the Mississippi Valley by settling there a population of loyal planters and laborers. But he also aimed to relieve the army of the expense and burden of supporting blacks in contraband camps and cities like Natchez and Vicksburg. Further, he sought to inculcate among blacks the habit (which army men believed they needed to be taught) of working for a living rather than relying upon the government for support. Essentially, Thomas offered able-bodied black men the choice of joining the army, working as military laborers, or signing plantation contracts. Black women, including the families of soldiers, were expected to go to work for Northern lessees or for Southern planters who renounced their allegiance to the Confederacy.[45]

By all accounts, the Northern men who leased plantations were "an unsavory lot," attracted by the quick profits seemingly guaranteed in wartime cotton production. In the scramble among army officers illegally engaged in cotton deals and Northern investors seeking to "pluck the golden goose" of the South, the rights of blacks received scant regard. In order to put the system into operation quickly, General Thomas set wages at a low level ($7 per month for men, $5 for women, minus the cost of medical attention and clothing). Even then, many lessees defrauded

44. Harrington, *Banks*, 105–106; John Eaton, *Grant, Lincoln and the Freedmen* (New York, 1907), 2–5, 207; Weymouth T. Jordan, "The Freedmen's Bureau in Tennessee," *ETHSP*, 11 (1939), 47–48; Steven J. Ross, "Freed Soil, Freed Labor, Freed Men: John Eaton and the Davis Bend Experiment," *JSH*, 44 (May 1978), 217; Gaines M. Foster, "The Limitations of Federal Health Care for Freedmen, 1862–1868," *JSH*, 48 (August 1982), 350–57.

45. James T. Currie, *Enclave: Vicksburg and Her Plantations, 1863–1870* (Jackson, 1980), 56–58; Gerteis, *From Contraband to Freedman*, 123–26; Noralee Frankel, "Workers, Wives, and Mothers: Black Women in Mississippi, 1860–1870" (unpub. diss., George Washington University, 1983), 31–39.

the freedmen of their earnings. In the winter of 1863–64, the Treasury Department briefly assumed control of the Mississippi Valley labor system, mandated a substantial increase in black wages, and contemplated leasing the plantations directly to the freedmen. The Treasury's regulations, complained Alexander Winchell, zoology professor at the University of Michigan were "framed in the exclusive interest of the negro and in the non-recognition of the moral sense and patriotism of the white man." (Professor Winchell had organized the Ann Arbor Cotton Company and sold stock to the university's president, whereupon he received a leave of absence to engage in cotton planting.) After a direct appeal to Lincoln, military authority was restored, whereupon Thomas reduced wages, even for those blacks who had already signed Treasury-designed contracts.

As in Louisiana, upon whose labor regulations Thomas now modeled his own, compulsory free labor did not produce the expected results. The lessees found themselves unable to obtain tools, wagons, or food, all of which were monopolized by the army. They stood helpless in the face of depredations by both Confederate raiders (over a dozen lessees were killed during 1864, along with hundreds of freedmen) and the army worm (an insect named for its resemblance to officers who found ways to appropriate nine tenths of the crop). Every payday, moreover, seemed to bring disputes with the labor force. The vast majority of the lessees returned home before the end of 1864.[46]

Despite local variations in policy, most army officials in charge of wartime labor relations assumed that the emancipated slaves should remain as plantation laborers. Only occasionally did glimmerings of an alternative point of view appear. In occupied Virginia and North Carolina, a few freedmen were settled on abandoned land, and several hundred managed to lease farms in the Mississippi Valley. Often, however, these were merely tiny garden plots (on Roanoke Island, North Carolina, one acre per family), to which blacks' legal title would prove anything but secure. The largest laboratory in black economic independence was Davis Bend, a peninsula formed by the tortuous course of the Mississippi River just south of Vicksburg, which contained the huge plantations of Confederate President Jefferson Davis and his brother Joseph. Davis Bend had already been the site of one utopian experiment before the Civil War. Influenced by Joseph's encounter with British socialist Robert Owen, the Davis brothers had attempted to establish a model slave community, with blacks far better fed and housed than elsewhere in the state and permitted an

46. Wiley, "Vicissitudes of Farming," 442, 446–50; Ludwell H. Johnson, "Northern Profit and Profiteers: The Cotton Rings of 1864–1865," *CWH*, 12 (June 1966), 101–15; David H. Overy, Jr., *Wisconsin Carpetbaggers in Dixie* (Madison, Wis., 1961), 16–20; Vernon L. Wharton, *The Negro in Mississippi 1865–1890* (Chapel Hill, 1947), 32–38; Currie, *Enclave*, 67, 79–80; Martha M. Bigelow, "Plantation Lessee Problems in 1864," *JSH*, 27 (August 1961), 354–67.

REHEARSALS FOR RECONSTRUCTION

extraordinary degree of self-government, including a slave jury system that enforced plantation discipline. Other planters mocked "Joe Davis's free negroes," but the system enhanced the family's reputation among blacks. After the war, one group of Mississippi freedmen pressed for Jefferson Davis's release from prison because "altho he tried hard to keep us all slaves . . . some of us well know of many kindness he shown his slaves on his plantation."[47]

The Civil War destroyed the "model" slave system at Davis Bend and exposed undercurrents of discontent of which the Davis brothers had been unaware. When Joseph Davis fled his plantation in 1862, the slaves not only refused to accompany him, but broke into his mansion and appropriated clothing and furniture. By the time Union troops arrived, the blacks were running the plantation. In 1863, General Grant decided that Davis Bend should become a "negro paradise" and directed John Eaton to lease land to the freedmen and establish a "home farm" for the black refugees who were crowding into the area. The following year the entire area was set aside for the exclusive settlement of freedmen, the land assigned collectively to groups of blacks who were to pay only for rations, mules, and tools belonging to the government. The leading spirit of the enterprise was Benjamin Montgomery, an embodiment of nascent black capitalism, who as a slave had been allowed to establish a plantation store and handle the Davis brothers' cotton transactions. By 1865 Davis Bend had become a remarkable example of self-reliance, whose laborers raised nearly 2,000 bales of cotton and earned a profit of $160,000. The community had its own system of government, complete with elected judges and sheriffs. From Davis Bend would emerge a number of black leaders of Mississippi Reconstruction, including Israel Shadd, speaker of the state's House of Representatives and husband of a Bend schoolteacher, Albert Johnson, who served in Mississippi's constitutional convention and legislature, and Thornton Montgomery, Benjamin's son, who became the first black to hold office in the state when Gen. E. O. C. Ord appointed him postmaster at Davis Bend in 1867.[48]

Under the direction of Benjamin Montgomery, Davis Bend demonstrated that not all blacks, if given the choice, would eschew the mar-

47. Gerteis, *From Contraband to Freedman*, 29, 33–37, 167–71, 184; Jerrell H. Shofner, *Nor Is It Over Yet: Florida in the Era of Reconstruction, 1863–1877* (Gainesville, Fla., 1974), 28; Edward Magdol, *A Right to the Land: Essays on the Freedmen's Community* (Westport, Conn., 1977), 93–94; Engs, *Black Hampton*, 41; Currie, *Enclave*, 75–76; Janet S. Hermann, *The Pursuit of a Dream* (New York, 1981), 13–34; "We the Colored People" to Benjamin G. Humphreys, December 3, 1865, F-41 1865, Registered Letters Received, Ser. 2052, Ms. Asst. Comr., RG 105, NA [FSSP A-9035].

48. Hermann, *Pursuit of a Dream*, 35–87, 121, 183–84, 196–97; Currie, *Enclave*, 92–144; Ross, "Freed Soil," 218–30.

ketplace and cotton. But it was not Davis Bend, any more than the South Carolina Sea Islands, that proved to be the true rehearsal for Reconstruction so far as labor relations were concerned. Despite their apparent failure, the Louisiana and Mississippi Valley experiments in free labor not only involved far larger numbers of blacks, but established a system of plantation agriculture based on yearly labor contracts, which would carry over into the postwar policies of the army and Freedmen's Bureau. Almost by default, and guided as much by wartime conditions as by any well-considered blueprint for the postwar South, military men had made crucial policy decisions that began to resolve one of the most complex problems to arise from the Civil War. And their labor policies exacerbated a developing split within the Republican party over the course of wartime Reconstruction and the implications of emancipation.

The Politics of Emancipation and the End of the War

By mid-1864, a white resident of Chattanooga noted, life in the Union-occupied South was "so different from what it used to be." Blacks were attending schools run by Yankee teachers, and hundreds of thousands were for the first time receiving wages for their labor. But instead of winning unqualified applause from Northern Republicans, the course of events in the South threatened to divide them. With emancipation now an article of party faith, political debate increasingly centered upon the freedmen's postwar status. Many Republicans were particularly troubled by developments in Louisiana, from which emerged disturbing reports concerning the state government and General Banks's labor system.

Already, abolitionists like Wendell Phillips insisted that Reconstruction could never be complete until blacks had been guaranteed education, access to land, and, most importantly, the ballot. But few in Congress echoed these demands; in 1864 the major concern even of Radical Republicans was equality before the law, not black suffrage, and the control of new Southern governments by genuine Unionists. On both counts, the new government of Louisiana appeared wanting. Banks's labor system and the overtly antiblack views expressed by his supporters convinced many Republicans that the freedmen could not receive equitable treatment from governments organized under Lincoln's 10 Percent Plan. And the plan's lenient pledge of future loyalty increasingly appeared a less adequate test of Unionism than the "ironclad oath" that one had never voluntarily aided the Confederacy. "The people of the North," Radical Sen. Jacob Howard told the Senate, "are not such fools as to fight through such a war as this, . . . and then turn around and say to the traitors, 'all you have to do is to come back into the councils of the nation and take an oath that henceforth you will be true to the Government'."[49]

The tangled threads of dissatisfaction with events in Louisiana, concern for the fate of the freedmen, and rival definitions of loyalty to the Union—with a dash of Republican Presidential politics thrown in—came together in July 1864 to produce the Wade-Davis bill. An alternative to Lincoln's 10 Percent Plan, this proposed to delay the start of Reconstruction until a majority of a state's white males had pledged to support the federal Constitution. After this had occurred, elections would be held for a constitutional convention, with suffrage restricted to those who could take the Ironclad Oath. The bill also contained guarantees of equality before the law, although not suffrage, for the freedmen. Fearing the measure would force him to repudiate the Louisiana regime, Lincoln pocket-vetoed it, adding that he had no objection if any Southern state chose to adopt the Wade-Davis plan voluntarily (hardly a likely occurrence). Whereupon the bill's Radical authors, Ohio Sen. Benjamin F. Wade and Maryland Congressman Henry Winter Davis, issued an intemperate "manifesto," accusing Lincoln of defying the judgement of Congress and exercising "dictatorial usurpation."[50]

Despite the harsh language of the Wade-Davis Manifesto, these events did not signal an irreparable breach between Lincoln and the Radical Republicans. The points of unity among Republicans, especially their commitment to winning the war and rendering emancipation unassailable, were far greater than their differences (even though many Radicals in 1864 preferred a different Presidential candidate). Lincoln enjoyed good personal relations with Radical leaders like Massachusetts Sen. Charles Sumner, and had cooperated with Radicals on issues ranging from war appropriations to the recruitment of black troops. The Wade-Davis bill, moreover, was not the work of a narrow faction, for it won almost unanimous support among Congressional Republicans, most of whom believed that Congress should have a larger voice in shaping Reconstruction policy and that greater care was needed before new state governments were created. As Massachusetts Congressman Henry L. Dawes later put it, moderates like himself had become convinced that "something more Radical" than the 10 Percent Plan was now in order.[51]

Nonetheless, the controversy underscored the party's genuine differ-

49. Lester C. Lamon, *Blacks in Tennessee 1791–1970* (Knoxville, 1981), 34; James M. McPherson, *The Struggle for Equality: Abolitionists and the Negro in the Civil War and Reconstruction* (Princeton, 1964), 239–43; Hyman, *Era of the Oath*, 23; *CG*, 38th Congress, 1st Session, 294–96.

50. Harold M. Hyman and William M. Wiecek, *Equal Justice Under Law: Constitutional Development 1835–1875* (New York, 1982), 269–74; Herman Belz, *A New Birth of Freedom: The Republican Party and Freedmen's Rights 1861–1866* (Westport, Conn., 1976), 57–62; Richardson, ed., *Messages and Papers*, 6:222–23.

51. Hans L. Trefousse, *The Radical Republicans: Lincoln's Vanguard for Racial Justice* (New York, 1969), 231, 250, 265; Donald, *Sumner*, 206–207; Herman Belz, *Emancipation and Equal Rights* (New York, 1978), 14, 25–30; *CG*, 41st Congress, 1st Session, 406.

ences concerning Reconstruction. Lincoln viewed Reconstruction as part of the effort to win the war and secure emancipation. His aim was to weaken the Confederacy by establishing state governments that could attract the broadest possible support, and for this purpose he defined as a Unionist virtually every white Southerner who took an oath pledging to uphold the Union and the abolition of slavery. To the Radicals, Reconstruction implied a far-reaching transformation in Southern society; as a result, they wished to delay the process until after the war and to limit participation to a smaller number of "iron-clad loyalists." As George S. Boutwell of Massachusetts told the House, the Wade-Davis bill contained "the germs of a new civilization for one half of a continent." Wade-Davis limited suffrage to loyal whites (to include blacks, said Senator Wade, would "sacrifice the bill"). But many Radicals were already convinced that black suffrage must come. In many states, Boutwell insisted, the freedmen "are almost the only people who are trustworthy supporters of the Union." Already in May 1864, a black meeting at Beaufort had elected sixteen delegates to the Republican National Convention, among them four blacks, including future Congressman Robert Smalls and Prince Rivers, then a sergeant in the Union Army and later a member of South Carolina's constitutional convention and legislature. But neither the statements of Radicals nor the appearance of Sea Island blacks at the convention that renominated Lincoln forced black suffrage onto the center stage of politics. That was accomplished by the political mobilization of New Orleans' free blacks, who compelled Congress and the President to grapple with the question as Louisiana sought readmission to the Union.[52]

Several developments during 1864 propelled these free blacks down a radical road. They were shocked by the racism so pervasive at Louisiana's constitutional convention and by the new lawmakers' complete indifference to their claims. The legislature rejected a Quadroon Bill, supported by General Banks, that would have given the vote to free men of color possessing three quarters white blood, made no appropriation for black education, and, in fact, did nothing "except provide for the pay of its members." Free blacks also discovered that "vagrancy" and curfew regulations issued in conjunction with the Banks labor system did not respect the distinction between themselves and the freedmen, placing severe limits upon their accustomed freedom of movement.[53]

For over a year, suffrage for men of free birth had been demanded by L'Union, a newspaper founded in 1862 by a group of wealthy free blacks

52. Michael L. Benedict, *A Compromise of Principle: Congressional Republicans and Reconstruction 1863–1869* (New York, 1974), 80–83; Cox, *Lincoln*, 36–41; Belz, *Reconstructing the Union*, 236–40; Hans L. Trefousse, *Benjamin Franklin Wade: Radical Republican from Ohio* (New York, 1963), 221; *CG*, 38th Congress, 1st Session, 2103; Rose, *Rehearsal*, 316.

53. Herman Belz, "Origins of Negro Suffrage During the Civil War," *SS*, 17 (Summer 1978), 123; Gerteis, *From Contraband to Freedman*, 100–102.

that appeared first in French and then bilingually. The first black newspaper to appear outside the Northern states, it enjoyed only limited influence and in 1864 suspended publication. But soon after the constitutional convention, the New Orleans *Tribune* was created as a rallying point for Louisiana Radicalism. The paper's founder and guiding spirit was Louis C. Roudanez, the wealthy son of a French merchant and a free woman of color, who had earned medical degrees at both the University of Paris and Dartmouth. In November 1864, Jean-Charles Houzeau, one of the most remarkable men to take part in the saga of Reconstruction, became the editor. Born to an aristocratic Belgian family, this journalist and astronomer had been converted to revolutionary ideas and thereupon lost his position at the Belgian Royal Observatory in 1849. He emigrated to Texas in 1858, sided with Unionists there early in the Civil War, and in 1864 arrived in Louisiana. In Houzeau, the *Tribune*'s proprietors found a man whose political outlook, like their own, had been shaped by the heritage of the Enlightenment and the French Revolution, and who identified the cause of American blacks as "only one chapter in the great universal fight of the oppressed of all colors and nations." In Dr. Roudanez and the others associated with the newspaper, Houzeau recognized "the vanguard of the African population of the United States."[54]

Even before Houzeau's arrival, the *Tribune* had moved beyond the earlier politics of the free black community. In August and September 1864, it condemned the Banks labor system as a reincarnation of slavery that allowed blacks to be "chained to the soil" and exploited by "avaricious adventurers from the North." "Every man," the *Tribune* insisted, "should own the land he tills." And it made the momentous decision to demand suffrage for the former slaves, the free blacks' *"dormant* partners." But it was Houzeau who transformed the *Tribune* into a journal widely respected in Northern Republican circles and known even in Europe (in 1865 it received a letter from Victor Hugo), and who broadened its message to encompass a coherent radical program embracing black suffrage, equality before the law, desegregation of Louisiana's schools, the opening of New Orleans streetcars to blacks, and division of the plantations among the freedmen. And it was he who made the alliance between free blacks and the freed the cornerstone of *Tribune* politics, as the only means of preventing the revolution unleashed by the Civil War from succumbing to reaction. As he wrote in December:

These two populations, equally rejected and deprived of their rights, cannot be well estranged from one another. The emancipated will find, in the old freemen, friends ready to guide them, to . . . teach them their duties as well

54. William P. Conner, "Reconstruction Rebels: The New Orleans *Tribune* in Post-Civil War Louisiana," *LaH*, 21 (Spring 1980), 161–65; Houzeau, *My Passage*, 2–5, 19–23, 75, 78.

as their rights. . . . The freemen will find in the recently liberated slaves a
mass to uphold them; and with this mass behind them they will command
the respect always bestowed to number and strength. . . .[55]

Leaving behind the heritage of white paternalism ("the age of guardianship is past forever," remarked the *Tribune*, "we now think for ourselves"), the free blacks clearly retained more than a hint of noblesse
oblige when it came to the freedmen. Not surprisingly, the light-
skinned, propertied, French-speaking, and Catholic free blacks believed that their role was to elevate and offer guidance to the former
slaves. The *Tribune*'s Religious Department lectured freedmen on the
sanctity of marriage and the importance of "correct deportment and
gentlemanly conduct" among ministers. But once the paper made its
commitment to suffrage for the freedmen, it did not turn back, and it
brought in tow the Radical party of Louisiana. In December 1864,
mass meetings in New Orleans heard white Radical Thomas J. Durant
and free blacks Oscar J. Dunn and James H. Ingraham (the latter recently returned from the national convention at Syracuse) demand
black suffrage. "We regard all black and colored men," Dunn declared,
"as fellow sufferers."[56]

In January 1865, with Ingraham as chairman, a convention of the Equal
Rights League assembled in New Orleans, its resolutions demanding
black suffrage and equal access to the city's streetcars, its speakers denouncing the state legislature for treating blacks with contempt. The
gathering, declared the *Tribune*, epitomized the new unity of Louisiana's
nonwhite population: "There, were seated side by side the rich and the
poor, the literate and educated man, and the country laborer, hardly
released from bondage." In fact, most delegates were New Orleans free
blacks, including a considerable number of army officers. More accurate
was another of the newspaper's assessments: "The speakers whom we
have seen rising to prominence in this Convention will be the champions
of their race." For although members of the free elite predominated,
those who spoke only French were obliged to take a secondary role. To
the fore now came English-speaking men of more humble origins, including Ingraham and Dunn, both destined to play major roles in Reconstruction.[57]

Despite the preponderance of free blacks, the convention revealed that
a new group, the emancipated slaves, had entered Louisiana's political
arena. Moreover, as the *Tribune* noted, "the country delegates were gen-

55. New Orleans *Tribune*, August 4, 11, 13, September 10, 24, December 29, 1864;
Houzeau, *My Passage*, 23–39, 79–80.
56. New Orleans *Tribune*, December 3, 27, 1864, January 20, 21, 1865; Tunnell, "Free
Negroes and Freedmen," 22–23; Vincent, *Black Legislators*, 30–31.
57. New Orleans *Tribune*, January 10–15, 1865; Houzeau, *My Passage*, 96–97; Tunnell,
"Free Negroes and Freedmen," 14–15.

erally more radical than most of the city delegates." One may surmise that the rural freedmen placed as much emphasis upon the interrelated land and labor questions as upon the suffrage. Certainly, during the early months of 1865 the New Orleans Radicals devoted increasing attention to economic issues. The *Tribune* now outlined plans for a cooperative economic order, probably influenced by Thomas J. Durant, a follower of the French utopian socialist Charles Fourier. The freedmen would be assisted in purchasing shares in "self-help banks," which, in turn, would acquire land to be operated by voluntary associations of laborers. Gone would be the compulsory contracts and vagrancy regulations of Banks's system. Military officials and the new state government remained unimpressed by this ferment within the black community. In March 1865 Gen. Stephen A. Hurlbut, Banks's successor, issued labor regulations for the coming year that reiterated the provisions the *Tribune* found most distasteful—compulsory yearly contracts, fixed wages, a pass system. A mass meeting in New Orleans denounced the "Banks oligarchy" and demanded the entire abolition of the army's Bureau of Free Labor, and the establishment of courts, composed in part of blacks, to hear labor disputes previously settled by provost marshals.[58]

Attuned to laissez-faire ideas and convinced by three years' experience that labor regulation in any form provided a cover for the exploitation of blacks, the Radicals had moved from a critique of the Banks system to a rejection of all government intervention in the labor market. General Hurlbut replied that the idea of allowing blacks a voice in arbitrating labor disputes was "impracticable" and went on to insist that without the army's presence, the freedmen would be left at the mercy of their former masters. This was a plausible contention, reiterated by the New Orleans *Black Republican*, a short-lived newspaper that appeared in April 1865 and claimed to speak for "the poor as well as the rich, the freedman as well as the freeman." Clearly inspired by Banks's supporters, the paper nonetheless reflected real social and ideological divisions among the city's black population. Its founders mostly came from outside the French, Catholic, free mulatto elite; the editor, S. W. Rogers, was a slave-born Protestant minister. The *Black Republican* insisted that the former slaves did not share the *Tribune*'s laissez-faire outlook: "We all know that not one in a hundred of our brethren on the plantations would ever receive his just earnings if the planter were left to himself." Here was a preview of differences in economic outlook between the free and the freed that would surface again during Louisiana's postwar Reconstruction.[59]

58. New Orleans *Tribune*, January 15, 29, February 18, 23, March 1, 31, 1865; Houzeau, *My Passage*, 37, 110n.

59. New Orleans *Black Republican*, April 15, 1865; David C. Rankin, "The Politics of Caste: Free Colored Leadership in New Orleans During the Civil War," in Robert R. Macdonald et al., eds., *Louisiana's Black Heritage* (New Orleans, 1979), 145.

Thus, as the Civil War drew to a close, the nation's most articulate and politicized black community was thoroughly estranged from the military and civil authorities of Free Louisiana. Within the state, they did not receive a hearing. But in Washington, their complaints against the Louisiana government found a sympathetic audience. Contact with this cultured, economically successful group challenged the racist assumptions widespread even in Republican circles and doubtless influenced Lincoln's own evolution toward a more egalitarian approach to Reconstruction. Because of Louisiana, black suffrage became a pressing issue in the session of Congress that assembled in December 1864, thwarting efforts to reach agreement on a plan of Reconstruction. Although the President assured key legislators that he would use his influence to secure the extension of the franchise to at least some Louisiana blacks, Congress refused to count the state's electoral vote of 1864, and Charles Sumner's filibuster blocked the seating of Louisiana's newly elected Senators. When Congress adjourned in March 1865, the issue of Reconstruction remained unresolved, and the Radicals prepared themselves for a massive effort to persuade the North of the necessity of black suffrage. "We need now a Tom Paine Common Sense pamphlet," wrote Edward Atkinson, "to cause public opinion to begin to crystallize on the questions of the future."[60]

Despite the Louisiana impasse, this second session of the Thirty-Eighth Congress was indeed a historic occasion. In 1864 the Senate had approved the Thirteenth Amendment, abolishing slavery throughout the Union, but it had failed to receive the required two-thirds majority in the House. On January 31, 1865, by a margin of 119 to 56, the Amendment won House approval, and was forwarded to the states for ratification. Lincoln, Secretary of State Seward, and the Radicals had expended considerable energy to secure this narrow victory. The vote set off wild cheering in the galleries, while Congressmen "joined in the shouting . . . [and] wept like children." The following morning Atkinson dated a letter, "Year 1 of American Independence."[61]

"The one question of the age is *settled,*" declared antislavery Congressman Cornelius Cole, but like so many other achievements of the Civil War, the Amendment closed one issue only to open a host of others. "What is freedom?" James A. Garfield later asked. "Is it the bare privilege of not being chained? . . . If this is all, then freedom is a bitter mockery, a cruel delusion." The meaning of the Amendment's second clause,

60. Cox, *Lincoln,* 103–27; Benedict, *Compromise of Principle,* 84–97; Belz, "Origins of Negro Suffrage," 126–30; Edward Atkinson to John Murray Forbes, February 17, 1865, Edward Atkinson Papers, MHS.
61. Cox, *Lincoln,* 18–19; Schafer, ed., *Intimate Letters of Schurz,* 315; "George W. Julian's Journal—The Assassination of Lincoln" *IndMH,* 11 (December 1915), 327; Edward Atkinson to John Murray Forbes, February 1, 1865, Atkinson Papers.

empowering Congress to enforce abolition with "appropriate legislation," was another open question. Already, many Republicans envisioned a nation where North and South, black and white, were ruled by "one law impartial over all." Once ratified, they believed, the Thirteenth Amendment would bring blacks within the purview of a national citizenship whose fundamental rights were protected by a beneficent federal government.[62]

Even the abolitionist movement could not decide whether the Amendment was an end or a beginning. "My vocation, as an Abolitionist, thank God, is ended," declared William Lloyd Garrison, urging the American Anti-Slavery Society at its May 1865 annual meeting, to dissolve in triumph. To which Frederick Douglass replied, "Slavery is not abolished until the black man has the ballot." After an ill-tempered debate, Garrison's proposal was defeated, Wendell Phillips replaced him as the society's president, and the *National Anti-Slavery Standard* appeared with a new motto on its masthead: "No Reconstruction Without Negro Suffrage." Moving beyond the abolitionists' traditional critique of slavery as essentially a moral condition, Phillips turned his attention to the balance of class power in a reconstructed South. Unless the freedmen were granted the vote, the planters' hegemony would be restored and the promise of emancipation undermined. "I do not believe," he wrote, "in an English freedom, that trusts the welfare of the dependent class to the good will and moral sense of the upper class."[63]

Lurking behind these debates was an even broader question suggested by the end of slavery: Should the freedmen be viewed as individuals ready to take their place as citizens and participants in the competitive marketplace, or did their unique historical experience oblige the federal government to take special action on their behalf? Although they had generally accepted the expansion of national authority during the war, many reformers did not abandon laissez-faire assumptions entirely. Assistance begets dependence, insisted Sea Island teacher William C. Gannett; the sooner blacks were "thrown upon themselves, the speedier will be their salvation." Even Frederick Douglass concluded that the persistent question "What shall we do with the Negro?" had only one answer: "Do nothing. . . . Give him a chance to stand on his own legs! Let him alone!" There was far more here than doctrinaire individualism. The injustices

62. Cornelius Cole, *Memoirs of Cornelius Cole* (New York, 1908), 220; Belz, *New Birth of Freedom*, 25–30, 118–25, 160; Burke A. Hinsdale, ed., *The Works of James Abram Garfield* (Boston, 1882–83), 1:86; Hyman and Wiecek, *Equal Justice Under Law*, 276–78; McPherson, *Struggle for Equality*, 221–22.

63. *National Anti-Slavery Standard*, May 20, 1865; Irving H. Bartlett, *Wendell Phillips: Brahmin Rebel* (Boston, 1961), 289–93; Lawrence J. Friedman, *Gregarious Saints: Self and Community in American Abolitionism 1830–1870* (New York, 1982), 267–70.

done to blacks by provost marshals and other agents of the federal government suggested that special action might result in limiting blacks' freedom—far better to empower blacks to defend their own interests via the vote. Douglass, moreover, knew all too well that the other face of benevolence is often paternalism, and feared that in a society founded, if only rhetorically, on the principle of equality, "special efforts" on the freedmen's behalf might "serve to keep up the very prejudices, which it is so desirable to banish," by promoting an image of blacks as privileged wards of the state.[64]

At the other end of this ideological spectrum stood Radicals advocating an act of federal intervention comparable in scope only to emancipation itself—the confiscation of planter lands and their division among the freedmen. The most persistent Congressional supporter of such a measure was George W. Julian, chairman of the House Committee on Public Lands. Long an enemy of "land monopoly," Julian insisted that without land reform, Southern society could not be remade according to the tenets of "radical democracy," and the freedmen would find themselves reduced to "a system of wages slavery . . . more galling than slavery itself." Julian led a fight to amend the Second Confiscation Act so as to authorize the permanent seizure of Confederates' land, and by the end of the war, both houses had approved separate versions, but no joint measure had been enacted. Nonetheless, the creation of the Freedmen's Bureau in March 1865 symbolized the widespread belief among Republicans that the federal government must shoulder broad responsibility for the emancipated slaves, including offering them some kind of access to land.[65]

The Freedmen's Bureau was the child of the American Freedmen's Inquiry Commission, created by the War Department in 1863 to suggest methods for dealing with the emancipated slaves. Its three members, Samuel Gridley Howe, James McKaye, and Robert Dale Owen, all longtime abolitionists and reformers, visited the Union-occupied South, heard testimony from whites and blacks, and wrote two joint reports and many pages of individual observations. The commission's recommendations reflected the tension between the laissez-faire and interventionist approaches to the aftermath of emancipation. They called for the creation of a Bureau of Emancipation to exercise a benevolent guardianship

64. Pearson, ed., *Letters From Port Royal*, 147; McPherson, *Struggle for Equality*, 397; Richard O. Curry, "The Abolitionists and Reconstruction: A Critical Appraisal," *JSH*, 34 (November 1968), 534–43; Philip S. Foner, ed., *The Life and Writings of Frederick Douglass* (New York, 1950–55), 3:189.

65. Patrick W. Riddleberger, *George Washington Julian: Radical Republican* (Indianapolis, 1966), 188–94; George W. Julian, *Speeches on Political Questions* (New York, 1872), 221–26; McPherson, *Struggle for Equality*, 247–56.

over the freedmen, but insisted that such a bureau must not become "a permanent institution" and that blacks should be placed on the road to self-reliance as quickly as possible. To ensure that federal guardianship need not be permanent, the commission endorsed civil and political equality. McKaye went even further, urging the confiscation and redistribution of the planters' land and a thorough "social reconstruction of the Southern states."[66]

Not until Congress stood on the verge of adjournment in March 1865, and after many months of disagreement between House and Senate over its provisions, was the bill establishing the Freedmen's Bureau adopted. (At the same time, Congress chartered the Freedman's Savings Bank to encourage habits of thrift among the former slaves.) The Bureau was to distribute clothing, food, and fuel to destitute freedmen and oversee "all subjects" relating to their condition in the South. Despite its unprecedented responsibilities and powers, the Bureau was clearly envisioned as a temporary expedient, for not only was its life span limited to one year, but, incredibly, no budget was appropriated—it would have to draw funds and staff from the War Department. Charles Sumner had proposed establishing the Bureau as a permanent agency with a secretary of Cabinet rank—an institutionalization of the nation's responsibility to the freedmen—but such an idea ran counter to strong inhibitions against long-term guardianship. Indeed, at the last moment, Congress redefined the Bureau's responsibilities so as to include Southern white refugees as well as freedmen, a vast expansion of its authority that aimed to counteract the impression of preferential treatment for blacks.[67]

In one respect, however, the Freedmen's Bureau appeared to promise a permanent transformation of the condition of the emancipated slaves. As suggested by its full title—Bureau of Refugees, Freedmen, and Abandoned Lands—it was authorized to divide abandoned and confiscated land into forty-acre plots, for rental to freedmen and loyal refugees and eventual sale with "such title as the United States can convey" (language that reflected the legal ambiguity surrounding the government's hold upon Southern land). Earlier drafts of the bill had envisioned the Bureau's operating plantations itself, with the freedmen as wage laborers. But while hardly a definitive commitment to land distribution, the fi-

66. John G. Sproat, "Blueprint for Radical Reconstruction," *JSH*, 23 (February 1957), 34–40; Harold M. Hyman, ed., *The Radical Republicans and Reconstruction 1861–1870* (Indianapolis, 1967), 113–16, 201; McPherson, *Struggle for Equality*, 178–87; McCrary, *Abraham Lincoln and Reconstruction*, 232.

67. Belz, *New Birth of Freedom*, 75–96; Carl R. Osthaus, *Freedmen, Philanthropy, and Fraud: A History of the Freedman's Savings Bank* (Urbana, Ill., 1976), 1–10; McPherson, *Struggle for Equality*, 189; LaWanda Cox, "The Promise of Land for the Freedmen," *MVHR*, 45 (December 1958), 438–39. The activities of the Freedmen's Bureau are discussed at length in Chapter 4.

nal version clearly anticipated at least some blacks becoming, with the government's assistance, independent farmers in a "free labor" South.[68]

While Congress deliberated, the gods of war, in the person of Gen. William T. Sherman and his 60,000-man army, dealt slavery its death blow in the heart of Georgia and added a new dimension to the already perplexing land question. Having captured Atlanta in September 1864, Sherman set out two months later on his March to the Sea. To Georgia's slaves the arrival of this avenging host seemed, as one federal officer put it, "the fulfillment of the millennial prophecies." By the thousands, men, women, and children abandoned the plantations to follow the Union army. They cheered the destruction of their owners' estates and refused to obey when the troops, following Sherman's orders, attempted to drive them away. "They flock to me, old and young," wrote Sherman from Savannah, "they pray and shout and mix up my name with Moses, and Simon . . . as well as 'Abram Linkom', the Great Messiah of 'Dis Jubilee'."[69]

On January 12, 1865, at the urging of Secretary of War Edwin M. Stanton, who had joined him in Savannah, Sherman gathered twenty leaders of the city's black community. Mostly Baptist and Methodist ministers, the majority had been born into slavery, although several had acquired their freedom before the Civil War. Among those present was James D. Porter, who for ten years had operated a clandestine school for black pupils "who have kept their secret with their studies, at home." The conversation revealed that these black leaders possessed a clear conception of the meaning of freedom. Asked what he understood by slavery, Garrison Frazier, a Baptist minister who had known bondage for sixty years before purchasing his freedom in 1857, responded that it meant one man's "receiving . . . the work of another man, and not by his consent." Freedom he defined as "placing us where we could reap the fruit of our own labor"; the best way to accomplish this was "to have land, and turn it and till it by our own labor." And in response to the question whether blacks preferred to live separately or among whites, he replied: "I would prefer to live by ourselves, for there is a prejudice against us in the South that will take years to get over." Four days later, Sherman issued Special Field Order No. 15, setting aside the Sea Islands and a portion of the low country rice coast south of Charleston, extending thirty miles inland, for the exclusive settlement of blacks. Each family would receive forty acres of land, and Sherman later provided that the army could assist them with the loan of mules. (Here, perhaps, lies the

68. Cox, "Promise of Land," 413–40; Belz, *New Birth of Freedom*, 100–105.

69. [William T. Sherman] *Memoirs of General W. T. Sherman, Written by Himself* (New York, 1891), 2:248–49; Paul D. Escott, "The Context of Freedom: Georgia's Slaves During the Civil War," *GaHQ*, 58 (Spring 1974), 92–93; John R. Dennett, *The South As It Is: 1865–1866*, edited by Henry M. Christman (New York, 1965), 177; M. A. De Wolfe Howe, ed., *Home Letters of General Sherman* (New York, 1909), 319.

origin of the phrase "forty acres and a mule" that would soon echo throughout the South.)[70]

Sherman was neither a humanitarian reformer nor a man with any particular concern for blacks. Instead of seeing Field Order 15 as a blueprint for the transformation of Southern society, he viewed it mainly as a way of relieving the immediate pressure caused by the large number of impoverished blacks following his army. The land grants, he later claimed, were intended only to make "temporary provisions for the freedmen and their families during the rest of the war," not to convey permanent possession. Understandably, however, the freedmen assumed that the land was to be theirs, especially after Gen. Rufus Saxton, assigned by Sherman to oversee the implementation of his order, informed a large gathering of blacks "that they were to be put in possession of lands, upon which they might locate their families and work out for themselves a living and respectability." Certainly, the freedmen hastened to take advantage of the Order. Baptist minister Ulysses Houston, one of the group that had met with Sherman, led 1,000 blacks to Skiddaway Island, Georgia, where they established a self-governing community with Houston as the "black governor." By June, in the region that had spawned one of the wealthiest segments of the planter class, some 40,000 freedmen had been settled on 400,000 acres of "Sherman land." Here in coastal South Carolina and Georgia, the prospect beckoned of a transformation of Southern society more radical even than the end of slavery.[71]

And now the war hastened inexorably to its close. Sherman's army moved into South Carolina, bringing in its wake, as one rice planter recorded in his journal, the "breath of Emancipation." On plantation after plantation, "perfect anarchy and rebellion" reigned, as the accumulated resentments of slavery burst forth in violence and in the conscious flouting of the planter aristocracy's authority and self-esteem. Planters' homes, smokehouses, and store rooms were plundered; an overseer was murdered; on one plantation blacks refused to listen any longer to the local white minister, but "would shout and sing after their own fashion." The magnificent plantation home at Middleton Place near Charleston was burned to the ground, the vaults of the family graveyard broken open and the bones scattered by the former slaves. Charles Mani-

70. Sherman, *Memoirs*, 2:245–52; Robert C. Morris, *Reading, 'Riting, and Reconstruction: The Education of Freedmen in the South 1861–1870* (Chicago, 1981), 124; "Colloquy With Colored Ministers," *JNH*, 16 (January 1931), 88–94; Vincent Harding, *There Is a River: The Black Struggle for Freedom in America* (New York, 1981), 261–65.

71. Gerteis, *From Contraband to Freedman*, 151; Sherman, *Memoirs*, 2:249–50; Howe, ed., *Sherman Letters*, 327–28; S. W. Magill to the AMA, February 3, 1865, AMA Archives, Amistad Research Center, Tulane University; Savannah *Daily Herald*, February 3, 1865; Magdol, *A Right to the Land*, 104–105; Claude F. Oubre, *Forty Acres and a Mule: The Freedmen's Bureau and Black Landownership* (Baton Rouge, 1978), 19.

gault, who had considered himself an indulgent master, was stunned by the "recklessness and ingratitude" of his slaves, "for they broke into our well furnished residences on each plantation and stole or destroyed everything therein." At one of Manigault's plantations, paintings taken from the walls of the big house were "hung up in their Negro houses, while some of the family portraits (as if to turn them into ridicule) they left out, night and day, exposed to the open air." On another rice plantation, "they divided out our land . . . and would obey no driver." What one planter called "the general wreck of our property" dealt the low country rice aristocracy a blow from which it never fully recovered.[72]

On February 18, 1865, Union forces entered Charleston, among them the black 54th Massachusetts Infantry singing "John Brown's Body." Five weeks later the city witnessed a "grand jubilee" of freedom, a vast outpouring of celebration and pride by the city's black community. Four thousand blacks took part in a massive parade—soldiers, fire companies, members of the various trades with their respective tools, schoolchildren carrying a banner with the inscription "We Know No Master but Ourselves." A mock funeral procession formed the centerpiece, complete with a coffin bearing the motto "Slavery Is Dead," followed by a long train of women "mourners." Then, on April 14, four years after the Civil War began in Charleston harbor, a band of Northern preachers, politicians, and abolitionists descended upon the city for the raising of the Union flag over Fort Sumter. Various notables delivered addresses, but the most affecting moment came when a black man stepped forward with his two small daughters to thank William Lloyd Garrison for his long labors on behalf of the slaves. Meanwhile, the flag-bedecked steamer *Planter*, captained by Robert Smalls and "perfectly black with colored people," traversed the harbor. (Smalls was so overcome by emotion that he lost his bearings, and the ship collided with another during the ceremony.) One white army officer was moved to tears by the raising of the standard "that now for the first time is the black man's as well as the white man's flag."[73]

72. Arney R. Childs, ed., *The Private Journal of Henry William Ravenel 1859–1887* (Chapel Hill, 1947), 217–19; Robert M. Myers, ed., *The Children of Pride* (New Haven, 1972), 1247; J. H. Easterby, ed., *The South Carolina Rice Plantation as Revealed in the Papers of Robert F. W. Allston* (Chicago, 1945), 206–11; Nicholas B. Wainwright, ed., *A Philadelphia Perspective: The Diary of Sidney George Fisher Covering the Years 1834–1871* (Philadelphia, 1967), 497–98; Charles Joyner, *Down by the Riverside: A South Carolina Slave Community* (Urbana, Ill., 1984), 228–29; "The Close of the War—The Negro, etc.," typescript copy of manuscript by Charles Manigault, USC; James M. Clifton, "A Half-Century of a Georgia Rice Plantation," *NCHR*, 47 (Autumn 1970), 401–405; George C. Rogers, *The History of Georgetown County, South Carolina* (Columbia, S.C., 1970), 416–22; Williams Middleton to Eliza Fisher, May 12, 1865, Cadwallader Collection, HSPa.

73. Williamson, *After Slavery*, 22; *New York Times*, March 30, April 4, 1865; *New York Tribune*, April 4, 1865; Sarah F. Hughes, ed., *Reminiscences of John Murray Forbes* (Boston, 1902), 2:356; Okon E. Uya, *From Slavery to Public Service: Robert Smalls 1839–1915* (New York, 1971), 27–29; S. Willard Saxton Journal, April 15, 1865, Rufus and S. Willard Saxton Papers, Yale University.

Farther north, similar scenes of jubilation occurred as Grant's army, on April 3, occupied Richmond. Amid a column of smoke from a fire set by retreating Southern troops to destroy stores of tobacco, cotton, and munitions, and with the sound of artillery shells still echoing in the distance, emancipation came to the capital of the Confederacy. Blacks thronged the streets, dancing, praying, singing "Slavery chain done broke at last," exulting particularly over the appearance of a unit of black soldiers whose band played "Year of Jubilee." The next day Lincoln, heedless of his own safety, walked the streets of Richmond accompanied only by a dozen sailors. At every step he was besieged by former slaves who hailed him as a "Messiah" and fell on their knees before the embarrassed President, who asked them to remain standing. "There was to be no more Marster and Mistress now," a Richmond black told his former owner, "all was equal. . . . All the land belongs to the Yankees now and they gwine divide it out among de coloured people." On April 9, Grant accepted Lee's surrender at Appomattox Court House.[74]

Amid this cavalcade of historic events, Reconstruction emerged as the central problem confronting the nation. But, as James G. Blaine later remarked, Lincoln did not turn to peacetime with a "fixed plan" of Reconstruction.[75] Different approaches had operated simultaneously in different parts of the South. Lincoln had approved the lenient policies of General Banks in Louisiana and the far more proscriptive acts of Andrew Johnson in Tennessee, all in an attempt to quicken Union victory and secure the abolition of slavery, rather than to fashion a blueprint for the postwar South. Policies such as the Mississippi Valley labor system had indeed been inaugurated that would influence the future course of Reconstruction, but these had been war measures, strongly affected by the military situation in particular theaters, and subject to debate and alteration according to the very different requirements of peace. Nor had wartime Reconstruction been particularly successful. Union governments had been created within the Confederacy in Virginia, Tennessee, Arkansas, and Louisiana, but none had attracted truly broad support and none had been recognized by Congress. Despite his political dexterity, Lincoln had failed to bring a single reconstructed state (except for West Virginia) into the Union. And while he had expended considerable political capital supporting the Louisiana regime, no one knew how firm his commitment was to the other governments or what terms would be laid down for dealing with the rest of the Confederacy.

In the period before Lee's surrender, in fact, Lincoln had displayed a shifting approach to policy that revealed either a desire to achieve peace

74. A. A. Hoehling and Mary Hoehling, *The Day Richmond Died* (San Diego, 1981), 202–207, 240–42; Michael B. Chesson, *Richmond After the War 1865–1890* (Richmond, 1981), 57–62; John T. O'Brien, "Reconstruction in Richmond: White Restoration and Black Protest, April-June 1865," *VaMHB*, 89 (July 1981), 261–63; Peter Randolph, *From Slave Cabin to the Pulpit* (Boston, 1894), 59.

75. James G. Blaine, *Twenty Years of Congress* (Norwich, Conn., 1884-86), 2:49.

as expeditiously as possible or the indecision of a man still searching for a coherent program for the postwar South. At a Cabinet meeting in February, he revived the discredited idea of compensated emancipation, but dropped it after unanimous opposition. Even the conservative Secretary of the Navy, Gideon Welles, laconically remarked that while desire for peace was admirable, "there may be such a thing as overdoing." Then early in April, Lincoln proposed to convene the Confederate legislature of Virginia in order to have it take the state out of the war, but was again dissuaded by the Cabinet. Such an act would have implied the repudiation of the Pierpont regime, previously recognized as the state's legitimate government. When the President, on April 10, met with Pierpont, he asked questions instead of providing answers. How many Unionists really existed in the South? Would they join the Republican party? If Pierpont is to be believed, the President remarked that he "had no plan for reorganization."[76]

The very next day, Lincoln delivered at the White House what came to be known as his "last speech" (a description that, while accurate enough, suggests a finality scarcely anticipated by the President). Essentially a defense of the wartime Louisiana government and a response to Radical demands for the enfranchisement of blacks, it marked the first occasion on which Lincoln (or any American President) publicly endorsed black suffrage, albeit in cautious and limited language: "I would myself prefer that [the vote] were now conferred on the very intelligent, and on those who serve our cause as soldiers." The speech was typical of Lincoln—an attempt to encourage Republicans to think of Reconstruction as a practical rather than a theoretical matter (the question of whether the South was "in or out of the Union," he remarked, was not only "practically immaterial" but downright "mischievous"), and to maintain public support and party unity by moving partway toward the Radical position on black suffrage, without promising to impose his "preference" on the defeated South. Chase, now Chief Justice of the Supreme Court, who had been bombarding Lincoln with appeals to enfranchise all "loyal citizens," viewed the speech as a step forward. Less kind was the Democratic New York *World:* "Mr Lincoln gropes . . . like a traveller in an unknown country without a map." Probably the most accurate assessment was that of the *New York Times*, which concluded that Lincoln had judged the time not yet ripe for "the statement of a settled reconstruction policy."[77]

On Good Friday, April 14, Secretary of War Stanton presented the

76. Howard K. Beale, ed., *Diary of Gideon Welles* (New York, 1960), 2:237, 279–80; Albert Mordell, ed., *Civil War and Reconstruction: Selected Essays by Gideon Welles* (New York, 1959), 186–87; Ambler, *Pierpont,* 254–58.

77. Basler, ed., *Lincoln Works,* 8:399–403; David Donald, ed., *Inside Lincoln's Cabinet: The Civil War Diaries of Salmon P. Chase* (New York, 1954), 264–66; New York *World,* April 13, 1865; *New York Times,* April 13, 1865.

Cabinet with a tentative plan to appoint military governors for Virginia and North Carolina. Little discussion ensued, and Lincoln urged his colleagues to devote their attention to "the great question now before us." Stanton was directed to redraft the proposal, for consideration at the next Cabinet meeting. That night, the President was mortally wounded by the actor John Wilkes Booth. He died the next day, mourned by blacks as a divinely appointed savior and by Northern whites as the man who had preserved the Union and emancipated the slaves.[78]

Years before the end of slavery, black abolitionist Charles L. Reason had predicted that emancipation would impose "severe trials" upon the freedmen: "The prejudice now felt against them for bearing on their persons the brand of slaves, cannot die out immediately." Even Lincoln's funeral illustrated the problem, for when his body reached New York, the city's municipal authorities sought to bar blacks from marching in the procession, only to be overruled by the War Department. Yet in a society resting on ideals of individual autonomy and competitive equality, could there be a permanent resting place between slavery and citizenship? The elusive concept of "freedom" differed substantially in the United States from its counterpart in societies accustomed to fixed social classes and historically defined gradations of civil and political rights. Here, emancipation led inexorably to demands for civil equality and the vote. For, as Frederick Douglass well appreciated:

> If I were in a monarchial government . . . where the few bore rule and the many were subject, there would be no special stigma resting upon me, because I did not exercise the elective franchise. . . . But here, where universal suffrage is the . . . fundamental idea of the Government, to rule us out is to make us an exception, to brand us with the stigma of inferiority.

And, on a more pragmatic level, without black suffrage could the old ruling class of the South be prevented from reestablishing its political hegemony? These questions had been repeatedly raised as the war neared its close, but the war did not provide an answer.[79]

It was becoming apparent, moreover, that whites would not debate these questions alone. In 1864 Isaac Brinckerhoff, a Northern Baptist missionary in Union-occupied Florida, delivered a series of lectures on the United States Constitution and the principles of republican government. The blacks, he recorded, "were deeply interested. Old Meredith, my colored servant, when asked how he was enjoying the lectures, spread both his hands and said, 'I grow bigger and bigger every time'." By the end of the war, small groups of freedmen were already learning their first

78. Benjamin P. Thomas and Harold M. Hyman, *Stanton: The Life and Times of Lincoln's Secretary of War* (New York, 1962), 357–58; Beale, ed., *Welles Diary*, 2:281.

79. Julia Griffiths, ed., *Autographs for Freedom* (2 Ser.: Auburn, N. Y., 1854), 14; New York *Tribune*, April 24, 28, 1865; Foner, ed., *Douglass*, 4:159; Donald, *Sumner*, 137–38.

lessons in political participation. At Mitchelville, in the South Carolina Sea Islands, blacks, under army supervision, had elected a mayor and city council, who controlled local schools and the administration of justice. On Amelia Island, Florida, blacks voted alongside whites in a local election. Other examples of local self-government could be found at Davis Bend and in contraband camps from Virginia to Mississippi. Here was a people just emerging onto the stage of politics, and they, too, would have a say in defining the meaning of freedom.[80]

As postwar Reconstruction commenced, Sidney George Fisher, a conservative Philadelphia lawyer, observed that the war had left fundamental issues unresolved:

> It seems our fate never to get rid of the Negro question. No sooner have we abolished slavery than a party, which seems [to] be growing in power, proposes Negro suffrage, so that the problem—What shall we do with the Negro?—seems as far from being settled as ever. In fact it is *incapable* of any solution that will satisfy both North and South.

Fisher understood all too clearly the inevitability of continuing conflict over the legacy of emancipation. "Verily," as Frederick Douglass put it, "the work does not end with the abolition of slavery, but only begins."[81]

---

80. Isaac Brinckerhoff, "Missionary Work Among the Freed Negroes. Beaufort, S. Carolina, St. Augustine, Georgia, 1862–1874," manuscript, American Baptist Historical Society, 127; James E. Sefton, "Chief Justice Chase as an Advisor on Presidential Reconstruction," *CWH*, 13 (September 1967), 260; Magdol, *Right to the Land*, 100–104; Martin Abbott, "Freedom's Cry: Negroes and Their Meetings in South Carolina, 1865–1869," *Phylon*, 20 (Fall 1959), 266.

81. Wainwright, ed., *Fisher Diary*, 499; Foner, ed., *Douglass*, 3:293.

CHAPTER 3

The Meaning of Freedom

FREEDOM came in different ways to different parts of the South. In large areas, slavery had disintegrated long before Lee's surrender, but elsewhere, far from the presence of federal troops, blacks did not learn of its irrevocable end until the spring of 1865. Despite the many disappointments that followed, this generation of blacks would always regard the moment when "de freedom sun shine out" as the great watershed of their lives. Houston H. Holloway, who had been sold three times before he reached the age of twenty in 1865, later recalled with vivid clarity the day emancipation came to his section of Georgia: "I felt like a bird out of a cage. Amen. Amen. Amen. I could hardly ask to feel any better than I did that day. . . . The week passed off in a blaze of glory." Six weeks later Holloway and his wife "received my free born son into the world."[1]

"Freedom," said a black minister, "burned in the black heart long before freedom was born." But what did "freedom" mean? "It is necessary to define that word," Freedmen's Bureau Commissioner O. O. Howard told a black audience in 1865, "for it is most apt to be misunderstood." Howard assumed a straightforward definition existed. But instead of a predetermined category or static concept, "freedom" itself became a terrain of conflict, its substance open to different and sometimes contradictory interpretations, its content changing for whites as well as blacks in the aftermath of the Civil War.[2]

Many Southern whites assumed that blacks confronted the demise of slavery entirely unprepared for the responsibilities of freedom. "The Negroes are to be pitied, . . ." wrote South Carolinian Julius J. Fleming, an educator, minister, and public official. "They do not understand the liberty which has been conferred upon them." In fact, blacks carried out

1. George P. Rawick, ed., *The American Slave: A Composite Autobiography* (Westport, Conn., 1972–79), Supplement, Ser. 2, 2:1945; Houston H. Holloway Autobiography, Miscellaneous Manuscript Collections, LC.
2. Quitman *Banner* in Savannah *Daily News and Herald*, July 15, 1867; New Orleans *Tribune*, November 6, 1865.

of bondage an understanding of their new condition shaped both by their experience as slaves and by observation of the free society around them. What one planter called their "wild notions of right and freedom" encompassed, first of all, an end to the myriad injustices associated with slavery. Like the Louisiana blacks interviewed by General Banks's agents during the Civil War, many former slaves saw freedom as an end to the separation of families, the abolition of punishment by the lash, and the opportunity to educate their children. Others, like black minister Henry M. Turner, stressed that freedom meant the enjoyment of "our rights in common with other men." "If I cannot do like a white man I am not free," Henry Adams told his former master in 1865. "I see how the poor white people do. I ought to do so too, or else I am a slave."[3]

But underpinning the specific aspirations lay a broader theme: a desire for independence from white control, for autonomy both as individuals and as members of a community itself being transformed as a result of emancipation. Before the war, free blacks had created a network of churches, schools, and mutual benefit societies, while slaves had forged a semiautonomous culture centered on the family and church. With freedom, these institutions were consolidated, expanded, and liberated from white supervision, and new ones—particularly political organizations— joined them as focal points of black life. In stabilizing their families, seizing control of their churches, greatly expanding their schools and benevolent societies, staking a claim to economic independence, and forging a distinctive political culture, blacks during Reconstruction laid the foundation for the modern black community, whose roots lay deep in slavery, but whose structure and values reflected the consequences of emancipation.

From Slavery to Freedom

Long after the end of the Civil War, the experience of bondage remained deeply etched in blacks' collective memory. As one white writer noted years later, blacks could not be shaken from the conviction "that the white race has barbarously oppressed them." They took particular offense at contentions that American slavery had been unusually benevolent and that "harmonious relations" had existed between master and slave. "All

3. John H. Moore, ed., *The Juhl Letters to the "Charleston Courier"* (Athens, Ga., 1974), 20; Will Martin to Benjamin G. Humphreys, December 5, 1865, Mississippi Governor's Papers, MDAH; Joseph P. Reidy, "Masters and Slaves, Planters and Freedmen: The Transition from Slavery to Freedom in Central Georgia, 1820–1880" (unpub. diss., Northern Illinois University, 1982), 162; 46th Congress, 2d Session, Senate Report 693, pt. 2:191. Most older historical accounts, and some more recent ones, echo the idea that the ex-slaves were completely unprepared for freedom. Walter L. Fleming, *Civil War and Reconstruction in Alabama* (New York, 1905), 270–71; Howard K. Beale, *The Critical Year* (New York, 1930), 188–89; William C. Harris, *Presidential Reconstruction in Mississippi* (Baton Rouge, 1967), 80–81.

of us know how happy we have been. . . ." declared one black orator. "Have these gentlemen forgotten so soon to what ills we have been subjected?" Fundamentally, however, blacks resented not only the incidents of slavery—the whippings, separations of families, and countless rituals of subordination—but the fact of having been held as slaves at all. During a visit to Richmond, Scottish minister David Macrae was surprised to hear a former slave complain of past mistreatment, while acknowledging he had never been whipped. "How were you cruelly treated then?" asked Macrae. "I was cruelly treated," answered the freedman, "because I was kept in slavery."[4]

In countless ways, the newly freed slaves sought to "throw off the badge of servitude," to overturn the real and symbolic authority whites had exercised over every aspect of their lives. Some took new names that reflected the lofty hopes inspired by emancipation. One Northern teacher in Savannah reported among her black pupils an Alexander Hamilton, a Franklin Pierce, even a General Joe E. Johnston; in Georgetown, South Carolina, former slaves' new names included Deliverance Berlin, Hope Mitchell, Chance Great, and Thomas Jefferson. Many blacks now demanded to be addressed by whites as Mr. or Mrs. rather than by their first name, as was conventional under slavery.[5]

Blacks relished opportunities to flaunt their liberation from the innumerable regulations, significant and trivial, associated with slavery. Freedmen held mass meetings and religious services unrestrained by white surveillance, acquired dogs, guns, and liquor (all barred to them under slavery), and refused to yield the sidewalks to whites. They dressed as they pleased, black women sometimes wearing gaudy finery, carrying parasols, and replacing the slave kerchief with colorful hats and veils. In the summer of 1865, Charleston saw freedmen occupying "some of the best residences," and promenading on King Street "arrayed in silks and satins of all the colors of the rainbow," while black schoolchildren sang " 'John Brown's Body' within ear-shot of Calhoun's tomb." Rural whites complained of "insolence" and "insubordination" among the freedmen, by which they meant any departure from the deference and obedience expected under slavery. On the Bradford plantation in Florida, one untoward incident followed another. First, the family cook told Mrs. Bradford "if she want any dinner she kin cook it herself." Then the former slaves went off to a meeting with Northern soldiers to discuss "our freedom."

4. Z. T. Filmore to History Company Publishers, March 2, 1887, Bancroft Library, University of California, Berkeley; undated manuscript speech, 1865 or 1866, Pinckney S. Pinchback Papers, Howard University; David Macrae, *The Americans at Home* (New York, 1952 [orig. pub. 1870]), 133.

5. Eliza F. Andrews, *The War-Time Journal of a Georgia Girl* (New York, 1908), 347; Sarah A. Jenness to Samuel Hunt, December 30, 1865, AMA Archives, Amistad Research Center, Tulane University; George C. Rogers, Jr., *The History of Georgetown County, South Carolina* (Columbia, S.C., 1970), 439–41.

Told that she and her daughter could not attend, one woman replied "they were now free and if she saw fit to take her daughter into that crowd it was nobody's business." "Never before had I a word of impudence from any of our black folk," recorded nineteen-year-old Susan Bradford, "but they are not ours any longer."[6]

The presence of black troops among the occupying Union army reinforced the freedmen's assertiveness and inspired constant complaint on the part of whites. Black soldiers acted, in the words of the New York *World*, as "apostles of black equality," spreading among the former slaves ideas of land ownership and civil and political equality. They intervened in plantation disputes and sometimes arrested whites. ("It is very hard . . ." wrote a Confederate veteran, "to see a white man taken under guard by one of those black scoundrels.") Black troops helped construct schools, churches, and orphanages, organized debating societies, and held political gatherings where "freedom songs" were sung and soldiers delivered "speeches of the most inflammatory kind." In Southern cities they demanded the right to travel on segregated streetcars, taunted white passersby with remarks like "We's all equal now," and advised freedmen in cities like Memphis that they need not obey military orders to return to the plantations.[7]

Among the most resented of slavery's restrictions were the rule that no black could travel without a pass and the patrols that enforced the pass system. With emancipation, it seemed that half the South's black population took to the roads. "Right off colored folks started on the move," a Texas slave later recalled. "They seemed to want to get closer to freedom, so they'd know what it was—like it was a place or a city." Blacks' previous treatment as slaves seemed to have little to do with the movement. "Every one of A. M. Dorman's negroes quit him," an Alabama planter reported. "They have always been as free and as much indulged as his children." The ability to come and go as they pleased would long remain a source of pride and excitement for former slaves. "The Negroes are literally crazy about traveling," wrote a white observer in 1877. "The

6. Vincent Harding, *There Is a River: The Black Struggle for Freedom in America* (New York, 1981), 278–81; Jacqueline Jones, *Labor of Love, Labor of Sorrow: Black Women, Work and the Family, from Slavery to the Present* (New York, 1985), 69; Joel Williamson, *After Slavery: The Negro in South Carolina During Reconstruction, 1861–1877* (Chapel Hill, 1965), 46–47; Leon F. Litwack, *Been in the Storm So Long: The Aftermath of Slavery* (New York, 1979), 359; Elias H. Dees to Anne Dees, August 12, 1865, Elias H. Deas Papers, USC; Joseph H. Mahaffey, ed., "Carl Schurz's Letters From the South," *GaHQ*, 35 (September 1951), 235; Susan B. Eppes, *Through Some Eventful Years* (Macon, 1926), 279–84, 294–95.

7. New York *World*, September 13, 1865; Ira Berlin et al., eds., *Freedom: A Documentary History of Emancipation 1861–1867* (New York, 1982–), Ser. 2, 733–39; Emile E. Delserier to Marguerite E. Williams, May 6, 1865, Marguerite E. Williams Papers, UNC; Jacob Schirmer Diary, June 1865, SCHS; *Weekly Anglo-African*, August 12, 1865; Charleston *South Carolina Leader*, March 31, 1866; Elizabeth A. Meriwether, *Recollections of 92 Years* (Nashville, 1958), 164–68; Bobby L. Lovett, "Memphis Riots: White Reactions to Blacks in Memphis, May 1865–July 1866," *THQ*, 38 (Spring 1979), 15–17.

railway officials are continually importuned by them to run extra trains, excursion trains, and so on, on all sorts of occasions: holidays, picnics, Sunday-school celebrations, church dedications."[8]

Undertaken in the face of determined opposition from planters, the army, and the Freedmen's Bureau, the massive population movement of early Reconstruction appeared to Southern whites, many Northerners, and subsequent historians as an "aimless migration," proof that blacks equated freedom with idleness and "vagabondage." In fact, a majority of freedmen did not abandon their home plantations in 1865, and those who did generally traveled only a few miles. Those blacks who did move usually had specific reasons for doing so. Henry Adams, for example, left his Louisiana plantation in 1865 to "see whether I am free by going without a pass." (This was not an idle exercise. A group of whites accosted Adams on the road, asked the name of his owner, and beat him when he replied that he belonged to no one.) Some blacks abandoned predominately white upcountry counties to seek the fellowship of their own race. One freedwoman left a Georgia farm saying "she couldn't live anywhere where there was no more negroes than here." The postwar "exodus" also reflected the massive displacement of the black population that had occurred during the Civil War. Thousands of slaves "refugeed" by their owners to Texas now returned to Mississippi and Louisiana, while in South Carolina blacks removed from the Sea Islands early in the war returned home, sometimes crossing paths with former mainland slaves who had escaped to the islands and were now traveling home. And considerable numbers, attracted by wages substantially higher than in the East, emigrated to Texas, Louisiana, and other southwestern states.[9]

For a variety of reasons, Southern towns and cities experienced an especially large influx of freedmen during and immediately after the Civil War. In the cities, many blacks believed, "freedom was free-er." Here were black social institutions—schools, churches, and fraternal societies—and here too, in spite of inequities in law enforcement, were the army (including black soldiers) and Freedmen's Bureau, offering protection from the violence so pervasive in much of the rural South. "People who get scared at others being beaten go to the cities," said a Georgia black legislator during Reconstruction. Between 1865 and 1870, the black population of the South's ten largest cities doubled, while the

8. Litwack, *Been in the Storm*, 292–301; Rawick, ed., *American Slave*, 4, pt. 2:133; Henry Watson, Jr., to James A. Wemyss, January 26, 1866, Henry Watson, Jr., Papers, DU; George B. Tindall, *South Carolina Negroes 1877–1900* (Columbia, S.C., 1952), 153.

9. Henderson H. Donald, *The Negro Freedman* (New York, 1952), 1 (a book which offers every conceivable stereotype concerning black behavior in 1865); Litwack, *Been in the Storm*, 31–33, 305–10, 322–26; 46th Congress, 2d Session, Senate Report 693, pt. 2:191; Laura Perry to Grant Perry, February 25, 1868, J. M. Perry Family Papers, Atlanta Historical Society; Williamson, *After Slavery*, 39–41; Robert P. Brooks, *The Agrarian Revolution in Georgia, 1865–1912* (Madison, Wis., 1914), 10.

number of white residents rose by only 10 percent. Smaller towns, from which blacks had often been excluded as slaves, experienced even more dramatic increases. The black population of Demopolis, Alabama, site of a regional Freedmen's Bureau office, grew from one individual in 1860 to nearly 1,000 ten years later.[10]

Black migrants who hoped to find an urban alternative to plantation labor and rural living conditions often encountered severe disappointment. The influx from the countryside flooded the labor market, undercutting the economic position of longtime city residents and consigning most urban blacks to low-wage, menial employment. Unable to obtain decent housing, black migrants lived in shanty towns that sprang up on the outskirts of Southern cities; in these districts of poverty, squalor, and periodic epidemics, the death rate far exceeded that among white city dwellers. The result was a striking change in Southern urban living patterns. Before the war, blacks and whites had lived scattered throughout Southern cities. Reconstruction witnessed the rise of a new, segregated, urban geography: "the main town, populated principally by whites, and containing the finest structures; and the 'free town' (which the whites often dub Liberia), consisting chiefly of wretched log cabins." For all these reasons, the urban migration slowed dramatically after 1870 and the proportion of Southern blacks living in cities stabilized at around 9 percent.[11]

Of all the motivations for black mobility, none was more poignant than the effort to reunite families separated during slavery. "In their eyes," wrote a Freedmen's Bureau agent, "the work of emancipation was incomplete until the families which had been dispersed by slavery were reunited." In September 1865, Northern reporter John Dennett encountered a freedman who had walked more than 600 miles from Georgia to North Carolina, searching for his wife and children from whom he had been separated by sale. Another freedman, writing from Texas, asked the aid of the Freedmen's Bureau in locating "my own dearest relatives," providing a long list of sisters, nieces, nephews, uncles, and in-laws, none of whom he had seen since his sale in Virginia twenty-four years before.

10. Litwack, Been in the Storm, 310–16; 42d Congress, 2d Session, House Report 22, Georgia, 7 (hereafter cited as KKK Hearings); Herbert A. Thomas, Jr., "Victims of Circumstance: Negroes in a Southern Town, 1865–1880," Register of the Kentucky Historical Society, 71 (July 1973), 253; Orville V. Burton, "The Rise and Fall of Afro-American Town Life: Town and Country in Reconstruction Edgefield, South Carolina," in Orville H. Burton and Robert C. McMath, Jr., eds., Toward a New South? Studies in Post-Civil War Southern Communities, (Westport, Conn., 1982), 152–53; Peter Kolchin, First Freedom: The Responses of Alabama's Blacks to Emancipation and Reconstruction (Westport, Conn., 1972), 10.

11. Constance M. Green, The Secret City: A History of Race Relations in the Nation's Capital (Princeton, 1967), 82; Jones, Labor of Love, 73–78, 110–12; Thomas, "Victims of Circumstance," 256–58; John Kellogg, "The Evolution of Black Residential Areas in Lexington, Kentucky, 1865–1887," JSH, 48 (February 1982), 21–52; [Belton O'Neill Townsend] "South Carolina Society," Atlantic Monthly, 39 (June 1877), 678; Robert Higgs, Competition and Coercion: Blacks in the American Economy, 1865–1914 (New York, 1977), 33.

"Family Record." A color lithograph, marketed to former slaves. (Library of Congress)

As late as the turn of the century, black newspapers carried advertisements that testified to the human tragedies that formed an everyday part of slavery. A typical plea for help appeared in the Nashville *Colored Tennessean:*

> During the year 1849, Thomas Sample carried away from this city, as his slaves, our daughter, Polly, and son. . . . We will give $100 each for them to any person who will assist them . . . to get to Nashville, or get word to us of their whereabouts.

Usually, such quests ended in failure, and others produced wrenching disappointment when spouses were located who had remarried. But few scenes were as affecting as the reunion of long-separated relatives. "I wish you could see this people as they step from slavery into freedom," a Union officer wrote his wife in May 1865. "Men are taking their wives and children, families which had been for a long time broken up are united and oh! such happiness. I am glad I am here."[12]

Strong family ties, it is clear, had existed under slavery, but had always been vulnerable to disruption. Emancipation allowed blacks to reaffirm and solidify their family connections, and most freedmen seized the opportunity with alacrity. During the Civil War, John Eaton, who, like many whites, believed that slavery had destroyed the sense of family obligation, was astonished by the eagerness with which former slaves in contraband camps legalized their marriage bonds. The same pattern was repeated when the Freedmen's Bureau and state governments made it possible to register and solemnize slave unions. Many families, in addition, adopted the children of deceased relatives and friends, rather than see them apprenticed to white masters or placed in Freedmen's Bureau orphanages. By 1870, a large majority of blacks lived in two-parent family households, a fact that can be gleaned from the manuscript census returns but also "quite incidentally" from the Congressional Ku Klux Klan hearings, which recorded countless instances of victims assaulted in their homes, "the husband and wife in bed, and . . . their little children beside them."[13]

But while emancipation thus made possible the stabilization and strengthening of the preexisting black family, it also transformed the roles of its members and relations among them. One common, significant

12. John W. DeForest, *A Union Officer in the Reconstruction,* edited by James H. Croushore and David M. Potter (New Haven, 1948), 36–37; John R. Dennett, *The South As It Is: 1865–1866,* edited by Henry M. Christman (New York, 1965), 130; Hawkins Wilson to "Chief of the Freedmen's Bureau at Richmond," May 11, 1867, Letters Received, Ser. 3892, Bowling Green Subasst. Comr., RG105, NA [FSSP A-8254]; Lester C. Lamon, *Blacks in Tennessee 1791–1970* (Knoxville, 1981), 43; John E. Bryant to Emma Bryant, May 29, 1865, John E. Bryant Papers, DU.

13. John Eaton, *Grant, Lincoln and the Freedmen* (New York, 1907), 34, 211; Herbert G. Gutman, *The Black Family in Slavery and Freedom, 1750–1925* (New York, 1976), 61–62, 141–42, 225–28, 417–20; KKK Hearings, Georgia, 817.

change was that slave families, separated much of the time because their members belonged to different owners, could now live together. More widely noticed by white observers in early Reconstruction was the withdrawal of black women from field labor. The nineteenth century's "cult of domesticity," which defined the home as a woman's proper sphere, was never thought to apply to blacks and certainly not to slaves. Although men performed the heaviest tasks, like plowing and splitting rails, on most plantations slave women regularly worked in the fields, and sometimes comprised a majority of the agricultural labor force. Among the slaves themselves, however, labor seems to have been divided along sexual lines, with men chopping wood, hunting, and assuming positions of leadership (such as driver and preacher), while women washed, sewed, cooked, gardened, and assumed primary responsibility for the care of children. Like free women, female slaves found that their responsibilities did not end when the "workday" was over.[14]

Beginning in 1865, and for years thereafter, whites throughout the South complained of the difficulty of obtaining female field laborers. Thus was lost, as a Georgian put it, "a very important per cent of the entire labor of the South." The editor of The Plantation lamented that black women would no longer "pick cotton, which is a woman's work. . . . They will merely take care of their own households and do but little or no work outdoors." In both cities and rural areas, black women also proved reluctant to labor as domestic servants in white homes, and those who did frequently refused to live in their employer's residence. "House servants are difficult to get out here," wrote a resident of upcountry Georgia. "Every negro woman wants to set up house keeping." Many contemporaries, who viewed white women who remained at home as paragons of the domestic ideal, saw their black counterparts as lazy and slightly ludicrous. Planters, Freedmen's Bureau officials, and Northern visitors all ridiculed the black "female aristocracy" for "acting the *lady*" or mimicking the family patterns of middle-class whites. White employers also resented their inability to force black children to labor in the fields, especially after the spread of schools in rural areas. "The freedmen," a Georgia newspaper reported in 1869, "have almost universally withdrawn their women and children from the fields, putting the first at housework and the latter at school."[15]

14. Jones, *Labor of Love*, 11–43; Michael P. Johnson, "Work, Culture, and the Slave Community: Slave Occupations in the Cotton Belt in 1860," *Labor History*, 27 (Summer 1986), 329; Charles Joyner, *Down by the Riverside: A South Carolina Slave Community* (Urbana, Ill., 1984), 60.

15. M. C. Fulton to Davis Tillson, April 17, 1866, Unregistered Letters Received, Ser. 632, Ga. Asst Comr., RG 105, NA [FSSP A-5379]; KKK Hearings, Georgia, 829; Thomas, "Victims of Circumstance," 263; John Kincaid to Mrs. E. K. Anderson, January 5, 1867, Kincaid-Anderson Papers, USC; Jones, *Labor of Love*, 58–59; DeForest, *Union Officer*, 94; *Southern Field and Factory*, 1 (April 1871), 149; Atlanta *Constitution*, July 10, 1869.

As these comments indicate, contemporaries appeared uncertain whether black women, black men, or both, were responsible for the withdrawal of females from agricultural labor. There is no question that many black men considered it a badge of honor to see their wives working at home and believed that, as head of the family, the man should decide how its labor was organized. In one part of Louisiana, where planters attempted to force black women into the fields, freedmen insisted that "whenever they wanted their wives to work they would tell them themselves; and if [they] could not rule [their] own domestic affairs on that place [they] would leave it." But all blacks resented the sexual exploitation that had been a regular feature of slave life, and shared the determination that the women no longer labor under the direct supervision of white men. And many black women independently desired to devote more time than had been possible under slavery to caring for their children and to domestic responsibilities like cooking, sewing, and laundering. Tasks like these, arduous enough in the days before electricity and running water in the rural South, were made even more time-consuming by the demise of the plantation slave quarter, where children were cared for by elderly slave women and household chores often done collectively.[16]

The shift in the locus of black female labor from the fields to the home proved, in large measure, a temporary phenomenon. The rise of renting and sharecropping, which made each family responsible for its own plot of land, placed a premium on the labor of all family members. "A man takes more or less land according to the number of his family," reported Northern journalist Charles Nordhoff after his trip across the South in 1875. "Where the negro works for wages, he tries to keep his wife at home. If he rents land, or plants on shares, the wife and children help him in the field." The dire poverty of many black families, deepened by the depression of the 1870s, made it essential for women as well as men to contribute to the family's income. Throughout this period, a far higher percentage of black than white women and children worked for wages outside their homes. Where women continued to concentrate on domestic tasks, and children attended school, they frequently engaged in seasonal field labor. This was the pattern at Davis Bend, Mississippi, where most black women listed their occupation as "Keeping House" or "At Home" in the 1870 census, but labored in the fields, often with their children, at cotton-picking time. Thus, emancipation did not eliminate wage labor by black females and children, but it fundamentally altered control over their labor. The family itself, rather than a white owner or

16. Rawick, ed., *American Slave*, 14:14, 306; Theodore Rosengarten, *All God's Dangers: The Life of Nate Shaw* (New York, 1974), 120, 266; 46th Congress, 2d Session, Senate Report 693, pt. 2:179; Jane Le Conte to Joseph Le Conte, December 13, 1865, Le Conte Family Papers, Bancroft Library, University of California, Berkeley; Jones, *Labor of Love*, 59–60.

overseer, now decided where and when black women and children would work.[17]

For blacks, liberating their families from the authority of whites was an indispensable element of freedom. But the family itself was in some ways transformed by emancipation. Although historians no longer view the slave family as "matriarchal," it is true that slave men did not function as economic breadwinners and that their authority within the household was ultimately inferior to that of their masters. In a sense, slavery had imposed upon black men and women the rough "equality" of powerlessness. With freedom came developments that strengthened patriarchy within the black family and institutionalized the notion that men and women should inhabit separate spheres.

Outside events strongly influenced this development. Service in the Union Army enabled black men to participate more directly than women in the struggle for freedom. The Freedmen's Bureau designated the husband as head of the black household, insisting that men sign contracts for the labor of their entire families and establishing wage scales that paid women less than men for identical plantation labor. The Freedmen's Bureau Act of 1865 spoke of assigning land to every "male" freedman and refugee; the Southern Homestead Act of 1866 allowed women to claim a portion of the public domain only if unmarried. Political developments further reinforced the distinction between the public sphere of men and the private world of women. In the early days of freedom both men and women took part in informal mass meetings, although from the start men alone served as delegates to organized black conventions. After 1867 black men could serve on juries, vote, hold office, and rise to leadership in the Republican party, while women, like their white counterparts, could not. Militia units and fraternal societies were likewise all-male, although they often had ladies' auxiliaries. And male leaders of the black community promoted a strongly patriarchal definition of the family and woman's role. Black preachers, editors, and politicians emphasized women's responsibility for making the home "a place of peace and comfort" for men, and urged them to submit to their husbands' authority. Militant Virginia political leader Thomas Bayne had a severely restricted definition of women's "rights": "It is a woman's right to raise and bear children, and to train them for their future duties in life."[18]

17. Charles Nordhoff, *The Cotton States in the Spring and Summer of 1875* (New York, 1876), 38; 43th Congress, 2d Session, House Report 265, 203; Edmund L. Drago, *Black Politicians and Reconstruction in Georgia* (Baton Rouge, 1983), 135–36; Claudia Goldin, "Female Labor Force Participation: The Origin of Black and White Differences, 1870 and 1880," *JEcH*, 37 (March 1977), 91–96; Sir George Campbell, *White and Black* (New York, 1879), 150; Janet S. Hermann, *The Pursuit of a Dream* (New York, 1981), 182.

18. Jones, *Labor of Love*, 62–68; Robert F. Engs, *Freedom's First Generation: Black Hampton, Virginia, 1861–1890* (Philadelphia, 1979), 41; Savannah *Colored Tribune*, January 15, 1876; Macon *American Union*, October 29, 1869; James O. Horton, "Freedom's Yoke: Gender Conventions Among Antebellum Free Blacks," *Feminist Studies*, 12 (Spring 1986), 51–76;

Not all black women placidly accepted the increasingly patriarchal quality of black family life. Indeed, many proved more than willing to bring family disputes before public authorities. The records of the Freedmen's Bureau contain hundreds of complaints by black women of beatings, infidelity, and lack of child support. "I notice that some of you have your husbands arrested, and the husbands have their wives arrested," declared Holland Thompson, one of Alabama's leading black politicians, in an 1867 speech. "All that is wrong—you can settle it among yourselves." Some black women objected to their husbands' signing labor contracts for them, demanded separate payment of their wages, and refused to be liable for their husbands' debts at country stores. And some women, married as well as single, opened individual accounts at the Freedman's Savings Bank. Yet if emancipation not only institutionalized the black family but also spawned tensions within it, black men and women shared a passionate commitment to the stability of family life as a badge of freedom and the solid foundation upon which a new black community could flourish.[19]

Building the Black Community

Second only to the family as a focal point of black life stood the church. And, as in the case of the family, Reconstruction was a time of consolidation and transformation for black religion. With the death of slavery, urban blacks seized control of their own churches, while the "invisible institution" of the rural slave church emerged into the full light of day. The creation of an independent black religious life proved to be a momentous and irreversible consequence of emancipation.

In antebellum Southern Protestant congregations, slaves and free blacks had enjoyed a kind of associate membership. Subject to the same rules and discipline as whites, they were required to sit in the back of the church or in the gallery during services, and excluded from Sabbath schools and a role in church governance. In the larger cities, the number of black members often justified the organization of wholly black congregations and the construction of separate churches. In 1860 Richmond boasted four black churches with a combined membership of over 4,000.

The Debates and Proceedings of the Constitutional Convention of the State of Virginia (Richmond, 1868), 524.

19. Noralee Frankel, "Workers, Wives, and Mothers: Black Women in Mississippi, 1860–1870" (unpub. diss., George Washington University, 1983), 177; Barry A. Crouch and Larry Madaras, "Reconstructing Black Families: Perspectives from the Texas Freedmen's Bureau Records," Prologue, 18 (Summer 1986), 117; Montgomery Daily Sentinel, July 25, 1867; J. L. Thorpe to John Tyler, April 30, 1867, Narrative Reports of Operations, Ser. 242, Ark. Asst. Comr. RG 105, NA [FSSP A-2486]; Bliss Perry, Life and Letters of Henry Lee Higginson (Boston, 1921), 1:257; Carl R. Osthaus, Freedmen, Philanthropy, and Fraud: A History of the Freedman's Savings Bank (Urbana, Ill., 1976), 84.

Many such institutions achieved a considerable degree of autonomy, even though the law required that the pastors be white. Some of these white ministers, like Rev. Robert Ryland of Richmond's First African Baptist Church, treated their black parishioners with genuine respect and allowed black deacons and class leaders elected by the members to exercise real authority; others seemed to have no broader concept of Christianity than the biblical injunction that servants obey their masters. ("The black people of this country hate that passage," one black minister remarked after the war, "and I cannot get my people to like it, even now.") Although generally constructed with funds contributed by blacks, church buildings, by law, belonged to white trustees. In the countryside, nearly every plantation had its black preacher, usually a "self-called" slave with a knowledge of the Bible and "some little smattering of theology." Their secret after-dark religious meetings provided a rare opportunity for slaves to congregate and express their sorrows and aspirations free from white surveillance.[20]

In the aftermath of emancipation, the wholesale withdrawal of blacks from biracial congregations redrew the religious map of the South. Two causes combined to produce the independent black church: the refusal of whites to offer blacks an equal place within their congregations and the black quest for self-determination. The end of slavery does not appear to have altered the views of many white clergymen as to the legitimacy of the peculiar institution or the desirability of preserving unaltered blacks' second-class status within biracial churches. The "whole doctrine" of the scriptural justification for slavery remained intact, declared the General Assembly of the Southern Presbyterian Church in December 1865; as late as the 1890s, Southern ecclesiastics were still denouncing the idea of the inherent sinfulness of slaveholding. While initially urging blacks to remain within their congregations, most white ministers insisted that the old inequalities—separate pews, the white monopoly on church governance—must continue.[21]

Given the alternatives of admitting blacks as equal members or acquiescing in the formation of separate black churches, most Southern whites took the second course. Some, like Richmond's Ryland, went even

20. Harvey K. Newman, "Piety and Segregation—White Protestant Attitudes Toward Blacks in Atlanta, 1865–1906," *GaHQ*, 63 (Summer 1979), 239–40; W. Harrison Daniel, "Southern Protestantism and the Negro, 1860–1865," *NCHR*, 41 (Summer 1964), 338–41; John T. O'Brien, "Factory, Church, and Community: Blacks in Antebellum Richmond," *JSH*, 44 (November 1978), 523–30; *Virginia Convention Debates*, 429; Robert Anderson, *From Slavery to Affluence: Memoirs of Robert Anderson, Ex-Slave* (Hemingford, Neb., n.d.), 22–24; *Weekly Anglo-African*, August 19, 1865.

21. John L. Bell, Jr., "The Presbyterian Church and the Negro in North Carolina During Reconstruction," *NCHR*, 40 (Winter, 1963), 15–27; H. Shelton Smith, *In His Image But . . . : Racism in Southern Religion, 1780–1910* (Durham, 1972), 209–13; W. Harrison Daniel, "Virginia Baptists and the Negro," *VaMHB*, 76 (July 1968), 340–43; *Christian Recorder*, June 3, 1865.

further; he resigned his pastorship of Richmond's First African Baptist Church and arranged for the deed to be transferred from the white trustees to the black members. Elsewhere, however, the ownership of church property provoked bitter controversy. A case in point is the dispute over control of the Front Street Methodist Church in Wilmington, North Carolina, whose congregation before the war numbered about 1,400, two thirds of them black. When Union soldiers occupied the city early in 1865, the black members informed Rev. L. S. Burkhead "that they did not require his services any longer as Pastor, . . . he being a rebel," and proceeded to elect a black minister in his place. Gen. John M. Schofield, emulating Solomon, ordered that the spiritual day be divided equally between the races, each with a minister of its own choosing. The conflict continued into 1866, with the Reverend Mr. Burkhead preaching in the old manner (although a few blacks, he complained, ostentatiously attempted to sit downstairs during his sermons). Eventually, the white minority regained control and most of the blacks left to form an independent congregation. In other similar disputes, however, blacks were able to win title to the church buildings they had constructed as slaves.[22]

Throughout the South, blacks emerging from slavery pooled their resources to purchase land and erect their own churches. Before the buildings were completed, they held services in structures as diverse as a railroad boxcar, where Atlanta's First Baptist Church gathered, or in an outdoor "bush arbor," where the First Baptist Church of Memphis congregated in 1865. The first new building to rise amid Charleston's ruins was a black church on Calhoun Street; by 1866 ten more had been constructed. In rural areas, former slave preachers and missionaries from the North spurred the creation of religious institutions. 1866 was "a year of revivals" in Georgia, and in 1867 North Carolina witnessed a series of immense outdoor meetings. Whites sometimes attended these gatherings along with blacks, generally sitting in separate sections and often listening to their own preachers, but at one "old fashioned camp meeting" in 1867 several hundred Georgia whites were moved to take seats "on the rude log seats in the midst of the negroes." In the countryside, construction of church buildings proceeded more slowly than in the cities. Often a community would build a single church, used in rotation by the various black denominations.[23]

22. John L. Bell, Jr., "Baptists and the Negro in North Carolina During Reconstruction," *NCHR*, 42 (Autumn 1965), 400–401; Howard N. Rabinowitz, *Race Relations in the Urban South 1865–1890* (New York, 1978), 199–200; L. S. Burkhead, "History of the Difficulties of the Pastorate of the Front Street Methodist Church, Wilmington, N. C., for the Year 1865," *Trinity College Historical Society Historical Papers*, 8 (1908–9), 35–118; W. M. Poisson to Andrew Johnson, April 18, 1866, P-53 1866, Letters Received, Ser. 2452, N. C. Asst. Comr., RG 105, NA [FSSP A-559].
23. Rabinowitz, *Race Relations*, 204; Armstead Robinson, "Plans Dat Comed from God: Institution Building and the Emergence of Black Leadership in Reconstruction Memphis," in Burton and McMath, eds., *Toward a New South* 73–75; C. P. Gadsden to Thomas Smythe,

By the end of Reconstruction in 1877, the vast majority of Southern blacks had withdrawn from churches dominated by whites. On the eve of the war, 42,000 black Methodists worshipped in biracial South Carolina churches; by the 1870s, only 600 remained. Cleveland County, North Carolina, counted 200 black members of biracial Methodist churches in 1860, ten in 1867, and none five years later. A partial exception to this pattern was the Catholic Church, which generally did not require black worshippers to sit in separate pews (although its parochial schools were segregated). Some freedmen abandoned Catholicism for black-controlled Protestant denominations, but others were attracted to it precisely because, a Northern teacher reported from Natchez, "they are treated on terms of equality, at least while they are in church." And Catholicism retained its hold on large numbers of New Orleans free blacks who, at least on Sunday, coexisted harmoniously with the city's French and Irish white Catholic population.[24]

Northern whites who ventured South to proselytize among the freedmen proved no more successful than Southerners in winning black converts, partly because of their ill-disguised contempt for uneducated black ministers and their emotional services. Teachers employed by the American Missionary Association used Bible classes to inveigh against "heathenish habits such as shouting" and "unchristian" behavior like that of the black funeral mourner who "clapped her hands, threw them over her head screaming 'glory to God' . . . dancing up and down in front of the pulpit." Most Northern missionaries believed the old-time slave preachers must be replaced by new men trained in theology. Only a few listened long enough to appreciate that these preachers often exhibited a remarkable command of the Bible and were capable of genuine eloquence. Army Chaplain George H. Hepworth recognized in one uneducated Louisiana minister a man of "genius," whose sermons employed phrases "epic in grandeur"; he spoke of "the rugged wood of the cross" to which Christ had been nailed, and how "the earth was unable to endure the tremendous sacrilege, and trembled." Baptist missionary Isaac Brinckerhoff lamented the ignorance of black churchmen and ridiculed the Georgia minister who opened Sabbath services by inviting the congregation "to tumble with him through the third chapter of John." Yet in South Carolina Brinckerhoff encountered a black preacher who caused him to feel "ashamed of myself":

October 27, 1865, Augustine T. Smythe Letters, USC; *Christian Recorder*, November 24, 1866; Griffin (Ga.) *Semi-Weekly Star*, undated clipping (1869), Bryant Papers; Griffin *American Union*, October 11, 1867.

24. Francis B. Simkins and Robert H. Woody, *South Carolina During Reconstruction* (Chapel Hill, 1932), 382–88; J. R. Davis, "Reconstruction in Cleveland County," *Trinity College Historical Society Historical Papers*, 10 (1914), 18; Katherine Schlosser to unknown, undated letter fragment (1869), Katherine S. Estabrook Letters, SHSW; Nordhoff, *Cotton States*, 73; New Orleans *Louisianian*, August 14, 1875.

He talked about Christ and his salvation as one who understood what he said. . . . Here was an unlearned man, one who could not read, telling of the love of Christ, of Christian faith and duty in a way which I have not learned.[25]

Blacks, Northern missionaries quickly learned, preferred to worship in churches with ministers of their own race. The African Methodist Episcopal Church gained ascendancy over its white-dominated rivals, both Northern and Southern, in the competition for the allegiance of black Methodists. But the black emissaries sent south by the AME encountered problems of their own, for many insisted upon the need for an educated ministry and demanded more sedate services than Southern blacks were accustomed to. "The old people were not anxious to see innovations introduced in religious worship," an AME leader later wrote, recalling how one black minister from the North with an undemonstrative preaching style was mocked as a "Presbyterian" by his Southern flock. For these and other reasons, Baptist churches attracted the largest number of freedmen. The Baptists' decentralized, democratic structure and the fervor of their services meant that slave preachers could establish churches without being beholden to bishops promoting an educated ministry, as in the AME, while the freedmen could worship as they pleased. By the end of Reconstruction, black Baptists outnumbered all the other denominations combined; taken together, the Baptist churches formed the largest black organization ever created in this country.[26]

The church was "the first social institution fully controlled by black men in America," and its multiple functions testified to its centrality in the black community. Places of worship, churches also housed schools, social events, and political gatherings. In rural areas, church picnics, festivals, and excursions often provided the only opportunity for fellowship and recreation. The church served as an "Ecclesiastical Court House," promoting moral values, adjudicating family disputes, and disciplining individuals for adultery and other illicit behavior. In every black community, ministers were among the most respected individuals, esteemed for their speaking ability, organizational talents, and good judgment on matters both public and private. "You know those who are the

25. Joe M. Richardson, "The Failure of the American Missionary Association to Expand Congregationalism Among Southern Blacks," SS, 18 (Spring 1979), 51–70; Timothy Lyman to George Whipple, February 1, 1865, Lyman to Michael E. Streiby, February 27, 1865, AMA Archives; Joe M. Richardson, ed., " 'We are Truly Doing Missionary Work': Letters from American Missionary Association Teachers in Florida, 1864–1874," FIHQ, 54 (October 1975), 191–92; George Hepworth, The Whip, Hoe, and Sword (Boston, 1864), 166–67; Isaac W. Brinckerhoff, "Missionary Work Among the Freed Negroes. Beaufort, South Carolina, St. Augustine, Florida, Savannah, Georgia," Manuscript, American Baptist Historical Society.

26. Clarence G. Walker, A Rock in a Weary Land: The African Methodist Episcopal Church during the Civil War and Reconstruction (Baton Rouge, 1982), 75–76, 83–107; Christian Recorder, December 15, 1866; L. J. Coppin, Unwritten History (Philadelphia, 1914), 165, 198–99; James M. Washington, Frustrated Fellowship: The Black Baptist Quest for Social Power (Macon, 1986), 83–105.

real leaders in every community of freedmen," wrote a white North Carolinian in 1868, "are religious exhorters." One visitor to the South remarked, however, that such men "are rather preachers because they are leaders than leaders because they are preachers," for, as one of the few available positions of power and prestige, the ministry inevitably attracted those with leadership potential.[27]

Inevitably, too, preachers came to play a central role in black politics during Reconstruction. Many agreed with AME minister Charles H. Pearce, who held several Reconstruction offices in Florida, that it was "impossible" to separate religion and politics: "A man in this State cannot do his whole duty as a minister except he looks out for the political interests of his people." Even those preachers who lacked ambition for political position sometimes found it thrust upon them. Often among the few literate blacks in a community, they were called upon to serve as election registrars and candidates for office. Over 100 black ministers, hailing from North and South, from free and slave backgrounds, and from every black denomination from AME to Primitive Baptist, would be elected to legislative seats during Reconstruction. And among the lay majority of black politicians, many built a political base in the church. Alabama legislator Holland Thompson, for example, had played a leading role in Montgomery's Baptist affairs since his days as a slave. Rare indeed was Frederick Douglass, the only prominent black political leader who not only lacked a tie to the black church but repudiated its mystical, evangelical rhetoric.[28]

Throughout Reconstruction, religious convictions profoundly affected the way blacks understood the momentous events around them, the very language in which they expressed aspirations for justice and autonomy. Blacks inherited from slavery a distinctive version of Christian faith, in which Jesus appeared as a personal redeemer offering solace in the face of misfortune, while the Old Testament suggested that they were a chosen people, analogous to the Jews in Egypt, whom God, in the fullness of time, would deliver from bondage. "There is no part of the Bible with which they are so familiar as the story of the deliverance of the Children of Israel," a white army chaplain reported from Alabama in 1866.[29]

27. W. E. B. Du Bois, "Reconstruction and Its Benefits," *AHR*, 15 (July 1910), 782; E. B. Rucker to "Editor of Republican," May 30, 1870, Bryant Papers; Rabinowitz, *Race Relations*, 210; Coppin, *Unwritten History*, 124–27; J. G. de Roulhac Hamilton and Max R. Williams, eds., *The Papers of William Alexander Graham* (Raleigh, 1957–), 7:514; Campbell, *White and Black*, 139.

28. KKK Hearings, Florida, 171; Walker, *Rock in a Weary Land*, 116–27; Drago, *Black Politicians*, 29–37; Howard N. Rabinowitz, "Holland Thompson and Black Political Participation," in Howard N. Rabinowitz, ed., *Southern Black Leaders of the Reconstruction Era*, (Urbana, Ill., 1982), 52; Robert L. Factor, *The Black Response to America: Men, Ideals, and Organization from Frederick Douglass to the NAACP* (Reading, Mass., 1970), 65–66, 102.

29. Walker, *A Rock in a Weary Land*, 125; 43d Congress, 2d Session, House Report 262, 779; W. G. Kephart to Lewis Tappan, May 9, 1866, AMA Archives.

Emancipation and the defeat of the Confederacy strongly reinforced this messianic vision of history. Blacks endowed these experiences with spiritual import, comprehending them in the language of Christian faith. "These are the times foretold by the Prophets, 'when a nation shall be born in a day'," declared the call for a black political gathering in 1865. A Tennessee newspaper commented in 1869 that freedmen habitually referred to slavery as Paul's Time, and Reconstruction as Isaiah's Time (referring perhaps to Paul's message of obedience and humility, and Isaiah's prophecy of cataclysmic change brought about by violence). God, who had "scourged America with war for her injustice to the black man," had allowed his agent Lincoln, like Moses, to glimpse the promised land of "universal freedom" and then mysteriously removed him before he "reached its blessed fruitions."[30]

Under some circumstances, such faith can produce quiescence, a belief that the contrivances of man are inadequate to bring about divine purposes. But during Reconstruction, black Christianity inspired not inaction but political commitment. When one speaker at a black political meeting complained of preachers who spoke more "of politics than of Christ," he was silenced by shouts of "Politics in Christ." Throughout Reconstruction, black republicanism was grounded in "the great Christian principle of the brotherhood of man." Even nonclerics used secular and religious vocabulary interchangeably, as in one 1867 speech recorded by a North Carolina justice of the peace:

> He said it was not now like it used to be, that . . . the negro was about to get his equal rights. . . . That the negroes owed their freedom to the courage of the negro soldiers and to God. . . . He made frequent references to the II and IV chapters of Joshua for a full accomplishment of the principles and destiny of the race. It was concluded that the race have a destiny in view similar to the Children of Israel.

Indeed, for black political leaders, the Bible—the one book with which they could assume familiarity among their largely illiterate constituents—served as a point of reference for understanding public events. When in 1870 North Carolina's House impeached Governor Holden, seventeen black legislators issued an address that began: "Know ye that since the time that Haman conspired to destroy all the Jews who dwelt in the Persian Dominions . . . no wickedness hath been devised that will bear

30. W. McKee Evans, *Ballots and Fence Rails: Reconstruction on the Lower Cape Fear* (Chapel Hill, 1967), 87; Stephen V. Ash, *Middle Tennessee Society Transformed, 1860–1870: War and Peace in the Upper South* (Baton Rouge, 1987), Chapter 9 [Isaiah 66:15: "The Lord will come with fire, and with his chariots like a whirlwind, to render his anger with fury, and his rebuke with flames of fire" and 66:22: "The new heavens and the new earth, which I will make, shall remain before me, saith the Lord."]; Walker, *Rock in a Weary Land*, 30–31; Savannah *Republican* in Raleigh *Journal of Freedom*, October 28, 1865; *Christian Recorder*, May 13, 1865, January 20, 1866.

any comparison with some of the measures proposed by the dominant party in the present General Assembly."[31]

The rise of the independent black church provides only the most striking example of the thriving institutional structure blacks created in the aftermath of emancipation. A host of fraternal, benevolent, and mutual-aid societies also sprang into existence. Even before the Civil War, free blacks had formed fraternal organizations, and secret societies of various kinds had existed among the slaves. In early Reconstruction, blacks created literally thousands of such organizations; a partial list includes burial societies, debating clubs, Masonic lodges, fire companies, drama societies, trade associations, temperance clubs, and equal rights leagues. Often spawned in black churches, they quickly took on lives of their own. By the 1870s, over 200 such organizations existed in Memphis, 400 in Richmond, and countless others were scattered across the rural South. Although their activities generally took place away from white observation, they appeared in public in the processions and celebrations that seemed ubiquitous, especially in Southern cities, during Reconstruction. Black parades commemorated special occasions like the ratification of the Fifteenth Amendment, but the largest celebrations were reserved for January 1 (the anniversary of the Emancipation Proclamation) and July 4—days on which Southern whites generally remained indoors.[32]

Offering social fellowship, sickness and funeral benefits and, most of all, a chance to manage their own affairs, these voluntary associations embodied a spirit of collective self-improvement. Robert G. Fitzgerald, who had been born free in Delaware, served in both the U.S. Army, and Navy, and came to Virginia to teach in 1866, was delighted to see rural blacks establishing churches, lyceums, and schools. "They tell me," he recorded in his diary, "before Mr. Lincoln made them free they had nothing to work for, to look up to, now they have everything, and will, by God's help, make the best of it." Linking blacks across lines of occupation, income, and prewar status, the societies offered the better-off the opportunity for wholesome and respectable association, provided the poor with a modicum of economic insurance, and opened positions of community leadership to men of modest backgrounds. (Among the leaders of Memphis' religious and benevolent societies, the majority were unskilled laborers.) Moreover, the spirit of mutual self-help extended

31. Richmond *Dispatch*, April 16, 1867; Charleston *Daily Republican*, August 25, 1870, Supplement; Eric Foner, "Reconstruction and the Black Political Tradition," in Richard L. McCormick, ed., *Political Parties and the Modern State* (New Brunswick, 1984), 62; *Address to the Colored People of North Carolina*, Broadside, Raleigh, December 19, 1870.

32. Robinson, "Plans," 78; Peter J. Rachleff, *Black Labor in the South: Richmond, Virginia, 1865–1890* (Philadelphia, 1984), 25; Rabinowitz, *Race Relations*, 228–29; *Christian Recorder*, January 27, 1866; Savannah *Freemen's Standard*, February 15, 1868; Jacob Schirmer Diary, July 4, 1871, January 1, 1874.

outward from the societies to embrace destitute nonmembers. In 1865 and 1866, blacks in Nashville, Jackson, New Orleans, and Atlanta, as well as in many rural areas, raised money to establish orphanages, soup kitchens, employment agencies, and poor relief funds. In some areas, such as a poverty-stricken corner of West Virginia, black organizations contributed money to aid suffering poor whites.[33]

Perhaps the most striking illustration of the freedmen's quest for self-improvement was their seemingly unquenchable thirst for education. Before the war, every Southern state except Tennessee had prohibited the instruction of slaves, and while many free blacks had attended school and a number of slaves became literate through their own efforts or the aid of sympathetic masters, over 90 percent of the South's adult black population was illiterate in 1860. Access to education for themselves and their children was, for blacks, central to the meaning of freedom, and white contemporaries were astonished by their "avidity for learning." A Mississippi Freedmen's Bureau agent reported in 1865 that when he informed a gathering of 3,000 freedmen that they "were to have the advantages of schools and education, their joy knew no bounds. They fairly jumped and shouted in gladness." The desire for learning led parents to migrate to towns and cities in search of education for their children, and plantation workers to make the establishment of a schoolhouse "an absolute condition" of signing labor contracts. (One 1867 Louisiana contract specified that the planter pay a "5 per cent tax" to support black education.) Adults as well as children thronged the schools established during and after the Civil War. A Northern teacher in Florida reported how one sixty-year-old woman, "just beginning to spell, seems as if she could not think of any thing but her book, says she spells her lesson all the evening, then she dreams about it, and wakes up thinking about it."[34]

For many adults, a craving "to read the word of God" provided the

33. Robert G. Fitzgerald Diary, July 27, 1867, Robert G. Fitzgerald Papers, Schomburg Center for Research in Black Culture; Robinson, "Plans," 81, 89–92; Paul D. Phillips, "A History of the Freedmen's Bureau in Tennessee" (unpub. diss., Vanderbilt University, 1964), 72–73; H. Gardner to A. W. Preston, September 30, 1866, G-119, Registered Letters Received, Ser. 2052, Miss. Asst. Comr., RG 105, NA [FSSP A-9109]; Charles Vincent, "Black Louisianians, During the Civil War and Reconstruction: Aspects of their Struggles and Achievements," in Robert R. Macdonald et al., eds., *Louisiana's Black Heritage*, (New Orleans, 1979), 92; Jerry Thornbery, "Northerners and the Atlanta Freedmen, 1865–69," *Prologue*, 6 (Winter 1974), 243–44; John E. Stealey III, ed., "Reports of Freedmen's Bureau Operations in West Virginia: Agents in the Eastern Panhandle," *WVaH*, 43 (Fall 1980–Winter 1981), 113.

34. William P. Vaughan, *Schools for All: The Blacks and Public Education in the South, 1865–1877* (Lexington, Ky., 1974), 1; Joseph Crosfield to Unknown, October 5, 1865 (typed copy), Society of Friends Library, Friends House, London; L. C. Hubbard to R. S. Donaldson, August 5, 1865, H-2 1865, Registered Letters Received, Ser. 2180, Jackson, Ms. Acting Asst. Comr., RG 105, NA [FSSP A-9304]; Whitelaw Reid, *After the War: A Southern Tour* (Cincinnati, 1866), 511; Hartford *Courant*, March 1, 1867; Dorothy Sterling, ed., *We Are Your Sisters: Black Women in the Nineteenth Century* (New York, 1984), 298–99; Richardson, ed., " 'We are Truly Doing Missionary Work'," 185.

immediate spur to learning. One elderly freedman sitting beside his grandchild in a Mobile school explained to a Northern reporter, "he wouldn't trouble the lady much, but he must learn to read the Bible and the Testament." Others recognized education as indispensable for economic advancement. "I gets almost discouraged, but I does want to learn to cipher so I can do business," an elderly Mississippi pupil told his teacher. But more generally, blacks' hunger for education arose from the same desire for autonomy and self-improvement that inspired so many activities in the aftermath of emancipation. As a member of a North Carolina education society put it in 1866, "he thought a school-house would be the first proof of their *independence.*"[35]

Northern benevolent societies, the Freedmen's Bureau, and, after 1868, state governments, provided most of the funding for black education during Reconstruction. But the initiative often lay with blacks themselves, a pattern established in the early days of the war. Mary Peake, the daughter of a free black mother and an English father, in 1861 established the first school for blacks in Hampton, Virginia, before the arrival of Northern teachers. When the Gideonites arrived in the Sea Islands in 1862, they found two schools already in operation, one of them taught by a black cabinetmaker who for years had conducted secret night classes for slaves. After the war, urban blacks took immediate steps to set up schools, sometimes holding classes temporarily in abandoned warehouses, billiards rooms, or, in New Orleans and Savannah, former slave markets. By the end of April 1865, less than a month after Union troops occupied the city, over 1,000 black children and seventy-five adults attended schools established by Richmond's black churches and the American Missionary Association. "Joy incomparable" was expressed by those who gathered for the opening. In rural areas, Freedmen's Bureau officials repeatedly expressed surprise at discovering classes organized by blacks already meeting in churches, basements, or private homes. Charles Hopkins, a freedman and Methodist preacher, in 1866 "obtained a room in a deserted hotel" in Greenville, South Carolina, "and began giving spelling and reading lessons." And then there was the seemingly ubiquitous learning that took place outside school—children teaching their parents the alphabet at home; laborers on lunch breaks "poring over the elementary pages"; the "wayside schools" described by a Bureau officer:

> A negro riding on a loaded wagon, or sitting on a hack waiting for a train, or by the cabin door, is often seen, book in hand delving after the rudiments of knowledge. A group on the platform of a depot, after carefully conning an old spelling book, resolves itself into a class.[36]

35. Dennett, *South As It Is,* 304; Katherine Schlosser to "the church in Milwaukee," December 25, 1868, Estabrook Papers; *National Freedman,* March 1866.
36. Joe M. Richardson, *Christian Reconstruction: The American Missionary Association and Southern Blacks, 1861–1890* (Athens, Ga., 1986), 4; Rupert S. Holland, ed., *Letters and Diary*

Throughout the South, blacks in 1865 and 1866 formed societies and raised money among themselves to purchase land, build schoolhouses, and pay teachers' salaries. Some communities voluntarily taxed themselves, while in others black schools charged tuition, although often a certain number of the poorest families were allowed to enroll their children free of charge. Robert G. Fitzgerald's salary was raised by a monthly tuition charge of twenty cents. However, he gave much of the money away to poor black families, and in 1869 abolished the fee altogether, to render his school accessible to "rich and poor, black and white, high and low." Black artisans donated their labor to construct schoolhouses, and black families offered room and board to teachers to supplement their salaries. The sums expended—$800 to purchase a lot in Georgetown, South Carolina, $2,000 to support fifty-six Georgia schools in November 1865— represented a genuine sacrifice for a largely impoverished community. Contemporaries could not but note the contrast between white families seemingly indifferent to education and blacks who "toil and strive, labour and endure in order that their children 'may have a schooling'." As one Northern educator remarked: "Is it not significant that after the lapse of one hundred and forty-four years since the settlement [of Beaufort, North Carolina], the Freedmen are building the first public school-house ever erected here."[37]

By 1870, blacks had expended over $1 million on education, a fact that long remained a point of collective pride. "Whoever may hereafter lay claim to the honor of 'establishing' . . . schools," wrote a black resident of Selma in 1867, "I trust the fact will never be ignored that Miss Lucy Lee, one of the emancipated, was the pioneer teacher of the colored children, . . . without the aid of Northern societies." But poverty undercut black educational efforts, forcing many schools to turn to the Freedmen's Bureau and Northern societies for aid. "We have plodded along this far,

of Laura M. Towne (Cambridge, Mass., 1912), 27; Lamon, Blacks in Tennessee, 38; Mobile Nationalist, July 19, 1866; Henry Swint, ed., Dear Ones at Home (Nashville, 1966), 157; unknown to Michael E. Strieby, April 30, 1865, AMA Archives; DeForest, Union Officer, 118–19; E. P. Breck to Miss Cochran, December 23, 1864, New England Freedmen's Aid Society Papers, SC; [Oliver O. Howard] Autobiography of Oliver Otis Howard (New York, 1907), 2:274–75; 39th Congress, 1st Session, House Report 30, pt 2:204; D. Burt to J. R. Lewis, October 1866, Tenn. Annual Reports of Asst. Comr., Ser. 32, Washington Headquarters, RG 105, NA [FSSP A-6000].
37. Kolchin, First Freedom, 84–86; Jane T. Shelton, Pines and Pioneers: A History of Lowndes County, Georgia, 1825–1900 (Atlanta, 1976), 167; Robert G. Fitzgerald Diary, February 8, March 5, 19, November 30, 1869, Fitzgerald Papers; Byron Porter to E. M. Wheelock, January 23, 1867, P-2 1867, Registered Letters Received, Ser. 3633, Tx. Supt. of Education, RG 105, NA [FSSP A-3485]; Jacqueline Jones, Soldiers of Light and Love: Northern Teachers and Georgia Blacks, 1865–1873 (Chapel Hill, 1980), 62–63; E. C. Rainey to J. K. Jillson, October 3, 1872, South Carolina Superintendent of Education Papers, SCDA; John E. Bryant, "Georgia Educational Movement," manuscript, January 1866, Bryant Papers; Ella Gertrude Thomas Journal, June 26, 1869, DU; H. S. Beals to Samuel Hunt, August 31, 1866, AMA Archives.

the best we could," wrote Florida freedman Emanuel Smith in April 1867, requesting the American Missionary Association to furnish and help pay the salary of a female teacher. "This is the first application that has been made to any source for help since we have been free." (Smith specified a woman because, as he explained, "I suppose they can be had on cheaper terms.")[38]

"Without help we can do nothing," a South Carolina freedman concluded in 1867. But with outside help came the prospect of outside control, as events in Savannah illustrate. When Sherman captured the city in December 1864, local black ministers immediately established the Savannah Educational Association, which by February had raised nearly $1,000, engaged fifteen black teachers, and enrolled 600 pupils in schools. Simultaneously, AMA missionaries from the North, headed by Rev. S. W. Magill, a white Georgian who had resided in Connecticut during the war, arrived in the city intent on educating the freedmen. Hoping to uplift blacks by providing them with "Christian education," Magill believed the black teachers incompetent and their school system "radically defective." "It will not do," he concluded, "to leave these people to themselves." The Bureau agents who reached Savannah in mid-1865 shared these assumptions, and withheld funds from the black school system. By 1866, unable to finance its schools, the Savannah Educational Association had no alternative but to turn them over to the AMA, which replaced the black teachers with its own white employees, retaining a few of the blacks as assistants. Ill-will generated by these events lingered. "There is jealousy of the superintendence of the white man in this matter," one Northern teacher remarked. "What they desire is assistance without control."[39]

Inevitably, the first black teachers appeared hopelessly incompetent in Northern eyes, for a smattering of education was enough to place an individual in front of a class. Acutely aware of their lack of preparation, some teachers worried about the poorly written reports they drafted for Freedmen's Bureau education officials. "I have no education only what I gave myself by chance so I ask you to excuse my unqualified address," one wrote; another poignantly explained, "I never had the chance of goen to school for I was a slave until freedom. . . . I am the only teacher because we can not doe better now." Yet even an imperfect literacy,

38. Arnold H. Taylor, *Travail and Triumph: Black Life and Culture in the South Since the Civil War* (Westport, Conn., 1976), 118; Mobile *Nationalist*, January 17, 1867; Richardson, ed., " 'We are Truly Doing Missionary Work'," 189–90; Herbert G. Gutman, "Schools for Freedom: Post-Emancipation Origins of Afro-American Education," in Herbert G. Gutman, *Power and Culture: Essays on the American Working Class*, edited by Ira Berlin (New York, 1987), 260–97.

39. Charleston *Advocate*, April 20, 1867; W. T. Richardson, "Mr. Richardson's Report of Doings at Savannah, Ga.," January 25, 1865, S. W. Magill to the AMA, February 3, 7, 26, 1865, AMA Archives; Alan Conway, *The Reconstruction of Georgia* (Minneapolis, 1966), 86–87; Jones, *Soldiers of Light and Love*, 71–76; *Freedmen's Record*, June 1865.

coupled with the courage often required to establish a rural school in the face of local white opposition, marked these teachers as community leaders. Black teachers played numerous roles apart from education, assisting freedmen in contract disputes, engaging in church work, and drafting petitions to the Freedmen's Bureau, state officials, and Congress. Robert Harris, a free black from Cleveland who had come to North Carolina to teach, passed up his usual summer trip to Ohio in 1869 because his family was "so connected with the educational, religious, social, and industrial affairs of the people that we cannot be spared." Like the ministry, teaching frequently became a springboard to political office. At least seventy black teachers served in state legislatures during Reconstruction. And many black politicians were linked in other ways to the quest for learning, like Alabama Congressman Benjamin S. Turner, an ex-slave "destitute of education," who personally funded a Selma school.[40]

Not surprisingly, the majority of black teachers who held political office during Reconstruction had been free before the Civil War. Indeed the schools, like the entire institutional structure established by blacks during Reconstruction, symbolized the emergence of a community that united the free and the freed, and Northern and Southern blacks. The process occurred most smoothly in the Upper South, where the cultural and economic gap between free blacks and slaves had always been less pronounced than in the coastal cities of the cotton states. While generally lighter in color than slaves, most Upper South free blacks were poor urban workers or farm laborers, often tied to the slave community through marriage and church membership. It was not uncommon after the Civil War to find free blacks like John Overton of Cedar Grove, North Carolina, heading educational societies composed mostly of freedmen. Many Northern-born blacks who ventured south after the end of the war also linked their fortunes to those of the former slaves. "I class myself with the freedmen," wrote Northern black teacher Virginia C. Green. "Though I have never known servitude they are in fact my people."[41]

In cities like New Orleans, Mobile, Savannah, and Charleston, however, affluent mulatto elites responded with deep ambivalence to the new situation created by emancipation. In 1866 Rev. Henry M. Turner expressed alarm at disputes within his church in which "the blacks were

40. Allan A. Williams to William H. Smith, June 12, 1867, Alabama Governor's Papers, ASDAH; Robert Alexander to J. R. Lewis, June 10, 1869, #95 1869, Letters Received, Ser. 657, Ga. Supt. of Education, RG 105, NA [FSSP A-5477]; Earle H. West, "The Harris Brothers: Black Northern Teachers in the Reconstruction South," *Journal of Negro Education*, 48 (Spring 1979), 132–33; Howard, *Autobiography*, 2:134.

41. Ira Berlin, *Slaves Without Masters: The Free Negro in the Antebellum South* (New York, 1974), 174–78; Engs, *Black Hampton*, 10–16; *National Freedman*, January 1866; Virginia C. Green to A. W. Preston, October 24, 1866 G-136 1866, Registered Letters Received, Ser. 2052, Ms. Asst. Comr., RG 105, NA [FSSP A-9110].

arrayed against the brown or mulattoes, and the mulattoes in turn against the blacks." Educated members of the free black elite often found the freedmen's religious practices excessively emotional, and were appalled by the anti-intellectualism of ex-slave preachers. (One sermon of the most famous "old-style" preacher, John Jaspar, entitled "The Sun Do Move," "disproved" the heliocentric theory of astronomers.) Free blacks welcomed the end of slavery, but many resented the elimination of their unique status and feared being submerged in a sea of freedmen. Even in New Orleans, where politically conscious free blacks had already moved to make common cause with the freedmen, a sense of exclusivity survived the end of slavery. The Freedmen's Bureau found many free blacks reluctant to send their children to school with former slaves. In Mobile, too, free mulattoes were divided between those who embraced the new order and emerged as social and political leaders during Reconstruction, and others who cultivated the goodwill of local whites in the hope of maintaining their elite position. The Mobile *Nationalist,* founded by a group of free blacks in 1865, excoriated the Creole Fire Company for acceding to white demands not to display an American flag during a parade, and for putting on airs of "supposed superiority." Yet the same paper advised the freedmen to "put away 'nigger' plays and songs" and adopt the "plays and amusements [of] . . . free men and women."[42]

After New Orleans, the South's largest and wealthiest community of free blacks resided in Charleston, although the free elite there was neither as rich nor as culturally distinct as its Louisiana counterpart. Before the war, no free person of color in Charleston owned as much property as the richest New Orleans mulattoes. The Charlestonians spoke English and worshipped at Protestant churches (although, unlike the slaves, they were mostly Episcopalians or Presbyterians), and they could not bear arms or testify in court against whites. Nonetheless, the Charleston free elite was no less conscious of the gap separating themselves not merely from the slaves, but from the city's poorer free blacks—a gap institutionalized in organizations like the Brown Fellowship Society, which excluded men with dark skins.[43]

Arriving in Charleston in November 1865, John R. Dennett found some members of the free elite cultivating the old spirit of exclusiveness. Others, however, had taken the lead in organizing assistance for destitute

42. *Christian Recorder,* January 20, 1866; Taylor, *Travail and Triumph,* 150; Thomas Holt, *Black over White: Negro Political Leadership in South Carolina During Reconstruction* (Urbana, Ill., 1977), 64; Robert C. Morris, *Reading, 'Riting, and Reconstruction: The Education of Freedmen in the South 1861–1870* (Chicago, 1981), 110; Kolchin, *First Freedom,* 140–42; Mobile *Nationalist,* March 8, April 26, May 3, 1866.

43. Holt, *Black over White,* 56–66; Robert L. Harris, Jr., "Charleston's Free Afro-American Elite: The Brown Fellowship Society and the Humane Brotherhood," *SCHM,* 82 (October 1981), 289–310.

freedmen and in teaching the former slaves. In June 1865, Francis L. Cardozo took charge of the AMA's largest Charleston school, which enrolled over 1,000 pupils. The son of a Jewish businessman who had married a free black woman, Cardozo had been educated in the city, left in 1858 to study at the University of Glasgow, and served as pastor of a New Haven Congregationalist church during the Civil War. Despite his privileged background, Cardozo made no distinction between free and freed children and ridiculed the idea that mulattoes learned more quickly than blacks. Other sons and daughters of prominent free families, mostly young people in their twenties, fanned out into the South Carolina countryside as teachers and missionaries (something that would have been impossible for the French-speaking New Orleans elite). Several thereby acquired positions of local political leadership, and later returned to Charleston as constitutional convention delegates and legislators. Thus the children of the Charleston elite cast their lot with the freedmen, bringing, as they saw it, modern culture to the former slaves. This encounter was not without its tensions. But in the long run it hastened the emergence of a black community stratified by class rather than color, in which the former free elite took its place as one element of a new black bourgeoisie, instead of existing as a separate caste as it had in the port cities of the antebellum Lower South.[44]

In the severing of ties that had bound black and white families and churches to one another under slavery, the coming together of blacks in an explosion of institution building, and the political and cultural fusion of former free blacks and former slaves, Reconstruction witnessed the birth of the modern black community. All in all, the months following the end of the Civil War were a period of remarkable accomplishment for Southern blacks. Looking back in January 1866, the Philadelphia-born black missionary Jonathan C. Gibbs could only exclaim: "we have progressed a century in a year."[45]

The Economics of Freedom

In no realm of Southern life did blacks' efforts to define the terms of their freedom have implications as explosive for the entire society as the economy. Blacks brought out of slavery a conception of themselves as a "Working Class of People" who had been unjustly deprived of the fruits of their labor. Reprimanded by a planter for laziness—"You lazy nigger,

44. Dennett, *South As It Is*, 218; Holt, *Black over White*, 70; Francis L. Cardozo to Samuel Hunt, December 13, 1865, AMA Archives; Morris, *Reading, 'Riting, and Reconstruction*, 86–87; Williamson, *After Slavery*, 364–66; Joel Williamson, *New People: Miscegenation and Mulattoes in the United States* (New York, 1980), 61–84.

45. Unidentified newspaper clipping, January 22, 1866, Rufus and S. Willard Saxton Papers, Yale University.

I am losing a whole day's labor by you"—a freedman responded, "Massa, how many day's labor have I lost by you?" Former slaves had no reluctance to express assertively their sense of having been wronged. "We have built up their houses and cultivated their lands. . . ." declared black minister Willis Hodges. "If they were to pay us but twenty-five cents on the dollar, they would all be very poor."[46]

For blacks, the abolition of slavery meant not an escape from all labor, but an end to unrequited toil. "We scorn and treat with contempt the allegation . . . that we understand Freedom to mean idleness and indolence," a mass meeting of Petersburg, Virginia, blacks resolved in June 1865. "But we do understand Freedom to mean industry and the enjoyment of the legitimate fruits thereof." To white predictions that they would not work, blacks responded that if any class could be characterized as "lazy," it was the planters, who had "lived in idleness all their lives on stolen labor." Blacks deeply resented incessant allegations of indolence and incapacity. "They say we will not work," complained a Virginia freedman. "He who makes that assertion asserts an untruth. We have been working all our lives, not only supporting ourselves, but we have supported our masters, many of them in idleness." It is certainly true that many blacks expected to labor less as free men and women than they had as slaves, an understandable aim in view of the conditions they had previously known. "Whence comes the assertion that the 'nigger won't work'?" asked an Alabama freedman. "It comes from this fact: . . . the freedman refuses to be driven out into the field two hours before day, and work until 9 or 10 o'clock in the night, as was the case in the days of slavery." As for predictions that they would be unable to care for themselves in freedom, one ex-slave responded: "We used to support ourselves and our masters too when we were slaves and I reckon we can take care of ourselves now."[47]

Yet freedom meant more than simply receiving wages. Freedmen wished to take control of the conditions under which they labored, free themselves from subordination to white authority, and carve out the greatest measure of economic autonomy. As in the case of their families, churches, and social life, economic emancipation meant freedom from white control. Probably the most ubiquitous example of this ambition was the widespread reluctance of freedmen to continue working in gangs under the direction of an overseer. Blacks "don't want any white man to

46. Edward Magdol, *A Right to the Land: Essays on the Freedmen's Community* (Westport, Conn., 1977), 273; Mahaffey, ed., "Schurz's Letters," 241; *Virginia Convention Debates*, 164–65.

47. New York *Tribune*, June 15, 1865; Berlin et al., eds., *Freedom*, Ser. 2, 582–83; *Proceedings of the Convention of the Colored People of Va., Held in the City of Alexandria* (Alexandria, 1865), 3; Mobile *Nationalist*, December 20, 1866; F. J. Massey to Orlando Brown, May 1, 1866, Monthly Reports, Ser. 4350, Yorktown Asst. Subasst. Comr., RG 105, NA [FSSP A-7889].

control them," a Bureau agent reported. On one Georgia plantation, the hiring of an overseer in 1865 "enraged the negroes so much" that they "ran off and went to Macon," with the result that the planter had to hire white laborers to harvest his crop.[48]

The desire to escape from white supervision and establish a modicum of economic independence profoundly shaped blacks' economic choices during Reconstruction, leading them to prefer tenancy to wage labor, and leasing land for a fixed rent to sharecropping. Above all, it inspired the quest for land of their own. Indeed, the same blacks arraigned for idleness sacrificed and saved in the attempt to acquire land, and those who succeeded clung to it with amazing tenacity. "They will almost starve and go naked before they will work for a white man," wrote a Georgia planter, "if they can get a patch of ground to live on, and get from under his control." Owning land, the freedmen believed, would "complete their independence." Without land, there could be no economic autonomy, for their labor would continue to be subject to exploitation by their former owners. "Gib us our own land and we take care ourselves," a Charleston black told Northern correspondent Whitelaw Reid, "but widout land, de ole masses can hire us or starve us, as dey please."[49]

To those familiar with the experience of other postemancipation societies, blacks' "mania for owning a small piece of land" did not appear unusual. Throughout the Western Hemisphere, the end of slavery was followed by a prolonged struggle over the control of labor and access to land. Freedmen in Haiti, the British and Spanish Caribbean, and Brazil all saw ownership of land as crucial to establishing their economic independence, and their efforts to avoid returning to plantation labor were strenuously resisted by the planter elite and local political authorities. Unlike freedmen in other countries, however, American blacks emerged from slavery convinced that the federal government had committed itself to land distribution. A millennial expectation of impending change swept through the South as the end of 1865 approached. A story circulated that the Freedmen's Bureau had received a "great document" bearing four seals, to be opened on January 1 to reveal the "final orders" of the federal government. Belief in an imminent division of land was most pervasive in the South Carolina and Georgia low country, with its tradition of black autonomy and the unique experience of Sherman's Field Order 15. But the idea was shared in other parts of the South as well, including counties

48. D. T. Corbin to H. W. Smith, February 1, 1866, Registered Letters Received, Ser. 2922, S. C. Asst. Comr., RG 105, NA [FSSP A-7093]; Carrie Kincaid to Mrs. E. K. Anderson, January 21, 1866, Kincaid-Anderson Papers.

49. Eric Foner, *Nothing But Freedom: Emancipation and Its Legacy* (Baton Rouge, 1983), 8–12, 18–19; *Southern Cultivator*, March 1867, 69; Eric Foner, *Politics and Ideology in the Age of the Civil War* (New York, 1980), 107; Reid, *After the War*, 59.

that had never been occupied by federal troops. For some blacks, more-over, land distribution seemed a logical consequence of emancipation. "If you had the right to take Master's niggers," one Virginia freedman told an army officer, "you had the right to take Master's land too." Others contended that "the land ought to belong to the man who (alone) could work it." Most often, however, blacks insisted that their past labor enti-tled them to at least a portion of their owners' estates. As an Alabama black convention put it: "The property which they hold was nearly all earned by the sweat of *our* brows."[50]

In its most sophisticated form, the claim to land rested on an apprecia-tion of the role blacks had played in the evolution of the American economy. When the army evicted blacks it had earlier settled on land near Yorktown, Virginia, freedman Bayley Wyat gave an impromptu speech protesting the injustice:

> We has a right to the land where we are located. For why? I tell you. Our wives, our children, our husbands, has been sold over and over again to purchase the lands we now locates upon; for that reason we have a divine right to the land. . . . And den didn't we clear the land, and raise de crops ob corn, ob cotton, ob tobacco, ob rice, ob sugar, ob everything. And den didn't dem large cities in de North grow up on de cotton and de sugars and de rice dat we made? . . . I say dey has grown rich, and my people is poor.[51]

In some parts of the South, blacks in 1865 did more than argue the merits of their case. Hundreds of freedmen refused either to sign labor contracts or to leave the plantations, insisting that the property belonged to them. A Virginia freedman informed his former owner that he was "entitled to a part of the farm after all the work he had done on it. The kitchen belonged to him because he had helped cut the timber to build it." On the property of a Tennessee planter, former slaves not only claimed to be "joint heirs" to the estate but, the owner complained, abandoned the slave quarters and took up residence "in the rooms of my house:"

> My foreman Sidney, having a wife and several children, . . . has brought them into my parlour. He claims the land from the lane to the river embracing all the houses and was so assured of his rights, that he dug up a nursery of young apple trees on my son's farm, and planted him an orchard. Randal,

50. Arney R. Childs, ed., *The Private Journal of Henry William Ravenel 1859–1887* (Co-lumbia, S.C. 1947), 272; Magdol, *A Right to the Land*, 140–41; Reid, *After the War*, 335; Dennett, *South As It Is*, 189, 229; 39th Congress, 1st Session, Senate Executive Document 27, 84; Randolph B. Campbell, *A Southern Community in Crisis: Harrison County, Texas, 1850–1880* (Austin, 1983), 252; W. A. McClure to Lewis E. Parsons, December 2, 1865, Alabama Governor's Papers; Robert E. Withers, *Autobiography of an Octogenarian* (Roanoke, Va., 1907), 229; Edward B. Heyward to Katherine Heyward, May 5, 1867, Heyward Family Papers, USC; Mobile *Nationalist*, May 16, 1867.
51. *A Freedman's Speech* (Philadelphia, 1867).

a boy I gave Mrs. Williams by his own consent, joined the Yankey army early in the war, has now returned, removed his wife and children . . . into my dining room. Jo and Andy two impudent whelps who rode boldly off from my home at Anderson with Stoneman's raiders . . . have gone to my farm and are in my private bed room.[52]

Few freedmen were able to maintain control of property seized in this manner, although, as will be related, the process of dispossession was prolonged and sometimes violent. A small number did, however, obtain land through other means, squatting on unoccupied real estate in sparsely populated states like Florida and Texas, buying tiny city plots, or cooperatively purchasing farms and plantations. Blacks in Hampton, Virginia, established Lincoln's Land Association, under the direction of a local Baptist minister, and acquired several hundred acres of land, worked collectively by a group of families. Two Texas freedmen purchased 4,000 acres on credit and, in the tradition of white land speculators and small-town boosters, went about advertising for settlers from the East. And a number of discharged black soldiers invested their bounties and back pay either in small farms or, collectively, in plantations. One regiment stationed in Louisiana accumulated $50,000 for this purpose.[53]

These, however, were isolated instances. The vast majority of blacks emerged from slavery lacking the ability to purchase land even at the depressed prices of early Reconstruction, and confronting a white community united in the refusal to advance credit or sell them property. Thus, they entered the world of free labor as wage or share workers on land owned by whites. The adjustment to a new social order in which their persons were removed from the market, but their labor was bought and sold like any other commodity, proved in many respects difficult. For it required the abandonment of some traditions inherited from slavery and the adaptation of others to the logic of the economic market, where the impersonal laws of supply and demand and the balance of power between employer and employee, rather than custom, justice, or personal dependency, determines a laborer's material circumstances.

Most freedmen welcomed the demise of the paternalist attitudes and mutual obligations of slavery—in this sense they eagerly embraced the free market, with its promise of individual mobility, personal autonomy, and freedom to choose among employers. But many, especially elderly

52. John T. O'Brien, "Reconstruction in Richmond: White Restoration and Black Protest, April–June 1865," *VaMHB*, 89 (July 1981), 262–63; O. R. Broyles to Thomas A. R. Nelson, August 15, 1865, Thomas A. R. Nelson Papers, McClung Collection, LML.
53. W. J. Purman to E. C. Woodruff, February 28, 1867, P-1 1867, Letters Received, Ser. 586, Fla. Asst. Comr., RG 105, NA [FSSP A-1392]; Mobile *Nationalist*, March 22, 1866; Engs, *Black Hampton*, 91; Anthony Blunt et al. to Commissioner, August 7, 1865, B-15 1865, Letters Received, Ser. 2452, N. C. Asst. Comr., RG 105, NA [FSSP A-509]; James Smallwood, "Perpetuation of Caste: Black Agricultural Workers in Reconstruction Texas," *Mid-America*, 61 (January 1979), 7; Claude F. Oubre, *Forty Acres and a Mule: The Freedmen's Bureau and Black Landownership* (Baton Rouge, 1978), 28.

blacks no longer able to work, insisted that their owners' responsibilities had not died with bondage. The oldest freedman on a South Carolina plantation indignantly informed his former master, "he was going to die on this place, and he was not going to do any work either." Other blacks saw no reason why emancipation should mean a diminution of either the privileges or the level of material well-being they had previously enjoyed. The slave, after all, possessed one customary "right" no free laborer could claim—the right to subsistence. "He that works we believe has a right to eat," a mass meeting of Petersburg blacks resolved in 1865. Harvard graduate Henry Lee Higginson, who with his wife and two friends purchased a Georgia plantation in 1865, found that the freedmen believed "they ought to get all their living and have wages besides, all extra." They did not, Mrs. Higginson concluded, understand "the value of work and wages." Other whites complained that freedmen had an "exorbitant" idea of what remuneration their labor ought to bring. "Each one seems to think his share will be worth a fortune," wrote the manager of a Mississippi plantation at the end of 1865. For their part, blacks resented being offered as annual wages sums far below what planters had paid before the Civil War to rent slaves for the year. ("The negro . . ." a Union army officer commented, "thought it strange he was not worth as much as before.")[54]

Beneath these "misunderstandings" lay the fact that blacks entered the new market in labor with their own purposes in view. Former slaves, to be sure, proved eager to enjoy the material amenities of freedom. They patronized the stores that sprang up throughout the rural South, purchasing "luxuries" ranging from sardines, cheese, and sugar, to new clothing—denims, shoes, handkerchiefs, and calico dresses. They saved money to build and support churches and educate their children. And they quickly learned to use and influence the market for their own ends. The early years of Reconstruction witnessed strikes or collective petitions for higher wages by black urban laborers, including Richmond factory workers, Jackson washerwomen, New Orleans and Savannah stevedores and mechanics in Columbus, Georgia. In rural areas, too, plantation freedmen sometimes bargained collectively over contract terms, organized strikes, and occasionally even attempted to establish wage schedules for an entire area. Late in 1866, after crop failures had left sharecroppers with only a meager return, South Carolina freedmen held mass meetings to consider terms for the coming year. At one such gathering, attended by over 1,000 blacks, speakers insisted "when their children were naked and starving, they could not work for so little, [and] claimed

54. Daniel E. Huger Smith et al., eds., *Mason Smith Family Letters 1860–1868* (Columbia, S.C., 1950), 235; New York *Tribune*, June 15, 1865; Perry, *Higginson*, 1:256; Wilmer Shields to William N. Mercer, December 19, 1865, William N. Mercer Papers, LSU; 39th Congress, 1st Session, House Report 30, pt. 2:12–13, pt. 3:6.

that one-half the crop should be their due." Blacks took full advantage of competition between planters and nonagricultural employers, seeking work on railroad construction crews, and at turpentine mills and other enterprises offering pay far higher than on the plantations. Freedmen used the market for their own benefit in additional ways. One group of Mississippi laborers no longer sold eggs and poultry raised on their own time to the planter, as under slavery, but marketed them to the highest bidder.[55]

Slavery, however, did not produce workers fully socialized to the virtues of economic accumulation. Despite the profits to be earned in early postwar cotton farming, many freedmen evinced a strong resistance to growing the "slave crop." "If ole massa want to grow cotton," said one Georgia freedman, "let him plant it himself." A. Warren Kelsey, dispatched by a group of Northern textile manufacturers to investigate prospects for the resumption of plantation agriculture, reported from upcountry South Carolina that cotton was "associated in the negroes' mind with memories of the overseer, the driver and the lash, in fact with the whole system of slavery." Moreover, another Northern visitor observed, blacks realized cotton would be sold by their employers and "pass out of their reach," while food crops could be consumed.[56]

Those blacks who managed to acquire farms or who simply squatted on unoccupied land often seemed content to pursue subsistence agriculture. "We never planted cotton, because we could not eat this," a Texas freedman recalled decades after emancipation. "I made bows and arrows to kill our meat. . . . We never came to the store for nothing." Such extreme self-sufficiency was rare indeed in nineteenth-century America. More typical were the freedmen in central Mississippi who, according to a local planter, preferred to live "by cultivating a small patch of corn and cotton and by raising a few hogs," or the Sea Island blacks who spent most of their time on food production, displaying interest in cotton only to "supply them with spending money." These blacks sought to farm in the manner of peasants in other parts of the world and white yeomen in the South—concentrating on food crops as a first priority, and only to a lesser extent on cotton or other staples to obtain ready cash. Rather than

55. Reid, *After the War*, 499; O'Brien, "Factory, Church, and Community," 511–35; Philip S. Foner and Ronald L. Lewis, eds., *The Black Worker: A Documentary History from Colonial Times to the Present* (Philadelphia, 1978–84), 1:345–46; Litwack, *Been in the Storm*, 441–42; J. T. Trowbridge, *The South: A Tour of Its Battle-Fields and Ruined Cities* (Hartford, Conn., 1866), 405; Savannah *Daily News and Herald*, January 26, 30, 31, February 26, July 3, 1867; Moore, ed., *Juhl Letters*, 65, 134; Wilmer Shields to William N. Mercer, November 28, 1866, Mercer Papers.

56. Barbara J. Fields, *Slavery and Freedom on the Middle Ground: Maryland During the Nineteenth Century* (New Haven, 1985), 157–58; *Christian Recorder*, August 19, 1865; Edward Atkinson to John Murray Forbes, May 8, 1865, A. Warren Kelsey to Edward Atkinson, September 2, 8, 1865, Edward Atkinson Papers, MHS; John Covode to Edwin M. Stanton, undated report [1865], John Covode Papers, LC.

choose irrevocably between self-sufficiency and farming for the market, they sought to avoid a complete dependence upon either while taking advantage of the opportunities each could offer. As "cotton detective" Kelsey shrewdly recognized, it was precisely the ability to choose, to organize their economic lives as independently as possible, that blacks most valued:

> The sole ambition of the freedman at the present time appears to be to become the owner of a little piece of land, there to erect a humble home, and to dwell in peace and security at his own free will and pleasure. If he wishes, to cultivate the ground in cotton on his own account, to be able to do so without anyone to dictate to him hours or system of labor, if he wishes instead to plant corn or sweet potatoes—to be able to do *that* free from any outside control. . . . That is their idea, their desire and their hope.[57]

Here was a definition of economic freedom that corresponded to the traditional republican ideal of a society of autonomous small producers. Thomas Jefferson would have fully appreciated this ambition to be master of one's own time, free from the coercion of either an arbitrary master or the impersonal marketplace.

Historical experience and modern scholarship suggest that acquiring small plots of land would hardly, by itself, have solved the economic plight of black families. The fate of the white yeomanry would soon demonstrate the precariousness of small farmers' hold on their land in the postwar South. Land is only one of the scarce resources of under-developed rural societies; where not accompanied by control of credit and access to markets, land reform can often be a "hollow victory." And where political power rests in hostile hands, small landowners often find themselves subjected to oppressive taxation and other state policies that severely limit their economic prospects. In such circumstances, the autonomy offered by land ownership tends to be defensive, rather than the springboard for sustained economic advancement. Yet while hardly an economic panacea, land distribution would have had profound consequences for Southern society, weakening the land-based economic and political power of the old ruling class, offering blacks a measure of choice as to whether, when, and under what circumstances to enter the labor market, and affecting the former slaves' conception of themselves. (Well into the twentieth century, blacks who did acquire land were more likely to register, vote, and run for office than other members of the rural

---

57. Rawick, ed., *American Slave*, Supplement, Ser. 2, 6:2152; George C. Osborn, "The Life of a Southern Plantation Owner During Reconstruction as Revealed in the Clay Sharkey Papers," *JMH*, 6 (April 1944), 105; Dennett, *South As It Is*, 211–14; Clarence L. Mohr, "Before Sherman: Georgia Blacks and the Union War Effort, 1861–1864," *JSH*, 45 (August 1979), 348; Gavin Wright, *The Political Economy of the Cotton South* (New York, 1978), 62–64; Frederick Cooper, "Peasants, Capitalists and Historians: A Review Article," *Journal of Southern African Studies*, 7 (April 1981), 289; A. Warren Kelsey to Edward Atkinson, September 8, 1865, Atkinson Papers.

community.) One might argue that immediate landownership would have encouraged blacks to lapse into self-sufficiency, with disastrous consequences for Southern economic development. Yet in the South, as in most parts of the hemisphere, the survival of the plantation system produced only economic stagnation, and as things turned out, blacks lacked even the partial shield against economic exploitation afforded by ownership of land.[58]

Blacks' quest for economic independence not only threatened the very foundations of the Southern political economy, but, as will be related, put the freedmen at odds with both former owners seeking to restore plantation labor discipline and Northerners committed to reinvigorating staple crop production. But as part of the broad quest for individual and collective autonomy, it remained central to the black community's effort to define the meaning of freedom. Indeed the fulfillment of blacks' "noneconomic" aspirations, from family autonomy to the creation of schools and churches, all depended in considerable measure on success in winning control of their working lives and gaining access to the economic resources of the South.

Origins of Black Politics

If the goal of autonomy inspired blacks to withdraw from religious and social institutions controlled by whites and to attempt to work out their economic destinies for themselves, in the polity, "freedom" meant inclusion rather than separation. Recognition of their equal rights as citizens quickly emerged as the animating impulse of Reconstruction black politics. In the spring and summer of 1865, blacks organized a seemingly unending series of mass meetings, parades, and petitions demanding civil equality and the suffrage as indispensable corollaries of emancipation. The most extensive mobilization occurred in areas that had been occupied by Union troops during the war, where political activity had begun even before 1865. Union Leagues and similar groups sprang up in low country South Carolina and Georgia, their meetings bringing together Freedmen's Bureau agents, black soldiers, and local freedmen, to demand the vote and the repeal of all laws discriminating against blacks. "By the Declaration of Independence," declared a gathering on St. Helena Island, "we believe these are rights which cannot justly be denied us."[59]

58. Kenneth Parsons, "Land Reform in the Postwar Era," *Land Economics*, 33 (August 1957), 215–16; Philip M. Raup, "Land Reform and Agricultural Development," in Herman W. Southwork and Bruce F. Johnson, eds., *Agricultural Development and Economic Growth*, (Ithaca, N.Y., 1967), 290–91; Foner, *Nothing But Freedom*, 21–26, 35–37; Leo McGee and Robert Boone, eds., *The Black Rural Landowner—Endangered Species* (Westport, Conn., 1979), xvii.

59. Salmon P. Chase to Andrew Johnson, May 21, 1865, J. G. Dodge to Johnson, June 20, 1865, Petition, June 1865, to Charles Sumner, Charles Sumner to Johnson,

Political mobilization also proceeded apace in Southern cities, where the flourishing network of churches and fraternal societies provided a springboard for organization, and the army and Freedmen's Bureau stood ready to offer protection. In Wilmington, North Carolina, freedmen in 1865 formed an Equal Rights League which, local officials reported, insisted upon "all the social and political rights of white citizens" and demanded that blacks be consulted in the selection of policemen, justices of the peace, and county commissioners. By midsummer, "secret political Radical Associations" had been formed in Virginia's major cities. Richmond blacks first organized politically to protest the army's rounding up of "vagrants" for plantation labor, but soon expanded their demands to include the right to vote and the removal of the "Rebel-controlled" local government. In Norfolk, occupied by the Union Army since 1862, blacks early in 1865 created the Union Monitor Club to press their claim to equal rights, and in May hundreds attempted to vote in a local election. A mass meeting endorsed a militant statement drafted by former fugitive slave Thomas Bayne: "Traitors shall not dictate or prescribe to us the terms or conditions of our citizenship."[60]

In Louisiana, where black politics had advanced furthest during the war, the New Orleans *Tribune* and its Radical allies continued to press the issue of black suffrage. A September 1865 convention composed of native white Radicals, Northerners like the young provost judge and future governor Henry C. Warmoth, and prominent members of the free black elite, voted to affiliate with the national Republican Party, called upon Congress to govern Louisiana as a territory, and demanded full legal and political equality for blacks. Meanwhile, mobilization penetrated the sugar country, with laborers, one planter complained, abandoning work at will to attend political gatherings. In November, as white Louisianans went to the polls, a Republican-sponsored "voluntary election" attracted some 20,000 voters, mostly blacks in New Orleans and the surrounding parishes, who "elected" Warmoth to serve as Louisiana's "Territorial delegate" to Congress. "The whole Parish was in an uproar" on election day, reported an army officer, with hundreds of freedmen abandoning the plantations, "stating that they were going to vote."[61]

June 30, 1865, Andrew Johnson Papers, LC; *Christian Recorder,* June 10, 1865; Holt, *Black over White,* 12; Herbert Aptheker, "South Carolina Negro Conventions, 1865," *JNH,* (January 1946), 93.

60. Jonathan Dawson, et al. to William W. Holden, July 12, 1865, North Carolina Governor's Papers, NCDAH; *Address. The Members of the Equal Rights League of Wilmington, N. C.,* printed circular, January 1866, AMA Archives; J. K. Van Fleet to Benjamin F. Butler, August 1, 1865, Benjamin F. Butler Papers, LC; Rachleff, *Black Labor,* 13–14, 35; O'Brien, "Reconstruction in Richmond," 274–80; *Equal Suffrage: Address from the Colored Citizens of Norfolk, Va., to the People of the United States* (New Bedford, Mass., 1865), 8–14.

61. New Orleans *Tribune,* September 17, 25, 27, 28, 29, 1865; J. Carlyle Sitterson, "The Transition from Slave to Free Economy on the William J. Minor Plantations," *AgH,* 17 (January 1943), 222; Henry C. Warmoth, *War, Politics and Reconstruction* (New York, 1930),

Statewide conventions held throughout the South in 1865 and early 1866 offered the most visible evidence of black political organization. Several hundred delegates attended these gatherings, some selected by local meetings occasionally marked by "animated debate," others by churches, fraternal societies, Union Leagues, and black army units, still others simply appointed by themselves. "Some bring credentials," observed North Carolina black leader James H. Harris, "others had as much as they could do to bring themselves, having to escape from their homes stealthily at night" to avoid white reprisal. Although little information survives about the majority of these individuals, certain patterns can be discerned from the fragmentary evidence. The delegates "ranged all colors and apparently all conditions," but urban free mulattoes took the most prominent roles, and former slaves were almost entirely absent from leadership positions. One speaker at the Tennessee gathering doubted it should be called a "Negro convention" at all, since its officers were "all mixed blood," some "as white as the editor of the New York *Herald.*" Charleston free blacks, along with six Northern-born newcomers, dominated South Carolina's gathering, and at Louisiana's Republican state convention nineteen of the twenty black delegates had been born free. But other groups also came to the fore in 1865. In Mississippi, a state with few free blacks before the war, ex-slave army veterans and their relatives comprised the majority of the delegates. Alabama and Georgia had a heavy representation of black ministers, and all the conventions included numerous skilled artisans. Many of the delegates, especially those born free, were relatively well-to-do, although the very richest blacks held aloof, too linked to whites economically and by kinship to risk taking an active role in politics.

The prominence of free blacks, ministers, artisans, and former soldiers in these early conventions established patterns that would characterize black politics for much of Reconstruction. From among these delegates would emerge such prominent officeholders as Alabama Congressman James T. Rapier and Mississippi Secretary of State James D. Lynch. The most remarkable continuity in black leadership occurred in South Carolina, for among the fifty-two delegates to the November 1865 convention sat four future Congressmen, thirteen legislators, and twelve delegates to the state's 1868 constitutional convention. In general, however, what is striking is how few of these early leaders went on to positions of prominence. Only two of Alabama's fifty-six delegates (William V. Turner and Holland Thompson) later played significant roles in Reconstruction politics, a pattern repeated in Virginia, North Carolina, Tennessee, Mississippi, Alabama, and Arkansas. In most states, black political

---

43–45; J. W. Greene to Charles W. Lowell, November 23, 1865, Letters Received, Ser. 1845, Provost Marshal, Dept. of La., RG 393, NA [FSSP A-808].

mobilization had advanced far more rapidly in cities and in rural areas occupied by federal troops during the war, than in the bulk of the plantation counties, where the majority of the former slaves lived. The free blacks of Louisiana and South Carolina who stepped to the fore in 1865 would remain at the helm of black politics throughout Reconstruction; elsewhere, however, a new group of leaders, many of them freedmen from the black belt, would soon supersede those who had taken the lead in 1865.[62]

The debates at these conventions illuminated conflicting currents of black public life in the immediate aftermath of emancipation. Tensions within the black community occasionally rose to the surface. One delegate remarked that he did not intend "to have the whip of slavery cracked over us by no slaveholder's son"; another voiced resentment that a Northerner (Pennsylvania-born James W. Hood) had been chosen president of North Carolina's convention; and the printed proceedings of the South Carolina convention included a discreet reference to the "spirited discussion" produced by a resolution referring to blacks' making "distinctions among ourselves." Relations between the races caused debate as well. A resolution urging blacks to employ, wherever possible, teachers of their own race, was tabled by the North Carolina delegates after considerable debate, and replaced by one thanking Northern societies for their efforts on behalf of the freedmen. By and large, however, the proceedings proved harmonious, the delegates devoting most of their time to issues that united blacks rather than dividing them. South Carolina's convention demanded the full gamut of opportunities and privileges enjoyed by whites, from access to education to the right to bear arms, serve on juries, establish newspapers, assemble peacefully, "enter upon all the avenues of agriculture, commerce, [and] trade," and "develop our whole being by all the appliances that belong to civilized society." Georgia's resolutions complained of violence inflicted on rural blacks, efforts to prevent freedmen from establishing schools, and attempts to keep from blacks the church property "paid for by our own earnings while we were in slavery."[63]

62. *Weekly Anglo-African*, August 19, 1865; *Christian Recorder*, October 28, 1865; Dennett, *South As It Is*, 148–50; Reid, *After the War*, 81; Nashville *Colored Tennessean*, August 12, 1865; Holt, *Black over White*, 15–16; Donald E. Everett, "Demands of the New Orleans Free Colored Population for Political Equality, 1862–1865," *LaHQ*, 38 (April 1955), 62–64; J. W. Blackwell to Andrew Johnson, November 24, 1865, Johnson Papers; Kolchin, *First Freedom*, 152–53; Drago, *Black Politicians*, 27–28; William C. Hine, "Charleston and Reconstruction: Black Political Leadership and the Republican Party, 1865–1877" (unpub. diss., Kent State University, 1978), 32–36. Conclusions as to delegates' origins and future careers are drawn from a file of biographical information about black Reconstruction officeholders compiled by the author.

63. Dennett, *South As It Is*, 152–53; Philip S. Foner and George E. Walker, eds., *Proceedings of the Black State Conventions, 1840–1865* (Philadelphia, 1979), 2:291, 302; *Proceedings of the Freedmen's Convention of Georgia* (Augusta, Ga., 1866), 17, 30.

The delegates' central preoccupation, however, was equality before the law and the suffrage. A number of New Orleans and Charleston free blacks, to be sure, still flirted with the idea of confining black suffrage to a privileged minority through some combination of property and educational qualifications, although they insisted that such requirements apply to both races. ("If the ignorant white man is allowed to vote," declared a petition of prominent Charleston free blacks, the "ignorant colored man" should be enfranchised as well.) Yet at the 1865 conventions, speaker after speaker echoed the view that universal manhood suffrage constituted "an essential and inseparable element of self-government." In justifying their demand for the vote, the delegates invoked America's republican traditions, especially the Declaration of Independence, "the broadest, the deepest, the most comprehensive and truthful definition of human freedom that was ever given to the world." "The colored people," Hood would declare in 1868, "had read the Declaration until it had become part of their natures." The North Carolina freedmen's convention he chaired in 1865 portrayed the Civil War and emancipation as chapters in the onward march of "progressive civilization," embodiments of "the fundamental truths laid down in the great charter of Republican liberty, the Declaration of Independence." Such language was not confined to the convention delegates. Eleven Alabama blacks, who complained of contract frauds, injustice before the courts, and other abuses, concluded their petition with a revealing masterpiece of understatement: "This is not the persuit of happiness."[64]

There was more here than merely familiar wording. Like Northern blacks steeped in the Great Tradition of prewar protest, the freedmen and Southern free blacks saw emancipation as enabling the nation to live up to the full implications of its republican creed—a goal that could be achieved only by abandoning racial proscription and absorbing blacks fully into the civil and political order. Isham Sweat, a slave-born barber who wrote the address of North Carolina's convention and went on to serve in the state legislature, told John R. Dennett that Congress should "declare that no state had a republican form of government if every free man in it was not equal before the law." Another 1865 speaker destined for Reconstruction prominence, Louisiana's Oscar J. Dunn, described the absence of "discrimination among men" and "hereditary distinctions" as the essence of America's political heritage. Continued proscription of blacks, Dunn warned, would jeopardize the republic's very future, open-

64. New Orleans *Tribune*, July 8, September 14, 1865; Aptheker, "Negro Conventions," 93; Holt, *Black over White*, 21-22, 67-68; Michael P. Johnson and James L. Roark, *Black Masters: A Free Family of Color in the Old South* (New York, 1984), 326; Nashville *Colored Tennessean*, August 12, 1865; John M. Langston, *Freedom and Citizenship* (Washington, D.C., 1883), 99-100, 110; Leonard Bernstein, "The Participation of Negro Delegates in the Constitutional Convention of 1868 in North Carolina," *JNH*, 34 (October 1949), 404; *Convention of the Freedmen of North Carolina* (Raleigh, 1865), 6; Foner, "Reconstruction," 60.

THE MEANING OF FREEDOM

ing "the door for the institution of aristocracy, nobility, and even monarchy."[65]

Like their Northern counterparts during the Civil War, Southern blacks proclaimed their identification with the nation's history, destiny, and political system. The very abundance of letters and petitions addressed by black gatherings and ordinary freedmen to officials of the army, Freedmen's Bureau, and state and federal authorities, as well as the decision of a number of conventions to send representatives to Washington to lobby for black rights, revealed a belief that the political order was at least partially open to their influence. "We are Americans," declared a meeting of Norfolk blacks, "we know no other country, we love the land of our birth." Their address reminded white Virginians that in 1619, "our fathers as well as yours were toiling in the plantations on James River" and that a black man, Crispus Attucks, had shed "the first blood" in the American Revolution. And, of course, blacks had fought and died to save the Union. America, resolved another Virginia meeting, was "now *our* country—made emphatically so by the blood of our brethren." "We stood by the government when it wanted help," a delegate to Mississippi's convention wrote President Johnson. "Now . . . will it stand by us?"[66]

Despite the insistent language of individual speeches, the convention resolutions and public addresses adopted a moderate tone, offering "the right hand of fellowship" to Southern whites. The Virginia convention proved an exception, for its address spoke of "injuries deeper and darker than the earth ever witnessed in the case of any other people." At one point, the Virginia delegates changed the wording of a public statement from "our former masters" to "our former oppressors." Elsewhere, however, a far more conciliatory approach prevailed. Leaders of North Carolina's convention advocated "equal rights, and a moderate conservative course in demanding them." One rural delegate who proposed that the assembly demand admission to the state's constitutional convention, then in session in Raleigh, was denounced as "absurd and foolish," and the gathering "respectfully and humbly" petitioned the state government for education and equality before the law, while avoiding reference to the suffrage. Georgia's delegates, divided between advocates of universal suffrage and those favoring a literacy or property test, compromised by claiming "at least conditional suffrage." Even the South Carolina conven-

65. Peter D. Klingman, *Josiah Walls* (Gainesville, Fla., 1976), 72–73; Dennett, *South As It Is*, 176; Sidney Andrews, *The South Since the War* (Boston, 1866), 125; *Proceedings of the Republican Party of Louisiana* (New Orleans, 1865), 4–5.

66. Peter D. Klingman, "Rascal or Representative? Joe Oates of Tallahassee and the 'Election' of 1866," *FlHQ*, 51 (July 1972), 52–57; *Equal Suffrage*, 1, 8; Joseph R. Johnson to O. O. Howard, August 4, 1865, Unregistered Letters Received, Ser. 457, D.C. Asst. Comr., RG 105, NA [FSSP A-9851]; J. W. Blackwell to Andrew Johnson, November 24, 1865, Johnson Papers.

tion, forthright in claiming civil and political equality and in identifying its demands with "the cause of millions of oppressed men" throughout the world, took pains to assure the state's white minority of blacks' "spirit of meekness," their consciousness of "your wealth and greatness, and our poverty and weakness."[67]

To some extent, this cautious tone reflected a realistic assessment of the political situation at a time when Southern whites had been restored to local power by President Johnson, and Congress had not yet launched its own Reconstruction policy. But the conventions' mixture of radicalism and conciliation also mirrored the indecision of an emerging class of black political leaders still finding their own voice in 1865 and 1866, and dominated by urban free blacks, ministers, and others who had in the past enjoyed harmonious relations with at least some local whites and did not always feel the bitter resentments of rural freedmen.

Nor did a coherent economic program emerge from these assemblies. Demands for land did surface at local meetings that chose convention delegates. One such gathering in Greensboro, Alabama, heard speakers call for "land or blood," while at Tarboro, North Carolina, where two candidates presented themselves to 1,500 blacks, the one who called for a division of the land was unanimously elected. Yet such views rarely found expression among the conventions' leadership. Virginia's delegates pointedly observed that the Freedmen's Bureau Act had promised blacks access to land, Georgia's petitioned Congress to validate the Sherman land grants, and South Carolina's requested Congress to place "a fair and impartial construction" upon the "pledges of government to us concerning the land question." But by and large, economic concerns figured only marginally in the proceedings, and the addresses and resolutions offered no economic program, apart from stressing the "mutual interest" of capital and labor, and urging self-improvement as the route to personal advancement. The Arkansas resolutions even remarked that blacks "are destined in the future, as in the past, to cultivate your cotton fields." A number of conventions chided idle freedmen for "vagrancy and pauperism," and urged them to remain on the land, labor diligently, and save money in order to purchase homesteads.[68]

67. Litwack, *Been in the Storm*, 505–509; Holt, *Black over White*, 16; *Proceedings of the Convention of the Colored People of Virginia* (Alexandria, 1865), 9–12, 20; *North Carolina Convention*, 5, 12–16; *Georgia Convention*, 19; Augusta *Loyal Georgian*, February 17, 1866; Foner and Walker, eds., *Black Conventions*, 2:298–300.

68. James S. Allen, *Reconstruction: The Battle for Democracy* (New York, 1937), 65; Foner and Walker, eds., *Black Conventions*, 2:272, 302; Manuel Gottlieb, "The Land Question in Georgia During Reconstruction," *Science and Society*, 3 (Summer 1939), 369; Philip S. Foner and George E. Walker, eds., *Proceedings of the Black National and State Conventions, 1865–1900* (Philadelphia, 1986–), 1:189–94; "Official Proceedings of the Colored Convention for the State of Mississippi, Vicksburg, November 22–25, 1865," manuscript, M-82 1866, Letters Received, Ser. 15, Washington Headquarters, RG 105, NA [FSSP A-9223]; *Georgia Convention*, 20, 30; Andrews, *South Since the War*, 126; Holt, *Black over White*, 18.

Thus, the ferment rippling through the Southern countryside found little echo at the state conventions of 1865—a reflection of the paucity of representation from the plantation counties and the prominence of political leaders more attuned to political equality and self-help formulas than to rural freedmen's thirst for land. Nor did the conventions' eloquent appeals for civil and political equality accomplish anything, for all were ignored by the intransigent state governments of Presidential Reconstruction. As a result, enthusiasm for such gatherings waned. Among the states of the Confederacy, only Georgia, Tennessee, North Carolina, and Texas witnessed black conventions in 1866. One delegate noted that his constituents believed "we do nothing but meet, pass resolutions, publish pamphlets, and incur expenses, without accomplishing good results."[69]

While understandable, this indictment was perhaps unfair, for these early black conventions both reflected and advanced the process of political mobilization. Some Tennessee delegates, for example, took to heart their convention's instruction to "look after the welfare" of their constituents. After returning home, they actively promoted black education, protested to civil authorities and the Freedmen's Bureau about violence and contract frauds, and struggled against unequal odds to secure blacks a modicum of justice in local courts. Chapters of the Georgia Equal Rights and Educational Association, established at the state's January 1866 convention, became "schools in which the colored citizens learn their rights." Spreading into fifty counties by the end of the year, the association's local meetings attracted as many as 2,000 freedmen, who listened to speeches on issues of the day and readings from Republican newspapers. And, although plagued by financial problems and the difficulty of reaching an overwhelmingly illiterate audience, the emerging black press also promoted the spread of political education. By 1866, nine (mostly short-lived) black newspapers had joined the New Orleans *Tribune*. Edited by two white Northerners, but owned and managed by a black board of directors, the Mobile *Nationalist* sent agents into the countryside to solicit subscriptions, report on local conditions, and urge freedmen "to stand up like men on behalf of [their] rights." Blacks able to read the *Nationalist*, one Alabama white complained, absorbed "the 'radicalism' it contains," became "*pugnacious*," and no longer exhibited proper respect for their former owners.[70]

69. Nashville *Daily Press and Times*, August 8, 1866.
70. Nashville *Daily Press and Times*, August 8, 9, 1866; Clinton B. Fisk to Frederick E. Trotter, August 9, 1866, F-97 1866, Registered Letters Received, Ser. 3379, Tenn. Asst. Comr., RG 105, NA [FSSP A-6251]; Henry Webb et al. to Clinton B. Fisk, March 26, 1866, M-89 1866, Registered Letters Received, Ser. 3379, Tenn. Asst. Comr., RG 105, NA [FSSP A-6294]; *Proceedings of the Convention of the Equal Rights and Educational Association of Georgia* (Augusta, Ga., 1866), 3–8; *Christian Recorder*, June 9, 1866; Henry L. Suggs, ed., *The Black Press in the South, 1865–1979* (Westport, Conn., 1983), 3–23; Dennett, *South As It Is*, 303; J. D. Williams to Robert M. Patton, September 11, 1866, Alabama Governor's Papers.

Although few in number, the statewide conventions of 1866 illustrated the results of this ongoing process of politicization. Twice as many counties were represented in the Georgia and North Carolina gatherings as the year before, reflecting how organization had penetrated the black belt. In Greene County, North Carolina, unrepresented at the first state convention, blacks in 1866 held an election to choose a delegate from between two candidates who conducted "a regular canvass." Former slaves now began to assume positions of prominence monopolized by the freeborn a year earlier, and the resolutions and speeches were noticeably more radical. North Carolina's delegates heard militant speeches chastising whites for violence against freedmen, injustice to black laborers, and opposition to black education. Their resolutions demanded equal suffrage (an issue sidestepped in 1865), praised Charles Sumner, Thaddeus Stevens, and other Radical Republicans as "beacon lights of our race," and urged blacks to combat economic inequalities by forming joint stock companies and patronizing, wherever possible, businessmen of their own race. The Tennessee convention called upon Congress to grant the state "a republican form of government" under which blacks could vote, bear arms, and educate their children. But even more striking than this new tone was the wholesale turnover in membership. Only a small minority of the 1865 delegates (seventeen of 106 in North Carolina, eighteen of 102 in Tennessee) reappeared in 1866. Even in South Carolina, with its continuity in black political leadership, Richard H. Cain observed that some early leaders, including prominent free blacks, had by 1866 "relapsed into secondary men; and the class who were hardly known" were stepping forward to assume prominent roles.[71]

All in all, the most striking characteristic of this initial phase of black political mobilization was its very unevenness. In some states, organization proceeded steadily in 1865 and 1866, in others, such as Mississippi, little activity occurred between an initial flurry in the summer of 1865 and the advent of black suffrage two years later. Large parts of the black belt remained untouched by organized politics, but many blacks were aware of Congressional debates on Reconstruction policy, and quickly employed on their own behalf the Civil Rights Act of 1866. "The negro of today," remarked a correspondent of the New Orleans *Tribune* in September 1866, "is not the same as he was six years ago. . . . He has been told of his rights, which have long been robbed." Only in 1867 would blacks enter the "political nation," but in organization, leadership, and an ideology that drew upon America's republican heritage to demand an equal

71. Roberta S. Alexander, *North Carolina Faces the Freedmen: Race Relations During Presidential Reconstruction, 1865–67* (Durham, 1985), 22, 81–90; *Minutes of Freedmen's Convention Held in the City of Raleigh* (Raleigh, 1866), 14, 21, 26, 29; Nashville *Daily Press and Times*, August 10, 1866; *Christian Recorder*, April 21, 1866.

place as citizens, the seeds that flowered then were planted in the first years of freedom.[72]

Violence and Everyday Life

The black community's religious, social, and political mobilization was all the more remarkable for occurring in the face of a wave of violence that raged almost unchecked in large parts of the postwar South. Although wartime conflicts between white Unionists and Confederates, as well as the economic destitution that inspired local bands to prey upon the property of others, contributed to the unsettled condition of Southern life, in the vast majority of cases freedmen were the victims and whites the aggressors.[73]

In some areas, violence against blacks reached staggering proportions in the immediate aftermath of the war. In Louisiana, reported a visitor from North Carolina in 1865, "they govern . . . by the pistol and the rifle." "I saw white men whipping colored men just the same as they did before the war," testified ex-slave Henry Adams, who claimed that "over two thousand colored people" were murdered in 1865 in the area around Shreveport, Louisiana. In Texas, where the army and Freedmen's Bureau proved entirely unable to establish order, blacks, according to a Bureau official, "are frequently beaten unmercifully, and shot down like wild beasts, without any provocation." Susan Merritt, a freedwoman from Rusk County, Texas, remembered seeing black bodies floating down the Sabine River, and said of local whites: "There sure are going to be lots of souls crying against them in Judgement." In some cases, whites wreaked horrible vengeance for offenses real or imagined. In 1866, after "some kind of dispute with some freedmen," a group near Pine Bluff, Arkansas, set fire to a black settlement and rounded up the inhabitants. A man who visited the scene the following morning found "a sight that apald me 24 Negro men woman and children were hanging to trees all round the Cabbins."[74]

72. William C. Harris, *The Day of the Carpetbagger: Republican Reconstruction in Mississippi* (Baton Rouge, 1979), 96; Joseph H. Catching to Benjamin G. Humphreys, August 24, 1866, Mississippi Governor's Papers; New Orleans *Tribune*, September 13, 1866; S. W. Laidler to Thaddeus Stevens, May 7, 1866, Thaddeus Stevens Papers, LC.

73. Dan T. Carter, *When the War Was Over: The Failure of Self-Reconstruction in the South, 1865–1867* (Baton Rouge, 1985), 6–23. The pervasiveness of violence undermines a crucial precept of the traditional interpretation of Reconstruction: that "the alienation of the races began" with the Reconstruction Acts of 1867 (as if race relations had been harmonious under slavery and during Presidential Reconstruction). Fleming, *Alabama*, 390.

74. Alfred Dockery to R. J. Powell, November 13, 1865, William H. Seward Papers, University of Rochester; 46th Congress, 2d Session, Senate Report 693, pt. 2:175–76, 191; 39th Congress, 1st Session, Senate Executive Document 27, 83; James W. Smallwood, *Time of Hope, Time of Despair: Black Texans During Reconstruction* (Port Washington, N.Y., 1981), 32; Ronnie C. Tyler and Lawrence R. Murphy, eds., *The Slave Narratives of Texas* (Austin, 1974), 121; William L. Mallet to Thaddeus Stevens, May 28, 1866, Stevens Papers.

The pervasiveness of violence reflected whites' determination to define in their own way the meaning of freedom and their determined resistance to blacks' efforts to establish their autonomy, whether in matters of family, church, labor, or personal demeanor. Georgia freedman James Jeter was beaten "for claiming the right of whipping his own child instead of allowing his employer and former master to do so." Black schools, churches, and political meetings also became targets. White students from the University of North Carolina twice in 1865 assaulted peaceful black meetings, one a gathering to select delegates to a statewide black convention, the second a meeting of a black "secret society" addressed by a speaker from the state capital.[75]

"Southern whites," a Freedmen's Bureau agent observed, "are quite indignant if they are not treated with the same deference that they were accustomed to" under slavery, and behavior that departed from the etiquette of antebellum race relations frequently provoked violence. Conduct deemed manly or dignified on the part of whites became examples of "insolence" and "insubordination" in the case of blacks. One North Carolina planter complained bitterly to a Union officer that a black soldier had "bowed to me and said good morning," insisting blacks must never address whites unless spoken to first. An Alabama overseer shot a black worker who "gave him sarse"; a white South Carolina minister "drew his pistol and shot [a freedman] thru the heart" after he objected to the expulsion of another black man from church services. In Texas, Bureau records listed the "reasons" for some of the 1,000 murders of blacks by whites between 1865 and 1868: One victim "did not remove his hat"; another "wouldn't give up his whiskey flask"; a white man "wanted to thin out the niggers a little"; another wanted "to see a d——d nigger kick." Gender offered no protection to black women—one was beaten by her employer for "using insolent language," another for refusing to "call him master," a third "for crying because he whipped my mother." The victims also included individuals who personified the ways freedmen had challenged customary racial mores. When delegates to the 1865 black conventions returned home, "many only found ashes and cinders." A group of Virginia whites beat a black veteran merely for stating that he was proud to have served in the Union Army. "As one of the disfranchised race," said a Louisiana black, "I would say to every colored soldier, 'Bring your gun home'."[76]

75. 40th Congress, 3d Session, House Miscellaneous Document 52, 129; Hope S. Chamberlain, *Old Days in Chapel Hill* (Chapel Hill, 1926), 117; H. C. Thompson to Benjamin S. Hedrick, September 14, 1865, Benjamin S. Hedrick Papers, DU.

76. 39th Congress, 1st Session, House Report 30, pt. 2:178; Swint, ed., *Dear Ones at Home*, 203; Moore, ed., *Juhl Letters*, 96–98; George W. Gile to J. Duval Greene, June 21, 1866, South Carolina Governor's Papers, SCDA; Claude Elliott, "The Freedmen's Bureau in Texas," *SWHQ*, 56 (July 1952), 6; Sterling, ed., *We Are Your Sisters*, 333–38; Mobile *Nationalist*, February 1, 1866; Litwack, *Been in the Storm*, 274.

Probably the largest number of violent acts stemmed from disputes arising from black efforts to assert their freedom from control by their former masters. Freedmen were assaulted and murdered for attempting to leave plantations, disputing contract settlements, not laboring in the manner desired by their employers, attempting to buy or rent land, and resisting whippings. One black who refused to be bound and whipped, asserting that "he was a freeman and he would not be tied like a slave," was shot dead by his employer, a prominent Texas lawyer. In parts of Tennessee, a Nashville newspaper reported early in 1867, "regulators . . . are riding about whipping, maiming and killing all negroes who do not obey the orders of their former masters, just as if slavery existed."[77]

In the face of this pervasive violence, local leaders of society and politics remained silent, reluctant to hold other whites responsible for crimes against blacks. A resident of southwestern Alabama wrote the governor of his shock at hearing "men of standing . . . countenance disorder and abuse of negroes" and their refusal to "restrain young men in their violence." John Wesley North, a Northerner who went to Knoxville after the war, in 1866 encountered a mob beating a freedman. When North intervened, the crowd dispersed, "evidently amazed that any person should venture to remonstrate against even the *murder* of a *black man.*" A local banker subsequently offered the Yankee this advice: "never in this country . . . interfere in behalf of a nigger."[78]

Considering the extent of white violence against blacks, it is remarkable in how few instances blacks attacked whites. Cases arising from assaults among blacks themselves appear not infrequently in the records of the Freedmen's Bureau and local courts, but violence or even threats against individual whites were all but unknown. On some occasions, freedmen did band together and take the law into their own hands to suppress crime. In 1866, a group of armed blacks apprehended and delivered to the county jail three whites who had been terrorizing Orangeburg, South Carolina, freedmen, and in Holly Springs, Mississippi, blacks formed a posse to hunt down a white man guilty of the cold-blooded murder of a freedwoman. But the obstacles to such actions were formidable indeed, for the slightest evidence of blacks holding secret meetings or arming themselves sufficed to set off waves of fear among Southern whites. In Tennessee, according to Gen. Clinton B. Fisk, a black man shooting squirrels in the woods inspired rumors that hundreds of armed blacks were preparing to "rise *en masse* and kill off the white people." As the year drew to a close, and talk of an impending division of lands by the federal

77. Barry A. Crouch, "A Spirit of Lawlessness: White Violence, Texas Blacks, 1865–1868," *JSocH,* 18 (Winter 1984), 218–20; Dennett, *South As It Is,* 125, 223; 39th Congress, 1st Session, Senate Executive Document 27, 23; Nashville *Press and Times* in Hartford *Courant,* February 6, 1867.

78. S. S. Houston to Lewis Parsons, August 19, 1865, Alabama Governor's Papers; Ann North to George and Mary Ann Loomis, July 8, 1866, North Family Correspondence, HL.

government circulated among blacks, an insurrection panic gripped much of the white South.[79]

New Year's Day, 1866, came and went with no sign of a black uprising. Federal army officers concluded that Southern whites had behaved like "frightened old women" overwhelmed by a pathological fear of their former slaves. But anxiety about black rebellion, so reminiscent of prewar insurrection panics, reflected more than an irrationality born of slavery. Touring the South in 1865, Carl Schurz predicted that if freedmen chose to behave "as free laborers in the North act every day without causing the least surprise," Southern whites would be seized by "a paroxysm of fright." A federal officer investigating reports of impending insurrection in Kingstree, South Carolina, concluded that exaggerated fears "spring from dread on the part of the planters of the freed people asserting their rights of manhood." Blacks bearing arms or, as at Kingstree, marching "with red colors flying" to demand better contract terms, symbolized the revolutionary transformation in social relations wrought by emancipation.

Indeed, like the pervasive violence, insurrection panics underscored what might be called the "politicization" of everyday life that followed the demise of slavery. A seemingly insignificant incident reported to the state's governor in 1869 by black North Carolinian A. D. Lewis graphically illustrates this development:

> Please allow me to call your kine attention to a transaction which occured to day between me and Dr. A. H. Jones. . . . I was in my field at my own work and this Jones came by me and drove up to a man's gate that live close by . . . and ordered my child to come there and open that gate for him . . . while there was children in the yard at the same time not more than twenty yards from him and jest because they were white and mine black he wood not call them to open the gate. . . . I spoke gently to him that [the white children] would open the gate. . . . He got out of his buggy . . . and walked nearly hundred yards rite into my field where I was at my own work and double his fist and strick me in the face three times and . . . cursed me [as] a dum old Radical. . . . Now governor I wants you to please rite to me how to bring this man to jestus.

No record exists of the disposition of this complaint, but Lewis' letter conveys worlds of meaning about Reconstruction: his powerful sense of place, his quiet dignity in the face of assault, his refusal to allow his son to be treated differently from white children or to let a stranger's authority be imposed on his family, the way an everyday encounter rapidly

79. Litwack, *Been in the Storm*, 288–89; J. W. Dukes to James L. Orr, December 3, 1866, South Carolina Governor's Papers; Kinlock Falconer to James M. Kennard, June 7, 1866, James M. Kennard Papers, MDAH; Wilmer Shields to William N. Mercer, December 19, 1865, Mercer Papers; 39th Congress, 1st Session, House Report 30, pt 3, 30; Dan T. Carter, "The Anatomy of Fear: The Christmas Day Insurrection Scare of 1865," *JSH*, 42 (August 1976), 345–64.

descended into violence and acquired political meaning, and Lewis' assumption (reflecting the situation after 1867) that blacks could expect justice from the government under which they lived. Most of all, it illustrates how day-to-day encounters between the races became infused with the tension inevitable when a social order, with its established power relations and commonly understood rules of conduct, has been swept away and a new one has not yet come into being. Only over time would the South's new system of social relations be worked out. As David L. Swain, former governor of North Carolina, remarked in 1865, "With reference to emancipation, we are at the beginning of the war."[80]

80. Carter, "Anatomy of Fear," 348; Mahaffey, "Schurz's Letters," 239; James W. Johnson to A. M. Crawford, December 17, 1866, South Carolina Governor's Papers; A. D. Lewis to William W. Holden, June 5, 1869, North Carolina Governor's Papers; Hamilton and Williams, eds., *Graham Papers*, 6:324.

CHAPTER 4

Ambiguities of Free Labor

NORTHERN journalists who hurried south at the end of the Civil War telegraphed back reports of a devastated society. Where the great armies had fought and marched, vast scenes of desolation greeted the observer. The Shenandoah Valley, Virginia's antebellum breadbasket, appeared "almost a desert," its barns and dwellings burned, bridges demolished, fences, tools, and livestock destroyed. Northern Alabama, having endured three years of fighting, and the state's central counties, which felt the wrath of the Union cavalry early in 1865, offered vistas of "absolute destitution"—houses razed, fields uncultivated, iron works and cotton gins burned. Along Sherman's track in Georgia and South Carolina, the scars of battle were everywhere. A white Georgian in August described in his diary a railroad journey through "a desolated land. Every village and station we stopped at presented an array of ruined walls and chimneys standing useless and solitary . . . thanks to that destroying vandal." Large portions of Southern cities lay in ruins; Columbia, South Carolina, presented "a melancholy sight," its center "a mass of blackened chimneys and crumbling walls." How distant seemed the active commerce, bustling factories, and bouyant optimism of the North. A South Carolina planter, on a visit to Baltimore, found it "hard to bear . . . this exulting, abounding, overrunning wealth of the North in contrast with the utter desolation of the unfortunate South."[1]

To be sure, large parts of the rural South, including most of Georgia and the entire state of Texas, had escaped physical devastation. Even apart from the disorganization caused by the end of slavery, however, the

1. 39th Congress, 1st Session, House Report 30, pt 2, 68; Robert H. McKenzie, "The Economic Impact of Federal Operations in Alabama During the Civil War," *A1HQ*, 38 (Spring 1976), 51–60; J. T. Trowbridge, *The South: A Tour of Its Battle-Fields and Ruined Cities* (Hartford, Conn., 1866), 440; Samuel P. Richards Diary, August 10, 1865, Atlanta Historical Society; *New York Times*, June 18, 1865; Sidney Andrews, *The South Since the War* (Boston, 1866), 1–2, 33; John R. Dennett, *The South As It Is: 1865–1866*, edited by Henry M. Christman (New York, 1965), 230; Whitelaw Reid, *After the War: A Southern Tour* (Cincinnati, 1866), 351; William H. Trescot to unknown, June 5, 1866 (copy), William H. Trescot Papers, USC.

widespread destruction of work animals, farm buildings, and machinery, and the deterioration of levees and canals, ensured that the revival of agriculture would be slow and painful. So too did the appalling loss of life, a disaster without parallel in the American experience. Thirty-seven thousand blacks, the great majority from the South, perished in the Union Army, as did tens of thousands more in contraband camps, on Confederate Army labor gangs, and in disease-ridden urban shanty-towns. Nearly 260,000 men died for the Confederacy—over one fifth of the South's adult white male population. Many more were wounded, some maimed for life. (Mississippi expended 20 percent of its revenue in 1865 on artificial limbs for Confederate veterans.) The region, moreover, was all but bankrupt, for the collapse of Confederate bonds and currency wiped out the savings of countless individuals and the resources and endowments of colleges, churches, and other institutions. Little money circulated and interest rates soared to exorbitant levels; even in an important railroad and manufacturing center like Chattanooga, many commodities in 1865 exchanged at barter. Moreover, capital continued to be drained from the region by the new national banking system and the federal tax on cotton, a wartime measure not repealed until 1868. "The only money here," a Charlestonian reported, "is in the hands of one or two Northern exchange brokers and bankers, . . . men entirely unknown to us."[2]

Agricultural statistics reveal the full extent of the economic disaster the South had suffered. Between 1860 and 1870, while farm output expanded in the rest of the nation, the South experienced precipitous declines in the value of farm land and the amount of acreage under cultivation. The number of horses fell by 29 percent, swine by 35 percent, and farm values by half. Georgia alone reported 1 million fewer swine, 50,000 fewer horses, and 200,000 fewer cattle, and had 3 million fewer acres under cultivation in 1870 than ten years earlier. For the South as a whole, the real value of all property, even discounting that represented by slaves, stood 30 percent lower than its prewar figure, and the output of the staple crops cotton, rice, sugar, and tobacco, and food crops like corn and potatoes, stood far below their antebellum levels. Confederate Gen. Braxton Bragg returned from the war to his "once prosperous" Alabama home to find "all, all was lost, except my debts." Bragg and his wife, a woman "raised in affluence," lived for a time in a slave cabin, "expecting," as the general wrote, "to be even deprived of the necessities of life."

2. William C. Harris, *Presidential Reconstruction in Mississippi* (Baton Rouge, 1967), 164; Francis B. Simkins and Robert H. Woody, *South Carolina During Reconstruction* (Chapel Hill, 1932), 11; James W. Livingood, "Chattanooga, Tennessee: Its Economic History in the Years Immediately Following Appomattox," *ETHSP*, 15 (1943), 39; George L. Anderson, "The South and Problems of Post-Civil War Finance," *JSH*, 9 (May 1943), 181–84; H. W. DeSaussere to Henry DeSaussere, June 12, 1865, DeSaussere Family Papers, SCHS.

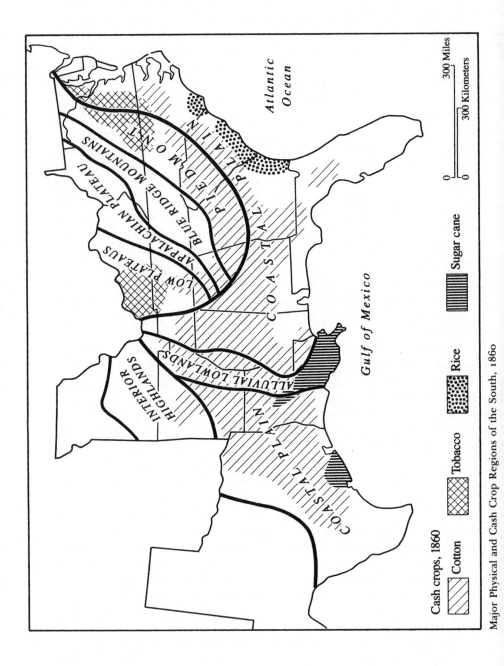

Major Physical and Cash Crop Regions of the South, 1860

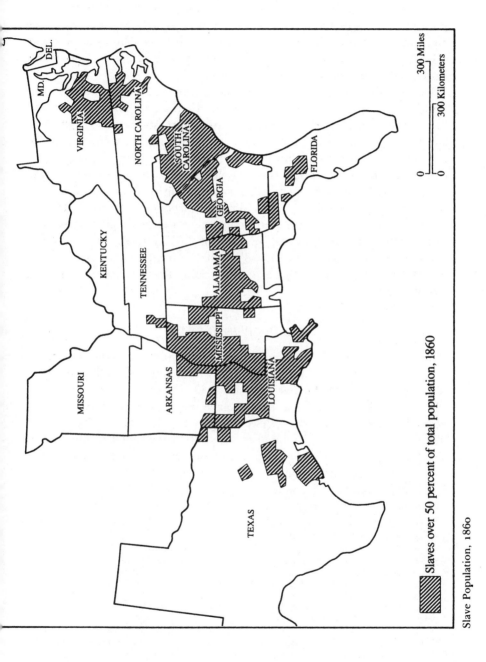

Slaves over 50 percent of total population, 1860

Slave Population, 1860

Bragg soon recouped his fortunes by securing employment as Alabama's Commissioner of Public Works and later with the Gulf, Colorado, & Santa Fe Railroad, but many others were not so fortunate. Thousands of poorer white farmers emerged from the war "wholly destitute of bread and meat." The widespread destruction of livestock dealt a particularly severe blow to upcountry yeomen, since the free ranging of hogs and cattle had been essential to their livelihood.[3]

Despite the grim reality of desolation and poverty, however, the South's economic recovery involved more than rebuilding shattered farms and repairing broken bridges. An entire social order had been swept away, and on its ruins a new one had to be constructed. "The revolution of our social fabric is too great, the entire upheaval and overthrow of all the foundations of our society too universal," wrote Dr. Richard D. Arnold, Savannah's wartime mayor, "not to affect every body and to place persons in an almost entirely new status." The process by which a new social and economic order replaced the old followed different paths in different parts of the South. But for black and white alike, the war's end ushered in what South Carolina planter William H. Trescot called "the perpetual trouble that belongs to a time of social change."[4]

Masters Without Slaves

Plantation slavery never dominated the entire South as it did, for example, most islands of the West Indies. But the plantation belt, containing the region's most productive land, the bulk of its economic wealth, and the majority of its slave population, gave rise to a ruling class that had shaped regional institutions, from the school and church to the state, in its own interests. A large landed estate specializing in the production of a staple crop for the world market, the plantation has historically required a disciplined, dependent labor force, since planters have found it nearly impossible to attract free laborers, especially where either land or alternative employment opportunities are available. Like their counterparts in

3. James L. Sellers, "The Economic Incidence of the Civil War in the South," *MVHR*, 14 (September 1927), 179–91; Willard Range, *A Century of Georgia Agriculture 1850–1950* (Athens, Ga., 1954), 66–67; Roger L. Ransom and Richard Sutch, *One Kind of Freedom: The Economic Consequences of Emancipation* (New York, 1977), 52–53; Eugene Lerner, "Southern Output and Agricultural Income, 1860–1880," *AgH*, 33 (July 1959), 117–25; Aaron M. Boom, ed., " 'We Sowed and We Have Reaped': A Postwar Letter from Braxton Bragg," *JSH*, 31 (February 1965), 74–79; L. M. Stiff et al. to Lewis E. Parsons, August 6, 1865, Alabama Governor's Papers, ASDAH; Grady McWhiney, "The Revolution in Nineteenth-Century Alabama Agriculture," *Alabama Review*, 31 (January 1978), 4–21.

4. Richard H. Shryock, ed., "Letters of Richard D. Arnold, M.D. 1808–1896," *Trinity College Historical Society Historical Papers*, 18–19 (1929), 128; William H. Trescot to unknown, February 9, 1866 (copy), Trescot Papers. The process of change among white yeomen is considered in Chapter 8.

other societies, American planters believed that the South's prosperity and their own survival as a class depended, as a Georgia newspaper put it, upon "one single condition"—*the ability of the planter to command labor.*" And the conflict between former masters attempting to re-create a disciplined labor force and blacks seeking to infuse meaning into their freedom by carving out autonomy in every aspect of their lives, profoundly affected the course of Reconstruction. For as Christopher G. Memminger, the Confederacy's Secretary of the Treasury, wrote in 1865, politics, race relations, and the very definition of "what is included in . . . emancipation" all turned "upon the decision which shall be made upon the mode of organizing the labor of the African race."[5]

Southern planters emerged from the Civil War in a state of shock. Their class had been devastated—physically, economically, and psychologically. Thousands of wealthy young men had heeded the Confederacy's call only to die in battle. The loss of the planters' slaves and life savings (to the extent that they had invested in Confederate bonds) wiped out the inheritance of generations. William Gilmore Simms, the South Carolina poet and novelist who had married a planter's daughter, lost "slaves, stock, furniture, books, pictures, horses . . . a property which was worth $150,000 in gold." In Dallas County, in the heart of Alabama's plantation belt, whites' per capita wealth fell from $19,000 in 1860 to one sixth that amount ten years later. Many slaveholding families faced the added indignity of the departure of their labor force. A Maryland Unionist described the plight of Gen. Tench Tilghman and his family:

> This family, one of our oldest and most respectable, once very wealthy, are now reduced to that state which is even worse in my estimation than actual poverty, large debts, large pride, large wants: small income, and small helpfulness. They are now without servants. . . . The young ladies on Wednesday and Thursday *milked* the cows, while their father the General held the umbrella over them to keep off the rain. . . . The general has to harness his own carriage horses and probably black his own boots.[6]

5. John F. Hart, "The Role of the Plantation in Southern Agriculture," *Proceedings, Tall Timbers Ecology and Management Conference*, 16 (1979), 1–20; Gavin Wright, " 'Economic Democracy' and the Concentration of Agricultural Wealth in the Cotton South, 1850–1860," *AgH*, 44 (January 1970), 84–85; Edgar Thompson, *Plantation Societies, Race Relations, and the South: The Regimentation of Populations* (Durham, 1975), xiii, 9, 55, 69; Augusta Transcript in *New York Times*, November 30, 1865; Elizabeth G. McPherson, ed., "Letters from North Carolina to Andrew Johnson," *NCHR*, 27 (July 1950), 477.

6. Frederic C. Jaher, *The Urban Establishment* (Urbana, Ill., 1982), 399–400; Michael Wayne, *The Reshaping of Plantation Society: The Natchez District, 1860–1880* (Baton Rouge, 1983), 31–38; Mary C. Oliphant et al, eds., *The Letters of William Gilmore Simms* (Columbia, S.C., 1952–56), 4:577; William L. Barney, "The Ambivalence of Change: From Old South to New in the Alabama Black Belt, 1850–1870," in Walter J. Fraser, Jr., and Winfred B. Moore, Jr., eds., *From the Old South to the New: Essays on the Transitional South* (Westport, Conn., 1981), 37–38; Charles L. Wagandt, ed., "The Civil War Journal of Dr. Samuel A. Harrison," *CWH*, 13 (June 1967), 142.

Bitterness and demoralization overwhelmed many a planter in 1865. Henry Watson, Jr., who spent the war years in Europe, returned to his Alabama plantation to find his neighbors determined to sell their land and leave the region. Facing "a joyless future of probable ignominy, poverty, and want," as many as 10,000 slaveowners abandoned their homes after the war, hoping to begin anew in the North or Europe or to reestablish themselves as planters in Mexico or Brazil. Others sought to "drown our troubles in a sea of gaiety," reviving the aristocratic social life of the antebellum years as if nothing had changed. Tournaments straight out of *Ivanhoe*, complete with knights adorned with lances and plumed helmets and ladies competing to be crowned Queen of Love and Beauty, made an incongruous reappearance—although in one North Carolina community, the cult of medieval nobility waned after blacks organized a Tournament Association of their own.[7]

Some Bourbons "clung to the dead body" of slavery, hoping for compensation for their slaves or even a Supreme Court challenge to the Emancipation Proclamation. Many others dreamed of a lily-white South in which blacks, like the Indians, had all "died out." But for the majority of planters, as for their former slaves, the Confederacy's defeat and the end of slavery ushered in a difficult period of adjustment to new forms of race and class relations and new ways of organizing labor.[8]

The first casualty of this transformation was the paternalist ethos of prewar planters. A sense of obligation based on mastership over an inferior, paternalism had no place in a social order in which labor relations were mediated by the impersonal market, and blacks aggressively pressed claims to autonomy and equality. Some former slaveholders, to be sure, continued to concern themselves with the fortunes of individual freedmen, assisting them in contract disputes with new employers, or petitioning governors to have them rescued from unfair court judgments. Among those genuinely concerned for the well-being of blacks directly dependent on them was Jonathan Worth, North Carolina's governor during Presidential Reconstruction. Worth presented one of his former slaves with a house as a wedding gift, served as security for another who purchased a horse and dray, and even assisted a black employee in obtaining the enlistment bounty owed to a son who had died in the Union Army.

7. James L. Roark, *Masters Without Slaves: Southern Planters in the Civil War and Reconstruction* (New York, 1977), 121–32; Susan B. Eppes, *Through Some Eventful Years* (Macon, 1926), 286; W. C. Nunn, *Escape from Reconstruction* (Fort Worth, 1956); William D. Henderson, *The Unredeemed City: Reconstruction in Petersburg, Virginia: 1865–1874* (Washington, D.C. 1977), 70–71; W. McKee Evans, *Ballots and Fence Rails: Reconstruction on the Lower Cape Fear* (Chapel Hill, 1967), 211–13.

8. I. W. Avery, *The History of the State of Georgia From 1850 to 1881* (New York, 1881), 342; J. D. Aiken to D. Wyatt Aiken, May 20, 1865, D. Wyatt Aiken Papers, USC; Billy D. Ledbetter, "White Texans' Attitudes Toward the Political Equality of Negroes, 1865–1870," *Phylon*, 40 (September 1979), 254–55; A. Warren Kelsey to Edward Atkinson, October 1, 1865, Edward Atkinson Papers, MHS.

These remnants of paternalism, however, impeded Worth's ability to accept the full implications of emancipation. While believing blacks entitled to the benevolent oversight of whites, he insisted that they were "incapable of permanent civilization" and must not be allowed to act independently to protect their own interests.[9]

"The Law which freed the negro," a Southern editor wrote in 1865, "at the same time freed the master, all obligations springing out of the relations of master and slave, except those of kindness, cease mutually to exist." And kindness proved all too rare in the aftermath of war and emancipation. In hundreds of cases, planters evicted from their plantations those blacks too old or infirm to labor, and transformed "rights" enjoyed by slaves—clothing, housing, access to garden plots—into commodities for which payment was due. "Some think," reported a Bureau agent, "that the only difference between freedom and slavery is that then the negroes were obliged to work for nothing; now they have to pay for what they used to have for nothing." One former slave later recalled how his master forced him to work for one full week to pay for "the clothes that I had when the government freed us." Robert Toombs, the prominent Georgia politician, gave each family among his former slaves land, a house, and a mule, but his generosity was indeed unusual. After the war, a Missouri freedman remembered, "I do know some of dem old slave owners to be nice enough to start der slaves off in freedom wid somethin' to live on . . . but dey wasn't in droves, I tell you."[10]

"The former relation has to be unlearnt by both parties," wrote one planter, but except for the obligations of paternalism, ideas inherited from slavery displayed remarkable resiliency. For those accustomed to the power of command, the normal give-and-take of employer and employee was difficult to accept. "It seems humiliating to be compelled to bargain and haggle with our own servants about wages," wrote Fanny Andrews, the daughter of a Georgia planter. Former slaveowners resented the very idea of having to negotiate with the freedmen. The employer, many planters believed, should be the sole judge of the value of his laborers' services. One white North Carolinian hired a freedman in the spring of 1865, promising to give him "whatever was right" after

9. Petition, February 12, 1868, South Carolina Governor's Papers, SCDA; Edward Sparrow to Benjamin G. Humphreys, December 3, 1866, Mississippi Governor's Papers, MDAH; Richard L. Zuber, *Jonathan Worth* (Chapel Hill, 1965), 291; Jonathan Worth to Benjamin S. Hedrick, December 30, 1866, Benjamin S. Hedrick Papers, DU; J. G. de Roulhac Hamilton, ed., *The Correspondence of Jonathan Worth* (Raleigh, 1909), 1:417, 421, 2:786, 804, 807, 875, 1094–95, 1154–56.

10. Wayne, *Reshaping of Plantation Society* 41; Roark, *Masters Without Slaves*, 144; Roberta S. Alexander, *North Carolina Faces the Freedmen: Race Relations During Presidential Reconstruction, 1865–67* (Durham, 1985), 96; Gerald D. Jaynes, *Branches Without Roots: Genesis of the Black Working Class in the American South, 1862–1882* (New York, 1986), 121–27; George P. Rawick, ed., *The American Slave: A Composite Autobiography* (Westport, Conn., 1972–79), 9, pt. 8:372, Supplement, Ser. 2, 9:3452.

the crop had been gathered; another, an ex-governor of the state, said he would pay wages "where I thought them earned, but this must be left to me." Behavior entirely normal in the North, such as a freedman leaving the employ of a Georgia farmer because "he thought he could do better," provoked cries of outrage and charges of ingratitude. A few planters did embrace the logic of the marketplace. South Carolina rice planter Edward B. Heyward boasted that he had no desire to bind blacks to his plantation: "Self interest *alone* retains my laborer. If he finds a better place, he can go." But in general, one historian's comment on the aftermath of Caribbean abolition is equally applicable to the United States: "Planters bitterly resisted the creation of a free labour market as implied by emancipation."[11]

Among white Southerners, the question "Will the free Negro work?" became the all-absorbing obsession of 1865 and 1866. "Nothing else is talked of," reported a Northern visitor. "Even the Reconstruction question has sunk into utter insignificance." It was an article of faith that the freedmen, naturally indolent, would work only under compulsion. Reports of blacks' incorrigible laziness filled the pages of planters' letters and Southern newspapers and magazines. "Our faults are daily published by the editors," observed a group of Mississippi blacks, "not a statement will you ever see in our favor." Indeed, the end of slavery produced among planters and overseers a nostalgia for the days when the lash could be freely used to compel slaves to labor. "I have come to the conclusion," wrote a Louisiana sugar planter, "that the great secret of our success was the great motive power contained in that little instrument."[12]

Carl Schurz and other Northerners who toured the South in 1865 concluded that white Southerners "do not know what free labor is." To which many planters replied that Northerners "do not understand the character of the negro." Free labor assumptions—economic rationality, internal self-discipline, responsiveness to the incentives of the market— could never, planters insisted, be applied to blacks. "They are improvident and reckless of the future," complained a Georgia newspaper; the "inborn nature of the negro," echoed a Louisiana planter, "cannot be

11. Arney R. Childs, ed., *The Private Journal of Henry William Ravenel 1859–1887* (Columbia, S.C., 1947), 269; Eliza F. Andrews, *The War-Time Journal of a Georgia Girl* (New York, 1908), 319; New Orleans *Times* in New Orleans *Tribune*, June 3, 1865; Eric Foner, *Politics and Ideology in the Age of the Civil War* (New York, 1980), 99; J. G. de Roulhac Hamilton and Max R. Williams, eds., *The Papers of William A. Graham* (Raleigh, 1957–), 6:311; Laura Perry to Grant Perry, February 3, 1869, J. M. Perry Family Papers, Atlanta Historical Society; Edward B. Heyward to Katherine Heyward, undated letter [1867], Heyward Family Papers, USC; Walter Rodney, "Slavery and Underdevelopment," *Historical Reflections*, 6 (Summer 1979), 284.

12. Sylvia H. Krebs, "Will the Freedmen Work? White Alabamans Adjust to Free Black Labor," *AlHQ*, 36 (Summer 1974), 151–63; A. Warren Kelsey to Edward Atkinson, September 2, 1865, Atkinson Papers; Foner, *Politics and Ideology*, 99; J. Carlyle Sitterson, *Sugar Country* (Lexington, Ky., 1953), 235.

changed by the offer of more or less money." Nor was another free labor axiom, the opportunity for social mobility, applicable in the South. "You must begin at the bottom of the ladder and climb up," Freedmen's Bureau Commissioner O. O. Howard informed a black audience in New Orleans. At least he offered the opportunity to climb. A Natchez newspaper at the same time was informing its readers: "The true station of the negro is that of a servant. The wants and state of our country demand that he should remain a servant." Or, as a Texan put it, the "destiny of the negro race" could be summarized "in one sentence—subordination to the white race."[13]

The conviction that preindustrial lower classes share an aversion to regular, disciplined toil had a long history in both Europe and America. In the Reconstruction South, this ideology took a racial form, and although racism, needless to say, was endemic throughout nineteenth-century America, the requirements of the plantation economy shaped its specific content in the aftermath of emancipation. Charges of "indolence" were often directed not against blacks unwilling to work at all, but at those who preferred to labor for themselves. "Want of ambition will be the devil of the race, I think," wrote Kemp P. Battle, a North Carolina planter and political leader, in 1866. "Some of my most sensible men say they have no other desire than to cultivate their own land in grain and raise bacon." On the face of it, such an aspiration appears ambitious enough, and hardly unusual in the nineteenth-century South. But in a plantation society, a black man seeking to work his way up the agricultural ladder to the status of self-sufficient farmer seemed not an admirable example of industriousness, but a demoralized freedman unwilling to work—work, that is, under white supervision on a plantation.[14]

The prior experience of abolition in the British West Indies reinforced Southern planters' certainty of the "disastrous" consequences of emancipation. Viewed through the lens of plantation agriculture, the West Indies taught an unmistakable lesson: Emancipation was a failure. Freedom had come to Haiti in the 1790s and to the British Caribbean in the 1830s, and in both settings former slaves had abandoned the sugar plantations in large numbers to establish themselves as subsistence-oriented small

13. Joseph H. Mahaffey, ed., "Carl Schurz's Letters from the South," *GaHQ*, 35 (September 1951), 227–28; Edgar A. Stewart, ed., "The Journal of James Mallory, 1834–1877," *Alabama Review*, 13 (July 1961), 228; Savannah *Daily Herald*, July 21, 1865; B. I. Wiley, "Vicissitudes of Early Reconstruction Farming in the Lower Mississippi Valley," *JSH*, 3 (November 1937), 450; New Orleans *Tribune*, November 6, 1865; Ross H. Moore, "Social and Economic Conditions in Mississippi During Reconstruction" (unpub. diss., Duke University, 1937), 41; Carl H. Moneyhon, *Republicanism in Reconstruction Texas* (Austin, 1980), 20.

14. A. W. Coats, "Changing Attitudes to Labour in the Mid-Eighteenth Century," *Economic History Review*, 2 Ser, 11 (August 1958), 35–51; Kemp P. Battle to Benjamin S. Hedrick, January 20, 1866, Hedrick Papers; Charles L. Flynn, Jr., *White Land, Black Labor: Caste and Class in Late Nineteenth-Century Georgia* (Baton Rouge, 1983), 24.

farmers. This was especially true in Haiti, where revolution had destroyed the planter class, and in Trinidad, Jamaica, and British Guiana, where large tracts of uncultivated land had been available to the former slaves. As a result, sugar production had plummeted; plantation agriculture never resumed in Haiti and in the British Caribbean, it survived only through the massive importation of indentured "coolies" from India and China. Caribbean emancipation stood as a symbol and a warning to the white South, a demonstration of the futility of all schemes to elevate blacks, and of the dire fate awaiting American planters in the aftermath of slavery. Most of all, it taught that the freedmen must be barred from access to land. Only on smaller islands like Barbados, where whites owned all the land "and the negro is unable to get possession of a foot of it," had plantation agriculture continued to flourish.[15]

In the United States, too, the questions of land and labor were intimately interrelated. Planters quickly concluded that their ability to control black labor rested upon maintaining their own privileged access to the productive land of the plantation belt. A. Warren Kelsey, the Northern "cotton detective," found planters convinced "that so long as they retain possession of their lands they can oblige the negroes to work on such terms as they please." Even if relatively few freedmen established themselves as independent farmers, plantation discipline would dissolve since, as William H. Trescot explained, "it will be utterly impossible for the owner to find laborers that will work contentedly for wages alongside of these free colonies." At public meetings in 1865, and in their private correspondence, planters resolved never to rent or sell land to freedmen and condemned those landowners heedless enough of the broader interests of their class to do so. In effect, they sought to impose upon blacks their own definition of freedom, one which repudiated the former slaves' equation of liberty and autonomy. "They have an idea that a hireling is not a freeman," Mississippi planter Samuel Agnew noted in his diary, and went on to observe:

> Our negroes have a fall, a tall fall ahead of them, in my humble opinion. They will learn that freedom and independence are different things. A man may be free and yet not independent.

Or, as a Kentucky newspaper succinctly put it, the former slave must be taught "that he is *free*, but free only to labor."[16]

Between the planters' need for a disciplined labor force and the freed-

---

15. Eric Foner, *Nothing But Freedom: Emancipation and Its Legacy* (Baton Rouge, 1983), 8–43; Vernon Burton, "Race and Reconstruction: Edgefield County, South Carolina," *JSocH*, 12 (Fall 1978), 36.

16. A. Warren Kelsey to Edward Atkinson, August 12, 1865, Atkinson Papers; William H. Trescot to Andrew Johnson, March 11, 1866 (copy), South Carolina Governor's Papers; Charleston *South Carolina Leader*, December 16, 1865; Jerrell H. Shofner, *Nor Is It Over Yet: Florida in the Era of Reconstruction, 1863–1877* (Gainesville, Fla., 1974), 49; Samuel Agnew Diary, November 3, December 15, 1865, UNC; Louisville *Democrat* in Columbia *Daily Phoenix*, August 3, 1866.

men's quest for autonomy, conflict was inevitable. Planters sought through written contracts to reestablish their authority over every aspect of their laborers' lives. "Let everything proceed as formerly," one advised, "the contractual relation being substituted for that of master and slave." These early contracts prescribed not only labor in gangs from sunup to sundown as in antebellum days, but complete subservience to the planter's will. One South Carolina planter required freedmen to obey the employer "and go by his direction the same as in slavery time." Many contracts not only specified modes of work and payment, but prohibited blacks from leaving plantations, entertaining visitors, or holding meetings without permission of the employer. Frequently, planters also drafted detailed provisions regulating blacks' personal demeanor. Contracts required employees "to establish a character for honesty, industry and thrift," "behave in a gentle manner," and avoid "impertinence, all disorderly or unseemly noises, all vulgar or indecent words, all profane swearing and all quarreling and fighting." One provided that disputes among the laborers were to be "settled by the employer." No detail of blacks' lives seemed exempt from outside control. One group of planters sought to compel laborers to cook communally, as under slavery, since preparation of meals by individual families involved "extravagance in wood and loss of time."[17]

Such provisions proved easier to compose than enforce. When one Alabama planter forbade a freedman to marry a certain woman, the employee responded that "he preferred to leave" rather than subject his personal life to outside dictation. Planters quickly learned that labor contracts could not by themselves create a submissive labor force. On the aptly named Vexation plantation in Texas, blacks in September 1865 were said to be "insolent and refusing to work." The employees of Louisiana's former Confederate governor, Thomas O. Moore, set their own pace of work, refused to plow when the ground was hard, and answered his complaints in a "disrespectful and annoying" manner. On the Alabama plantation of Henry Watson, Jr., freedmen refused to agree to contracts but remained on the plantation, "idle and doing nothing." Eventually, Watson rented the plantation to his overseer, who leased individual plots to black families.[18]

17. Roark, *Masters Without Slaves*, 141–42; Wayne, *Reshaping of Plantation Society*, 41–42; Robert P. Brooks, *The Agrarian Revolution in Georgia, 1865–1912* (Madison, Wis., 1914), 18; Labor Contract, George Wise, January 1866, USC; Labor Contract, T. T. Tredway, January 1, 1866, Contracts and Indentures, Ser. 3975, Farmville, Va., Subasst Comr., RG 105, NA [FSSP A-8134]; Lewis C. Chartock, "A History and Analysis of Labor Contracts Administered by the Bureau of Refugees, Freedmen, and Abandoned Lands in Edgefield, Abbeville, and Anderson Counties in South Carolina, 1865–1868" (unpub. diss., Bryn Mawr College, Graduate School of Social Work and Social Research, 1973), 138–52; Jaynes, *Branches Without Roots*, 111–12.

18. Labor Contract, January 1, 1866, James A. Gillespie Papers, LSU; W. Stanley Hoole, ed., "The Diary of Dr. Basil Manley, 1858–1867," *Alabama Review*, 5 (April 1952), 146–47; John Q. Anderson, ed., *Brockenburn: The Journal of Kate Stone, 1861–1868* (Baton Rouge,

Watson was not the only planter to abandon the struggle to control free black labor. Paul C. Cameron, North Carolina's largest slaveholder before the Civil War, found it impossible to govern his employees in 1865; the following year he rented his plantation lands to white farmers, each of whom hired his own workers. "If any one want the negro . . . take him!" Cameron wrote. Most planters persevered, in the hope, as a Georgian put it, "that capital will at someday control labor." But their correspondence and the Southern press abounded with lamentations of failure. One planter offered a $50 reward for "information that will enable me to make a living . . . by the use of Negro labor." Another, whose black employees set their own slow pace of work "out of mere spite," longingly recalled the antebellum world: "Oh, for one of the days of '60."[19]

Clearly, the "labor problem" involved more than questions of wages and hours. Planters' inability to establish their authority arose from the clash between their determination to preserve the old forms of domination and the freedmen's desire to carve out the greatest possible independence for themselves and their families. Conflict was endemic on plantations throughout the South. Blacks, planters complained, insisted on setting their own hours of labor and demanded extra compensation for, or refused to do, work not directly related to the growing crop but necessary for the plantation's upkeep. Freedmen, wrote a Mississippi planter, "seem to think . . . that a horse never gets hungry or thirsty, and that the farming utensils will never be needed after the present crop is completed." When a South Carolina planter ordered his employees to repair a fence, they asked, "How it was they were doing this work for nothing?" Other blacks refused to weed cotton fields in the rain. Still others would not perform the essential but hated "mud work" of a rice plantation—dredging canals and repairing dikes—forcing some rice planters "to hire Irishmen to do the ditching." House servants, too, had their own ideas of where their obligations began and ended. Butlers refused to cook or polish brass, domestics would not black the boots of plantation guests, chambermaids declared that it was not their duty to answer the front door, serving girls insisted on the right to entertain male visitors in their rooms.[20]

1955), 362; Bell I. Wiley, *Southern Negroes 1861–1865* (New Haven, 1938), 257; Peter Kolchin, *First Freedom: The Responses of Alabama's Blacks to Emancipation and Reconstruction* (Westport, Conn., 1972), 32.

19. Robert C. Kenzer, "Portrait of a Southern Community, 1849–1881: Family, Kinship, and Neighborhood in Orange County, North Carolina" (unpub. diss., Harvard University, 1982), 138; John A. Cobb to Mrs. Howell Cobb, February 14, 1867, Howell Cobb Papers, UGa; John A. Moore, ed., *The Juhl Letters to the "Charleston Courier"* (Athens, Ga., 1974), 53; Wiley, *Southern Negroes*, 258; Frank B. Conner to Lemuel Conner, November 30, 1866, Lemuel P. Conner Family Papers, LSU.

20. Flavellus G. Nicholson Diary, August 23, 1866, March 15, 1867, MDAH; F. W. Leidtke to Robert K. Scott, August 22, 1866, R. Y. Dwight Papers, USC; John S. Strickland, " 'No More Mud Work': The Struggle for the Control of Labor and Production in Low Country

Southern whites were not the only ones to encounter difficulty disciplining the former slaves. During and immediately after the war, a new element joined the South's planter class: Northerners who purchased land, leased plantations, or formed partnerships with Southern planters. These newcomers were a varied, ambitious group, mostly former soldiers eager to invest their savings in this promising new frontier, and civilians lured south by press reports of "the fabulous sums of money to be made in the South in raising cotton." Although few had been farmers at home, Massachusetts teacher-turned-planter Henry Warren later recalled, most succumbed to the " 'get rich quick' enterprise of raising cotton." Joined with the quest for profit, however, was a reforming spirit, a vision of themselves as agents of sectional reconciliation and the South's "economic regeneration." Accustomed to viewing Southerners, black and white, as devoid of economic initiative and self-discipline, they believed that only "Northern capital and energy" could bring "the blessings of a free labor system" to the region. Garth James, an officer of the black 54th Massachusetts Infantry during the Civil War who, with his brother Robertson, purchased $40,000 worth of Florida real estate in 1866, hoped to "vindicate the principle . . . that the freed Negro under decent and just treatment can be worked to profit to employer and employee." An Illinois man farming in Texas agreed: "I am going to introduce new ideas here in the farming line," he wrote, "and show the beauties of free over slave labor."[21]

Later reviled as "carpetbaggers," these Northerners generally received a warm welcome in 1865. The Southern press assured them of their safety, and capital-starved businessmen welcomed their participation in commission houses, banks, and planting partnerships. Despite instances of violent hostility or ostracism, most Southern planters recognized that Northern investment, ironically, was raising land prices and rescuing many former slaveholders from debt—in a word, stabilizing their class. What most annoyed Southern whites was the newcomers' sublime confidence that they knew better than former slaveowners how to supervise

South Carolina, 1863–1880," in Walter J. Fraser, Jr., and Winfred B. Moore, Jr., eds., *The Southern Enigma: Essays on Race, Class, and Folk Culture*, (Westport, Conn., 1983), 48–51; Audley Couper to Francis P. Corbin, March 15, 1867, Francis P. Corbin Papers, NYPL; Daniel E. Sutherland, "A Special Kind of Problem: The Response of Household Slaves and Their Masters to Freedom," SS, 20 (Summer 1981), 158–62; Lucy M. Cohen, *Chinese in the Post-Civil War South: A People Without a History* (Baton Rouge, 1984), 61.

21. Lawrence N. Powell, *New Masters: Northern Planters During the Civil War and Reconstruction* (New Haven, 1980); Reid, *After the War*, 371–72, 405, 481; David H. Overy, Jr., *Wisconsin Carpetbaggers in Dixie* (Madison, Wis., 1961), 11–14, 25–26, 56; Henry W. Warren, *Reminiscences of a Mississippi Carpetbagger* (Holden, Mass., 1914), 9–11; Andrews, *South Since the War*, 4; Indiana *True Republican* (Richmond), July 27, 1865; [George C. Benham] *A Year of Wreck* (New York, 1880), 136; Jane Maher, *Biography of Broken Fortunes: Wilkie and Bob, Brothers of William, Henry, and Alice James* (Hamden, Conn., 1986), 77–86; W. C. Wagley to J. W. Shaffer, November 21, 1866, Elihu B. Washburne Papers, LC.

free black labor. "They believed, in their supercilious folly," one planter later wrote, "that *they* could get along with the free negroes as laborers, and teach the Southerners how to manage them."[22]

Southern planters predicted that the newcomers would soon complain about the character of black labor, and they were not far wrong. The very "scientific" methods Northerners hoped to introduce, involving closely supervised work and changes in customary plantation routines, challenged the more irregular pace of work preferred by blacks, and their desire to direct their own labor. Harriet Beecher Stowe, who lived on a Florida plantation in the early 1870s, described the freedmen as "great conservatives," who clung to old ways of working and rejected routines Northerners thought more efficient. One former slave demanded of his new Northern employer, "What the use of being free was, if he had to work harder than when he was a slave?" Increasingly, Northerners found that "the free labor system can be managed much more easily in the lecture room or an editor's sanctum *theoretically* north, than in the field practically here." Like their Southern counterparts, they began to complain of blacks' "shiftlessness" and unreliability. Henry Lee Higginson, who with his wife and two friends invested $65,000 in a Georgia cotton plantation, became convinced that uplifting the freedmen would be a "long, long struggle against ignorance, prejudice and laziness. . . . It is discouraging to see how . . . much more hopeful they appear at a distance than near to."[23]

As time passed, the Northern planters sounded and acted more and more like Southern. Some sought to restore corporal punishment, only to find that the freedmen would not stand for it. Perhaps the problem arose from the fact that, like Southern whites, most of the newcomers did not believe blacks had emerged from slavery capable of "self-directed labor." If the freedmen were to become productive free laborers, said the *New York Times* with unintended irony, "it must be done by giving them new masters." Blacks, however, wanted to be their own masters. And, against employers both Southern and Northern, they used whatever weapons they could find in the chaotic economic conditions of the postwar South, to influence the conditions of their labor.[24]

Blacks did, indeed, enjoy considerable bargaining power because of the "labor shortage" that followed the end of slavery. Particularly acute in sparsely populated Florida and the expanding cotton empire of the

22. Powell, *New Masters*, 35–50; Robert F. Futrell, "Efforts of Mississippians to Encourage Immigration, 1865–1880," *JMH*, 20 (April 1958), 61–62; D. E. Huger Smith, *A Charlestonian's Recollections 1846–1913* (Charleston, 1950), 131.

23. Powell, *New Masters*, 80, 97–98, 101–09, 120–21; Harriet Beecher Stowe, *Palmetto-Leaves* (Boston, 1873), 290; E. T. Wright to H. B. Clitz, October 6, 1865, B-69 1865, Letters Received, Ser. 4109, Dept. of S. C., RG 393, pt. 1, NA [FSSP C-1361]; Bliss Perry, *Life and Letters of Henry Lee Higginson* (Boston, 1921), 1:265–66.

24. Mahaffey, "Schurz's Letters," 230; *New York Times*, June 18, 1865.

Southwest, competition for labor affected planters throughout the South. With cotton prices still high in 1866 and many blacks refusing to sign labor contracts because they expected to receive land from the federal government, the competition for workers was "frightful and the planters are literally cutting one another's throats." "Laborers," a white Alabaman complained to the state's governor, "without any cause or provocation are leaving their employers and are encouraged to do so for the reason that they find no difficulty in getting employment." The same situation prevailed in many areas at the start of 1867. "The struggle seems to be who will get the negro at any price," lamented Texas planter Frank B. Conner. Planters, he concluded, must band together to "establish some maximum figure," stop "enticing" each others' workers, and agree that anyone "breaking the established custom should be driven from the community."[25]

To attract laborers, many planters in 1866 and 1867 found it necessary to raise wages, promise additional pay for harvest work, and offer land free of charge for garden plots. Some employers adopted more novel methods, like the Tennessee planter who drove a wagonload of blacks to a barbecue, and the Floridian who "offered to take all the Negroes confined in the jail at Tallahassee, pay any charges that may be against them, regardless of the offenses they had committed . . . and pay good wages." The labor shortage made planters reluctant to enforce the disciplinary provisions of their contracts, for fear blacks would abandon their employ. An upcountry South Carolina planter complained to President Johnson in 1866 that his laborers "absent themselves when they please, and lounge lazily about," and that any attempt to reduce their wages would lead to his being "left without any operative at all." And those planters with a reputation for having abused or cheated their laborers sometimes found it difficult to hire new ones at year's end. Dissatisfied with their treatment the previous year, Howell Cobb's former slaves not only refused to sign contracts in January 1867, but "have been busy spreading tales about our overseers and trying to prevent new hands from coming to us." "They feel no gratitude towards us," Cobb added, "and I now feel no obligations to them."[26]

The scarcity of labor was no mirage. Measured in hours worked per capita, the supply of black labor dropped by about one third after the

25. Roark, *Masters Without Slaves*, 135; John S. Harris to Robert M. Patton, April 21, 1866, Alabama Governor's Papers; Frank B. Conner to Lemuel P. Conner, September 26, 1866, February 3, 1867, Conner Family Papers.

26. Kolchin, *First Freedom*, 39; John G. Chance to E. L. Deane, February 6, 1867, Registered Letters Received, Ser. 3202, Georgetown Subasst. Comr., RG 105, NA [FSSP A-7196]; *Christian Recorder*, May 26, 1866; Clifton Paisley, *From Cotton to Quail: An Agricultural Chronicle of Leon County, Florida 1860–1967* (Gainesville, Fla., 1968), 25; C. W. Dudley to Andrew Johnson, July 9, 1866, Andrew Johnson Papers, LC; Howell Cobb to Mrs. Howell Cobb, December, 1866, January 1, 1867, Cobb-Erwin-Lamar Papers, UGa.

Civil War, largely because all former slaves were determined to work fewer hours than under slavery, and many women and children withdrew altogether from the fields. Moreover, as a Georgia newspaper put it, "the negroes engaged do not work as they formerly did." But the "labor shortage" was a question not only of numbers, but of power. It arose from black families' determination to use the rights resulting from emancipation to establish the conditions, rhythms, and compensation of their work, and to create time to pursue the personal and community goals that have earlier been described. Events on the Laurel Hill plantation of Natchez nabob William N. Mercer illustrate how black employees took advantage of competition among local planters to press demands for higher pay, community advancement, and day-to-day independence. As 1866 drew to a close, Laurel Hill freedmen insisted on the right to leave the plantation, access to garden plots, education for their children, and the right to set their own pace of work. Mercer's problems were compounded by his neighbors' efforts to hire his laborers, "offering every inducement they can to get them away," as overseer Wilmer Shields reported. One promised blacks the use of teams and horses to ride to Natchez on Saturdays, there, Shields suspected, to sell stolen produce. Another offered "plenty of whiskey and every latitude and liberty to do as they please." A third simply gave his plantation over to freedmen to operate, "and quietly awaits *his* half." Shields did agree to establish a school, but otherwise did nothing to counteract local blacks' impression "that we are too strict and do not pay enough." In January 1867, nearly all of Mercer's employees abandoned Laurel Hill "and we cannot blame them, in view of the tempting offers made to them."[27]

If the shortage of labor enhanced blacks' bargaining power, successive postwar crop failures seriously undermined it. Spurred by the high cotton prices of 1865, Southern planters and Northern newcomers plunged into cotton production, only to find that wartime destruction of tools and animals, the use of old seed that failed to produce vigorous crops, and persistent conflicts over labor discipline combined to produce a disappointing harvest. In 1866 nature compounded the problems of Southern farmers. The weather alternated between periods of drought and heavy rain, stunting the crops. Battered by the war, levees on the Mississippi, Red, and Arkansas rivers gave way, flooding prime cotton lands and drowning cattle and mules. Finally, the dreaded army worm attacked what

27. Ransom and Sutch, *One Kind of Freedom*, 6; Quitman *Banner* in Savannah *Daily News and Herald*, April 30, 1865; Wilmer Shields to William N. Mercer, September 21, November 28, December 1, 12, 1866, January 1, 9, May 22, 1867, William N. Mercer Papers, LSU; Ronald F. Davis, *Good and Faithful Labor: From Slavery to Sharecropping in the Natchez District, 1860–1890* (Westport, Conn., 1982), 90–97.

cotton had survived. "Planters of thirty years' experience pronounce this the worst crop season within their knowledge," wrote a South Carolinian. In 1867, many of the same conditions, especially flooding and the army worm, returned, and the crop again proved disappointing.[28]

Many planters suffered devastating losses in these years. In black belt Alabama, Henry Watson, Jr., surveyed the condition of his neighbors at the beginning of 1867. Of fourteen plantations, only one had turned a profit (to the amount of $294), while over $100,000 had been lost. Large numbers of Northern planters and investors suffered the same fate. Whitelaw Reid lost $20,000, mostly borrowed from a Washington bank, and decided to return north to make his living as a journalist. Abram P. Andrew, a twenty-three-year-old Civil War veteran from Indiana, lost the same amount and also set his sights for home: "I am disgusted with cotton, Carroll Parish, La., and the South generally, and shall not honor them (cotton included) with my presence any longer than I have to." Henry Lee Higginson in May 1867 left Georgia for Boston, there to launch a career as a banker and philanthropist (among his accomplishments would be the founding of the Boston Symphony Orchestra in 1881). Garth James, ruined by crop failures, sold out in 1870, moving to Wisconsin and the comparative security of a job with the Chicago, Milwaukee, & St. Paul Railroad. Not a few Northerners who remained in the South would shortly turn to politics in quest of the livelihood that had eluded them in the cotton fields.[29]

Despite successive crop failures, the price of cotton never rebounded to the artificially high levels of 1865. The South thus faced the worst of both worlds—a poor crop coupled with declining prices. "When cotton falls from $1.25 per lb. to 20 cents or less somebody must be bitten thereby," observed financier Jay Cooke. As the poorest part of the region's population, blacks suffered the most. On the South Carolina Sea Islands, where the corn and cotton crops had failed, "the negroes soon were almost starving," and in the state as a whole, the Freedmen's Bureau estimated, not one freedman in ten who had labored on the plantations in 1866 "realized enough to support their families." The crop failures of 1866 and 1867 ruined those few blacks who had managed to set up independently and, by exerting an irresistible downward pressure on

28. William E. Highsmith, "Some Aspects of Reconstruction in the Heart of Louisiana," *JSH*, 13 (November 1947), 467; C. J. Barrow to Anne E. Barrow, July 21, 1866, William Barrow Family Papers, LSU; Frank B. Conner to Lemuel P. Conner, August 9, 1866, Conner Family Papers; Savannah *Daily News and Herald*, May 26, September 24, 1866; Moore, ed., *Juhl Letters*, 115.

29. Henry Watson, Jr., "Cost of Cultivation and Yield of 14 Plantations," manuscript, January 16, 1867, Edward Atkinson Papers; Bingham Duncan, *Whitelaw Reid: Journalist, Politician, Diplomat* (Athens, Ga., 1975), 34–37; Andrew Gray, ed., "The Carpetbagger's Letters," *LaH*, 20 (Fall 1979), 445–49; Powell, *New Masters*, 145–49; Maher, *Biography of Broken Fortunes*, 100–14.

wages, all but eliminated whatever chance others may have had to move up the "agricultural ladder."[30]

"If Providence had smiled on this region in 1866, by giving it a reasonable crop, . . ." Northern planter George Benham later remarked, "injustice to the negro and the new-comer, bitterness of heart and hatred of the government would all have disappeared. In the absence of a good crop . . . all these were intensified." Benham may have been overly sanguine about the prospects for social harmony in the postwar South, but there is no doubt that crop failures further embittered relations between planters and their black employees. Those working for a share of the crop discovered that poor harvests reduced their income virtually to nothing; others, laboring for wages, found planters, themselves impoverished, unable to meet their obligations. Thousands of blacks were evicted from the plantations without pay as soon as crops had been harvested. Decades later, freedwoman Ella Wilson would recall how her employer drove her family from a Louisiana plantation: "We didn't get no half. We didn't git nothin'. . . . We hadn't done nothin' to him. He just wanted all the crop for hisself, and he run us off. That's all."[31]

The "Misrepresented Bureau"

Despite the intensity of their conflict, neither former master nor former slave possessed the power to define the South's new system of labor. A third protagonist, the victorious North, also attempted to shape the transition from slavery to freedom. To the Freedmen's Bureau, more than any other institution, fell the task of assisting at the birth of a free labor society. The Bureau's commissioner was Gen. Oliver Otis Howard, a graduate of Bowdoin College and a veteran of the Civil War, whose close ties to the freedmen's aid societies had earned him the sobriquet "Christian General." Despite being universally regarded as temporary, Howard's agency was an experiment in social policy that, a modern scholar writes, "did not belong to the America of its day." Its responsibilities can only be described as daunting; they included introducing a workable system of free labor in the South, establishing schools for freedmen, providing aid to the destitute, aged, ill, and insane, adjudicating disputes among blacks and between the races, and attempting to secure for blacks and white Unionists equal justice from the state and local governments established during Presi-

30. Wayne, *Reshaping of Plantation Society,* 61–68; Jay Cooke to William E. Chandler, May 20, 1867, William E. Chandler Papers, LC; "Life and Recollections of Joseph W. Barnwell," 207–10, manuscript, SCHS; Vernon L. Wharton, *The Negro in Mississippi 1865–1890* (Chapel Hill, 1947), 61; 40th Congress, 2d Session, House Executive Document 1, 669; *Report of the Commissioner of Agriculture for the Year 1876* (Washington, D.C., 1877), 130–31.

31. Benham, *Year of Wreck,* 402–403; Leon F. Litwack, *Been in the Storm So Long: The Aftermath of Slavery* (New York, 1979), 420–25; Rawick, ed., *American Slave,* 11, pt. 7:205.

dential Reconstruction. In turn diplomat, marriage counselor, educator, supervisor of labor contracts, sheriff, judge, and jury, the local Bureau agent was expected to win the confidence of blacks and whites alike in a situation where race and labor relations had been poisoned by mutual distrust and conflicting interests. "It is not . . . in your power to fulfill one tenth of the expectations of those who framed the Bureau," Gen. William T. Sherman advised Howard. "I fear you have Hercules' task."[32]

Since Congress initially made no appropriation for the Bureau, Howard drew its personnel from the army. Some appointees, including Rufus Saxton and Thomas W. Conway, the Bureau's first assistant commissioners in South Carolina and Louisiana, came from antislavery backgrounds and were genuinely committed to winning justice for the freedmen. Others, like Brevet Brig. Gen. Ralph Ely, who operated five plantations in South Carolina while neglecting his Bureau duties, were wholly unfit for the job. The majority brought to their posts a combination of paternalist assumptions about race and sensitivity to the plight of blacks, the precise mixture varying with the individual. The Bureau's employees also included a handful of blacks, among them John M. Langston, working out of Howard's Washington office as inspector of schools, and Martin R. Delany, agent at Hilton Head, South Carolina. In a sense, the agents' greatest handicap was inadequate manpower and resources. Perennially underfinanced, the Bureau employed, at its peak, no more than 900 agents in the entire South. Only a dozen served in Mississippi in 1866, and the largest contingent in Alabama at any time comprised twenty. In Colleton District, South Carolina, one agent had responsibility for 40,000 freedmen.[33]

At first glance, the Bureau's activities appear as a welter of contradictions, reflecting differences among individual agents in interpreting general policies laid down in Washington, and themselves evolving. Given the chaotic conditions in the postwar South, agents spent most of their time coping with day-to-day crises, and did so under adverse circumstances and with resources unequal to the task. But unifying the Bureau's activities was the endeavor to lay the foundation for a free labor society— one in which blacks labored voluntarily, having internalized the values of the marketplace, while planters and civil authorities accorded them the

32. John H. Cox and LaWanda Cox, "General O. O. Howard and the 'Misrepresented Bureau'," *JSH*, 19 (November 1953), 432–33, 441–42; Martin Abbott, *The Freedmen's Bureau in South Carolina, 1865–1872* (Chapel Hill, 1967), 133; Donald G. Nieman, *To Set the Law in Motion: The Freedmen's Bureau and the Legal Rights of Blacks, 1865–1868* (Millwood, N.Y., 1979), xv.

33. William S. McFeely, *Yankee Stepfather: General O. O. Howard and the Freedmen* (New Haven, 1968), 66–69, 289; Kenneth B. White, "Wager Swayne—Racist or Realist?" *Alabama Review*, 31 (April 1978), 94; Donald G. Nieman, "The Freedmen's Bureau and the Mississippi Black Code," *JMH*, 40 (May 1978), 114; Elizabeth Bethel, "The Freedmen's Bureau in Alabama," *JSH*, 14 (February 1948), 63; Abbott, *Freedmen's Bureau*, 11–14, 20–23.

rights and treatment enjoyed by Northern workers. To the extent that this meant putting freedmen back to work on plantations, the Bureau's policies coincided with the interests of the planters. To the extent that it prohibited coercive labor discipline, took up the burden of black education, sought to protect blacks against violence, and promoted the removal of legal barriers to blacks' advancement, the Bureau reinforced the freedmen's aspirations. In the end, the Bureau's career exposed the ambiguities and inadequacies of the free labor ideology itself. But simultaneously, the former slaves seized the opportunity offered by the Bureau's imperfect efforts on their behalf, to bolster their own quest for self-improvement and autonomy.

Education, for General Howard, was the foundation upon which all efforts to assist the freedmen rested, and the encouragement and oversight of schools for blacks occupied a significant portion of local agents' time. Because of its limited resources, however, the Bureau did not establish schools itself, but coordinated the activities of Northern societies committed to black education. By 1869 nearly 3,000 schools, serving over 150,000 pupils, reported to the Bureau, and these figures did not include many evening and private schools operated by the missionary societies and by blacks themselves. Plagued by financial difficulties and inadequate facilities, and more successful in reaching black youngsters in towns and cities than in rural areas, Bureau schools nonetheless helped lay the foundation for Southern public education. Education probably represented the agency's greatest success in the postwar South.

In a number of states local whites, some with blatantly racist attitudes, staffed the freedmen's schools, but most of the teachers were middle-class white women, the majority from New England, sent south by the Northern aid societies. Most had received some higher education, either in normal schools or in the few colleges, like Oberlin, that admitted women, and nearly all carried with them a commitment to uplifting the freedmen. "I feel that it is a precious privilege," one wrote, "to be allowed to do something for these poor people." Their letters of application invoked the example of husbands and brothers who had fought for the Union, and spoke of teaching as a way women could enlist as "soldiers" in a peacetime campaign to fulfill the promise of emancipation.[34]

Such commitment was essential if Northern teachers were to persevere in the face of the hardships they encountered. In cities, the teachers often resided in comfortable homes rented by sponsoring societies or provided by the Bureau, but in rural areas, ostracized by local whites, they had to

34. Ransom and Sutch, *One Kind of Freedom*, 24; Kenneth B. White, "The Alabama Freedmen's Bureau and Black Education: The Myth of Opportunity," *Alabama Review*, 34 (April 1981), 114; Jacqueline Jones, *Soldiers of Light and Love: Northern Teachers and Georgia Blacks, 1865–1873* (Chapel Hill, 1980); Esther W. Douglass to Samuel Hunt, February 1, 1866, AMA Archives, Amistad Research Center, Tulane University.

live alone or board with black families, accustoming themselves to impoverished surroundings and an unfamiliar diet. Other difficulties included the primitive condition of many schools ("we shut out the pigs, goats, and cows . . . by a very modest partly broken fence," wrote a teacher in Georgia), lack of sufficient books and materials, and classes of 100 or more children whose attendance fluctuated daily. Equally disheartening, the male-dominated aid societies denied women teachers a role in decision-making and expected them to meet their own travel and living expenses on salaries lower than they could have earned by remaining in the North. Some female teachers resigned in protest over rules that barred them from becoming school principals and superintendents. Overall, the Northern teachers' average tenure in the South did not exceed two years, but a few remained to teach blacks long after the end of Reconstruction.[35]

The Northern societies urged teachers to "dispossess [their] thoughts of the vulgar prejudice against color" and treat blacks in every respect "as we would if they were white." The American Missionary Association dismissed one teacher for having "frequently used the vulgar phrase 'niggers' when speaking of the colored people." But the realities of life in the postwar South often forced compromises upon the societies and their teachers. In principle, the AMA and American Freedmen's Union Commission, along with many Bureau education officials, favored integrated schools but, extremely sensitive to charges of fostering "social equality" between the races, they in fact established only a handful. And while the AMA attempted to train black instructors as quickly as possible, it proved reluctant to increase local hostility by boarding them together with white teachers. Three Northern black women who arrived to teach in Natchez complained early in 1866 that the local AMA superintendent had barred them from living in the society's residence unless they agreed to "room with the domestics." Such incidents embarrassed the Northern societies but did not deter them from adding blacks to the roster of instructors. With assistance from the Freedmen's Bureau, they founded and staffed the first black colleges in the South, including Berea, Fisk, Hampton, and Tougaloo, all initially designed to train black teachers. By 1869, among the approximately 3,000 freedmen's teachers in the South, blacks for the first time outnumbered whites.[36]

35. *National Freedman*, January, March 1866; John A. Rockwell to "My Dear Kate," October 22, 1868, John A. Rockwell Papers, UGa; Sandra E. Small, "The Yankee Schoolmarm in Freedmen's Schools: An Analysis of Attitudes," *JSH*, 45 (August 1979), 393; Jacqueline Jones, "Women Who Were More Than Men: Sex and Status in Freedmen's Teaching," *History of Education Quarterly*, 19 (Spring 1979), 47–50.

36. Lewis Tappan, *Caste: A Letter to a Teacher Among the Freedmen* (New York, 1867), 10–11; *National Freedman*, March 1866; Howard N. Rabinowitz, *Race Relations in the Urban South 1865–1890* (New York, 1978), 154, 162; Joe M. Richardson, *The Negro in the Reconstruction of Florida, 1865–1877* (Tallahassee, 1965), 108; C. W. Buckley to George Whipple, March 13, 1866, Blanche V. Harris to Whipple, January 23, March 10, 1866, Palmer Litts to Whipple, April 27, 1866, Samuel S. Ashley to Samuel Hunt, November 2, 1865, AMA

A typical nineteenth-century amalgam of benevolent uplift and social control, freedmen's education aimed simultaneously to equip the freedmen to take full advantage of citizenship, and to remake the culture blacks had inherited from slavery, by inculcating qualities of self-reliance and self-discipline. (In this sense, its aims did not differ substantially from common school education in the North.) Few Northerners involved in black education could rise above the conviction that slavery had produced a "degraded" people, in dire need of instruction in frugality, temperance, honesty, and the dignity of labor. Rare indeed was *The Freedmen's Book*, a primer written by abolitionist Lydia Maria Child, that sought to develop a sense of racial pride through brief biographies of black figures from Benjamin Banneker and Frederick Douglass to Toussaint L'Ouverture. In classrooms, alphabet drills and multiplication tables alternated with exhortations to piety, cleanliness, and punctuality. After school hours, teachers and Bureau officials tried to become involved in the freedmen's private lives, organizing temperance societies and delivering lectures on the obligation of husbands to provide for their families.[37]

At their most conservative, teachers sought to inculcate, in the words of Samuel S. Ashley, an AMA education official and later North Carolina's superintendent of education, "obedience to law and respect for the rights and property of others, and reverence for those in authority." With the cooperation of the Bureau and AMA, Gen. Samuel C. Armstrong in 1868 founded Hampton Institute, a citadel of industrial education whose black pupils were advised to eschew political involvement and concentrate on character development and economic self-help. Social control, however, did not exhaust the purposes of freedmen's education. The effort to acculturate former slaves to the workings of the market carried with it a complex and sometimes contradictory set of implications. Reflecting the aims of the Northern common school movement from which many of them sprang, Bureau educators viewed schooling as a solvent of class lines that equipped individuals to advance themselves economically while enabling the society at large to avoid class conflict. Demonstrating the complex of motives that inspired efforts to educate the former slaves, AMA President Rev. Edward N. Kirk excoriated those who sought to set class against class or resort to "mob violence" or strikes, while insisting that no barriers, "legislative, or conventional" should prevent laborers, black or white, from "becoming the equal of the capitalist." The educated

Archives; Robert C. Morris, *Reading, 'Riting, and Reconstruction: The Education of Freedmen in the South 1861–1870* (Chicago, 1981), 58.

37. Joe M. Richardson, *Christian Reconstruction: The American Missionary Association and Southern Blacks, 1861–1890* (Athens, Ga., 1986), 40–44, 242; Carl F. Kaestle, *Pillars of the Republic: Common Schools and American Society, 1780–1860* (New York, 1983), 104–35; Jones, *Soldiers of Light and Love*, 9–22, 49–50, 68–69, 123–26; Lydia Maria Child, *The Freedmen's Book* (Boston, 1865); Small, "Yankee Schoomarm," 383–90; Clinton B. Fisk, *Plain Counsels for Freedmen: In Sixteen Brief Lectures* (Boston, 1866), 26, 32–34; Fisk P. Brewer to Edward P. Smith, April 7, 1868, AMA Archives.

laborer, declared Kirk, "knows what his rights are, and will make no contracts that ignore these rights."[38]

Far from attempting to blunt blacks' ambitions, educators hoped to expand the freedmen's material aspirations, the "wants" that classical economists viewed as the most effective spur to self-directed labor. The Northern preference for a system in which educated laborers worked without coercion to satisfy ever-expanding wants, generating an endless spiral of prosperity for capital and labor alike, was strikingly expressed by Maine-born Bureau agent John N. Bryant:

> Formerly, you were obliged to work or submit to punishment, now you must be induced to work, not compelled to do it. . . . You will be better laborers if educated. . . . That man who has the most wants will usually labor with the greatest industry. . . . The more intelligent men are the more wants they have, hence it is for the interest of all that the laborers shall be educated.

As an added bonus, an educated labor force with expanding "wants" would create in the South a vast new market for Northern goods. It is not, perhaps, coincidental that among the officers of the freedmen's aid societies were railroad men, merchants, and manufacturers.[39]

In the broadest sense, the schools established by the Freedmen's Bureau and Northern aid societies, to quote the American Freedmen's Union Commission, aimed "to plant a genuine republicanism in the southern States." The divorce of schooling from ideals of equal citizenship would come only after the end of Reconstruction when, under the auspices of Armstrong's pupil Booker T. Washington and the South's white "Redeemers," the Hampton philosophy gained ascendancy in Southern black education. Before then, most teachers favored equality before the law and black suffrage, and many promoted black political organization. Even the most paternalist educators viewed their efforts to reshape the character of blacks as a way of equipping them to make decisions for themselves. Perhaps the purpose of most teachers was summarized in an incident related by Nellie Morton, an AMA teacher in Mississippi: "One told me yesterday, 'We've no chance—the white people's arms are longer than ours.' What we want to do is lengthen the colored people's arms till they can reach as far as their old masters." Blacks, moreover, proved perfectly capable of resisting both interference in their personal lives and efforts to "reform" their religious worship. There is no reason to believe that Northern teachers were any more successful in inculcating unwanted values, or that the notions of self-discipline, temperance, and thrift lacked utility and appeal for

38. Morris, *Reading, 'Riting, Reconstruction,* 148–61; Donald Spivey, *Schooling for the New Slavery: Black Industrial Education, 1868–1915* (Westport, Conn., 1978) 9–22; Jones, *Soldiers of Light and Love,* 25–26, 140–42; E. N. Kirk, *Educated Labor* (New York, 1868), 2–7.

39. "Address of Capt. J. E. Bryant to Freedmen's Convention of Georgia, January 13, 1866," manuscript, John E. Bryant Papers, DU; Henry L. Swint, "Northern Interest in the Shoeless Southerner," *JSH,* 16 (November 1950), 471.

a black community seeking to carve out an independent existence.[40]

The Bureau activity most celebrated in the North, education was also the only one intended to leave permanent institutions in the South. The Bureau regarded the other aspects of its work as temporary, needed only until local authorities recognized blacks as equals and the "laws" of the "free market" came to govern economic relations. Even as his agency created an extensive network of courts, hospitals, charitable institutions, and labor regulations, Howard sought to limit their life span, convinced that blacks would benefit more by recognition as equal citizens than from being treated as a special class permanently dependent upon federal assistance and protection. Yet in all these realms, the assumption of its own impermanence severely limited the scope and effectiveness of Bureau policies. Nowhere was this more evident than in the most pressing task confronting beleaguered Bureau agents—preventing the freedmen from being "treated with greater inhumanity and brutality than when in bondage," and assuring them the protection of the law.[41]

While insisting that, as citizens, the former slaves enjoyed the right to bear arms, Freedmen's Bureau officers strongly discountenanced any talk of self-defense or retaliation by blacks against violence. They advised freedmen to rely instead on local and federal protection. But many agents found suppressing violence a difficult and frustrating task. "A freedman is now standing at my door," one agent wrote in 1866, "his tattered clothes bespattered with blood from his head caused by blows inflicted by a white man with a stick and we can do nothing for him. . . . Yet these people flee to us for protection as if we could give it." Gen. E. O. C. Ord, the military commander in Arkansas and Mississippi, believed that without troops, agents of the Bureau were "worse than useless," but very few had military force at their disposal. Despite Southern complaints of bayonet rule, the Union Army rapidly demobilized after the war. Numbering 1 million on May 1, 1865, the roster of men under arms fell to 152,000 by the end of the year, and only 38,000, many of them stationed on the Indian frontier, by the fall of 1866. In most cases, therefore, Bureau agents had to rely upon their own powers of persuasion, and the cooperation of local officials, in attempting to protect the freedmen.[42]

A bewildering array of federal, state, and local authorities confronted

40. *National Freedman,* March 1866; August Meier, *Negro Thought in America 1880–1915* (Ann Arbor, 1963), 89–93; Morris, *Reading, 'Riting, and Reconstruction,* 180–84; John A. Rockwell to Alfred Rockwell, January 28, 1869, Rockwell Papers; Nellie M. Morton to Gustavus D. Pike, April 7, 1871, AMA Archives.

41. Charles L. Price, "John C. Barrett, Freedmen's Bureau Agent in North Carolina," *East Carolina University Papers in History,* 5 (1981), 59.

42. Davis Tillson to H. F. Horne et al., September 11, 1866, Press Copies of Letters Sent, Ser. 625, Ga. Asst. Comr., RG 105, NA [FSSP A-5403]; A. E. Niles to H. W. Smith, May 2, 1866, South Carolina Governor's Papers; E. O. C. Ord to O. O. Howard, January 27, 1867, E. O. C. Ord Papers, Bancroft Library, University of California, Berkeley; *AC,* 1865, 33, 1866, 30–32.

those freedmen who sought redress of grievances. State governments created under Presidential Reconstruction moved quickly to fill law enforcement posts and establish local courts. The army tried under martial law crimes having a military cast, such as depredations by armed bands of whites, and set up provost courts to deal with less serious offenses. The Freedmen's Bureau asserted its own judicial authority, with local agents individually adjudicating a wide variety of complaints and in some areas setting up temporary three-man courts to hear individual contract disputes, with planters, freedmen, and the Bureau each choosing a member. Although neither army nor Bureau courts recognized racial distinctions in legal rights, blacks often received less than equal justice. In South Carolina, where the system of military tribunals became most extensive, blacks consistently received sentences more severe than those meted out to whites convicted of the same crimes. Bureau courts generally treated blacks more equitably. In Georgia and Alabama, however, where Assistant Commissioners Davis Tillson and Wager Swayne authorized local magistrates to act as agents in judicial proceedings, blacks' rights often went unprotected. Tillson and Swayne believed the freedmen could only benefit if prominent whites were enlisted in the cause of equality before the law. By all accounts, however, the experiment failed, for these officials, mostly former slaveholders who had held office under the Confederacy, frequently inflicted severe punishments on freedmen charged with vagrancy, contract violations, or "insolence" (a crime that did not exist for whites).[43]

From Howard himself to the local agents, most Bureau officials remained deeply committed to the idea of equality before the law. But instead of strengthening its own court system, the Bureau devoted its energy to persuading the Southern states to recognize racial equality in their own judicial proceedings. After initial hesitation, especially on the matter of allowing blacks to testify against whites, one state after another revised its judicial proceedings in accordance with Bureau guidelines, precisely to rid themselves of Bureau courts. By the end of 1866, local courts had regained jurisdiction over cases involving freedmen. Disbanding their own courts, Bureau agents now monitored state and local judicial proceedings on behalf of blacks, defending freedmen, prosecuting whites, and retaining the authority to overturn discriminatory decisions. But it quickly became clear that the formal trappings of equality could not guarantee blacks substantive justice. The basic problem, concluded Col. Samuel Thomas, who directed the Bureau in Mississippi in 1865, was that

43. James E. Sefton, *The United States Army and Reconstruction 1865–1877* (Baton Rouge, 1967), 30–32, 44; Thomas D. Morris, "Equality, 'Extraordinary Justice', and Criminal Justice: The South Carolina Experience, 1865–1866," *SCHM*, 83 (January 1982), 16–22; C. Mildred Thompson, *Reconstruction in Georgia* (New York, 1915), 46; 40th Congress, 2d Session, House Executive Document 1, 674; Bethel, "Freedmen's Bureau," 51–54.

white public opinion could not "conceive of the negro having any rights at all":

> Men, who are honorable in their dealings with their white neighbors, will cheat a negro without feeling a single twinge of their honor; to kill a negro they do not deem murder; to debauch a negro woman they do not think fornication; to take property away from a negro they do not deem robbery. . . . They still have the ingrained feeling that the black people at large belong to the whites at large.[44]

In this climate, it is hardly surprising that blacks preferred to bring legal cases to the Freedmen's Bureau and, where possible, to maximize their own influence on its courts. General Howard assumed that the freedmen would select "an intelligent white man who has always seemed to be their friend" to represent them on the three-member tribunals that reviewed contract disputes. He seemed genuinely surprised when they chose black representatives instead. On the grounds that no white Virginian would sit on a court with a black, the Bureau twice overruled Yorktown, Virginia, freedmen who insisted upon selecting Dr. Daniel M. Norton, an escaped slave who had practiced medicine in New York before the war. The freedmen responded that "they were now citizens and could take care of themselves," but agents replaced Norton with a white Virginian.[45]

Despite such conflicts, blacks recognized in the agents and their courts their best hope for impartial justice in the postwar South. They brought to the Bureau problems trivial and substantial, including many involving family and personal disputes among themselves. Most cases, however, arose from black complaints against whites, especially violence, nonpayment of wages, and unfair division of crops. Freedmen appreciated that whatever the outcome of individual cases, the very existence of these tribunals represented a challenge both to notions of local autonomy and to inherited ideas of racial domination. For their part, Southern whites perceived the necessity of answering charges brought by former slaves as an indignity. "He listened to the slightest complaint of the negroes, and dragged prominent white citizens before his court upon the mere accusation of a dissatisfied negro," whites complained about one Mississippi

44. [Oliver O. Howard] *Autobiography of Oliver Otis Howard* (New York, 1907), 2:279; Wager Swayne to John Sherman, December 26, 1865, John Sherman Papers, LC; Donald G. Nieman, "Andrew Johnson, the Freedmen's Bureau, and the Problem of Equal Rights, 1865–1866," *JSH*, 44 (August 1978), 399–420; James Oakes, "A Failure of Vision: The Collapse of the Freedmen's Bureau Courts," *CWH*, 25 (March 1979), 66–76; William F. Mugleston, ed., "The Freedmen's Bureau and Reconstruction in Virginia: The Diary of Marcus Sterling Hopkins, a Union Officer," *VaMHB*, 86 (January 1978), 73; Samuel Thomas to O. O. Howard, September 6, 1865, M-5 1865, Letters Received, Ser. 15, Washington Headquarters, RG 105, NA [FSSP A-9206].

45. Howard, *Autobiography*, 2:252; Robert F. Engs, *Freedom's First Generation: Black Hampton, Virginia, 1861–1890* (Philadelphia, 1979), 89, 104–105, 130–31.

agent. A Georgian considered it outrageous that blacks had "white men arrested and carried to the Freedmen's court . . . where their testimony is taken as equal to a white man's."[46]

A similar combination of limited resources, free labor precepts, a commitment to assisting the freedmen, and the conviction that civil authorities must assume responsibility for the equitable treatment of blacks, shaped another set of Bureau activities—its efforts to alleviate illness and destitution among the black population. In a region suffering from the devastation of war and recurrent crop failures and where the aged and infirm had been evicted from countless plantations, tens of thousands of blacks stood in dire need of medical care and economic assistance. Since local authorities generally refused to appropriate money for black health facilities and many white doctors would not treat blacks unless paid cash, the only medical attention available to blacks was that provided by employers under the terms of labor contracts, or by their own "root doctors" and "conjure men." The remedies of neither proved effective in the face of smallpox, yellow fever, and cholera epidemics that swept the black shantytowns of Vicksburg, New Orleans, Memphis, and other Southern cities, and then made their way into the plantation belt.[47]

The Freedmen's Bureau made heroic but limited efforts to remedy what can only be termed the postemancipation crisis of health among the former slaves. Assuming operation of hospitals established by the army during the war, it expanded the system into areas not previously under military control. Larger cities like Richmond and New Orleans had hospitals with professional staffs; elsewhere the Bureau established dispensaries providing medical care and drugs at a nominal cost or free of charge. Plagued by inadequate funding, a shortage of hospital beds, and a lack of facilities in the rural areas where most blacks lived, the Bureau nonetheless managed in the early years of Reconstruction to treat an estimated half million suffering freedmen, as well as a smaller but significant number of whites. Howard, however, viewed the medical program as provisional and devoted considerable effort to persuading local authorities to shoulder the responsibility for black health care, as well as for the insane, blind, deaf, and dumb. Several Southern cities did establish public, segregated medical facilities and asylums for blacks, and by 1867, the Bureau's medical system, like its courts, had virtually been disman-

46. J. Thomas May, "The Freedmen's Bureau at the Local Level: A Study of a Louisiana Agent," *LaH*, 9 (Winter 1968), 14–15; Barry A. Crouch, "Black Dreams and White Justice," *Prologue*, 6 (Winter 1974), 356–63; Ruth Watkins, "Reconstruction in Marshall County," *PMHS*, 12 (1912), 170; Jane Le Conte to Joseph Le Conte, December 13, 1865, Le Conte Family Papers, Bancroft Library, University of California, Berkeley.

47. Rabinowitz, *Race Relations*, 128–32; James A. Bullock to Dr. H. C. Vogell, May 3, 1869, North Carolina Governor's Papers; D. E. Cadwallader and F. J. Wilson, "Folklore Medicine Among Georgia's Piedmont Negroes After the Civil War," *GaHQ*, 49 (June 1965), 217–18; Ronnie C. Tyler and Lawrence R. Murphy, eds., *The Slave Narratives of Texas* (Austin, 1974), 88–91; Shryock, ed., "Arnold Letters," 131.

tled. To Howard, establishing the principle that public authorities must treat the freedmen as part of the "people" was an essential step in promoting a broad acceptance of black citizenship. But since the Bureau's medical department neither trained black physicians and medical assistants nor undertook a general program of health education, its activities did little to alter the fact that control of medicine remained firmly in white hands.[48]

Howard also believed that local governments must learn to assume the responsibility for indigent blacks but would never do so "while the general government furnishes assistance." Equally important in shaping the Bureau's relief policy was his fear of lending credence to persistent charges that the Bureau sustained able-bodied blacks in idleness. "A man who can work has no right to support by government. No really respectable person wishes to be supported by others." This injunction, issued in 1865 by a Mississippi Bureau agent, epitomized the contemporary Northern view that poor relief undermined manly independence. "A man can scarcely be called free," remarked a Tennessee Bureau official, "who is the recipient of public charity."[49]

Howard himself considered public assistance to the poor "abnormal to our system of government." As the Bureau brought its activities to a close in 1869, he recalled with pride that his had not been a "pauperizing agency," since "so few" had actually been assisted. In many cases, in fact, the Bureau regarded economic assistance as a loan rather than an outright grant. The Bureau's "war on dependency" led it to close with unseemly haste camps for fugitive and homeless blacks that had been established during the war, and to restrict direct assistance to children, the aged, and those unable to work. Economic realities, however, undermined its premature efforts to eliminate relief altogether. In the first fifteen months following the war, the Bureau issued over 13 million rations (a supply of corn meal, flour, and sugar sufficient to feed a person for one week), two thirds to blacks. Then, in the fall of 1866, Howard ordered such relief discontinued, except for hospital patients and inmates of orphan asylums—a draconian measure impossible to implement in the wake of the 1866 crop failure and ensuing destitution.[50]

48. Marshall S. Legan, "Disease and the Freedmen in Mississippi During Reconstruction," *Journal of the History of Medicine and Allied Sciences*, 28 (July 1973), 257–67; Todd L. Savitt, "Politics in Medicine: The Georgia Freedmen's Bureau and the Organization of Health Care," *CWH*, 28 (March 1982), 45–64; Howard A. White, *The Freedmen's Bureau in Louisiana* (Baton Rouge, 1970), 89–100; Gaines M. Foster, "The Limitations of Federal Health Care for Freedmen, 1862–1868," *JSH*, 48 (August 1982), 349–72.

49. 40th Congress, 3d Session, House Executive Document 1, 1058; 39th Congress, 1st Session, House Executive Document 70, 155; Paul D. Phillips, "A History of the Freedmen's Bureau in Tennessee" (unpub. diss, Vanderbilt University, 1964), 125.

50. Howard, *Autobiography*, 2:226; Robert H. Bremner, *The Public Good: Philanthropy and Welfare in the Civil War Era* (New York 1980), 125; Phillips, "Freedmen's Bureau in Tennessee," 67–68; Jerry Thornbery, "Northerners and the Atlanta Freedmen, 1865–69," *Prologue*,

Throughout its existence, the Bureau regarded poor relief as a temptation to idleness. Blacks, declared Virginia Bureau head Orlando Brown, must "feel the spur of necessity, if it be needed to make them self-reliant, industrious and provident." Clearly, this position reflected not only attitudes toward blacks, but a more general Northern belief in the dangers of encouraging dependency among the lower classes. Yet the Bureau's assumption that blacks *wished* to be dependent on the government persisted in the face of evidence that the black community itself, wherever possible, shouldered the task of caring for orphans, the aged, and the destitute, or the fact that in many localities more whites than blacks received Bureau aid. In Mobile, Whitelaw Reid observed, "a stranger might have concluded that it was the white race that was going to prove unable to take care of itself, instead of the emancipated slaves."[51]

Early in 1865, as Congress considered the bill establishing the Bureau, the American Missionary Association warned against empowering the agency to use charity as a weapon against the freedmen: "If the commissioner can withhold rations from those who will not work, then he can compel them to work when and as he will." Yet its relief policy was only one part of a larger contradiction in the Bureau's purposes. The aim of revitalizing the South's production of agricultural staples in many ways undercut that of guaranteeing the freedmen's rights. Nowhere was this inherent ambiguity in its mission more evident than in the Bureau's efforts to supervise the transition from slave to free labor in the postwar South.[52]

The Freedmen's Bureau, Land, and Labor

In the war's immediate aftermath, federal policy regarding black labor was established by the army. And after welcoming Union soldiers as emissaries of liberation, many freedmen quickly became disillusioned with military occupation. For along with the experience of wartime experiments in "free labor," two convictions influenced military authorities: that only the revival of the South's shattered economy could avert starvation and lighten the army's responsibilities, and that former slaves preferred to live on government charity. The course the army adopted seemed to have only one object in view—to compel freedmen to return to work on the plantations.

In the spring and early summer of 1865, military commanders issued

6 (Winter 1974), 240–43; 39th Congress, 2d Session, House Executive Document 1, 713; 40th Congress, 2d Session, House Executive Document 1, 639–40.

51. William T. Alderson, "The Influence of Military Rule and the Freedmen's Bureau on Reconstruction in Virginia, 1865–1870" (unpub. diss., Vanderbilt University, 1952), 38; Reid, *After the War*, 221.

52. *Objections to the Adoption of the Report of the Committee of Conference on Freedmen's Affairs*, printed circular (February, 1865), AMA Archives.

stringent orders aimed at stemming the influx of freedmen into Southern cities. Gen. Godfrey Weitzel, Union commander at Richmond immediately after the Confederate surrender, was deluged by both whites and blacks seeking economic relief. He ordered that before receiving assistance, whites take an oath of loyalty and blacks sign labor contracts—an indication of how he understood the faults of each group. Soon afterward, Department of Virginia commander Gen. E. O. C. Ord barred rural freedmen seeking employment, family members, and protection against violence from entering the city. In early June, soldiers and local Richmond police arrested several hundred blacks and shipped them to the countryside. Elsewhere, military regulations forbade blacks to travel without passes from their employers or be on the streets at night, and prohibited "insubordination" on their part. Col. William Gurney, the federal commander at Charleston, ordered freedmen to leave the city to take up rural labor, an absurd directive, observed the resident British consul, since in large parts of the countryside, blacks were exposed to violence by "owners of plantations . . . filled with revenge."[53]

Although Southern whites generally resented the presence of Union soldiers, many came to regard them as a salutary influence upon the freedmen. The army's "only care," complained Sea Island teacher Laura Towne, was "to make the blacks work." Towne exaggerated, since army officers also issued orders forbidding corporal punishment, disallowing flagrantly discriminatory vagrancy laws, and requiring that blacks be allowed access to education. But overall, military policy reflected little concern for the freedmen's rights. Some Union officers even inflicted upon recalcitrant black laborers brutal penalties such as hanging by one's thumbs, or even punishments used under slavery. In June 1865, Capt. Randolph Stoop, an army provost marshal at Columbia, Virginia, resorted to the "wooden horse" to punish a black couple who had resisted a whipping. Afterward, the two traveled fifty-six miles to complain to Orlando Brown, who had just assumed direction of the Freedmen's Bureau in the state.[54]

The New Orleans *Tribune* indignantly exposed the fallacy behind the

53. John T. O'Brien, "Reconstruction in Richmond: White Restoration and Black Protest, April–June 1865," *VaMHB*, 89 (July 1981), 266–72; Michael B. Chesson, *Richmond After the War 1865–1890* (Richmond, 1981), 73–74; Bernard Cressap, *Appomattox Commander: The Story of General E. O. C. Ord* (San Diego, 1981), 230–34; Bobby L. Lovett, "Memphis Riots: White Reaction to Blacks in Memphis, May 1865–July 1866," *THQ*, 28 (Spring 1979), 14–15; James W. Smallwood, *Time of Hope, Time of Despair: Black Texans During Reconstruction* (Port Washington, N.Y., 1981), 38–39; Thompson, *Reconstruction in Georgia*, 49; H. Pinckney Walker to Earl Russell, April 7, 1865, F. O. 115/443/12–15, Public Record Office, London.

54. Rupert S. Holland, ed., *Letters and Diary of Laura M. Towne* (Cambridge, Mass., 1912), 171; John R. Kirkland, "Federal Troops in the South Atlantic States During Reconstruction, 1865–1877" (unpub. diss., University of North Carolina, 1967), 336–41; Statement of Frederick Nichols and Meiner Poindexter and endorsement by Col. Orlando Brown, June 28, 1865, Robert Brock Collection, HL.

army's regulations: "We were and still are oppressed; we are not demoralized criminals." In several cities, postwar black political organization began with protests against army policies. A group of Memphis free blacks condemned the rounding up of "vagrants" for plantation labor: "It seems the great slave trade is revived again in our city." From Richmond, several "colored men" drafted an eloquent letter to the New York *Tribune* complaining of the "mounted patrol, with their sabers drawn, whose business is the hunting of colored people. . . . All that is needed to restore slavery in full, is the auction block as it used to be." On June 10, 3,000 Richmond blacks gathered at a Baptist church to approve a formal remonstrance. It was delivered to President Johnson by a delegation headed by Fields Cook, a free black barber and local church leader who would take a prominent part in black politics between 1865 and 1867. Partly as a result of the Richmond protest, Ord was replaced by Gen. Alfred Terry, who struck down the city's discriminatory vagrancy laws, announcing that the army would treat all inhabitants as equals before the law. Secretary of War Stanton at the end of July instructed Southern commanders to discontinue pass requirements and forbade the army to hinder in any way blacks' freedom of movement. But the assumption underpinning military policy—that the interests of the South, the nation, and the freedmen themselves would best be served by blacks' return to plantation labor—remained intact as the Freedmen's Bureau assumed command of the transition from slave to free labor.[55]

The idea of free labor, wrote a Tennessee agent, was "the noblest principle on earth." Gen. Robert K. Scott, the Bureau's chief officer in South Carolina, informed Gov. James L. Orr that the state could not hope to escape "the fixed principles which govern [free labor] all over the world." "To the establishment of these principles," he added, "the Bureau is committed." Like Northern Republicans more generally, Bureau officers held what in retrospect appear as amazingly utopian assumptions about the ease with which Southern labor relations could be recast in the free labor mold. Blacks and whites merely had to abandon attitudes toward labor, and toward each other, inherited from slavery, and the market would do the rest. "Let it be understood that a fair day's wages will be paid for a fair day's work," Scott announced, "and the planter will not want for reliable and faithful laborers." With the Bureau acting as midwife at its birth, the free market would quickly assume its role as arbiter of the South's economic destinies, honing those qualities that distinguished free labor from slave—efficiency, productivity, and eco-

55. New Orleans *Tribune*, August 31, 1865; Anthony Motley to Clinton B. Fisk, September 28, 1865, M-84 1865, Registered Letters Received, Ser. 3379, Tenn. Asst. Comr., RG 105, NA [FSSP A-6172]; New York *Tribune*, June 12, 17, 1865; Peter J. Rachleff, *Black Labor in the South: Richmond, Virginia 1865–1890* (Philadelphia, 1984), 14–15, 36; Cressap, *Ord*, 230–34.

nomic rationality—and insuring equitable wages and working conditions. The result would be an enterprising and progressive South whose material rewards both races shared.[56]

In fact, this social vision was to a large extent irrelevant to the social realities the Bureau confronted. The effort to create a free labor South would expose tensions and ambiguities within free labor thought, as well as the free labor ideology's doubtful applicability in a former slave economy organized around the plantation. The ideology rested on a theory of universal economic rationality and the conviction that all classes in a free labor society shared the same interests. In reality, former masters and former slaves inherited from slavery work habits and attitudes at odds with free labor assumptions, and both recognized, more clearly than the Bureau, the irreconcilability of their respective interests and aspirations. The free labor social order, moreover, ostensibly guaranteed the ambitious worker the opportunity for economic mobility, the ability to move from wage labor to independence through the acquisition of productive property. "There is not of necessity," Lincoln had declared at the outset of the Civil War, "any such thing as a free hired laborer being fixed to that condition for life." Yet what became of this axiom in an impoverished society where even the highest agricultural wages remained pitiably low, and whose white population was determined to employ every means at its disposal to prevent blacks from acquiring land or any other means of economic independence?[57]

Establishing themselves in the South in the summer and fall of 1865, Bureau agents hoped to induce Southerners to "give the system a fair and honest trial." To planters' desire for a disciplined labor force governed by the lash, agents responded that *"bodily coercion* fell as an incident of slavery." To the contention that blacks would never work voluntarily or respond to market incentives, they replied that the problem of economic readjustment should be viewed through the prism of labor, rather than race. As agent John E. Bryant explained:

> *No* man loves work naturally. Interest or necessity induces him to labor. . . . Why does the *white man* labor? That he may acquire property and the means of purchasing the comforts and luxuries of life. The *colored man* will labor for the same reason.

Justice to the laborer formed the basis of a free labor economy—this was the unwelcome message the Bureau brought to Southern planters. "To make free labor successful," Bryant declared, "it is necessary that the

56. Phillips, "Freedmen's Bureau in Tennessee," 138; Robert K. Scott to James L. Orr, December 13, 1866, South Carolina Governor's Papers; Robert K. Scott, "Circular to Landlords and Laborers of South Carolina, December 1866," unidentified newspaper clipping, Reconstruction Scrapbook, USC.

57. James D. Richardson, ed., *A Compilation of the Messages and Papers of the Presidents 1789–1897* (Washington, D.C., 1896–99), 6:57–58.

laborer shall be treated fairly. . . . Give to the freedman justice, *impartial justice*, and we believe, he will work better as a freedman than he did as a slave."[58]

The "two evils" against which the Bureau had to contend, an army officer observed in July 1865, were "cruelty on the part of the employer and shirking on the part of the negroes." Yet despite its efforts to instruct planters in free labor principles, the Bureau, like the army, seemed to consider black reluctance to labor the greater threat to its economic mission. Tennessee Bureau head Gen. Clinton B. Fisk published *Plain Counsels for Freedmen*, a collection of sixteen lectures that sketched a rather spartan portrait of freedom in which blacks eschewed "fine cigars" and "useless dress and ornaments," and lived by the maxim "time is money." In some areas, agents continued the military's urban pass systems and vagrancy patrols, as well as the practice of rounding up unemployed laborers for shipment to plantations. When conservative Joseph S. Fullerton assumed command of the Louisiana Bureau in October 1865, he closed the black orphan asylum, apprenticed the inmates to white masters, and ordered the arrest as "idlers and vagrants" of all New Orleans blacks who lacked written evidence of employment. In Florida, Bureau head Thomas W. Osborn ordered agents to "take every precaution in their power to prevent freedmen from collecting about towns, military posts, railroad depots, or in isolated communities, with an apparent intention to escape labor on the plantations." And in Memphis, blacks were regularly rounded up in the fall of 1865 to meet the labor needs of the surrounding countryside. Freedmen who had come to the city, believed Gen. Davis Tillson, were "lazy, worthless vagrants," and his patrols accosted children with schoolbooks in their hands, informing them they should be out picking cotton. Meanwhile Bureau courts dispatched impoverished blacks convicted of crimes to labor for whites who would pay their fines. "What a mockery to call those 'Freedmen' who are still subjected to such things," commented a local minister.[59]

United as to the glories of free labor, Bureau officials, like Northerners generally, differed among themselves about the ultimate social implications of the free labor ideology. Some believed the freedmen would remain a permanent plantation labor force; others insisted they should

58. J. L. Haynes to B. F. Morey, July 8, 1865, H-17 1865, Registered Letters Received, Ser. 2052, Ms. Asst. Comr., RG 105, NA [FSSP A-9044]; 39th Congress, 1st Session, Senate Executive Document 27, 28; Trowbridge, *The South*, 369; Augusta *Loyal Georgian*, January 20, 1866.

59. J. L. Haynes to B. F. Morey, July 8, 1865, H-17 1865, Registered Letters Received, Ser. 2052, Ms. Asst. Comr., RG 105, NA [FSSP A-9044]; Fisk, *Plain Counsels*, 10–23; Litwack, *Been in the Storm*, 379–82; New Orleans *Tribune*, October 28, 1865; J. S. Fullerton, *Report of the Administration of Freedmen's Affairs in Louisiana* (Washington, D.C., 1865), 3–6; 39th Congress, 1st Session, House Executive Document 70, 80; Lovett, "Memphis Riots," 15–17; T. E. Bliss to N. A. M. Dudley, November 3, 1865, C-93 1865, Registered Letters Received, Ser. 3379, Tenn. Asst. Comr., RG 105, NA [FSSP A-6100].

enjoy the same opportunity to make their way up the social ladder to independent proprietorship as Northern workers; still others hoped the federal government would assist at least some blacks in acquiring their own farms. At their most conservative, agents appeared to share the planters' limited definition of the ex-slaves' freedom. Charles Soule, wartime officer of a black regiment, supervised contracts for the Bureau at Orangeburg, South Carolina. In June 1865, he informed local freedmen:

> There must be a head man everywhere, and on a plantation the head man who gives all the orders, is the owner of the place. Whatever he tells you to do you must do at once, and cheerfully. . . . There are different kinds of work. . . . Every man has his own place . . . and he must stick to it. . . . Some people must be rich, to pay the others, and they have the right to do no work except to look out after their property.

Soule, however, hardly typified the views of Bureau officials. Indeed, Commissioner Howard reminded him that agents must avoid lending their authority to the planters' desire for "absolute power" over their labor force.[60] Like Soule, Howard believed most freedmen must return to plantation labor, but under conditions that allowed them the opportunity to work their way out of the wage-earning class. At the same time, he took seriously the provision in the act establishing his agency that authorized it to settle freedmen on confiscated and abandoned lands. In 1865, Howard and a group of sympathetic Bureau officials attempted to breathe life into this alternative vision of a free labor South.

Even though the Lincoln Administration had left the 1862 Confiscation Act virtually unenforced, the Bureau controlled over 850,000 acres of abandoned land in 1865, hardly enough to accommodate all the former slaves but sufficient to make a start toward creating a black yeomanry. Howard's subordinates included men sincerely committed to settling freedmen on farms of their own and protecting the rights of those (mostly on the "Sherman reservation") who already occupied land. In Tennessee, General Fisk began locating blacks on the 65,000 acres under his control. In Louisiana, Thomas Conway invited applications from freedmen who wished to "procure land for their own use," and leased over 60,000 acres to blacks (including a plantation owned by the son of former President Zachary Taylor). Orlando Brown, an advocate of *"extensive confiscation,"* urged Howard to "take possession of all the abandoned and confiscated land we require, and permit the negroes to work it on their own behalf." Most dedicated of all to the idea of black landownership was Gen. Rufus Saxton, a prewar abolitionist who directed the Bureau in South Carolina, Georgia, and Florida during the summer of 1865. Saxton had already overseen the settlement of thousands of blacks on lands

60. Ira Berlin et al., eds., "The Terrain of Freedom: The Struggle over the Meaning of Free Labor in the U. S. South," *History Workshop*, 22 (Autumn 1986), 117–23, 130n.

reserved for them under General Sherman's Field Order 15. In June 1865, he announced his intention to use the property under Bureau control to provide freedmen with forty-acre homesteads "where by faithful industry they can readily achieve an independence." Market-oriented farming was Saxton's ideal. "Put in all the cotton and rice you can," he advised black farmers, "for these are the crops which will pay the best. . . . Let the world see ere long the fields of South Carolina, Georgia and Florida white with [cotton]." In this way, he argued, blacks tilling their own lands could demonstrate the superiority of free labor to slave.[61]

Initially, Howard himself shared the radical aims of Conway, Brown, and Saxton. "He says he will give the freedmen *protection, land and schools,* as far and as fast as he can," a friend reported in March 1865. But regarding land, this policy was not to be. For reasons that will be related below, President Johnson during the summer and fall issued a rash of special pardons, restoring the property of former Confederates. Johnson's actions threw into question the status of confiscated and abandoned land, including the Sherman reservation. At the end of July, without consulting the President, Howard issued Circular 13, which instructed Bureau agents to "set aside" forty-acre tracts for the freedmen as rapidly as possible. Presidential pardons, he insisted, did not carry with them the restoration of land that had been settled by freedmen in accordance with the law establishing the Bureau. Johnson, however, soon directed Howard to rescind his order. A new policy, drafted in the White House and issued in September as Howard's Circular 15, ordered the restoration to pardoned owners of all land except the small amount that had already been sold under a court decree. Soon thereafter, the government suspended land sales scheduled in Virginia and South Carolina. Once growing crops had been harvested, virtually all the land in Bureau hands would revert to its former owners.[62]

Most of the land occupied by blacks in the summer and fall of 1865 lay within the Sherman reservation, where 40,000 freedmen had been settled. "Could a just Government," Saxton asked, "drive out these loyal men?" To Howard fell the task of informing the freedmen that the land would be restored to their former owners, and that they must either agree

61. Foner, *Politics and Ideology,* 131–32; Weymouth T. Jordan, "The Freedmen's Bureau in Tennessee," *ETHSP,* 11 (1939), 54–55; Claude F. Oubre, *Forty Acres and a Mule: The Freedmen's Bureau and Black Landownership* (Baton Rouge, 1978), 24–25, 33–34; C. Peter Ripley, *Slaves and Freedmen in Civil War Louisiana* (Baton Rouge, 1976), 84, Orlando Brown to O. O. Howard, June 6, 1865, B-190 1865, Orlando Brown to O. O. Howard, July 6, 1865, B-197 1865, Letters Received, Ser. 15, Washington Headquarters, RG 105, NA [FSSP A-7465]; Reid, *After the War,* 92, 117, 269; Abbott, *Freedmen's Bureau,* 54; 39th Congress, 1st Session, House Executive Document 70, 92.

62. McFeely, *Yankee Stepfather,* 16, 97–98, 103–105, 126–34; Howard, *Autobiography,* 2:234–36; 39th Congress, 1st Session, House Executive Document 70, 111–15; Hope S. Chamberlain, *Old Days in Chapel Hill* (Chapel Hill, 1926), 115. For the evolution of Johnson's policies, see Chapter 5.

to work for the planters or be evicted. In October, he traveled to low-country South Carolina, hoping to "ease the shock as much as possible, of depriving the freedmen of the ownership of the lands." On Edisto Island occurred one of the most poignant confrontations of the era. Blacks here had been holding weekly meetings, where issues of "general interest" were discussed and Republican newspapers read aloud. They fully anticipated Howard's message, and when he rose to speak to more than 2,000 freedmen gathered at a local church, "dissatisfaction and sorrow were manifested from every part of the assembly." Finally, a "sweet-voiced negro woman" quieted the crowd by leading it in singing spirituals: "Nobody Knows the Trouble I Seen" and "Wandering in the Wilderness of Sorrow and Gloom." When the freedmen fell silent, Howard begged them to "lay aside their bitter feelings, and to become reconciled to their old masters." He was continually interrupted by members of the audience: "No, never," "Can't do it," "Why, General Howard, do you take away our lands?"[63]

Howard requested the assembled freedmen to appoint a three-man committee to consider the fairest way of restoring the planters to ownership. The committee's eloquent response did not augur well for a tranquil settlement:

> General, we want Homesteads, we were promised Homesteads by the government. If it does not carry out the promises its agents made to us, if the government haveing concluded to befriend its late enemies and to neglect to observe the principles of common faith between its self and us its allies in the war you said was over, now takes away from them all right to the soil they stand upon save such as they can get by again working for *your* late and their *all time* enemies . . . we are left in a more unpleasant condition than our former. . . . You will see this is not the condition of really freemen.
>
> You ask us to forgive the land owners of our island. *You* only lost your right arm in war and might forgive them. The man who tied me to a tree and gave me 39 lashes and who stripped and flogged my mother and my sister and who will not let me stay in his empty hut except I will do his planting and be satisfied with his price and who combines with others to keep away land from me well knowing I would not have anything to do with him if I had land of my own—that man, I cannot well forgive. Does it look as if he has forgiven me, seeing how he tries to keep me in a condition of helplessness?

"The condition of really free men," "I cannot well forgive," "their *all time* enemies," "the war you said was over." In these words, the committee expressed with simple dignity convictions that freedmen throughout the South had come to share—land was the foundation of freedom, the

63. Abbott, *Freedmen's Bureau*, 55; S. Willard Saxton Journal, October 22, 1865, Rufus and S. Willard Saxton Papers, Yale University; Charleston *South Carolina Leader*, December 9, 1865; Henry E. Tremain, *Two Days of War: A Gettysburg Narrative and Other Excursions* (New York, 1905), 270; Mary Ames, *From a New England Woman's Diary in Dixie in 1865* (Springfield, Mass., 1906), 96–97; Howard, *Autobiography*, 2:237–39.

evils of slavery could not be quickly forgotten, the interests of former master and former slave were fundamentally irreconcilable. The freedmen pleaded with Howard for the right to rent or purchase land, and displayed "the greatest aversion" to signing labor contracts with the landowners, equating such agreements, a Union officer noted, with "a practical return to slavery."[64]

Genuinely moved by this experience, Howard pledged to renew the fight for land when Congress reconvened in December. Meanwhile, planters complained of the hostile attitude of General Saxton. So long as Saxton headed the Bureau in South Carolina, William H. Trescot informed the President, the land would never be restored. Early in 1866, Johnson ordered Saxton removed. On his departure, thousands of South Carolina freedmen contributed "pennies and five cent pieces" to present "two or three handsome pieces of silver" to their staunchest friend in the Freedmen's Bureau.[65]

As Trescot noted, Johnson's actions, especially Circular 15, "changed radically the character of the Bureau." Gone was the idea of settling any considerable portion of the freedmen on land, and gone too, by 1866, were officials like Saxton and Conway, most committed to such a policy. In subsequent years, individual agents would encourage blacks to rent or purchase land, sometimes advancing rations—as a loan against the future crop—to independent black farmers. But in November and December 1865, the Bureau had no choice but to implement the President's orders. Agents made a major effort to persuade freedmen throughout the South to sign labor contracts for the coming year, and to disabuse them of the idea that the government intended to divide land among them. Even men like Col. Samuel Thomas, the assistant commissioner in Mississippi until the spring of 1866, who believed blacks should be working their own lands, threatened those freedmen who refused to sign contracts with arrest. By mid-1866, half the land in Bureau hands had been restored to its former owners, and more was returned in subsequent years. Although in some areas, the process of restoration was delayed by court challenges into the 1870s, in the end the amount of land that came into the possession of blacks proved to be minuscule. Johnson had in effect abrogated the Confiscation Act and unilaterally amended the law creating the Bureau. The idea of a Freedmen's Bureau actively promoting black landownership had come to an abrupt end.[66]

64. Howard, *Autobiography*, 2:239; Andrews, *South Since the War*, 1866, 212; Berlin et al., eds., "Terrain of Freedom," 127–28; Tremain, *Two Days of War*, 267–76. Howard had lost an arm at the Civil War battle of Fair Oaks.

65. McFeely, *Yankee Stepfather*, 139–43, 196, 226–28; William H. Trescot to Andrew Johnson, December 1, 1865 (copy), Saxton Papers; Francis L. Cardozo to George Whipple, January 27, 1866, AMA Archives.

66. Columbia *Daily Phoenix*, January 10, 1866; 40th Congress, 3d Session, House Executive Document 1, 1041–42; Oubre, *Forty Acres*, 159; Circular Letter, Freedmen's Bureau,

The restoration of land required the displacement of tens of thousands of freedmen throughout the South. The army evicted most of the 20,000 blacks settled on confiscated and abandoned property in southeastern Virginia, as well as freedmen near Wilmington, North Carolina, cultivating land assigned them by Gen. Joseph Hawley. The 62,000 acres farmed by Louisiana blacks were restored to their former owners; as the wife of a New Orleans editor observed, Gen. Joseph S. Fullerton, who succeeded Conway, "can't seem to hustle out fast enough the occupants of confiscated property." Even Davis Bend, Mississippi, where black farmers cultivated 10,000 acres in 1865, reverted to Joseph Davis. Unlike most planters, however, Davis thereupon sold the land on long-term credit to Benjamin Montgomery, the most prominent leader of the Bend's black community. Davis died in 1870, leaving a will instructing his heirs to "extend a liberal indulgence" with respect to Montgomery's payments. But in 1878, after the $300,000 purchase price had fallen due, a Mississippi court ordered the land restored to the Davis family—a melancholy end to the dream of establishing a "Negro paradise" at Davis Bend.[67]

Nowhere, however, was the restoration process so disruptive as in the Georgia and South Carolina lowcountry. Even after General Howard's visit in the fall of 1865, blacks had no intention of abandoning their claim to the land. On more than one occasion, freedmen armed themselves, barricaded plantations, and drove off owners attempting to dispossess them. Black squatters told one party of Edisto Island landlords in February 1866, "you have better go back to Charleston, and go to work there, and if you can do nothing else, you can pick oysters and earn your living as the loyal people have done—by the sweat of their brows." Bureau agents, black and white, bent every effort to induce lowcountry freedmen to sign contracts with their former owners, while federal troops forcibly evicted those who refused.[68]

The confrontation on Jehossee Island, South Carolina, where former governor William Aiken owned the South's largest rice plantation, typified scenes throughout the lowcountry. In late December 1865, accompanied by a number of soldiers, Aiken returned to Jehossee, only to discover that his former slaves, some of whom had received forty-acre land grants from General Saxton, "utterly refused" to work for him. On

November 11, 1865, J. B. DeBow Papers, DU; 39th Congress, 1st Session, Senate Executive Document 27, 36; Oubre, *Forty Acres*, 37; Foner, *Politics and Ideology*, 138.

67. Alderson, "Military Rule and Freedmen's Bureau," 51–52; Evans, *Ballots and Fence Rails*, 54–59, 68–70; William F. Messner, *Freedmen and the Ideology of Free Labor: Louisiana 1862–1865* (Lafayette, La., 1978), 185; Dolly Dill to Eleanor Le Conte, November 10, 1865, Le Conte Family Papers; Janet S. Hermann, *The Pursuit of a Dream* (New York, 1981), 61–218.

68. 39th Congress, 2d Session, House Executive Document 1, 708; Myrta L. Avery, *Dixie After the War* (New York, 1906), 344–45; Litwack, *Been in the Storm*, 406–407; Dorothy Sterling, ed., *The Trouble They Seen* (Garden City, N.Y. 1976), 39–40; Trowbridge, *The South*, 539.

January 5, Aiken and Maj. James P. Roy returned to the island and the former governor—a paternalist who before the war had constructed two hospitals and a "very neat commodious church" for his slaves—offered a contract that he correctly described as more generous that those proposed by other planters: half the crop of rice, cotton, and provisions, support for a school, and "a comfortable home and a piece of land" for freedmen too aged or infirm to work. Major Roy attempted to persuade the blacks to agree but "found it uphill work for I could not dispossess them of the idea that they would eventually own the land." In the end, the freedmen unanimously rejected Aiken's offer; no matter how generous the terms, they refused to work as hired laborers on land they claimed as their own. "Several remarked," Major Roy reported, "that they knew Congress was in session and would provide for them."[69]

That freedmen on an isolated South Carolina Sea Island were following the course of Congressional deliberations illustrates the extent of blacks' politicization in the early years of Reconstruction. Congress, however, did not fulfill their expectations. The question of land restoration became embroiled in the broader national debate over Reconstruction that will be examined below. It is sufficient here to note that President Johnson, working in close consultation with William H. Trescot, who represented South Carolina Governor Orr in Washington, successfully blocked efforts to validate the Sherman land titles. In February 1866, Johnson's veto killed a bill extending the life of the Bureau that would have granted three-year "possessory titles" to freedmen on the Sherman land and authorized Howard to provide forty acres to those dispossessed. When Congress, in July, finally enacted a Freedmen's Bureau law (over another veto), the blacks holding land warrants from General Sherman received only the right to lease or purchase twenty-acre plots on government land.[70]

Even this did not end resistance among lowcountry blacks. At Delta plantation on the South Carolina side of the Savannah River, armed blacks early in 1867 refused to sign contracts to work as wage laborers, leave the premises, or exchange their Sherman warrants for government land as Congress had provided. "They would die where they stood before they would surrender their claims to the land," they told a reporter. Among their leaders was Aaron A. Bradley, a South Carolina-born slave who had escaped during the 1830s, studied law in Boston, and in 1865 returned to the lowcountry, where he urged freedmen to resist orders to vacate restored land. After Bradley's arrest, troops evicted those unwill-

69. James M. Clifton, "Jehossee Island: The Antebellum South's Largest Rice Plantation," *AgH*, 69 (January 1985), 59–65; *Liberator*, December 29, 1865; James P. Roy to W. L. M. Burger, February 1, 1866, Letters Received (unentered), Ser. 4109, Dept. of S. C., RG 393, pt. 1, NA [FSSP C-1385].
70. William H. Trescot to James L. Orr, February 4, 6, March 31, 1866, South Carolina Governor's Papers; McFeely, *Yankee Stepfather*, 205–37; Oubre, *Forty Acres*, 67–69.

ing to sign labor contracts. In the end, only about 2,000 South Carolina and Georgia freedmen actually received the land they had been promised in 1865.[71]

The events of 1865 and 1866 kindled a deep sense of betrayal among freedmen throughout the South. Land enough existed, wrote former Mississippi slave Merrimon Howard, for every "man and woman to have as much as they could work." Yet blacks had been left with

> "no *land*, no *house*, not so much as place to lay our head. . . . Despised by the world, hated by the country that gives us birth, denied of all our writs as a people, we were friends on the march, . . . brothers on the battlefield, but in the peaceful pursuits of life it seems that we are strangers."

Long after the end of slavery, the memory of this injustice lingered. "De slaves," a Mississippi black would recall, "spected a heap from freedom dey didn't git. . . . Dey promised us a mule an' forty acres o' lan'." "Yes sir," agreed a Tennessee freedman, "they should have give us part of Maser's land as us poor old slaves we made what our Masers had."[72]

Thus, by 1866 the Bureau's definition of "free labor" had been significantly transformed. Instead of carrying out a two-pronged labor policy, in which some blacks farmed independently, while others worked as hired laborers for white employers, the Bureau found itself with no alternative but to encourage virtually all freedmen to sign annual contracts to work on the plantations. Its hopes for long-term black advancement and Southern economic prosperity now came to focus exclusively upon the labor contract itself.

"The laws of contract," a Northern Republican wrote from New Orleans in 1874, "are the foundation of civilization." The contract ostensibly reconciled free choice and social order, and epitomized the principle that legitimate systems of authority must rest upon consent rather than coercion. By voluntarily signing and adhering to contracts, both planters and freedmen would develop the habits of a free labor economy, and come to understand their fundamental harmony of interests. In retrospect, the burdens placed upon the labor contract appear hopelessly unrealistic, partly because in nearly all societies the idea of free choice central to contract ideology masks an unequal distribution of economic power, and partly because of the enormous obstacles to placing a workable contract system into operation in the postwar South. When the Bureau assumed control of labor relations in 1865, large numbers of

71. Savannah *Weekly Republican*, January 26, 1867; Savannah *Daily News and Herald*, January 21, 23, 27, February 1, 1867; Joseph P. Reidy, "Aaron A. Bradley: Voice of Black Labor in the Georgia Lowcountry," in Howard N. Rabinowitz, ed., *Southern Black Leaders of the Reconstruction Era* (Urbana, Ill., 1982), 281–308; Oubre, *Forty Acres*, 69–70; 40th Congress, 2d Session, House Executive Document 1, 672; Howard C. Westwood, "Sherman Marched—and Proclaimed 'Land for the Landless'," *SCHM*, 85 (January 1984), 49–50.

72. Merrimon Howard to O. O. Howard, April 7, 1866, H-104 1866, Letters Received, Ser. 15, Washington Headquarters, RG 105, NA [FSSP A-9113]; Rawick, ed., *American Slave*, 7, pt. 2:147, Supplement 2, 3:877.

blacks were already at work under highly disadvantageous agreements approved by army officers, with wages, apart from food or shelter, extremely low or nonexistent, and planters allowed to regulate their employees' personal lives. Then, at the beginning of 1866, agents found themselves required to perform a nearly impossible balancing act. Spending much of their time disabusing blacks of the idea that they would soon obtain land from the government, and threatening to arrest those who refused to sign a contract or leave the plantations, agents simultaneously insisted upon blacks' right to bargain freely for employment and attempted to secure more advantageous contracts than had prevailed in 1865. Bureau officials also moved to stabilize the labor market by transporting former slaves (generally at the freedmen's own expense) to parts of the South where wages were the highest.[73]

The extent to which the Bureau intervened to alter the provisions of contracts varied from state to state and from agent to agent. The law did not require that all contracts be submitted to the Bureau for approval; in some areas, few received an agent's attention, in others, agents did all they could to encourage planters to submit contracts for review. Some Bureau officers approved agreements in which the laborer would receive nothing at all if the crop failed and could incur fines for such vaguely defined offenses as failure to do satisfactory work or "impudent, profane or indecent language." More conscientious agents revoked contract provisions regulating blacks' day-to-day lives, and insisted that laborers who left plantations before the harvest must be paid for their work up to the date of departure. One Bureau official lectured a North Carolina planter who desired to bar blacks from leaving the plantation without his permission: "Contracts of this nature when the landowner undertakes to control the personal liberty of the laborers, are utterly foreign to free institutions." And virtually all agents insisted that planters acknowledge that their power to employ physical coercion had come to an end. Convinced that "free men should find their worth in the free competition of the marketplace," Howard at first instructed agents not to fix wages. But in Georgia, with a labor surplus and widespread rural violence, the "free market" in 1865 produced contracts with wage rates so low—one tenth or one twentieth of the crop or two to four dollars per month—that Bureau chief Davis Tillson was moved to establish a minimum wage.[74]

73. George W. Welch to Benjamin F. Butler, May 19, 1874, Benjamin F. Butler Papers, LC; Mahaffey, ed., "Schurz's Letters," 231; Richardson, *Negro in Florida*, 58–60; George D. Humphrey, "The Failure of the Mississippi Freedmen's Bureau in Black Labor Relations, 1865–1867," *JMH*, 45 (February 1983), 27–28; White, *Freedmen's Bureau*, 122; Robert K. Scott to James L. Orr, January 13, 1866, South Carolina Governor's Papers; Phillips, "Freedmen's Bureau in Tennessee," 113–18; William Cohen, "Black Immobility and Free Labor: The Freedmen's Bureau and the Relocation of Black Labor, 1865–1868," *CWH*, 30 (September 1984), 222–27.

74. McFeely, *Yankee Stepfather*, 151; Columbia *Daily Phoenix*, July 15, 1865; Edmund L. Drago, *Black Politicians and Reconstruction in Georgia* (Baton Rouge, 1983), 113–14; Paul A.

Then there was the thorny problem of ensuring that freedmen actually received the payment. In theory, wages due under a labor contract would be paid promptly and honestly, or else the courts would enforce the laborers' claims. In fact, poverty made it impossible for blacks either to hire a lawyer or to wait the twelve to eighteen months that might elapse before a court ruling, and at any rate, local juries could not be relied upon to deal fairly with the freedmen. To remedy this situation, the Bureau recognized claims for wages as a first lien on the crop, and in many instances agents impounded the harvest, preventing its sale by the planter or its seizure by his creditors until laborers received their share. "I now hold nearly one hundred bags of cotton in the warehouses of this city," a Georgia agent reported in 1868, "which I have seized for the payment of the hands that produced it."[75]

The Bureau's role in supervising labor relations reached its peak in 1866 and 1867; thereafter, federal authorities intervened less and less frequently to oversee contracts or settle plantation disputes. To the extent that the contract system had been intended to promote stability in labor relations in the chaotic aftermath of war and allow commercial agriculture to resume, it could be deemed a success. But in other ways, the system failed. Howard and many of his subordinates assumed that working for wages afforded the Northern laborer the opportunity to accumulate savings and move into the ranks of independent artisans and farmers. The same, they sincerely believed, would be true for black agricultural workers in the South. But even wages considered high by planters proved barely enough for subsistence, let alone accumulation. "With labor at fifteen dollars a month. . . ." one Bureau agent observed in 1866, "it is one endless struggle to beat back poverty."[76]

Even more fundamentally, the entire contract system in some ways violated the principles of free labor. Agreements, Howard announced soon after assuming office, "should be free, *bona fide* acts." Yet how "voluntary" were labor contracts agreed to by blacks when they were denied access to land, coerced by troops and Bureau agents if they refused to sign, and fined or imprisoned if they struck for higher wages? Propertyless individuals in the North, to be sure, were compelled to labor for wages, but the compulsion was supplied by necessity, not by public

Cimbala, "The 'Talisman Power': Davis Tillson, The Freedmen's Bureau, and Free Labor in Reconstruction Georgia, 1865–1866," *CWH*, 28 (June 1982), 160.

75. LaWanda Cox and John H. Cox, eds., *Reconstruction, the Negro, and the New South* (Columbia, S.C., 1973), 348–49; Bethel, "Freedmen's Bureau," 55–56, 76; Thornbery, "Northerners and the Atlanta Freedmen," 244; Jaynes, *Branches Without Roots*, 141–57; 40th Congress, 2d Session, House Executive Document 1, 675; *CG*, 40th Congress, 2d Session, 181.

76. Davis, *Good and Faithful Labor*, 76–77; May, "Freedmen's Bureau at the Local Level," 5–20; John E. Stealey III, ed., "Reports of Freedmen's Bureau Operations in West Virginia: Agents in the Eastern Panhandle," *WVaH*, 42 (Fall 1980–Winter 1981), 105.

officials, and contracts did not prevent them from leaving work whenever they chose. Why, asked the New Orleans *Tribune* again and again, did the Bureau require blacks to sign year-long labor contracts when "laborers throughout the civilized world"—including agricultural laborers in the North—could leave their employment at any time? The contract system, charged the *Tribune*, appeared to be based on the idea "that agricultural pursuits need the binding of the laborer to the plantation." To which one may add that even the most sympathetic Bureau officials assumed that blacks would constitute the rural labor force, at least until the natural workings of the market divided the great plantations into small farms. "There are today as many houseless, homeless, poor wandering, idle white men in the South, as there are Negroes," noted Mississippi Bureau head Samuel Thomas in 1865. But "idle white men" were never required to sign labor contracts or ordered to leave Southern cities for the country-side—a fact that made a mockery of the Bureau's professed goal of equal treatment for the freedmen.[77]

Howard always believed that the Bureau's policies, viewed as a whole, benefited the freedmen more than their employers, especially since civil authorities offered blacks no protection against violence or fraud and the courts provided no justice to those who sought legal redress. He viewed the system of annual labor contracts as a temporary expedient, which would disappear once free labor obtained a "permanent foothold" in the South "under its necessary protection of equal and just laws properly executed." Eventually, as in the North, the market would regulate em-ployment. "This," contended Wager Swayne, the Bureau's assistant com-missioner in Alabama, "is more and better than all laws." Yet in the early years of Reconstruction, operating within the constraints of the free labor ideology, adverse crop and market conditions, the desire to restore pro-duction of the South's staple crops, and presidential policy, Bureau deci-sions conceived as temporary exerted a powerful influence on the emer-gence of new economic and social relations, closing off some options for blacks, shifting the balance of power in favor of employers, and helping to stabilize the beleaguered planter class.[78]

Partly because of variations in policy among assistant commissioners and local agents, contemporaries differed widely in assessing the Bureau. Planters hastened to take advantage of those aspects of Bureau policy aimed at compelling blacks to work—the denial of rations, the insistence that they sign labor contracts—and, where they deemed local agents sympathetic, called upon them to settle plantation disputes. When

77. Howard, *Autobiography*, 2:222–23; Harold D. Woodman, "Post-Civil War Southern Agriculture and the Law," *AgH*, 53 (January 1979), 319–22; Smallwood, *Time of Hope, Time of Despair*, 53; New Orleans *Tribune*, November 19, December 17, 1865; Jaynes, *Branches Without Roots*, 71–74.

78. Howard, *Autobiography*, 2:221, 247, 307; 39th Congress, 1st Session, Senate Executive Document 27, 60; Nieman, *To Set the Law in Motion*, 53–62, 160–70.

Charles Rauschenberg took up his job as agent at Cuthbert, Georgia, in 1867, he found that his predecessor had been regarded by local whites as "a substitute for overseers and drivers." Taking advantage of the Bureau, however, did not necessarily mean desiring that its presence in the South continue. "Of course everyone abuses the Freedmen's Bureau," the British ambassador reported after a visit to Virginia in early 1866, precisely when agents were exerting their greatest effort to induce blacks to sign labor contracts. Indeed, whatever the policies of individual agents, most Southern whites resented the Bureau as a symbol of Confederate defeat and a barrier to the authority reminiscent of slavery that planters hoped to impose upon the freedmen. Even if, in individual cases, the Bureau's intervention enhanced the power of the employer, the very act of calling upon a third and ostensibly disinterested party served to undermine his standing, by making evident to the freedmen that the planter's authority was not absolute. John F. Couts, member of a prominent Middle Tennessee planter family, confirmed that for many whites the Bureau's presence was a humiliation:

> The Agent of the Bureau . . . *requires* citizens (former owners) to make and enter into *written contracts* for the hire of their *own* negroes. . . . When a negro is not *properly* paid or fairly dealt with and *reports* the facts, then a squad of Negro soldiers is sent after the *offender,* who is *escorted* to town to be dealt with as per the negro testimony. In the name of God how long is such things to last?

Again and again the Bureau's critics repeated this theme—that by treating blacks with a semblance of equality, the Bureau "demoralized" plantation labor. "The anxiety of the people to get rid of the United States troops and the Freedmen's Bureau," wrote an AMA missionary in Mississippi, "ought to open the eyes of President Johnson to the necessity of keeping them here."[79]

For their part, blacks sought to avail themselves of the assistance the Bureau offered, while limiting the damage done by policies and agents more attuned to the planters' interests than their own. On occasion, freedmen requested the removal of hostile officials, recommended the appointment of others known to be sympathetic to their aspirations, and sometimes demanded the right to select Bureau agents themselves. The New Orleans *Tribune* persistently castigated Louisiana agents as "the planter's guards, and nothing else." Yet for the same reasons that Southern whites demanded its removal, most blacks remained "doggedly

79. Cox and Cox, eds., *Reconstruction*, 339–40; Wilbur D. Jones, ed., "A British Report on Postwar Virginia," *VaMHB*, 69 (July 1961), 351; Roark, *Masters Without Slaves*, 154; John F. Couts to Cave Johnson Couts, January 12, 1866, Cave Johnson Couts Papers, HL; Audley Couper to Francis P. Corbin, September 18, 1866, Corbin Papers; John P. Bardwell to Michael E. Strieby, November 20, 1865, AMA Archives.

loyal" to the Bureau. "Many Negroes will tell you, 'they are as free as you are'," John F. Couts reported, "and if you fool with me I will order you to the 'Bureau'."[80]

To the very end of Reconstruction, blacks would insist that "those who freed them shall protect that freedom." The strength of their commitment to this principle, and to the Bureau as an embodiment of the nation's responsibility, became clear in 1866 when President Johnson sent generals John Steedman and Joseph S. Fullerton on an inspection tour of the South. Johnson hoped to elicit enough complaints to discredit the agency, but in city after city, blacks rallied to the Bureau's support. At Norfolk, a delegation, including three Civil War veterans, informed the generals that it would be folly to remove the Bureau "while we are not recognized as men by the whites of the South." In Wilmington, North Carolina, 800 blacks crowded into the Brick Church to voice support. "If the Freedman Bureau was removed," one speaker insisted, "a colored man would have better sense than to speak a word in behalf of the colored man's rights, for fear of his life." Somewhat taken aback, General Steedman asked the assemblage if the army or the Freedmen's Bureau had to be withdrawn, which they would prefer to have remain in the South. From all parts of the church came the reply, "The Bureau."[81]

Col. W. H. H. Beadle, a native of Michigan and Bureau agent at Wilmington, most likely attended the meeting in Brick Church. If so, he may have learned a lesson in the realities of Southern life. Scarcely a month earlier, in testifying before the Joint Congressional Committee on Reconstruction, Beadle had lamented the fact that Southerners did not comprehend a crucial axiom of the free labor ideology—"the mutual dependence of labor and capital." Even as planters sought to limit the freedom of the former slaves in every way possible, and blacks understood that they lived "in the midst of enemies to our race," Bureau agents held fast to the conviction that "the interests of capital and labor are identical." This belief severely compromised the Bureau's understanding of the society in which it had been placed. Northerners in the South, observed reporter Sidney Andrews, found themselves "sorely perplexed" by the "direct antagonism" of former master and former slave, their "positive hostility" to one another. Bureau agents were no exception. Even those most sympathetic to the freedmen viewed the struggle on the plantations as an

80. Sterling, ed., *Trouble They Seen*, 77–78; James M. Smith to Davis Tillson, July 4, 1866, Unregistered Letters Received, Ser. 632, Ga. Asst. Comr., RG 105, NA [FSSP A-5426]; New Orleans *Tribune*, October 31, 1867; *Christian Recorder*, December 1, 1866; John W. Blassingame, ed., *Slave Testimony* (Baton Rouge, 1977), 737–38; McFeely, *Yankee Stepfather*, 3; John F. Couts to Cave Johnson Couts, April 28, 1866, Couts Papers.

81. *Christian Recorder*, May 26, 1866; *National Anti-Slavery Standard*, May 16, 1866; Samuel S. Ashley to George Whipple, May 16, 1866, AMA Archives; 39th Congress, 1st Session, House Executive Document 120, 39–43.

irrational legacy of slavery that would disappear as soon as planters and freedmen absorbed free labor principles. Thus, instead of seeing itself as the champion of one side or the other in an ongoing conflict, the Bureau, in Howard's words, sought to stand "between the two classes" and make both aware of the common interests they shared, if only they would recognize it. In fact, however, the South's "labor problem" arose not from misunderstanding, but from the irreconcilable interests of former masters and former slaves as each sought to define the meaning of emancipation. Perhaps the greatest failing of the Freedmen's Bureau was that it never quite comprehended the depths of racial antagonism and class conflict in the postwar South.[82]

Beginnings of Economic Reconstruction

Thus began the forging of a new class structure to replace the shattered world of slavery—an economic transformation that would culminate, long after the end of Reconstruction, in the consolidation of a rural proletariat composed of the descendants of former slaves and white yeomen, and of a new owning class of planters and merchants, itself subordinate to Northern financiers and industrialists. The historian, however, must avoid telescoping the actual course of events into a predetermined, linear progress. Both planters and freedmen viewed labor relations as a shifting conflict on a terrain where victories and defeats remained provisional, and trial and error altered each group's perceptions of its own self-interest. A new set of labor arrangements did not spring up overnight, and there was no preordained outcome to the workings of what a federal Treasury agent described as "the new *system* of labor if system it may be called, when there is endless confusion, and absurd contradiction."[83]

In some parts of the South, planters in the early postwar years found it almost impossible to resume production, notably in the sugar and rice kingdoms, where the estates of exceptionally wealthy aristocracies lay in ruins. The war had devastated the expensive grinding and threshing mills and the elaborate systems of dikes, irrigation canals, and levees, and thoroughly disrupted the labor system. Only a handful of Louisiana's

82. 39th Congress, 1st Session, House Report 30, pt. 2:266–67; Jacqueline B. Walker, "Blacks in North Carolina During Reconstruction" (unpub. diss., Duke University, 1979), 58; Robert K. Scott, "Circular to Landlords and Laborers of South Carolina, December, 1866," unidentified newspaper clipping, Reconstruction Scrapbook, USC; Sidney Andrews, "Three Months Among the Reconstructionists," *Atlantic Monthly*, 16 (February 1866), 243; 39th Congress, 1st Session, House Executive Document 70, 157; Howard, *Autobiography*, 2:212–14, 250, 312–13, 423.

83. Barbara J. Fields, "The Nineteenth-Century American South: History and Theory," *Plantation Society in the Americas*, 2 (April 1983), 22–25; Irving H. Bartlett, ed., "New Light on Wendell Phillips: The Community of Reform 1840–1880," *Perspectives in American History*, 12 (1979), 165.

sugar plantations operated at all in 1865; the rest stood idle, overgrown with weeds, and the crop amounted to only one tenth of that raised in 1861. In the rice region, "labor was in a disorganized and chaotic state, production had ceased, and . . . the power to compel laborers to go into the rice swamp utterly broken." Thousands of black families had been settled on abandoned rice plantations in accordance with General Sherman's Field Order 15, and countless others occupied estates deserted by their owners. President Johnson's restoration of planters to control of their land failed to end the "undeclared war" between former masters and former slaves that consumed the lowcountry.[84]

Where agricultural production did resume, a variety of arrangements often coexisted in the same area, sometimes on the same plantation. On twenty estates near Natchez, Northern traveler J. T. Trowbridge encountered ten different kinds of labor contract. In the early years of Reconstruction, payments included cash wages, paid monthly or at year's end, a share of the crop, divided collectively among the entire labor force or among smaller groups of workers, various combinations of wage and share payments, time-sharing plans in which freedmen worked part of the week for the planter and part on their own land, wages in kind, and cash wages for specific tasks. Well could Wilmot G. DeSaussere, scion of a distinguished South Carolina family, remark in 1865, "at present a state of transition is going on, in what to result who can tell."[85]

Beneath the welter of contradictions, however, certain broad patterns may be discerned. In much of the South, especially the cotton and sugar regions, planters initially attempted to revive the antebellum gang system, with closely supervised labor and a system of fines and forfeitures to ensure steady work. Although many planters preferred to pay cash wages so as to be able to dismiss recalcitrant workers immediately, the shortage of currency and credit meant that payment usually could not be made until the crop had been harvested and sold. Most planters, moreover, preferred to delay payment to the end of the year in order to prevent laborers from leaving during the crop season. Thus, in 1865 and 1866, a majority of labor contracts involved agreements between planters and large groups of freedmen. Payment was generally either in "standing wages" withheld until year's end, or, more frequently, "share wages"—a share of the crop sometimes paid collectively to the workers and divided among themselves, and sometimes allocated according to their working capacity. "Full" hands—the ablest adult men—received more than did

84. Walter Prichard, "The Effects of the Civil War on the Louisiana Sugar Industry," *JSH*, 5 (August 1939), 319–31; Sitterson, *Sugar Country*, 233–34; George A. Rogers and R. Frank Saunders, Jr., *Swamp Water and Wiregrass: Historical Sketches of Coastal Georgia* (Macon, 1984), 162; Strickland, " 'No More Mud Work'," 51.

85. Trowbridge, *The South*, 391; Cox and Cox, eds., *Reconstruction*, xxvii–xxviii; Ralph Shlomowitz, "The Origins of Southern Sharecropping," *AgH*, 53 (July 1979), 558–62; Wilmot G. DeSaussere to Henry DeSaussere, July 2, 1865, DeSaussere Family Papers.

"half" or "three quarters" workers, and women were invariably paid less than men. Skilled artisans such as blacksmiths and carpenters often signed individual contracts, either for monthly wages or a share of the crop larger than that of field hands. Laborers also received housing, often in the old slave quarters, and food, although these were sometimes charged against the postharvest payments. In the contracts of 1865, the shares paid to the freedmen were usually extremely low, sometimes as little as one tenth of the crop.[86]

In effect, the postponement of payment to the end of the year represented an interest-free extension of credit from employee to employer, as well as a shifting of part of the risk of farming to the freedmen. The practice not only left share workers penniless in the event of a poor crop, but offered numerous opportunities for fraud on the part of planters, some of whom deducted excessive fines for poor work or other infractions, or presented freedmen with bills for rations that more than equaled the wages due them. The laws of Presidential Reconstruction, moreover, awarded factors, bankers, and others who advanced credit to the planter a first lien on the entire crop, often leaving the laborers with a reduced share or nothing at all. Not surprisingly, blacks wanted "*all* the cotton sold to pay them," as Mississippi overseer Wilmer Shields reported. Share-wage agreements, a Freedmen's Bureau agent in Georgia observed, also gave rise to other "complications, misapprehensions . . . and all kinds of trouble." The laborers insisted that along with part of the risk of farming, they had assumed part of the authority for decision making— determining the mix of crops, the use of fertilizer, the manner of dividing and ginning the cotton, and so forth. Planters maintained that they alone should decide how the estates were to be run.[87]

"Even with our former absolute authority over the negro," observed a Southern newspaper in 1865, "compromise between the two races was necessary." And in 1866 and 1867, the freedmen's demand for an improvement in their economic condition and a greater degree of autonomy in their working lives set in motion a train of events that fundamentally transformed the plantation labor system. Aware that the prospect of land distribution had been all but eliminated, freedmen reluctantly agreed to new labor contracts, but often demanded a change in the mode of payment. Where wages had been low or withheld, they pressed for share

86. Labor Contracts, John M. DeSaussere Papers, USC; Shlomowitz, "Origins of Share-cropping," 566–71; Cox and Cox, eds., *Reconstruction*, 332; Flavellus G. Nicholson Diary, July 16, 1865; Jonathan M. Wiener, *Social Origins of the New South: Alabama 1860–1885* (Baton Rouge, 1978), 35–39; Range, *Century of Georgia Agriculture*, 85.

87. Jaynes, *Branches Without Roots*, 24–56, 141–223; Wilmer Shields to William N. Mercer, December 12, 1866, Mercer Papers; 40th Congress, 3d Session, House Miscellaneous Document 52, 122; Lacy Ford, "Labor and Ideology in the South Carolina Up-Country: The Transition to Free Labor Agriculture," in Walter J. Fraser, Jr., and Winfred B. Moore, Jr., eds., *The Southern Enigma: Essays on Race, Class, and Folk Culture*, (Westport, Conn., 1983), 32.

payments; those who had received shares in the poor crop year of 1866 often insisted upon money wages in subsequent negotiations. Sometimes, the vote of a plantation's laborers decided between wages and shares. On one Mississippi plantation, blacks in February 1867 "by a majority decided to take the money instead of a share in the crop." In most contracts, the freedmen's portion of the crop rose from the extremely low figures of 1865 to one third or even one half, and the new contracts generally contained fewer onerous regulations concerning their private lives and personal demeanor.[88]

Simultaneously, blacks' desire for greater autonomy in the day-to-day organization of work produced a trend toward the subdivision of the labor force. Gang labor for wages persisted where planters had access to outside capital and thus could offer high monthly wages, promptly paid. Thanks to an influx of Northern investment, this was the case on sugar plantations that managed to resume production. On many cotton plantations in 1866 and 1867, however, squads of a dozen or fewer freedmen replaced the gangs so reminiscent of slavery. Generally organized by the blacks themselves, these squads sometimes consisted entirely of members of a single family, but more often included unrelated men. In 1866, Linton Stephens contracted with four squads of freedmen on his black belt Georgia plantation, each receiving half of the crop it raised, minus deductions for seed, implements, and supplies. On Stephens' plantation the squads worked under an overseer's supervision, but elsewhere a driver or "headman" chosen from among themselves directed the laborers. By 1867 the gang system was disappearing from the cotton fields. "You do not see as large gangs together as of old times," reported an Alabama newspaper, "but more frequently squads of five or ten in a place."[89]

The final stage in the decentralization of plantation agriculture was the emergence of sharecropping. Unlike the earlier share-wage system, with which it is often confused, in sharecropping individual families (rather than large groups of freedmen) signed contracts with the landowner and became responsible for a specified piece of land (instead of working in gangs). Generally, sharecroppers retained one third of the year's crop if the planter provided implements, fertilizer, work animals, and seed, and half if they supplied their own. The transition to sharecropping occurred

88. Columbia *Daily Phoenix*, September 15, 1865; Paisley, *From Cotton to Quail*, 26; Wilmer Shields to William N. Mercer, February 13, 1867, Mercer Papers; Brooks, *Agrarian Revolution*, 22–26; Howell Cobb, Jr., to Howell Cobb, January 3, 1866, Cobb-Erwin-Lamar Papers.

89. Ralph Shlomowitz, "The Squad System on Postbellum Cotton Plantations," in Orville V. Burton and Robert C. McMath, eds., *Toward a New South? Studies in Post-Civil War Southern Communities* (Westport, Conn., 1982), 265–80; Linton Stephens to Alexander H. Stephens, December 29, 1865, Alexander H. Stephens Papers, Manhattanville College; Charles Hill to Benjamin G. Humphreys, May 21, 1867, Mississippi Governor's Papers; Kolchin, *First Freedom*, 46.

at different rates on different plantations and continued well into the 1870s, but the arrangement appeared in some areas soon after the Civil War. One Southern newspaper claimed in 1867 that this "new system" of labor originated in Mississippi, and it seems to have developed faster on the smaller farms of the Southern Piedmont than on plantations where employers clung to the hope of retaining gang labor.[90]

To blacks, sharecropping offered an escape from gang labor and day-to-day white supervision. For planters, the system provided a way to reduce the cost and difficulty of labor supervision, share risk with tenants, and circumvent the chronic shortage of cash and credit. Most important of all, it stabilized the work force, for sharecroppers utilized the labor of all members of the family and had a vested interest in remaining until the crop had been gathered. Yet whatever its economic rationale, many planters resisted sharecropping as a threat to their overall authority and inefficient besides (since they believed blacks would not work without direct white supervision). A compromise not fully satisfactory to either party, the system's precise outlines remained a point of conflict. Planters insisted sharecroppers were wage laborers, who must obey the orders of their employer and possessed no property right in the crop until they received their share at the end of the year. But sharecroppers, a planter complained in 1866, considered themselves "partners in the crop," who insisted on farming according to their own dictates and would not brook white supervision. Only a system of wages, payable at the end of the year, he concluded, would allow whites to "work them in accordance with our former management." But precisely because it seemed so far removed from "our former management," blacks came to prefer the sharecropping system.[91]

If freedmen in the cotton fields sought to escape the gang labor associated with bondage, those in the rice swamps struggled to retain and strengthen the familiar task system, the foundation of the partial autonomy they had enjoyed as slaves. "We want to work just as we have always worked," declared a group of freedmen in South Carolina's rice region, and to attract labor, rice planters found themselves obliged to let the blacks "work . . . as they choose without any overseer." Some employers agreed in writing that labor would be organized according to "the customary tasks," while those who attempted to establish close supervision of their employees found the effort both futile and dangerous. Out of the wreck of the rice economy and blacks' insistence upon autonomy

90. Jaynes, *Branches Without Roots*, 141–90; Shlomowitz, "Origins of Sharecropping," 563; Columbia *Daily Phoenix*, May 9, 1867; Ronald L. F. Davis, "Labor Dependency Among Freedmen, 1865–1880," in Fraser and Moore, eds., *From Old South to New*, 157; Ford, "Labor and Ideology," 29.

91. Joseph D. Reid, Jr., "Sharecropping as an Understandable Market Response—The Post-Bellum South," *JEcH*, 33 (March 1973), 106–30; Wiener, *Social Origins*, 66–69; "Fairfield" to James L. Orr, December 14, 1866, South Carolina Governor's Papers.

emerged an unusual set of labor relations. Some planters simply rented their plantations to blacks for a share of the crop, or divided the land among groups of freedmen to cultivate as they saw fit. As a consequence, lamented Mary Jones, widow of a leading lowcountry planter and clergyman, "the negro is virtually master of the soil." Others agreed to a system of labor sharing, in which freedmen worked for two days on the plantation in exchange for an allotment of land on which to grow their own crops. On nearly all rice plantations, blacks continued to labor under the traditional task system, often supplementing their income by hunting, fishing, and growing crops on their own time.[92]

Thus, the struggles of early Reconstruction planted the seeds of new labor systems in the rural South. The precise manner in which these seeds matured would be worked out not only on Southern farms and plantations, but on the Reconstruction battlefields of local, state, and national politics.

92. John S. Strickland, "Traditional Culture and Moral Economy: Social and Economic Change in the South Carolina Low Country, 1865–1910," in Steven Hahn and Jonathan Prude, eds., *The Countryside in the Age of Capitalist Transformation* (Chapel Hill, 1985), 144–46; Labor Contract, May 29, 1867, Edward M. Stoeber Papers, USC; Mary Jones to Charles Colcock Jones, Jr., January 19, 1866, Charles Colcock Jones, Jr., Collection, UGa; Robert M. Myers, ed., *The Children of Pride* (New Haven, 1972), 1366; Foner, *Nothing But Freedom*, 82–87. For the subsequent development of the South's labor systems, see Chapter 8.

CHAPTER 5

The Failure of Presidential Reconstruction

Andrew Johnson and Reconstruction

AT first glance, the man who succeeded Abraham Lincoln seemed remarkably similar to his martyred predecessor. Born within two months of each other, both knew poverty in early life, neither enjoyed much formal schooling, and in both deprivation sparked a powerful desire for notoriety and worldly success. Lincoln's "ambition was a little engine that knew no rest," remarked his law partner, William Herndon, and one who knew Andrew Johnson well described his life as "one intense, unceasing, desperate upward struggle." During the prewar decades, both achieved material comfort, Lincoln as an Illinois corporation lawyer, Johnson rising from tailor's apprentice to become a prosperous landowner and investor in railroad stock. And for both, antebellum politics became a path to power and respect.

In terms of sheer political experience, few men have seemed more qualified for the Presidency than Andrew Johnson. Beginning as a Greenville, Tennessee, alderman in 1829, he held office almost continuously, rising to the state legislature and Congress, serving two terms as governor, and in 1857 entering the Senate. Even more than Lincoln, Johnson gloried in the role of tribune of the common man. Both in Washington and the rough-and-tumble world of Tennessee stump speaking, his speeches lauded "honest yeomen" and thundered against the "slaveocracy"—a "pampered, bloated, corrupted aristocracy." The issues most closely identified with Johnson's prewar career were tax-supported public education, a reform enacted into law during his term as governor, and homestead legislation, which he promoted tirelessly in the Senate.[1]

Apart from the education law, however, Johnson's political career was

1. Richard Hofstadter, *The American Political Tradition* (New York, 1948), 93; Oliver P. Temple, *Notable Men of Tennessee from 1833 to 1875* (New York, 1912), 466; Leroy P. Graf and Ralph W. Haskins, eds., *The Papers of Andrew Johnson* (Knoxville, 1967–), 1:xix–xxix, 2:xviii–xxix, 3:xvii–xxi; James E. Sefton, *Andrew Johnson and the Uses of Constitutional Power* (Boston, 1980), 1–57, 104.

remarkably devoid of substantive accomplishment, especially in light of his long tenure in various offices. In part, this failure stemmed from traits that would go a long way toward destroying his Presidency. If in Lincoln, early poverty and the struggle for success somehow produced wit, political dexterity, and sensitivity to the views of others, Johnson's personality turned in upon itself. An accomplished public orator, privately Johnson was a self-absorbed, lonely man, who had few friends and confided in no one. Gideon Welles, who as Secretary of the Navy staunchly upheld Johnson's Reconstruction policies, identified isolation as the President's greatest weakness: "He has no confidants and seeks none," and his major decisions appeared to have been made without consultation with "anyone whatever." No one could doubt Johnson's personal courage, yet early in his career other less commendable qualities had also become apparent, among them stubbornness, intolerance of the views of others, and an inability to compromise. As governor, Johnson failed to work effectively with his legislature; as military governor he proved unable to elicit popular support for his administration. Hardly a political novice, he found himself, as President, thrust into a role that required tact, flexibility, and sensitivity to the nuances of public opinion—qualities Lincoln possessed in abundance, but that Johnson lacked.[2]

"Andrew Johnson was the queerest character that ever occupied the White House," one contemporary remarked, and it seems safe to predict that his career as President will always remain something of an enigma. When he assumed office on April 15, 1865, his past career led many to expect a Reconstruction policy that envisioned far-reaching change in the defeated South. "Treason must be made odious, and traitors must be punished and impoverished," Johnson had declared in 1864; in the same year he offered himself as a "Moses," leading blacks to a promised land of freedom. "It was supposed," John Sherman later recalled, "that President Johnson would err, if at all, in imposing too harsh terms upon these states."[3]

Long dissatisfied with their partly cooperative, partly antagonistic relationship with Lincoln, Radical Republicans initially viewed Johnson's accession as a godsend. Hardly had Lincoln died, but a group of Radical lawmakers met with the new President. "Johnson, we have faith in you," declared Senator Benjamin F. Wade. "By the Gods, there will be no trouble now in running the government." To which the President replied: "I hold this: . . . *treason* is a crime, and *crime* must be punished." The Radicals were overjoyed. "I believe," one declared, "that the Al-

2. Howard K. Beale, ed., *Diary of Gideon Welles* (New York, 1960), 3:190; Graf and Haskins, eds., *Johnson Papers*, 1: xxi; 2: xxvi–xviii; 5: xxiv; Albert Castel, *The Presidency of Andrew Johnson* (Lawrence, Kans., 1979), 2–6; Eric L. McKitrick, *Andrew Johnson and Reconstruction* (Chicago, 1960), 85–89.

3. Shelby M. Cullom, *Fifty Years of Public Service* (Chicago, 1911), 143; John Sherman, *Recollections of Forty Years in the House, Senate and Cabinet* (Chicago, 1895), 1:359.

mighty continued Mr. Lincoln in office as long as he was useful, and then substituted a better man to finish the work."[4]

Although men like Wade applauded Johnson's talk of punishing traitors (which the President repeated with monotonous regularity to delegations visiting the White House in April), neither vindictiveness toward the South nor a carefully worked-out plan of Reconstruction distinguished Radicalism in the spring of 1865. Radicalism when Lincoln died was defined above all by an insistence upon black suffrage as the sine qua non of Reconstruction. In the weeks following the assassination, leading Radicals met frequently with the new President to press this issue. Most persistent was Charles Sumner, who "waited upon" Johnson almost daily between mid-April and mid-May, reiterating the theme with which his political career was identified—"justice to the colored race." Sumner became convinced that he and Johnson essentially agreed on black suffrage. As Sumner understood it, Chief Justice Salmon P. Chase's trip south early in May, undertaken after a conference with the President, was intended to convince white Southerners of the necessity of enfranchising blacks. Chase's tour became a one-man campaign for black suffrage, although, when advocating the policy publicly, he hastened to add that he did not speak for "the Government." Nonetheless, his mostly black audiences were duly impressed. " 'Tisn't only what he says, but it's de man what says it. . . ." commented one listener, "his words hab weight." (He might have been forgiven for wondering why the Chief Justice of the United States did not comprise part of "the government.")[5]

Well into May, Radicals considered Johnson, in Sumner's words, "the sincere friend of the negro and ready to act for him decisively." Yet in reality, Johnson shared neither the Radicals' expansive conception of federal power nor their commitment to political equality for blacks. Despite his own vigorous exercise of authority as military governor, Johnson had always believed in limited government and a strict construction of the Constitution. In Congress, he had moved to reduce the salaries of government workers, voted against aid to famine-stricken Ireland, and even opposed appropriations to pave Washington's muddy streets. His fervent nationalism in no way contradicted his respect for the rights of the states. These rights did not, in his view, include secession, and Johnson had insisted in 1861 that no state could leave the Union. He clung to this

4. "George W. Julian's Journal—The Assassination of Lincoln," *IndMH*, 11 (December 1915), 335; George W. Julian, *Political Recollections, 1840 to 1872* (Chicago, 1884), 256–57; Zachariah Chandler to Mrs. Chandler, April 23, 1865, Zachariah Chandler Papers, LC.

5. Graf and Haskins, eds., *Johnson Papers*, 7:583, 610–14, 630–32, 655; Julian, *Political Recollections*, 263; Edward L. Pierce, ed., *Memoir and Letters of Charles Sumner* (Boston, 1877–93), 4:241–46; Charles Sumner to Henry L. Dawes, June 25, 1865, Henry L. Dawes Papers, LC; James E. Sefton, ed., "Chief Justice Chase as an Advisor on Presidential Reconstruction," *CWH*, 13 (September 1967), 242–64; Whitelaw Reid, *After the War: A Southern Tour* (Cincinnati, 1866), 83.

position with characteristic tenacity. Individual "traitors" should be punished, and punished severely, but the states had never, legally, seceded, or surrendered their right to govern their own affairs.[6]

Logically, as Carl Schurz later commented, Johnson had "a pretty plausible case"—secession had been null and void, the states remained intact, and Reconstruction meant enabling them to resume their full constitutional rights as quickly as possible. The situation actually confronting the nation, however, bore little resemblance to Johnson's neat syllogism. "To say because they had no right to go out therefore they could not," declared California railroad magnate Leland Stanford, "does not seem to me more reasonable than to say that because a man has no right to commit murder therefore he cannot. A man does commit murder and that is a fact which no reasoning can refute." Eleven Southern states had levied war against the Union, a situation, as Schurz pointed out to the President, "not foreseen in the Constitution." Like Lincoln before him, Johnson planned to impose conditions, including the irrevocable abolition of slavery, for their return. But did this not imply that the states had sacrificed some of their accustomed rights? For, as Thaddeus Stevens remarked, "How the executive can remodel states *in the Union* is past my comprehension." And, Radicals insisted, if Johnson did possess the power to appoint provisional governors and lay down terms for reunion, he could also prescribe voting qualifications. In this sense, the Radicals and Johnson disagreed less on a constitutional issue in the spring of 1865 than on a matter of policy: whether black suffrage should be made a requirement for the South's readmission.[7]

Nothing in Johnson's career had prepared him to take so dramatic and unprecedented a step. The owner of five slaves before the war, Johnson had sincerely embraced emancipation as military governor, but his speeches condemning slavery dwelled almost obsessively on racial miscegenation as the institution's main evil, and he made no commitment to civil equality or a political role for the freedmen. Worse, as a perceptive constituent warned Illinois Congressman Elihu B. Washburne in 1865, "I have grounds to fear President Johnson may hold almost unconquerable prejudices against the African race." Three years later, the President's private secretary, Col. William G. Moore, recorded in his diary that Johnson "has at times exhibited a morbid distress and feeling against the negroes." Finding "half a dozen stout negroes" at work on White House

6. Charles Sumner to John Bright, April 24, 1865, Add. Mss. 43,390, f. 222, John Bright Papers, British Museum; Michael L. Benedict, *The Impeachment and Trial of Andrew Johnson* (New York, 1973), 4; Sefton, *Johnson*, 105.

7. Carl Schurz, *The Reminiscences of Carl Schurz* (New York, 1907–08), 3:219–20; Leland Stanford to Cornelius Cole, February 9, 1867, Cornelius Cole Papers, University of California, Los Angeles; Carl Schurz to Andrew Johnson, June 6, 1865, Thaddeus Stevens to Johnson, May 16, 1865, Andrew Johnson Papers, LC; Samuel S. Cox, *Three Decades of Federal Legislation* (Providence, 1885), 379–80.

grounds, the President "at once wanted to know if all the white men had been discharged." In his December 1867 annual message to Congress, Johnson insisted that blacks possessed less "capacity for government than any other race of people. No independent government of any form has ever been successful in their hands. On the contrary, wherever they have been left to their own devices they have shown a constant tendency to relapse into barbarism." This was probably the most blatantly racist pronouncement ever to appear in an official state paper of an American President. Johnson went on to warn that black suffrage would result in "a tyranny such as this continent has never yet witnessed," although it is difficult to imagine what regime blacks might impose more tyrannical than chattel slavery.[8]

In 1865, as racial policy emerged as a focus of Reconstruction controversy, Johnson took steps to counteract the image of being indifferent to the freedmen's fate. In August, he advised Mississippi Gov. William L. Sharkey to extend the franchise to literate and propertied blacks; as a result, "the radicals, who are wild upon negro suffrage," would be "completely foiled." The contrast with Lincoln's 1864 suggestion that Louisiana partially enfranchise blacks could not have been more striking. Lincoln pressed the justice of the case and worked behind the scenes to bolster those supporting his views; Johnson's recommendation, as one Southern Unionist remarked, "was only a policy." In October, in an interview with Massachusetts abolitionist George L. Stearns, the President went as far as he ever would on the issue, stating that were he in Tennessee, he would favor the vote for the literate and propertied as well as army veterans. But Johnson never wavered from the conviction that the federal government lacked the authority to impose such a policy upon the states, and that the status of blacks must not become an obstacle to the speedy completion of Reconstruction.[9]

"White men alone must manage the South," the President remarked to California Senator John Conness in 1865. Johnson's prejudices are often ascribed to his upbringing as a "poor white" and his self-defined role as spokesman for the South's yeomanry. So long as it does not suggest that the poor harbored racism more deeply than other classes of white Southerners, this assessment contains considerable truth. Johnson

8. John Cimprich, "Military Governor Johnson and Tennessee Blacks, 1862–65," *THQ,* 39 (Winter 1980), 469; Graf and Haskins, eds., *Johnson Papers,* 7:liii, 251–53, 281–82; J. Weldon to Elihu B. Washburne, August 21, 1865, Elihu B. Washburne Papers, LC; Col. William G. Moore Diary (transcript), April 9, 1868, Johnson Papers; James D. Richardson, ed., *A Compilation of the Messages and Papers of the Presidents 1789–1897* (Washington, D.C. 1896–99), 6:564–66.

9. Edward McPherson, *The Political History of the United States of America During the Period of Reconstruction* (Washington, D.C., 1875), 19–20, 49–50; Daniel R. Goodloe, "History of Provisional Governments of 1865," manuscript, Daniel R. Goodloe Papers, UNC; Dan T. Carter, *When the War Was Over: The Failure of Self-Reconstruction in the South, 1865–1867* (Baton Rouge, 1985), 29–30.

had long combated the planter aristocracy of Tennessee and the South, and believed they had dragooned a reluctant yeomanry into secession. In the Tennessee legislature, he had favored a "white basis" of representation so that masters would not benefit politically from slave property, and had once advocated separate statehood for East Tennessee, to liberate yeomen from the Slave Power's yoke. He seems to have assumed that the Confederacy's defeat had shattered the power of the "slaveocracy" and made possible the political ascendancy of loyal white yeomen. The freedmen had no role to play in his vision of a reconstructed South. When a delegation of blacks visited him at the White House in February 1866, Johnson proposed that their people emigrate to some other country.[10]

Throughout his Presidency, Johnson held the view—not uncommon among Southern yeomen—that slaves had in some way joined forces with their owners to oppress nonslaveholding whites. "The colored man and his master combined kept [the poor white] in slavery," he told the black delegation, "by depriving him of a fair participation in the labor and productions of the rich land of the country." Blacks, he believed, identified with their former masters and looked down upon poor whites, while the latter feared and despised the freedmen. The most likely result of black enfranchisement would therefore be an alliance of blacks and planters, restoring the Slave Power's hegemony and effectively excluding the yeomanry from political power. As Johnson put it in his interview with Stearns, "the negro will vote with the late master, whom he does not hate, rather than with the non-slaveholding white, whom he does hate." (At the February White House interview, Frederick Douglass proposed a different political scenario—a union of yeomen and freedmen in "a party . . . among the poor," but such a possibility does not seem to have interested the President.)[11]

Not until the end of May 1865 did Johnson's plan of Reconstruction become known. On April 14, it will be recalled, Secretary of War Stanton had brought before the Cabinet a proclamation establishing military rule in parts of the South. Two days later, with Lincoln dead and Johnson in the White House, the Cabinet briefly discussed a modified version of the plan. That night, a group of influential Republicans met with Stanton. Charles Sumner and Speaker of the House Schuyler Colfax insisted that without suffrage, freedom for blacks "is a mockery." The Secretary responded that agitation of the issue would split the Republican party but apparently left the meeting convinced, for he revised his proclamation to provide that when the time came to form civil governments, all "loyal citizens" would enjoy the franchise. On May 8, Reconstruction again

10. Charles Nordhoff to William Cullen Bryant, February 21, 1867, Bryant-Godwin Papers, NYPL (recounting a conversation of 1865); Graf and Haskins, eds., *Johnson Papers*, 1: xxvi; McPherson, *Political History*, 49–56.
11. Philadelphia *Press*, July 18, 1865; McPherson, *Political History*, 49–56.

came before the Cabinet, and black suffrage quickly emerged as the only major point of disagreement. For two days the debate continued; in the end, the Cabinet divided evenly, with Welles, Interior Secretary John P. Usher, and Secretary of the Treasury Hugh McCulloch in favor of limiting suffrage to whites, while Stanton, Attorney General James Speed, and Postmaster General William Dennison supported black voting. The most important member, Secretary of State Seward, was absent, recuperating from an attempt upon his own life the night Lincoln was killed.[12]

Throughout this discussion, Johnson remained silent. But on the very day of the Cabinet vote, he extended recognition to the Southern governments created under the Lincoln administration (Arkansas, Louisiana, Tennessee, and Virginia), none of which had enfranchised blacks. Even more troubling to Radicals was Johnson's response to events in Louisiana. Here, Michael Hahn, who resigned as governor in March upon his election to the Senate, had been succeeded by James Madison Wells, a prosperous sugar and cotton planter who had owned nearly 100 slaves before the war. Wells's loyalty to the Union was above reproach—he had organized armed bands to harass Confederate forces in Louisiana—and he entered office with a reputation as a Radical. Yet, taking stock of the political situation as the war drew to a close, he realized that returning Confederate soldiers and their supporters vastly outnumbered the white Louisianans loyal to his regime. Moving swiftly to broaden his political base, Wells began the wholesale removal of officials appointed by General Banks, replacing them with Conservative Unionists and Confederate veterans who wished to overturn the free state constitution of 1864.[13]

Wells proposed, a close ally later explained, to replace the Free State party with a coalition of "representative men of the old parties of the pre-rebellion times; men of irreproachable integrity, of suitable age, social importance, and proper educational qualifications." To many Unionists, this seemed nothing less than the restoration of the old order. Upon his return to Louisiana late in April, Banks countermanded Wells's appointments, warning that the governor's course would "re-establish in power men of the old system of slavery." One member of the legislature wrote the President that Wells had become too close to "the aristocratic leaders" who had "gulled . . . the poor men into rebellion" and intended to "have slavery revived by another name." Johnson proved indifferent to such appeals. On May 17, he removed Banks from his command and

12. Beale, ed., *Welles Diary*, 2:291; Benjamin P. Thomas and Harold M. Hyman, *Stanton: The Life and Times of Lincoln's Secretary of War* (New York, 1962), 402–404; Albert Mordell, ed., *Civil War and Reconstruction: Selected Essays by Gideon Welles* (New York, 1959), 194–95, 200–201, 212–13; Charles Sumner to Henry L. Dawes, June 25, 1865, Dawes Papers.

13. Thaddeus Stevens to Charles Sumner, May 10, 1865, Charles Sumner Papers, HU; Walter McG. Lowrey, "The Political Career of James Madison Wells," *LaHQ,* 31 (October 1948), 995–1034; Peyton McCrary, *Abraham Lincoln and Reconstruction: The Louisiana Experiment* (Princeton, 1978), 309–12, 354–55.

confirmed Wells's appointive power. The infant free state of Louisiana created by Lincoln and Banks had been virtually dismantled.[14]

The definitive announcement of Johnson's plan of Reconstruction came in two proclamations issued on May 29, 1865. The first conferred amnesty and pardon, including restoration of all property rights except for slaves, upon participants in the rebellion who took an oath pledging loyalty to the Union and support for emancipation. Fourteen classes of Southerners, however, most notably major Confederate officials and owners of taxable property valued at more than $20,000, were required to apply individually for Presidential pardons. Simultaneously, Johnson appointed William W. Holden provisional governor of North Carolina, instructing him to call a convention to amend the state's prewar constitution so as to create a "republican form of government" that would entitle North Carolina to its rights within the Union. All those who had not been pardoned under the terms of the first proclamation were excluded from voting for delegates, but otherwise, voter qualifications in effect immediately before secession (when the franchise, of course, was limited to whites) would apply. Identical edicts for the rest of the Confederacy soon followed.

The May proclamations appeared to reflect Johnson's long-time aim of breaking the political and economic hegemony of the "slaveocracy" and establishing the ascendancy of the South's Unionist yeomanry. Indeed, while Johnson always claimed that his Reconstruction policy merely continued Lincoln's, in crucial respects it was very much his own. On the one hand, Lincoln, at the end of his life, had made clear that he favored a limited suffrage for Southern blacks; on the other, he had never suggested exemptions to Presidential amnesty as sweeping as those contained in Johnson's proclamation. The $20,000 clause, excluding the Confederacy's economic elite from a voice in Reconstruction, gave Johnson's proclamations an aura of sternness quite unlike any of Lincoln's Reconstruction statements.[15]

In his "Moses" speech of 1864, Johnson had mentioned the possibility of confiscating the large estates and their division and sale among "honest farmers." Shortly before assuming the Presidency, he had spoken of the need to "punish and impoverish" the leaders of the rebellion, break their "social power," and remunerate Union men for wartime losses from the property of "wealthy traitors." The New York *World* predicted that because of this "monomania," Johnson would vigorously enforce the confiscation laws, break up the plantations, and "create communities of small freeholders." Whether or not Johnson actually contemplated such

---

14. Hugh Kennedy to Andrew Johnson, September 7, 16, 1865, Johnson Papers; LaWanda Cox, *Lincoln and Black Freedom: A Study in Presidential Leadership* (Columbia, Mo., 1981), 135–39; Nathaniel P. Banks to Montgomery Blair, May 6, 1865, Blair Family Papers, LC; R. L. Brooks to Johnson, May 5, 1865, Johnson Papers.
15. Richardson, ed., *Messages and Papers*, 6:310–14; Castel, *Johnson*, 27–29.

a policy, many in May 1865 believed he intended the $20,000 clause to "keep these people out in the cold," enabling spokesmen for the yeomanry to shape Reconstruction. Others, however, believed Johnson planned to use the requirement for individual pardons not to punish the "aristocracy," but to force it to endorse his terms of Reconstruction. Once in receipt of a pardon and with it the all-important recognition of their property rights, wealthy Southerners would be expected to use their political influence in favor of the administration. The full implications of Johnson's plan would depend on whether the $20,000 clause was a prelude to a full-fledged assault upon the "slaveocracy," or a means of building political support for the President among the South's old elite.[16]

The latter course had its attractions, especially since it would contribute to Johnson's own reelection, a consideration that could not have been far from the mind of so intensely ambitious a man. The experience of the two previous Vice Presidents who entered the White House accidentally hardly inspired optimism. John Tyler alienated the Whig party that had elected him, threw his support to the Democrats, and sacrificed his political influence; his name became a byword for treachery and ineptitude. Millard Fillmore failed to win renomination. Johnson's own situation was complicated by the fact that before the war, the Republican party had been virtually nonexistent in the Southern states. Politically, Johnson would have to retain the support of the majority of Northern Republicans, win over moderate Democrats, and build up a following in the South. Were Presidential Reconstruction successful, Johnson would have created an unassailable political coalition capable of reuniting the Union, winning him a triumphant reelection, and determining the contours of American politics for a generation or more.

Blacks, of course, would remain forever outside the bounds of citizenship. A Tennessee Unionist driven from his home during the war pointed out the contradiction: "You say you believe in democratic government, or *consent* of loyal people. Yet you *dare not* avow with practical effect the right of the colored man to vote. Are you honest?" By the end of his life, Lincoln would have opened the door to future change by recognizing some blacks as part of the political nation. Johnson's suggestion that in the absence of Presidential dictation, individual states might take the initiative was certainly disingenuous, for not a single state, North or South, had expanded the political rights of blacks since the founding of the republic. For the freedmen, it already seemed clear that, as one recalled years afterwards, "things was hurt by Mr. Lincoln gettin' kilt."[17]

16. Graf and Haskins, eds., *Johnson Papers*, 7:251–53, 543–46; William D. Foulke, *Life of Oliver P. Morton* (Indianapolis, 1899), 1:440–41; New York *World*, April 25, 1865; [William W. Holden] *Memoirs of W. W. Holden* (Durham, 1911), 55; Raleigh *Standard*, June 5, 1865; James G. Blaine, *Twenty Years of Congress* (Norwich, Conn., 1884–86), 2:73–75.

17. Joseph Noxon to Andrew Johnson, May 27, 1865, Johnson Papers; George P. Rawick, ed., *The American Slave: A Composite Autobiography* (Westport, Conn., 1972–79), Supplement, Ser. 1, 1:257.

Launching the South's New Governments

Whatever their differences, all Northern proposals for Reconstruction took for granted that political power in the South must rest in the hands of loyal men. But what constituted genuine loyalty? Legally, at least, the Ironclad Oath, an affirmation that an individual had never voluntarily aided the Confederacy, defined devotion to the Union. And "unconditional" Unionists who could meet this stringent requirement, and had often suffered for their convictions, assumed they would reap the political benefits of Reconstruction. Already, such men had come to power in Maryland, West Virginia, Missouri, Arkansas, and Johnson's own Tennessee. Yet only in mountain areas like western North Carolina and some parts of the Upper Piedmont did a large group of unconditional Unionists stand ready to assume political power; elsewhere, they comprised a small faction, despised by the white majority as "Tories" and traitors. "There is almost no such thing as loyalty here, as that word is understood in the North," a Union officer reported from Louisiana in the summer of 1865. Whether deluded by the Slave Power or not, three quarters of the South's white males aged eighteen to forty-five had served in the Confederate armies at one time or another. As Whitelaw Reid observed during a tour of the South that convinced him a "strong Union party" simply did not exist, "it remains to be seen how long a minority, however loyal, can govern in a republican country."[18]

An alternative definition of Unionism focused on an individual's position during the secession crisis. A large number of white Southerners had opposed disunion in the winter of 1860–61, but "went with their states" with the coming of war. "I was an original Union man," declared E. F. Keen, "and fought secession as long, perhaps, as he who fought longest in Virginia." But when his state seceded, Keen, and thousands like him, entered the Confederate army. Such individuals demanded recognition as "Union men," and indignantly repudiated the labels secessionist or traitor. Even Alexander H. Stephens, the South's wartime Vice President, classed himself as part of Georgia's "Union element." By this definition—opposition to secession and willingness to "accept the situation" at war's end, with intervening actions conveniently forgotten—nearly everyone in the South qualified as loyal, for an "original secessionist" proved difficult to find in 1865. One federal official wondered how Florida had in fact seceded, since everyone claimed to have been a Unionist who merely "followed the state." The fact that "Unionism" possessed such divergent meanings in 1865 produced considerable misunderstanding between North and South. White Southerners believed that electing to office

18. Harold M. Hyman, *A More Perfect Union: The Impact of the Civil War and Reconstruction on the Constitution* (New York, 1973), 176–77; Sarah W. Wiggins, *The Scalawag in Alabama Politics, 1865–1881* (University, Ala., 1977), 9; C. E. Lippincott to Lyman Trumbull, August 29, 1865, Lyman Trumbull Papers, LC; Emory M. Thomas, *The Confederate Nation 1861–1865* (New York, 1979), 155; Reid, *After the War*, 20, 298–99.

former opponents of secession, even though they had served the Confederacy in high military and civilian posts, complied with the demand for "loyal government." To unconditional Unionists and Northern Republicans, such individuals were no less rebels than the architects of secession.[19]

Former Whigs comprised the majority of antisecession Southerners claiming the Unionist mantle. Although long moribund as a political organization, Whiggery survived in the hearts of many Southerners who, as one later recalled, "expected to take control of affairs" at war's end. The idea of absorbing a revived Whiggery into the Republican party had influenced Lincoln's Reconstruction policies and would beguile Northern politicians well into the 1870s. But the actual extent of "persistent Whiggery" remains open to question. The party had, in fact, been killed by the slavery issue. Tennessee's fiercely partisan Whig William G. Brownlow declared in 1860 that he would only join the Democrats "when the sun shines at midnight . . . when flowers lose their odor . . . when Queen Victoria consents to be divorced by a county court in Kansas." But by that year, most Southern Whig leaders had already moved into the Democratic camp. Their opposition to secession proved "remarkably feeble," and nearly all cast their lot with the Confederacy.[20]

One thing, however, was plain: In 1865, Southern Unionism, whether unconditional, Old Line Whig, or simply anti-Confederate, did not necessarily imply a willingness to extend civil and political equality to the freedmen. Events in Maryland, Tennessee, Louisiana, and Missouri had already revealed the limited support for black suffrage among wartime Unionists. For most, Reconstruction meant the proscription of "rebels," not recognition of the rights of blacks. Resolutions adopted by a series of Unionist meetings in North Carolina immediately after the war concentrated almost entirely on demands that men "who always *have been*" loyal be appointed to office and leading Confederates be punished. Jealous of their local autonomy, upcountry Unionists resented the presence of black troops and Freedmen's Bureau agents. They also shared President Johnson's assumption that blacks, if enfranchised, would vote with their former owners and seemed eager to rid their states of the freedmen alto-

19. 39th Congress, 1st Session, House Report 30, pt. 2:165; *The Debates and Proceedings of the Constitutional Convention of the State of Virginia* (Richmond, 1868), 256; Myrta L. Avary, ed., *Recollections of Alexander H. Stephens* (New York, 1910), 524; 39th Congress, 1st Session, House Report 30, pt. 2:212; Joe M. Richardson, *The Negro in the Reconstruction of Florida 1865–1877* (Tallahassee, 1965), 2.

20. Thomas B. Alexander, "Persistent Whiggery in the Confederate South, 1860–1877," *JSH*, 27 (August 1961), 305–29; 42d Congress, 2d Session, House Report 22, Georgia, 761–63; James L. Bliss to Lewis E. Parsons, July 24, 1865, Alabama Governor's Papers, ASDAH; E. Merton Coulter, *William G. Brownlow, Fighting Parson of the Southern Highlands* (Chapel Hill, 1937), 127–28; Daniel W. Howe, *The Political Culture of the American Whigs* (Chicago, 1980), 238; John V. Mering, "Persistent Whiggery in the Confederate South: A Reconsideration," *SAQ*, 69 (Winter 1970), 124–43.

gether. "It is hard to tell which they hate most," Brownlow commented to a visitor, "the Rebels, or the negroes."[21]

As for Old Line Whigs claiming the Unionist mantle, many were merchants, bankers, and wealthy black belt planters—confirmed elitists who had never accepted the democratizing trends of the antebellum era. At the Virginia and Georgia secession conventions, such men had supported efforts to disenfranchise propertyless whites and create a "patriarchal republic" insulated from the popular will. It was hardly likely that those who believed too many whites enjoyed the franchise would favor extending it to blacks. If the upcountry feared black voters would be controlled by their former masters, many former Whigs believed universal suffrage would call forth "reckless demagogues" from among the less affluent white population, to rally the freedmen with radical proposals for the division of property. Thus, whatever combination of Unionists emerged to direct state affairs, blacks had little basis for optimism at the outset of Presidential Reconstruction.[22]

For a man bent on making treason odious and displacing the South's traditional leadership, Andrew Johnson displayed remarkable forbearance in choosing provisional governors to launch the reconstruction process. Two appointments did appear provocative to many white Southerners: Andrew J. Hamilton of Texas, who had served in the Union army and been appointed his state's military governor by Lincoln, and William W. Holden, outspoken champion of North Carolina's yeomanry and leader of the 1864 peace movement. Each was cordially hated by his state's elite. Elsewhere, however, Johnson passed over unconditional loyalists to select men acceptable to a broader segment of white public opinion. In Georgia, the President chose James Johnson, an obscure former Whig Congressman who sat out the war without taking sides. In Alabama, Johnson ignored the upcountry Unionist opposition and selected Lewis E. Parsons, a former Whig Congressman who served as a "peace party" member of the wartime Alabama legislature and enjoyed close ties to the state's mercantile and railroad interests. Mississippi's new governor was William L. Sharkey, a prominent Whig planter who retired from political affairs after secession; Florida's was William Marvin, a New York-born businessman who spent most of the war as a judge behind federal lines. The South Carolina post, of great symbolic importance in

21. Raleigh *Daily Record*, June 3, July 19, 1865; Raleigh *Standard*, April 24, July 21, 1865; Willis D. Boyd, "Negro Colonization in the Reconstruction Era, 1865–1870," *GaHQ*, 40 (December 1956), 369; Knoxville *Whig*, June 28, August 2, September 27, October 11, December 13, 1865; J. T. Trowbridge, *The South: A Tour of Its Battle-Fields and Ruined Cities* (Hartford, Conn., 1866), 284.

22. Michael Perman, *The Road to Redemption: Southern Politics, 1869–1879* (Chapel Hill, 1984), 94; Michael P. Johnson, *Toward a Patriarchal Republic: The Secession of Georgia* (Baton Rouge, 1977), 166–78; T. Harry Williams, "An Analysis of Some Reconstruction Attitudes," *JSH*, 12 (November 1946), 476–77; Raleigh *Standard*, August 29, 1865.

Northern eyes, went to Benjamin F. Perry, who "wept like a child" at secession, but sat in the state legislature during the war. His main qualification, apart from prewar Unionism, was that he lived in the upcountry and had long opposed planters' domination of the state's politics.

Taken together, these men were loyal by almost any standard, although not all could take the Ironclad Oath. A majority would eventually make their way into the Republican party. Holden, Johnson, Perry, and Hamilton might be said to represent the upcountry constituencies so important to Johnson, while Sharkey, Marvin, and Parsons epitomized the Whiggish planters and entrepreneurs who had opposed secession. All, however, faced the identical task: building political support for themselves and the President in the aftermath of Johnson's proclamations. With both black suffrage and widespread white disenfranchisement apparently out of the question, the governors had little choice but to conciliate the majority of voters who had aided the Confederacy.[23]

In nineteenth-century America, patronage oiled the machinery of politics, and Johnson's governors possessed unprecedented patronage powers. Every state and local office stood vacant, subject to gubernatorial appointment pending new elections. By mid-August, Holden alone had named over 4,000 officials, ranging from mayors to judges and constables. Rather than fill these positions with unconditional Union men (whose numbers in some states did not suffice to staff all these offices), the governors used patronage to attract the support of a portion of the South's antebellum and Confederate political leadership—a course Wells had already embarked upon in Louisiana. Sharkey, Parsons, and Perry allowed local Confederate officeholders to remain at their posts. ("Monroe County," one Mississippian complained, "is in the hands of the vilest seceshionist in the state.") Perry and Johnson, each with a political base in the upcountry, sought via local appointments to placate the lowcountry aristocracy. Even Hamilton, who relied more heavily on wartime Unionists, still appointed prominent pro-Confederate citizens in plantation counties rather than attempt to bring new men to power. And Holden used patronage primarily to reward political friends, punish longtime antagonists, and expand his personal following. Whether an individual was a planter or yeoman, or had supported the Union or Confederacy, appeared less important than whether he had opposed Holden in the past

23. Carter, *When the War Was Over*, 25–26; Michael Perman, *Reunion Without Compromise: The South and Reconstruction 1865–1868* (New York, 1973), 62–69; Carl H. Moneyhon, *Republicanism in Reconstruction Texas* (Austin, 1980), 20–24; Harvey M. Watterson to Andrew Johnson, October 20, 1865, Johnson Papers; Otto H. Olsen, "Reconsidering the Scalawags," *CWH*, 12 (December 1966), 307; Olive H. Shadgett, "James Johnson, Provisional Governor of Georgia," *GaHQ*, 36 (March 1952), 1–3; Lillian A. Pereyra, *James Lusk Alcorn: Persistent Whig* (Baton Rouge, 1966), 13, 40–42; Jerrell H. Shofner, *Nor Is It Over Yet: Florida in the Era of Reconstruction, 1863–1877* (Gainesville, Fla., 1974), 34–36; Benjamin F. Perry to Benson J. Lossing, September 2, 1866, Benjamin F. Perry Papers, USC.

or seemed likely to threaten his reelection. All in all, the new governors' appointment policies sounded the death knell of wartime Unionists' hopes that Reconstruction would bring to power "a new class of politicians from the *plain* people."[24]

At the same time, the new governors moved to reassure their white constituents that emancipation did not imply any further change in the freedmen's status. Holden's newspaper listed "unqualified opposition to what is called negro suffrage" as one of six principles of Southern Unionism and, charged a North Carolina antislavery veteran, used his power as governor "to defeat every man who favored allowing any rights to the Negroes" whatever. Florida Governor Marvin insisted blacks must not "delude themselves" into believing that their fate had been central to the Civil War or that emancipation implied civil equality or the vote. He advised freedmen to return to the plantations, labor diligently, and "call your old Master—'Master'."[25]

To the bulk of white Southerners, these policies came as an unexpected tonic. In the immediate aftermath of defeat, a considerable number had been prepared to acquiesce in whatever directives emerged from Washington. Northern correspondent Whitelaw Reid probed the white South's mood in May and concluded that any conditions for reunion specified by the President, even black suffrage, would be "promptly accepted." By June, as Johnson's policy unfolded, Reid discerned a change in the Southern spirit. Now, relief at the mildness of Johnson's terms for reunion mingled with defiant talk of state rights and resistance to black suffrage. By midsummer, prominent whites realized that Johnson, once despised as "a renegade, demagogue, and drunkard," had left the Southern states in "the undisputed management of their own internal affairs." "Every political right which the State possessed under the Federal Constitution," declared Alabama Governor Parsons in July, "is hers today, with the single exception relating to slavery." White Southerners appreciated that Johnson's Reconstruction empowered them to shape the transition from slavery to freedom and define blacks' civil status without Northern interference. Harvey M. Watterson, a Tennessee Unionist dispatched in June on a Southern tour by the President, assured the region's leaders that Johnson would never abandon the position that suffrage belonged entirely to the states. He found the implications of Johnson's policies well

24. William W. Holden to Andrew Johnson, August 26, 1865, North Carolina Governor's Papers, NCDAH; Perman, *Reunion Without Compromise*, 115–19; unknown to William L. Sharkey, July 20, 1865, Mississippi Governor's Papers; Randolph B. Campbell, *A Southern Community in Crisis: Harrison County, Texas, 1850–1880* (Austin, 1983), 248–59; Carter, *When the War Was Over*, 32, 49–53; Thomas J. Rayner to Thaddeus Stevens, January 30, 1866, Thaddeus Stevens Papers, LC.

25. Raleigh *Standard*, August 3, 1865; Benjamin S. Hedrick to John Covode, June 22, 1868, Benjamin S. Hedrick Papers, DU; Thomas Wagstaff, "Call Your Old Master—'Master' ": Southern Political Leaders and Negro Labor During Presidential Reconstruction," *Labor History*, 10 (Summer 1969), 323–24.

understood. The President favored "a white man's government," and, reported Watterson, "more than anything else," Johnson's position "in regard to the negro suffrage," had inspired Southern whites of nearly all political persuasions to rally to his support.[26]

Events in the summer and fall of 1865 further encouraged white Southerners to look upon the President as their ally and protector. In August, as we have seen, Johnson overruled Freedmen's Bureau Commissioner Howard and ordered the return of confiscated and abandoned land to pardoned Southerners. In the same month, he sided with Mississippi Governor Sharkey in a dispute over the raising of a state militia. Fearing the force would be composed of Confederate veterans who could not be trusted to deal fairly with freedmen and Unionists, Maj. Gen. Henry W. Slocum prohibited its formation, only to see his order countermanded in Washington. In the fall, Johnson yielded to complaints by Louisiana whites and removed Bureau Assistant Commissioner Thomas W. Conway. And the President acquiesced in pleas for the removal of black troops, whose presence, "besides being a painful humiliation," was said to destroy plantation discipline. In Mississippi, one planter contrasted a neighboring county garrisoned by white troops, where "the negroes are made to work and behave," with his own where, because of the presence of black soldiers, "they do much less and their deportment is intolerable." In August, Johnson announced that black units would be removed from the South and within two years nearly all had been mustered out of the service.[27]

Johnson's pardon policy also reinforced his emerging image as the white South's champion. Despite talk of punishing traitors, the President embarked on a course of amazing leniency. No mass arrests followed the collapse of the Confederacy; only Henry Wirtz, commandant of Andersonville prison camp, paid the ultimate penalty for treason. Jefferson Davis spent two years in federal prison but was never put on trial and lived to his eighty-second year; his Vice President, Alexander H. Stephens, served a brief imprisonment, returned to Congress in 1873, and died ten years later as governor of Georgia. Almost from the outset of Presidential Reconstruction, moreover, the administration ignored the Ironclad Oath in making such politically important appointments as reve-

26. Reid, *After the War*, 18–19, 154–55, 219, 296–97; David Schenck Diary (typescript), May 1865, UNC; Joseph H. Parks, *Joseph E. Brown of Georgia* (Baton Rouge, 1976), 334–35; *Journal of the Proceedings of the Convention of the State of Alabama, Held in the City of Montgomery* . . . (Montgomery, 1865), 4; Martin Abbott, ed., "A Southerner Views the South, 1865: Letters of Harvey M. Watterson," *VaMHB*, 68 (October 1960), 478–89; Harvey M. Watterson to Andrew Johnson, June 20, 27, July 8, 1865, Johnson Papers.

27. William C. Harris, *Presidential Reconstruction in Mississippi* (Baton Rouge, 1967), 72–75; William S. McFeely, *Yankee Stepfather: O. O. Howard and the Freedmen* (New Haven, 1968), 172–75; Richard Jones and fifty-nine others to Lewis E. Parsons, September 26, 1865, Alabama Governor's Papers; "Bolivar" to William L. Sharkey, September 4, 1865, Mississippi Governor's Papers; Ira Berlin et al., eds., *Freedom: A Documentary History of Emancipation* (Washington, D.C. 1982–), Ser. 2, 733–37, 747–54.

Emancipated Negroes Celebrating the Emancipation Proclamation of President Lincoln": a scene in orthern Virginia near Winchester. (*Le Monde Illustré*, March 21, 1863)

Robert G. Fitzgerald in His Navy Uniform, 1863. After serving in both the Union Army and Navy, Fitzgerald became a school teacher in Virginia and North Carolina. (Estate of Pauli Murray)

nidentified Civil War Sergeant (Chicago Histori-
al Society)

Black Refugees Crossing the Rappahannock River, Virginia, 1862. (Library of Congress)

"Secret Meeting of Southern Unionists." (*Harper's Weekly*, August 4, 1866)

The Riots in New York: The Mob Lynching a Negro in Clarkson Street." (*Illustrated London News*, August 8, 1863)

Richmond Ruins, 1865. (Library of Congress)

PILLARS OF THE BLACK COMMUNITY: SCHOOL,
CHURCH, AND FAMILY

Laura M. Towne and Pupils, 1866. One of the origina "Gideonites," Towne taught on the South Carolina Sea Islands until her death in 1901. (New York Publi Library, Schomburg Center for Research in Black Cul ture)

"Prayer Meeting." (*Harper's Weekly*, February 2, 1867)

Unidentified Black Family. (New-York Historical Society)

TWO IMAGES OF THE FREEDMEN'S BUREAU

"The Freedmen's Bureau": the Bureau as promoter of racial peace in the postwar South. (*Harper's Weekly*, July 25, 1868)

Democratic Broadside, from Pennsylvania's Congressional and gubernatorial campaign of 1866. (Library of Congress)

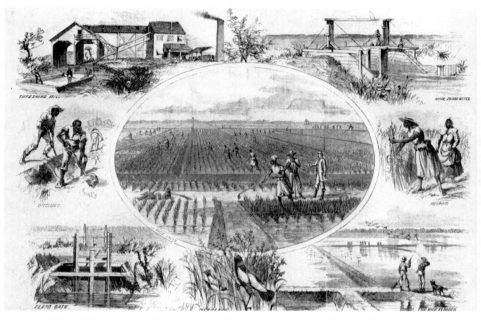

"Rice Culture on the Ogeechee, near Savannah, Georgia." (*Harper's Weekly*, January 5, 1867)

"The Sugar Harvest in Louisiana." (*Harper's Weekly*, October 30, 1875)

Cotton Pickers. (New-York Historical Society)

An Upcountry Family, Dressed in Homespun, Cedar Mountain, Virginia. (Library of Congress)

Andrew Johnson, Seventeenth President of the United States. (Library of Congress)

Thaddeus Stevens, Congressman from Pennsylvania, floor leader of House Republicans, and outspoken Radical. (Library of Congress)

Lyman Trumbull, Senator from Illinois, Chairman of the Judiciary Committee, and author of the Civil Rights and Freedmen's Bureau Bills of 1866. (Library of Congress)

Charles Sumner, Senator from Massachusetts and eloquent proponent of equality before the law. (Library of Congress)

nue assessor, tax collector, and postmaster. Simultaneously, Johnson's initial policy of excluding Confederate leaders and disloyal planters from political affairs unraveled. Some 15,000 Southerners, a majority barred from the general amnesty because of their wealth, filed applications for individual pardons. At first, the President granted pardons cautiously, but by September they were being issued wholesale, sometimes hundreds in a single day. By 1866, over 7,000 Southerners excluded from amnesty under the $20,000 clause had received individual pardons.[28]

Why the President so quickly abandoned the idea of depriving the prewar elite of its political and economic hegemony has always been something of a mystery. Johnson, a hostile Republican Congressman would comment in 1866, "is no poor white trash now. They have taken him on their platform at last, and he is wonderfully elevated and elated." Did flattery by planters and their wives play upon the Presidential ego and lead Johnson to abandon a planned assault upon the Confederacy's elite? More likely, he came to view cooperation with the planters as indispensable to two interrelated goals—white supremacy in the South and his own reelection as President. Blacks' unexpected militancy in 1865 may well have hardened Johnson's prejudices, convincing him that the freedmen were liable to descend into idleness, crime, and dissipation, and simultaneously leading him to reevaluate his traditional hostility to the planter class. This was the portrait he sketched to the British Ambassador, Sir Frederick Bruce, who, after conversations with Johnson and Seward, recorded their belief that blacks needed to be kept "in order" while receiving "the care and civilizing influence of dependence on the white man." Only planters could supervise and control the black population, but once entrusted with this responsibility, they could hardly be barred from a political role within society at large. Then too, the support of prominent Southerners seemed essential if Johnson were to fulfill his dream of a second term in the White House, especially since Unionist yeomen did not seem to be stepping forward to claim political power. Already, Southern leaders openly predicted a breach between Johnson— the "standard bearer of the South"—and the Radicals on the issue of black suffrage, and called for the formation of a new party to "rally around" the President "and sweep away everything that opposes [him]."[29]

The white South's identification with Johnson as a protector against the

28. Roy F. Nichols, "United States vs. Jefferson Davis," *AHR*, 31 (January 1926), 266–84; Thomas and Hyman, *Stanton*, 447–52; Beale, ed., *Welles Diary*, 2:357–58; Jonathan T. Dorris, *Pardon and Amnesty Under Lincoln and Johnson* (Chapel Hill, 1953), 135–46; McKitrick, *Andrew Johnson*, 142–48; 39th Congress, 1st Session, House Executive Document 99, 16.

29. Cornelius Cole to Olive Cole, March 6, 1866, Cole Papers; Eric Foner, ed., "Andrew Johnson and Reconstruction: A British View," *JSH*, 41 (August 1975), 389; Harvey M. Watterson to Andrew Johnson, July 9, 1865, Johnson Papers; J. G. de Roulhac Hamilton and Max R. Williams, eds., *The Papers of William A. Graham* (Raleigh, 1957–), 6:324; Raleigh *Standard*, October 12, 1865.

"ultra fanatics" of the North quickly rendered serious discussion of alternatives to a "white man's government" all but impossible. A few prominent Southerners departed from the regional consensus to advocate some form of limited black suffrage based upon property and educational qualifications. Antebellum newspaper editor and Confederate army officer Alfred M. Waddell of North Carolina pointed out that abolition had made blacks part of the "people"; to exclude them from a voice in public affairs would be "a bitter satire upon our 'free institutions'." While confined in a Boston prison, Alexander H. Stephens developed a plan whereby qualified blacks on a separate voting list would elect their own representatives to the legislature.[30]

Most of these suggestions, however, remained private, for Southerners who publicly advocated any form of black voting found themselves subject to tremendous abuse. Imprisoned Confederate Postmaster General John H. Reagan's public letter urging limited black suffrage created a situation in which, as a former governor of the state informed him, "every man in Texas who expects to be a candidate for anything from governor to constable seems to regard it as his duty to denounce you morning, noon, and night." Had Johnson lent his support, the opinions of men like Reagan might have carried considerable weight; in the absence of Presidential backing, their suggestions went unheeded. Years later, Christopher G. Memminger admitted that the South ought to have adopted a "different course as to the negroes" in 1865, but, he explained, Andrew Johnson "held up before us the hope of a 'white man's government,' and . . . it was natural that we should yield to our old prejudices."[31]

The rapid growth of white support for his Presidency ought to have given Johnson the power to shape Southern events. Instead, his identification with the new governments became so close that failings on their part inevitably reflected upon his leadership. Disturbing reports about his governors' actions, such as a letter by a Presidential emissary charging that South Carolina Governor Perry had "put upon their legs a set of men who . . . like the Bourbons have learned nothing and forgotten nothing," evoked no reevaluation of policy. Lincoln had always insisted his Reconstruction policies were experiments that did not rule out other approaches; he would never have allowed his own position to be compromised by shortcomings of the South. Yet as Southern constitutional conventions assembled in the summer and fall of 1865, Johnson, instead of master of developments in the South, increasingly appeared to be their prisoner.[32]

30. Solomon Cohen to James Johnson, September 11, 1865, Georgia Governor's Papers, UGa; Perman, *Reunion Without Compromise*, 152; James L. Lancaster, "The Scalawags of North Carolina, 1850–1868" (unpub. diss., Princeton University, 1974), 80; Alexander H. Stephens to Linton Stephens, July 4, 1865, Alexander H. Stephens Papers, Manhattanville College; Avary, ed, *Recollections of Stephens*, 268–73.

31. John H. Reagan, *Memoirs* (New York, 1906), 227–34; Frederic Bancroft, ed., *Speeches, Correspondence and Political Papers of Carl Schurz* (New York, 1913), 2:256.

32. William W. Boyce to Francis P. Blair, October 7, 1865, Johnson Papers.

If Presidential Reconstruction were to bring to power a new Unionist political leadership, the election of convention delegates in the summer of 1865 provided the opportunity. Few high Confederate officials or men of wealth had yet received individual pardons, and the politically discredited architects of secession did not seek election. As a result, over two thirds of those elected had opposed secession in 1860. Most were former Whigs, many of whom had held office before the war, but the top level of the antebellum political leadership was conspicuous by its absence. Alabama's convention, according to one member, contained few men "of real ability" and "a good many chuckle heads." In northern Alabama, western North Carolina, and loyalist enclaves like Jones County, Mississippi, unconditional Unionists who could take the Ironclad Oath, some of them Union army veterans, supplanted the prewar leadership. Such delegates, however, were far outnumbered by Southern Whigs who had served the Confederacy in one capacity or another. Nor did the delegates, mostly planters, professionals, and businessmen, differ substantially in occupation from their prewar counterparts. If the elections repudiated the South's prewar secessionist leadership, they did not herald the coming to power of those who had actively opposed the Confederacy, or of previously subordinate social classes.[33]

For the Unionist Whigs who dominated the conventions, Johnson's conditions for Reconstruction ought to have appeared mild indeed. Initially, delegates had only to acknowledge the abolition of slavery and repudiate secession; in October, the President also directed them to void state debts incurred in aid of the Confederacy. These measures merely recognized inescapable consequences of Confederate defeat; no one realistically believed slavery or secession could be revived or that Confederate money and bonds would regain their value. ("We cannot now, if left alone, reassert dominion over the slave," observed a prominent Southern jurist.) Yet, although conscious that their every action received careful scrutiny in the North, the conventions became embroiled in "petty and rancorous" disputes that undermined confidence in the President's policy and cast doubt on the willingness even of self-styled Unionists to abandon prewar beliefs and prejudices.[34]

First to assemble, in mid-August, was Mississippi's convention, composed almost entirely of former Whigs. It immediately embarked upon what one delegate called "ceaseless wrangling over an immaterial issue"—the precise wording of a constitutional amendment prohibiting slavery. Some delegates, having campaigned on platforms calling for

33. Carter, *When the War Was Over*, 65; Joshua B. Moore Diary (typescript), September 17, 1865, ASDAH; Winbourne M. Drake, "The Mississippi Reconstruction Convention of 1865," *JMH*, 21 (October 1959), 231–34; Malcolm C. McMillan, *Constitutional Development in Alabama 1798–1901: A Study in Politics, the Negro, and Sectionalism* (Chapel Hill, 1955), 92–93; C. Mildred Thompson, *Reconstruction in Georgia* (New York, 1915), 148–49.

34. Perman, *Reunion Without Compromise*, 70–76; Hamilton and Williams, eds., *Graham Papers*, 6:326; Carter, *When the War Was Over*, 63.

federal compensation for the South's slaves, opposed any language that might appear to surrender this claim. Others insisted that since the state had never legally been out of the Union, it had no obligation to accede to Presidential dictation. Even the more prudent majority, who favored unconditional acceptance of the President's requirements, did not hesitate to denounce Northern Radicalism and predict the formation of new political coalition uniting white Southerners with "the great conservative party of the North," and powerful enough to "control the next Presidential election." In the end, the convention adopted a simple acknowledgment of abolition that sidestepped the question of responsibility: "The institution of slavery having been destroyed," it would no longer exist in Mississippi. Subsequent conventions, some after debates as damaging, in Northern eyes, as Mississippi's, adopted similar wording, although Georgia specified that abolition did not imply the relinquishment of slaveowners' claims to compensation.[35]

The repudiation of secession and state debts sparked similar debates. Many unconditional Unionists insisted that the Confederacy had never enjoyed legal standing and wanted the secession ordinances declared "null and void." Some added that all actions of wartime state governments should be deprived of legal validity, a course that would have thrown into question thousands of state actions having nothing to do with secession, including court judgments, property transfers, and acts of incorporation. Other delegates argued for the "repeal" of secession ordinances; to act otherwise would cast aspersion on the "intelligence and patriotism" of prewar Southern leaders. In the end, most conventions adopted language declaring secession null and void, although in Mississippi "repeal" failed by only two votes. As for the repudiation of Confederate state debts, totaling some $54 million, wartime Unionists rather than the President initially raised this demand. Only when South Carolina took no action on its debt and North Carolina strongly resisted this "humiliating act" did Johnson publicly require the voiding of "every dollar." (North Carolina, remarked diarist Cornelia Phillips Spencer, seemed "always in the rear"—last to support secession, it now lagged "in progress toward reconciliation or reunion.")[36]

If the debates on abolition, secession, and state debts revealed a reluctance to confront forthrightly the consequences of the Civil War, in other ways the conventions seized the opportunities for change created by the Confederacy's defeat. Upcountry delegates, in particular, pressed for long-desired changes in the region's political structure. They were most successful in South Carolina, under whose antiquated political system

35. *Journal of the Proceedings and Debates in the Constitutional Convention of the State of Mississippi* (Jackson, 1865), 44–70, 105–108, 165; Drake, "Mississippi Convention," 229–31; Carter, *When the War Was Over*, 83; McPherson, *Political History*, 20.

36. *Journal of Mississippi Convention*, 173–79, 219–20; McMillan, *Constitutional Development*, 95–98; Carter, *When the War Was Over*, 68–74; Hope S. Chamberlain, *Old Days in Chapel Hill* (Chapel Hill, 1926), 118.

state officials and Presidential electors had been chosen not by popular vote, but by a legislature dominated by the coastal parishes. For thirty years, lowcountry planters had resisted demands for democratic reform; now, the upcountry insisted, the time had arrived to "lop off all the aristocratic features of the State government, and make it a people's state." At the urging of Governor Perry and with the support of the President, the convention provided for the popular election of the governor, abolished property qualifications for membership in the legislature, and adjusted the system of representation so as to "give the power to the upper counties almost entirely." In Alabama, too, the convention adopted the "white basis" of legislative apportionment, a victory for the upcountry in its long-standing campaign to reduce the political power of the plantation region. The black belt, declared an upcountry Unionist, had had its "day" in state politics: "Let us have ours." Elsewhere, the forces of change proved less successful. Florida's constitution continued to use the federal ratio (whites plus three fifths of blacks) as the basis of apportionment, while Georgia adopted a complicated formula that for the first time counted the entire black population, thus enhancing the power of the plantation counties.[37]

If nothing else, these debates revealed that long-standing divisions in the Southern polity had survived the Civil War. Yet when it came to the status of the freedmen, there appeared to be little difference between the views of upcountry and lowcountry, Democrat and Whig. A handful of delegates, such as Texas Unionist Edward Degener, called for extending the suffrage to literate blacks, but in general the idea was scarcely contemplated. Even among unconditional Unionists, the demand for democratic reform meant enhancing the political power of those counties where whites predominated, a goal that would be fatally subverted were blacks included as part of the political nation. In his message to the South Carolina convention, Governor Perry cited the Dred Scott decision to demonstrate that blacks could not be citizens, railed against the presence of black troops, and insisted black suffrage, like representation based on total population, would throw political power into the hands of "the man of wealth and large landed possessions." Black belt representatives, however, did not seem to agree that it would serve their interests to enfranchise the freedmen. A Mississippi delegate expressed the prevailing opinion: "'tis nature's law that the superior race must rule and rule they will."[38]

37. William W. Boyce to Andrew Johnson, June 23, 1865, Johnson Papers; Sidney Andrews, *The South Since the War* (Boston, 1866), 47–83; Daniel E. Huger Smith et al., eds., *Mason Smith Family Letters 1860–1868* (Columbia, S.C., 1950), 235; McMillan, *Constitutional Development*, 104–105; Jeremiah Clements to Andrew Johnson, April 21, 1865, Johnson Papers; Shofner, *Nor Is It Over*, 42; Francis L. Thorpe, ed., *The Federal and State Constitutions* (Washington, D.C., 1909), 2:809–22.

38. Charles W. Ramsdell, *Reconstruction in Texas* (New York, 1910), 101; Columbia *Daily Phoenix*, September 4, 1865; Lillian A. Kibler, *Benjamin F. Perry: South Carolina Unionist*

With Johnson's requirements fulfilled, the South in the fall of 1865 proceeded to elect legislators, governors, and members of Congress. In a majority of the states, former Whigs who had opposed secession swept to victory. Of seven Southern governors elected in 1865, six had been antisecessionist Whigs, and the same group dominated the new legislatures and Congressional delegations. "The old Unionists are now having it all their own way," a North Carolinian remarked that summer, and Southerners believed they had met the last prerequisite for reunion, choosing loyal men to direct their state governments and represent them in Washington. A closer look at the 1865 elections, however, discloses a striking difference between results in Upper South states where wartime Reconstruction governments had survived into 1865, and those that had experienced only Presidential Reconstruction. Of twenty-five men sent to Congress from Arkansas, Tennessee, and Virginia, five had served in the Union Army, many others had aided the federal war effort, and nearly all could take the Ironclad Oath. Even here, however, former Confederates secured many local offices. Most disconcerting, perhaps, was the election in Nashville, which ousted Johnson's wartime city government, the mayor he had appointed receiving only 10 percent of the vote.[39]

Farther South, although former Whigs controlled the elections, service to the Confederacy emerged as a virtual prerequisite for victory. The vast majority of the new Senators and Congressmen had opposed secession, yet nearly all had followed their states into the rebellion. Apart from western North Carolina, which sent federal army veteran Alexander H. Jones to Congress, and a few upcountry counties in Alabama and Mississippi, men who had actively aided the Union were resoundingly defeated. In South Carolina, James L. Orr, a longtime upcountry critic of the state's planter-dominated political system, only narrowly defeated a write-in vote for Confederate Gen. Wade Hampton, who had refused to be a candidate. Georgia's legislature, dominated by Whig opponents of secession, disregarded the President's wishes and sent Confederate Vice President Alexander H. Stephens to the Senate.[40]

Probably the most closely watched contest occurred in North Carolina, where Jonathan Worth, a Unionist Whig and Confederate state treasurer,

(Durham, 1946), 408–409; Samuel Matthews to Robert Matthews, November 19, 1865, James and Samuel Matthews Papers, MDAH.

39. Carter, *When the War Was Over*, 94, 229–30; David Schenck Diary (typescript), August 16, 1865, UNC; Alan G. Bromberg, "The Virginia Congressional Elections of 1865: A Test of Southern Loyalty," *VaMHB*, 84 (January 1976), 76–98; Peter Maslowski, *Treason Must Be Made Odious: Military Occupation and Wartime Reconstruction in Nashville, Tennessee, 1862–1865* (Millwood, N.Y., 1978), 147–48.

40. Raleigh *Standard*, December 12, 1865; 39th Congress, 1st Session, House Report 30, pt. 3:8; W. J. Brantley to Benjamin G. Humphreys, August 8, 1866, Mississippi Governor's Papers; Perman, *Reunion Without Compromise*, 158–64.

defeated Governor Holden. Despite Worth's liabilities as a candidate (he had made his living before the war as a lawyer collecting debts due to merchants and businessmen), and Holden's frantic efforts to identify his own candidacy with the President while painting his opponent as a stalking horse for secessionists, Worth swept the eastern and middle counties of the state, overcoming Holden's large majority in the west. Once in office, he quickly restored the old elite, whose power Holden had to some extent challenged, to control of local affairs. The result confirmed the continued dominance of wartime political leadership in a state with one of the South's largest populations of nonslaveholding yeomen. Even more ominous was the outcome in Louisiana, where Governor Wells won reelection as the candidate of both Democrats and Conservative Unionists. Here, where New Orleans free blacks and Radical Unionists had for months been advocating black suffrage, secessionist Democrats assumed political power on a platform declaring that Louisiana government should be "perpetuated for the exclusive benefit of the white race." In the face of black demands for political rights, "persistent Whiggery" had entirely disappeared.[41]

All in all, the 1865 elections threw into question the future of Presidential Reconstruction. Johnson himself sensed that something had gone awry: "There seems, in many of the elections," he wrote at the end of November, "something like defiance, which is all out of place at this time." The stark truth was that outside the Unionist mountains, Johnson's policies had failed to create a new political leadership to replace the prewar "slaveocracy"—partly because the President himself had so quickly aligned with portions of the old elite. If the architects of secession had been repudiated, the South's affairs would still be directed by men who, while Unionist in 1860, formed part of the antebellum political establishment. Their actions would go a long way toward determining the fate of Johnson's Reconstruction experiment. For, as James H. Bell, an unsuccessful candidate for the Texas constitutional convention, prophetically observed, if these governments "refuse to do what reasonably may be demanded as a condition of restoration . . . the people of the North may easily be driven to act upon those views which are now considered radical."[42]

---

41. Richard L. Zuber, *Jonathan Worth* (Chapel Hill, 1965), 78; Jonathan Worth to Benjamin S. Hedrick, October 18, 1865, Hedrick Papers; J. G. de Roulhac Hamilton, *Reconstruction in North Carolina* (New York, 1914), 136–41; Paul D. Escott, *Many Excellent People: Power and Privilege in North Carolina, 1850–1900* (Chapel Hill, 1985), 101; Mark W. Summers, "The Moderates' Last Chance: The Louisiana Election of 1865," *LaH*, 24 (Winter 1983), 56–69; Lowrey, "Wells," 1056.
42. Perman, *Reunion Without Compromise*, 107; Carter, *When the War Was Over*, 271; John P. Carrier, "A Political History of Texas During the Reconstruction, 1865–1874" (unpub. diss., Vanderbilt University, 1971), 30–31.

The Anatomy of Presidential Reconstruction

Although Southern whites faced many challenges in the aftermath of the Civil War, one problem took precedence as the new legislatures assembled. As William H. Trescot explained to South Carolina's governor in December 1865, "you will find that this question of the control of labor underlies every other question of state interest." The ferment in the countryside, the history of other societies that had experienced emancipation, and ideologies and prejudices inherited from slavery, combined to convince the white South that only through some form of coerced labor could the production of plantation staples be resumed. The transformation of blacks from property to prospective citizens, however, required a fundamental change in the way their labor was regulated. "Our little sovereignties and Feudal arrangements are all levelled to the ground," bemoaned one former slaveowner, and with the collapse of their personal authority over blacks, planters turned to the state to reestablish labor discipline. Laws regarding labor, property rights, taxation, the administration of justice, and education were of one piece: All formed part of a broad effort to employ the state's power to shape the new social relations arising from slavery. Thus began the effort, as one early historian of Reconstruction described it, "to put the state much in the place of the former master."[43]

Virtually from the moment the Civil War ended, the search began for legal means of subordinating a volatile black population that regarded economic independence as a corollary of freedom and the old labor discipline as a badge of slavery. Many localities in the summer of 1865 adopted ordinances limiting black freedom of movement, prescribing severe penalties for vagrancy, and restricting blacks' right to rent or purchase real estate and engage in skilled urban jobs. Opelousas, Louisiana, even established a pass system and curfew for blacks and barred them from living in town except as servants—regulations that attracted national attention after a local free black dispatched a copy to the New Orleans *Tribune*. Labor policy, however, could hardly be settled by a series of local measures. As the gathering of the new legislatures neared, the Southern press and the private correspondence of planters resounded with calls for what a New Orleans newspaper called "a new labor system . . . prescribed and enforced by the State." "I have but little faith in Negro labour unless our states can make sufficient laws to require the negro to remain at one place and labour," wrote a Mississippi planter. Another proposed that each plantation be declared a town, "with the planter as Judge of Police," in effect resurrecting the slaveholder's per-

43. William H. Trescot to James L. Orr, December 13, 1865, South Carolina Governor's Papers; D. B. McLurin to N. R. Middleton, September 26, 1867, Middleton Papers, Langdon Cheves Collection, SCHS; William W. Davis, *The Civil War and Reconstruction in Florida* (New York, 1913), 425.

sonal authority under the aegis of the law. Others suggested driving into the countryside freedmen who congregated in towns and cities. But whatever the proposal, Southern whites seemed convinced that, if blacks remained averse to plantation labor, "we can control by wise laws."[44]

Andrew Johnson had by now guaranteed the white South a virtual free hand in regulating the region's internal affairs. And throughout this period, as a Georgia plantation agent wrote, state rights included "the right and power to govern our population in our own way." Several of the new legislatures proved reluctant to ratify the Thirteenth Amendment abolishing slavery, for fear it opened the door to national legislation concerning "the *negro* question." Mississippi rejected the Amendment entirely, claiming it would provide a field day for "radicals and demagogues," while several states ratified with the "understanding" that Congress lacked the power to determine the future of the former slaves.[45]

The initial response to demands for a legislative solution to the "labor problem" was embodied in the Black Codes, a series of state laws that would play a crucial role in the undoing of Presidential Reconstruction. Intended to define the freedmen's new rights and responsibilities, the codes authorized blacks to acquire and own property, marry, make contracts, sue and be sued, and testify in court in cases involving persons of their own color. But their centerpiece was the attempt to stabilize the black work force and limit its economic options apart from plantation labor. Henceforth, the state would enforce labor agreements and plantation discipline, punish those who refused to contract, and prevent whites from competing among themselves for black workers. The codes amply fulfilled Radical Benjamin F. Flanders' prediction as Louisiana's legislature assembled: "Their whole thought and time will be given to plans for getting things back as near to slavery as possible."[46]

Mississippi and South Carolina enacted the first and most severe Black Codes toward the end of 1865. Mississippi required all blacks to possess, each January, written evidence of employment for the coming year. Laborers leaving their jobs before the contract expired would forfeit wages already earned, and, as under slavery, be subject to arrest by any white citizen. A person offering work to a laborer already under contract risked imprisonment or a fine of $500. To limit the freedmen's economic oppor-

44. Leon F. Litwack, *Been in the Storm So Long: The Aftermath of Slavery* (New York, 1979), 319–22; Edward Jackson to Henry C. Warmoth, April 3, 1867, Henry C. Warmoth Papers, UNC; *CG*, 39th Congress, 1st Session, 517; William E. Highsmith, "Louisiana Landholding During War and Reconstruction," *LaHQ*, 38 (January 1955), 44; John O. Grisham to Colonel and Mrs. Joseph J. Norton, October 7, 1865, Joseph J. Norton Papers, USC; I. Pierce to William L. Sharkey, July 15, 1865, Mississippi Governor's Papers; Raleigh *Semi-Weekly Record*, August 23, 26, 1865.

45. Audley Couper to Francis P. Corbin, July 28, 1866, Francis P. Corbin Papers, NYPL; Samuel L. M. Barlow to Montgomery Blair, November 13, 1865, Samuel L. M. Barlow Papers, HL; *AC*, 1865, 18–19; McPherson, *Political History*, 22–25.

46. Benjamin F. Flanders to Henry C. Warmoth, November 23, 1865, Warmoth Papers.

tunities, they were forbidden to rent land in urban areas. Vagrancy—a crime whose definition included the idle, disorderly, and those who "misspend what they earn"—could be punished by fines or involuntary plantation labor; other criminal offenses included "insulting" gestures or language, "malicious mischief," and preaching the Gospel without a license. In case anything had been overlooked, the legislature declared all penal codes defining crimes by slaves and free blacks "in full force" unless specifically altered by law.[47]

South Carolina's Code was in some respects even more discriminatory, although it contained provisions, such as prohibiting the expulsion of aged freedmen from plantations, designed to reinvigorate paternalism and clothe it with the force of law. It did not forbid blacks to rent land, but barred them from following any occupation other than farmer or servant except by paying an annual tax ranging from $10 to $100 (a severe blow to the free black community of Charleston and to former slave artisans). The law required blacks to sign annual contracts and included elaborate provisions regulating relations between "servants" and their "masters," including labor from sunup to sundown and a ban on leaving the plantation, or entertaining guests upon it, without permission of the employer. A vagrancy law applied to unemployed blacks, "persons who lead idle or disorderly lives," and even traveling circuses, fortune tellers, and thespians.[48]

The uproar caused in the North by these laws led other Southern states to modify the language, if not the underlying purpose, of early legislation regarding the freedmen. Virtually all the former Confederate states enacted sweeping vagrancy and labor contract laws, supplemented by "antienticement" measures punishing anyone offering higher wages to an employee already under contract. Virginia attempted to outlaw collective action for higher pay by including within the definition of vagrancy those who refused to work for "the usual and common wages given to other laborers." Florida's code, drawn up by a commission whose report praised slavery as a "benign" institution deficient only in its inadequate regulation of black sexual behavior, made disobedience, impudence, and even "disrespect" to the employer a crime. Blacks who broke labor contracts could be whipped, placed in the pillory, and sold for up to one year's labor, while whites who violated contracts faced only the threat of civil suits. Louisiana and Texas, seeking to counteract the withdrawal of black women from field labor, mandated that contracts "shall embrace the labor of all the members of the family able to work." Louisiana also provided that all disputes between the employer and his laborers "shall

47. Harris, *Presidential Reconstruction*, 99–100, 112–15, 130–31; 39th Congress, 2d Session, Senate Executive Document 6, 192–95.
48. Francis B. Simkins and Robert H. Woody, *South Carolina During Reconstruction* (Chapel Hill, 1932), 48–50; 39th Congress, 2d Session, Senate Executive Document 6, 218–19.

be settled . . . by the former." Unlike the Mississippi and South Carolina codes, many subsequent laws made no reference to race, to avoid the appearance of discrimination and comply with the federal Civil Rights Act of 1866. But it was well understood, as Alabama planter and Democratic politico John W. DuBois later remarked, that "the vagrant contemplated was the plantation negro."[49]

Although all of these measures inspired protest from blacks, the most bitter complaints centered on apprenticeship laws, which seized upon the consequences of slavery—the separation of families and the freedmen's poverty—as an excuse for providing planters with the unpaid labor of black minors. Generally, these laws allowed judges to bind to white employers black orphans and those whose parents were deemed unable to support them. The former owner usually had first preference, the consent of the parents was not required, and the law permitted "moderate corporal chastisement." Although apprenticeship had a venerable history in Europe and America, these arrangements bore little resemblance to the traditional notion of training youths in a skilled trade. Indeed, in some areas, courts bound out individuals for uncompensated labor who could hardly be considered minors; one tenth of the apprentices in one North Carolina county exceeded the age of sixteen, including an "orphan" working at a turpentine mill and supporting his wife and child. To blacks, such apprenticeships represented nothing less than a continuation of slavery. In Maryland and North Carolina, courts bound out thousands to white "guardians" without the consent, sometimes without the knowledge, of their parents. Blacks deluged the Freedmen's Bureau with demands for help in releasing their own children or those of deceased relatives. "Surely the law does not Call for Children to be bond out," wrote one freedman, "when their peopel is Abel to Keep them." "I think very hard of the former oners," declared another, "for Trying to keep My blood when I kno that Slavery is dead." As late as the end of 1867, Bureau agents and local justices of the peace were still releasing black children from court-ordered apprenticeships.[50]

49. William Cohen, "Negro Involuntary Servitude in the South, 1865–1940: A Preliminary Analysis," *JSH*, 42 (February 1976), 35–50; Theodore B. Wilson, *The Black Codes of the South* (University, Ala., 1965), 96–100; 39th Congress, 2d Session, Senate Executive Document 6, 172–77, 180–83, 222–26; John T. O'Brien, "From Bondage to Citizenship: The Richmond Black Community, 1865–1867" (unpub. diss., University of Rochester, 1974), 304; Michael Wayne, *The Reshaping of Plantation Society: The Natchez District, 1860–1880* (Baton Rouge, 1983), 46–47; John W. DuBose, *Alabama's Tragic Decade*, edited by James K. Greer (Birmingham, 1940), 55.

50. 39th Congress, 2d Session, Senate Executive Document 6, 172–74, 180–81, 190, 209–10; Roberta S. Alexander, *North Carolina Faces the Freedmen: Race Relations During Presidential Reconstruction, 1865–67* (Durham, 1985), 115–17; W. McKee Evans, *Ballots and Fence Rails: Reconstruction on the Lower Cape Fear* (Chapel Hill, 1967), 74–75; Joseph P. Reidy, "Masters and Slaves, Planters and Freedmen: The Transition from Slavery to Freedom in Central Georgia, 1820–1880" (unpub. diss., Northern Illinois University, 1982), 216; Charles M. Hooper to Wager Swayne, April 20, 1867, Wager Swayne Papers, ASDAH;

The statutes regulating labor and apprenticeship, in the words of Northern reporter Sidney Andrews, "acknowledge the overthrow of the special servitude of man to man, but seek . . . to establish the general servitude of man to the commonwealth." The same was true of new criminal laws designed to enforce the property rights of landowners against the claims of their former slaves. Under slavery the boundary between public and private authority had been indefinite; crimes like theft, looked upon as labor troubles, were generally settled by planters themselves. "I ain't never heard nothin' 'bout no jails in slavery time," a Georgia freedman later remarked. "What dey done den was 'most beat de life out of de niggers to make 'em behave." Abolition obviously required a restructuring of enforcement machinery. As George A. Trenholm, a prominent South Carolina merchant, explained soon after the end of the Civil War: "Hitherto these depredations were either overlooked, or the culprit punished lightly and restored to favor. Now it must necessarily be different. Theft is no longer an offense against his master, but a crime against the State."[51]

In the aftermath of abolition, as under slavery, planters complained of widespread theft by blacks. Plantation laborers, wrote an Alabama planter's son, "are errant rogues and steal everything they can. Horses and jewelry, poultry and pork, vegetables and fuel and every conceivable kind of property, are subject to their thefts." One man's theft, however, may be another's "right," especially where the former's claim to his property is open to question. One black preacher, imprisoned for larceny, had been "known to say from his pulpit that it was no harm to steal from white people, that his hearers would only be getting back what belonged to them." For their part, legislators moved to bolster whites' hold upon their property by sharply increasing the penalty for petty larceny. Virginia and Georgia in 1866 made the theft of a horse or mule a capital crime. South Carolina required blacks employed in agriculture to present written authorization from their "masters" before selling farm produce. And North Carolina, at the urging of former Gov. William A. Graham, made "the *intent* to steal" a crime and decreed that all attempted thefts, even if unsuccessful, should be prosecuted as larceny.[52]

Rebecca Scott, "The Battle over the Child: Child Apprenticeship and the Freedmen's Bureau in North Carolina," *Prologue*, 10 (Summer 1978), 101–13.

51. Sidney Andrews, "Three Months Among the Reconstructionists," *Atlantic Monthly*, 17 (February 1866), 244; George A. Trenholm to Alfred G. Trenholm, October 7, 1865, George A. Trenholm Papers, USC.

52. Eric Foner, *Nothing But Freedom: Emancipation and Its Legacy* (Baton Rouge, 1983), 57–58; John Sanford to John W. A. Sanford, February 25, 1866, John W. A. Sanford Papers, ASDAH; D. M. Carter to Tod R. Caldwell, May 19, 1872, North Carolina Governor's

Simultaneously, Southern lawmakers moved to limit blacks' independent access to economic resources. Rights such as hunting, fishing, and
the free grazing of livestock, which whites took for granted and many
blacks had enjoyed as slaves, were now, in some areas, transformed into
crimes. Planters resented hunting as a way blacks could obtain subsistence while avoiding plantation labor; it also often involved trespass, thus
flouting whites' property rights. Several states now made it illegal for
blacks to own weapons, or imposed taxes on their dogs and guns. Georgia
in 1866 outlawed hunting on Sundays in counties with large black populations, and forbade the taking of timber, berries, fruit, or anything "of any
value whatever" from private property. Meanwhile, the first steps were
taken to restrict livestock from ranging freely on unfenced land, a tradition deeply valued both by upcountry yeomen and the freedmen. Although laws requiring livestock owners to fence in their animals (thus
making it impossible for the landless to own pigs or cattle) remained a
point of controversy into the twentieth century, the initial steps toward
closing the Southern range came during Presidential Reconstruction.
Often, however, these early fence laws applied only to black belt counties.
Clearly, these measures served more than one purpose—the ban on
owning guns reflected, in part, white fears of black insurrection, and
agricultural reformers had long complained of the expense of fencing
large tracts of land to protect crops from roaming animals. But by limiting
hunting and fishing and requiring fencing, the law made it more difficult
for blacks to obtain food or income without working on plantations.[53]

The entire complex of labor regulations and criminal laws was enforced by a police apparatus and judicial system in which blacks enjoyed
virtually no voice whatever. Whites staffed urban police forces as well as
state militias, intended, as a Mississippi white put it in 1865, to "keep
good order and discipline amongst the negro population." Although
disorder was hardly confined to blacks, virtually all the militiamen patrolled black belt counties. Often composed of Confederate veterans still
wearing their gray uniforms, they frequently terrorized the black population, ransacking their homes to seize shotguns and other property and
abusing those who refused to sign plantation labor contracts. Louisiana

nor's Papers; Alrutheus A. Taylor, *The Negro in the Reconstruction of Virginia* (Washington,
D.C., 1926), 21; *Georgia Acts 1865–66*, 232; *South Carolina Acts 1865*, 274; Alexander, *North
Carolina*, 48.

53. Foner, *Nothing But Freedom*, 62–66, 115–17; Charles L. Flynn, *White Land, Black Labor:
Caste and Class in Late Nineteenth-Century Georgia* (Baton Rouge, 1983), 123–25; Steven H.
Hahn, *The Roots of Southern Populism: Yeoman Farmers and the Transformation of the Georgia
Upcountry 1850–1890* (New York, 1983), 58–61; 39th Congress, 2d Session, Senate Execu-
tive Document 6, 174, 183–84, 195–96; *Fur, Fin and Feather: A Compilation of the Game Laws
of the Principal States and Provinces of the United States and Canada* (New York, 1872 ed.), 142–50;
J. Crawford King, "The Closing of the Southern Range: An Exploratory Study," *JSH,* 48
(February 1982), 53–70.

blacks called the militia the "patrol," a reminder of slavery days, and could not understand "why men who but a few months since were in armed rebellion against the government should now have arms put in their hands."[54]

Nor did the courts offer hope of impartial justice. The initial prohibitions on blacks testifying in cases involving whites were soon rescinded, although for reasons having little to do with equity. When Alabama Governor Parsons in 1865 requested a group of distinguished lawyers to report on the advisability of allowing blacks to testify against whites, the group described such a step as "distasteful, injurious, and humiliating," but recommended its adoption to rid the state of military and Freedmen's Bureau courts and to appease Northern public opinion. By mid-1866, most of the Southern states allowed blacks to testify on the same terms as whites. But, as a British barrister noted after observing Richmond's courts early in 1867, with blacks barred from jury service, "the verdicts are always for the white man and against the colored man." Some states attempted to discourage blacks from going to court at all. North Carolina awarded landlords double settlements in cases in which a tenant unsuccessfully appealed a local court decision, and Mississippi provided that a black who "falsely and maliciously" brought legal charges against a white could be fined, imprisoned, and hired out to labor.[55]

Sheriffs, justices of the peace, and other local officials proved extremely reluctant to prosecute whites accused of crimes against blacks. To do so, said a Georgia sheriff, would be "unpopular" and dangerous, while an Arkansas counterpart told a Bureau agent that to take action against a planter who had defrauded freedmen "would defeat him in the coming fall election." When civil authorities or Bureau agents did bring such cases to court, "it seldom results in anything but the acquittal of the criminal," complained South Carolina Bureau head Robert K. Scott. If convictions did follow, judges imposed sentences far more lenient than blacks received for the same crimes. Texas courts indicted some 500 white men for the murder of blacks in 1865 and 1866, but not one was convicted. "No white man in that state has been punished for murder

54. Howard N. Rabinowitz, "The Conflict Between Blacks and the Police in the Urban South, 1865–1900," *Historian*, 39 (November 1976), 63–64; Carter, *When the War Was Over*, 220; George C. Rable, *But There Was No Peace: The Role of Violence in the Politics of Reconstruction* (Athens, Ga., 1984), 21–28; Clinton A. Reilly to J. W. Ames, August 14, 1865, J. R. Ellis to William W. Holden, June 25, 1865, North Carolina Governor's Papers; Thomas J. Durant to Henry C. Warmoth, January 13, 1866, Warmoth Papers; Thomas Kanady to Z. K. Wood, December 23, 1865, L-896 1865, Letters Received, Ser. 1757, Dept. of La., RG 393, pt. 1, NA [FSSP C-655].

55. T. Reavis et al. to Lewis E. Parsons, August 11, 1865, Alabama Governor's Papers; Wharton, *Negro in Mississippi*, 136–37; Henry Latham, *Black and White* (London, 1867), 105; Roberta D. Miller, "Of Freedom and Freedmen: Racial Attitudes of White Elites in North Carolina During Reconstruction, 1865–1877" (unpub. diss., University of North Carolina, 1976), 252–53; 39th Congress, 2d Session, Senate Executive Document 6, 196–97.

since it revolted from Mexico," commented a Northern visitor. "Murder is considered one of their inalienable state rights."[56]

In much of the South, the courts of Presidential Reconstruction appeared more interested in disciplining the black population and forcing it to labor than in dispensing justice. "The fact is," one of Florida's few white Radicals commented shrewdly early in 1867, "custom with them [local whites] has become the law." If employers could no longer subject blacks to corporal punishment, courts could mandate whipping as a punishment for vagrancy or petty theft. If individual whites could no longer hold blacks in involuntary servitude, courts could sentence freedmen to long prison terms, force them to labor without compensation on public works, or bind them out to white employers who would pay their fines. The convict lease system, moreover, which had originated on a small scale before the war, was expanded so as to provide employers with a supply of cheap labor. In Texas in 1867, blacks constituted about one third of the convicts confined to the state penitentiary, but nearly 90 percent of those leased out for railroad labor.[57]

Arrested by white sheriffs, tried before white judges and juries, blacks understandably had little confidence in the courts of Presidential Reconstruction. There, a group of Charleston blacks complained early in 1867, "justice is mocked and injustice is clothed in the garb of righteousness." Blacks searched for alternatives to the Southern court system, preferring the Freedmen's Bureau or army to adjudicate disputes and punish crimes. Abram Colby, leader of the Greene County, Georgia, Equal Rights Association, asked the Bureau to bestow upon blacks the "right of tribunal to settle business between colored and others" since "we have no chance of justice before the courts." Blacks, a Bureau official concluded, "would be *just as well* off with no law at all or no Government," as with the legal system of Presidential Reconstruction.[58]

Taxation provided yet another example of the inequitable turn taken by public policy. Before the war, landed property in most Southern states

56. John A. Carpenter, "Atrocities in the Reconstruction Period," *JNH*, 47 (October 1962), 238; W. A. Inman to Dennis H. Williams, June 30, 1866, Narrative Reports of Operations, Ser. 242, Ark. Asst. Comr., RG 105, NA; [FSSP A-2458]; Robert K. Scott to James L. Orr, December 13, 1866, South Carolina Governor's Papers; James W. Smallwood, *Time of Hope, Time of Despair: Black Texans During Reconstruction* (Port Washington, N.Y., 1981), 33; O. D. Bartlett to Benjamin F. Wade, July 7, 1867, Benjamin F. Wade Papers, LC.
57. Shofner, *Nor Is It Over Yet*, 105; 39th Congress, 1st Session, House Report 30, pt. 2:272; Edward L. Ayers, *Vengeance and Justice: Crime and Punishment in the Nineteenth-Century American South* (New York, 1984), 165; J. Thorsten Sellin, *Slavery and the Penal System* (New York, 1976), 145–62; A. Elizabeth Taylor, "The Origins and Development of the Convict Lease System in Georgia," *GaHQ*, 26 (June 1942), 113–15; James P. Butler to J. T. Kirkman, May 13, 1867, Box 1, B13, Fifth Military District, RG 393, NA.
58. *CG*, 39th Congress, 2d Session, 1709; Abram Colby et al. to Davis Tillson, [August 1866], Unregistered Letters Received, Ser. 632, Ga. Asst. Comr, RG 105, NA [FSSP A-5349]; Nelson G. Gill to O. O. Howard, October 14, 1866, G-310 1866, Letters Received, Ser. 15, Washington Headquarters, RG 105, NA [FSSP A-9528].

had gone virtually untaxed, while poll taxes and levies on slaves, luxuries, commercial activities, and professions provided the bulk of revenue. As a result, white yeomen paid few taxes, planters paid more, although hardly an amount commensurate with their wealth and income (partly because they generally assessed the value of their own holdings), and urban and commercial interests complained of an excessive tax burden. War and emancipation inevitably increased the demands upon state revenues while the region's poverty made even low tax rates seem a hardship. Under these circumstances, the tax system became a battleground where the competing claims of various classes of Southerners were fought out.[59]

Throughout the world, political elites have employed poll or "head" taxes, which require individuals to obtain cash, to prod self-sufficient peasants to enter the labor market. In Presidential Reconstruction, tax policy was intended, in part, to reinforce the planter's position vis-à-vis labor. Heavy poll taxes were levied on freedmen, encouraging them to work for wages, while those unable to pay were deemed vagrants, who could be hired out to anyone meeting the tax bill. Meanwhile, levies on landed property remained extremely low (one tenth of 1 percent in Mississippi, for example), shielding planters and yeomen from the burden of rising government expenditures. As a result, "the man with his two thousand acres paid less tax than any one of the scores of hands he may have had in his employ who owned not a dollar's worth of property." He also paid less than town craftsmen, whose earnings were taxed at rates far higher than real estate. In Mississippi's Warren County, the three largest landowners each paid less than $200 in taxes, while the owner of a livery stable paid nearly $700, a butcher over $200, and a shoemaker $75. In addition, localities added poll taxes of their own, sometimes, in black belt counties, raising the bulk of their revenue in this manner. Mobile levied a special tax of $5 on every adult male "and if the tax is not paid," reported the city's black newspaper, "the chain-gang is the punishment." With state, county, and local levies, blacks might find themselves paying $15 in poll taxes alone.[60]

Not surprisingly, blacks resented a highly regressive revenue system

59. Foner, *Nothing But Freedom*, 67–68; J. Mills Thornton III, "Fiscal Policy and the Failure of Radical Reconstruction in the Lower South," in J. Morgan Kousser and James M. McPherson, eds., *Region, Race, and Reconstruction: Essays in Honor of C. Vann Woodward*, (New York, 1982), 351–60; J. M. Hollander, ed., *Studies in State Taxation* (Baltimore, 1900), 85–92, 187. Poll taxes are taxes levied directly on individuals. They have no necessary connection with suffrage, but payment has sometimes been made a requirement for the right to vote.

60. Murdo J. MacLeod, "The Sociological Theory of Taxation and the Peasant," *Peasant Studies Newsletter*, 4 (July 1975), 2–6; Jackson *Daily Mississippi Pilot*, August 29, 1870, August 24, 1871; Jackson *Weekly Mississippi Pilot*, January 23, 1875; 42d Congress, 2d Session, House Report 22, Mississippi, 728 (hereafter cited as KKK Hearings); A. T. Morgan, *Yazoo: or, On the Picket Line of Freedom in the South* (Washington, D.C., 1884), 262–63; Mobile *Nationalist*, March 22, 1866; 40th Congress, 3d Session, House Miscellaneous Document 52, pt. 2:274–87.

from whose proceeds, as a North Carolina Bureau agent reported, "they state, and with truth, that they derive no benefit whatever." The freedmen did not object to paying taxes per se; they often taxed themselves to raise school funds. But they did insist on receiving some return from the revenue. The governments of Presidential Reconstruction, however, sought to avoid any responsibility for the black population (except for compelling it to labor). A few states in these years took the first steps toward establishing a legally enforced system of racial segregation, imposing barriers between the races that had not existed under slavery. Mississippi and Texas barred blacks (except servants traveling with their employers) from first-class railroad cars, and Florida made it illegal for any black to "intrude himself" into religious services, public meetings, or transportation facilities set aside for whites. These measures, however, were only a small harbinger of the later system of de jure segregation. Instead of segregation—which implied providing separate facilities of some kind for blacks—the Southern states attempted to exclude them altogether from public institutions.[61]

Even though taxes on blacks as well as whites helped fill their coffers, states and municipalities barred blacks from poor relief, orphanages, parks, schools, and other public facilities, insisting that the Freedmen's Bureau should provide whatever services blacks required. Georgia in 1866 appropriated $200,000 to aid children and widows of Confederate (but not Union) soldiers and other "aged or infirm white persons." Not until 1867 did Richmond vote any funds for black paupers. Where states did attempt to provide for black needs, funds were generally derived from special taxes levied on blacks themselves, rather than from general revenues. Texas and Florida, the only states to make any arrangements for black education during Presidential Reconstruction, financed these schools by a separate black poll tax, supplemented by tuition charges. Louisiana, on the other hand, whose school superintendent Robert M. Lusher believed education should "vindicate the honor and supremacy of the Caucasian race," dismantled the thriving black school system General Banks had established.[62]

The fate of public education in North Carolina illustrates the lengths to which the leaders of Presidential Reconstruction were prepared to go to avoid recognizing blacks as part of their common constituency. Gov.

61. Allan Rutherford to J. T. Chur, October 29, 1866, Annual Reports of Operations, Ser. 2463, N.C. Asst. Comr., RG 105, NA [FSSP A-692]; 39th Congress, 2d Session, Senate Executive Document 174, 79; Smallwood, *Time of Hope*, 123; Richardson, *Negro in Florida*, 43.

62. Howard N. Rabinowitz, "From Exclusion to Segregation: Southern Race Relations, 1865–1890," *JAH*, 63 (September 1976), 326–27; Peter Wallenstein, *From Slave South to New South: Public Policy in Nineteenth-Century Georgia* (Chapel Hill, 1987), 138; O'Brien, "From Bondage to Citizenship," 313; William P. Vaughan, *Schools for All: The Blacks and Public Education in the South, 1865–1877* (Lexington, Ky., 1974), 52–54; Roger A. Fischer, *The Segregation Struggle in Louisiana 1862–77* (Urbana, Ill., 1974), 27–28.

Jonathan Worth, elected in 1865, had earlier in his career steered to passage the bill establishing public education in North Carolina, but he now persuaded the legislature to abolish the state school system altogether. Although the school fund had been "swept away by the Avalanche," financial considerations did not determine Worth's attitude. Rather, the governor feared that if white children were educated at public expense, "we will be required to educate the negroes in like manner." To avoid having to expend public monies on black education, Worth and his legislature authorized localities to establish tax-supported private academies, risking, as one ally warned, "the entire alienation of the poorer class" of whites, and destroying the South's only extensive system of public education.[63]

Many Southern whites believed the work of the Presidential Reconstruction legislatures constituted a good faith effort to confront the aftermath of emancipation. Some laws of 1865 and 1866, they pointed out, had been modeled on army and Freedmen's Bureau labor regulations, while vagrancy statutes existed in the North and free blacks there often lacked the rights to serve on juries, join militias, and attend public schools. Army and Bureau regulations, however, were temporary expedients, not a permanent blueprint for the Southern labor system, and the North's racial proscription, while allowing Southerners to charge their critics with hypocrisy, affected a tiny minority, not the large bulk of the laboring population. Northern courts tended to view those without work as unfortunates rather than criminals, usually employing vagrancy laws only to discipline prostitutes and petty thieves. Nor did a parallel exist in the North for laws requiring a person to have an employer at the beginning of each year, imposing criminal penalties for breach of contract, or barring the enticement of a servant—the cornerstones of Southern labor legislation. Except for sailors who jumped ship, contract violations were civil, rather than criminal matters. "I have never known a master to go to court" against a departing employee, New York iron magnate Abram Hewitt declared in 1867, adding that he viewed criminal laws punishing breach of contract as "very undesirable."[64]

By creating a large class barred from equal access to the courts and full participation in the marketplace, Southern laws violated the fundamental principles of the free labor ideology. For this reason, the most flagrant

63. Daniel J. Whitener, "Public Education in North Carolina During Reconstruction, 1865–1876," in Fletcher M. Green, ed., *Essays in Southern History* (Chapel Hill, 1949), 66–73; Charles Phillips to unknown, March 19, 1866, Cornelia Phillips Spencer Papers, UNC; J. G. de Roulhac Hamilton, ed., *The Correspondence of Jonathan Worth* (Raleigh, 1909), 1:467; Hamilton and Williams, eds., *Graham Papers*, 7:10, 41.

64. Wilson, *Black Codes*, 36–39, 57–59; Allen R. Steinberg, "The Criminal Courts and the Transformation of Criminal Justice in Philadelphia, 1815–1874" (unpub. diss., Columbia University, 1983), Chapter 3; Richard B. Morris, "Labor Controls in Maryland in the Nineteenth Century," *JSH* 14 (August 1948), 393–96; *U. K. Parliamentary Session Papers*, 1867, 32, c. 3913–28.

provisions of the Black Codes never went into effect. Gen. Daniel E. Sickles, who insisted "all laws shall be applicable alike to all inhabitants," suspended South Carolina's code, and Gen. Alfred H. Terry overturned Virginia's vagrancy law as an attempt to reestablish "slavery in all but its name." Alarmed by reaction to the early codes, Mississippi Governor Sharkey declared the measure barring blacks from renting real estate "palpably in violation of the constitution, and therefore void," while in Alabama, Governor Patton vetoed Alabama's labor contract and apprenticeship bills. By the end of 1866, most Southern states had repealed those laws applying only to blacks. Yet Southern courts continued to enforce vagrancy, breach of contract, and apprenticeship statutes that made no direct reference to race, and tax policies, the militia system, and the all-white judiciary remained unchanged.[65]

Historians hostile to Radical Reconstruction later defended the legislation of 1865 and 1866 as an expansion of black rights when compared with slavery, or attributed its oppressive quality to "the antipathy of the lower-class whites to the negroes." Yet few legislatures of Presidential Reconstruction contained many poor whites (and Tennessee and Arkansas, which did, failed to enact new vagrancy and antienticement laws). Nor were these laws the work of extreme secessionists; they were drafted by distinguished lawyers and conservative Whig Unionists. Eminent jurists David L. Wardlaw and Armistead Burt, the former "the ablest lawyer the state ever produced," drew up South Carolina's Black Code, and prominent sugar planter Duncan F. Kenner prepared Louisiana's. As Daniel R. Goodloe, a North Carolina-born abolitionist, later observed, it seemed "little short of madness" for Southern leaders to believe "the triumphant North . . . would tolerate this new slave code." But obsessed by the idea that blacks would never work voluntarily, and alarmed by the freedmen's unexpected militancy, Southern lawmakers strove, as one newspaper put it, "to do what was best for the State and for society; ever remembering that it was a white man's State [they were] legislating for."[66]

No one can claim that the complex structure of labor, property, and tax laws enacted in 1865 and 1866 succeeded fully in controlling the black

65. Wilson, *Black Codes*, 70–75; Jack P. Maddex, Jr., *The Virginia Conservatives, 1867–1879* (Chapel Hill, 1970), 39; William L. Sharkey to William H. Seward, December 30, 1865, William H. Seward Papers, University of Rochester; McPherson, *Political History*, 21; Donald G. Nieman, *To Set the Law in Motion: The Freedmen's Bureau and the Legal Rights of Blacks, 1865–1868* (Millwood, N. Y., 1979), 73–96.

66. Wilson, *Black Codes*, 196–97; William A. Dunning, *Reconstruction, Political and Economic 1865–1877* (New York, 1907), 57–58; interview with Joseph W. Barnwell, Southern Notebook "C," Frederic Bancroft Papers, Columbia University; *Ex-Governor Hahn on Louisiana Legislation Relating to Freedmen* (n.p., 1866), 3; Daniel R. Goodloe, "History of Southern Provisional Governments of Louisiana," manuscript, Goodloe Papers; Ross H. Moore, "Social and Economic Conditions in Mississippi During Reconstruction" (unpub. diss., Duke University, 1937), 41.

laborer or shaping the evolution of the Southern economy. The "labor shortage" persisted, as did black efforts to resist plantation discipline. The law is an inefficient mechanism for compelling people to work in a disciplined manner. As a South Carolina plantation physician put it, "they can be forced by law *to contract,* but how to enforce their labor is not yet determined." Nonetheless, the legal system of Presidential Reconstruction had profound consequences, limiting the options open to blacks, reinforcing whites' privileged access to economic resources, shielding planters from the full implications of emancipation, and inhibiting the development of a free market in land and labor.[67]

The aim of resurrecting as nearly as possible the old order with regard to black labor, moreover, contradicted a second purpose of the new governments: reshaping the economy so as to create a New South. Before the war, a small band of modernizers had urged the South to reorient its economy by building railroads and factories alongside the plantations. Although few of these reformers challenged slavery's status as the central feature of the Southern economy, their plans had foundered on the opposition of the planter class. But with abolition accomplished and King Cotton apparently dethroned, the prospect beckoned of a South more fully attuned to nineteenth-century "progress." Northern investment would spur the growth of railroads and factories, immigration would introduce a new spirit of enterprise, and farmers would no longer see their capital frozen in the labor force. The Southern press extolled the idea of expanding the small prewar textile industry so as to utilize cotton locally and provide employment for the thousands widowed and orphaned by the war. Criticisms of the plantation system itself, a minor theme before the war, now blossomed, with calls for a more scientific agriculture based on small farms and diversified crops. "Our large plantations," declared a South Carolina newspaper in 1866, "must be carved up into respectable farms; our water power must be made available in the erection of manufactories; . . . our young men must learn to work." One New South enthusiast, himself a planter, went so far as to advocate a heavy tax on unimproved land, to force owners to cultivate their holdings or place them on the market.[68]

This comprehensive vision of economic change never commanded majority support during Presidential Reconstruction. But governors like James L. Orr, James Johnson, and Robert Patton extolled the virtues of a New South, and found a receptive audience among the former Whigs who dominated politics and eagerly embraced the idea of state-promoted

67. J. H. Easterby, ed., *The South Carolina Rice Plantation as Revealed in the Papers of Robert F. W. Allston* (Chicago, 1945), 224; Foner, *Nothing But Freedom,* 71.

68. Eugene D. Genovese, *The Political Economy of Slavery* (New York, 1965), 124–53; Carter, *When the War Was Over,* 106–16; Patrick J. Hearnden, *Independence and Empire: The New South's Cotton Mill Campaign, 1865–1901* (DeKalb, Ill., 1982), 3–13; Columbia *Daily Phoenix,* March 9, 1866; Aubrey L. Brooks and Hugh T. Lefler, eds., *The Papers of Walter Clark* (Chapel Hill, 1948–50), 1:145–46, 158.

railroad and industrial development as the key to economic modernization. Before the war, most Southern states had assisted railroad development in one way or another, even, in the case of Georgia, building (with
government-owned slaves) a railroad itself. Yet the South lagged far
behind the rest of the nation in railroad mileage, and most Southern
roads remained small, undercapitalized ventures, their primary purpose
to transport plantation staples to the coast. Now, a veritable "railroad
fever" swept over the region, as states and localities invested funds and
expectations in the iron horse. Alabamans believed railroads would un-
lock the state's mineral resources and create the "identity of feeling"
never before shared by the northern and southern parts of the state.
Commercially threatened cities like Charleston and Vicksburg saw railroads as panaceas for economic stagnation, while to upcountry towns
they offered a share of national commerce, a means of bypassing older
port cities and trading directly with the North. Railroads, declared a
Mississippi newspaper, would "revive the energies of the people, open up
the resources of the State, and put us in the way of growth and general
prosperity."[69]

Although the policy of lending the state's credit to promote railroad
construction is usually associated with Radical Reconstruction, it in fact
originated under the Johnson governments. An act of February 1867,
authorizing the state to endorse railroad bonds at the rate of $12,000 for
each mile built, formed the basis of subsequent Alabama aid. Texas,
whose governor, James Throckmorton, was a longtime railroad devel-
oper, chartered sixteen new roads and adopted an aid program even
more generous than Alabama's. Simultaneously, legislatures chartered
manufacturing, mining, banking, and insurance corporations. And to
promote investment in agriculture, states gave the force of law to credit
arrangements guaranteeing a first lien on the crop to persons advancing
loans or supplies for farming. These early laws avoided the vexing question of priority among creditors by declaring anyone who made advances
equally entitled to a lien. Blacks complained that by failing to establish
a laborer's lien on the crop, these laws allowed an indebted planter's
entire harvest to be seized by creditors, leaving nothing to pay his employees. But the laws' primary purpose, as a former governor of Mississippi explained, was to attract funds to the South by "inspiring confidence on the part of capitalists."[70]

69. Paul M. Gaston, *The New South Creed* (New York, 1970), 23–30; *Address of the Hon. James L. Orr Governor Elect, Delivered Before the Legislature of South Carolina* (Columbia, S.C., 1865), 6; Carter Goodrich, *Government Promotion of American Canals and Railroads 1800–1890* (New York, 1960), 112–19; John F. Stover, *Iron Roads to the West: American Railroads in the 1850s* (New York, 1978), 89, 116; Simkins and Woody, *South Carolina*, 186–91; Huntsville *Advocate*, July 12, 1865; Jamie W. Moore, "The Lowcountry in Economic Transition: Charleston since 1865," *SCHM*, 80 (April 1979), 156–58; Harris, *Presidential Reconstruction*, 214.

70. A. B. Moore, "Railroad Building in Alabama During the Reconstruction Period," *JSH*, 1 (February 1935), 422–24; Carrier, "Political History of Texas," 75–76, 122, 144–45; Carter Goodrich, "Public Aid to Railroads in the Reconstruction South," *Political Science*

The South's governors also sought, with varying degrees of success, to resist growing demands for debtor relief, fearing "stay laws" would discourage outside investment. The question of debt provoked more division within the white South than any other issue of Presidential Reconstruction, although the class and sectional alignments hardly conformed to a simple pattern. Not a few indebted planters found stay laws appealing, as did other men of property seeking to avoid repaying prewar loans from Northern merchants and financiers. (As a Texas political leader later recalled, "the idea to take it against Northern debts was all right and a part of the war.") Still others believed stay laws would protect planters from employees suing for the payment of wages. Generally, however, the demand for laws "staying" the collection of debts or repudiating them altogether came from upcountry yeomen whose economic independence was threatened by wartime devastation and postwar crop failures.[71]

From this complex situation emerged a series of laws, varying from state to state, that postponed the collection of debts, "scaled" old debts to take account of the drastic decline in property values, put off court sessions, and exempted a certain amount of land and personal property from seizure. Some represented real victories for debtors, others merely attempted to head off more radical measures like complete repudiation. All, however, evoked considerable protest in the North—no one would invest, wrote a Northern correspondent of Georgia Gov. Charles Jenkins, where "stay laws are made to prevent the collection of debts"—and top state officials moved swiftly to limit their effectiveness. State courts voided debtor relief laws in Georgia, Mississippi, and South Carolina. Alabama Governor Patton turned a deaf ear to upcountry demands that he cancel court sessions after the disastrous crop failure of 1866, and North Carolina's Governor Worth denounced stay laws as "pernicious alike to debtor and creditor," although he was unable to prevent the enactment of several relief measures.[72]

The willingness of courts and governors to risk "considerable dissatis-

Quarterly, 71 (September 1956), 411–25; Harold D. Woodman, "Post Civil War Southern Agriculture and the Law," AgH, 53 (January 1979), 319–37; William C. Carson and forty others to George C. Meade, January 1868, C-55 1868, Letters Received, Third Military District, RG 393, NA; William McWillie to Benjamin G. Humphreys, December 31, 1866, Mississippi Governor's Papers.

71. Carter, When the War Was Over, 77–79; John Murphy to John O'Neil, January 15, 1867, John O'Neil Papers, USC; "Lawabiding Citizen" to James L. Orr, January 13, 1866, South Carolina Governor's Papers; John B. Stiles to William H. Seward, December 9, 1865, Seward Papers; Lucy A. Erath, "Memoirs of Major General Bernard Erath," SWHQ, 27 (October 1923), 161.

72. Carter, When the War Was Over, 135–45; AC, 1866, 13; A. W. Spies to Charles Jenkins, November 30, 1866, Georgia Governor's Papers; Simkins and Woody, South Carolina, 46–47; Harris, Presidential Reconstruction, 157–59, 174–76; Hamilton, ed., Worth Correspondence, 4:48.

faction among the masses" in order to reassure prospective Northern investors illustrates the depth of their commitment to economic modernization. In nearly every respect, however, the economic policies of Presidential Reconstruction failed. The programs of railroad aid accomplished virtually nothing—in the eleven states of the Confederacy, only 422 miles of track were laid in 1866 and 1867, none at all in Louisiana, South Carolina, and Mississippi. Despite the appointment of commissioners of immigration, nearly all the immigrants who entered the country remained in the North and West—the number of foreign-born residents of the eleven Confederate states was lower in 1870 than in 1860. Industrial development remained insignificant. A few establishments, like Richmond's Tredegar Iron Works, attracted enough Northern investment to resume production, but most Southern entrepreneurs who ventured north in search of capital returned home empty-handed. With lucrative opportunities available in the West, investors had no desire to risk their funds in the South's unstable political climate. "Very few persons are willing to trust their investments in the South," concluded former Massachusetts Gov. John Andrew. "It is easier to sell an imaginary copper mine in Jupiter, than it is to hire ten per cent on the best lands in the South, on the northern market."[73]

The stillbirth of this early New South program had many causes, some far beyond the power of Southern politicians to affect. Neither the disastrous economic consequences of the Civil War nor the legacy of decades of plantation dominance could be erased in two short years. But in some ways, the failure reflected the divided mind and contradictory aims of those advocating economic change. The South's inability to attract immigration illustrates the problem. Some reformers looked upon newcomers as prospective landowners, whose presence would facilitate the breakup of the plantation system and the rise of modern, market-oriented small farming. Others, however, intended immigration not to undermine the plantation but to preserve it. "Immigration," a prominent North Carolina lawyer wrote in 1865, "would, doubtless, be a blessing to us, provided we could always control it, and make it entirely subservient to our wants." Not surprisingly, European immigrants did not relish the idea of taking the place of blacks as plantation laborers. One Alabama planter brought in thirty Swedes in 1866, housed them in slave cabins and fed them the usual rations. Within a week the laborers had departed, informing him "they were not slaves." To attract immigrants, observed A. B. Cooper, another Alabaman, the South must abandon the idea of labor associated

73. Benjamin H. Wilson to James L. Orr, June 22, 1866, South Carolina Governor's Papers; Goodrich, "Public Aid to Railroads," 436; Simkins and Woody, *South Carolina*, 244–46, 289; U.S. Census Office, *Ninth Census, 1870, Population*, 1:299; Charles B. Dew, *Ironmaker to the Confederacy: Joseph R. Anderson and the Tredegar Iron Works* (New Haven, 1966), 318; Broadus Mitchell, *The Rise of Cotton Mills in the South* (Baltimore, 1921), 63; John Andrew to Montgomery Blair, January 7, 1867, Letterbook, John Andrew Papers, MHS.

with the plantation: "We must divest ourselves of the idea that we can *command, control* the laborer. We must be prepared to receive him as a free man, an equal, and treat him as such."[74]

Cooper grasped an important fact. Just as the realities of a political economy dominated by slavery frustrated efforts at agricultural reform before the war, so genuine postwar modernization would have required an assault on the plantation. Throughout the world, plantation societies are characterized by persistent economic backwardness. Geared to producing agricultural staples for the world market, they have weak internal markets, and planter classes use their political power to prevent the emergence of alternative economic enterprises that might threaten their control of the labor force.[75] Because they failed to come to grips with the plantation itself, the leaders of Presidential Reconstruction lacked a coherent vision of Southern progress. They wanted the trappings of economic development without accepting its full implications—an agrarian revolution and a free labor market. Newspapers that called for breaking up the plantations in the same breath demanded strict legislation immobilizing black labor. Laws intended to modernize the Southern economy coincided with measures, enacted by the same legislatures, to discipline the plantation labor force. Taxes on property remained so low that the establishment of public schools or other forward-looking social services became impossible.

At least the planter class possessed the virtue of consistency; it had no intention of presiding over its own dissolution. It would support railroads, factories, and Northern investment so long as these supplemented and invigorated the plantation and did not threaten the stability of the black labor force. Those who spoke of dismantling the plantations had no idea what would become of the black population. The entire New South program, in fact, assumed that substitutes would be found for black labor. Scientific agriculture and the introduction of machinery would enable large estates to "dispense with the services of freedmen" altogether. Family labor would suffice for small farms. Reformers spoke of factories employing white laborers, and of small farms tilled by white newcomers replacing black belt plantations, without making any provision for the former slaves, apart from morbid predictions that they would conveniently "die out." Certainly, spokesmen for a New South had no intention of seeing the finest land in the region fall into black hands.[76]

74. Highsmith, "Louisiana Landholding," 44–45; J. G. de Roulhac Hamilton, ed., *The Papers of Thomas Ruffin* (Raleigh, 1918–20), 4:45; Mobile *Nationalist,* May 9, 1867; A. B. Cooper to Robert M. Patton, May 25, 1867, Alabama Governor's Papers.

75. George L. Beckford, *Persistent Poverty: Underdevelopment in Plantation Economies of the Third World* (New York, 1972).

76. Wiener, *Social Origins,* 156–57; Wagstaff, " 'Call Your Old Master'," 337–38; Richmond *Dispatch,* April 19, 1867; Brooks and Lefler, eds., *Clark Papers,* 1:154–56; Highsmith, "Louisiana Landholding," 48–49.

The entire experience of Presidential Reconstruction reveals how profoundly attitudes toward the emancipated slaves and their place in the new social order affected efforts to reshape the Southern polity and economy. The unwillingness of unconditional Unionists to broaden their political base to include the freedmen necessitated the massive disenfranchisement of former Confederates, exacerbating hatreds generated by the Civil War and tainting border governments with illegitimacy. Andrew Johnson's obsession with keeping blacks in order led inevitably to abandonment of the idea of destroying planters' economic and political hegemony. And the inability of the leaders of the governments he created to conceive of blacks as anything but plantation laborers doomed the idea of real economic reform. In the end, their policies envisioned less a New South than an improved version of the old.

The outcome typified the failure of vision that marked the South's attempt at "self-reconstruction" from beginning to end. As Presidential Reconstruction drew to a close, Southern whites recognized that an opportunity had slipped away. Lawmakers found themselves castigated by the press in language later turned upon Radical governments: They were inept, lazy, and had failed to deal effectively with the region's problems. "Probably the best thing the Legislature can do," remarked a correspondent of South Carolina Governor Orr, "will be to *go home.*" Thoughtful observers would later acknowledge that the white South brought Radical Reconstruction upon itself. "We had, in 1865, a white man's government in Alabama. . . ." declared Johnson's Provisional Gov. Lewis E. Parsons, "but we lost it." The "great blunder" was not to "have at once taken the negro right under the protection of the laws."[77]

Among its many consequences, Presidential Reconstruction profoundly affected black political behavior. In postemancipation Jamaica, the tiny white population had attempted, with considerable success, to win political allies among the mulatto elite. But the laws of Presidential Reconstruction applied indiscriminately to free and freed. By eliminating the middle ground between black and white that some free mulattoes had hoped to preserve, white policies propelled free blacks into a political alliance with the freedmen. "As far as the law is concerned," declared the Mobile *Nationalist,* "the Creole and freedman stand upon the same level. . . . They must, in the future, rise or fall together." Invocations of the Black Codes as "a disgrace to civilization" quickly became a staple of black political rhetoric. "If you call this Freedom," wrote one black veteran, "what do you call Slavery?" The "undisputed history" of these years, black Congressman Josiah Walls later observed, explained why Southern blacks refused to cast Democratic ballots, and stood as a warn-

77. Thomas Holt, *Black over White: Negro Political Leadership in South Carolina During Reconstruction* (Urbana, Ill., 1977), 24; William E. Earle to James L. Orr, December 15, 1866, South Carolina Governor's Papers; KKK Hearings, Alabama, 93.

ing "as to what they will do if they should again obtain control of this Government." Most of all, Presidential Reconstruction reinforced blacks' identification with federal authority. Only outside intervention could assure the freedmen a modicum of justice.[78]

The North's Response

When first announced, Andrew Johnson's Reconstruction policy enjoyed overwhelming Northern support. Many Republicans would have welcomed a declaration of black suffrage (for the South), but few, apart from the Radicals, seemed anxious to make this a party issue or saw it as a reason to repudiate the President. Early in May 1865, Connecticut Sen. James Dixon commented on the "warm and generous disposition" Northerners manifested toward Johnson: "It seems to be a restoration of the 'era of good feeling'." The Reconstruction proclamations at the end of the month did little to alter this public mood. A broad coalition embraced Johnson's program and sought to influence its implementation. Along with numerous Northerners who, for one reason or another, favored the rapid restoration of the Southern states to the Union, it included Democrats, who hoped to revive their party's fortunes after its "most disastrous epoch," Republicans of Democratic antecedents who shared Johnson's state rights orientation and racial prejudices, and Republicans who saw an opportunity to enhance their position within the party by identifying themselves with the new President.[79]

Northern Democrats quickly realized that Johnson's plan embraced ideas central to the party's ideology and essential for its political revival—control of local affairs by the individual states, white supremacy, and the quick resumption of the South's place within the Union. First to appreciate the benefits the party might derive from an identification with the President were its leaders in New York State. The "cockpit" of nineteenth-century Democratic politics, New York supplied seven of the party's presidential candidates between 1836 and 1900 and boasted in August Belmont its leading financier, and in the New York *World* its premier national organ. The *World*'s co-owner, Samuel L. M. Barlow, was involved in all sorts of railroad, mining, and commercial ventures, including, during Reconstruction, investments in iron mines and rice plantations in the South. His first love, however, was politics, and he devoted himself in 1865 to convincing the President that his political future lay

78. *New York Times*, September 8, 1868; *CG*, 43d Congress, 2d Session, Appendix, 169; Henry Mars to Edwin M. Stanton, May 14, 1866, M-240 1866, Letters Received, Ser. 15, Washington Headquarters, RG 105, NA [FSSP A-6032].

79. Michael L. Benedict, *A Compromise of Principle: Congressional Republicans and Reconstruction 1863–1869* (New York, 1974), 43; James Dixon to Andrew Johnson, May 5, 1865, Johnson Papers; Irving Katz, *August Belmont: A Political Biography* (New York, 1968), 91.

in placing himself at the head of the Democracy. Once certain that "the Southern people are universally pleased with Mr. Johnson's course," the *World* endorsed the President's program with enthusiasm.[80]

As in the South, close identification with Johnson stifled consideration of alternative scenarios for Reconstruction. Early in May, in a major departure, the *World* had announced that the freedmen could not be "permanently excluded from the elective franchise" and insisted that Southern states not pass laws "discriminating against men of color." But once Johnson announced his program, the paper quickly reverted to its accustomed racism, calling blacks prone to "licentiousness" and the Freedmen's Bureau a "department of pauperism." Because they lacked a sense of family, enfranchising the freedmen would lead to a Mormon ascendancy and the legalization of polygamy! Throughout the North, support for the President and opposition to enfranchising blacks became the Democracy's rallying cries. State platforms insisted black suffrage would undermine the republic by "bringing the right to vote into disgrace," and party speakers described Republicans as "nigger adorers" unconcerned with the white race. The suffrage issue, Democratic Congressman Samuel S. Cox believed, not only formed the "basis" of Reconstruction, but would teach Johnson "upon whom to rely . . . and [whose] support he needs."[81]

"The whole party is today a Johnson party," Barlow announced in July, and Democrats deluged the President with calls to endorse the party's candidates in the Northern fall elections and remove Seward and Stanton from the Cabinet. Their plans were seconded by a group of Republicans whose political antecedents, like Johnson's, lay in the prewar Democracy. Democratic-Republicans had long resented what they considered former Whigs' domination of the wartime Republican party. They saw in Johnson's accession an opportunity to reshape political alignments to their own benefit. A "big row," wrote Assistant Postmaster General Alexander W. Randall, was on the horizon. "The whole radical portion of our party" would be pushed aside in favor of the "Democratic element." Democratic-Republicans like Secretary of the Navy Welles viewed the Radicals as lineal descendants of "old blue light federalism," advocates of "a strong central government with supreme

80. Edward L. Gambill, *Conservative Ordeal: Northern Democrats and Reconstruction, 1865–1868* (Ames, Iowa, 1981), 20–33; George T. McJimsey, *Genteel Partisan: Manton Marble, 1834–1917* (Ames, Iowa, 1971), 60–63; Robert F. Hohe to Samuel L. M. Barlow, December 8, 1865, P. M. Nightingale to Barlow, August 30, 1867, Barlow Papers; New York *World*, June 29, 1865.

81. New York *World*, May 3, 13, June 23, 29, July 29, August 21, October 12, 1865; *AC*, 1865, 685; Kent A. Peterson, "New Jersey Politics and National Policy-Making, 1865–1868" (unpub. diss., Princeton University, 1970), 110–11; Samuel S. Cox to Montgomery Blair, May 16, 1865, Blair Family Papers, LC.

power over the states and people." Like Johnson, Welles had a penchant for reducing complex political questions to matters of supposedly "plain and unmistakable" constitutional interpretation. The South, he insisted, retained complete power over its "local and municipal affairs," including the suffrage; to act on any other assumption would "change the character of the government."[82]

Among Democratic-Republicans, the Blair family came to exert the greatest influence upon the President. Like Johnson, the Blairs were border-state Unionists who had strongly supported the war effort but had no love for Radicalism or blacks. Francis P. Blair, the family patriarch, had been a trusted adviser of Johnson's hero, Andrew Jackson. Frank, Jr., helped build the Republican party in Missouri and rose to major general in the Union Army; his brother Montgomery played an active role in Maryland Republican politics and had served Lincoln as Postmaster General until Radical pressure forced his dismissal. Johnson's ties with the family were solidified in March 1865. Suffering from typhoid fever at the time of his inauguration as Vice President, Johnson fortified himself with strong whiskey. Apparently, he imbibed too much, for his speech became "a rambling and strange harangue, which was listened to with pain and mortification." While the press ridiculed him as a "drunken clown," the Blairs hurried Johnson off to the family estate in Silver Spring for recuperation.[83]

Johnson, the Blairs insisted, should "dispense with the support of the Radicals" and rely upon "his natural supporters"—Unionist Democrats and Democratic-Republicans like themselves. The key to postwar politics, they believed, lay in changing the focus of debate from slavery to race. The Blair family had long been obsessed with the specter of racial "amalgamation," and as with Johnson, prejudice led them down a path of illogic in which blacks and advocates of black suffrage were somehow equated with slavery and the Slave Power. For the Blairs, this was because slavery had led to racial intermixing and tendencies toward a "hybrid race" and black voting would produce the same results. Moreover, the race issue was good politics. "If we can dispose of the slave question," Montgomery Blair predicted, "we shall have the miscegenationists [his term for anyone favoring black voting] in a party to themselves and can

82. Samuel L. M. Barlow to Montgomery Blair, July 24, 1865, Johnson Papers; Eric Foner, *Free Soil, Free Labor, Free Men: The Ideology of the Republican Party Before the Civil War* (New York, 1970), 149–85; Richard N. Current, *The Civil War Era, 1848–1873* (Madison, Wis., 1976), 575; James R. Doolittle to Lucius Fairchild, September 16, 1865, Lucius Fairchild Papers, SHSW; untitled manuscript, ca. 1867, Gideon Welles to Andrew Johnson, July 27, 1869, Gideon Welles Papers, HL; Gideon Welles to Charles Sumner, June 30, 1865, Sumner Papers; Gideon Welles to Mark Howard, January 6, 1866, Mark Howard Papers, Connecticut Historical Society.

83. George F. Milton, *The Age of Hate: Andrew Johnson and the Radicals* (New York, 1930), 145–48; Castel, *Johnson*, 9–10.

beat them easily." The Blairs strongly endorsed Johnson's intention to leave suffrage—"the only question of any real importance"—in the hands of Southern whites. So did other Democratic-Republicans, who, as a group, had always been the most racist element in the Republican coalition.[84]

"The ascendancy is with the Blairs," wrote Charles Sumner in June, but Johnson was too experienced a politician to commit himself completely to a single faction of the Republican party, or to a Democracy still identified in many minds with treason. He refused to throw administration support to Democratic candidates in the North's fall elections, which the Republicans carried handily. One index of the President's independence was his refusal to bow to pressure to remove Secretary of State Seward, the old political nemesis of Democrats and Democratic-Republicans. His earlier radicalism having long since faded, Seward emerged from the war with a philosophical attitude toward the fate of the freedmen: They must "take their level" as laborers in the cotton fields, and not stand in the way of reestablishing national harmony. He rose from his sickbed to argue against the $20,000 exclusion, but otherwise seems to have played little role in Johnson's early political deliberations. But as the summer progressed his influence waxed, until, by the end of 1865, many saw him as the guiding spirit of the administration. Seward still commanded broad respect within the Republican party. Certainly, so long as he remained in the Cabinet, it seemed implausible to contend that Johnson had "gone over" to the Democrats.[85]

Throughout 1865, in the words of a New York newspaper, the belief spread "that the old parties must soon give way to new combinations." But political realignment meant different things to different supporters of the President, although all sang the siren song of reelection. If the Blairs envisioned a new organization with loyal Democrats and ex-Democrats at the helm, Seward and his New York political lieutenant Thurlow Weed saw Johnson's Reconstruction policy as a means of redefining power relations within the Republican party. Democrats were welcome, but only as junior partners. In New York, the support of the President and moderate Democrats would enable Seward and Weed to regain ascendancy from Radicals led by Gov. Reuben Fenton. In the South, Seward

84. LaWanda Cox and John H. Cox, *Politics, Principle, and Prejudice 1865–1866* (New York, 1963), 54–56; William E. Smith, *The Francis Preston Blair Family in Politics* (New York, 1933), 2:328–29; Montgomery Blair to Samuel L. M. Barlow, January 12, June 14, July 16, November 12, 1865, Barlow Papers; Jean H. Baker, *The Politics of Continuity: Maryland Political Parties from 1858 to 1870* (Baltimore, 1973), 91–101, 143–47; Francis P. Blair to Andrew Johnson, August 1, 1865, Johnson Papers; Foner, *Free Soil, Free Labor, Free Men,* 267–70.

85. Castel, *Johnson,* 43; Sefton, *Johnson,* 114; S. L. M. Barlow to Montgomery Blair, November 8, 1865, Barlow Papers; Glyndon G. Van Deusen, *William Henry Seward* (New York, 1967), 389, 415–35, 452–53; Montgomery Blair to Samuel L. M. Barlow, September 14, 1867, Barlow Papers; Blaine, *Twenty Years,* 2:68–82.

looked to former Whigs, including planters, as political allies. Seward and his supporters sought to impress upon Johnson the necessity of reaching out to his erstwhile enemies in the planter class and avoiding the politically fatal impression that he was being "captured" by the Democrats.[86]

Also lending their backing to the President's program was an influential body of Northerners who believed the speedy revival of cotton production essential for the nation's economic health. King Cotton may have been dethroned, but as the nation's leading export it remained, as the *New York Times* put it, "a magnate of the very first rank." The trade in the "white gold" was crucial to the wealthy merchants who dominated the economic life of Boston, Philadelphia, New York, and other commercial centers, and to a wide range of other businessmen and professionals such as lawyers, bankers, insurance brokers, and ship owners. Unless cotton production were speedily revived, such men believed, Southerners could never repay their prewar debts, New England textile factories would have to close, and the nation would be unable to earn enough foreign exchange to resume specie payments and pay its overseas indebtedness. Without cotton, declared Rhode Island textile manufacturer and Republican Sen. William Sprague, America would be "bankrupt in every particular."[87]

These economic realities had produced, long before the war, a community of interest among Eastern merchants, financiers, cotton manufacturers, and Southern slaveholders—an alliance, notorious to abolitionists, of the Lords of the Loom and Lords of the Lash. "Wall Street," a Radical editor would remark in 1868, "is Copperhead." In fact, most substantial financiers and merchant capitalists had supported and helped finance the war effort. But they tended to be either moderate Democrats or conservative Republicans who deprecated political agitation that might upset the revival of the Southern economy. Black suffrage, they feared, would disrupt the Southern labor force, and they welcomed policies coercing freedmen into signing labor contracts. Thus, powerful Northern economic interests had a stake in speedy reunion and the resumption of staple agriculture, and believed Johnson's policies could accomplish these goals. "All the sober, substantial men" in New York, Washington, and St. Louis, reported General Grant's aide-de-camp Adam Badeau in October 1865, endorsed the President's program. So did leading urban newspapers tied to commercial interests, like New York's *Times* and *Evening Post*, Philadelphia's *Public Ledger*, and the bulk of the business press.

86. Homer A. Stebbins, *A Political History of the State of New York 1865–1869* (New York, 1913), 65; Thurlow W. Barnes, ed., *Memoir of Thurlow Weed* (Boston, 1884), 451; Cox and Cox, *Politics, Principle, Prejudice*, 45–51; Van Deusen, *Seward*, 422.
87. *New York Times*, August 3, 1865; George L. Anderson, "The South and Problems of Post-Civil War Finance," *JSH*, 9 (May 1943), 186–88; George R. Woolfolk, *The Cotton Regency: The Northern Merchants and Reconstruction, 1865–1880* (New York, 1958), 41–60; *CG*, 39th Congress, 2d Session, 1929.

"If the entire interests of the colored race," declared the New York *Journal of Commerce*, "were remanded where they belong, to the several states, there would be . . . vastly more productive labor." Even businessmen less directly dependent upon Southern raw materials and trade viewed controversy over Reconstruction essentially as an annoyance. The nation, wrote banker Jay Cooke, needed a "plain, simple, common-sense plan" that would enable it to turn its attention to economic concerns.[88]

Against the natural tendency among Northerners to support the new President and the range of interests that united behind him, only one group openly opposed Johnson's program. Radical Republicans were stunned by the May proclamations, believing that on the question of black suffrage, Johnson had misled them. *"We must broaden our front,"* Charles Sumner wrote in June, "so that when Congress meets it will be ready to declare the true doctrine." During the summer and fall of 1865, Radicals and abolitionists embarked on a campaign to convince Northern public opinion that suffrage was "the logical sequence of negro emancipation." Some grounded their appeal upon elemental fairness to the freedman. "There is no party in the South," wrote Wendell Phillips, "that is friendly to the negro or disposed to do him justice." Without the vote, blacks would be doomed to "a century of serfdom." Others adopted arguments intended to appeal to a broader spectrum of Northern opinion. Maryland Radical Henry Winter Davis, avoiding "vague generalities" like "justice and humanity," insisted that only black votes could destroy "the power of those who rebelled. . . . It is a question of power, not of right." To some Radicals, black suffrage formed only one part of a broader program of federal intervention that would remake Southern society; others found it appealing because it offered an alternative to permanent national responsibility for the freedmen. Once blacks had the vote, declared *Harper's Weekly*, "the 'Negro question' will take care of itself." These differences of emphasis were omens of later divisions over Reconstruction policy and the eventual breakup of the Radical coalition. In 1865, however, all Radicals could unite on the principle that without black suffrage there could be no Reconstruction.[89]

88. Joseph Medill to John Sherman, January 7, 1868, John Sherman Papers, LC; David Montgomery, *Beyond Equality: Labor and the Radical Republicans 1862–1872* (New York, 1967), 59–67; Adam Badeau to Elihu B. Washburne, October 20, 1865, Washburne Papers; Peter Kolchin, "The Business Press and Reconstruction," *JSH*, 33 (May 1967), 183–86; Dorothy Dodd, *Henry J. Raymond and The New York "Times" during Reconstruction* (Chicago, 1936), 24–27; Philadelphia *Public Ledger*, January 23, 1866; New York *Journal of Commerce*, December 28, 1865; Ellis P. Oberholzer, *Jay Cooke, Financier of the Civil War* (New York, 1907), 2:22.

89. Pierce, ed., *Memoir and Letters of Sumner*, 4:253–54; Charles Sumner to Carl Schurz, June 15, 1865, Carl Schurz Papers, LC; *The Works of Charles Sumner* (Boston, 1870–83), 9:424–25, 441–77; Philip S. Foner, ed., *The Life and Writings of Frederick Douglass* (New York, 1950–55), 4:158–65; *National Anti-Slavery Standard*, June 3, 1865; Henry Winter Davis to Charles Sumner, June 20, 1865, Sumner Papers; Henry Winter Davis, *Speeches and Addresses* (New York, 1867), 580–83; McKitrick, *Andrew Johnson*, 53–54; *Harper's Weekly*, May 20, 1865.

Throughout these months letters passed back and forth between leading Radicals, lamenting Johnson's policies and promising to organize against them. But an unmistakable note of gloom pervaded this correspondence. "I hope you will do all that can be done for the protection of the poor negroes," Sen. Henry Wilson wrote Freedmen's Bureau Commissioner Howard, since "this nation seems about to abandon them to their disloyal masters." "All is wrong," wrote Ben Butler in July. "We are losing the just results of this four years struggle." Outside one or two strongholds like Massachusetts, the Radicals were on the defensive. Most Republicans appeared to agree with the new Secretary of the Interior, James Harlan, that a battle with Johnson over black suffrage "would lay the foundation of a certain and inevitable division of the Union organization, which I am inclined to think would result in the triumph of the President's policy."[90]

The question of black suffrage, commented New York diarist George Templeton Strong, was "full of difficulties and conflicting rights. No statesman ever had a more knotty problem set him by destiny." Despite the easing of some racial proscriptions in 1864 and 1865 and the agitation of Northern blacks for suffrage, only five states, all in New England, allowed blacks to vote on the same terms as whites. The majority of Republicans were not Radicals but moderates and conservatives who resented the "element that seem to have the negro on the brain all the time" and feared the issue of black rights would prove fatal to the party's electoral prospects. "It is apparent to any sensible man," an Ohio Republican wrote Senator John Sherman, "that its agitation will do harm."[91]

Just how much harm quickly became apparent as, in state after state, black rights emerged as a divisive and potentially damaging issue within the Republican party. Outside New England, the results were scarcely encouraging for the Radicals. New York's Republican convention, controlled by Thurlow Weed, endorsed the President's policies. Pennsylvania's, which supported Thaddeus Stevens' call for the confiscation of Confederate planters' lands, passed over the question of black voting rights as "heavy and premature." In Indiana, Gov. Oliver P. Morton seized upon opposition to immediate black suffrage as a weapon in his long-standing campaign to reduce Radical influence. Wisconsin's state convention adopted a plank drafted by Sen. James R. Doolittle strongly supporting Johnson's policy of leaving suffrage to the individual states. Gen. Jacob Cox, Republican gubernatorial candidate in Ohio, endorsed

90. Thaddeus Stevens to Charles Sumner, June 3, 14, 1865, Sumner Papers; Charles Sumner to Thaddeus Stevens, June 19, 1865, Stevens Papers; McFeely, *Yankee Stepfather*, 91; Benjamin F. Butler to Benjamin F. Wade, June 26, 1865, Wade Papers; James Harlan to Charles Sumner, June 15, 1865, Sumner Papers.

91. Allan Nevins and Milton H. Thomas, eds., *The Diary of George Templeton Strong* (New York, 1952), 4:62; A. L. Brewer to John Sherman, June 11, 1865, Joseph A. Geiger to Sherman, October 19, 1865, Sherman Papers.

Johnson's program and proposed the colonization of blacks in a separate territory where they could "work out their own salvation." Cox considered his stance the only way to protect the party against Democratic charges of favoring racial equality. "On that issue, if made," he told Radical Congressman James A. Garfield, "you will be beaten."[92]

To judge by three 1865 referenda on extending the franchise to the North's tiny black population, Cox was an accurate judge of the public temper. In Minnesota, where the Republican convention endorsed voting rights for the state's few hundred black residents, a constitutional amendment failed by 2,600 votes even though a Republican was elected governor. Wisconsin gubernatorial candidate Gen. Lucius Fairchild, who personally favored black voting, refused to commit himself on the issue until late in the campaign, preferring, as he put it privately, defeat on the suffrage question to "losing the ticket and the amendment." Fairchild won a 10,000-vote majority, but the amendment fell 9,000 votes short of passage. Equally disheartening for the Radicals was the outcome in Connecticut, where black suffrage was defeated by 6,000 votes.[93]

If they gazed long enough, Radicals could discern a silver lining in these results. Black suffrage had attracted 43 percent of the vote in Connecticut, 45 in Minnesota, and 47 in Wisconsin, figures that far exceeded anything achieved in prewar referenda on the issue. Moreover, while nearly all Democrats opposed the policy, a large majority of Republicans voted in favor, an indication that the party's attitude toward black rights had indeed changed during the Civil War. Yet so long as a minority of Republicans remained ready to side with Democrats on the question, black suffrage stood little chance of passage. Subsequent referenda in Kansas, Ohio, New York, and Nebraska Territory produced the same outcome; during these years, only Iowa and Minnesota (both in 1868) would vote to enfranchise their black residents. The 1865 referenda emboldened Democrats and the white South, strengthened Johnson's hand, and weakened the Radicals. They helped convince the President's supporters that his critics formed a tiny "radical and fanatic element," and deepened Johnson's commitment to his own course. As Assistant Secretary of the Treasury William E. Chandler explained to one Radical, Johnson believed the party "could not carry [black suffrage] as a national issue; and the result in Connecticut proves that he is right." Many Radi-

92. Stebbins, *Political History of New York*, 58–62; Thaddeus Stevens to Charles Sumner, August 26, 1865, Sumner Papers; Foulke, *Morton*, 1:446–52; James R. Doolittle to Andrew Johnson, September 9, 1865, Johnson Papers; George H. Porter, *Ohio Politics During the Civil War Period* (New York, 1911), 206–18; Wilbert H. Ahearn, "The Cox Plan of Reconstruction: A Case Study in Ideology and Race Relations," *CWH*, 16 (December 1970), 293–308; Jacob B. Cox to James A. Garfield, July 21, 1865, James A. Garfield Papers, LC.

93. Martin Ridge, *Ignatius Donnelly: The Portrait of a Politician* (Chicago, 1962), 94–102; Lucius Fairchild to H. B. Harshaw, September 15, 1865, Fairchild to "My Dear General," September 22, 1865, Letterbook, Fairchild Papers; Sam Ross, *The Empty Sleeve: A Biography of Lucius Fairchild* (Madison, Wis., 1964), 72–75.

cals concluded that a tactical retreat was in order. "If we shall not be able to maintain the fight on the suffrage question," wrote Garfield, "should we not make a preliminary resistance to immediate restoration and thereby gain time?" When Congress reassembled in December, the issue of black suffrage was, for the moment, politically dead.[94]

Yet beneath the apparent triumph of Johnson's policies, a certain uneasiness was evident by the fall of 1865 among broad sectors of Northern public opinion and influential Republican leaders. Desire for a return to normal did not mean Northerners lacked criteria for judging whether Presidential Reconstruction had succeeded. Northern Republicans demanded concrete evidence that the South genuinely accepted the consequences of the Civil War—the defeat of secession, emancipation of the slaves, and ascendancy of the Republican party. At a minimum, this meant the repudiation of the leaders of the Confederacy, security for Unionists and Northerners in the South, and the protection of the freedmen's basic rights. More indefinite, but no less important, was the anticipation that Southerners would exhibit, in their mood, statements, and behavior, genuine regret over secession and a willingness to "accept the situation" created by their defeat. The *"spirit"* of the white South, the *New York Times* declared, even more than its "formal acts" would determine the Northern assessment of Presidential Reconstruction. If the South did not appear to make a good-faith effort to conform to these requirements, both concrete and symbolic, Republicans, whatever their opinions on black suffrage, might well conclude that Johnson's policies required modification. "The most vivid hope I have," wrote Ben Butler in July, "is that the rebels will behave so outrageously as to awaken the Government and the North once more."[95]

Throughout the summer and fall, Northerners carefully scrutinized every piece of evidence for clues to the "temper and disposition of the Southern people." A host of newspaper reporters and individual fact finders sent back intelligence from the defeated Confederacy, and their reports were supplemented by the private correspondence of soldiers, federal officials, and other Northerners, as well as letters from Southerners, white and black. Few accounts, of course, were entirely objective. "I notice," observed William E. Chandler, "that everyone who goes South, whether Radical or Conservative, comes back confirmed in his previous

94. William Gillette, *The Right to Vote* (Baltimore, 1969 ed.), 25–28; G. Galin Berrier, "The Negro Suffrage Issue in Iowa—1865–1868," *Annals of Iowa*, Ser. 3, 39 (Spring 1968), 241–60; John H. Moore, ed., *The Juhl Letters to the "Charleston Courier"* (Athens, Ga., 1974), 59–60, 72; William E. Chandler to George S. Boutwell, October 7, 1865, William E. Chandler Papers, LC; Richard H. Abbott, *Cobbler in Congress: The Life of Henry Wilson, 1812–1875* (Lexington, Ky., 1972), 166–69; James A. Garfield to Salmon P. Chase, October 4, 1865, Salmon P. Chase Papers, LC. Wisconsin blacks obtained the vote as the result of a state Supreme Court decision in 1866.
95. McKitrick, *Andrew Johnson*, 16–19, 28–30; *New York Times*, December 2, 1865; Benjamin F. Butler to Benjamin F. Wade, July 26, 1865, Wade Papers.

opinion." Yet taken together, these reports eroded support for Johnson's policies. For, with some exceptions, they spoke of the white South's inability to adjust to the end of slavery, the widespread mistreatment of blacks, Unionists, and Northerners, and a pervasive spirit of disloyalty. By the end of November reports from the South had convinced many Northerners that, as a New Yorker put it, Presidential Reconstruction "is no reconstruction at all."[96]

News of violence against the freedmen and the passage of the Black Codes aroused an indignation that spread far beyond Radical circles. Virtually all Republicans agreed, as Edward Atkinson put it, that the Civil War had been "a war for the establishment of free labor, call it by whatever other name you will." Thus, efforts to "restore all of slavery but its name" by state legislation were anathema. "Whoever assented to the President's plan of Reconstruction," Congressman James G. Blaine later observed, "assented to these laws, and, beyond that, assented to the full right of the rebellious states to continue legislation of this odious type." Johnson would never quite understand that, to mainstream Republicans like Blaine, the freedmen had earned a claim upon the conscience of the nation. Partly because of the service of black troops, their fate had become identified with the outcome of the Civil War. Many Northerners who did not share the Radicals' commitment to black political rights insisted the freedmen's personal liberty and ability to compete as free laborers must be guaranteed or emancipation would be little more than a mockery.[97]

Reports also circulated of insults to Northern travelers—of hotels and restaurants refusing to serve them, steamboats denying them passage. The "great majority" of Southern whites, a Rochester resident related to Secretary of State Seward after a trip through four Southern states, "are exceedingly sore towards the North and northern people, however much they may sometimes attempt to cover it up." But probably the most damaging accounts were those describing a revival of "rebel" political hegemony and the proscription of Southern Unionists. "As for negro suffrage," declared Chicago editor Charles A. Dana in September, "the mass of the Union men in the Northwest do not care a great deal. What scares them is the idea that the rebels are all to be let back . . . and made a power in the government again, just as though there had been no rebellion." The fact that so many of the South's new Senators and Congressmen had opposed secession in 1860 seemed less significant than that they had gone on to serve the Confederacy—including ten as gener-

96. Perman, *Reunion Without Compromise*, 13–14; Leon B. Richardson, *William E. Chandler, Republican* (New York, 1940), 66; J. Michael Quill, *Prelude to the Radicals: The North and Reconstruction During 1865* (Washington, D.C., 1980), 127–30; S. F. Wetmore to John Andrew, November 20, 1865, Andrew Papers.

97. Edward Atkinson, *On Cotton* (Boston, 1865), 40; Blaine, *Twenty Years*, II, 92–93, 105; Eugene H. Berwanger, *The West and Reconstruction* (Urbana, Ill., 1981), 43–53.

als. "Obnoxious traitors," the moderate Indianapolis *Journal* called them. From Mississippi, former slaveowner Robert Flournoy advised Stevens that Johnson's course had destroyed "whatever genuine Union sentiment was forming and would in time have grown up." He urged Congress not to seat the Southern Congressmen: "Let these rebellious states know and feel that there is a power left that can reach and punish treason."[98]

Coupled with praise Northern Democrats lavished upon the President—a display many Republicans found disquieting—such reports had a profound cumulative impact. Even Gideon Welles concluded that the white South had pursued precisely the course to strengthen the Radicals and "embarrass those who are willing to befriend them." Southern events slowly reshaped the thinking of such influential moderates as Sen. Lyman Trumbull of Illinois, who had at first strongly supported Johnson's policies. By no means an advocate of black suffrage, Trumbull became convinced that further federal measures to protect blacks' civil rights, encourage Southern Unionism, and suppress violence were necessary before the South could resume its place in national life. In mid-November, Speaker of the House Schuyler Colfax delivered an impromptu speech in Washington, demanding that the South meet additional requirements before Reconstruction could be deemed complete. Blacks must be granted equality before the law, Congressmen must be elected who had never voluntarily aided the Confederacy, and Southerners must "give evidence of their earnest and cheerful loyalty." He made no mention of black suffrage. The speech won an enthusiastic response from the Republican press, including the pro-Johnson *New York Times*. Here, wrote a Washington reporter, was "the key-note of action of the Republicans in Congress." Colfax believed he had outlined a platform upon which the party "could sweep the country from ocean to ocean next fall."[99]

When the Thirty-Ninth Congress gathered early in December, Johnson's position remained impressive. The President sincerely believed he had created a new political order in the South, controlled by men loyal to the Union who genuinely accepted the end of slavery. He simply could not believe, one suspects, that Northern Republicans would jettison his program over so quixotic an issue as the freedmen's rights. Indeed, few

98. Powell, *New Masters*, 57–63; Benjamin F. Sanford to William H. Seward, November 23, 1865, Seward Papers; McKitrick, *Andrew Johnson*, 36; Charles A. Dana to Isaac Sherman, September 17, 1865, Isaac Sherman Papers, Private Collection; Carter, *When the War Was Over*, 229–30; Charles Roll, "Indiana's Part in Reconstruction," in *Studies in American History Dedicated to James Albert Woodburn* (Bloomington, Ind., 1926), 308; Robert W. Flournoy to Thaddeus Stevens, November 20, 1865, Stevens Papers.

99. McKitrick, *Andrew Johnson*, 60–61; Beale, ed., *Welles Diary*, 2:420; Gideon Welles to Mark Howard, December 12, 1865, Howard Papers; Mark M. Krug, *Lyman Trumbull: Conservative Radical* (New York, 1965), 231–32; O. J. Hollister, *Life of Schuyler Colfax* (New York, 1887), 270–73; Kenneth L. Tomlinson, "Indiana Republicans and the Negro Suffrage Issue, 1865–1867" (unpub. diss., Ball State University, 1971), 88; Quill, *Prelude*, 133.

Republicans seemed eager for a conflict that would split the party and perhaps throw the federal government into Democratic hands. "We ought to do all in our power to avoid a break with him," Radical Sen. Jacob Howard told Sumner in November, "and to keep him with us." Even Massachusetts Sen. Henry Wilson, who had been denouncing Johnson's policies all summer, informed Seward that the question confronting Congress was not black voting, "but the annulment of all laws against the freedmen and their full liberty." He pleaded for administration support in this endeavor. The door stood open for Johnson to embrace the emerging consensus within the Republican party that the freedmen were entitled to civil equality short of suffrage, and that wartime Unionists must play a more prominent role in Southern politics. To oppose these changes in his Reconstruction policy would force the bulk of Republicans to choose between abandoning their principles and constituents and abandoning the President, and might well bring about the very outcome Johnson most feared. For, as a West Virginia editor had a few months earlier observed after a tour of the North, "abstractly the American people are not in favor of Negroes voting, but as against rebeldom ruling in Congress or even in the South, they are."[100]

Those close to Johnson, however, knew he was not prone to compromise. Indeed, they relished the prospect of a political battle over Reconstruction. "A fight between the Radicals and the Executive is inevitable," declared Harvey Watterson. "Let it come. The sooner the better for the whole country."[101]

100. W. R. Brock, *An American Crisis* (London, 1963), 196–98; Jacob Howard to Charles Sumner, November 12, 1865, Sumner Papers; Henry Wilson to William H. Seward, November 20, 1865, Seward Papers; Richard O. Curry, ed., "A Note on the Motives of Three Radical Republicans," *JNH*, 47 (October 1962), 277.

101. Harvey M. Watterson to Joseph E. Brown, December 13, 1865, Hargrett Collection, UGa.

CHAPTER 6

The Making of Radical Reconstruction

IT was a peculiarity of nineteenth-century politics that more than a year elapsed between the election of a Congress and its initial meeting. The Thirty-Ninth Congress, elected in the midst of war, assembled in December 1865 to confront the crucial issues of Reconstruction: Who would control the South? Who would rule the nation? What was to be the status of the emancipated slave? In both Houses, Republicans outnumbered Democrats by better than three to one. Clearly, a united party would have no difficulty enacting a Reconstruction policy and, if necessary, overriding Presidential vetoes. But American parties are not tightly organized, ideologically unified political machines, and the interaction between the Republican party's distinctive factions would go a long way toward determining the contours of Congressional policy.

The Radical Republicans

On the party's left stood the Radical Republicans, a self-conscious political generation with a common set of experiences and commitments, a grass-roots constituency, a moral sensibility, and a program for Reconstruction. At the core of Congressional Radicalism were men whose careers had been shaped, well before the Civil War, by the slavery controversy: Charles Sumner, Ben Wade, and Henry Wilson in the Senate; Thaddeus Stevens, George W. Julian, and James M. Ashley in the House. With the exception of Stevens they represented constituencies centered in New England and the belt of New England migration that stretched across the rural North through upstate New York, Ohio's Western Reserve, northern Illinois, and the Upper Northwest. Here lay rapidly growing communities of family farms and small towns, where the superiority of the free labor system appeared self-evident, antebellum reform had flourished, and the Republican party, from the moment of its birth, commanded overwhelming majorities.

Bringing to politics the moral sensibility of abolitionism, Radicals had long insisted that slavery and the rights of blacks must take precedence over other political questions. Most, observed young Georges Clemenceau, reporting American events for a French newspaper, "embarked on the abolitionist sea without any clear idea of where their course would lead." Yet they had sailed that sea for years and had learned to navigate its waters. Their very political longevity gave them a sense of common purpose and enhanced their influence at a time of remarkably high turnover between Congresses. In a House half of whose members were in their first terms, it mattered that a man like Stevens had a career stretching back before the war, and could recall the days "when the mighty Toombs [the powerful antebellum Congressman from Georgia], with his shaggy locks, headed a gang who, with shouts of defiance, rendered this a hell of legislation."[1]

The preeminent Radical leaders, Thaddeus Stevens and Charles Sumner, differed in personality and political style. The recognized floor leader of House Republicans, Stevens was a master of Congressional infighting, parliamentary tactics, and blunt speaking. One contemporary called him "a perfect political brigand, . . . a rude jouster in political and personal warfare," and his quick tongue and sarcastic wit were legendary. "When it was first proposed to free the slaves and arm the blacks," he remarked in 1865, "did not half the nation tremble? The prim conservatives, the snobs, and the male waiting maids in Congress, were in hysterics." Even those who disagreed with his policies could not avoid a grudging admiration for the man and his honesty, idealism, and indifference to praise and criticism—qualities not altogether common among politicians. Seventy-three in 1865, Stevens impressed Clemenceau as the "Robespierre" of "the second American Revolution." Yet Stevens knew, too, the limits of his influence. "No man was oftener outvoted," a Boston newspaper commented after his death, and Stevens proved more than willing to seize immediate political advantages, and to compromise when necessary.[2]

Like Stevens, Sumner was "a man absolutely convinced, and in a sense

1. Eric Foner, *Free Soil, Free Labor, Free Men: The Ideology of the Republican Party Before the Civil War* (New York, 1970), 105–106; Hans L. Trefousse, *The Radical Republicans: Lincoln's Vanguard for Racial Justice* (New York, 1969), 49–55; Margaret Shortreed, "The Anti-Slavery Radicals: From Crusade to Revolution 1840–1868," *Past and Present*, 16 (November 1959), 65–87; Georges Clemenceau, *American Reconstruction*, edited by Fernand Baldensperger, translated by Margaret MacVeagh (New York, 1928), 278; Morton Keller, *Affairs of State: Public Life in Late Nineteenth Century America* (Cambridge, Mass., 1977), 61; CG, 39th Congress, 1st Session, 2544.

2. S. W. Bowman to John Sherman, March 1, 1866, John Sherman Papers, LC; *Reconstruction. Speech of the Hon. Thaddeus Stevens Delivered in the City of Lancaster, September 7, 1865* (Lancaster, Pa., 1865), 7; Springfield *Republican*, August 15, 1868; Clemenceau, *American Reconstruction*, 77–79; Boston *Advertiser*, August 13, 1868.

rightly, that he and history were for the moment in perfect step." Disliked by Senate colleagues for egotism, self-righteousness, and stubborn refusal to compromise, Sumner acted as though he were the voice, the embodiment, of the New England conscience. And, in a sense, he was. Unconcerned with the details of committee work and legislative maneuvering, Sumner's forte lay in lengthy, erudite speeches in which he expounded the recurrent theme of his political career: the principle of equality before the law. His speeches rarely had much impact on the details of legislation, but in the crisis of war and Reconstruction, Sumner's ideas acquired an increasing following in Republican circles. Sumner's uncompromising stance on black rights, a Californian wrote early in 1866, had caused many Republicans to reassess their own opinions: "You have hundreds of believers in your doctrine in this State where you had not one four years ago." Unlike Stevens, who never enjoyed close connections with moral reform circles, Sumner corresponded frequently with abolitionists, who considered him *their* politician. So too did ordinary blacks, North and South, who deluged him with requests for advice and accounts of their grievances. "Your name," wrote a black army veteran in 1869, "shall live in our hearts for ever."[3]

Uniting Stevens, Sumner, and the other Radicals in 1865 was the conviction that the Civil War constituted a "golden moment," an opportunity for far-reaching change that, if allowed to pass, "will have escaped for years, if not forever." While some constituents (including the aptly named Massachusetts man Henry A. Coffin), demanded the execution of Southern leaders as punishment for treason, only a handful of Radical leaders echoed these calls.[4] Rather than vengeance, the driving force of Radical ideology was the utopian vision of a nation whose citizens enjoyed equality of civil and political rights, secured by a powerful and beneficent national state. For decades, long before any conceivable political benefit derived from its advocacy, Stevens, Sumner, and other Radicals had defended the unpopular cause of black suffrage and castigated the idea that America was a "white man's government" (a doctrine, Stevens remarked, "that damned the late Chief Justice [Roger B. Taney] to everlasting fame; and, I fear, to everlasting fire"). Although Stevens and Sumner were, by any standard, racial egalitarians, many Radicals

3. Eric L. McKitrick, *Andrew Johnson and Reconstruction* (Chicago, 1960), 268; David Donald, *Charles Sumner and the Rights of Man* (New York, 1970); Asa Hatch to Charles Sumner, February 16, 1866, Parker Pillsbury to Sumner, March 16, 1866, James P. Lee and fourteen others to Sumner, February 12, 1866, G. B. Thomas to Sumner, April 3, 1869, Charles Sumner Papers, HU.

4. *CG*, 39th Congress, 1st Session, 1016; Henry A. Coffin to Charles Sumner, February 22, 1866, Sumner Papers. Sumner, Wade, and Stevens opposed any executions for treason, and Henry Wilson interceded with President Johnson on behalf of Jefferson Davis. Charles Sumner to John Bright, April 18, 1865, Add. Mss 43,390, f. 216, John Bright Papers, British Museum; Trefousse, *Radical Republicans*, 342.

could not free themselves entirely from the prejudices so pervasive in their society. Many accepted the common stereotypes of the manly and vigorous Anglo-Saxon, the religious, "feminine," black, and the intelligent mulatto, and assumed blacks were naturally suited to warmer climes. Yet even those who harbored doubts about blacks' innate capabilities insisted that to limit on racial grounds the egalitarian commitments central to American political culture made a mockery of republican institutions. There was no room for a legally and politically submerged class in the "perfect republic" that must emerge from the Civil War.[5]

"I believe in equality among citizens—equality in the broadest and most comprehensive democratic sense," declared Massachusetts Sen. Henry Wilson. But the implications of the elusive term "equality" were anything but clear in 1865. At the outset of Reconstruction most Republicans still adhered to a political vocabulary inherited from the antebellum era, which distinguished sharply between natural, civil, political, and social rights. The first could not legitimately be circumscribed by government; slavery had been wrong, fundamentally, because it violated the natural rights—life, liberty, and the pursuit of happiness—common to all humanity. Equality in civil rights—equal treatment by the courts and civil and criminal laws—most Republicans now deemed nearly as essential, for an individual's natural rights could not be secured without it. Although Radicals insisted black suffrage must be part of Reconstruction, the vote was commonly considered a "privilege" rather than a right; requirements varied from state to state, and unequal treatment or even complete exclusion did not compromise one's standing as a citizen. And social relations—the choice of business and personal associates—most Americans deemed a personal matter, outside the purview of government. Throughout Reconstruction, indeed, the term "social equality" conjured up fantastic images of blacks forcing their way into whites' private clubs, homes, and bedrooms. "Negro equality, . . ." Stevens assured the House, "does not mean that a negro shall sit on the same seat or eat at the same table with a white man. That is a matter of taste which every man must decide for himself."[6]

To this egalitarianism, the Civil War wedded a new conception of the powers and potentialities of the national state. More fully than other Republicans, the Radicals embraced the wartime expansion of federal authority, carrying into Reconstruction the conviction that federalism and state rights must not obstruct a sweeping national effort to define and

5. Foner, *Free Soil, Free Labor, Free Men*, 149, 281–92; *Nation*, September 14, 1865; George S. Boutwell, *Speeches and Papers* (Boston, 1867), 472; Hans L. Trefousse, "Ben Wade and the Negro," *OHQ*, 68 (April 1959), 161–76; *CG*, 39th Congress, 1st Session, 75, 258; 2d Session, 251.

6. *CG*, 40th Congress, 3d Session, 1326; Foner, *Free Soil, Free Labor, Free Men*, 291–97; *CG*, 39th Congress, 2d Session, 252.

protect the rights of citizens. The "demon of Caste," Sumner declared in 1865, must be destroyed. "The same national authority that destroyed slavery must see that this other pretension is not permitted to survive."[7]

For Stevens, the war had created its own logic and imperatives. "We are making a nation," he told the House, and obsolete "technical scruples" must not be allowed to stand in the way. The Southern states had seceded, waged war against the Union, and been vanquished; having sacrificed their constitutional standing, they could be treated as conquered provinces, governed according to the will of Congress. This position had the advantage that it accorded with reality. Whatever the metaphysical reasoning of legal theorists, the South had in fact been subjugated on the battlefield. Yet Stevens displayed too little regard for constitutional niceties to command broad support. Other Radicals sought to locate authority for their political program elsewhere than in the barrel of the conquerer's gun. Sumner insisted that the Declaration of Independence must be recognized as a "fundamental law" coequal with the Constitution and authorizing federal action to secure equality before the law. He also advanced the notion of "state suicide"— while not legally seceding, the Southern states had, in effect, reverted to the condition of territories, and were now subject to Congressional authority.[8]

Increasingly, however, Radicals turned to another reservoir of federal power, the Constitution's clause guaranteeing to each state a republican form of government. Originally designed to permit the federal government to suppress insurrections and prevent a rebirth of monarchial and aristocratic institutions, the guarantee clause lay dormant until the Civil War. But Radical Republicans found it attractive precisely because it might clothe federal intervention on behalf of citizens' rights with constitutional legitimacy. This was "the jewel of the Constitution," said Illinois Sen. Richard Yates; Sumner called it "a sleeping giant . . . never until this recent war awakened, but now it comes forward with a giant's power. There is no clause in the Constitution like it. There is no other clause which gives to Congress such supreme power over the states." The guarantee clause, however, had an enigmatic, "almost Delphic" quality. What was a republican form of government? How should the United States guarantee it? "The time has come," Sumner announced, "to fix a meaning on these words." Having read "everything on the subject" of republicanism "from Plato to the last French pamphlet," he delivered a typically erudite speech in February 1866 on the "foundations of the Republics of the world." It was also typically long-winded, lasting two full days and

7. Allan G. Bogue, *The Earnest Men: Republicans of the Civil War Senate* (Ithaca, N.Y., 1981), 313–25; *The Works of Charles Sumner* (Boston, 1870–83), 9:424–25.

8. *CG*, 40th Congress, 2d Session, 1966; 39th Congress, 1st Session, 697, 1310; *Reconstruction. Stevens Speech*, 3; McKitrick, *Andrew Johnson*, 110; *Works of Sumner*, 12:191–92.

filling forty-one fine-print columns of the *Congressional Globe*. But its point was straightforward: A government that denied any of its citizens equality before the law and did not rest fully on the consent of the governed could not be considered republican. (Critics responded that, according to this definition, "there never has been a republican government on the face of the earth.")[9]

Like the republicanism of the American Revolution, Reconstruction Radicalism was first and foremost a civic ideology, grounded in a definition of American citizenship. On the economic issues of the day—the tariff, currency, railroad aid—no distinctive or unified Radical position existed. Stevens, himself a small iron manufacturer, favored an economic program geared to the needs of small producers and aspiring businessmen, including tariff protection, low interest rates, plentiful greenback currency, and government promotion of internal improvements. For Stevens, his Pennsylvania colleague William D. Kelley, and Ben Butler of Massachusetts, these policies went hand in hand with hostility to middlemen, financiers, and other "nonproducers," and their program of aggressive economic nationalism won support from rising entrepreneurs, including those in expanding war-related industries. On the other hand, Radicals like Charles Sumner and *Nation* editor E. L. Godkin, men attuned to orthodox laissez-faire economic theory, favored a low tariff, the swift resumption of specie payments, and minimal government involvement in the economy, policies supported by bankers, commercial capitalists, and well-established northeastern manufacturers.[10]

Some Radical lawmakers, like prosperous Michigan merchant and land speculator Zachariah Chandler, and John B. Alley, the wealthiest man in the shoe manufacturing city of Lynn, Massachusetts, were themselves successful capitalist entrepreneurs; others, like Iowa Congressman William B. Allison, were closely tied to local railroad interests. But generally, Congressional Radicals viewed economic issues as secondary to those of Reconstruction. "No question of finance, or banks, or currency, or tariffs," declared Richard Yates, "can obscure this mighty moral question of the age." James Ashley's Republican opponents charged that by devot-

9. William M. Wiecek, *The Guarantee Clause of the U.S. Constitution* (Ithaca, N.Y., 1972), 1–2 and passim; Charles O. Lerche, Jr., "Congressional Interpretations of the Guarantee of a Republican Form of Government During Reconstruction," *JSH*, 15 (May 1949), 192–211; Edward L. Pierce, ed., *Memoir and Letters of Charles Sumner* (Boston, 1877–93), 3:258–59; *CG*, 39th Congress, 1st Session, 673–87; 40th Congress, 1st Session, 614; 2d Session, 1961; 41st Congress, 2d Session, 1327.

10. Glenn M. Linden, " 'Radicals' and Economic Policies: The Senate, 1861–1873," *JSH*, 32 (May 1966), 189–99; Glenn M. Linden, " 'Radicals' and Economic Policies: The House of Representatives, 1861–1873," *CWH*, 13 (March 1967), 51–65; Irwin Unger, *The Greenback Era: A Social and Political History of American Finance 1865–1879* (Princeton, 1964), 85–86; David Montgomery, "Radical Republicanism in Pennsylvania, 1866–1873," *PaMHB*, 85 (October 1961), 439–57; Pierce, ed., *Sumner*, 4:355–60; *Nation*, March 22, 1866; Amos A. Lawrence to Charles Sumner, January 25, 1867, July 18, 1868, Sumner Papers; Edward Atkinson to Henry Wilson, July 9, 1866, Letterbooks, Edward Atkinson Papers, MHS.

ing all his attention to Reconstruction, he neglected the commercial interests of his Toledo district. Sumner, who once remarked that he knew nothing of finance, said little on economic matters until 1868 when, hoping to gain the support of Boston's "mercantile classes" in his bid for reelection, he maladroitly proposed the immediate resumption of specie payments—a plan so out of touch with fiscal reality that businessmen told him it would prove economically disastrous. As for Stevens, one Philadelphia businessman complained, "He seems to oppose any measure that will not benefit the *nigger.*"[11]

For their part, some prominent Northern entrepreneurs were strongly attracted to the Radical ideology. Bostonian John Murray Forbes, a leading investor in midwestern railroads and a "rare Republican" among Boston's commercial elite, viewed black suffrage as the only way of creating the political conditions necessary for Northern investment in a reconstructed South. Radicals also won support among entrepreneurs in the small-town North and manufacturers who believed blacks, if allowed to rise in the social scale, would form a new market for their products. But other businessmen, especially those tied directly or indirectly to the cotton trade or who hoped to invest in the South, feared Radical policies would "disrupt the cheap Southern labor force" and interfere with the resumption of cotton production. Clearly, capitalists did not agree among themselves on Reconstruction. Even California's renowned railroad magnates divided. E. B. Crocker and Leland Stanford supported the Radicals, while Collis P. Huntington denounced the idea of black suffrage as a product of "French Philosophy . . . better fitted to ornament the hall of some Social Science meeting than the halls of legislation."[12]

Radical Republicanism did possess a social and economic content, but one that derived from the free labor ideology rather than from the interests of any particular set of businessmen. The North's victory, declared Horace Greeley, marked "the triumph of republicanism and free labor."

11. W. R. Brock, *An American Crisis* (London, 1963), 82–84; Alan Dawley, *Class and Community: The Industrial Revolution in Lynn* (Cambridge, Mass., 1976), 100; Leland L. Sage, *William Boyd Allison: A Study in Practical Politics* (Iowa City, 1956), 65–67, 77–78, 85–86; CG, 40th Congress, 2d Session, Appendix, 351; Maxine B. Kahn, "Congressman Ashley in the Post-Civil War Years," *Northwest Ohio Quarterly*, 36 (Autumn 1964), 206–7; John G. Sproat, *"The Best Men": Liberal Reformers in the Gilded Age* (New York, 1968), 188; Edward L. Pierce to Charles Sumner, July 24, 31, 1868, John Murray Forbes to Sumner, November 16, 1868, Edward Atkinson to Sumner, December 4, 1868, Amasa Walker to Sumner, December 25, 1868, Sumner Papers; J. Williamson to Thomas A. Jenckes, February 16, 1866, Thomas A. Jenckes Papers, LC.

12. Stanley Coben, "Northeastern Business and Radical Reconstruction: A Re-Examination," *MVHR*, 46 (June 1959), 67–90; Dale Baum, *The Civil War Party System: The Case of Massachusetts, 1848–1876* (Chapel Hill, 1984), 75; Sarah F. Hughes, ed., *Letters (Supplementary) of John Murray Forbes* (Boston, 1905), 3:25–26, 43–44; Mark Howard to Mrs. Howard, July 12, 1865, Mark Howard Papers, Connecticut Historical Society; E. B. Crocker to Cornelius Cole, April 12, 1865, Leland Stanford to Cole, February 9, 1867, Cornelius Cole Papers, University of California, Los Angeles; Collis P. Huntington to Andrew Johnson, February 27, 1866, Andrew Johnson Papers, LC.

And just as Radicals extended the notion of civil and political equality to embrace blacks, so they insisted the freedmen were entitled to the same economic opportunities enjoyed by white laborers. A correspondent of Sumner's, describing how no New York City hotel would offer his black servant accommodations, gave striking expression to the Radical ideal of equal opportunity regardless of race:

> Is not this state of things a disgrace to America, as a land of liberty and freedom? Must the black man—as free—be insulted and humiliated at every step? . . . The white servant is deemed not on an equality with his employer— yet recognized the right to rise to that equality. Neither is the black servant on an equality with his employer—yet has an equal right with the white servant to gain it.

If the once-popular description of Radical Republicanism as an expression of the interests of Northern business is untenable when couched in terms of specific issues like the tariff, it remains true that Radicals hoped to reshape Southern society in the image of the small-scale competitive capitalism of the North. As Carl Schurz put it, "a free labor society must be established and built up on the ruins of the slave labor society." While many moderates shared this goal, few wished to hold Reconstruction hostage to such a transformation. To Radicals, however, the South's political and social "regeneration" formed two sides of the same coin. "My dream," one explained in 1866, "is of a model republic, extending equal protection and rights to all men. . . . The wilderness shall vanish, the church and school-house will appear; . . . the whole land will revive under the magic touch of free labor."[13]

At the outer limits of Radical Republicanism, the idea of remaking Southern society produced a plan for national action to overturn the plantation system and provide the former slaves with homesteads. In a speech to Pennsylvania's Republican convention in September 1865, Stevens called for the seizure of the 400 million acres belonging to the wealthiest 10 percent of Southerners. Forty acres would be granted to each adult freedman and the remainder—some 90 percent of the total— sold "to the highest bidder" in plots, he later added, no larger than 500 acres. The proceeds would enable the federal government to finance pensions for Civil War veterans, compensate loyal men who had suffered losses during the war (not a few of whom lived in Stevens' southern Pennsylvania district), and retire the bulk of the national debt. As was typical of its author, the plan combined idealism, expediency, and Northern self-interest, all in the service of a far-reaching social revolution:

13. Horace Greeley to Thomas Dixon, June 2, 1865, Horace Greeley Papers, NYPL; William C. Jewell to Charles Sumner, February 19, 1866, Sumner Papers; Howard K. Beale, "The Tariff and Reconstruction," *AHR*, 35 (January 1930), 276–94; Joseph Schafer, ed., *Intimate Letters of Carl Schurz 1841–1869* (Madison, Wis., 1928), 340–41; *CG*, 39th Congress, 2d Session, 118, Appendix, 78.

The whole fabric of southern society *must* be changed, and never can it be done if this opportunity is lost. Without this, this Government can never be, as it has never been, a true republic. . . . How can republican institutions, free schools, free churches, free social intercourse exist in a mingled community of nabobs and serfs? If the South is ever to be made a safe republic let her lands be cultivated by the toil of the owners or the free labor of intelligent citizens.[14]

As this statement suggests, Stevens' aims were political in the broadest sense rather than strictly economic. As Kelley said shortly after Stevens' death: "He knew that a landed aristocracy and a landless class are alike dangerous in a republic, and by a single act of justice he would abolish both." Confiscation, Stevens believed, would break the power of the South's traditional ruling class, transform the Southern social structure, and create the basis for a triumphant Southern Republican party composed of black and white yeomen and Northern purchasers of planter land. Even among abolitionists and Radical Republicans, however, only a handful stressed the land question as persistently and forcefully as Stevens. Throughout the 1860s, George W. Julian carried on a relentless Congressional struggle against "land monopoly," advocating confiscation measures in aid of blacks and condemning land grants to railroads and other claims upon the public domain that endangered the homestead principle. Wendell Phillips and, occasionally, Benjamin F. Butler and Charles Sumner, also advocated a federal land distribution policy. Yet to most Radicals, land for the freedmen, though a commendable idea, was not nearly as crucial to Reconstruction as black suffrage. "If I were a black man, with the chains just stricken from my limbs . . . " said James M. Ashley, "and you should offer me the ballot, or a cabin and forty acres of cotton land, I would take the ballot."[15]

The response to the land question illuminated long-standing divisions, even among Radicals, over the definition of freedom and the economic future of the postemancipation South. It is hardly surprising that many Radicals proved reluctant to support a program that so contravened the sanctity of property as confiscation; what is striking is how few suggested an alternative, other than holding out the prospect of individual advancement in accordance with the free labor ideology. Stevens, in a sense, echoed eighteenth-century republican thought in his concern for the social basis of representative institutions and in identifying freedom with ownership of productive property. "The system of labor for wages," said

14. Eric Foner, *Politics and Ideology in the Age of the Civil War* (New York, 1980), 128–49; *Reconstruction. Stevens Speech*, 5–7; *CG*, 40th Congress, 1st Session, 203.
15. *CG*, 40th Congress, 3d Session, 133–34; Patrick W. Riddleberger, "George W. Julian: Abolitionist Land Reformer," *AgH*, 29 (July 1955), 108–15; Foner, *Politics and Ideology*, 134; Benjamin W. Arnett, ed., *Duplicate Copy of the Souvenir from the Afro-American League of Tennessee to the Hon. James M. Ashley of Ohio* (Philadelphia, 1894), 408.

Kelley, "is not the freedom of which he . . . dreamed." And Stevens appreciated that only governmental action of a radical kind could prevent blacks from emerging from slavery as landless laborers. But most Radicals believed that in a free labor South, with civil and political equality secured, black and white would find their own level, and as Ben Wade put it, "finally occupy a platform according to their merits." The key was that all must be given "a perfectly fair chance." Moreover, contended Massachusetts financier and cotton manufacturer Edward Atkinson, a self-described "business radical," market forces themselves would inevitably produce the demise of the plantation system, once high cotton prices and competition for their labor enabled blacks to earn good wages. Atkinson anticipated the eventual rise of a black yeomanry. But confiscation, he insisted, would "ruin the freedmen" by leading them to believe they could acquire land without "*working* for it and purchasing."[16]

The debate over the land issue illuminated both divisions within Radical ranks and the limits of the Radical ideology. Radicals' expansive talk of equality alarmed conservative Republicans like Pennsylvania Sen. Edgar Cowan, who thought the social structure, "after all is said and done, is pretty well arranged." "Who would black boots and curry the horses, who would do the menial offices of the world," Cowan asked, if Sumner's doctrines became reality? Yet Sumner made it clear that equality before the law did not rule out large disparities of wealth, status, and power. He would prohibit railroads, he told the Senate in 1871, from maintaining separate cars for black and white passengers, but not first, second, and third class accommodations: "That is simply a matter of price." Beyond equality, in other words, lay questions of class relations crucial to the freedmen and glimpsed in debates over confiscation, but lying beyond the purview of Radical Republicanism.[17]

Yet whatever Radicals' indecision as to the economic future of the postwar South, the core of the ideology—the idea of a powerful national state guaranteeing blacks equal standing in the polity and equal opportunity in a free labor economy—called for a striking departure in American public life. As Congress assembled, no one knew how many Republicans were ready to advance this far. Even Radical Congressman John M. Broomall of Pennsylvania, writing of the need to bring together all House members favoring black suffrage, could add, "after the first dozen who

16. *CG*, 40th Congress, 3d Session, 133–34; 39th Congress, 2d Session, 63; Frederic Bancroft, ed., *Speeches, Correspondence and Political Papers of Carl Schurz* (New York, 1913), 1:370–71; Edward Atkinson to John Andrew, December 9, 1865, Atkinson to Hugh McCulloch, January 2, 1866, Atkinson to Henry Ward Beecher, June 25, 1867, Atkinson to J. C. Peters, June 26, 1867, Letterbooks, Atkinson Papers; Atkinson to Charles Sumner, July 8, 1867, Sumner Papers.

17. *CG*, 39th Congress, 1st Session, 342; *Works of Sumner*, 14:362; David Montgomery, *Beyond Equality: Labor and the Radical Republicans 1862–1872* (New York, 1967).

can tell who they are?" Retrospective analysis reveals that the Radicals, while hardly "in control" of Congress, enjoyed substantial power, constituting nearly half the Republican members of the House and a lesser but still significant portion of the Senate. Yet many Republicans defied precise classification, and many others shifted position as the crisis over Reconstruction deepened. Indeed, the crisis itself shaped in contradictory ways the life of the Thirty-Ninth Congress. Placing a premium on party unity and political moderation, since two thirds of Congress was necessary to override a veto, it simultaneously created a political climate that allowed Radicalism to flourish.[18]

The growing perception of white Southern intransigence and President Johnson's inability to satisfy Congressional concern for the rights of blacks and white loyalists would help propel the party's center of gravity to the left. Radicalism, however, possessed a dynamic of its own, based first of all on the reality that in a time of crisis, Radicals alone seemed to have a coherent sense of purpose. The "one body of men who had any positive affirmative ideas," Texas Senator-elect Oran M. Roberts discovered upon arriving in Washington, were "the vanguard of the radical party. They knew exactly what they wanted to do, and were determined to do it." Then too, the Radicals represented a significant strand of Northern opinion, an essential part of the Republican constituency. "Without the votes of the moderate men we cannot win," observed Henry Wilson. But by the same token, if Radicals could not generally "make" policy, their acquiescence in policy was indispensable. Any settlement unacceptable to the Radicals might well shatter the Republican party; certainly it would guarantee a continuation of the long agitation over slavery and its legacy. Here, indeed, lay the Republican dilemma. On the one hand, black suffrage was a dangerous issue—as Wilson told the Senate in 1869, not "a square mile of the United States" existed where identification with the rights of blacks did not cost the party white votes. Yet, another Radical responded, the loyalty of the party's most committed followers depended on its reformist image; should Republicans abandon the "rights of the colored men . . . the party itself would not be worth preserving."[19]

Radicals, a New York *Tribune* correspondent observed, could lead Congress only "in the direction of its own convictions." Yet in a period of intense ideological conflict, those convictions were in a state of rapid flux. Time and again, Radicals had staked out unpopular positions, only to be

18. John M. Broomall to Thaddeus Stevens, October 27, 1866, Thaddeus Stevens Papers, LC; Michael L. Benedict, *A Compromise of Principle: Congressional Republicans and Reconstruction 1863–1869* (New York, 1974), 25–26, 348–57; Edward L. Gambill, "Who Were the Senate Radicals?," *CWH*, 11 (September 1965), 237–44.

19. O. M. Roberts, "The Experience of an Unrecognized Senator," *Texas State Historical Association Quarterly*, 12 (October 1908), 132; Henry Wilson to Henry C. Carey, April 16, 1866, Henry C. Carey Papers, Edward Carey Gardiner Collection, HSPa; *CG*, 40th Congress, 3d Session, 672, 708.

vindicated by events. Uncompromising opposition to slavery's expansion, emancipation, the arming of black troops—all enjoyed little support when first proposed, yet all had come to be embraced by the mainstream of Republican opinion. "These are no times of ordinary politics," declared Wendell Phillips. "These are formative hours: the national purpose and thought grows and ripens in thirty days as much as ordinary years bring it forward." Everyone knew the Radicals were prepared for renewed agitation. Whatever the merits of legal and political equality for blacks, a correspondent of moderate Ohio Sen. John Sherman noted, "if you reconstruct upon any principle short of this, you . . . cause a continuous political strife which will last until the thing is obtained."[20]

Origins of Civil Rights

From the day the Thirty-Ninth Congress assembled, it was clear the Republican majority harbored misgivings about what Johnson had accomplished. Clerk of the House Edward McPherson omitted the names of newly elected Southern Congressmen as he called the roll, and the two Houses proceeded to establish a Joint Committee on Reconstruction to investigate conditions in the Southern states and report whether any were entitled to representation. (They also denied Southern claimants living expenses, leaving them, one remarked, the alternatives "go home or starve.") To Johnson, these decisions in effect admitted that the South had actually left the Union, and many of his supporters spoke darkly of a Radical coup. Yet the Republican caucus had approved the Southerners' exclusion by a nearly unanimous vote, and the committee's membership was carefully balanced among the party's factions. Moderate Sen. William Pitt Fessenden of Maine occupied the chair, while Sumner, considered "too ultra," was left off entirely.[21]

Some of Johnson's supporters considered these steps a direct challenge to Presidential authority, but Johnson's annual message to Congress took a conciliatory approach. Describing secession as, from the beginning, "null and void," it acknowledged that state authority in the South had fallen into "abeyance," and reviewed the process by which new governments had been established and the Thirteenth Amendment ratified. For the freedmen, Johnson expressed paternal concern; while arguing that he had no right to impose black voting upon the South, he held out the prospect of future state action to expand the suffrage, and

20. New York Tribune, February 17, 1866; Wendell Phillips to Charles Sumner, March 24, 1866, Sumner Papers; Thomas Richmond to John Sherman, February 27, 1866, Sherman Papers.
21. McKitrick, Andrew Johnson, 258; N. G. Taylor to Thomas A. R. Nelson, December 27, 1865, Thomas A. R. Nelson Papers, McClung Collection, LML; James E. Sefton, Andrew Johnson and the Uses of Constitutional Power (Boston, 1980), 121; New York Times, December 3, 1865.

indirectly criticized Southern legislatures by insisting that blacks must be secure in "the right to claim the just reward of their labor." Essentially, however, "the work of restoration" was now complete—all that remained was for Congress to admit Southern members. Carefully crafted to appeal to the broadest spectrum of opinion, North and South, Johnson's message invited diverse, even contradictory interpretations. The use of the term "restoration" rather than "reconstruction" suggested that on the rights of the states, Johnson was likely to be unyielding. On the other hand, the President conceded that Congress had the right to determine the qualifications of its members, apparently offering it some role in judging Reconstruction's progress. Most Republicans appear to have agreed with the New York *Tribune:* "We doubt whether any former Message has, on the whole, contained so much that will be generally and justly approved, with so little that will or should provoke dissent."[22]

"Whatever one may think of the policies of Mr. Johnson. . . ." Georges Clemenceau reported at the end of December, "he seems to have won out . . . over the leaders of the extreme radicals." The response to the Radicals' initial efforts to shape Congressional policy reinforced this judgment. In the Senate, Ben Wade demanded that no Southern state denying blacks the right to vote be admitted to representation, Sumner introduced a series of resolutions calling for civil and political equality, and Timothy Howe of Wisconsin moved to establish provisional governments in the "districts" formerly occupied by the Confederacy. Stevens ridiculed the South's governments before the House—"aggregations of white-washed rebels," he called them—and demanded that Congress create territorial regimes as a prelude to a social transformation involving black suffrage and land distribution. But proposals to overturn the Johnson governments and commit Congress to black suffrage fell on deaf ears. "No party, however strong, could stand a year on this platform," one Republican newspaper commented. The only Radical proposal to meet with even partial success was a bill to enfranchise blacks in the District of Columbia, which passed the House in January only to die in a Senate committee. (Hoping to forestall Congressional action, the District in December 1865 held a referendum among white voters. The result: 35 in favor of black suffrage, 6,951 against.)[23]

22. James D. Richardson, ed., *A Compilation of the Messages and Papers of the Presidents, 1789–1897* (Washington, D.C., 1896–99), 6:355–59; LaWanda Cox and John H. Cox, *Politics, Principle, and Prejudice 1865–1866* (New York, 1963), 129–39; New York *Tribune,* December 6, 1865.

23. Clemenceau, *American Reconstruction,* 63; Hans L. Trefousse, *Benjamin Franklin Wade: Radical Republican from Ohio* (New York, 1963), 263; Donald, *Sumner,* 239–41; *CG* 39th Congress, 1st Session, 74, Appendix, 217; Benedict, *Compromise of Principle,* 141–46; Constance M. Green, *The Secret City: A History of Race Relations in the Nation's Capital* (Princeton, 1967), 75–77.

With the failure of the Radical initiative, political leadership in Congress passed to the moderates. Nearly all Republicans shared a commitment to "loyal" government and a free labor economy in the South, protection of the freedmen's basic rights, and the integrity and ascendancy of the party that had saved the Union. But politically, ideologically, and temperamentally, moderate leaders like James G. Blaine and John A. Bingham in the House, and Lyman Trumbull, John Sherman, and William Pitt Fessenden in the Senate, differed markedly from their Radical colleagues. Moderates generally represented districts outside the belt of New England migration where Radicalism flourished, or states of the Lower North internally divided (like the nation itself) into northern and southern sections, each with distinct political traditions. Occupying the middle ground in politics was the key to victory in states like Illinois, Indiana, and Ohio, and well before 1865, moderates had sought to limit Radicalism's influence there. Men like Sherman believed the country had "suffered too much from ultraists and wranglers," and hoped Reconstruction could be settled quickly so the nation could turn its attention to pressing economic concerns. While fully embracing the changes brought about by the Civil War, moderate Republicans viewed Reconstruction as a practical problem, not an opportunity to impose an open-ended social revolution upon the South. Nor did they believe a break with Johnson inevitable or desirable. If "Sumner and Stevens, and a few other such men do not embroil us with the President," Fessenden insisted, "matters can be satisfactorily arranged—satisfactorily, I mean, to the great bulk of Union men throughout the States."[24]

Nor were moderates enthusiastic about the prospect of black suffrage, either in the North, where it represented a political liability, or the South, where it seemed less likely to provide a stable basis for a new Republican party than a political alliance with forward-looking white Southerners. Many agreed with John Andrew, whose January 1866 address, on leaving the Massachusetts governorship, argued that the white South must be judged according to "present loyal purpose" rather than "past disloyalty." His experience with the American Land Company and Agency, a venture designed to buy Southern land for resale to Northern settlers and former slaves while turning a profit for Northern investors, had convinced Andrew that many planters honestly accepted the defeat of secession and the principles of free labor and could be trusted to treat the freedmen fairly. "They see a new order of society, to which they must conform." At any rate, such men, "who know how to *lead*," could not be

24. John Sherman, *Recollections of Forty Years in the House, Senate and Cabinet* (Chicago, 1895), 1:366–67; William Sheffield to J. C. Hall, January 13, 1866, C. J. Albright to John Sherman, February 27, 1866, Sherman Papers; McKitrick, *Andrew Johnson*, 76–78; Charles A. Jellison, *Fessenden of Maine* (Syracuse, 1962), 198–201.

permanently excluded from power in the South. Black suffrage, Andrew believed, would be more secure voluntarily granted by the South's "natural leaders" than simply imposed by the North.[25]

Nonetheless, moderate Republicans believed Johnson's Reconstruction policies required modification. Alarmed by the large number of "rebels" holding office in the South, they insisted upon further guarantees of "loyalty" and hoped Johnson would repudiate talk of party realignment and stop meeting so openly with "obnoxious Democrats." Equally important, while rejecting black suffrage, mainstream Republicans had by now embraced civil equality for blacks, a commitment strengthened by continuing reports of violations of free labor precepts in the South. The North, Illinois Republican leader Jesse Fell insisted, had an obligation "to adopt measures for the safety and elevation of the African race. Their present nominal freedom is nothing but a *mockery.*" Like Lincoln, moderates were practical men of affairs, to whom debates over the South's precise legal status appeared pointless. "Whether by the acts of secession the rebel states are technically out of the Union or not, I do not, for one, regard as of much practical consequence," a constituent wrote Trumbull. "Congress has power enough to prescribe the terms and conditions upon which they shall be restored to their full power."[26]

This is not to say that moderate Republicans shared Radicalism's expansive view of federal authority. They accepted the enhancement of national power resulting from the Civil War, but did not believe the legitimate rights of the states had been destroyed, or the traditional principles of federalism eradicated. Nor did they agree that the Constitution's guarantee clause authorized endless national intervention in state affairs. These differences affected issues having nothing to do with Reconstruction. In a May 1866 debate over protecting Americans against cholera introduced from abroad, Sumner called "the national power" best equipped to handle the problem, while moderate Senators responded that public health matters fell "within the police power of the States exclusively." The moderates' dilemma was that most of the rights they sought to guarantee for blacks had always been state concerns. Federal action to secure these rights raised the specter of an undue

25. Jacob Cox to James A. Garfield, January 1, 1866, James A. Garfield Papers, LC; Henry G. Pearson, *The Life of John Andrew* (Boston, 1904), 2:273–83; John Andrew to Herman Bokum, November 30, 1865, Letterbook, Andrew Papers; Lawrence N. Powell, "The American Land Company and Agency: John A. Andrew and the Northernization of the South," *CWH*, 21 (December 1975), 294–98, 302–303.

26. M. Stone to John Sherman, December 27, 1865, Sherman Papers; Ransom Balcom to William H. Seward, December 27, 1865, William H. Seward Papers, University of Rochester; Jellison, *Fessenden*, 203; Cox and Cox, *Politics, Principle, and Prejudice*, 207; Jesse W. Fell to Lyman Trumbull, December 26, 1865, J. G. Wilson to Trumbull, January 21, 1866, Lyman Trumbull Papers, LC.

"centralization" of power. Rejecting talk of "conquered provinces" and states reverting to territories, moderates adopted a constitutional position not unlike the President's. The states remained indestructible, but they had forfeited some of their rights by attempting secession; they could be held, temporarily, in the "grasp of war" while the war's outcome was secured. Johnson had used essentially this reasoning to appoint provisional governors and require states to ratify the Thirteenth Amendment. Moderates believed the same logic empowered Congress to withhold representation from the South until the essential rights of the freedmen had been guaranteed.[27]

Two bills reported to the Senate soon after the New Year by Lyman Trumbull, chairman of the Judiciary Committee, embodied the moderates' policy. One of the most influential men in Congress, Trumbull believed the Radicals far too hasty in condemning the President. He met with Johnson just before Christmas and emerged convinced that "the President wishes no issue with Congress and if our friends would be reasonable we would all get along harmoniously." He also spent time studying Freedmen's Bureau operations and reports from the South. His conversations with Commissioner Howard strongly influenced the first of Trumbull's proposals, a measure extending the Bureau's life, providing the first direct funding for its activities, and authorizing Bureau agents to take jurisdiction of cases involving blacks and punish state officials denying blacks the "civil rights belonging to white persons."[28]

In normal times, the bill would have represented a radical departure in federal policy, but as Trumbull assured the Senate, the Bureau was "not intended as a permanent institution." More far-reaching was his second measure, the Civil Rights bill, which Henry J. Raymond, editor of the *New York Times* and a Congressman from New York, called "one of the most important bills ever presented to this House for its action." This defined all persons born in the United States (except Indians) as national citizens and spelled out rights they were to enjoy equally without regard to race—making contracts, bringing lawsuits, and enjoying "full and equal benefit of all laws and proceedings for the security of person and property." No state law or custom could deprive any citizen of what Trumbull called these "fundamental rights belonging to every man as a free man." The bill went on to authorize federal district attorneys, marshals, and Bureau officials to bring suit against violations; allowed these

27. Herman Belz, *Emancipation and Equal Rights* (New York, 1978), 77–78; *Works of Sumner*, 10:435–49; Richard Henry Dana, Jr., *Speeches in Stirring Times*, edited by Richard Henry Dana III (Boston, 1910), 243–48; Benedict, *Compromise of Principle*, 124–27.

28. Lyman Trumbull to Dr. William Jayne, December 24, 1865, Dr. William Jayne Papers, Illinois State Historical Library; [Oliver O. Howard] *Autobiography of Oliver Otis Howard* (New York, 1907), 2:280; Donald G. Nieman, *To Set the Law in Motion: The Freedmen's Bureau and the Legal Rights of Blacks, 1865–1868* (Millwood, N. Y., 1979), 106–109.

cases to be heard in federal courts; and made all persons, including local officials, who deprived a citizen of civil rights under color of state law liable to fine or imprisonment.[29]

In constitutional terms, the Civil Rights bill represented the first attempt to give meaning to the Thirteenth Amendment, to define in legislative terms the essence of freedom. Again and again during the debate on Trumbull's bills, Congressmen spoke of the national government's responsibility to protect the "fundamental rights" of American citizens. But as to the precise content of these rights, uncertainty prevailed. To Radicals, equality before the law was an expansive doctrine embracing nearly every phase of public life. Moderates had in mind a narrower definition, focusing on those rights essential for blacks to enter the world of contract, to compete on equal terms as free laborers. The bill proposed, one Congressman declared, "to secure to a poor, weak class of laborers the right to make contracts for their labor, the power to enforce the payment of their wages, and the means of holding and enjoying the proceeds of their toil." If states could deny blacks these rights, another Republican remarked, "then I demand to know, of what practical value is the amendment abolishing slavery?" But, beyond these specific rights, moderates, like the Radicals, rejected the entire idea of laws differentiating between black and white in access to the courts and penalties for crimes. The shadow of the Black Codes hung over these debates, and Trumbull began his discussion of the Civil Rights Bill with a reference to recent laws of Mississippi and South Carolina, declaring his intention "to destroy all these discriminations."[30]

As the first statutory definition of the rights of American citizenship, the Civil Rights Bill embodied a profound change in federal-state relations and reflected how ideas once considered Radical had been adopted by the party's mainstream. Before the Civil War, James G. Blaine later remarked, only "the wildest fancy of a distempered brain" could envision an Act of Congress conferring upon blacks "all the civil rights pertaining to a white man." Moreover, the bill was not limited to the South or to blacks. Although primarily intended to benefit the freedman, it invalidated many discriminatory laws in the North. And, Republicans hoped, the ability to remove cases from state to federal courts (where judges and jurors were required to take the Ironclad Oath of past loyalty) would protect Southern loyalists and federal officials in the South from

29. CG, 39th Congress, 1st Session, 319, 474, 1266; Harold M. Hyman, A More Perfect Union: The Impact of the Civil War and Reconstruction on the Constitution (New York, 1973), 461–62. Bingham believed the Civil Rights Bill unconstitutional, but every other influential Republican considered it warranted by the second clause of the Thirteenth Amendment.

30. Harold M. Hyman and William M. Wiecek, Equal Justice Under Law: Constitutional Development 1835–1875 (New York, 1982), 406–407; CG, 39th Congress, 1st Session, 42–43, 474, 1151–55, 1159.

damage suits and other forms of legal harassment. Reflecting the conviction, born of the Civil War, that the federal government possessed the authority to define and protect citizens' rights, the bill represented a striking departure in American jurisprudence. "I admit," said Maine Sen. Lot M. Morrill, "that this species of legislation is absolutely revolutionary. But are we not in the midst of a revolution?"[31]

In fact, however, the bill combined elements of continuity and change, reflecting the state of Republican opinion in early 1866. Instead of envisioning continuous federal intervention in local affairs, it honored the traditional presumption that the primary responsibility for law enforcement lay with the states, while creating a latent federal presence, to be triggered by discriminatory state laws. "When I reflect how easy it is for the states to avoid the operations of this bill . . ." declared Nevada Sen. William Stewart, "I think that it is robbed of its coercive features." Nor did Congress create a national police force or permanent military presence to protect the rights of citizens. Instead, it placed an unprecedented—and unrealistic—burden of enforcement on the federal courts. Once states enacted color-blind laws, these courts, despite their expanded jurisdiction, would probably find it difficult to prove discrimination by local officials. And despite its intriguing reference to "customs" that deprived blacks of legal equality, the Civil Rights Bill was primarily directed against public, not private, acts of injustice. To Trumbull and other moderates, the greatest threat to blacks' rights lay in discriminatory state laws, a questionable assumption at a time when the freedmen faced rampant violence as well as unequal treatment by sheriffs, judges, and juries, often under color of laws that did not in fact mention race. And, as Trumbull insisted, the bill contained nothing "about the political rights of the Negro." It thus remained true to the tradition that voting was a privilege, not an essential attribute of citizenship. Civil rights, short of the suffrage, was "as far as the country will go at the present time."[32]

Nor were Republicans willing to move very far toward the goal of land for the freedmen, as became evident during House debates on the Freedmen's Bureau Bill. Trumbull, as we have seen, included a provision confirming for three years blacks' right to lands set aside for them by General Sherman. The bill also allotted 3 million acres of public land in the South for homesteading by blacks, and authorized the Bureau to purchase additional land for resale. When the bill came before the House early in February, Thaddeus Stevens moved to add "forfeited estates of

31. Robert J. Kaczorowski, *The Politics of Judicial Interpretation: The Federal Courts, Department of Justice and Civil Rights, 1866–1876* (New York, 1985), 1–4; James G. Blaine, *Twenty Years of Congress* (Norwich, Conn., 1884–86), 2:179; *CG*, 39th Congress, 1st Session, 570, 1291.
32. Hyman and Wiecek, *Equal Justice Under Law*, 416; *CG*, 39th Congress, 1st Session, 319, 599–600, 1785; Nieman, *To Set the Law in Motion*, 110–14; Lyman Trumbull to Dr. William Jayne, January 11, 1866, Jayne Papers.

the enemy" to the land available to blacks. The proposal was overwhelmingly defeated, with a large majority of Republicans, including such Radicals as Ashley and Kelley, voting in opposition. "Thus we see . . . the real strength of the Jacobins in the House," exulted the pro-Johnson New York *Herald*. (Only one group of former masters was compelled by Congress to provide their former slaves with land in these years—slaveholding Indians who had sided with the Confederacy.)[33]

As if to demonstrate the land question's complexities, only two days after this vote and with virtually unanimous Republican support, the House passed Julian's Southern Homestead Act, opening public land in the South to settlement and giving blacks and loyal whites preferential access until 1867. Republicans were quite willing to offer freedmen the same opportunity to acquire land as whites already enjoyed under the Homestead Act of 1862, but not to interfere with planters' property rights. Despite extravagant hopes that it would "break down land monopoly" in the South, Julian's bill proved a dismal failure. Plantations monopolized the best land in the South; public land—swampy, timbered, far from transportation—was markedly inferior. The freedmen, moreover, entirely lacked capital, and federal land offices were few and poorly managed. By 1869 only 4,000 black families had even attempted to take advantage of the act, three quarters of them in sparsely populated Florida, and many of these subsequently lost their land. By far the largest acreage claimed under the law went to whites, often acting as agents for lumber companies.[34]

Thus, by February 1866, Republicans had united upon Trumbull's Freedmen's Bureau and Civil Rights bills as necessary amendments to Presidential Reconstruction. Radicals viewed them as first steps toward more fundamental change, moderates as a prelude to readmitting the South to Congressional representation. Meanwhile, the persistent complaints of persecution forwarded to Washington by Southern blacks and white loyalists altered the mood in Congress by eroding the plausibility of Johnson's central assumption—that the Southern states could be trusted to manage their own affairs without federal oversight. Particularly alarming was the testimony being gathered by the Joint Committee on Reconstruction. Although witnesses differed on many points (former Confederate Vice President Alexander H. Stephens even reaffirmed the

33. William S. McFeely, *Yankee Stepfather: General O. O. Howard and the Freedmen* (New Haven, 1968), 211–36; *CG*, 39th Congress, 1st Session, 299, 655, 688; New York *Herald*, February 8, 1866; M. Thomas Bailey, *Reconstruction in Indian Territory* (Port Washington, N.Y., 1972), 60–72.

34. *CG*, 39th Congress, 1st Session, 748; Boston *Daily Advertiser*, June 23, 1866; Christie F. Pope, "Southern Homesteads for Negroes," *AgH*, 44 (April 1970), 201–12; Claude F. Oubre, *Forty Acres and a Mule: The Freedmen's Bureau and Black Landownership* (Baton Rouge, 1978), 90–92, 109–16, 137–42, 156, 188; Paul W. Gates, "Federal Land Policy in the South, 1866–1888," *JSH*, 6 (August 1940), 307–10.

right of secession), army officers, Bureau agents, and Southern Unionists repeated tales of injustice against blacks, loyal whites, and Northerners. Speaker after speaker criticized Johnson's amnesty policies for encouraging white intransigence. The few blacks called before the committee agreed. "If [Southern] representatives were received in Congress," one told the committee, "the condition of the freedmen would be very little better than that of the slaves." Early in February, North Carolina Senator-elect John Pool concluded that Southern members would not gain admission for some time, and that the South would have to "submit to conditions that would never have been thought of, if a more prudent and wise course had been adopted" by the Johnson governments.[35]

Virtually all Republicans assumed Johnson would sign the Freedmen's Bureau and Civil Rights bills. The first passed Congress in February with nearly unanimous Republican support, including the votes of such outspoken supporters of the President as Senators Doolittle and Dixon. After separate meetings with Johnson, influential moderate Senators Fessenden, Trumbull, and James W. Grimes of Iowa all emerged convinced of his approval. "A veto at that time," Illinois Congressman Shelby Cullom later recalled, "was almost unheard of." If Johnson signed these measures, a local Republican official wrote Trumbull from Illinois, "then no one will care except a few, how soon the Senators and members of the H. R. are admitted to seats," and "no one will ever hear of the old democratic party . . . again."[36]

To the utter surprise of Congress, the President vetoed the Freedmen's Bureau Bill. Moreover, rejecting a conciliatory draft written by Seward, which criticized the bill's specifics while acknowledging a federal responsibility for the freedmen, his message repudiated the Bureau entirely, deriding it as an "immense patronage" unwarranted by the Constitution and unaffordable given "the condition of our fiscal affairs." Congress, he pointed out, had never felt called upon to provide economic relief, establish schools, or purchase land for "our own people"; such aid, moreover, would injure the "character" and "prospects" of the freedmen by implying that they did not have to work for a living. These matters, Johnson went on, should not be decided while eleven states remained unrepresented, and at any rate the President—"chosen by the people of all the States"—had a broader view of the national interest than members of Congress, elected "from a single district."

35. John A. Bidwell to George A. Gillespie, January 29, 1866, George A. Gillespie Papers, HL; Benjamin B. Kendrick, *The Journal of the Joint Committee of Fifteen on Reconstruction* (New York, 1914), 264–67; 39th Congress, 1st Session, House Report 30, pt. 2:30–31, 55–56; John Pool to Thomas Settle, February 4, 1866, Thomas Settle Papers, UNC.

36. Edward McPherson, *The Political History of the United States of America During the Period of Reconstruction* (Washington, D.C., 1875), 74; John Niven, *Gideon Welles* (New York, 1973), 518–19; Shelby M. Cullom, *Fifty Years of Public Service* (Chicago, 1911), 150; D. L. Phillips to Lyman Trumbull, January 7, 1866, Trumbull Papers.

This was, to say the least, a remarkable document. In appealing to fiscal conservatism, raising the specter of an immense federal bureaucracy trampling upon citizen's rights, and insisting self-help, not dependence upon outside assistance, offered the surest road to economic advancement, Johnson voiced themes that to this day have sustained opposition to federal intervention on behalf of blacks. At the same time, he misrepresented the aims of Congress—calling the Bureau "a permanent branch of the public administration," which it was not—and avoided any expression of sympathy for the freedmen's plight. As for Johnson's exalting himself above Congress, this, one Republican remarked, "is modest for a man . . . made President by an assassin." The veto virtually ensured a bitter political struggle between Congress and the President, for, as Fessenden accurately predicted, according to its logic, "he will and must . . . veto every other bill we pass" concerning Reconstruction.[37]

Why did Johnson choose this path? First, he sincerely believed the Bureau unauthorized by the Constitution and feared it would encourage blacks to lead a "life of indolence." He had already removed assistant commissioners he considered too sympathetic to the freedmen, and throughout 1866 would seek to discredit the agency and undermine its legal authority. The idea of political realignment, moreover, still hung in the air. Johnson had been remarkably successful in retaining support among Northerners and Southerners, Republicans and Democrats, but the Freedmen's Bureau Bill forced him to begin choosing among his diverse allies. An obscure North Carolina legislator summarized the situation: "If the President vetoes . . . then the fuss commences between him and the radicals; if he signs . . . all will go on well with them but will raise a muss in the south." Johnson knew of the Bureau's unpopularity among Southern whites, and of Northern Democrats' clamor for its destruction. He seems to have interpreted moderate Republican efforts to avoid a split as evidence that they feared an open breach in the party. And he had become convinced that the Radicals were conspiring against him, perhaps even planning to remove him from office. Advice from Gideon Welles reinforced his unwillingness to compromise, for the doctrinaire Secretary of the Navy insisted that the President must "meet this question squarely, and have a square and probably a fierce fight with these men."[38]

Johnson, reported William H. Trescot, hoped to provoke the Radicals

37. Richardson, ed., *Messages and Papers*, 6:399–405; John H. Cox and LaWanda Cox, "Andrew Johnson and His Ghost Writers," *JAH*, 48 (December 1961), 460–79; McKitrick, *Andrew Johnson*, 284–297. In July a second bill extending the Bureau's life passed over the President's veto.

38. Richardson, ed., *Messages and Papers*, 6:425; George R. Bentley, *A History of the Freedmen's Bureau* (Philadelphia, 1955), 125–30; Otto H. Olsen and Ellen Z. McGrew, "Prelude to Reconstruction: The Correspondence of State Senator Leander Sams Gash, 1866–67, Part I," *NCHR*, 60 (January 1983), 58; Cox and Cox, *Politics, Principle, and Prejudice*, 178–80; Howard K. Beale, ed., *Diary of Gideon Welles* (New York, 1960), 2:409–34.

into opposing him, so that they could be isolated and destroyed, while the Republican mainstream would "form a new party with the President." Unfortunately for this strategy, Johnson's belief that concern for the freedmen's rights was confined to the Radicals caused him to misconstrue the lines of division within Republican ranks. The Senate vote on overriding his veto ought to have given him pause, for while the bill fell two votes short of the necessary two thirds, thirty of thirty-eight Republicans voted for repassage. Trescot now recognized that Republicans might well unite against the President, inaugurating "a fight this fall such as has never been seen." But Johnson could not believe the majority of Republicans would contest him on the issue of federal protection for the freedmen. The day after the Senate vote, the President continued his assault upon the Radicals. In an impromptu Washington's Birthday speech, he equated Stevens, Sumner, and Wendell Phillips with Confederate leaders, since all were "opposed to the fundamental principles of this Government." He even implied that they were plotting his assassination.[39]

The Washington's Birthday speech displayed Johnson at his worst—self-absorbed (in a speech one hour long he referred to himself over 200 times), intolerant of criticism, and out of touch with political reality. But, coming on the heels of the Freedmen's Bureau veto, it thrilled those who hoped to benefit from division in Republican ranks. Parts of the speech, a Connecticut Democratic leader admitted, were in "questionable" taste, but "he says just what you, I and others have been saying for the last quarter of a century." Johnson also won support from conservative Northern business interests. A mass meeting at New York's Cooper Institute to endorse the veto attracted some of the city's most prominent bankers and merchants, who disparaged the Bureau for interfering with the plantation discipline essential for a revival of cotton production. "There is great clamor to coerce black men to work," wrote one New Yorker of this gathering, "but I hear none to make those who have heretofore lived upon the black man's labor do likewise."[40]

Many Republicans considered the veto a declaration of war against the party and the freedmen. Johnson's name, wrote a resident of Detroit, would soon be "as infamous as that of John Tyler's or even Benedict Arnold's." Yet moderate party leaders warned against reading Johnson out of the party. Connecticut held its gubernatorial campaign during these weeks, and the party convention managed to side simultaneously with Congress and the President—"a very skillful piece of equestrianism, . . . where two horses are ridden at once," commented one official.

39. William H. Trescot to James L. Orr, February 28, 1866, South Carolina Governor's Papers, SCDA; McPherson, *Political History*, 59–61, 74.

40. McKitrick, *Andrew Johnson*, 292–93; Whitelaw Reid, *After the War: A Southern Tour* (Cincinnati, 1866), 574; C. M. Ingersoll to Thomas H. Seymour, undated [February 1866], Thomas H. Seymour Papers, Connecticut Historical Society; New York *Tribune*, February 23, 1866; Paul Babcock to Charles Sumner, February 24, 1866, Sumner Papers.

The same circus trick was performed by state Republican conventions in Indiana and California. Attention now turned to the Civil Rights Bill, which passed Congress with nearly unanimous Republican support. "We all feel, that the *most important interests are at stake, . . .*" a member of the Ohio Senate wrote Sherman. "If the President vetoes the Civil Rights bill, I believe we shall be obliged to draw our swords for a fight and throw away the scabbards." Republican opinion, Johnson's supporters within the party frankly informed him, insisted that the freedmen must have "the same rights of property and person" as whites. They urged him to sign the bill, even if, as Ohio Governor Cox wrote, it became necessary to "*strain a point* in order to meet the popular spirit." Every member of the Cabinet except Welles and Seward hoped Johnson would sign the bill, and the Secretary of State advised him to be as conciliatory as possible in his veto, and explicitly endorse the principle of black citizenship. But this principle Johnson was unwilling to accept.[41]

Like his rejection of the Freedmen's Bureau Bill, Johnson's veto message repudiated not merely the specific terms of the Civil Rights Bill, but the entire principle behind it. Federal protection of blacks' civil rights and the broad conception of national power that lay behind it, he insisted, violated "all our experience as a people" and constituted a "stride towards centralization, and the concentration of all legislative powers in the national Government." Yet what was most striking about the message was its blatant racism; what had been muted in the Freedmen's Bureau veto now became explicit. Somehow, the President had convinced himself that clothing blacks with the privileges of citizenship discriminated against whites—"the distinction of race and color is by the bill made to operate in favor of the colored and against the white race." He also presented the curious argument that immigrants from abroad were more deserving of citizenship than blacks, because they knew more about "the nature and character of our institutions." Johnson even invoked the specter of racial intermarriage as the logical consequence of Congressional policy.

For Republican moderates, the Civil Rights veto ended all hope of cooperation with the President. In a biting speech, Trumbull dissected Johnson's logic, especially the notion that guaranteeing blacks civil equality impaired the rights of whites. Underscoring the intensity of Republican feeling, the Senate expelled Democratic Sen. John P. Stockton shortly before the vote on repassage, on the questionable grounds that the New Jersey legislature in 1865 had illegally altered its rules in order to elect him. Early in April, for the first time in American history, Congress

41. Joseph Warren to William H. Seward, February 22, 1866, Seward Papers; *CG*, 39th Congress, 1st Session, Appendix, 128–32; William Faxon to Mark Howard, February 27, 1866, Howard Papers; Russel M. Seeds, *History of the Republican Party of Indiana* (Indianapolis, 1899), 35–36; Winfield J. Davis, *History of Political Conventions in California, 1849–1892* (Sacramento, 1893), 234; E. B. Sadler to John Sherman, March 25, 1866, Sherman Papers; Jacob D. Cox to Andrew Johnson, March 22, 1866, Johnson Papers; Niven, *Welles*, 523.

enacted a major piece of legislation over a President's veto. A headline in one Republican newspaper summed up the political situation: "The Separation Complete."[42]

Johnson's rejection of the Civil Rights Bill has always been viewed as a major blunder, the most disastrous miscalculation of his political career. If the President aimed to isolate the Radicals and build up a new political coalition around himself, he could not have failed more miserably. Moderates now concluded that Johnson's policies "would wreck the Republican party." They also believed the Civil Rights Bill, as Sherman put it, was "clearly right." Whatever their differences, virtually all Republicans by now endorsed the view expressed by the Springfield *Republican* after the veto: Protection of the freedmen's civil rights "follows from the suppression of the rebellion. . . . The party is nothing, if it does not do this—the nation is dishonored if it hesitates in this."[43]

Yet despite the veto's outcome, Johnson's course cannot be explained simply in terms of insensitivity to Northern public opinion. Not only was racism deeply embedded in Northern as well as Southern public life, but, as Frederick Douglass observed around this time, no "political idea" was "more deeply rooted in the minds of men of all sections of the country [than] the right of each State to control its own local affairs." Given the Civil Rights Act's astonishing expansion of federal authority and blacks' rights, it is not surprising that Johnson considered it a Radical measure and believed he could mobilize voters against it. When, during one April speech, Johnson asked rhetorically, "What does the veto mean?" a voice from the crowd shouted: "It is keeping the nigger down." Johnson chose the issue on which to fight—federal protection for blacks' civil rights— and it was an issue on which he did not expect to lose.[44]

The Fourteenth Amendment

As relations with the President moved toward a rupture, Republicans grappled with the task of fixing in the Constitution, beyond the reach of Presidential vetoes and shifting political majorities, their understanding of the fruits of the Civil War. At one point in January, no fewer than

---

42. Richardson, ed., *Messages and Papers*, 6:406–13; *CG*, 39th Congress, 1st Session, 1755–60; Kent A. Peterson, "New Jersey Politics and National Policy-Making, 1866–1868" (unpub. diss., Princeton University, 1970), 97–99, 117–18; McKitrick, *Andrew Johnson*, 323; George M. Blackburn, "Radical Republican Motivation: A Case Study," *JNH*, 54 (April 1969), 126.

43. Cox and Cox, *Politics, Principle, and Prejudice*, 203; Albert Castel, *The Presidency of Andrew Johnson* (Lawrence, Kans., 1979), 71–73; Cullom, *Fifty Years*, 146; Rachel S. Thorndike, ed., *The Sherman Letters* (New York, 1894), 270; Stephen J. Arcanti, "To Secure the Party: Henry L. Dawes and the Politics of Reconstruction," *Historical Journal of Western Massachusetts*, 5 (Spring 1977), 42.

44. Philip S. Foner, ed., *The Life and Writings of Frederick Douglass* (New York, 1950–55), 4:199; Brock, *An American Crisis*, 121n.

seventy constitutional amendments had been introduced. What James
G. Blaine called a "somewhat startling result" of emancipation was the
first constitutional issue to confront the Joint Committee on Reconstruction. Before the war, three fifths of the South's slaves had been
included in calculating Congressional representation. Now, as free persons, all would be counted, significantly enhancing Southern power in
the House of Representatives and the Electoral College. Left unchanged, predicted one Congressman, the prewar system would allow
"unrepentant . . . traitors," in alliance with Northern Democrats, to
gain control of Congress, compensate slaveowners for emancipation,
and elect Robert E. Lee President in 1868. Since the Radicals' solution—giving the vote to Southern blacks—did not command majority
support, the search began for alternatives. The simplest was to base
representation on the number of qualified voters, a plan which had the
advantage of leaving voting requirements to the states, while indirectly
promoting black suffrage. In a rare instance of agreement, Thaddeus
Stevens introduced such a measure, and Johnson at the end of January
indicated his approval.

Simple it may have been, but the plan quickly succumbed to political
and demographic realities. Such an amendment, Blaine observed, would
set off an "unseemly scramble" for voters. Because of the westward
migration of young males, New England contained a higher proportion
of women than the Western states; it would face a choice between a
decline in its political power and female suffrage. Large numbers of
unnaturalized foreigners also resided in the North; states like Rhode
Island would either have to forfeit representation based on these persons
or eliminate literacy and naturalization requirements for voting, in which
case, Blaine continued, the ballot "would be demoralized and disgraced."
Toward the end of January, the Joint Committee reported a new proposal—an amendment declaring that when a state denied any citizen the
right to vote because of race, all members of that race would be excluded
from enumeration. Here was a compromise ingeniously designed to
allow the North to continue to bar women, aliens, and the illiterate
without penalty, while preventing the South from benefiting politically
from its disenfranchised black population. A "half-way measure," it offered blacks little real protection. Southern whites, as a Virginia legislator
candidly told the Joint Committee, would adopt the "obvious policy" of
enacting nonracial literacy or property qualifications that would "give us
the benefit of the negro race in counting our population, and under which
white people would do all the voting." Or, they could accept a reduced
national role and continue to rule blacks as they pleased. "It is a compromise," declared one Senator, "that the whites may govern the blacks in
the . . . South and wring from them sweat and tears, provided the north-

ern and eastern states are permitted to control the national Government."[45]

Nonetheless, the amendment passed the House on January 31. But in the Senate it encountered the formidable opposition of Charles Sumner, who condemned it as a "compromise with wrong" for acknowledging that states were entitled to limit suffrage on racial grounds. On March 9, five Radicals including Sumner joined with Democrats and Johnson supporters to block the required two thirds majority. The amendment, said Stevens, had been "mortally wounded in the house of its friends . . . slaughtered by a puerile and pedantic criticism." A second proposal from the Joint Committee fared no better. This was an amendment drafted by Ohio Congressman John A. Bingham granting Congress the authority to secure the "privileges and immunities" and "equal protection of life, liberty, and property" of all citizens. Most Republicans believed Congress already possessed this power, as they demonstrated in passing the Civil Rights Bill. Early in March, the House tabled Bingham's proposal.[46]

"The President has gone over to the enemy," wrote one Congressman, "and our friends are all split up among themselves." Nevada Sen. William Stewart sought to break the impasse with a plan for "universal amnesty and universal suffrage," but neither part of this formula commanded majority support. Not until the end of April, after a series of further proposals, votes, and reconsiderations, did the committee report an amendment to Congress. This time it placed all the issues under consideration in a single proposition. Its first clause (which will be discussed in detail shortly) prohibited the states from abridging equality before the law. The second provided for a reduction in representation in proportion to the number of male citizens denied the suffrage. (This prevented the South from using literacy and property tests to bar blacks from voting while retaining its full representation; it also raised the prospect that some Northern states might face a reduction in political power.) The third clause excluded those who had voluntarily aided the Confederacy from voting in national elections until 1870. The amendment also prohibited payment of the Confederate debt, and empowered Congress to enforce its provisions through "appropriate" legislation. Wrote Senator Grimes: "It is not exactly what any of us wanted, but we were each compelled to surrender some of our individual preferences in order to secure anything."[47]

45. Blaine, *Twenty Years*, 2:189; *CG*, 39th Congress, 1st Session, 141, 342, 353, 407, 2540, Appendix, 115; Kendrick, *Journal of the Joint Committee*, 41; Charles R. Williams, ed., *Diary and Letters of Rutherford Birchard Hayes* (Columbus, Ohio, 1922–26), 3:16; 39th Congress, 1st Session, House Report 30, pt. 2:158.

46. *Works of Sumner*, 10:119–237; *CG*, 39th Congress, 1st Session, 1224–33, 1284–89, 2459; Joseph B. James, *The Framing of the Fourteenth Amendment* (Urbana, Ill., 1956), 81–86.

47. Benedict, *Compromise of Principle*, 160–61; McKitrick, *Andrew Johnson*, 341–42; Kendrick, *Journal of the Joint Committee*, 83–117; James, *Fourteenth Amendment*, 112–17.

Early in May, Stevens opened the House debate. The first section, he declared, established the principle that state laws "shall operate *equally* upon all." This was the meaning of the Civil Rights Act, and "I can hardly believe that any person can be found who will not admit that . . . [it] is just." The representation clause, he continued, would either compel the South to enfranchise blacks or "keep [it] forever in a hopeless minority in the national Government." Disenfranchisement of Confederates until 1870 was "the mildest of all punishments ever inflicted on traitors," and as for repudiating the Confederate debt, "none dare object . . . who is not himself a rebel." Of the amendment's five sections, most Republicans now considered all but the third unexceptionable. "Give us the third section or give us nothing," Stevens implored, but a majority of Republicans considered disenfranchisement vindictive, undemocratic, and likely to arouse opposition in the North. William H. Trescot shrewdly analyzed the situation: Moderate Republicans believed they could "go to the country safely" on the amendment once disenfranchisement had been deleted. House Democrats, "evidently fearing the truth of this opinion," and urged on by the President, joined with Stevens and his supporters to block its removal. The amendment passed the House with the third section intact, but Senate Republicans substituted a provision that made no reference to voting, but barred from national and state office men who had taken an oath of allegiance to the Constitution and then aided the Confederacy. Two thirds of Congress could remove this disability. Even the pro-Johnson *New York Times* voiced approval: "The exclusion from office of men who added perjury to treason is certainly not severe."[48]

Finally, on June 13, the House accepted the Senate's changes and, without a Democrat in favor or Republican opposed, gave the amendment final approval. Stevens' speech just before passage was an eloquent statement of his political creed:

> In my youth, in my manhood, in my old age, I had fondly dreamed that when any fortunate chance should have broken up for awhile the foundation of our institutions, and released us from obligations the most tyrannical that ever man imposed in the name of freedom, that the intelligent, pure and just men of this Republic . . . would have so remodeled all our institutions as to have freed them from every vestige of human oppression, of inequality of rights, of the recognized degradation of the poor, and the superior caste of the rich. . . . This bright dream has vanished 'like the baseless fabric of a dream.' I find that we shall be obliged to be content with patching up the worst portions of the ancient edifice, and leaving it, in many of its parts, to be swept through by . . . the storms of despotism.
>
> Do you inquire why, holding these views and possessing some will of my

48. *CG*, 39th Congress, 1st Session, 2459, 2544, 2869; William H. Trescot to James L. Orr, May 12, 1866, South Carolina Governor's Papers; *New York Times*, May 31, 1866.

own, I accept so imperfect a proposition? I answer, because I live among men and not among angels.[49]

Because it implicitly acknowledged the right of states to limit voting because of race, Wendell Phillips denounced the amendment as a "fatal and total surrender." He was not the only reformer to condemn it. Susan B. Anthony, Elizabeth Cady Stanton, and other leaders of the women's suffrage movement also felt betrayed, because the second clause for the first time introduced the word "male" into the Constitution. Alone among suffrage restrictions, those founded on sex would not reduce a state's representation.

Ideologically and politically, nineteenth-century feminism had been tied to abolition; most advocates of women's suffrage were dedicated foes of slavery, and many male abolitionists had long been active in the movement for women's rights. During the war, the women's movement had put aside the suffrage issue to join in the crusade for Union and emancipation. Yet at the same time, it shared in the flowering of egalitarian and nationalist ideology promoted by the conflict. "The contest with the South which destroyed slavery," commented diarist Sidney George Fisher, "has caused an immense increase to the popular passion for liberty and equality," a passion women shared. Feminist leaders now turned Radical ideology back upon Congress. If "special claims for special classes" were illegitimate and unrepublican, how could the denial of women's rights be justified? Should not sex, like race, be deemed an unacceptable basis for legal distinctions among citizens? Confronted by the argument that Reconstruction was "the negro's hour," they defined it, instead, as the hour for change, an opportunity that must be seized or another generation might pass "ere the constitutional door will again be opened."[50]

In politics, it was indeed the "Negro's hour." A Civil War had not been fought over the status of women, nor had thirty years of prior agitation awakened public consciousness on the issue. Yet the dispute over the Fourteenth Amendment marked a turning point in nineteenth-century reform. Leaving feminist leaders with a deep sense of betrayal, it convinced them, as Stanton put it, that woman "must not put her trust in man" in seeking her own rights. Women's leaders now embarked on a course that severed their historic alliance with abolitionism and created an independent feminist movement, seeking a new constituency outside

49. CG, 39th Congress, 1st Session, 3148.
50. Wendell Phillips to Thaddeus Stevens, April 30, 1866, Stevens Papers; Ellen C. DuBois, *Feminism and Suffrage: The Emergence of an Independent Women's Movement in America, 1848–1869* (Ithaca, N.Y., 1978), 36–37, 50–51, 60–75; Nicholas B. Wainwright, ed., *A Philadelphia Perspective: The Diary of Sidney George Fisher Covering the Years 1834–1871* (Philadelphia, 1967), 523; Ida H. Harper, *The Life and Work of Susan B. Anthony* (Indianapolis, 1898), 1:260; Elizabeth Cady Stanton, *Eighty Years and More: Reminiscences 1815–1897* (New York, 1898), 256.

the reform milieu. At the same time, the breach allowed previously unarticulated social and racial prejudices to surface among middle-class feminist leaders. Even as support for black (male) suffrage solidified among her former Radical allies, Stanton spoke not only of extending the franchise to women, but of limiting it on the basis of "intelligence and education." The black woman, she contended, would be better off as "the slave of an educated white man, than of a degraded, ignorant black one." For their part, abolitionists and Radicals, who had always prided themselves on a willingness to stake out a principled position and brook no compromise, now heeded the siren song of political expediency, illustrating how closely the war had tied reformers to the Republican party and the national state. Sumner's broad egalitarianism, Fessenden remarked, logically applied to women as well as men, "but I noticed that the honorable Senator dodged that part of the proposition very carefully." Sumner could only explain that women's suffrage, while not germane at present, was "the great question of the future."[51]

The Fourteenth Amendment, one Republican newspaper observed, repudiated the two axioms on which the Radicals "started to make their fight last December: dead States and equal suffrage."[52] Yet it clothed with constitutional authority the principle Radicals had fought a lonely battle to vindicate: equality before the law, overseen by the national government. For its heart was the first section, which, in its final form, declared all persons born or naturalized in the United States both national and state citizens and prohibited the states from abridging their "privileges and immunities," depriving any person of life, liberty, or property without "due process of law," or denying them "equal protection of the laws." For more than a century, politicians, judges, lawyers, and scholars have debated the meaning of this elusive language. The problem of establishing the Amendment's "original intent" is complicated by the fact that its final wording resulted from a series of extremely narrow votes in the Joint Committee and subsequent alteration on the floor of Congress. To complicate matters further, the broadest statements of the Amendment's purposes originated with opponents, who, in an attempt to discredit the measure, claimed it would destroy the powers of the states, establish equality "in every respect" between black and white, and empower Congress to legislate on any local matter it chose—an interpretation Republicans vehemently denied.

Whether the courts *should* be bound by the "original intent" of a constitutional amendment is a political, not historical question. But the

---

51. Ellen C. DuBois, ed., *Elizabeth Cady Stanton, Susan B. Anthony: Correspondence, Writings, Speeches* (New York, 1981), 90–92; Theodore Stanton and Harriot Stanton Blatch, eds., *Elizabeth Cady Stanton* (New York, 1922), 2:108–10; *CG*, 39th Congress, 2d Session, 40, 55–62, 84; *Works of Sumner*, 10:238.

52. James, *Fourteenth Amendment*, 145.

aims of the Fourteenth Amendment can only be understood within the political and ideological context of 1866: the break with the President, the need to find a measure upon which all Republicans could unite, and the growing consensus within the party around the need for strong federal action to protect the freedmen's rights, short of the suffrage. Despite the many drafts, changes, and deletions, the Amendment's central principle remained constant: a national guarantee of equality before the law. Some critics, indeed, played on racial fears to charge that no state could henceforth regulate the rights not merely of blacks, but of Indians, Chinese, even Gypsies. ("I have lived in the United States now for many a year," declared one Republican Senator, "and really I have heard more about Gypsies within the past two or three months than I have heard before in my life.") But the principle of equal civil rights was now so widely accepted in Republican circles, and had already been so fully discussed, that compared with the now-forgotten disqualification and representation clauses, the first section inspired relatively little discussion. It was "so just," a moderate Congressman declared, "that no member of this House can seriously object to it."[53]

Republicans did not deny one Democrat's description of the Amendment as "open to ambiguity and . . . conflicting constructions." The debate abounded in generalities such as "the fundamental rights of citizens," and Republicans rejected calls to define these with precision. Unlike the Civil Rights Act, which listed numerous rights a state could not abridge, the Amendment used only the broadest language. Clearly, Republicans proposed to abrogate the Black Codes and eliminate any doubts as to the constitutionality of the Civil Rights Act. Yet to reduce their aims to this is to misconstrue the difference between a statute and a constitutional amendment. Some amendments, dealing with narrow, immediate concerns, can be thought of as statutes writ large; altering one aspect of national life, they leave the larger structure intact. Others are broad statements of principle, giving constitutional form to the resolution of national crises, and permanently altering American nationality. The Fourteenth Amendment was a measure of this kind. In language that transcended race and region, it challenged legal discrimination through-

---

53. Robert J. Kaczorowski, "Searching for the Intent of the Framers of the Fourteenth Amendment," *Connecticut Law Review*, 5 (Winter 1972–73), 368–98; *CG*, 39th Congress, 1st Session, 2510, 2530. As Kaczorowski points out, too many attempts by legal scholars to ascertain the "original intent" of the Fourteenth Amendment rely on a handful of selected quotations from Congressional debates rather than the full historical context of Republican ideology and its evolution in the Civil War era. They often rest, in addition, on a now outdated interpretation of Reconstruction, which saw the Congressional program as lacking broad support among the Northern electorate and reflecting little genuine concern for the rights of blacks. Charles Fairman, "Does the Fourteenth Amendment Incorporate the Bill of Rights?: The Original Understanding," *Stanford Law Review*, 2 (1949), 5–139, and Raoul Berger, *Government By Judiciary: The Transformation of the Fourteenth Amendment* (Cambridge, Mass., 1977), are influential examples of such accounts.

out the nation and changed and broadened the meaning of freedom for all Americans.

On the precise definition of equality before the law, Republicans differed among themselves. Even moderates, however, understood Reconstruction as a dynamic process, in which phrases like "privileges and immunities" were subject to changing interpretation. They preferred to allow both Congress and the federal courts maximum flexibility in implementing the Amendment's provisions and combating the multitude of injustices that confronted blacks in many parts of the South. The final version, it should be noted, was far stronger than Bingham's earlier proposal directly granting national lawmakers the power to enforce civil rights, for this would become a dead letter if Democrats regained control of the House or Senate. Now, discriminatory state laws could be overturned by the federal courts regardless of which party dominated Congress. (Indeed, as in the Civil Rights Act, Congress placed great reliance on an activist federal judiciary for civil rights enforcement—a mechanism that appeared preferable to maintaining indefinitely a standing army in the South, or establishing a permanent national bureaucracy empowered to oversee Reconstruction.)[54]

In establishing the primacy of a national citizenship whose common rights the states could not abridge, Republicans carried forward the state-building process born of the Civil War. "It is a singular fact," Wendell Phillips declared as Congress deliberated, "that, unlike all other nations, this nation has yet a question as to what makes or constitutes a citizen." The Amendment remedied this situation; as a result, observed John A. Bingham, "the powers of the States have been limited and the powers of Congress extended." The states, declared Michigan Sen. Jacob Howard, who guided the Amendment to passage in the Senate, could no longer infringe upon the liberties the Bill of Rights had secured against federal violation; henceforth, they must respect "the personal rights guaranteed and secured by the first eight Amendments." Bingham said much the same thing in the House. Some portions of the Bill of Rights were of little moment in 1866 (no one was threatening to quarter soldiers in a home without consent of the owner). But it is abundantly clear that Republicans wished to give constitutional sanction to states' obligation to respect such key provisions as freedom of speech, the right to bear arms, trial by impartial jury, and protection against cruel and unusual punishment and unreasonable search and seizure. The Freedmen's Bureau had already taken steps to protect these rights, and the Amendment was deemed necessary, in

54. CG, 39th Congress, 1st Session, 2467, 2537, 2765, 2890; Daniel A. Farber and John E. Muench, "The Ideological Origins of the Fourteenth Amendment," *Constitutional Commentary*, 1 (Summer 1984), 269; Jacobus TenBroek, *Equal Under Law* (New York, 1965), 207.

part, precisely because every one of them was being systematically violated in the South in 1866.[55]

Yet if a degree of federal intervention in state affairs scarcely conceivable before 1860 now became possible, few Republicans wished to break completely with the principles of federalism. Only if state governments failed to protect citizens' rights would federal action become necessary. "It takes from no State," declared Bingham, "any right, . . . but it imposes a limitation upon the States to correct their abuses of power."[56] Most Republicans assumed the states would retain the largest authority over local affairs—in this, they followed the path marked out by the Civil Rights Act. But there was an important difference. The Fourteenth Amendment can only be understood as a whole, for while respecting federalism, it intervened directly in Southern politics, seeking to conjure into being a new political leadership that would respect the principle of equality before the law.

Some Republicans still believed the prospect of reduced representation would lead the South to embrace black suffrage. Without it, one newspaper calculated, the region would sacrifice one third of its House membership. Yet in 1866, control of local affairs concerned Southern political leaders far more than the size of their Congressional delegation. A. O. P. Nicholson, a prominent Tennessee jurist and politician, declared himself "perfectly willing" to waive national representation entirely for ten years, "if the North would agree to let the Southern states try the experiment of self-government." Republicans, however, did not intend to leave "the basis of political society in the southern states untouched." Upper South loyalists had already grappled with the dilemma of promoting political change while avoiding black suffrage, and the original third section, temporarily disenfranchising all Confederates, extended to the entire South a policy already adopted in Tennessee, Missouri, and other states. Border Republican Congressmen, who had displayed little interest in measures aiding blacks, hailed this as "the only salvation" for Union men. Yet the final version of the Amendment, barring from office Confederates who before the war had taken an oath of allegiance (required of officials ranging from President down to postmaster), although seemingly more lenient, in some ways had broader implications. The original provision had applied only to national elections, leaving the structure of state politics intact, while now, although not depriving "rebels" of the vote, the Amendment made virtually the entire political leadership of the South ineligible for office. Congressional Republicans were deeply disturbed by

---

55. *National Anti-Slavery Standard,* June 3, 1865; *CG,* 42d Congress, 1st Session, Appendix, 83; 39th Congress, 1st Session, 1033, 1088, 2765; Michael K. Curtis, "The Fourteenth Amendment and the Bill of Rights," *Connecticut Law Review,* 14 (Winter 1982), 237–306.
56. James, *Fourteenth Amendment,* 160–61.

the drumbeat of complaints from blacks and Unionists that unrepentant Confederates controlled Johnson's governments. Without saddling the party with the burden of black suffrage, Section 3 aimed to promote a sweeping transformation of Southern public life.[57]

Last but not least, the Fourteenth Amendment was framed with the elections of 1866 very much in mind. Republicans, declared Speaker of the House Colfax, must not "take on any loads of popular prejudice that can be avoided." The Amendment supplied Republicans with a platform for the fall campaign, while leaving to the future the issue of black suffrage. And the Report of the Joint Committee, released in June, provided an official explanation and defense of the Amendment upon which, as *The Nation* declared, "a great party must take its stand." Carefully side-stepping differences between Radicals and moderates (it called the South's precise constitutional status a "profitless abstraction"), the Report insisted the freedmen could not be abandoned "as free men and citizens," and defended the representation clause as a compromise that left voting requirements to the states yet would lead "at no distant day" to black suffrage. Having elevated "rebels" to power and abused the freedmen and loyal whites, it insisted, the Southern states were not ready for readmission.[58]

Some moderates still entertained hope that Johnson would endorse the Fourteenth Amendment, or at least encourage the South to accept it. But, having held aloof from Congressional deliberations, the President quickly repudiated the entire process. Instead, he moved to convene a national gathering of his supporters, to launch the fall campaign. Drafted by Senator Doolittle, a call for a National Union Convention was released to the press on June 25. Here, at last, was the long-anticipated effort at political realignment. By now, however, Johnson had few supporters in Republican ranks, a fact embarrassingly clear in the list of men who signed the call—a handful of Senators and a few Cabinet members, none of whom commanded real political influence. (Three members of the Cabinet resigned rather than endorse the movement, while Stanton, who also opposed it, remained in office.) It was not surprising that in Radical areas like upstate New York, "not a single Union man" could be found who endorsed the President's policy. But even a conservative like William M. Evarts—the kind of man indispensable for the success of a new party— held aloof, concluding its only result would be to strengthen the Democracy. One sign of the times was that the New York *Herald*, with the largest

57. New York *Commercial Journal* in Raleigh *Weekly North Carolina Standard*, June 27, 1866; A. O. P. Nicholson to Marius C. Church, June 7, 1866, Seward Papers; *CG*, 39th Congress, 1st Session, 2504–5, 2532; Philip J. Avillo, Jr., "Ballots for the Faithful: The Oath and the Emergence of Slave State Republican Congressmen, 1861–1867," *CWH*, 22 (June 1976), 172.

58. O. J. Hollister, *Life of Schuyler Colfax* (New York, 1887), 285; *Nation*, June 12, 1866; 39th Congress, 1st Session, House Report 30, x–xviii.

circulation of any newspaper in the country, abandoned the President. Its editor James Gordon Bennett, according to one *Herald* writer, "goes with . . . the party [he] believes to be the strongest." Could Johnson win a contest centered on the Fourteenth Amendment and civil rights for blacks? The President believed so, but Trescot, surveying the political scene, concluded differently: "From all that I can learn I think not."[59]

When Congress adjourned in July, two divisive questions had been left unresolved. One was what precisely the Southern states had to do to achieve readmission. Tennessee quickly ratified the Fourteenth Amendment and regained its right to representation, but without an explicit Congressional acknowledgment that this established a binding precedent. Congress, pleaded Chicago *Tribune* editor Joseph Medill, could not dissolve without presenting to the country a final plan of Reconstruction and "offering some terms of admission to the South." But that is exactly what it did. And, for the moment, the vexed question of black voting rights had been laid aside. Henry M. Turner, the black minister and political organizer who had been sent to Washington to lobby for black rights by Georgia's statewide black convention, reported home: "Several Congressmen tell me, 'the negro must vote,' but the issue must be avoided now so as 'to keep up a two thirds power in Congress'." Viewing the Amendment as a temporary resting place, not a final settlement, the Radicals believed their star would soon be in the ascendant. "If I was ever Conservative, I am Radical now," wrote Congressman John A. Bidwell of moderate California, and in Kansas, party leaders decided to "run in the radical groove." Even conservative Republican Sen. John B. Henderson of Missouri believed black suffrage was destined to come: "It will not be five years from today before this body will vote for it. You cannot get along without it."[60]

The Campaign of 1866

On May 1, 1866, two horse-drawn hacks, one driven by a white man, the other by a black, collided on the streets of Memphis. When police arrested the black driver, a group of recently discharged black veterans intervened, and a white crowd began to gather. From this incident fol-

59. Lyman Trumbull to Julia Trumbull, May 15, 1866, Trumbull Family Papers, Illinois State Historical Library; James L. Sellers, "James R. Doolittle," *WMH*, 18 (September 1934), 30; Niven, *Welles*, 530; Ransom Balsom to William H. Seward, July 13, 1866, Seward Papers; William M. Evarts to John A. Dix, July 25, 1866, John A. Dix Papers, Columbia University; W. B. Phillips to Andrew Johnson, May 20, 1866, Johnson Papers; William H. Trescot to James L. Orr, May 12, 1866, South Carolina Governor's Papers.

60. McKitrick, *Andrew Johnson*, 333; Joseph Medill to Lyman Trumbull, July 17, 1866, Trumbull Papers; Henry M. Turner to John E. Bryant, April 13, 1866, John E. Bryant Papers, DU; John A. Bidwell to George A. Gillespie, July 31, 1866, Gillespie Papers; Mark A. Plummer, *Frontier Governor: Samuel J. Crawford of Kansas* (Lawrence, Kans., 1971), 58–59; *CG*, 39th Congress, 1st Session, 3033.

lowed three days of racial violence, with white mobs, composed in large part of the mostly Irish policemen and firemen, assaulting blacks on the streets and invading South Memphis, an area that included a shantytown housing families of black soldiers stationed in nearby Fort Pickering. Before the rioting subsided, at least forty-eight persons (all but two of them black) lay dead, five black women had been raped, and hundreds of black dwellings, churches, and schools were pillaged or destroyed by fire.

One of the bloodiest outbreaks in this violent era, the Memphis riot had its roots in tensions that had gripped the city for over a year. The black population had more than quadrupled during the war, and signs of change abounded, from the black soldiers to Freedmen's Bureau hospitals and schools. It is difficult to say which proved more threatening to local whites—the large number of impoverished rural freedmen who thronged the streets in search of employment, or the considerable group that managed to achieve modest economic success (many of the black victims were robbed of cash, watches, tools, and furniture). Racial altercations were frequent, and the city's press constantly abused black residents. "Would to God they were back in Africa, or some other seaport town," declared the Memphis *Argus* two days before the riot began, "anywhere but here." For black and white alike, the riot taught many lessons, not least the impotence of federal authorities, for the Bureau proved unable to aid those who crowded its offices seeking protection, and the army commander refused to dispatch troops of either race to assist the families of black soldiers. The violence exposed divisions within the black community, for in its aftermath, a group of better-off free men of color urged the removal of rural freedmen from the city, to reduce racial tensions. And it underscored the limits of Tennessee Unionists' policy of proscription. The law disenfranchising former Confederates had, as intended, brought new men to power, for recently arrived Irish immigrants now dominated city government. But they proved no more sympathetic to blacks than the old elite. Most of all, the violence discredited Johnson's policies. As one newspaper declared, "if anything could reveal, in light as clear as day, the demoniac spirit of the southern whites toward the freedmen, . . . it is such an event as this."[61]

Even more damaging was the outbreak twelve weeks later in New Orleans. Once again, local whites were the aggressors and blacks the victims, but this time the violence arose directly from Reconstruction politics. The growing power of former Confederates under the adminis-

61. Bobby L. Lovett, "Memphis Riots: White Reaction to Blacks in Memphis, May 1865–July 1866," *THQ*, 38 (Spring 1979), 9–33; James G. Ryan, "The Memphis Riot of 1866: Terror in a Black Community During Reconstruction," *JNH*, 62 (July 1977), 243–57; Altina L. Waller, "Community, Class and Race in the Memphis Riot of 1866," *JSocH*, 18 (Winter 1984), 233–46; 39th Congress, 1st Session, House Report 101; Cleveland *Leader*, June 4, 1866.

tration of Governor Wells had long dismayed the city's Radicals, and eventually alarmed Wells himself. Over his veto, the state legislature early in 1866 mandated new municipal elections, which returned to power the city's Confederate mayor. Wells now endorsed a Radical plan to reconvene the constitutional convention of 1864 in order to enfranchise blacks, prohibit "rebels" from voting, and establish a new state government. For weeks, Wells's opponents agonized over the prospect that Louisiana would be "revolutionized," and it appears certain that some members of the city police, made up largely of Confederate veterans, conspired to disperse the gathering by force. On the appointed day, July 30, only twenty-five delegates in fact assembled, soon joined by a procession of some 200 black supporters, mostly former soldiers. Fighting broke out in the streets, police converged on the area, and the scene quickly degenerated into what Gen. Philip H. Sheridan later called "an absolute massacre," with blacks assaulted indiscriminately and the delegates and their supporters besieged in the convention hall and shot down when they fled, despite hoisting white flags of surrender. By the time federal troops arrived, thirty-four blacks and three white Radicals had been killed, and well over 100 persons injured. The son of former Vice President Hannibal Hamlin, a veteran of the Civil War, wrote that "the wholesale slaughter and the little regard paid to human life I witnessed here" surpassed anything he had seen on the battlefield.[62]

Even more than the Memphis riot, the events in New Orleans discredited Presidential Reconstruction, even though Lincoln, not Johnson had established the two state governments and blame for the New Orleans violence was widely shared. Reconvening the constitutional convention two years after it had adjourned certainly appeared irregular, the local army commander cabled Washington about the danger of violence but did nothing to prepare for it. Secretary of War Stanton failed to forward this telegram to the President, and Johnson informed Louisiana's lieutenant governor the convention could be dispersed. But the stark fact remained that nearly all the victims had been blacks and convention delegates, and that the police, far from preserving order, had joined in the assault. Many Northerners agreed with Gen. Joseph Holt that Johnson's leniency had unleashed "the barbarism of the rebellion in its renaissance."[63]

62. Gilles Vandal, *The New Orleans Riot of 1866: Anatomy of a Tragedy* (Lafayette, La., 1983); Walter McG. Lowrey, "The Political Career of James Madison Wells," *LaHQ*, 31 (October 1948), 1055; Duncan Kenner to James M. Wells, July 8, 1866, Sloo Collection, Louisiana State Museum; Ted Tunnell, *Crucible of Reconstruction: War, Radicalism, and Race in Louisiana 1862–1877* (Baton Rouge, 1984), 103–107; Cyrus Hamlin to Hannibal Hamlin, August 19, 1866, Hannibal Hamlin Papers, University of Maine, Orono.

63. Joseph G. Dawson III, *Army Generals and Reconstruction: Louisiana, 1862–1877* (Baton Rouge, 1982), 40–41; Benjamin P. Thomas and Harold M. Hyman, *Stanton: The Life and Times of Lincoln's Secretary of War* (New York, 1962), 496; Benedict, *Compromise of Principle*, 206; Joseph Holt to Henry C. Warmoth, August 1, 1866, Henry C. Warmoth Papers, UNC.

From Johnson's point of view, the New Orleans riot could not have occurred at a worse time—only two weeks before the National Union Convention assembled in Philadelphia. Already, however, the goal of a grand coalition of moderates and conservatives had foundered on internal dissension. Conservative Republicans like Secretary Seward and his New York allies, Thurlow Weed and Henry J. Raymond, envisioned a new alliance within the Republican party, controlled by themselves. Doolittle and Welles maintained that no realignment could take place so long as Seward had a prominent place in the movement. Democrats, for their part, had no intention of serving as the tail on a Republican dog. They were happy to cooperate with the President, but on their own terms and for their own purposes. Samuel L. M. Barlow dismissed the idea of a "great new party" as nothing but "a wild dream," but hoped the movement might draw off enough Republican votes to "distract and ultimately destroy the Radical party and give us the control in nearly every Northern State."[64]

On the surface, harmony prevailed at the Philadelphia Convention. The 7,000 spectators cheered wildly as South Carolina's massive Gov. James L. Orr marched down the main aisle arm-in-arm with the diminutive Gen. Darius N. Crouch of Massachusetts, leading a procession of the delegates. (A less pleasant reception awaited Judge David L. Wardlaw, principal author of South Carolina's Black Code, whose wallet containing $500 was stolen by a pickpocket.) Notorious Peace Democrats like Ohio's Clement Vallandigham remained away, along with known secessionists. Yet behind the scenes, dissension reigned. Raymond had been persuaded to deliver the convention's main address, but his draft of the platform included guarded praise of the Fourteenth Amendment and oblique criticism of slavery. This proved too much for the Resolutions Committee, which omitted the offending passages. In the end, the convention did not try to establish a new national party, but called for the election of Congressmen who would support Johnson's policies.[65]

The President now decided to take his case to the Northern people. On August 28, accompanied by Welles, Grant, Adm. David Farragut, and other notables, he embarked on the "swing around the circle," an unprecedented speaking tour aimed at influencing the coming elections. At first, things went well, for New York and Philadelphia men of commerce

64. Dorothy Dodd, *Henry J. Raymond and the New York "Times" during Reconstruction* (Chicago, 1936), 51–57; James R. Doolittle to Mary Doolittle, July 1, 1866, James R. Doolittle Papers, SHSW; Beale, ed., *Welles Diary*, 2:533, 540–41; Samuel L. M. Barlow to Montgomery Blair, August 9, 1866, Barlow to Richard Taylor, July 31, 1866, Samuel L. M. Barlow Papers, HL.

65. Thomas Wagstaff, "The Arm-in-Arm Convention," *CWH*, 14 (June 1968), 101–19; Michael Perman, *Reunion Without Compromise: The South and Reconstruction 1865–1868* (New York, 1973), 220–21; Maurice Baxter, *Orville H. Browning: Lincoln's Friend and Critic* (Bloomington, Ind., 1957), 187–89.

and finance welcomed him with enthusiasm. Then the party traveled through upstate New York, and on to the West. Again and again Johnson delivered essentially the same speech, calling for reconciliation between North and South, affirming the loyalty of Southern whites, and arguing that suffrage requirements should be left to the states. He never mentioned the Fourteenth Amendment. When the party reached Ohio, Johnson, interrupted by hecklers, could not resist responding in kind. At Cleveland, when a member of the audience yelled "hang Jeff Davis," the President replied, "Why not hang Thad Stevens and Wendell Phillips?" Johnson also indulged his unique blend of self-aggrandizement and self-pity. On one occasion, he intimated that Providence had removed Lincoln to elevate Johnson himself to the White House. At St. Louis, he blamed Congress for instigating the New Orleans riot and unleashed a "muddled tirade" against his opponents: "I have been traduced, I have been slandered, I have been maligned. I have been called Judas Iscariot. . . . Who has been my Christ that I have played the Judas with? Was it Thad Stevens?"[66]

Americans, more often than not, choose mediocre Presidents, but require of them a decorum foreign to other aspects of their political life. This was not the Greenville debating society, where Johnson had excelled as a youth, or the rough-hewn world of the Tennessee stump. In crossing the line between political debate and uncontrolled harangue, Johnson surrendered the dignity his office had thrust upon him. Even his partisans were mortified. "Thoroughly reprehensible," exclaimed the New York *Journal of Commerce*, while former Georgia Gov. Herschel V. Johnson declared Johnson had sacrificed "the moral power of his position, and done great damage to the cause of Constitutional reorganization." In mid-September, the President returned to Washington from what one admirer called "a tour it were better had never been made."[67]

The "swing around the circle," claimed Doolittle, cost the President's supporters 1 million Northern votes. But it was hardly the only factor that sealed Johnson's political doom. The President's effort to use his formidable patronage power against his political enemies proved equally unsuccessful. Throughout the spring, Democrats and Johnson Republicans had crowded the White House, urging the President to make a clean sweep of federal officeholders and build up a party loyal to himself. "If

66. Montgomery, "Radical Republicanism in Pennsylvania," 444–45; Montgomery, *Beyond Equality*, 70; Gregg Phifer, "Andrew Johnson Argues a Case," *THQ*, 11 (June 1952), 148–70; Gregg Phifer, "Andrew Johnson Delivers His Argument," *THQ*, 11 (September 1952), 212–34; McPherson, *Political History*, 132–37; Carl Schurz, *The Reminiscences of Carl Schurz* (New York, 1907–8), 3:243.

67. Michael L. Benedict, *The Impeachment and Trial of Andrew Johnson* (New York, 1973), 3–4; New York *Journal of Commerce*, September 11, 1866; Herschel V. Johnson to James L. Orr, September 25, 1866, South Carolina Governor's Papers; John Erskine to Joseph E. Brown, September 21, 1866, Hargrett Collection, UGa.

we had a good energetic body of appointments of the right stamp," wrote one supporter from Ohio, "we would as a balance of power be feared and felt, as it is we are despised by both sides." Johnson, however, remained, as the New York *Herald* put it, a man with "two coattails," each of which distrusted the other. To replace existing appointees with Democrats would unite all Republicans against him. Not until the final split with Congress did the Administration send out circulars demanding that officeholders support the Philadelphia convention, and heads soon began to roll. Doolittle received control of Wisconsin patronage and began victimizing supporters of Sen. Timothy Howe; in Illinois, Orville H. Browning, the new Secretary of the Interior, removed appointees loyal to Trumbull. Postmaster General Alexander Randall oversaw a wholesale purge of his department; by the fall, over 1,600 postmasters had been removed. Friends of moderate Congressmen like Henry L. Dawes and Columbus Delano fared no better than those who owed their posts to Radicals.[68]

The patronage weapon, however, failed to produce the desired results. Some officeholders prudently sided with the President, but a remarkable number courted removal rather than support him. In Wisconsin, reported Howe, "the victims of the guillotine are full of defiance." Hannibal Hamlin resigned his lucrative position as collector of customs at Boston rather than endorse Johnson's policies. Patronage, the President discovered, can solidify one's position within an existing party, but cannot magically create a new one. Rather than a conservative coalition overseen by Johnson Republicans, the National Union movement quickly became little more than the Democratic party in a new guise. In nominating conventions, Welles complained, Democrats pressed forward candidates "whom good Union men cannot willingly support." By the fall, most conservative Republicans had returned to the party fold. Even Raymond, whose *New York Times* lost one third of its readership because of his support for the President, in October endorsed the state's Republican ticket. Thus, instead of witnessing the birth of a new party, the 1866 campaign pitted familiar adversaries against one other: the Democracy, claiming loyalty to Johnson, and Republicans, united in opposition to the President.[69]

68. Lewis D. Campbell to Andrew Johnson, June 22, 1866, Johnson Papers; Joseph Geiger to James R. Doolittle, June 25, 1866, Doolittle Papers; Cox and Cox, *Politics, Principle, and Prejudice*, 107; Wagstaff, "Arm-in-Arm Convention," 106–107; Sellers, "Doolittle," 27, 33–34; William H. Russell, "Timothy O. Howe, Stalwart Republican," *WMH*, 35 (Winter 1951), 95; James G. Randall, ed., *The Diary of Orville Hickman Browning* (Springfield, Mass., 1925–33), 2:77–79; Mark M. Krug, *Lyman Trumbull: Conservative Radical* (New York, 1965), 244–45; Benedict, *Impeachment and Trial*, 48–51; Arcanti, "Dawes," 42–43; Columbus Delano to William E. Chandler, August 30, 1866, William E. Chandler Papers, LC.
69. Timothy O. Howe to Grace Howe, June 2, 1866 (copy), Timothy O. Howe Papers, SHSW; Hannibal Hamlin to Andrew Johnson, August 28, 1866 (copy), Hamlin Papers; McKitrick, *Andrew Johnson*, 377–82, 392–94; Beale, ed., *Welles Diary*, 2:590, 603; Homer A. Stebbins, *A Political History of the State of New York, 1865–1869* (New York, 1913), 111.

For the first time in American history, civil rights for blacks played a central part in a major party's national campaign. But on the issue of black suffrage, Republicans remained divided. In moderate areas, while Democrats pressed the question, Republicans remained silent or denied any intention of forcing black voting upon the South. Radicals, however, used the campaign "to educate the public mind up to the standard of universal suffrage." Public opinion was in flux and, as Ohio's party chairman reported, the "general average" of Republican sentiment seemed impossible to determine: "In the [Western] Reserve counties, some of our speakers have openly advocated impartial suffrage, while in other places it was thought necessary . . . to oppose it."[70]

Like most campaigns of the period, that of 1866 did not lack for colorful, scurrilous political rhetoric. "The Democratic party," proclaimed Indiana Gov. Oliver P. Morton, "may be described as a common sewer and loathsome receptacle, into which is emptied every element of treason, North and South." (For their part, Democrats assailed the physically infirm and semi-paralyzed Morton as "a fetid excrescence on the body politic," "a wretch accursed of God and enjoying a foretaste of hell on earth.") Johnson's supporters, and later historians sympathetic to the President, would contend that a small band of fanatics secured the Republican victory by stirring up Northern opinion against "rebels" and "Copperheads" and obscuring the real issues at stake. Alternatively, they chastized the President for not engaging Republicans on economic issues like the tariff, on which a majority could ostensibly have been forged. Yet both parties remained divided on economic questions, and voters displayed little interest in them in 1866. "The readers of the [Chicago] *Tribune,*" editor Horace White had reported earlier in the year, "have lost nearly all interest in the Free Trade vs. Protection controversy. . . . The tariff question will not and cannot become a vital one while the reconstruction question remains unsettled." More than anything else, the election became a referendum on the Fourteenth Amendment. Seldom, declared the *New York Times,* had a political contest been conducted "with so exclusive reference to a single issue." And the result was a disastrous defeat for the President. Defying the usual pattern whereby the party in power loses strength in off-year elections, voters confirmed the massive Congressional majority Republicans had achieved in 1864. In the next Congress, Republicans would outnumber Democrats and Johnson conservatives by well above the two-thirds majority required to override a veto.[71]

70. *New York Times,* October 24, 1866; Leslie H. Fishel, Jr., "Northern Prejudice and Negro Suffrage," *JNH,* 39 (January 1954), 17; Blaine, *Twenty Years,* 2:243–44; B. R. Cowen to Salmon P. Chase, October 12, 1866, Salmon P. Chase Papers, LC.

71. Foulke, *Morton,* 474–77; Beale, ed., *Welles Diary,* 2:616–17; Howard K. Beale, *The Critical Year: A Study of Andrew Johnson and Reconstruction* (New York, 1930), 5–7, 142, 300, 385; Joseph Logsdon, *Horace White: Nineteenth Century Liberal* (Westport, Conn., 1971), 120–23; *New York Times,* October 11, 1866.

"This is the most decisive and emphatic victory ever seen in American politics," exclaimed *The Nation*. In its aftermath, the course of prudence seemed plain. The South, warned the *Times*, must ratify and comply with the Fourteenth Amendment and the President must cease to oppose it; otherwise, black suffrage could not be avoided. An editorial writer on Bennett's *Herald* agreed: For the President to heed "the voice of the people," now abundantly clear, would be an act of "the highest statesmanship." Johnson, however, refused to alter his opposition to the Amendment. Northern Democrats adopted the same position. The South, an Ohio party leader advised Georgia Governor Jenkins, should "let the Jacobins run their course, spurn and indignantly reject all their degrading and humiliating terms of constitutional amendment and universal or impartial suffrage." More than stubbornness lay behind this counsel. Johnson's close advisers still believed Congress could never muster the votes to enact black suffrage over a veto, and that any attempt to do so would "present the issue squarely . . . upon [which] we can beat them at the next Presidential election." Northern Democrats, too, welcome the prospect of a national campaign centered on blacks' right to vote.[72]

Southern political leaders might well have pondered whether the advice of those who had so often proved wrong ought to be trusted. From Washington, North Carolinian Benjamin S. Hedrick warned that the South would "gain nothing by fighting the battles of defeated Northern demagogues." But most white Southern leaders believed their fate inextricably tied to the President's. Southern newspapers, moreover, consistently misinformed their readers about Northern politics, overestimating the strength of the National Union movement, portraying Johnson's opponents as a band of Radical fanatics who lacked broad popular support, and predicting Congress could not possibly do things it then proceeded to do. The election returns came as a shock, but produced no political reassessment.[73]

White public opinion, Alabama editor John Forsyth observed, was "very unanimous against adopting the Amendment." Although many objected to the representation clause as an opening wedge for black suffrage, the section barring from office what one newspaper called "the best portion of our citizens" aroused the strongest opposition. South

72. *Nation*, November 15, 1866; *New York Times*, October 11, 1866; W. B. Phillips to Andrew Johnson, October 7, November 8, 1866, Johnson Papers; Alexander Long to Charles J. Jenkins, November 19, 1866, Alexander Long Papers, Cincinnati Historical Society; James R. Doolittle to Orville H. Browning, November 8, 1866 (copy), Doolittle Papers; George T. McJimsey, *Genteel Partisan: Manton Marble, 1834–1917* (Ames, Iowa, 1971), 111–13.

73. J. G. de Roulhac Hamilton, ed., *The Correspondence of Jonathan Worth* (Raleigh, 1909), 2:784; Dan T. Carter, *When the War Was Over: The Failure of Self-Reconstruction in the South, 1865–1867* (Baton Rouge, 1985), 241; Perman, *Reunion Without Compromise*, 238–48.

Carolina Governor Orr declared Southerners were being asked to "concede more to the will of their conquerors" than any people in history. Between October 1866 and the following January, all ten Southern legislatures that considered the Amendment repudiated it by overwhelming majorities. It made no difference whether governors urged rejection or, as in Louisiana, Virginia, and Arkansas, favored ratification. Louisiana's vote was unanimous: "A Governor without a single supporter in the legislature is without precedent in the political annals of this country," commented a New Orleans newspaper. All told, only thirty-three Southern lawmakers braved public opposition to vote for ratification. "Are we not," wondered one South Carolinian, "actually inviting our own destruction?"[74]

For one group of white Southerners, however, the Amendment held out hope for a political revolution that would sweep them into power. In most of the South, wartime Unionists remained a beleaguered minority, confronting "rebel" officials and a hostile public opinion. Only in a few areas were they numerous enough to organize in opposition to the Johnson governments. In the mountains of Alabama, wartime Unionists refused to pay taxes to the "rebel State," prevented judges appointed by the governor from holding court, and harassed prominent Confederates. But generally, they did little but look to the federal government for assistance. Increasingly, they sided with Congress in the dispute over Reconstruction. Among those gravitating to the Republicans were Governors Wells and Pierpont, who remained in office but had "parted with all power," their efforts to create a political middle ground in Louisiana and Virginia overwhelmed by the realities of postwar Southern politics. They were joined by former Presidential Reconstruction governors Andrew J. Hamilton, whose handpicked successor Texas voters had defeated, and William W. Holden, whom the North Carolina electorate had turned out of office.[75]

The same disability clause that made the Fourteenth Amendment unacceptable to the South's officialdom won the enthusiastic support of wartime Unionists. In the Upper South, few expressed much interest in the section guaranteeing blacks equal protection of the law, and still fewer endorsed black suffrage. As one Union veteran from Virginia declared in

74. John Forsyth to Manton Marble, October 13, 1866, Marton Marble Papers, LC; Perman, *Reunion Without Compromise*, 236, 252–59; *Message No. 1 of His Excellency Gov. J. L. Orr* (Columbia, S.C., 1866), 19; McPherson, *Political History*, 194; C. W. Dudley to James L. Orr, December 31, 1866, South Carolina Governor's Papers.

75. J. J. Grier to A. Chester, September 17, 1866, G-294 1866, Letters Received, Ser. 15, Washington Headquarters, RG 105, NA [FSSP, A-2210]; W. B. Wood to Robert M. Patton, November 3, 1866, Alabama Governor's Papers, ASDAH; Alfred F. Terry to Joseph R. Hawley, February 1, 1866, Joseph R. Hawley Papers, LC; Jack P. Maddex, Jr., *The Virginia Conservatives 1867–1879* (Chapel Hill, 1970), 41–43; Carter, *When the War Was Over*, 256; William W. Holden to Andrew Johnson, July 11, 1866, Johnson Papers; Perrin Busbee to Benjamin S. Hedrick, September 21, 1866, Benjamin S. Hedrick Papers, DU.

August, he had become "one of Thad Stevens' friends . . . except the nigger portion of his policy." But in the Deep South, with far fewer white loyalists, Union men, albeit reluctantly, increasingly favored enfranchising blacks. "The vilest traitors" dominated state government, a South Carolina man wrote Stevens. "I have slowly but deliberately come to the conclusion that negro suffrage is our only hope."[76]

Just how deeply this issue divided the South's Unionists became clear in September, when the Southern Loyalists Convention assembled in Philadelphia. Most of the nearly 500 delegates hailed from the Upper South and Louisiana; elsewhere, public hostility made attendance nearly impossible. Georgia's Joshua Hill explained he could not take part, since to do so would "call down upon one's head an avalanche of ridicule and contempt, . . . [and] would have driven away nearly every associate of one's family." Brotherly love seemed absent from the city that week, for the delegates included only one black, P. B. Randolph of Louisiana, and when Frederick Douglass appeared, he was urged not to embarrass the gathering by taking part. The convention readily endorsed the Fourteenth Amendment, but failed to agree on much else. Delegates from the Deep South insisted that their salvation lay in black votes. But the majority, representing border states with Republican regimes, still believed they could retain power through disenfranchising former Confederates and defeated all calls for extending the ballot to blacks. Northerns present, moreover, lobbied strenuously against any endorsement of black suffrage, fearing its impact on their own upcoming elections. Only after most of the border men had departed did the remaining participants approve a platform that denounced the Johnson governments as instruments of a rebel aristocracy and advocated black suffrage as "the one all-sufficient remedy" for the plight of loyal Southerners.[77]

The disastrous results of the 1866 elections converted even Upper South Unionists to black suffrage. In March, opponents of the Brownlow regime had made a "clean sweep" of local elections in Middle and West Tennessee. In the fall, defying the Northern pattern, Democrats ousted wartime Unionists from control of the legislatures of Arkansas (whose state courts had voided the 1864 disenfranchisement law), and Maryland. "*Rebels* [must be] disfranchised . . . and the negro enfranchised," Union-

76. Rutherford *Star*, October 3, 1866; George H. Berger to Leonidas C. Houk, August 4, 1866, Leonidas C. Houk Papers, McClung Collection, LML; James A. Baggett, "Birth of the Texas Republican Party," *SWHQ*, 78 (July 1974), 11–12; Simeon J. Corley to Thaddeus Stevens, February 6, 1866, Stevens Papers.

77. Joshua Hill to Horace Greeley, September 22, 1866, Greeley Papers; [Frederick Douglass] *Life and Times of Frederick Douglass* (New York, 1962 ed.), 388–90; *The Southern Loyalists Convention* (New York, 1866); Richard H. Abbott, *The Republican Party and the South, 1855–1877: The First Southern Strategy* (Chapel Hill, 1986), 70–71.

ists in these states now concluded. Soon afterward, the Southern Republican Association, meeting in Washington, demanded the replacement of the Johnson governments by new ones with "all men, except rebels," eligible to vote. The "true Unionists" of western North Carolina now echoed this demand, as did black leader James H. Harris, who arrived in Washington in December hoping to convince Congress that the planters would never "control the votes of our people." In January, insisting this was the only way to keep Tennessee out of "disloyal hands," Brownlow won enactment of a bill enfranchising blacks (but still excluding them from office and jury service). Residents of the mountains, wrote one Unionist, were not "afraid of negro equality," if "rebel superiority" was the alternative.[78]

Thus, as Congress reassembled, black suffrage was once again on the political agenda, thanks to Radical Republicans, Southern Unionists, and the freedmen themselves. Their efforts had been unintentionally abetted by Johnson and the governments he had established. Not for the first time, Southern intransigence played into the Radicals' hands. For, as Benjamin S. Hedrick had warned, "If the Northern people are forced by the South to follow Thad Stevens or the Copperheads, I believe they will prefer the former."[79]

The Coming of Black Suffrage

The Republicans who gathered in December 1866 for the second session of the Thirty-Ninth Congress considered themselves "masters of the situation." Grimes, who a year earlier had bent every effort to finding a modus vivendi between Congress and the White House, expressed the prevailing mood: "The President has no power to control or influence anybody and legislation will be carried on entirely regardless of his opinions or wishes." Johnson's annual message, pleading for the immediate restoration of the "now unrepresented States," was ignored. The President, declared the New York *Herald,* his erstwhile supporter, "forgets that we have passed through the fiery ordeal of a mighty revolution, and that

78. Abbott, *Republican Party and the South,* 72–76; William G. Brownlow to Oliver P. Temple, March 8, 1866, Oliver P. Temple Papers, University of Tennessee; Paige Mulhollen, "The Arkansas General Assembly of 1866 and its Effect on Reconstruction," *ArkHQ,* 20 (Winter 1961), 333; Charles Wagandt, "Redemption of Reaction?—Maryland in the Post-Civil War Years," in Richard O. Curry, ed., *Radicalism, Racism, and Party Realignment: The Border States During Reconstruction* (Baltimore, 1969), 162–76; H. B. Allis to Richard Yates, November 13, December 25, 1866, Richard Yates Papers, Illinois State Historical Library; Raleigh *Tri-Weekly Standard,* January 1, June 4, 1867; Alexander H. Jones to Thaddeus Stevens, January 4, 1867, William B. Downey to Stevens, January 7, 1867, Stevens Papers; Knoxville *Whig,* January 2, 1867.
79. Benjamin S. Hedrick to Jonathan Worth, April 3, 1866, Jonathan Worth Papers, NCDAH.

the pre-existing order of things is gone and can return no more—that a great work of reconstruction is before us, and that we cannot escape it."[80]

Yet the triumphant Republicans faced difficulties of their own, not least two decisions now announced by the Supreme Court. In *Ex parte Milligan*, the Court voided the wartime conviction of an Indiana man by a military tribunal on the grounds that civilian courts had been functioning at the time of his trial. Although Justice David Davis, who wrote the opinion, insisted it had nothing to do with the South, the case threw into question the legality of martial law and Freedmen's Bureau courts. In *Cummings* v. *Missouri*, the Court overturned Missouri's constitutional requirement that lawyers, ministers, and others swear to a prescribed loyalty oath. Ironically, as the first time a state constitution had been taken under federal judicial review, the decision reflected the expansion of national power arising from the Civil War. But, along with *Milligan*, it suggested that Republicans might soon face a hostile Supreme Court.[81]

As Congress deliberated, the "spirit" of Benjamin Franklin appeared before one political observer, predicting the imminent passage of a measure "disfranchising the rebels in the South, and enfranchising the negroes." One did not have to be clairvoyant, however, to know that black suffrage was on the horizon. In mid-December, citing the Constitution's guarantee clause, Trumbull told the Senate that Congress possessed the authority to "enter these States and hurl from power the disloyal element which controls and governs them," an important announcement that moderates were now prepared to overturn the Johnson governments. In January 1867, the bill enfranchising blacks in the District of Columbia became law over the President's veto. (A month later, the city held an election and, for the first time in living memory, Democrats chose not to parade with banners "in regard to *niggers, miscegenation* and similar matters.") Then, Congress extended manhood suffrage to the territories. Three points appeared "fully settled" among Republican leaders, John Pool reported from Washington: existing Southern governments should be superseded, "rebels" should hold no place in the new regimes, and "the negroes should vote." Even more radical proposals were in the air, including widespread disenfranchisement, martial law for the South, confiscation, the impeachment of the President. A New York *Herald* editorial writer apologized to Johnson for the paper's advocacy of

80. Willard H. Smith, *Schuyler Colfax: The Changing Fortunes of a Political Idol* (Indianapolis, 1952), 247; James W. Grimes to Charles H. Ray, December 2, 1866, Charles H. Ray Papers, HL; Richardson, ed., *Messages and Papers*, 6:446–48; Montgomery, *Beyond Equality*, 72.

81. Willard L. King, *Lincoln's Manager: David Davis* (Cambridge, Mass., 1960), 245–56; Charles Fairman, *Reconstruction and Reunion 1864–68: Part One* (New York, 1971), 232–36; Thomas S. Barclay, *The Liberal Republican Movement in Missouri 1865–1871* (Columbia, Mo., 1926), 58–59, 114. *Ex Parte Milligan* had been decided in April, but the full opinion was only released in December.

his removal: Its editor, James Gordon Bennett, always went with the political tide, and the tide now flowed toward the Radicals.[82]

With moderates lacking a program since the South's rejection of the Fourteenth Amendment, Radicals moved to seize the legislative initiative. Stevens introduced a bill requiring the existing Southern governments to hold constitutional conventions, elected by manhood suffrage with the exception of former Confederates, who would be deprived of citizenship for five years. Then, James Ashley proposed a substitute, sweeping away the Johnson governments entirely. From Bingham came the moderate counterattack. Stevens' bill, he declared, treated white Southerners as "alien enemies" and failed to address the immediate plight of blacks and loyalists—"it gives no protection to anybody." Ashley's substitute he termed "a bill of anarchy, not restoration." On January 16, Bingham moved to refer both measures to the Joint Committee on Reconstruction. Against this proposal, Stevens fought a desperate rearguard action, persuading Ashley to withdraw his substitute, promising to open his own bill to amendment from the floor, and warning that committal meant "the death of the measure." But on January 28, by a vote of 88 to 65, the House approved Bingham's motion. Stevens carried a majority of his party, and Democratic votes provided the margin of moderate victory.[83]

Once again, Congress found itself without a program, a situation not all Republicans regretted. On the same day as Bingham's success, Julian warned against precipitous action. What the South needed was not "hasty restoration" or oaths that invited men to commit perjury, but *"government, the strong arm of power, outstretched from the central authority here in Washington."* Only a prolonged period of federal control would enable loyal public opinion to sink deep roots and permit "Northern capital and labor, Northern energy and enterprise" to venture South, there to establish "a Christian civilization and a living democracy." The South, he proposed, should be governed directly from Washington and only readmitted at "some indefinite future time" when its "political and social elements" had been thoroughly transformed.[84]

Julian's speech struck a chord in Congress. The Joint Committee quickly approved a bill to impose military rule on the South. One Congressman described it as turning the political clock back to "the point where Grant left off the work, at Appomattox Court-House." Even conservative Republican M. Russell Thayer of Pennsylvania, acknowledging the measure was "not a reconstruction bill" at all, welcomed it as a way

82. L. M. Smith to Horace Greeley, January 19, 1867, Greeley Papers; *CG*, 39th Congress, 2d Session, 159–60; McPherson, *Political History*, 160, 184; Hamilton, ed., *Worth Correspondence*, 2:902–903; Richard L. Zuber, *Jonathan Worth* (Chapel Hill, 1965), 245–46; W. B. Phillips to Andrew Johnson, January 25, 1867, Johnson Papers.

83. Benedict, *Compromise of Principle*, 214–20; *CG*, 39th Congress, 2d Session, 250–54, 500–504, 781–82, 815–17.

84. *CG*, 39th Congress, 2d Session, Appendix, 78.

of offering immediate protection to "the down-trodden and persecuted Union men of the South." But many moderates feared the bill portended an endless prolongation of the crisis; they accepted military rule as a temporary expedient, but insisted the means by which the South could establish new civil governments and regain its standing within the Union must be clearly specified. On February 12, Bingham proposed to readmit the Southern states once they had ratified the Fourteenth Amendment and established black suffrage, to which Blaine added that they should also be permitted to disenfranchise those who had participated in the rebellion. An indication of how the political climate had shifted since 1866 was that debate centered not on extending the franchise to blacks, but restricting it among whites.[85]

The disenfranchisement of former Confederates was incorporated in a bill reported from a committee investigating the New Orleans riot, which provided for the establishment of a new civil government in Louisiana, elected by blacks and loyal whites. On February 12, this measure passed the House and attention turned again to the military bill. The next day was one of high drama on the House floor. Blaine moved to recommit the bill to the Joint Committee, with instructions that it be reported back immediately, incorporating provisions for Southern readmission. By a narrow margin, the House agreed to an immediate vote on this motion. Stevens then made a supreme effort to avert defeat. In a voice so weak his colleagues had to crowd around to hear him, he addressed the "burning crisis" of Reconstruction, pleading with the House not to bind itself in advance to terms of readmission when so much needed to be accomplished. The speech was "one of the few ever delivered in Congress that have resulted in the changing of votes." The House defeated Blaine's motion and passed the unamended military bill. Stevens was jubilant, but had twenty-three Democrats who voted with him supported Blaine instead, he would not have triumphed. Although their party loathed even temporary military rule as "the death-knell of civil liberty," many Democrats now hoped for a Radical triumph that might split Republicans and alienate the electorate. "Let the Radicals go on. . . ." wrote a Connecticut Democratic leader. "The more extreme the measures . . . the sooner they will be whipped."[86]

The next day, the Louisiana and military bills came before the Senate. "Each is excellent," declared Sumner. "One is the beginning of a true reconstruction; the other is the beginning of a true protection." But others had their doubts, and a hail of amendments greeted the two measures. To prevent Democrats from determining the outcome of close votes, Republican Senators retired to a caucus, which appointed a seven-

85. CG, 39th Congress, 2d Session, 1097, 1176–82, 1211, Appendix, 175.

86. CG, 39th Congress, 2d Session, 1077, 1129, 1175, 1213–15; David Donald, The Politics of Reconstruction, 1863–1867 (Baton Rouge, 1965), 70–75; McKitrick, Andrew Johnson, 464.

member committee, chaired by Sherman, to fashion an acceptable measure. Essentially, the committee combined the two bills into one providing both military rule and conditions of readmission for the entire South. The main point of dispute concerned black suffrage. All agreed it must operate in elections for constitutional conventions, but not whether the new constitutions should be required to incorporate it as well. To Sherman, this was a matter of "detail." To Sumner, it was crucial, and when the committee failed to bind the new documents to black suffrage, he appealed to the full caucus. The question of black voting, said Sumner, must be settled, or "every State and village between here and the Rio Grande would be agitated by it." By a margin of two, the Republican caucus overturned the committee's decision. Exclaimed Henry Wilson: "This is the greatest vote that has been taken on this continent."[87]

And so Republicans decided that blacks must enter the South's body politic. But further controversy lay ahead before a Reconstruction policy could be settled. When the amended bill returned to the House, it touched off a storm of Radical protest. Rebels, charged Boutwell, had been handed "the chief places in the work of reconstruction," for while establishing military rule, the bill left the Johnson governments in place and failed to disenfranchise former Confederates. Border and Upper South Republicans were especially furious. The bill, declared William B. Stokes of Tennessee, was a "death-blow" to Southern Unionists: "Pass this bill and where are your loyal men? White and black, they go under. Yes, sir, I repeat, they go under." With Democrats again providing the margin of victory, the House refused to accept the Senate amendments. But Senate moderates, angered by "the union of a few extreme Radicals and the whole Democratic party *en masse*," declined to discuss further changes.[88]

The Radicals, as Stevens remarked, now realized "we had to take that or nothing." Two amendments, intended to place the Reconstruction process in the hands of loyal men, made the bill more palatable to its critics. The first barred anyone disqualified from office under the Fourteenth Amendment from electing, or serving as, constitutional convention delegates. The second declared the Johnson governments subject to modification or abolition at any time, and prohibited individuals disqualified under the Fourteenth Amendment from voting or holding office under them. As many contemporaries pointed out, these disabilities contained a muddle of contradictions. They temporarily disenfranchised antebellum officeholders who had served the Confederacy, including Unionists who had assumed minor positions to avoid military service, but

87. CG, 39th Congress, 2d Session, 1303; 41st Congress, 2d Session, 1177–82; Pierce, ed., *Memoir and Letters of Sumner*, 4:319–20; Brock, *An American Crisis*, 194.
88. CG, 39th Congress, 2d Session, 1315–16, 1340, 1557; *New York Times*, February 20, 1867.

not Southern leaders who had occupied no prewar office, or the vast majority of Confederate soldiers. No one knew how many persons these eleventh-hour changes affected: Sherman guessed 10,000 to 15,000 "leading rebels," others offered different estimates. But for Southern Unionists, they represented a major victory. The larger part of a political generation, men of local influence ranging from prewar postmasters and justices of the peace to legislators and Congressmen, had been temporarily excluded from office and voting. "This Amendment . . . will prove of vital importance in the work of reconstruction. . . ." declared Holden's Raleigh *Standard.* "We rejoice that there is to be an end to rebel rule."[89]

Throughout these deliberations, Johnson remained silent. Toward the end of February, New York *Evening Post* editor Charles Nordhoff visited the White House. He found the President "much excited," certain "the people of the South . . . were to be trodden under foot 'to protect niggers'." Nordhoff had once admired the President; now he judged him a "pig-headed man" with only one idea: "bitter opposition to universal suffrage." Gone was the vision of a reconstructed South controlled by loyal yeomen. "The old southern leaders . . ." declared the man who had once railed against the Slave Power, "must rule the South." When the Reconstruction bill reached his desk on March 2, Johnson returned it with a veto, which Congress promptly overrode. Maryland Sen. Reverdy Johnson, a man, like Stevens, born during Washington's Presidency, was the only member to break party ranks. Whatever its flaws, he declared, the bill offered the South a path back into the Union, and the President should have long since acceded to the plainly expressed will of the people rather than persist in his intransigence. Reverdy Johnson's was the only Democratic vote cast in favor of any of the Reconstruction measures of 1866–67.[90]

In its final form, the Reconstruction Act of 1867 divided the eleven Confederate states, except Tennessee, into five military districts under commanders empowered to employ the army to protect life and property. And without immediately replacing the Johnson regimes, it laid out the steps by which new state governments could be created and recognized by Congress—essentially the writing of new constitutions providing for manhood suffrage, their approval by a majority of registered voters, and ratification of the Fourteenth Amendment. (A precedent existed for requiring a state to ratify an amendment to gain representation in Congress, for Johnson had done precisely the same thing with regard to the Thirteenth.) The act contained no mechanism for beginning the process

89. New York *Herald,* July 8, 1867; *CG,* 39th Congress, 2d Session, 1399–1400, 1564, 1625–26; Brock, *An American Crisis,* 197–98; N. G. Foster to John Sherman, April 5, 1867, Sherman Papers; Raleigh *Tri-Weekly Standard,* February 23, 26, 1867.

90. Charles Nordhoff to William Cullen Bryant, February 21, 1867, Bryant-Godwin Papers, NYPL; Richardson, ed., *Messages and Papers,* 6:507–509; *CG,* 39th Congress, 2d Session, 1733, 1796, 1972.

of change, an oversight soon remedied by a supplemental measure authorizing military commanders to register voters and hold elections. Simultaneously, and with little debate, Congress passed the Habeas Corpus Act, which greatly expanded citizens' ability to remove cases to federal courts, and a law abolishing peonage (holding a person to labor until a debt is paid). It also called the Fortieth Congress into special session on March 4.[91]

The astonishingly rapid evolution of Congressional attitudes that culminated in black suffrage arose both from the crisis created by the obstinacy of Johnson and the white South, and the determination of Radicals, blacks, and eventually Southern Unionists not to accept a Reconstruction program that stopped short of this demand. And like all the decisions of the Thirty-Ninth Congress, the Reconstruction Act contained a somewhat incongruous mixture of idealism and political expediency. Every Radical element appeared balanced by a moderate one. The bill established military rule, but only as a temporary measure to keep the peace, with the states relatively quickly regaining their place within the Union. It looked to a new political order for the South, but failed to place Southern Unionists in immediate control. It made no economic provisions for the freedmen. Even the commitment to black suffrage applied only to the defeated Confederacy, not the nation as a whole.

In all these ways, the Reconstruction Act reflected the circumstances of its creation, especially the necessity of finding a program upon which two thirds of Congress could agree and the Northern electorate would support, and the deeply held beliefs and prejudices that placed limits upon Congressional action. The free labor ideology simultaneously inspired efforts to guarantee civil and political equality—essential attributes of autonomous citizenship in a competitive society—and inhibited efforts to provide an economic underpinning for blacks' new freedom. Indeed, it suggested that once accorded equal rights, the freedmen would find their social level and assume responsibility for their own fate. Federalism also affected Republican policy-making. For if the Civil War created the national state and Reconstruction added the idea of a national citizenry whose common rights no state could abridge, most Republicans still believed the states retained rights beyond the scope of federal intervention, and expected the relatively rapid return of the Southern states as equal members of the Union. Nor did Reconstruction create new bureaucratic agencies institutionally committed to protecting black rights. The Freedmen's Bureau had always been conceived as temporary, and the long-term burden of overseeing the local administration of justice would fall to the overworked, understaffed Justice Department and to the federal judiciary with its newly expanded jurisdiction. By the mid-

91. McPherson, *Political History*, 191–92; William M. Wiecek, "The Great Writ and Reconstruction: The Habeas Corpus Act of 1867," *JSH*, 36 (November 1970), 530–48.

1870s, virtually every case involving blacks, white loyalists, and federal officials in the South could be removed to federal courts.[92]

Republicanism, too, helped shape Congressional Reconstruction. It was somehow fitting that as Congress deliberated in February 1867, the last surviving veteran of the American Revolution died. For like the Revolution, Reconstruction was an era when the foundations of public life were thrown open for discussion. Republicanism offered a potent argument for black suffrage, but ruled out the massive disenfranchisement of Southern whites. "If we exclude from voting the rebels of the South . . ." asked Sherman, "what becomes of the republican doctrine that all governments must be founded on the consent of the governed?" It also required that Southerners pass judgment on the governments that would rule them, rather than simply have them imposed from outside. From the submission of the Fourteenth Amendment for Southern approval to the Reconstruction Act's insistence that new constitutions be ratified by majority vote, the consent of an electorate that included large numbers of "rebels" remained central to Congressional policy. Now, however, the definition of the voting population included black men as well as white.[93]

Black suffrage, of course, was the most radical element of Congressional Reconstruction, but this too derived from a variety of motives and calculations. For Radicals, it represented the culmination of a lifetime of reform. For others, it seemed less the fulfillment of an idealistic creed than an alternative to prolonged federal intervention in the South, a means of enabling blacks to defend themselves against abuse, while relieving the nation of that responsibility. Many Republicans placed utopian burdens upon the right to vote. "The ballot," Radical Sen. Richard Yates exclaimed, "will finish the negro question; it will settle everything connected with this question. . . . We need no vast expenditures, we need no standing army. . . . Sir, the ballot is the freedman's Moses." When such expectations proved unrealistic, disillusionment was certain to follow.[94]

Despite all its limitations, Congressional Reconstruction was indeed a radical departure, a stunning and unprecedented experiment in interracial democracy. In America, the ballot did more than identify who could vote—it defined a collective national identity (as women's suffrage advocates so tirelessly pointed out). Democrats had fought black suffrage on precisely these grounds. "Without reference to the question of equality,"

92. Michael L. Benedict, "Preserving Federalism: Reconstruction and the Waite Court," *Supreme Court Review*, 1978, 40–48; Brock, *An American Crisis*, 301; William M. Wiecek, "The Reconstruction of Federal Judicial Power, 1863–1875," *American Journal of Legal History*, 13 (October 1969), 333–34.

93. *CG*, 39th Congress, 2d Session, 1348, 1563–64; 41st Congress, 2d Session, 565; Perman, *Reunion Without Compromise*, 5–6, 272.

94. Eugene H. Berwanger, *The West and Reconstruction* (Urbana, Ill., 1981), 112–13; Brock, *An American Crisis*, 268; *CG*, 39th Congress, 1st Session, Appendix, 103.

declared Indiana Sen. Thomas Hendricks, "I say we are not of the same race; we are so different that we ought not to compose one political community." The enfranchisement of blacks marked a powerful repudiation of such thinking. In some ways it was an astonishing leap of faith. Were the mass of freedmen truly prepared for political rights? Gen. E.O.C. Ord, federal commander in Arkansas, believed them "so servile and accustomed to submit" to white dictation that they would "not dare to present themselves at the polls." Even some Radicals harbored inner doubts, fearing the black vote would be swayed by "demagogues" or controlled by their former masters. Other contemporaries warned that political rights would prove meaningless without economic independence.[95]

In the course of Reconstruction, the freedmen would disprove these and other forecasts. They would demonstrate political shrewdness and independence, and the ability to use the ballot to affect the conditions of their freedom. However inadequate as a response to the legacy of slavery, it remains a tragedy that the lofty goals of civil and political equality were not permanently achieved. And the end of Reconstruction would come not because propertyless blacks succumbed to economic coercion, but because a politically tenacious black community, abandoned by the nation, fell victim to violence and fraud.

Alone among the nations that abolished slavery in the nineteenth century, the United States, within a few years of emancipation, clothed its former slaves with citizenship rights equal to those of whites. Henceforth, the central political battleground of Reconstruction would shift from Washington to the South. And here the enfranchisement of blacks posed an immense challenge to traditional white prerogatives and the inherited structure of society. "It touches every portion of the social fabric," the *Nation* would later observe, "from foundation to apex." For good reason, landed elites in plantation economies fear political democracy, for their control of the scarce resources of these societies rests, in considerable measure, on their local political hegemony. If black suffrage was not accompanied by an economic program such as that envisioned by Stevens, it did raise the prospect of further change within the South. It was a revolution, lamented North Carolina Gov. Jonathan Worth, and "nobody can anticipate the action of revolutions."[96]

As if reflecting the death throes of an old order, on March 1, with the Reconstruction Act's final passage imminent, the last North Carolina

95. *CG*, 39th Congress, 1st Session, 880; E.O.C. Ord to O. O. Howard, January 27, 1867, E.O.C. Ord Papers, Bancroft Library, University of California, Berkeley; Springfield (Mass.) *Weekly Republican*, August 10, 1867; *CG*, 39th Congress, 2d Session, 1316, 1323; 40th Congress, 1st Session, 114; New York *World*, January 28, 1867.

96. *Nation*, September 7, 1876; Eric Foner, *Nothing But Freedom: Emancipation and Its Legacy* (Baton Rouge, 1983), 37; Jeffrey M. Paige, *Agrarian Revolution* (New York, 1975), 17–21; Hamilton, ed., *Worth Correspondence*, 2:914.

legislature elected solely by whites dissolved into a drunken frolic. "Some of the rich old colts from the East," recounted one member, "gave a general treat on the East Portico of the capital furnishing the very best liquor. . . . The whole Capitol was in an uproar." Three days later the lawmakers adjourned; some of their places would soon be occupied by emancipated slaves. "We have cut loose from the whole dead past," wrote Senator Howe, "and have cast our anchor out a hundred years." His colleague, Waitman T. Willey of West Virginia, adopted a more cautious tone: "The legislation of the last two years will mark a great page of history for good or evil—I hope the former. The crisis, however, is not yet past."[97]

97. Otto H. Olsen and Ellen Z. McGrew, "Prelude to Reconstruction: The Correspondence of State Senator Leander Sams Gash, 1866–67, Part III," *NCHR*, 60 (July 1983), 360; Timothy O. Howe to Grace Howe, February 26, 1867 (copy), Howe Papers; Waitman T. Willey Diary, March 5, 1867, Waitman T. Willey Papers, University of West Virginia.

CHAPTER 7

Blueprints for a Republican South

The Political Mobilization of the Black Community

LIKE emancipation, the passage of the Reconstruction Act inspired blacks with a millennial sense of living at the dawn of a new era. Former slaves now stood on an equal footing with whites, a black speaker told a Savannah mass meeting; before them lay "a field, too vast for contemplation." As in 1865, blacks found countless ways of pursuing aspirations for autonomy and equality, and seizing the opportunity to press for further change. Workers on John B. Gordon's Georgia rice plantation drove off the overseer, an act the Confederate general attributed to the advent of Radical Reconstruction. Strikes broke out among black longshoremen in Charleston, Savannah, Mobile, Richmond, and New Orleans, and quickly spread to other workers, including Richmond coopers and Selma restaurant waiters. Hundreds of South Carolina blacks refused to pay taxes to the existing state government, and crowds rescued companions arrested by the all-white Richmond police. At Jacksonville, a speaker told a mass meeting that "the bone and sinew of the colored man" had built the city; "thus, we have an equal title to enjoy and govern it."[1]

Horse-drawn urban streetcars, which relegated blacks, except for servants accompanying their employers or escorting white children, to outside platforms or separate cars, became a particular target of protest. Three blacks refused to leave a whites-only vehicle in Richmond, and

1. Savannah *Daily News and Herald*, March 19, 1867; Allen P. Tankersley, *John B. Gordon: A Study in Gallantry* (Atlanta, 1955), 234; Philip S. Foner and Ronald L. Lewis, eds., *The Black Worker: A Documentary History from Colonial Times to the Present* (Philadelphia, 1978–84), 1:352–53; William C. Hine, "Black Organized Labor in Reconstruction Charleston," *Labor History*, 25 (Fall 1984), 506; Peter J. Rachleff, *Black Labor in the South: Richmond, Virginia, 1865–1900* (Philadelphia, 1984), 42–43; Joseph G. Dawson III, *Army Generals and Reconstruction: Louisiana, 1862–1877* (Baton Rouge, 1982), 52; Thomas H. Wade to James L. Orr, April 16, 1867, South Carolina Governor's Papers, SCDA; Barbara A. Richardson, "A History of Blacks in Jacksonville, Florida, 1860–1895: A Socio-Economic and Political Study" (unpub. diss., Carnegie-Mellon University, 1975), 184.

crowds rushed to the scene shouting, "Let's have our rights." Several
Charleston blacks "sat in" on a streetcar, and in New Orleans groups
forcibly commandeered cars and drove them around the city in triumph.
By midsummer, integrated transportation had come to these and other
Southern cities.[2]

But in 1867, politics emerged as the principal focus of black aspirations. In that annus mirabilis, the impending demise of the structure of
civil authority opened the door for political mobilization to sweep across
the black belt. Itinerant lecturers, black and white, brought the Republican message into the heart of the rural South. A black Baptist minister
calling himself Professor J. W. Toer journeyed through parts of Georgia
and Florida with a "magic lantern" exhibiting "the progress of reconstruction. . . . He has a scene, which he calls 'before the proclamation,'
another 'after the proclamation' and then '22nd Regt. U. S. C[olored]
T[roops] Duncan's Brigade'." Voting registrars instructed freedmen in
American history and government and "the individual benefits of citizenship." In Monroe County, Alabama, where no black political meeting had
occurred before 1867, freedmen crowded around the speaker shouting,
"God bless you," "Bless God for this." Throughout the South, planters
complained of blacks neglecting their labor. Once a week during the
summer of 1867, "the negroes from the entire county" quit work and
flocked to Waco, Texas, for political rallies. In Alabama, "they stop at any
time and go off to Greensboro" for the same purpose. On August 1,
Richmond's tobacco factories were forced to close because so many black
laborers attended the Republican state convention.[3]

So great was the enthusiasm that, as one ex-slave minister later wrote,
"Politics got in our midst and our revival or religious work for a while
began to wane." The offices of the black-controlled St. Landry (Louisiana) *Progress*, where several hundred freedmen gathered each Sunday to
hear the weekly issue read aloud, temporarily displaced the church as a
community meeting place. More typically, the church, and indeed every
other black institution, became politicized. Every AME preacher in
Georgia was said to be actively engaged in Republican organizing, and
political materials were read aloud at "churches, societies, leagues, clubs,

2. Michael B. Chesson, *Richmond After the War 1865–1890* (Richmond, 1981), 102; Ra-
chleff, *Black Labor*, 42; William C. Hine, "The 1867 Charleston Streetcar Sit-Ins: A Case of
Successful Black Protest," *SCHM*, 77 (April 1976), 110–14; Roger A. Fischer, "A Pioneer
Protest: The New Orleans Street Car Controversy of 1867," *JNH*, 53 (July 1968), 219–33;
Mobile *Nationalist*, July 25, 1867.

3. Elias Yulee to C. C. Sibley, April 23, 1867, Y-552 1867, Letters Received, Ser. 631,
Ga. Asst. Comr., RG 105, NA [FSSP A-193]; Robert L. Hall, "Tallahassee's Black Churches,
1865–1885," *Florida Historical Quarterly*, 58 (October 1979), 194; Samuel S. Gardner to O.
D. Kinsman, July 23, 1867, Wager Swayne Papers, ASDAH; Frank B. Conner to Lemuel P.
Conner, May 16, 1867, Lemuel P. Conner Family Papers, LSU; George W. Hagins to Henry
Watson, Jr., September 8, 1867, Henry Watson, Jr., Papers, DU; Edward Magdol, *A Right
to the Land: Essays on the Freedmen's Community* (Westport, Conn., 1977), 42.

balls, picnics, and all other gatherings." One plantation manager summed up the situation: "You never saw a people more excited on the subject of politics than are the negroes of the south. They are perfectly wild."[4]

The meteoric rise of the Union League reflected and channeled this political mobilization. Having originated as a middle-class patriotic club in the Civil War North, the league now emerged as the political voice of impoverished freedmen. Even before 1867, local Union Leagues had sprung up among blacks in some parts of the South, and the order had spread rapidly during and after the war among Unionist whites in the Southern hill country. Now, as freedmen poured into the league, "the negro question" disrupted some upcountry branches, leading many white members to withdraw altogether or retreat into segregated branches. Many local leagues, however, achieved a remarkable degree of interracial harmony. In North Carolina, one racially mixed league composed of freedmen, white Unionists, and Confederate Army deserters, met "in old fields, or in some out of the way house, and elect candidates to be received into their body."[5]

By the end of 1867, it seemed, virtually every black voter in the South had enrolled in the Union League or some equivalent local political organization. Although the league's national leadership urged that meetings be held in "a commodious and pleasant room," this often proved impossible; branches convened in black churches, schools, and homes, and also, when necessary, in woods or fields. Usually, a Bible, a copy of the Declaration of Independence, and an anvil or some other emblem of labor lay on a table, a minister opened the meeting with a prayer, new members took an initiation oath, and pledges followed to uphold the Republican party and the principle of equal rights, and "to stick to one another." Armed black sentinels—"a thing unheard of in South Carolina history," according to one alarmed white—guarded many meetings. Indeed, informal self-defense organizations sprang up around the leagues,

4. Houston H. Holloway Autobiography, Miscellaneous Manuscript Collections, LC; Geraldine McTigue, "Forms of Racial Interaction in Louisiana, 1860–1880" (unpub. diss., Yale University, 1975), 175–78, 271; Henry M. Turner to Thomas L. Tullock, July 8, 23, 1867 (copies), Robert B. Schenck Papers, Hayes Memorial Library; John H. Parrish to Henry Watson, Jr., August 6, 1867, Watson Papers.

5. Henry W. Bellows, *Historical Sketch of the Union League Club of New York* (New York, 1879); Edmund L. Drago, *Black Politicians and Reconstruction in Georgia* (Baton Rouge, 1982), 76–77; Rutherford *Star*, September 17, 1870; J. M. Hare and A. H. Merrill to Robert M. Patton, May 29, 1867, Alabama Governor's Papers, ASDAH; Henry W. Warren, *Reminiscences of a Mississippi Carpetbagger* (Holden, Mass., 1914), 44; Susie L. Owens, "The Union League of America: Political Attitudes in Tennessee, the Carolinas, and Virginia, 1865–1870" (unpub. diss., New York University, 1943), 89. Earlier scholars portrayed the League as little more than a means by which white Republicans manipulated gullible freedmen attracted to its meetings by secret passwords and colorful initiation rites. Roberta F. Cason, "The Loyal League in Georgia," *GaHQ*, 20 (June 1936), 125–53; Walter L. Fleming, *Civil War and Reconstruction in Alabama* (New York, 1905), 553–59.

and reports of blacks drilling with weapons, sometimes under men with self-appointed "military titles," aroused considerable white apprehension.[6]

The leagues' main function, however, was political education. "We just went there," explained an illiterate North Carolina black member, "and we talked a little; made speeches on one question and another." Republican newspapers were read aloud, issues of the day debated, candidates nominated for office, and banners with slogans like "Colored Troops Fought Nobly" prepared for rallies, parades, and barbecues. One racially mixed North Carolina league on various occasions discussed the organization of a July 4 celebration, cooperation with the Heroes of America (itself experiencing a revival among wartime Unionists in 1867), and questions like disenfranchisement, debtor relief, and public education likely to arise at the state's constitutional convention. A York County, South Carolina, league "frequently read and discussed" the Black Code, a reminder of injustices in the days of Presidential Reconstruction.[7]

The detailed minute book of the Union League of Maryville, Tennessee, a mountain community with a long-standing antislavery tradition, offers a rare glimpse of the league's inner workings. It records frequent discussions of such issues as the national debt and the impeachment of President Johnson, as well as broader questions: "Is the education of the Female as important as that of the male?" "Should students pay corporation tax?" "Should East Tennessee be a separate state?" Although composed largely of white loyalists—mainly small farmers, agricultural laborers, and town businessmen, many of them Union Army veterans—and located in a county only one-tenth black, the Maryville league chose a number of black officers, called upon Tennessee to send at least one black to Congress, and in 1868 nominated a black justice of the peace and four black city commissioners, all of whom won election.[8]

The local leagues' multifaceted activities, however, far transcended electoral politics. Often growing out of the institutions blacks had created

6. J. M. Edmunds to Thomas W. Conway, April 9, 1867, Loring Moody Papers, Boston Public Library; 42d Congress, 2d Session, House Report 22, South Carolina, 320–21, 805–806, 949–52 (hereafter cited as KKK Hearings); Vernon L. Wharton, *The Negro in Mississippi 1865–1890* (Chapel Hill, 1947), 165–66; C. Smith to James L. Orr, April 25, 1867, E. S. Canby to Orr, November 25, 1867, W. C. Bennett to Robert K. Scott, July 26, 1868, Milledge L. Bonham to Scott, August 19, 1868, South Carolina Governor's Papers; E. A. Brady to Wager Swayne, April 7, 1867, Swayne Papers.
7. *Trial of William Woods Holden* (Raleigh, 1871), 3:1198–99; Benjamin G. Humphreys to Lucius Q. C. Lamar, January 3, 1875, Lamar-Mayes Papers, MDAH; Journal of Hamburg Lodge, Union League of America, John M. Brown Papers, UNC; S. B. Hall, *A Shell in the Radical Camp* (Charleston, 1873), 22–23.
8. Maryville Union League Minute Book, McClung Collection, LML; Ralph W. Lloyd, *Maryville College: A History of 150 Years:1819–1969* (Maryville, Tenn., 1969), 11–14, 202–206. Inez E. Burns, *History of Blount County, Tennessee* (Maryville, Tenn., 1957), contains information concerning many League members, which I have supplemented with Knoxville *Whig*, January 30, 1867, and the manuscript United States census for 1870.

in 1865 and 1866, they promoted the building of schools and churches and collected funds "to see to the sick." League members drafted petitions protesting the exclusion of blacks from local juries and demanding the arrest of white criminals. In one instance, in Bullock County, Alabama, they organized their own "negro government" with a code of laws, sheriff, and courts. (The army imprisoned its leader, former slave George Shorter.)[9]

The league also served the freedmen's economic interests, not least by making members, as many whites complained, "impudent" and assertive in dealing with their employers. "I have all the rights that you or any other man has, and I shall not suffer them abridged," one laborer told future Alabama Gov. George S. Houston after being evicted for attending a league meeting. In the Alabama black belt, league organizer George W. Cox found himself besieged by freedmen requesting information about suing their employers, avoiding fines for attending political meetings, and ensuring a fair division of crops at harvest time. Here and in South Carolina, local leagues engaged in strikes for higher wages and encouraged members not to contract for less than half the crop, while some Texas leagues demanded back wages for blacks held in slavery after the Emancipation Proclamation. One North Carolina league official—a minister describing himself as "a poor Colord man"—proposed that the organization "stand as gardians" for freedmen who "don't know how to make a bargain . . . and see that they get the money." Whites, too, used the Union League for economic purposes. The Maryville league's leaders warned that "public sentiment" would retaliate against landowners who proposed "new contracts with their renters more onerous than any custom of the country has ever been."[10]

This hothouse atmosphere of political mobilization made possible a vast expansion of the black political leadership (mostly, it will be recalled, freeborn urban mulattoes) that had emerged between 1864 and 1867. Some, like the Charleston free blacks who fanned out into the black belt spreading Republican doctrine and organizing Union Leagues, did have years of political activism behind them. Others were among the more

9. KKK Hearings, Mississippi, 569; Georgia, 661–62; *New York Times*, July 1, 1867; New York *Tribune*, December 5, 23, 1867; C. H. Smith to E.O.C. Ord, September 16, 1867, Letters Received, Office of Civil Affairs, Box 1, A232, Fourth Military District, RG 393, NA.

10. W. D. Wood, *Reminiscences of Reconstruction in Texas* (n.p., 1902), 14; Burnet Houston to George S. Houston, August 3, 1867, George S. Houston Papers, DU; Michael W. Fitzgerald, "The Union League Movement in Alabama and Mississippi: Politics and Agricultural Change in the Deep South during Reconstruction" (unpub. diss., University of California at Los Angeles, 1986), Chapters 4–5; J.C.A. Stagg, "The Problem of Klan Violence: The South Carolina Up-Country, 1868–1871," *Journal of American Studies*, 8 (December 1974), 310; John P. Carrier, "A Political History of Texas During the Reconstruction, 1865–1874" (unpub. diss., Vanderbilt University, 1971), 215–16; Samuel Lewis to William W. Holden, January 4, 1869, North Carolina Governor's Papers, NCDAH; Knoxville *Whig*, May 27, 1868.

than eighty "colored itinerant lecturers" financed by the Republican Congressional Committee—men like William U. Saunders, a Baltimore barber and Union Army veteran, James Lynch, who left the editorship of the *Christian Recorder* to organize Republican meetings in Mississippi, and even James H. Jones, former "body servant" of Jefferson Davis. Of the black speakers who crisscrossed the South in 1867 and 1868, Lynch was widely regarded as the greatest orator. "Fluent and graceful, he stirred the audience as no other man did or could do," and his eloquence held gatherings of 3,000 freedmen or more spellbound for hours at a time.[11]

Not a few of the blacks who plunged into politics in 1867 had been born or raised in the North. Even in South Carolina, with its well-established native leadership, Northern blacks assumed a conspicuous role. One white participant in the state's first Republican convention, "astonished" by "the amount of intelligence and ability shown by the colored men," singled out Ohio-born William N. Viney, a young veteran (he was twenty-five in 1867) who had purchased land in the low country and, after the passage of the Reconstruction Act, organized political meetings throughout the region at his own expense. Many Northern blacks, like Viney, had come south with the army; others had served with the Freedmen's Bureau, or as teachers and ministers employed by black churches and Northern missionary societies. Still others were black veterans of the Northern antislavery crusade, fugitive slaves returning home, or the children of well-to-do Southern free blacks who had been sent north for the education (often at Oberlin College) and economic opportunities denied them at home. Reconstruction was one of the few times in American history that the South offered black men of talent and ambition not only the prospect of serving their race, but greater possibilities for personal advancement than existed in the North. And as long as it survived, the southward migration continued. As late as 1875, twenty-two year old D. B. Colton came to South Carolina from Ohio and promptly won a position as election manager. As a consequence, Northern black communities were drained of men of political ambition and of lawyers and other professionals. Having known discrimination in the North—Jonathan C. Gibbs had been "refused admittance to eighteen colleges" before finding a place at Dartmouth—black migrants carried with them a determination that Reconstruction must sweep away racial distinctions in every aspect of American life.[12]

11. Thomas Holt, *Black over White: Negro Political Leadership in South Carolina During Reconstruction* (Urbana, Ill., 1977), 28; Thomas Haughey to Loring Moody, March 22, 1867, Moody Papers; "Names of Speakers and Organizers employed or aided by the Republican Congressional Committee," undated manuscript [September 1867], Schenck Papers (a list of 118 speakers, eighty-three of them black); Joe M. Richardson, *The Negro in the Reconstruction of Florida, 1865–1877* (Tallahassee, 1965), 141–44; W. H. Hardy, "Recollections of Reconstruction in East and Southeast Mississippi," *PMHS*, 4 (1901), 126.

12. Richard H. Abbott, ed., "A Yankee Views the Organization of the Republican Party in South Carolina, July 1867," *SCHM*, 85 (July 1984), 247; Joel Williamson, *New People: Miscegenation and Mulattoes in the United States* (New York, 1980), 432; Russell Duncan,

Even more remarkable than the prominence of Northern blacks was the rapid emergence of indigenous leadership in the black belt. Here, where few free blacks had lived before the war, and political mobilization had proceeded extremely unevenly before 1867, local leaders tended to be ex-slaves of modest circumstances who had never before "had the privilege" of expressing political opinions "in public." Many were teachers, preachers, or individuals who possessed other skills of use to the community. Former slave Thomas Allen, a Union League organizer who would soon win election to the Georgia legislature, was a propertyless Baptist preacher, shoemaker, and farmer. But what established him as a leader was literacy: "In my county the colored people came to me for instructions, and I gave them the best instructions I could. I took the New York Tribune and other papers, and in that way I found out a great deal, and I told them whatever I thought was right." In occupation, the largest number of local activists appear to have been artisans. Comprising 5 percent or less of the rural black population, artisans were men whose skill and independence set them apart from ordinary laborers, but who remained deeply embedded in the life of the freedmen's community. Many had already established their prominence as slaves, like Emanuel Fortune, whose son, editor T. Thomas Fortune, later recalled: "It was natural for [him] to take the leadership in any independent movement of the Negroes. During and before the Civil War he had commanded his time as a tanner and expert shoe and bootmaker. In such life as the slaves were allowed and in church work, he took the leader's part." The Union League catapulted others into positions of importance. James T. Alston, an Alabama shoemaker and musician and the former slave of Confederate Gen. Cullen A. Battle, had "a stronger influence over the minds of the colored men in Macon county" than any other individual, a standing he attributed to the commission he received in 1867 to organize a local Union League.[13]

And there were other men, respected for personal qualities—good sense, oratorical ability, having served in the army, or, like South Carolina Republican organizer Alfred Wright, being "an active person in my principles." Calvin Rogers, a Florida black constable, was described by another freedman as "a thorough-going man; he was a stump speaker, and tried to excite the colored people to do the right thing. . . . He would work

Freedom's Shore: Tunis Campbell and the Georgia Freedmen (Athens, Ga., 1986), 13–15; 44th Congress, 2d Session, Senate Miscellaneous Document 48, 2:535–36; Carter G. Woodson, *A Century of Negro Migration* (New York, 1918), 123–25; Jonathan Gibbs to O. O. Howard, January 14, 1873, O. O. Howard Papers, Bowdoin College.

13. Charleston *Advocate*, April 20, 1867; KKK Hearings, Georgia, 607–14; Michael P. Johnson, "Work, Culture, and the Slave Community: Slave Occupations in the Cotton Belt in 1860," *Labor History*, 27 (Summer 1986), 331; Peter Kolchin, *First Freedom: The Responses of Alabama's Blacks to Emancipation and Reconstruction* (Westport, Conn., 1972), 163–66; 180; Magdol, *A Right to the Land*, 113–25, 136; Dorothy Sterling, ed., *The Trouble They Seen* (Garden City, N.Y., 1976), 111; KKK Hearings, Alabama, 1016–29.

for a man and make him pay him." Such attributes seemed more important in 1867 than education or political experience. "You can teach me
the law," wrote one black Texan, "but you cannot [teach] me what justice
is." Nor, in a region that erected nearly insuperable barriers against black
achievement, did high social status appear necessary for political distinction. "All colored people of this country understand," a black writer later
noted, "that what a man does, is no indication of what he is."[14]

In Union Leagues, Republican gatherings, and impromptu local meetings, ordinary blacks in 1867 and 1868 staked their claim to equal citizenship in the American republic. Like Northern blacks schooled in the Great
Tradition of protest, and the urban freemen who had dominated the state
conventions of 1865 and 1866, former slaves identified themselves with
the heritage of the Declaration of Independence, and insisted America
live up to its professed ideals. In insistent language far removed from the
conciliatory tones of 1865, an Alabama convention affirmed its understanding of equal citizenship:

> We claim exactly *the same rights, privileges and immunities as are enjoyed by white
> men*—we ask nothing more and will be content with nothing less. . . . The
> law no longer knows white nor black, but simply men, and consequently we
> are entitled to ride in public conveyances, hold office, sit on juries and do
> everything else which we have in the past been prevented from doing solely
> on the ground of color.

At their most utopian, blacks in Reconstruction envisioned a society
purged of all racial distinctions. This does not mean they lacked a sense
of racial identity, for they remained proud of the accomplishments of
black soldiers, and preferred black teachers for their children and black
churches in which to worship. But in the polity, those who had so long
been proscribed because of color defined equality as color-blind. "I
heard a white man say," black teacher Robert G. Fitzgerald recorded
in his diary, "today is the black man's day; tomorrow will be the white
man's. I thought, poor man, those days of distinction between colors is
about over in this (now) free country." Indeed, black politicians sometimes found their listeners unreceptive to the rhetoric of racial self-
consciousness. Martin R. Delany, the "father of black nationalism," and
a South Carolina Republican organizer, found it "dangerous to go into the
country and speak of color in any manner whatever, without the angry
rejoinder, 'we don't want to hear that; we are all one color now'."[15]

Nor did blacks during Radical Reconstruction evince much interest in

14. KKK Hearings, South Carolina, 1173; Florida, 113–14; J. M. Donalson to Charles
Griffith, July 16, 1867, Letters Received, Office of Civil Affairs, Box 1, D22, Fifth Military
District, RG 393, NA; *Christian Recorder*, November 23, 1876.
15. Montgomery *Alabama State Sentinel*, May 21, 1867; Robert G. Fitzgerald Diary, April
22, 1868, Robert G. Fitzgerald Papers, Schomburg Center for Research in Black Culture;
New National Era, August 31, 1871.

emigration. Over 1,000, "tired of the unprovoked scorn and prejudice we daily and hourly suffer," had sailed from Georgia and South Carolina for Liberia in 1866 and early 1867 under American Colonization Society auspices. But the optimism kindled by the advent of black suffrage brought the movement to an abrupt halt. "You could not get one of them to think of going to Liberia now," wrote a white advocate of colonization. Blacks probably considered themselves more fully American than at any time in the nineteenth century; some even echoed the exuberant nationalism of what one speaker called "our civilization." Throughout Reconstruction, blacks took pride in parading on July 4, "the day," a Charleston diarist observed, "the Niggers now celebrate, and the whites stay home and work." As late as 1876, a speaker at a black convention aroused "positive signs of disapproval" by mentioning emigration. "Damn Africa," declared one delegate. "If Smith wants to go let him; we'll stay in America."[16]

As was true throughout Reconstruction, politics in 1867 was intimately tied to the freedmen's economic aspirations. Many Northern blacks, it is true, carried south the ideology of free labor, with its respect for private property and individual initiative. Leading "black carpetbaggers" believed the interests of capital and labor identical and the freedmen entitled to little more than an "honest chance in the race of life." Such views were echoed by the Southern free black elite, many of whom opposed talk of confiscation and insisted that political equality did not imply the end of class distinctions: "We do not ask that the ignorant and degraded shall be put on a social equality with the refined and intelligent." 1867, however, was an inopportune moment to preach individual self-help to black belt freedmen, for successive crop failures had left those on share contracts with little or no income and caused a precipitous decline in cash wages. "We have tried [plantation labor] three years," wrote an Alabama black, "and are worse off than when we started. . . . We cannot accumulate enough to get a home." Drawing on widespread dissatisfaction with a contract system that appeared to consign them permanently to poverty and dependence, rural blacks raised, once again, the demand for land.[17]

16. Francis B. Simkins, "The Problems of South Carolina Agriculture After the Civil War," *NCHR*, 7 (January 1930), 54–55; Wyatt Moore to William Coppinger, July 5, 1866, E. M. Pendleton to Coppinger, March 18, 1867, American Colonization Society Papers, LC; P. Sterling Stuckey, "The Spell of Africa: The Development of Black Nationalist Theory, 1829–1945" (unpub. diss., Northwestern University, 1973), 91–92; *Proceedings of the Southern States Convention of Colored Men, Held in Columbia, South Carolina* (Columbia, S.C., 1871), 99–100; Jacob Schirmer Diary, July 4, 1867, July 4, 1872, SCHS; Cincinnati *Commercial*, April 10, 1876.

17. *Christian Recorder*, January 7, August 5, September 9, 16, 1865; *CR*, 43d Congress, 2d Session, 982; Mobile *Nationalist*, April 25, 1867; 40th Congress, 3d Session, House Executive Document 1, 1040; Mobile *Nationalist*, October 24, 1867.

The land issue animated grass-roots black politics in 1867. The Reconstruction Act rekindled the belief that the federal government intended to provide freedmen with homesteads and, as in 1865, December saw many refusing to sign labor contracts for the coming year. Talk of confiscation, reported an AMA official from North Carolina, "has much greater prominence . . . here than our friends at the North seem to be aware. The expectations of the poor have been very generally excited." A mass meeting in Savannah, attended by armed low country freedmen, heard Aaron A. Bradley call for the division among black families of lands belonging to "rich whites." Having grown to adulthood as a slave before escaping to Boston, Bradley emerged as one of the few black leaders from the North to become actively involved in the freedmen's land struggles. The city's best-known black politicians repudiated his program, but Bradley's ideas won support among freedmen in the rice kingdom, and some blacks working white-owned land on the city's outskirts began refusing to pay rent. In Alabama, freedmen delivered "inflammatory" speeches asserting that "all the wealth of the white man had been made by negro labor, and that the negroes were entitled to their fair share of all these accumulations." "Didn't you clear the white folks' land?" asked one orator. "Yes," voices answered from the crowd, "and we have a right to it!" There seemed "a great deal more danger," wrote former South Carolina Gov. Benjamin F. Perry, "of 'Cuffee' than Thad Stevens taking over lands."[18]

By mid-1867, William H. Trescot observed, blacks had become convinced that membership in the Union League "will in some way, they do not exactly know how, secure them the possession of the land." Yet this was only one among the many goals blacks sought to achieve through Reconstruction politics. In a society marked by vast economic disparities and a growing racial separation in social and religious life, politics became the only arena where black and white encountered each other on a basis of equality. And although elective office and the vote remained male preserves, black women shared in the political mobilization. They took part in rallies and parades, and, to the consternation of some male participants, voted on resolutions at mass meetings. During the 1868 campaign, Yazoo County, Mississippi, whites found their homes invaded

18. Michael Wayne, *The Reshaping of Plantation Society: The Natchez District, 1860–1880* (Baton Rouge, 1983), 122; Fisk Brewer to E. P. Smith, May 27, 1867, AMA Archives, Amistad Research Center, Tulane University; Joseph P. Reidy, "Aaron A. Bradley: Voice of Black Labor in the Georgia Lowcountry," in Howard N. Rabinowitz, ed., *Southern Black Leaders of the Reconstruction Era* (Urbana, Ill., 1982), 281–308; Manuel Gottlieb, "The Land Question in Georgia During Reconstruction," *Science and Society*, 3 (Summer 1939), 373–77; *Savannah Daily News and Herald*, April 8, 1868; KKK Hearings, Alabama, 976; James S. Allen, *Reconstruction: The Battle for Democracy* (New York, 1937), 124; Benjamin F. Perry to F. Marion Nye, May 25, 1867, Letterbook, Benjamin F. Perry Papers, UNC.

by buttons depicting General Grant, defiantly worn on the clothing of black maids and cooks.[19]

Throughout Reconstruction, blacks remained "irrepressible democrats." "Negroes all crazy on politics again," noted a Mississippi plantation manager in the fall of 1873. "Every tenth negro a candidate for some office." And the Republican party—the party of emancipation and black voting rights—became an institution as central to the black community as the church and school. The few black Democrats (mostly individuals dependent for a livelihood on white patronage) were considered "enemies to our people." When not deterred by violence, blacks eagerly attended political gatherings, and voted in extraordinary numbers; their turnout in many elections approached 90 percent. "It is the hardest thing in the world to keep a negro away from the polls," commented an Alabama white, "that is the one thing he will do, to vote." Long after they had been stripped of the franchise, blacks would recall the act of voting as a defiance of inherited norms of white superiority, and regard "the loss of suffrage as being the loss of freedom."[20]

Early in 1868, a Northerner reporting on Alabama's election day captured the sense of possibility with which Radical Reconstruction began. "In defiance of fatigue, hardship, hunger, and threats of employers," blacks had come en masse to the polls. Not one in fifty wore an "unpatched garment," few possessed a pair of shoes, yet for hours they stood on line in a "pitiless storm." Why? "The hunger to have the same chances as the white men they feel and comprehend. . . . That is what brings them here."[21] Rarely has a community invested so many hopes in politics as did blacks during Radical Reconstruction.

The Republican Coalition

As political mobilization swept the black belt, the South's traditional leaders appeared stunned. "All society stands now like a cone on its Apex," wrote prewar Congressman Francis W. Pickens. "The least jostle is likely to make a total overthrow of the whole structure." Many feared even worse events lay on the horizon. "I hope to plant 250 acres more next year," Edward B. Heyward wrote in July, "if the Blacks don't take

19. "Letter of William Henry Trescot on Reconstruction in South Carolina, 1867," *AHR*, 15 (April 1910), 575–76; Richmond *Dispatch*, August 2, 1867; A. T. Morgan, *Yazoo: or, On the Picket Line of Freedom in the South* (Washington, 1884), 231–33, 293.

20. [Belton O'Neall Townsend], "The Political Condition of South Carolina," *Atlantic Monthly*, 39 (February 1877), 192; A. D. Grambling to Stephen Duncan, August 10, 1873, Stephen Duncan Papers, Natchez Trace Collection, UTx; 44th Congress, 2d Session, Senate Miscellaneous Document 48, 2:227; KKK Hearings, Mississippi, 725; Melinda M. Hennessey, "Reconstruction Politics and the Military: The Eufaula Riot of 1874," *AlHQ*, 38 (Summer 1976), 116; Paul D. Escott, *Slavery Remembered* (Chapel Hill, 1979), 153–54; 41st Congress, 2d Session, House Miscellaneous Document 154, 181–82.

21. Cincinnati *Commercial* in *American Freedman*, February 1868, 373.

possession and eject the proprietors, which is quite possible." Many whites seemed paralyzed by indecision. "Shall we keep aloof from the negro," wondered a North Carolina editor, "or enter the field for his support?" Ironically, many of those who advocated the latter course, dubbing themselves "cooperationists," were secessionists who had dropped out of the political limelight in 1865. By bowing to the inevitable, they argued, the white South could secure itself against confiscation, while waiting for men of "influence, character [and] substantial wealth" to resume their accustomed political hegemony.[22]

Initially, cooperationists professed optimism about winning blacks' political support. One Tennessee politico boasted he would "carry the Negro vote with a banjo and a jug of whiskey." Others adopted a more sophisticated approach, endeavoring to convince the freedmen that Southern whites were their "best friends." "We have been generous masters," declared an Alabama speaker, "we have grown up together; some of you have our blood in your veins." Some flattered free blacks with talk of their superiority to rural freedmen, urging them to "separate themselves from the black people." In Alexandria, Louisiana, whites seeking black votes for the first time integrated the town's ballroom. But to the cooperationists' surprise, their appeals fell on deaf ears. The fact is that cooperationist speeches were not well calculated to win black support. "There was a general cry that they were our best friends," observed Savannah black leader James Simms, "but in the next paragraph below you will see us abused as badly as any people can be." Gen. Wade Hampton's attempt to "direct the negro vote" provides a case in point. More concerned about black control of local affairs than their role in national politics, Hampton declared his willingness to "send negroes to Congress" if "they will let us have the state." Yet while addressing blacks as "southern men" who shared with local whites a community of interest against "strangers," Hampton made clear that he abhorred the Reconstruction Act and did not really accept the principle of manhood suffrage.[23]

Cooperationists actively promoted the idea, widely accepted in the

22. J. H. Easterby, ed., *The South Carolina Rice Plantation as Revealed in the Papers of Robert F. W. Allston* (Chicago, 1945), 236; Edward B. Heyward to Katherine M. C. Heyward, July 23, 1867, Heyward Family Papers, USC; J. A. Engelhard to Manton Marble, May 10, 1867, Manton Marble Papers, LC; Michael Perman, *Reunion Without Compromise: The South and Reconstruction 1865–1868* (New York, 1973), 274–90; Herschel V. Johnson to Alexander H. Stephens, March 29, 1867, Alexander H. Stephens Papers, LC; William H. Trescot to Daniel E. Sickels, August 8, 1867, Miscellaneous Manuscripts, New-York Historical Society.

23. Manuscript newspaper article by "Alpha," May 27, 1867, Leonidas Houk Papers, McClung Collection, LML; Thomas M. Peters to Wager Swayne, May 29, 1867, Swayne Papers; Mobile *Times*, July 7, 1867; Mobile *Nationalist*, July 11, 18, 1867; William E. Highsmith, "Some Aspects of Reconstruction in the Heart of Louisiana," *JSH*, 13 (November 1947), 484; 40th Congress, 3d Session, House Miscellaneous Document 52, 10; Wade Hampton to James Conner, March 24, April 9, 1867, Hampton to unknown, March 31, 1867, Hampton Family Papers, USC; Charleston *Daily Courier*, March 23, 1867.

contemporary North and by many subsequent historians, that wealthier Southerners were more disposed to treat blacks equitably than lower-class whites. "Tolerance toward the negro broadens with the planter's acres," commented a Connecticut man after a Southern tour. Certain Northern blacks, privileged slaves, and freeborn artisans (who knew that unions of white craftsmen had long sought "to crush out the colored mechanics" of the South) also saw things this way. Few ordinary freedmen seem to have agreed. "Our old masters have become our enemies," a black delegation told Andrew Johnson shortly after the war ended, a conviction Presidential Reconstruction and day-to-day conflict on the plantations only reinforced. When Gen. John Steedman in 1866 asked a Wilmington, North Carolina, mass meeting which class of whites "display the worst feeling," one black spokesman replied, "The ruling class."[24]

Most of all, blacks identified the old political leaders with slavery. After Georgia's antebellum governor, Herschel V. Johnson, addressed a freedmen's meeting at Augusta, a black preacher replied, to cheers from the crowd, "that old ship, the institution of slavery is dead, and I am glad of it. Shall I employ its captain or its manager to bear me through the ocean again?" Indeed, as freedmen stepped onto the stage of politics in 1867, recollections and images of slavery emerged as a staple of black political rhetoric. James Lynch (who had been born free) "paint[ed] the horrors of slavery (as they existed in his imagination) in pathetic tones of sympathy until . . . every negro in the audience would be weeping." Just such a speech by a Northern Republican convinced William H. Trescot of the futility of efforts to win over the black vote:

> He stepped forward to the edge of the stage and looking around exclaimed: . . . 'Am I dreaming—is this Charleston where I came ten years ago to see human beings sold at auction? Are you all here for sale. . . . I will put you all at auction. . . . Going! Going!' and then he suddenly stopped and looking up to the sky, pointed upward and went on. 'Look yonder—do you see who is bidding for you—the soul of Abraham Lincoln. . . . Lincoln takes the whole lot—gone!' You could feel the effect of this speech all through the crowd. The old woman in front of me shouted: 'Yes, you blessed old man, Hallelujah.' . . . What do you suppose we can say in reply to that?[25]

Many prominent Southern whites attacked the cooperationist program from the beginning. Having prided themselves before 1867 on political moderation, former South Carolina Governor Perry, Georgia planter and lawyer Benjamin H. Hill, and North Carolina Whig leader William A.

---

24. Stephen Powers, *Afoot and Alone: A Walk from Sea to Sea* (Hartford, Conn., 1872), 28; Whitelaw Reid, *After the War: A Southern Tour* (Cincinnati, 1866), 51; E. Merton Coulter, *The South During Reconstruction 1865–1877* (Baton Rouge 1947), 164; *Christian Recorder*, December 2, 1865, June 9, 1866; New York *Tribune*, June 17, 1865; 39th Congress, 1st Session, House Executive Document 120, 42.

25. Drago, *Black Politicians*, 33; Hardy, "Recollections," 126; William H. Trescot to unknown (copy), April 7, 1867, William H. Trescot Papers, USC.

Graham now castigated black suffrage in the most extreme language. Reconstruction, declared Perry, would throw control of the South into the hands of "ignorant, stupid, demi-savage paupers." Graham believed enfranchising blacks would "roll back the tide of civilization two centuries at least." As it became evident that efforts to control the black vote were futile, men like Hampton pulled back, urging white voters to oppose the calling of constitutional conventions. The Southern establishment's growing unity in opposition to Reconstruction further solidified black support for the Republican party. "I do not in my heart wonder that the negroes vote the radical ticket. . . ." observed Southern diarist Ella Clanton Thomas. "Think of it, the right to vote . . . is within their *very grasp*. . . . I should think twice before I voted to have it taken away from me."[26]

Throughout Reconstruction, blacks constituted a large majority of Southern Republicanism. But, one Northerner had observed in 1866, "a party sustained only by black votes will not grow old." Of the eleven states of the old Confederacy, only South Carolina, Mississippi, and Louisiana contained a black majority; blacks constituted roughly one quarter of the population of Texas, Tennessee, and Arkansas, 40 percent in Virginia and North Carolina, and a bit less than half in Alabama, Florida, and Georgia. Even in states with black majorities, however, Southern Republicanism had to attract white support. And "carpetbaggers" (migrants from the North) and "scalawags" (native Southerners who cast their lot politically with the freedmen) found themselves subjected to a torrent of abuse by their Democratic opponents, an odium that persisted in the morality play of traditional Reconstruction historiography.[27]

Political, regional, and class prejudices combined to produce the image of the carpetbagger as a member of "the lowest class" of the Northern population. Able to pack "all his earthly belongings" in his carpetbag, he supposedly journeyed south after the passage of the Reconstruction Act "to fatten on our misfortunes," in the process poisoning the allegedly harmonious race relations of 1865–67. In fact, far from the dregs of Northern society, carpetbaggers tended to be well educated and middle class in origin. Not a few had been lawyers, businessmen, newspaper editors, and other pillars of Northern communities. The majority (includ-

26. Lillian A. Kibler, *Benjamin F. Perry: South Carolina Unionist* (Durham, 1946), 449–60; Columbia *Daily Phoenix*, May 30, 1867; Benjamin H. Hill, Jr., *Senator Benjamin H. Hill of Georgia: His Life, Speeches and Writings* (Atlanta, 1893), 730–911; J. G. de Roulhac Hamilton and Max R. Williams, eds., *The Papers of William Alexander Graham* (Raleigh, 1957–), 7:358; Charleston *Daily Courier*, August 29, 1867; Ella Gertrude Thomas Journal, November 2, 1868, DU.

27. A. B. Butler to John Sherman, February 27, 1866, John Sherman Papers, LC; Samuel D. Smith, *The Negro in Congress 1870–1901* (Chapel Hill, 1940), 2. The disenfranchisement of Confederates who had held office before the war, and white apathy or revulsion against Reconstruction, left blacks in 1867 with a temporary voting majority in Alabama and Florida as well.

ing fifty-two of the sixty who served in Congress during Reconstruction) were veterans of the Union Army, and their ranks also included teachers, Freedmen's Bureau agents, and men who had invested tens of thousands of dollars in cotton plantations. Nearly all had come South before 1867, when blacks lacked the franchise and the prospect of office appeared remote. Illinois-born Henry C. Warmoth, Louisiana's first Republican governor, plunged into politics almost from the moment he arrived in Louisiana with the army in 1864, but most carpetbaggers did not move to the South seeking political position.[28]

Talented and ambitious men in their twenties and early thirties (Ohio-born Emerson Bentley, a newspaper editor and schoolteacher in St. Landry Parish, Louisiana, was all of eighteen in 1867), carpetbaggers viewed the South as so many nineteenth-century Americans did the West—as a field for personal advancement. Some had already exhibited a restlessness typical of the period. John Wesley North, who came to Tennessee in 1866, had previously been involved in Western railroad building, mining, and land speculation. John B. Callis of Wisconsin had prospected for gold in California and traveled in Central America before enlisting in the army. He later worked for the Freedmen's Bureau at Huntsville, Alabama (where local blacks had a watch engraved with a picture of Callis restraining a whip-wielding planter), and went on to serve a term in Congress. Such men saw no need to apologize for taking up residence in the South. A majority of the members of Congress, one pointed out in 1871, represented states other than those of their birth, but only Northern-born Southern Republicans bore the "opprobrious epithet" "carpetbagger."[29]

A variety of motives and experiences propelled these Northerners into Southern Republican politics in 1867. Some were typical nineteenth-century men on the make for whom politics offered the opportunity for a quick profit. New York-born George Spencer came to Alabama with the army and quickly seized the "chances of making a fortune," engaging in a "big speculation" in depreciated railroad stocks and another in contraband cotton. He eventually made his way to the United States Senate. Others entered politics because they had earned the freedmen's goodwill

28. Cleveland *Leader,* June 16, 1877; Hardy, "Recollections," 109; Raleigh *Daily Sentinel,* July 10, 1868; Richard N. Current, "Carpetbaggers Reconsidered," in David H. Pinkney and Theodore Ropp, eds., *A Festschrift for Frederick B. Artz* (Durham, 1964), 139–57; Terry L. Seip, *The South Returns to Congress: Men, Economic Measures, and Intersectional Relationships, 1868–1879* (Baton Rouge, 1983), 58; Ted Tunnell, *Crucible of Reconstruction: War, Radicalism, and Race in Louisiana* (Baton Rouge, 1984), 136–50. The terms "carpetbagger" and "scalawag" have become so unavoidable a part of the lexicon of Reconstruction that I have continued to employ them, without intending to accept their pejorative implications.

29. Emerson Bentley to Henry C. Warmoth, June 12, 1868, Henry C. Warmoth Papers, UNC; Merlin Stonehouse, *John Wesley North and the Reform Frontier* (Minneapolis, 1965); David H. Overy, Jr., *Wisconsin Carpetbaggers in Dixie* (Madison, Wis., 1961), 46; *CG,* 42d Congress, 1st Session, Appendix, 265.

as Bureau agents or, in the case of Louisiana Congressman Joseph P. Newsham, a carpetbagger from Illinois, as the only lawyer in their new communities who would serve black clients. Still others, having lost their life's savings in unsuccessful cotton planting, saw in politics a means of earning a living. Henry Warren, a Massachusetts-born Yale graduate, and Albert T. Morgan, a young army veteran from Ohio, had both failed in the Mississippi cotton fields. Each seized with alacrity a political opportunity that arose in 1867. Military authorities asked Warren to serve as voter registrar, and local blacks invited Morgan to run for the constitutional convention, thus launching careers that took both to the legislature. And some carpetbaggers were idealists for whom Southern Republicanism seemed a fulfillment of abolitionist ideals, or who, like future Mississippi Gov. Adelbert Ames, became convinced that he "had a Mission with a large M" to assist the former slaves.[30]

Most carpetbaggers probably combined the desire for personal gain with a commitment to taking part in an effort "to substitute the civilization of freedom for that of slavery." To such men, Northern economic investment and Northerners' engagement in Republican politics formed two parts of a single effort to reform the "unprogressive" South by "establishing free institutions, free schools, and the system of free labor." Their commitment to far-reaching changes in Southern life created a bond of interest between carpetbaggers and blacks. Carpetbaggers generally supported measures aimed at democratizing and modernizing the South—civil rights legislation, aid to economic development, the establishment of public school systems. Many sympathized with blacks' quest for land, although few supported confiscation. Lands, Alabama carpetbagger C. W. Dustan announced, "cannot be owned without being earned, and they cannot be earned without labor." (Dustan did not fully adhere to this maxim in his own life, for he acquired a sizable holding by marrying the daughter of a Demopolis planter.)[31]

Generally representing black belt constituencies, carpetbaggers garnered a major share of Reconstruction offices, especially in states like Florida, South Carolina, and Louisiana, with relatively few native-born white Republicans. Northerners, however, could hardly provide a voting base for Southern Republicanism for in no state did they constitute even

30. Sarah V. Woolfolk, "George E. Spencer: A Carpetbagger in Alabama," *Alabama Review*, 19 (January 1966), 41–52; Seip, *South Returns to Congress*, 62–63; Lawrence N. Powell, "The Politics of Livelihood: Carpetbaggers in the Deep South," in J. Morgan Kousser and James M. McPherson, eds., *Region, Race, and Reconstruction: Essays in Honor of C. Vann Woodward* (New York, 1982), 317–21; Warren, *Reminiscences*, 95–97; Morgan, *Yazoo*, 124–38; Adelbert Ames to James W. Garner, January 17, 1900, James W. Garner Papers, MDAH.
31. William C. Harris, "The Creed of the Carpetbaggers: The Case of Mississippi," *JSH*, 40 (May 1974), 199–224; *CG*, 42d Congress, 1st Session, Appendix, 264; J. P. Rexford to Charles Sumner, August 3, 1872, Charles Sumner Papers, HU; Otto H. Olsen, *Carpetbagger's Crusade: The Life of Albion Winegar Tourgée* (Baltimore, 1965); C. W. Dustan, "Political Speech to Negroes," undated manuscript (ca. 1867), C. W. Dustan Papers, ASDAH.

2 percent of the total population. Far more numerous were Southern-born white Republicans, or "scalawags." To Democrats, a scalawag was "the local leper of the community," even more reprehensible than the hated carpetbaggers. "We can appreciate a man who lived north, and ... even fought against us," declared a former governor of North Carolina, "but a traitor to his own home cannot be trusted or respected." In no Southern state did Republicans attract a majority of the white vote. But given the twin legacies of slavery and defeat, and the opprobrium attached to those who attended "Negro conventions" and took unorthodox political positions, what is remarkable is not how few whites supported the party but, in some states, how many. "It costs nothing for a northern man, to be a Union man," wrote one Georgia Republican, "but the rebuff and persecution received here ... tells a horrible tale." Yet for a time, "scalawags" dominated the governments of Alabama, Georgia, Tennessee, Texas, and North Carolina, and played a major role in Mississippi.[32]

Castigated by their opponents as "white negroes" who had betrayed their region in the quest for office, scalawags had even more diverse backgrounds and motivations than Northern-born Republicans. They included men of prominence and rank outsiders, wartime Unionists and advocates of secession, enterpreneurs advocating a modernized New South and yeomen seeking to preserve semisubsistence agriculture. Their common characteristic was the conviction that they stood a greater chance of advancing their interests in a Republican South than by casting their lot with Reconstruction's opponents.

Some scalawags were men of stature, including a few members of the South's wealthiest families. Charles Hays, "one of the largest planters in Alabama," supported secession and fought in the Confederate Army, but joined the Republicans in 1867 and went on to serve in the legislature and Congress. His social prominence did not shield him from vilification; Hayes, commented one editor, had "sounded a depth of infamy" unknown even to carpetbaggers, and the election of "the worst negro that was ever flogged at the whipping post" would be far preferable. G. B. Burnett, Georgia's largest slaveholder north of Atlanta, joined the Republicans, as did Daniel L. Russell, son of the largest landholder in North Carolina. And although few Confederate leaders, either military or civilian, did likewise, the party did gain the support of Gen. James Longstreet, whose example inspired some Confederate veterans to follow in his footsteps. One was Alabama-born Albert R. Parsons, a descendant of

32. Maurice M. Vance, "Northerners in Late Nineteenth Century Florida: Carpetbaggers or Settlers?" *FlHQ*, 38 (July 1959), 3; Sarah W. Wiggins, *The Scalawag in Alabama Politics, 1865–1881* (University, Ala., 1977), 1; J. G. de Roulhac Hamilton, ed., *The Correspondence of Jonathan Worth* (Raleigh, 1909), 944; Caleb Tompkins to Benjamin F. Butler, July 16, 1867, Benjamin F. Butler Papers, LC.

Mayflower Pilgrims, who in 1867 established a Waco newspaper advocating black rights, and stumped central Texas for the Republicans.[33]

Many scalawags possessed considerable political experience; their ranks included prewar Congressmen, judges, and local officials. Most such men were former Whigs who viewed the Republican party as the "legitimate successor" to Whiggery. James L. Alcorn, owner of the largest plantation in the Yazoo-Mississippi delta, was probably the most prominent Old Whig scalawag. Having demonstrated his political sagacity by advocating limited black suffrage in 1865 and favoring the ratification of the Fourteenth Amendment in 1866, Alcorn now addressed some "blunt speaking" to Mississippi's whites. The black electoral majority was an unavoidable fact of life and only if men like himself took the lead in Reconstruction could "a harnessed revolution" take place in which the freedmen's civil and political rights were guaranteed, but whites retained control of state government. Moreover, only the Republican Congress could provide the capital essential for the state's economic recovery. Alcorn's aim, complained Radical carpetbagger Albert T. Morgan, was to see "the old civilization of the South *modernized,*" rather than an entirely new order constructed. Precisely for this reason, a number of black belt Whig planters followed him into the Republican party, assuming they could "lead and direct the colored vote." So long as blacks remained junior partners in a white-dominated coalition, these Old Whig scalawags remained, but when blacks refused any longer to submit to white dictation, Alcorn and his followers would abandon the Republican party.[34]

Yet another group of scalawags were modernizers who saw in Republicanism a "progressive party" that could bring a long-overdue social and

---

33. Hamilton and Williams, eds., *Graham Papers*, 7:533; William W. Rogers, " 'Politics is Mighty Uncertain': Charles Hays Goes to Congress," *Alabama Review*, 30 (July 1977), 163–90; KKK Hearings, Georgia, 950–54; Jeffrey J. Crow and Robert F. Durden, *Maverick Republican in the Old North State: A Political Biography of Daniel L. Russell* (Baton Rouge, 1977), 1–18; William B. Hesseltine, *Confederate Leaders in the New South* (Baton Rouge, 1950), 23, 103–11; William L. Richter, "James Longstreet: From Rebel to Scalawag," *LaH*, 11 (Summer 1970), 215–30; Paul Avrich, *The Haymarket Tragedy* (Princeton, 1984), 3–10. Longstreet's decision to join the Republican party made him an "object of hatred" among Southern Democrats for the remainder of his life. When he died in 1903, the United Daughters of the Confederacy voted not to send flowers to his funeral, and unlike other Confederate generals, no statues of Longstreet graced the Southern landscape. Thomas L. Connelly and Barbara L. Bellows, *God and General Longstreet: The Lost Cause and the Southern Mind* (Baton Rouge, 1982), 34–37.

34. James A. Baggett, "Origins of Early Texas Republican Party Leadership," *JSH*, 40 (August 1974), 454; J. R. G. Pitkin, *To the Whigs* (n. p., 1876), 2–3; Lillian A. Pereyra, *James Lusk Alcorn: Persistent Whig* (Baton Rouge, 1966); James L. Acorn to Elihu B. Washburne, June 29, 1868, Elihu B. Washburne Papers, LC; *Views of the Hon. J. L. Alcorn on the Political Situation of Mississippi* (Friar's Point, Miss., 1867), 3–5; *Address of J. L. Alcorn to the People of Mississippi* (n.p., 1869), 13; Albert T. Morgan to Charles Sumner, February 4, 1870, Sumner Papers; Warren A. Ellem, "Who Were the Mississippi Scalawags?," *JSH*, 38 (May 1972), 217–40; *CG*, 41st Congress, 2d Session, 2721; David H. Donald, "The Scalawag in Mississippi Reconstruction," *JSH*, 10 (November 1944), 447–60.

economic revolution to the South. Prewar "ideas and feelings," declared Thomas Settle of North Carolina, must be buried "a thousand fathoms deep. . . . Yankees and Yankee notions are just what we want in this country. We want their capital to build factories and workshops. We want their intelligence, their energy and enterprise." Georgia's Joseph E. Brown, a political chameleon whose coloration changed over time from advocacy of secession to obstruction of the war effort as Georgia's Confederate governor, to Reconstruction Republicanism and finally a stint in the U. S. Senate under the Redeemers, sounded the same theme. Brown urged white voters to support the Republican party as the only way to attract the outside capital needed for "the development of our state." Like Alcorn, he displayed little concern for black rights and fully expected whites to control Reconstruction politics: "They have superior numbers, they have the wealth, what is left, the education, and the intelligence." While denounced by most of the state's prewar leadership as a latter-day Benedict Arnold, Brown won the backing of a group of urban business promoters who embraced his vision of a "new era," complete with railroads, mines, machine shops, and cotton mills, all financed from the North. His allies included Georgia's first Republican governor, Rufus Bullock, president of the Macon & Augusta Railroad and a director of Augusta's first national bank, to whom "economic questions and material concerns" far overshadowed other Reconstruction issues.[35]

Republicans also attracted a number of urban and small town artisans and others to whom Reconstruction offered opportunities denied under the plantation regime. South Carolina tailor Simeon Corley, in antebellum days "hated and despised" as a Unionist, temperance advocate, and critic of slavery, served a term as a Republican Congressman. Among the South's foreign-born urban workingmen, radical views imported from Europe led some into the Republican party, including P. J. Coogan, a Charleston Irish nationalist and the only member of the state legislature to support the Fourteenth Amendment in 1866, and Frederic Carie, a "Red Republican" exile from Second Empire France, who served as sheriff of Terrebone Parish, Louisiana. The Germans of southwestern Texas comprised the largest bloc of immigrant Southern Republicans, helping to send to Congress Edward Degener, a San Antonio grocer who had taken part in the revolution of 1848, been imprisoned by Confederate authorities, and seen his two sons executed for treason.[36]

35. Raleigh *Tri-Weekly Standard*, July 25, 1867; Joseph H. Parks, *Joseph E. Brown of Georgia* (Baton Rouge, 1976), 369–410; Joseph E. Brown to William D. Kelley, March 18, 1868, Joseph E. Brown Letterbook, Hargrett Collection, UGa; Atlanta *Daily New Era*, November 17, 1866, February 26, 1867; Rufus B. Bullock, "Reconstruction in Georgia, 1865–70," *The Independent*, 55 (March 1903), 671.

36. Simeon Corley to Charles Lanman, July 28, 1868, USC; *Proceedings of the Constitutional Convention of South Carolina* (Charleston, 1868), 1:200–201; Simeon Corley, *To the Voters of the Third Congressional District of South Carolina*, broadside, April 4, 1868, USC; Ira Berlin and

The most extensive concentration of white Republicans, however, lay in the upcountry bastions of wartime Unionism. "Genuine Republicanism, Republicanism from conviction and principle," wrote Texas party leader James P. Newcomb, "exists among but a small portion of the white people of the South, except in East Tennessee, Missouri, North Carolina, West Virginia, and West Texas." (He might have added northwestern Arkansas and northern Alabama.) All these areas became party strongholds, and some counties, like Winston in Alabama, with only seventeen black residents in 1880, remained so into the twentieth century. North Carolina's "banner county" for white Republicanism was Wilkes, in the western part of the state, hundreds of whose residents "wore the Blue during the war." Such areas produced Republican governors William H. Smith of Alabama and Edmund J. Davis of Texas, both of whom had been driven from their homes after secession and fought in the federal army. In these regions, the party attracted what one member called "the virtuest, honest and poorest portion of our people." In Rutherford County, North Carolina, in the foothills of the Blue Ridge mountains, Republicans included "a large majority of the small farmers, tenants, laborers," and small town artisans. In 1870, every candidate for county office was "a true representative of the farm or workshop," but the party attracted "not a leading white gentleman . . . not a single lawyer, nor a single physician."[37]

To upcountry Republicans the party represented, first and foremost, the inheritor of wartime loyalism. Bitter memories of persecution cemented their attachment, and more than other Republicans, upcountry scalawags insisted on the sweeping proscription of "rebels." The children of Confederates, declared Heroes of America President William F. Henderson, "ought to say, 'My father was disfranchised on the ground of endeavoring to destroy the best government that ever the sun from high Heaven looked down upon.' That is the kind of Republican I am." But if wartime Unionism was the seedbed of yeoman Republicanism, 1867 was a year of political mobilization and expanding ideological horizons in upcountry communities. "A great many men" accustomed to

Herbert G. Gutman, "Natives and Immigrants, Free Men and Slaves: Urban Workingmen in the Antebellum South," *AHR*, 88 (December 1983), 1182–87; H. Pinkney Walker to Lord Stanley, December 22, 1866, F. O. 5/1078/192–3, Public Record Office, London; William Downing to Charles Sumner, April 26, 1869, Sumner Papers; Philip J. Avillo, Jr., "Phantom Radicals: Texas Republicans in Congress, 1870–1873," *SWHQ*, 77 (April 1974), 433.

37. "An Appeal in Behalf of the Republicans of Texas," manuscript, March 19, 1869, James P. Newcomb Papers, UTx; Allen W. Trelease, "Who Were the Scalawags?," *JSH*, 29 (November 1963), 456–63; U. S. Census Office, *Compendium of the Tenth Census* (Washington, D.C., 1883), 1:336; W. B. Siegrist to William E. Chandler, October 16, 1868, William E. Chandler Papers, LC; 39th Congress, 1st Session, House Report 30, pt. 3:10–11; Baggett, "Origins of Early Republican Leadership," 441; J. Pace to William W. Holden, August 20, 1868, North Carolina Governor's Papers; J. G. de Roulhac Hamilton, ed., *The Papers of Randolph Abbott Shotwell* (Raleigh, 1929–39), 2:280, 293; Rutherford *Star*, June 18, 1870; New York *Herald*, June 13, 1871.

taking political guidance from "slaveholders of the aristocratic stamp," declared an East Tennessee scalawag, "now can and will meet these slavery patriots in political debate." In meetings of Union Leagues and the Heroes of America, long-standing class antagonisms and intrastate sectional rivalries were transmuted into the coin of nineteenth-century radicalism. Traditional regional demands, like an end to property qualifications for members of the North Carolina legislature, now merged with a retrospective denunciation of slavery as the foundation of a social order that had held "poor whites . . . as much under bondage" as blacks. Inspired by the hope that "the reign of the would be aristocracy is near at a close," upcountry white Republicanism took on a reforming fervor. "Now is the time," wrote a resident of the Georgia mountains, "for every man to come out and speak his principles publickly and vote for liberty as we have been in bondage long enough."[38]

To some upcountry scalawags, Reconstruction promised an end to state policies that had favored plantation counties at the expense of their own. One Republican convention in the Georgia mountains, gathering on July 4, 1867, called for railroad construction, the establishment of a free common school system "without distinction of race or color," and encouragement to immigrants and capital to enter the state. Economic boosters like Tennessee Gov. William G. Brownlow had long looked to a statewide railroad system ("at whatever cost its completion may require") and infusions of Northern capital (with suitable "guarantees for its protection") to bring East Tennessee the benefits of capitalist development. Many small farmers, however, had more immediate economic concerns, especially in upcountry communities where wartime devastation had been succeeded by disastrous crop failures. In the early days of Radical Reconstruction, many Georgia and South Carolina yeomen stood "on the eve of starvation," and "scenes of the Irish famine" were reported in hill country Alabama. Families who carried prewar debts into Reconstruction, or who had recently mortgaged their farms to replace livestock and implements destroyed during the war, now confronted the possibility of losing their homes to creditors. Many looked to the Republican party to rescue "the poor . . . from this great privation" by providing the debtor relief Presidential Reconstruction had failed to supply. In North Carolina, the Heroes of America demanded that no one be required to pay more than 10 percent of the value of past debts.[39]

38. *The Great Republic* (Washington, D. C.), July 30, 1867; Rutherford *Star*, June 8, August 17, 1867; Orangeburg *Carolina Times*, May 27, 1867; W. H. D. McAteer to Robert K. Scott, October 28, 1871, South Carolina Governor's Papers; B. S. Lindsey to George E. Meade, April 1, 1868, Letters Received, L39 (1868), Third Military District, RG 393, NA.

39. *The Great Republic* (Washington, D. C.), July 4, 1867; E. Merton Coulter, *William G. Brownlow: Fighting Parson of the Southern Highlands* (Chapel Hill, 1937), 87–89, 374; Knoxville *Whig*, April 4, 1866, December 23, 1868; James B. Campbell, "East Tennessee During the Radical Regime, 1865–1869," *ETHSP*, 20 (1948), 97; Petition (ninety-two signatures), May

Here lay the seeds of future conflict over Republican economic policy. Debt-ridden yeomen were less than enthusiastic about ambitious economic programs that would inevitably raise taxes and offer creditors security so as to attract outside capital. The platform of J. M. Wilhite, a Republican sent by Winston County to the Alabama constitutional convention, called for disenfranchising Confederates, the establishment of a public school system, and "low taxes." Some upcountry scalawags even joined blacks in demanding a redistribution of planter land, a program economic boosters viewed as certain to antagonize potential Northern investors. The land issue agitated white as well as black Republicans in 1867; were confiscation submitted to the people, claimed one Northern correspondent, "a majority [of] both . . . would vote for it." Confiscation appealed to upcountry radicals anxious to reduce a "rich nabob aristocracy" to "splitting rails for a living," indemnify loyalists driven from their homes during the Civil War, and assist poor Union men "destitute of land." Generally, radical scalawags appeared more concerned with confiscation's benefits to the white than the black poor. But one group of Georgia yeomen, calling upon Congress to grant homesteads both to black families and to improverished white Unionists, revealed how black and white political language intermingled in 1867. If blacks absorbed the republican discourse of the larger political culture, these petitioners embraced the secular and religious underpinning of the freedmen's claim to land:

> Appropriate out of the vast amount of the surplus lands of the wealthy, a comfortable home for the helpless and dependent black man whose arduous labor for the last two hundred years justly entitles him to such inheritance. . . .
> We believe the freedman is just as much entitled to a home out of the lands of the secession party who tried to dissolve the Union in order to perpetuate slavery as the children of Israel were to the promised land.[40]

Such genuine identification with blacks' aspirations remained rare in the upcountry, where colonization sentiment still flourished and, as one local party leader put it, "our people are more radical against rebels than in favor of negroes." Certainly, scalawags almost unani-

---

30, 1867, South Carolina Governor's Papers; Robert A. Gilmour, "The Other Emancipation: Studies in the Society and Economy of Alabama Whites During Reconstruction" (unpub. diss., Johns Hopkins University, 1972), 63–68, 114, 130–44; Columbia *Daily Phoenix*, October 30, 1867; John Conklin to William W. Holden, July 20, 1868, North Carolina Governor's Papers; Raleigh *Tri-Weekly Standard*, November 9, 1867.

40. Wesley S. Thompson, *The Free State of Winston: A History of Winston County, Alabama* (Winfield, Ala., 1968), 108; Philadelphia *Press* in *New York Times*, May 30, 1867; James Mullins et al. to Thaddeus Stevens, March 10, 1867, Thaddeus Stevens Papers, LC; William Birthright to John M. Broomall, July 14, 1867, John M. Broomall Papers, HSPa; C. C. Ewing to Benjamin F. Butler, August 1, 1867, Butler Papers; J. Robert and ten others to John Sherman, May 1, 1867, Sherman Papers. The signers of the May 1 petition were located in the 1870 manuscript federal census for Meriwether County, Georgia.

mously rejected the idea that black political rights implied "social equality" of any sort. Robert W. Flournoy, a wealthy former slave-owner who earned a reputation as "the William Lloyd Garrison of Mississippi," in his newspaper, the Pontotoc *Equal Rights,* advocated racially integrated education from elementary schools to the state university, but his was indeed a lonely voice. Most scalawags agreed with the North Carolinian who wrote, "There is not the slightest reason why blacks and whites should sit in the same benches, in Churches, school houses, or Hotels. Each can have the equal protection and benefits of the law without these."[41]

For most scalawags, the alliance with blacks remained a marriage of convenience. Yet whatever its origins, this partnership carried with it a further commitment, entirely unprecedented for men reared in a slave society, to defend blacks' political and civil equality. Unionists, declared a North Carolina Republican newspaper, must choose "between salvation at the hand of the negro or destruction at the hand of the rebels. . . . If any union man is now too good or too white to cooperate with the negro, on a basis of complete and absolute political equality . . . let him go over to the rebels." Moreover, contact with blacks in Union Leagues and other political meetings, as an East Tennessee Republican reported, subtly weakened "the cradle-nurtured, unfounded prejudice against the Negro." One organizer of North Carolina's Heroes of America, after hearing several blacks speak at a Republican convention, concluded "there is but little if any difference in the talents of the two races and I am willing to give them all an equal start in the race. I am for 'Liberty, Union, and political equality'." It would certainly be inaccurate to suggest that the heritage of racism had suddenly vanished. But the fact that so many upcountry whites stood ready "to affiliate . . . politically with the negro element" underscored the extent of the political revolution that swept across the South in 1867.[42]

Thus, Southern Republicanism attracted a broad coalition of supporters with overlapping but distinct political agendas. Providing unity and inspiration were the party's commitment to civil and political equality, and its self-image as a "party of progress, and civilization" that would infuse the region with new social ideas and open "the avenues to success and promotion" to black and white, rich and poor. "This," a North Carolina Republican declared, "is the great revolution, which God grant, may soon come with the success of the Republican party." The sense of

41. KKK Hearings, Alabama, 1765; W. P. Hubbard to William E. Chandler, September 7, 1868, Chandler Papers; Huntsville *Advocate,* August 23, 1867; M. G. Abney, "Reconstruction in Pontotoc County," *PMHS,* 11 (1910), 254; *New National Era,* July 10, 1873; Hamilton and Williams, eds., *Graham Papers,* 7:349.

42. Raleigh *Register* in Rutherford *Star,* March 23 1867; *The Great Republic* (Washington, D.C.), July 30, 1867; Joel Ashworth to Nathan H. Hill, April 15, 1867, Nathan H. Hill Papers, DU; Perman, *Reunion Without Compromise,* 284–85.

living in a new era of progress animated the party as its first statewide conventions gathered in the spring and summer of 1867. Yet these meetings also revealed the inner tensions that would plague Southern Republicanism throughout its brief and stormy career. Virtually every party convention found itself divided between "confiscation radicals"—generally blacks but in some cases radical whites—and moderates committed to white control of the party and a policy of economic development that offered more to outside investors and native promoters than to impoverished freedmen and upcountry yeomen.[43]

A particularly divisive situation emerged in Virginia, where James W. Hunnicutt, a newspaper editor who had fled to the North during the war, emerged as Radicalism's leading voice. Establishing the Richmond *New Nation* in 1866, Hunnicutt had won the respect of the city's blacks by defending their claims to suffrage, speaking at their political meetings, and promoting the vision of a "revolutionized" Virginia with common schools, "manufacturing towns and industries," and the division of the lands of "the negro-oligarchs." Hunnicutt and his supporters dominated the first Republican state convention, two thirds of whose delegates were black. Meeting outdoors in Richmond in April 1867, the gathering resembled a mass rally more than a traditional party convention, for the thousands of black spectators made their preferences clear when delegates disagreed. "Almost to a man," the black delegates, both urban and rural, free and freed, supported the idea of confiscation, and while the platform eschewed an explicit call for land distribution, it dedicated the party to uplifting the "poor and humble" of the state.[44]

Although it endorsed a number of Whiggish economic planks—statesponsored internal improvements, and a modification of usury laws "to induce foreign capital to seek investment in the State"—the platform failed to mollify moderates alarmed by the delegates' "defiant speeches" and radical rhetoric. Some Radicals had coupled support for economic development with calls for state regulation of railroad rates, millers' fees, and bank interest to prevent excess profits, and Hunnicutt, to "tremendous cheers," had denounced the existing system of plantation labor as "robbery." The prospect of a radical, black-dominated party alarmed "intelligent and respectable white citizens" hoping to control Virginia Republicanism. Aided by a delegation dispatched by the Union Leagues of Boston, New York, and Philadelphia, moderate

43. Rutherford *Star*, June 16, 1868; Montgomery *Alabama State Sentinel*, August 1, 1867; Raleigh *Tri-Weekly Standard*, July 11, 1867; *New York Times*, June 12, 1867.

44. Richmond *New Nation*, April 12, 1866, March 7, 1867; *The Debates and Proceedings of the Constitutional Convention of the State of Virginia* (Richmond, 1868), 691–93; Alrutheus A. Taylor, *The Negro in the Reconstruction of Virginia* (Washington, D.C., 1926), 210; Rachleff, *Black Labor*, 41–42; *New York Times*, April 19, 1867; Edward McPherson, *The Political History of the United States of America During the Period of Reconstruction* (Washington, D.C., 1875), 253–54.

Whigs succeeded in arranging a second state convention. But when this gathered in Richmond, blacks crowded into the hall, Hunnicutt denounced the moderates for having previously opposed black suffrage and the Fourteenth Amendment, and the meeting adjourned to a public square where it turned into a mass rally and reaffirmed the April platform. Radicals, one black delegate noted, had worked "so rapidly that the conservatives stood aghast not knowing how to take hold of the matter." For the moment, Hunnicutt and his supporters controlled the Virginia party.[45]

Similar divisions occurred in North Carolina, where local party meetings had heard demands for land from black Union Leaguers and loyalist whites. "The people of this state," one delegate told the state convention, which met in Raleigh in September, "have a hope in confiscation, and if that is taken away the Republican party give away the power they have gained." The convention divided over a proposed plank terming confiscation "repugnant"; most blacks opposed the statement, but it won the support of a majority of whites as well as of prominent black leaders Abraham H. Galloway, James H. Harris, and James W. Hood (all of whom had lived in the North before the war). Eventually, a compromise was reached—a pledge to abide by the decision of Congress on the subject. The balance between party factions remained precarious. Moderates dominated the state committee, but the convention defeated Old Whig resolutions offering amnesty to former Confederates and proposing to limit taxes on large landed estates, and at the grass roots, demands for land and debtor relief continued to flourish.[46]

In South Carolina, radicalism was triumphant in 1867. In March, a Charleston Republican convention dominated by the city's freeborn blacks adopted the year's most radical platform, a program for sweeping change in nearly every aspect of the state's life. To a call for a "liberal system" of internal improvements (with blacks and whites enjoying "an equal and fair share" in the awarding of contracts), the platform wedded more far-reaching reforms: an integrated common school system, the abolition of corporal punishment and imprisonment for debt, protection of the "poor man's homestead" against seizure for debt, government responsibility for "the aged, infirm and helpless poor," and heavy taxation of uncultivated land to weaken the system of "large land monopolies" and promote "the division and sale of unoccupied lands among the poorer classes." In July, the state convention adopted essentially the

45. Richmond *Dispatch*, April 18, August 1–3, 1867; William D. Henderson, *The Unredeemed City: Reconstruction in Petersburg, Virginia: 1865–1874* (Washington, D.C., 1977), 165; Charles H. Lewis to Henry Wilson, November 19, 1867, Henry Wilson Papers, LC; Union League Club of New York, *Report of the Proceedings of the Conference at Richmond* (New York, 1867), 6–12; Robert G. Fitzgerald Diary, August 1, 2, 1867, Fitzgerald Papers.

46. Raleigh *Tri-Weekly Standard*, September 7, 10, 14, 1867; New York *Tribune*, April 24, September 9, 1867; Olsen, *Tourgée*, 83; *CG*, 42d Congress, 1st Session, Appendix, 169.

same platform, including its "mild confiscation" provision regarding land.[47]

If in South Carolina and Virginia, free blacks in 1867 shared the radical demands of rural freedmen, the situation proved very different in Louisiana, a state where "innumerable petty antagonisms" afflicted the Republican party from its birth. Carpetbaggers sought to attract the support of native whites by a program combining economic development and the protection of sugar interests with the promise that blacks would not dominate a Republican state government. Simultaneously, they sought to take advantage of "the apparent jealousy existing between free colored people and freedmen" to assert political leadership among rural blacks. For its part, the New Orleans *Tribune,* speaking for the "old free population," denounced "newcomers" for lacking genuine commitment to racial equality, and insisted that only the free elite could "take the recently emancipated black man by the hand and show them the path that leads to liberty and equality."[48]

The carefully balanced platform adopted by the party convention of June reflected the concerns of both carpetbaggers and freeborn blacks. Dominated by New Orleans delegates, the meeting endorsed a program of Mississippi River levee construction, invited immigrants and capital to enter the state, and, to appeal to urban mechanics, supported the eight-hour day. The resolutions defined "perfect equality before the law" to include equal access to jury service, travel, places of entertainment, and schools, and pledged that nominations for public office would be divided equally between the races. (The platform added, however, that, among blacks, "no distinction shall be made, whether said nominees or appointees were born free or not." Free blacks thus insisted upon proportional representation by color for the state as a whole, while rejecting the same principle within the black community.) The only delegates left unsatisfied were rural former slaves, for carpetbaggers, scalawags, and free blacks united in opposition to land distribution. "All the freedmen, *save one,*" reported a Republican newspaper, "were in favor of confiscation, and the measure would have been adopted . . . had it not been for the energetic exertions of the white and free born colored members." The furthest the convention would go was to favor the division of rural Louisiana into small farms "as far as practicable," and to urge Congress to seize the land of employers who discharged black laborers for political reasons.[49]

There is no need to delineate in detail the inner conflicts and diverse

47. Charleston *Daily Courier*, March 22, 1867; *New York Times*, July 31, 1867.

48. Richard H. Abbott, *The Republican Party and the South, 1855–1877: The First Southern Strategy* (Chapel Hill, 1986), 94–99; Thomas W. Conway to J. M. Edmunds, May 7, 1867, Moody Papers; Richter, "Longstreet," 221–22; J. H. Sypher to Henry C. Warmoth, August 24, 1867, Warmoth Papers; New Orleans *Tribune*, May 8, 19, 1867.

49. New Orleans *Tribune*, June 14, 15, 18, 1867; St. Landry (La.) *Progress*, September 14, 1867.

political strategies in other states. Georgia Republicans downplayed blacks' demands in an attempt to appeal to the "wool-hat boys"—upcountry yeomen—on the issue of debtor relief. In Texas, moderate Republicans, interested in reconciliation with former Confederates, a vigorous program of railroad building, and a secondary role for blacks, held the upper hand, but were challenged by Radicals demanding the proscription of rebels, an aggressive defense of black rights, and an "antimonopoly" system of internal improvements, with safeguards against abuses by government-aided railroads. Such factionalism was hardly unusual in American politics. Some contemporaries, however, wondered whether a new party confronted by opponents far superior in finances, education, and political experience, could afford not to place a higher premium on internal unity. More than the struggle for personal advantage explains these divisions. At stake were competing visions of the party's social composition and of the extent of change that would result from Reconstruction.[50]

Despite these internal divisions, the Republican party appeared in remarkably good health for an institution that six months earlier had scarcely existed on Southern soil. Indeed, if anything, its supporters were far too optimistic in 1867 about their prospects for enduring success. "There is a sense of security displayed by our people," warned P. B. S. Pinchback, "that is really alarming. They seem to think that . . . the Great Battle has been fought and the victory won." More realistic was Tennessee Governor Brownlow: "Never was such a conflict witnessed as we are to have."[51]

The North and Radical Reconstruction

Even as Southern Republicans debated among themselves the character of their region's new social order, outside forces attempted to turn Reconstruction in a moderate direction. Of the military commanders appointed by President Johnson, only Gen. Philip H. Sheridan, commander in Texas and Louisiana, actively assisted the formation of Union Leagues and the Republican party, and used his powers to remove numerous officeholders, including governors James W. Throckmorton and James M. Wells, as well as the entire New Orleans Board of Aldermen. The President, however, soon replaced Sheridan with the far more conservative Gen. Winfield S. Hancock, who announced that henceforth the military would remain subordinate to civilian authori-

50. Theophilus G. Steward, *Fifty Years in the Gospel Ministry* (Philadelphia, 1921), 128; Elizabeth S. Nathans, *Losing the Peace: Georgia Republicans and Reconstruction, 1865–1871* (Baton Rouge, 1968), 40–43; C. D. Lincoln to William P. Fessenden, September 20, 1867, William P. Fessenden Papers, LC; Carrier, "Texas During Reconstruction," 168–219.

51. James Haskins, *Pinckney Benton Stewart Pinchback* (New York, 1973), 52; William G. Brownlow to unknown, May 8, 1867 (copy), Moody Papers.

ties. Gen. John M. Schofield, the commander in Virginia, who viewed the Fourteenth Amendment as "unjust and unwise," aided "respectable Republicans as against the lower class of men who have acquired control over the mass of colored voters." In Mississippi and Arkansas, Gen. E. O. C. Ord not only opposed the Reconstruction Act "in toto," but urged freedmen not to neglect their labor "to engage in political discussions." With most commanders reluctant to take on civil responsibilities, the state and local structures of Presidential Reconstruction generally remained intact. A handful of blacks did receive minor offices—the first appears to have been Pascal M. Tourne, appointed inspector of hay for New Orleans in April. But during the period of military rule, civil courts controlled by local oligarchs continued to function, their decisions only occasionally overturned by the army, and federal patronage often remained in the hands of Johnson supporters. "With a Copperhead General commanding, a Rebel governor and rebels filling every office in the state, . . ." complained a Mississippi Republican, "we have much to contend with."[52]

Ultimately, however, Reconstruction policy was established by Congress, even though Johnson retained considerable influence as military commander-in-chief. And from the opening of the Fortieth Congress in March 1867, it became clear that mainstream Republicans, while determined to defend what had been accomplished, were in no mood to advance further. Despite Radical pressure, Congress decided not to remain in session throughout the summer to oversee Johnson's actions. And when Sumner introduced resolutions to broaden the scope of Reconstruction by immediately abolishing existing Southern governments, requiring the establishment of integrated public school systems, and providing the freedmen with homesteads, even Radical Henry Wilson objected, declaring "the terms we have made are hard enough." To moderates, such proposals seemed a form of what Sen. James W. Grimes called "class legislation"—singling out one group of citizens for special government favors. "That is more than we do for white men," declared Fessenden in opposing the land proposal. (To which Sumner responded, "White men have never been in slavery.") In the House, Stevens, too ill

52. Dawson, *Army Generals and Reconstruction*, 46–84; Carl Moneyhon, *Republicanism in Reconstruction Texas* (Austin, 1980), 67–70; James E. Sefton, ed., "Aristotle in Blue and Braid: General John M. Schofield's Essays on Reconstruction," *CWH*, 17 (March 1971), 45–57; John M. Schofield to Ulysses S. Grant, April 2, 1868, Ulysses S. Grant Papers, LC; Bernard Cresap, *Appomattox Commander: The Story of General E.O.C. Ord* (San Diego, 1981), 239–61, 293; New Orleans *Tribune*, April 19, 1867; Donald G. Nieman, *To Set the Law in Motion: The Freedmen's Bureau and the Legal Rights of Blacks, 1865–1868* (Millwood, N. Y., 1979), 200–208; A. C. Fish to William E. Chandler, July 24, 1868, Chandler Papers; Charles W. Clarke to John Covode, August 31, 1867, John Covode Papers, LC. In 1867, 16,066 soldiers were stationed in the entire South, of whom one quarter were in Texas and mainly concerned with Indian warfare. 40th Congress, 2d Session, House Executive Document 1, 462–63.

to speak, had a colleague read a long speech and a bill providing forty acres to freedmen from confiscated land. "This is the finishing work of reconstruction," wrote one of his admirers.[53]

In the South, Stevens' speech contributed to the upsurge of land agitation, and copies were read aloud at Union League meetings. "Confiscation must come or all past efforts for the better condition of the colored man will prove abortive," wrote a Virginia scalawag. "I would to God we had one hundred Ben Butlers and Thad Stevens." Moderate Northern leaders, however, had no intention of being "run over by the car of Revolution." The course of events since Appomattox had slowly created a party consensus in favor of civil and political rights for Southern blacks. No such consensus existed on the land question, and a barrage of criticism greeted Stevens' proposal. Confiscation, said Sherman, was a "fearful" proposition that would "disorganize and revolutionize society in the southern states." The Springfield *Republican* insisted that for the government to give blacks land would be an act of "mistaken kindness" that would prevent them from learning "the habits of free workingmen." More conservative Republicans denounced Stevens for adding to "the distrust which already deters capitalists" from investing in the South. The *New York Times* printed alarming reports that "fear of confiscation" had paralyzed Southern business, and that if the issue remained alive, "neither capital nor emigration will flow this way."[54]

Even more threatening, in the *Times*'s view, was the prospect that the precedent set by confiscation "would not be confined to the South." The North, it warned, had its own "extremists" eager to destroy "the inviolability of property rights." Indeed, other "schemes for interference with property or business" agitated the political scene in 1867. In Kansas, Ben Wade delivered a widely publicized speech declaring that with the slavery issue settled, the relations of capital and labor would now move to the center stage of politics. "Property," he declared, "is not equally divided, and a more equal distribution of capital must be wrought out." In London, Wade's speech so impressed Karl Marx that he mentioned it as a "sign of the times" in the first German edition of *Capital*, published that year. Few Northern Republicans shared his enthusiasm. The *Times* lambasted the Radicals for desiring "a war on property . . . to succeed the war on Slavery," and *The Nation* linked Wade's speech with the confisca-

53. Michael L. Benedict, *A Compromise of Principle: Congressional Republicans and Reconstruction 1863–1869* (New York, 1974), 246–55; *CG*, 40th Congress, 1st Session, 15, 51, 55, 79, 114, 147, 203–208, 304–308, 463; C. J. Baylor to Benjamin F. Butler, March 26, 1867, Butler Papers.

54. Raleigh *Tri-Weekly Standard*, April 4, 1867; Gottlieb, "Land Question," 371; Thomas L. Gale to Butler, June 16, 1867, Butler Papers; Kemp P. Battle to Benjamin S. Hedrick, July 5, 1867, Benjamin S. Hedrick Papers, DU; *CG*, 40th Congress, 1st Session, 52; Springfield *Weekly Republican*, June 15, 1867; *New York Times*, February 19, March 10, April 10, June 27, 1867.

tion issue, organized labor's demand for an eight-hour day, and the rise of Fenianism among Irish-Americans as illustrations of how "demagogues" had abandoned "true radicalism"—equality before the law and equal economic opportunity—in favor of "special favors for special classes of people."[55]

Such views inevitably affected how Northerners responded to developments within the South. Indeed, their positions on the land issue closely paralleled Northern Republicans' attitudes toward their new Southern wing. Men like Stevens believed confiscation could cement a political alliance between blacks and poorer whites. Reconstruction, declared one Radical newspaper, was fundamentally a question "of labor, and not of race"; the road to success lay in "making the sentiments of Charles Sumner and Thaddeus Stevens as popular in Texas as they are supposed to be in Massachusetts." This, of course, was precisely what moderate Republicans feared. They clung to the vision of a Southern coalition uniting enlightened planters, urban business interests, and black voters, with white propertied elements firmly in control. Encouraged by assurances from Whiggish Southerners that they would join the party once "their rights of property" had been guaranteed, moderates insisted that the land issue be dropped, and denounced "demagogues" like Virginia's Hunnicutt for driving from the party "the property holders of the South." Early in 1868, Clerk of the House McPherson killed Hunnicutt's newspaper by rescinding its government printing contract.[56]

To counteract proconfiscation influence, a group of Republican orators visited the South in the late spring of 1867 to address gatherings of freedmen. Horace Greeley told a large meeting at Richmond's African Church, "you are more likely to earn a home than get one by any form of confiscation." Henry Wilson, convinced by conversations with Southern moderates that many whites could be induced to cooperate in Reconstruction, brought the same message to Virginia and South Carolina. And William D. Kelley, traveling as an emissary of both the free labor gospel and a Northern combine preparing Southern investments, praised landownership as the basis of "a manly and honorable independence," but "discouraged the idea of confiscation." Many blacks, to be sure, rejected the message of those Stevens called the "Republican meteors" pursuing an "erratic . . . course" through the South. Recalling one of Wilson's speeches, a black clergyman later remarked: "The kind Senator did not leave them without some good advice. They were not looking for advice,

55. *New York Times*, June 14, July 9, 1867; William F. Zornow, " 'Bluff Ben' Wade in Lawrence, Kansas: The Issue of Class Conflict," *OHQ*, 66 (January 1956), 44–52; Karl Marx, *Capital* (New York, 1967 ed.), 1:10; *Nation*, June 27, July 18, 1867.

56. Eric Foner, *Politics and Ideology in the Age of the Civil War* (New York, 1980), 145–47; Washington *Daily Morning Chronicle*, April 30, 1867; W. M. Herbert to John Sherman, July 30, 1867, Sherman Papers; *Nation*, October 31, 1867; Springfield *Weekly Republican*, May 11, 1867; Abbott, *Republican Party and the South*, 110–11, 125–36.

however, but land to plant corn and potatoes, for their wives and children."[57]

Clearly, whatever Southern Republicans desired, the party in the North remained unwilling to embrace the land issue. And one reason was that Northern Republicans in 1867 faced problems of their own at home. With the apparent resolution of the Southern question, new issues posing new hazards now entered the political arena. One was the demand of Westerners of both parties for an inflation of the currency, or at least a halt to the policy of retiring greenbacks from circulation, inaugurated in 1865 by Secretary of the Treasury Hugh McCulloch in order to pave the way for a resumption of specie payments. Strongly supported by Eastern bankers with a vested interest in tight currency, the contraction policy drew sharp criticism from debtors, entrepreneurs desiring low interest rates, and Western farmers and businessmen whose region was plagued by "too little currency," not too much. Although Democrats hoped to profit from the money issue, its political implications were ambiguous at best, for the contraction policy stirred increasing opposition in both parties. Democrats had a better chance of making political capital, at least in the West, from the issue of the national debt. During the war, Congress had provided for the payment in gold of interest on war bonds, but had neglected to specify how the principal of one series—the 5-20s—would be redeemed. In 1867, former Congressman George H. Pendleton proposed what came to be known as the Ohio Idea—repaying the bonds in greenbacks.[58]

A few Republicans, including Stevens and Butler, enthusiastically supported Pendleton's plan. Most, however, viewed the sanctity of the national debt as a moral legacy of the war second only to emancipation itself. "Every greenback," said Sumner, "is red with the blood of fellow-citizens." Such Republicans equated the Ohio Idea with repudiation, although it was President Johnson, not Pendleton, who in his last annual message made the novel suggestion that future interest payments should be counted as reducing the bonds' principal. Although the issue did not adhere to party lines, since Wall Street Democrats like August Belmont strenuously opposed Pendleton's formula, it did win Democrats Western support in 1867. "We are in favor of negro suffrage," wrote one Kansas

57. New York *Tribune*, May 17, 27, 1867; Richard H. Abbott, *Cobbler in Congress: The Life of Henry Wilson, 1812–1875* (Lexington, Ky., 1972), 182–89; Raleigh *Tri-Weekly Standard*, May 2, 9, 1867; *CG*, 42d Congress, 1st Session, 341; William D. Kelley, *Speeches, Addresses and Letters on Industrial and Financial Questions* (Philadelphia, 1872), 167–69; *New York Times*, May 29, 1867; Peter Randolph, *From Slave Cabin to the Pulpit* (Boston, 1893), 69.

58. Herbert S. Schell, "Hugh McCulloch and the Treasury Department, 1865–1869," *MVHR*, 17 (December 1930), 404–21; Edward Atkinson to William B. Allison, December 3, 1867, Letterbook, Edward Atkinson Papers, MHS; Winslow S. Pierce to Elihu B. Washburne, October 24, 1868, Washburne Papers; Robert P. Sharkey, *Money, Class, and Party: An Economic Study of Civil War and Reconstruction* (Baltimore, 1967 ed.), 59–82, 110–17.

"The Reconstruction Policy of Congress, As Illustrated in California." In this Democratic cartoon from the election campaign of 1867, Republican candidate for governor George C. Gorham's support of manhood suffrage is portrayed as undermining American government by admitting blacks, Chinese immigrants, and Indians to the right to vote. (Library of Congress)

Republican, "but we are not in favor of the grand system of financial robbery and plunder of the producing classes which the Republican party has inaugurated."[59]

This writer did not mention his opinion of female suffrage, but, in Kansas at least, that issue also gave politics a new turn in 1867. The Republican-controlled legislature placed two referenda on the fall ballot, enfranchising, respectively, blacks and women. That summer, Elizabeth Cady Stanton and Susan B. Anthony arrived, hoping to campaign for both. They found Kansas Republicans endorsing black suffrage, but unwilling to risk an identification with votes for women. Whereupon Stanton and Anthony turned to the Democratic party, enlisting the help of George Francis Train, an eccentric merchant and financier given to anti-black diatribes, Irish nationalism, and flamboyant adventure (it was he who in 1870 traveled around the world in eighty days). Train financed Stanton and Anthony's speaking tour across the Kansas prairies and appeared with them each night, in full evening dress. To Radicals and abolitionists, cooperation with Train was anathema; William Lloyd Garrison described him as a "crack-brained harlequin and semi-lunatic." But Stanton and Anthony insisted on their right to seek political allies wherever they could find them.[60]

Currency, bonds, and women's rights offered glimpses of political issues of the future, but in most Northern states the familiar questions of race and Reconstruction dominated the fall elections. For the first time, Republicans went before the voters united in support of black suffrage, at least for the South. "There is not a single argument against negro suffrage," insisted *The Nation*, "which is not based on prejudice." But Democrats seized upon that prejudice with a vengeance. Republicans, the New York *World* declared, had made voters of "the lazy, licentious, brutalized elements" of the South, a race given to such "strange, wild, fantastic and savage practices" as witchcraft and exorcism. In Ohio, where a measure enfranchising blacks was on the fall ballot, Democratic gubernatorial candidate Allan G. Thurman pledged to save the state from "the thralldom of niggerism." On the West Coast, Democrats added anti-Chinese appeals, arguing that the Republican doctrine of "universal equality for all races, in all things" would lead to an "Asiatic" influx and

59. Benjamin F. Butler to S. P. Cummings, January 22, 1868, endorsement on Cummings to Butler, January 18, 1868, Butler Papers; Morton Keller, *Affairs of State: Public Life in Late Nineteenth Century America* (Cambridge, Mass., 1977), 191; James D. Richardson, ed., *A Compilation of the Messages and Papers of the Presidents 1789–1897* (Washington, D.C., 1896–99), 6: 678; A. J. Groves to Butler, October 20, 1867, Butler Papers.

60. Ellen C. DuBois, *Feminism and Suffrage: The Emergence of an Independent Women's Movement in America, 1848–1869* (Ithaca, N.Y., 1978), 62–103; Theodore Stanton and Harriot Stanton Blatch, eds., *Elizabeth Cady Stanton* (New York, 1922), 1:212–13; Robert E. Riegel, "The Split of the Feminist Movement in 1869," *MVHR*, 49 (December 1962), 490; Walter M. Merrill and Louis Ruchames, eds., *The Letters of William Lloyd Garrison* (Cambridge, Mass., 1971–81), 6:2.

control of the state by an alliance of "the Mongolian and Indian and African."[61]

For their part, many Republicans tried to sidestep the race issue. "Both sides," an Ohio politician observed, "are making their strongest appeal to prejudice—the one harping on the 'nigger' and the other on the 'copperhead'." Yet the crisis that had shifted the center of gravity of national politics to the left had also affected grass-roots Republicans. A remarkable number of non-Radicals now endorsed political equality for Northern blacks. Ohio gubernatorial candidate Rutherford B. Hayes supported black suffrage, as did New York's Republican state platform—"a sacrifice of political strength on the altar of consistency," one newspaper called it. Even in conservative New Jersey, the party now committed itself to black voting, and California gubernatorial candidate George C. Gorham repudiated the "anticoolie" movement, insisting "the same God created both Europeans and Asiatics."[62]

Two sets of elections occurred in the fall of 1867, producing sharply divergent outcomes. In the South, where voters considered calls for constitutional conventions and selected delegates, the result in every state was a Republican triumph. "The negroes," declared a white Republican in Alabama's Tennessee Valley, "voted their entire walking strength—no one staying at home that was able to come to the polls." Black turnout ranged from about 70 percent in Georgia to nearly 90 percent in Virginia, and their vote was all but unanimous (three states failed to record a single black ballot against holding a convention). Among whites, apathy and the hope that abstention would prevent conventions from achieving the necessary majority of registered voters kept turnout far lower, and the Republican campaign to attract white voters achieved only partial success. In North Carolina, Georgia, Alabama, and Arkansas, more than one white voter in five supported the calling of a convention, but in others, white Republicans remained a tiny minority. The party made encouraging inroads in the upcountry, but elsewhere the electorate remained polarized along racial lines.[63]

Counterbalancing victory in the South was a serious setback in the

61. *Nation*, May 22, 1866; New York *World*, September 30, November 14, 1867; Reginald McGrane, *William Allen: A Study in Western Democracy* (Columbus, Ohio, 1925), 179; *Speech of H. H. Haight, Esq. Democratic Candidate for Governor* (n.p., 1867), 3; Alexander Saxton, *The Indispensable Enemy: Labor and the Anti-Chinese Movement in California* (Berkeley, 1971), 81.

62. Mary L. Hinsdale, ed., *Garfield-Hinsdale Letters* (Ann Arbor, 1949), 96–97; Kenneth E. Davison, *The Presidency of Rutherford B. Hayes* (Westport, Conn., 1972), 10–11; Homer A. Stebbins, *A Political History of the State of New York, 1865–1869* (New York, 1913), 169, 263; Kent A. Peterson, "New Jersey Politics and National Policy-Making, 1865–1868" (unpub. diss., Princeton University, 1970), 145–54, 166–67; *Speech Delivered by George C. Gorham of San Francisco* (San Francisco, 1867), 13.

63. H. H. Russell to William H. Smith, October 19, 1867, Swayne Papers; Perman, *Reunion Without Compromise*, appendix; Martin E. Mantell, *Johnson, Grant, and the Politics of Reconstruction* (New York, 1973), 47–49; Abbott, *Republican Party and the South*, 137.

North. From Maine to California, Democrats gained ground dramatically, winning in New York by more than 50,000 votes, coming within 3,000 of electing Ohio's governor and gaining control of the legislature (thus precluding Ben Wade's reelection to the Senate), sweeping California, and, according to *The Nation*'s calculation, reducing Republicans' massive majorities of 1866 by three quarters. Meanwhile, black suffrage went down to defeat in Minnesota, Ohio, and Kansas (as did women's suffrage in the latter state, receiving slightly fewer votes). Since few major offices were at stake, the result did not severely injure the Republican party. Optimists could even take comfort in the facts that a large majority of Republican voters had supported black suffrage and that Ohio had cast more ballots in its favor than any state had ever given. (Georges Clemenceau considered this a remarkable outcome, in view of the difficulty of dislodging prejudice from "one of the out of the way corners of the tight Anglo-Saxon brain.") Moreover, local issues—prohibition in Massachusetts and New York, hostility to railroad monopolies in California, Ohio's Pendleton Plan—all influenced the outcome. Local questions, however, could hardly explain a consistent regional pattern, and both parties concluded that the race issue had produced the stunning result. "It would be vain to deny," declared *The Nation*, "that the fidelity of the Republican party to the cause of equal rights . . . has been one of the chief causes of its heavy losses."[64]

The results had a major impact on the balance of power within the party, convincing moderates that issues like disenfranchisement, black voting in the North, and impeachment must be avoided at all costs. Even in Massachusetts, reported one Republican editor, "extreme radical measures are becoming more and more unpopular." Elsewhere, too, the forthright egalitarianism of 1867 receded. "Judicious" men, a New Yorker wrote House Speaker Schuyler Colfax, must inform "the extreme radicals, thus far have we gone with you, but we cannot go any further . . . you see the disasters which have happened to our cause in the fall elections, from our adopting your views." Business-oriented Republicans like Henry D. Cooke, an Ohio banker and party leader, and James G. Blaine, moreover, associated the greenbackism and "repudiation" policies of "ultra infidelic radicals" like Stevens and Butler with Radical positions on Reconstruction. All, they believed, must be "put down." Looming on the horizon was a new Republican self-image, in which devotion to the Union and fiscal responsibility would overshadow civil and political equality. Although few Republicans considered abandoning

64. Michael L. Benedict, "The Rout of Radicalism: Republicans and the Election of 1867," *CWH*, 18 (December 1972), 334–44; Eugene H. Berwanger, *The West and Reconstruction* (Urbana, Ill., 1981), 163–67; Georges Clemenceau, *American Reconstruction*, edited by Fernand Baldensperger, translated by Margaret MacVeagh (New York, 1928), 116; *Nation*, November 21, 1867.

Reconstruction, many agreed with the Ohio politico who observed, "The Negro will be less prominent for some time to come." Certainly, the election results marked the end of any hope that Northern Republicans would embrace a program of land distribution. "Let confiscation be . . . an unspoken word in your state," Henry Wilson advised North Carolina black leader James H. Harris in November. "It has no meaning here." Henceforth, it appeared certain, the party would strive to bring Reconstruction to a successful conclusion rather than press it forward. And this could not but affect the balance of power within the South.[65]

The Constitutional Conventions

With most antebellum officials barred from membership and blacks and carpetbaggers representing the plantation belt, the Southern constitutional conventions of 1867–69, as a British visitor remarked, mirrored "the mighty revolution that had taken place in America." The delegates, hostile newspapers contended, possessed little political experience and virtually "no property at all." And, of course, blacks for the first time sat alongside whites as lawmakers, a fact that riveted the attention of Southern freedmen. In parts of South Carolina, blacks refused to sign labor contracts for 1868, "expecting the convention to give them lands." When important issues came to the floor in Virginia, blacks crowded the galleries, tobacco factories reported mass absenteeism, and white households found themselves forced by a lack of servants "to cook their own dinners, or content themselves with a cold lunch." Nor were blacks alone in observing the proceedings with "intense interest." From Congress came appeals for "wise and judicious action" and warnings that the completed documents must meet the approval of "the enlightened judgment of the country." To Northern officeholders, complained carpetbagger Albion W. Tourgée, it seemed to make "no difference what may be the needs of the people in these States. The Republican party and its interests are paramount."[66]

Since most opponents of Reconstruction had abstained from voting,

65. Benedict, *Compromise of Principle*, 256, 275–76; Dale Baum, *The Civil War Party System: The Case of Massachusetts, 1848–1876* (Chapel Hill, 1984), 123–25; Peterson, "New Jersey Politics," 170–75; John Binney to Schuyler Colfax, November 2, 1867 (copy), Fessenden Papers; H. D. Cooke to John Sherman, October 12, 1867, Sherman Papers; Ellis P. Oberholzer, *Jay Cooke, Financier of the Civil War* (New York, 1907), 2:28; Hinsdale, ed., *Garfield-Hinsdale Letters*, 112; Raleigh *Tri-Weekly Standard*, November 14, 1867.

66. David Macrae, *The Americans at Home* (New York, 1952 [orig. pub. 1870]), 138; Malcolm C. McMillan, *Constitutional Development in Alabama 1798–1901: A Study in Politics, The Negro, and Sectionalism* (Chapel Hill, 1955), 114–17; William M. Jenkins to James L. Orr, January 31, 1868, South Carolina Governor's Papers; Arney R. Childs, ed., *The Private Journal of Henry William Ravenel, 1859–1887* (Columbia, S.C., 1947), 317; Rachleff, *Black Labor*, 45–48; Schuyler Colfax to John C. Underwood, January 7, 1868, Elihu B. Washburne to Underwood, December 7, 1867, John C. Underwood Papers, LC; *National Anti-Slavery Standard*, January 4, 1868.

the total of just over 1,000 delegates included few Democrats or Conservatives, and a high rate of absenteeism further reduced their influence. Generally, in the words of one black delegate, they contented themselves with assailing Republican principles "with all that bitter resentment and revenge so characteristic of a once ruling, but now waning aristocracy." Some Conservatives defended slavery and the Dred Scott decision, others warned that political equality would give freedmen the right "to marry their daughters, and, if necessary, hug their wives." Their speeches won praise from much of the Southern press, which ridiculed the "Bones and Banjo Conventions" and the former slaves who believed themselves "competent to frame a code of laws."[67]

As the first large group of elected Southern Republicans, the delegates epitomized the party's social composition. About one sixth were carpetbaggers, most elected from black belt districts. Generally veterans of the Union army, carpetbaggers were the best-educated Republican delegates, numbering many lawyers, physicians, and other professionals. Talented, ambitious, and youthful (their average age was thirty-six), carpetbaggers usually chaired the key committees and drafted the most important provisions of the new constitutions.[68] Southern white Republicans formed the largest group of delegates; they were especially numerous in North Carolina, Georgia, Arkansas, Alabama, and Texas. Those of established standing, like Benjamin F. Saffold, son of Alabama's former chief justice, were far outnumbered by upcountry farmers and small-town merchants, artisans, and professionals, few of whom had ever held political office. Nearly all had opposed secession, and many had served in the federal army or been imprisoned for Unionist sentiments. Like the carpetbaggers, scalawag delegates believed the new constitutions must build a "nobler and more enduring civilization" on the ruins of the Old South, but more immediate concerns also commanded their attention, especially the proscription of Confederates and relief for debt-ridden small farmers. On racial issues, they were sharply divided. One representative of the Arkansas mountains cited Scripture and the Declaration of Independence to support his conviction that "all men are created equal," and others had worked closely with blacks in the Union League. But many scalawags, an opponent of Reconstruction noted, had been

67. Richard L. Hume, "The 'Black and Tan' Constitutional Conventions of 1867–1869 in Ten Former Confederate States: A Study of Their Membership" (unpub. diss., University of Washington, 1969); Autobiography of George Teamoh, Carter G. Woodson Papers, LC; *Debates and Proceedings of the Convention Which Assembled at Little Rock . . .* (Little Rock, 1868), 88–89, 431, 637 (hereafter cited as *Arkansas Convention Debates*); Cal M. Logue, "Racist Reporting During Reconstruction," *Journal of Black Studies*, 9 (March 1979), 335–50.

68. Richard L. Hume, "Carpetbaggers in the Reconstruction South: A Group Portrait of Outside Whites in the 'Black and Tan' Constitutional Conventions," *JAH*, 64 (September 1977), 313–30; *CG*, 41st Congress, 1st Session, 431; Olsen, *Tourgée*, xii; Jack B. Scroggs, "Carpetbagger Constitutional Reform in the South Atlantic States, 1867–1868," *JSH*, 27 (November 1961), 475–77; *Arkansas Convention Debates*, 642–43.

"elected with feelings opposed to the negro, however Republican they may be."[69]

Although accounting for a total of 265 delegates, blacks remained substantially underrepresented in most states. They formed a majority of the Louisiana and South Carolina conventions and nearly 40 percent in Florida, but only about one fifth in Alabama, Georgia, Mississippi, and Virginia, and 10 percent in Arkansas, North Carolina, and Texas. Of those for whom biographical information is available, 107 had been born into slavery (of whom nineteen had become free before the Civil War) and eighty-one free. Twenty-eight had spent all or most of their lives in the North or, in the case of two South Carolina delegates, the West Indies. At least forty black delegates had served in the Union Army, and the largest occupational groups were ministers, artisans (mostly carpenters, shoemakers, blacksmiths, and barbers), farmers, and teachers. Only a handful were field hands or common laborers. A few black delegates owned substantial amounts of property, but even in South Carolina the large majority paid no state taxes (apart from poll taxes) whatever. Nearly all would go on to hold other Reconstruction offices, including 147 elected to state legislatures, and nine to Congress.[70]

These figures, however, obscure local patterns that reflected the nature of black politics and the structure of black society in different parts of the South. Nearly half the black delegates, and a majority of those born free, served in South Carolina and Louisiana, where political organizing, led by the free urban elite, had the longest history. In Louisiana, indeed, the freeborn enjoyed a virtual monopoly of black positions. Georgia, with a small free population and few blacks who had served in the army or come to the state with the Freedmen's Bureau (since General Sherman had refused black recruits and the Bureau had hired virtually none), saw the church provide the bulk of leadership—the twenty-two black delegates included no fewer than seventeen ministers. In Virginia, reflecting the greater opportunity for manumission and escape enjoyed by Upper South slaves, and the close ties between free and freed, over one third of the delegates born as slaves had gained their freedom before 1860. One freeborn Virginia delegate, Willis Hodges, had before the war de-

69. Richard L. Hume, "Scalawags and the Beginnings of Congressional Reconstruction in the South" (unpub. paper, annual meeting of American Historical Association, 1978); McMillan, *Constitutional Development in Alabama,* 121; Olsen, *Tourgée,* 93; Raleigh *Daily Standard,* February 24, 1868; *Arkansas Convention Debates,* 661–62; Richard G. Lowe, ed. "Virginia's Reconstruction Convention: General Schofield Rates the Delegates," *VaMHB,* 80 (July 1972), 347, 352; M. J. Keith to Henry Watson, Jr., October 2, 1867, Watson Papers.

70. Richard L. Hume, "Negro Delegates to the State Constitutional Conventions of 1867–69," in Rabinowitz, ed., *Southern Black Leaders,* 129–53; Holt, *Black over White,* appendix; KKK Hearings, South Carolina, 1241–44. My own count differs slightly from Hume's. As can best be determined, the number of black delegates was Alabama 17, Arkansas 8, Florida 19, Georgia 36, Louisiana 50, Mississippi 16, North Carolina 14, South Carolina 71, Texas 10, Virginia 24.

scribed blacks, free and slave, as "one man of sorrow" and assisted fugitives escaping from bondage.[71]

The educated, articulate, and politically experienced freeborn delegates of South Carolina and Louisiana played major roles in their conventions, dominating debate and often outmaneuvering white participants. Other states produced articulate individuals who took an active part in proceedings—men like William H. Grey of Arkansas, who as the free servant of Virginia Gov. Henry A. Wise had attended sessions of Congress before the war, George T. Ruby of Texas, a Northerner in his midtwenties who had organized schools for the Freedmen's Bureau and risen to head the state's Union League, and fugitive slave Dr. Thomas Bayne, a veteran of Virginia black politics since 1865, and a man with "a good flow of speech, a vast amount of general knowledge, a fund of apposite and humorous anecdotes . . . and withal a good deal of common sense."[72]

Outside Louisiana and South Carolina, however, most black delegates, lacking formal education, found "agricultural degrees and brick yard diplomas" poor preparation for the complex proceedings of a constitutional convention. They had "little to say" during debates, and sometimes allowed white delegates to take advantage of their inexperience. Henry M. Turner later admitted that Georgia's black delegates had been misled into believing that a poll tax to support education could not be employed to limit black voting. Yet while generally remaining silent, black delegates proved perfectly capable of judging political and constitutional questions and promoting the interests of their constituents. On issues of civil rights and access to education, blacks in every state formed a unified bloc; on disenfranchisement and economic policy, black delegates, like whites, divided, reflecting diverse social currents and political strategies within the black community.[73]

South Carolina Gov. James L. Orr, speaking for many Southern whites, thought gatherings in which the region's "intelligence, refinement and wealth" were unrepresented could never frame constitutions attuned to the needs of the postwar South. In fact, most of the conventions produced modern, democratic documents, "magnificent," wrote New Orleans *Tribune* editor Houzeau, for their "liberal principles." The constitutions established the South's first state-funded systems of free public education overseen by central commissioners of education, and in South

71. Tunnell, *Crucible of Reconstruction*, 231–33; Drago, *Black Politicians*, 37–39; Willard B. Gatewood, ed., *Free Man of Color: The Autobiography of Willis Augustus Hodges* (Knoxville, 1982), xxxv, 3–4.

72. Holt, *Black over White*, 35–37; Joseph M. St. Hilaire, "The Negro Delegates in the Arkansas Constitutional Convention," *ArkHQ* 33 (Spring 1974), 43, 61; Carl H. Moneyhon, "George T. Ruby and the Politics of Expediency in Texas," in Rabinowitz, ed., *Southern Black Leaders*, 363–68; *New York Times*, January 11, 1868.

73. George Teamoh Autobiography, Woodson Papers; KKK Hearings, Georgia, 1041.

Carolina and Texas made school attendance compulsory, a provision strongly supported by black delegates. Clauses mandating the establishment of penitentiaries, orphan asylums, homes for the insane, and, in some cases, the provision of poor relief, further expanded public responsibilities.

All the constitutions guaranteed blacks' civil and political rights, completing, as a Texas newspaper put it, the "equal rights revolution." They abolished holdovers from the old regime resented by blacks and whites alike—whipping as a punishment for crime, property qualifications for office and jury service, viva voce voting, and imprisonment for debt (the last termed by one scalawag a "barbarous relic of a feudal age"). Florida even granted Seminole Indians two representatives in the state legislature. The constitutions reduced the number of capital crimes and in three cases reorganized local government along the lines of the New England township, to overthrow local oligarchies centered in unelected county courts. Nine of the ten states recognized a married woman's separate property rights (although more to protect families against a husband's creditors than as a gesture to feminism). And South Carolina for the first time in its history authorized the granting of divorces. Antebellum bills of rights were rewritten to include language from the Declaration of Independence declaring all men created equal and recognizing a citizen's paramount loyalty to the federal government. Such provisions brought Southern government into line with changes already in place elsewhere. "We want these . . . constitutions to be like our constitutions," Henry Wilson had declared, and except for black rights (guaranteed in few Northern states), they were, with many articles copied directly from the North.[74]

Although the conventions clearly embodied Southern Republicans' commitment to equal rights and a New South, they also revealed the party's inner divisions. Republicans achieved far greater agreement on general principles than their actual implementation: on public education but not whether schools should be racially integrated; civil and political rights for blacks but not "social equality"; the expansion of democracy but not black domination of local or state governments; loyal rule but not the disenfranchisement of rebels; economic modernization, but not how to balance the need for outside capital with white farmers' demands for

74. *South Carolina Convention Proceedings*, 1:47–50; Jean-Charles Houzeau, *My Passage at the New Orleans "Tribune": A Memoir of the Civil War Era*, edited by David C. Rankin, translated by Gerard F. Denault (Baton Rouge, 1984), 143; *The Campaign Speech of Hon. Foster Blodgett . . .* (Atlanta, 1870), 6; Suzanne Lebsock, "Radical Reconstruction and the Property Rights of Southern Women," *JSH*, 43 (May 1977), 201–207; Nelson M. Blake, *The Road to Reno: A History of Divorce in the United States* (New York, 1962), 63, 234; *CG*, 40th Congress, 1st Session, 144; McMillan, *Constitutional Development in Alabama*, 142–44. The texts of the Reconstruction constitutions may be found in Francis N. Thorpe, ed., *The Federal and State Constitutions*, 7 vols. (Washington, D.C., 1909).

debtor relief and blacks' for land. The outcome of these debates illuminated the balance of power within individual states and differing perceptions of whether the party should try to forge a political majority by striving to serve the interests of blacks and poorer whites, or by making its main concerns the attraction of "respectable" whites to its ranks and outside capital to the South.

To Conservatives, the issue of race relations offered the best means of embarrassing their foes and disrupting the Republican coalition. At every turn, they sought to place Republicans on record on such questions as interracial marriage and separate schools for black and white. Many Republicans hoped to avoid these divisive questions entirely, but opponents of Reconstruction succeeded in embroiling a number of conventions in lengthy discussions of interracial marriage. (One Georgian moved that any minister officiating at such a wedding be imprisoned for up to twenty years or deported to Liberia.) Black delegates expressed little interest in marrying white women, but some felt constrained to point out that the "purity of blood" lauded by their opponents had "already been somewhat interfered with" by planters assaulting or cohabiting with female slaves.[75]

Conservatives, however, were not the only delegates to raise questions of "social equality." Freeborn Southerners and blacks from the North, many of them men of some means who had experienced "considerable discomfiture" when traveling or seeking admission to hotels and restaurants, led the demand for equal access to transportation and public accommodations. James W. D. Bland, a freeborn carpenter, raised the issue in Virginia, Tunis G. Campbell, a former New York hotel steward, in Georgia, Ovide Gregory, the "acknowledged leader" of Mobile's free community, in Alabama, and James White, an Indiana-born minister and army veteran, in Arkansas. Both free and freed black delegates, however, supported such measures, but with carpetbaggers and scalawags divided or opposed, most conventions evaded the question, neither guaranteeing equal access to public facilities nor mandating segregation. South Carolina and Mississippi enacted vague provisions barring "distinction" on the basis of color, and only Louisiana, where freeborn blacks dominated the proceedings, explicitly required equal treatment in transportation and by licensed businesses.[76]

Even more charged was the subject of integration in education. No state actually required separate schools, an omission that led thirteen

75. *Journal of the Constitutional Convention of the State of North Carolina at Its Session 1868* (Raleigh, 1868), 473; *Journal of the Proceedings of the Constitutional Convention of the People of Georgia* (Augusta, 1868), 143; *Arkansas Convention Debates*, 363, 491–99, 501.

76. *Virginia Convention Debates*, 154; Drago, *Black Politicians*, 40–41; *Official Journal of the Constitutional Convention of the State of Alabama* (Montgomery, 1868), 15; St. Hilaire, "Negro Delegates," 64; *Official Journal of the Proceedings of the Convention for Framing a Constitution for the State of Louisiana* (New Orleans, 1868), 125.

white Alabama delegates to resign from the Republican party. But only Louisiana and South Carolina explicitly forbade them. Most blacks appeared more concerned with educational opportunities for their children and employment for black teachers than with the remote prospect of racially mixed education. Only a narrow majority of Virginia's black delegates supported Dr. Thomas Bayne's unsuccessful measure mandating such integration. Even in South Carolina, the same black delegates who praised the school integration clause as "laying the foundation of a new structure" of society acknowledged that both races preferred separate education. But in every state, blacks objected to constitutional language *requiring* racial segregation. James W. Hood, later North Carolina's assistant superintendent of education, favored separate schools because nearly all white teachers, "educated as they necessarily are in this country," viewed black children as "naturally inferior." But he adamantly opposed writing segregation into the constitution: "Make this distinction in your organic law and in many places the white children will have good schools . . . while the colored people will have none." Only the threat of integration would force states to provide blacks with "good schools" of their own.[77]

Republicans also differed among themselves over the extent, and implications, of the democratization of Southern politics. With black suffrage, in most states, came legislative apportionment based either on total population or registered voters. The end of the "white basis" of representation marked the final defeat of the upcountry's long battle to reduce the power of plantation counties, and led some scalawags to seek ways of limiting blacks' statewide power. Moreover, demands by black delegates and reform-minded whites for popular election of state and local officials were countered by fears of Democratic majorities in white counties and concern among Whiggish scalawags who hoped they, not blacks, would control local affairs in the plantation belt. Torn between the desire to expand popular control of government and uncertainty about the breadth of their white support, the conventions adopted contradictory policies, in some cases greatly enhancing democracy, in others actually limiting it.

In North Carolina, where upcountry resentment at the undemocratic structure of state and local politics had a long history and the Republican party's biracial coalition appeared secure, the convention produced what Henry Wilson called "the most republican constitution in the land." This was an apt remark, for language evoking revolutionary-era republicanism echoed in the debates. One scalawag proposed to abolish the state senate

77. McMillan, *Constitutional Development in Alabama*, 152; *Louisiana Convention Journal*, 186; Hume, "Negro Delegates," 144; *Journal of the Constitutional Convention of the State of Virginia* (Richmond, 1867 [actually 1868]), 333–40; *South Carolina Convention Proceedings*, 2:889–901; Raleigh *Daily Standard*, March 7, 1868.

as an example of "special advantage and protection of a particular class of citizens." Another denounced the state's property qualifications for office and its antiquated system of legislative representation as deriving from "the old fallacy that the people are incapable of self-government." Although the senate survived, the delegates dismantled the old state Executive Council and replaced county courts appointed by the General Assembly with state and local officials chosen by popular vote. And they made judges, from justices of the peace to the state supreme court, elective.[78]

At the other end of the political spectrum stood Georgia and Florida, whose conventions, dominated by moderate scalawags, adopted ingenious methods of making Reconstruction palatable to white voters and minimizing what Georgia's first Republican governor called "the danger of negro suffrage." Georgia's county-based system of legislative apportionment restricted the influence of the geographically concentrated black population, while provisions for legislative appointment of all state officials except the governor, and gubernatorial selection of judges above the level of justice of the peace, seemed to promise that blacks would occupy few of these positions. The constitution also omitted any mention of blacks' right to hold office and required that jurors be "worthy and intelligent" citizens, thus giving local officials ample authority to exclude freedmen. Florida's convention, controlled after a series of complex maneuvers by a coalition of business-oriented white Republicans and Whiggish Conservatives, likewise skewed legislative representation in favor of white counties, gave the governor "imperial" powers of appointment, and authorized the legislature to establish an educational qualification for voting. Designed to attract white voters to a moderate Republican party devoted to Florida's economic development, the constitution, commented *The Nation*, "surpasses in conservatism that of any State in the Union."[79]

Another issue pitting commitment to democracy against party survival was the disenfranchisement of former Confederates. Many Republicans could not reconcile their party's democratic rhetoric with proposals to strip large numbers of "rebels" of the vote. Although upcountry scalawags, especially those who had suffered for their Unionist beliefs or hailed from areas devastated by the South's internal civil war, supported

78. *CG*, 40th Congress, 2d Session, 2691; "Resolution of Mr. Congleton, February 14, 1868," Secretary of State Papers, Constitutional Convention, NCDAH; Raleigh *Daily Standard*, February 14, 1868; Paul D. Escott, *Many Excellent People: Power and Privilege in North Carolina, 1850–1900* (Chapel Hill, 1985), 142.

79. Scroggs, "Carpetbagger Constitutional Reform," 485–88; Bullock, "Reconstruction in Georgia," 672; Drago, *Black Politicians*, 40–44; Richard L. Hume, "Membership of the Florida Constitutional Convention of 1868: A Case Study of Republican Factionalism in the Reconstruction South," *FIHQ*, 51 (July 1972), 5–7, 15–16; Samuel Walker to Elihu B. Washburne, June 12, 1868, Washburne Papers; *Nation*, May 21, 1868.

disenfranchisement most vehemently, the issue followed no simple pattern, for it became embroiled in Republican divisions over the party's prospects of attracting white voters. Five states disenfranchised few or no Confederates: Georgia, Florida, and Texas, where moderates committed to luring white Conservatives into the party controlled the proceedings; South Carolina, with its overwhelming black voting majority; and North Carolina, where the party's white base appeared firm. (North Carolina's mountain delegates, however, objected to their constitution's leniency.) Alabama and Arkansas barred from voting men disqualified from office under the Fourteenth Amendment as well as those who had "violated the rules of civilized warfare" during the Civil War, and required all voters to take an oath acknowledging black civil and political equality. Even this was not enough for one delegate from the strife-torn Arkansas upcountry, who "would have disfranchised every one of them." Louisiana, where the likelihood of white support appeared bleak, disenfranchised Confederates from newspaper editors and ministers who had advocated disunion to those who had voted for the secession ordinance, but exempted men willing to swear to an oath favoring Radical Reconstruction. Mississippi and Virginia, to the chagrin of Whiggish Republicans, also barred considerable numbers of "rebels" from voting.[80]

A hallmark of upcountry Republicanism, disenfranchisement generated less interest among black delegates, many of whom seemed uncomfortable with a policy that appeared to undermine the party's commitment to manhood suffrage. "I have no desire to take away the rights of the white man," declared former slave Thomas Lee, an Alabama delegate. "All I want is equal rights in the court house and equal rights when I go to vote." Others feared the policy might embarrass the party nationally. Virginia black delegate James W. Bland inquired urgently of Congressman Elihu Washburne: "Is it policy to further disfranchise *Rebels*? . . . Give me your advice immediately." Generally, black voting on the issue paralleled that of the white Republican leadership, favoring leniency in Georgia and North Carolina, and following Hunnicutt in supporting proscription in Virginia. Louisiana's black majority found itself divided by the issue. The New Orleans *Tribune* had long opposed disenfranchisement, arguing, "If we refuse the franchise to any class, it can as well be withheld from us." Yet a Northern reporter found black delegates more proscriptive than their white counterparts, and a solid majority, including the few identified as former slaves, voted in favor of the provision.[81]

80. *Alabama Convention Journal*, 30–34; *Arkansas Convention Debates*, 320–21, 658, 673; *Virginia Convention Debates*, 528–33; William A. Russ, Jr., "Disfranchisement in Louisiana (1862–1870)," *LaHQ*, 18 (July 1935), 575–76; Hamilton J. Eckenrode, *The Political History of Virginia During Reconstruction* (Baltimore, 1904), 98–102; *Journal of the Proceedings of the Constitutional Convention of the State of Mississippi, 1868* (Jackson, 1871), 732.

81. McMillan, *Constitutional Development in Alabama*, 129; James W. D. Bland to Elihu B. Washburne, March 15, 1868, Washburne Papers; *Georgia Convention Journal*, 299–300; *North*

On economic matters, the developmental spirit prevailed. With a modern constitution, declared Mississippi Union League president Allston Mygatt, "large landed estates shall melt away into small divisions, . . . mechanism [will] flourish, agriculture become scientific, internal improvements be pushed on." The desire to promote economic growth led Alabama's convention to establish a Bureau of Industrial Resources, and Georgia's to move the state capital from the sleepy village of Milledgeville to the thriving commercial entrepôt of Atlanta (where, Joseph Brown explained, "the State can build a splendid granite Capitol, hewn out of the Stone Mountain, with convict labor, at a very light cost"). The new constitutions allowed extensive public aid to railroads and other ventures, although sometimes adding safeguards against abuse. (Mississippi and Virginia barred the loan of state credit to corporations, and Texas prohibited railroad land grants, but all three allowed direct financial subsidies.) North Carolina's delegates, unwilling to wait for legislative action, voted $2 million in immediate railroad aid. Several states authorized the granting of charters under general incorporation laws, and some for the first time established limited liability for corporate stockholders. Many voided usury laws entirely or dramatically increased the legal ceiling on interest rates.[82]

The desire to attract outside capital doomed Radicals' demand for an ab initio Reconstruction that would wipe away existing state debts and all laws, including corporate charters and railroad land grants, dating from the Confederacy. In every state, carpetbaggers emerged as the strongest proponents of economic modernization; upcountry scalawags and blacks, especially former slaves, were more cautious. An attempt by Alabama Radical Daniel Bingham to levy a $50,000 tax on railroads receiving state aid won the support of upcountry white Republicans and a majority of blacks, but was defeated by a coalition of carpetbaggers, Conservatives, and black belt scalawags. North Carolina's carpetbaggers joined with blacks and Conservatives to validate prewar state bonds, held in large amounts by railroad companies, but an attempt to establish a commissioner of immigration fell before the opposition of Conservatives, scalawags, and blacks. (One black delegate chided those hoping to use state funds to assist immigrants—"if anything was to be distributed, the poor black should have it.") Georgia had the most cautious black delegates, who unanimously helped carpetbaggers and urban scalawags defeat a clause, favored by many upcountry Republicans, authorizing the legisla-

---

Carolina Convention Journal, 251; Virginia Convention Journal, 221, 239–40, 271, 283–84, 295; New Orleans Tribune, November 25, 1866; New York Times, February 1, 1868.
   82. Mississippi Convention Journal, 4; Atlanta Daily New Era, March 11, 1868; Mark W. Summers, Railroads, Reconstruction, and the Gospel of Prosperity: Aid Under the Radical Republicans, 1865–1877 (Princeton, 1984), 25; Carter Goodrich, "The Revulsion Against Internal Improvements," JEcH, 10 (November 1950), 158–61; Jonathan M. Wiener, Social Origins of the New South 1860–1885 (Baton Rouge, 1978), 148–51.

ture to repeal or modify corporate charters. And in South Carolina, free blacks from the North, some already involved in railroad ventures, opposed limiting aid to internal improvements. "This is a progressive age," declared Pennsylvania-born Jonathan J. Wright, and the government should be left free "to do those things for the public good which the public good requires."[83]

Even more complex debates and alignments surrounded debtor relief, an issue that had won Republicans considerable support in the North Carolina, Georgia, and Alabama upcountry. Facing the loss of their land, many indebted small farmers viewed the relief question in stark class terms. "This is a strife between capital and labor," declared a Georgia scalawag delegate, "between the wealthy aristocrats and the great mass of the people." Actually, "wealthy aristocrats" were among those clamoring for relief. "The whole South is now bankrupt," a planter's wife lamented, and throughout the region estates were being advertised for sale to liquidate debts. Whatever their political differences, upcountry farmers and black belt planters shared a common interest in retaining control of their land. Others, however, viewed debtor relief less enthusiastically. Merchants forced to meet the demands of out-of-state suppliers would be ruined if local clients could avoid repaying advances. Economic promoters feared Northerners would never lend capital to railroads and other enterprises if barred from taking recalcitrant debtors to court. And many blacks agreed with the Georgian who told his former owner, "The freedmen cannot be benefited by this measure. They owe no debts." Blacks, moreover, knew that stay laws had enabled employers to avoid obligations to their laborers and had no desire to help rescue the planter class from liquidation. Staying the collection of debts, James W. Hood told North Carolina's convention, would benefit "those who now hold lands" and "prevent the poor people from ever getting land."[84]

Only in Georgia, which abrogated all debts dating from before 1865, did a substantial number of black delegates support stay measures, subordinating their constituents' desire for land to the attempt to woo white voters. As an alternative to debtor relief, a majority of Mississippi's black delegates urged the army to establish a public works program for the destitute of both races (a proposal in which military authorities displayed

83. Moneyhon, *Republicanism in Reconstruction Texas*, 72–76, 87; *Alabama Convention Journal*, 137; *North Carolina Convention Journal*, 204, 214, 310; Olsen, *Tourgée*, 110; Leonard Bernstein, "The Participation of Negro Delegates in the Constitutional Convention of 1868 in North Carolina," *JNH*, 34 (October 1949), 408; *Georgia Convention Journal*, 336–37; *South Carolina Convention Proceedings*, 1:249–50, 2:659–62.

84. Atlanta *Daily New Era*, February 2, 1868; Mrs. W. A. Kincaid to Mrs. E. K. Anderson, August 20, 1867, Kincaid-Anderson Papers, USC; Henry Watson, Jr., to James Dixon, December 20, 1867, Watson Papers; Jerrell H. Shofner, "A Merchant Planter in the Reconstruction South," *AgH*, 44 (April 1972), 291–96; A. W. Spies to Charles Jenkins, November 30, 1866, Georgia Governor's Papers, UGa; Joshua Hill to John Sherman, January 10, 1868, Sherman Papers; Raleigh *Daily Standard*, February 3, March 6, 1868.

no interest). In South Carolina, freeborn Charleston tailor Robert G. DeLarge and Northerner William J. Whipper urged delegates to heed "the voice of the impoverished people of the state." But Francis L. Cardozo opposed any step that might prevent the breakup of "the infernal plantation system." By a narrow margin, the delegates called upon the army to suspend the collection of debts; scalawags overwhelmingly favored the measure, with blacks, free and slave, strongly opposed.[85]

In preference to stay ordinances, which benefited rich and poor debtors indiscriminately, blacks generally favored the homestead exemption, a more selective form of relief. Measures shielding a specified amount of real and personal property from seizure by creditors guaranteed that an indebted small farmer or artisan would not lose his land, tools, or household furniture. "The petitions of the $20,000-men have been heard by Andrew Johnson," declared an interracial group of Virginia Republicans, calling on their state's convention to secure the homes of "the $1,000 or $500-men" against foreclosure. Every constitution except Louisiana's incorporated a homestead exemption—ranging from $1,500 worth of property in North and South Carolina to $5,000 in Arkansas and Texas— while barring its use against laborers' claims for wages and, in some cases, tax liabilities. Since Southern land values had fallen sharply since the war, the larger exemptions effectively protected the vast majority of white Southerners. "There are not many men," one correspondent remarked, "whose farms and residences . . . would bring $5,000 in cash were they sold tomorrow." Blacks generally favored low exemptions, hoping to benefit small farmers while compelling those with "fifty or eighty thousand acres" to part with some of their land if they fell into debt.[86]

If debtor relief temporarily stabilized Southern landholding patterns, tax reform seemed to offer an opportunity to reshape class relations. At Virginia's Radical-dominated convention, delegates black and white expressed bitter resentment against the antebellum tax system and the inordinately high state and local poll taxes of Presidential Reconstruction. "The poor people have to bear all the burdens of taxation in this State. . . ." declared Thomas Bayne, noting that freedmen paid poll taxes of $3 to $4, while "vacant lots worth thousands of dollars were taxed but fifty cents." Hunnicutt pointed out that low land taxes encouraged planters and speculators to accumulate large uncultivated tracts, which

85. *Georgia Convention Journal*, 198–99, 252, 458–59; William C. Harris, *The Day of the Carpetbagger: Republican Reconstruction in Mississippi* (Baton Rouge, 1979), 134–35; *South Carolina Convention Proceedings*, 1:108–14, 125–30, 148. The breakdown of South Carolina votes on the motion (which passed 57–52) calling upon the Army to suspend the collection of debts was carpetbaggers 6–4, scalawags 27–6, Northern blacks 6–8, freedmen 13–22, Southern free blacks 4–11, unidentified 1–1.

86. Olsen, *Tourgée*, 129–35; *Virginia Convention Debates*, 87–88; J. H. Thomas, "Homestead and Exemption Laws of the Southern States," *American Law Register*, 19 (1871), 1–17, 137–50; *New York Times*, February 5, 1868; *North Carolina Convention Journal*, 348–49; *South Carolina Convention Proceedings*, 1:137, 141.

a heavy land tax might force onto the market. "If we do not tax the land," declared Frank Moss, a black delegate who described himself as "a working man," "we might just as well not have come here to make a Constitution."

Some blacks proposed to make the tax structure progressive, through an extra levy on uncultivated land or, as one Louisiana delegate suggested, exempting farms of fewer than sixty acres from taxation altogether. The only such measure adopted, however, was one in Virginia, authorizing a special levy on annual incomes of over $600. Generally, the new constitutions rested state revenues on a general property tax, a principle already well-established in the North, but a striking departure in Southern fiscal policy. All property—land, personal possessions, stocks and bonds—would henceforth be taxed equally, drastically increasing the burdens of landowners from planter to small farmer, while benefiting propertyless freedmen as well as commercial interests, artisans, and professionals previously burdened by heavy license fees. Despite upcountry opposition, the constitutions also authorized modest poll taxes, with the revenue earmarked for the new school systems. Blacks generally supported these imposts, convinced that public education financed entirely by property taxes would quickly lose white support.[87]

When the constitutional conventions assembled, many Conservatives, in the words of a New Orleans newspaper, feared a policy of "unadulterated agrarianism" (the nineteenth-century term for attacks on private property). With "Negroes and Tories" in control, a Georgian warned, "a general division of lands" would follow, along with "laws regulating the price of labor and the rent of lands—all to benefit the negro and the poor." In one form or another, the interrelated questions of land and labor came before a majority of the conventions, but the results hardly justified either the hopes or fears kindled in 1867. Several conventions awarded mechanics and laborers liens for wages on the property of their employers, and Texas outlawed the establishment of "any system of peonage," but free labor assumptions doomed other attempts to improve the position of black workers. An attempt by South Carolina Confederate Army deserter James M. Allen to set one half of the crop as a maximum rent failed, as did a Virginia carpetbagger's proposal to establish an eight-hour day for "hired labor." Louisiana's constitution included a provision, strongly supported by carpetbaggers and Conservatives, barring the legislature from "fixing the price of manual labor." Wealthy free black employers like sugar planters Pierre G. Deslonde and Auguste

87. *Virginia Convention Debates*, 104, 129–30, 138, 487, 652, 686–87, 695, 713–15, 722; Louis Post, "A Carpet-Bagger in South Carolina," *JNH*, 10 (January 1925), 31; *Louisiana Convention Journal*, 112; J. Mills Thornton III, "Fiscal Policy and the Failure of Radical Reconstruction in the Lower South," in Kousser and McPherson, eds., *Region, Race, and Reconstruction*, 349–94; Olsen, *Tourgée*, 108.

Donato favored the clause, as did prominent New Orleans free blacks to whom any interference in the labor market recalled the hated contract regulations of General Banks. Other New Orleans delegates, black and white, opposed the provision, hoping to satisfy the eight-hour demands of the city's labor movement. Both ex-slaves voting opposed the provision guaranteeing a free market in labor, possibly anticipating future state regulations favoring employees.[88]

Both black and white Radicals spoke of the need to provide freedmen with land and encourage the breakup of the plantation system. "I have gone through the country," reported Richard H. Cain in South Carolina, "and on every side I was besieged with questions: How are we to get homesteads?" A few constitutions took modest steps toward meeting this demand. Texas offered free homesteads to settlers on the state's vast public domain, and Mississippi provided that land seized by the state to satisfy tax claims would be sold in tracts of no more than 160 acres. Louisiana set an even lower limit of fifty acres, but defeated a proposal to limit the amount any individual could purchase at such sales. (New Orleans free blacks favored the first clause but unanimously opposed the second, possibly hoping to acquire sizable holdings themselves.) The most substantial action was taken by South Carolina's convention, which authorized the legislature to establish a state commission to purchase land for resale on long-term credit. Soon after the convention opened, the delegates killed a motion by black carpetbagger Landon S. Langley, a veteran of the famous 54th Massachusetts Infantry, that confiscation and disenfranchisement be "forever abandoned." Sixty percent of white Republicans favored Langley's proposal, but a large majority of both Southern free blacks and freedmen opposed it. The most significant black support came from Northerners like Langley, carriers of free labor ideas. But confiscation never came to the floor, partly because the delegates feared Congress would reject a constitution containing such a provision.[89]

With each state producing its own mixture of radical and moderate elements, the new constitutions, taken together, failed to satisfy the economic aspirations that had animated much of the grass-roots organizing of 1867. But they introduced changes in the region's political structure that appeared dangerously radical to those wedded to the traditions of

88. New Orleans *Picayune*, February 20, 1868; Alan Conway, *The Reconstruction of Georgia* (Minneapolis, 1966), 148; *South Carolina Convention Proceedings*, 1:194; *Virginia Convention Journal*, 297; *Louisiana Convention Journal*, 26, 120–21. Blacks divided evenly, 21–21, on the Louisiana clause, which was adopted 43–34.

89. *South Carolina Convention Proceedings*, 1:380; *Journal of the Reconstruction Convention, Which Met at Austin, Texas* (Austin, 1870), 1:895; *Mississippi Convention Journal*, 739; *Louisiana Convention Journal*, 266–67, 1465. The vote on a motion to take Langley's motion from the table, which failed 46–61 (*South Carolina Convention Proceedings*, 1:43) was scalawags 20–10, carpetbaggers 6–5, freedmen 8–25, Northern blacks 6–8, Southern free blacks 5–11, unidentified 1–2.

the Old South and Presidential Reconstruction. Democrat and Whig alike, the bulk of the region's traditional leadership now mobilized in opposition. Their campaigns rested on forthright appeals to race. Other issues were "trivial" compared to one overriding question: "Shall this country be ruled by the whites or the niggers?" On Alabama's Democratic Executive Committee, only one member favored attempting to win the black vote, the remainder preferring "the battle cry of white man's government." North Carolina Conservatives harped upon the specter of integration in the new public schools, where white children would "take in all the base and lowly instincts of the African."[90]

Racial appeals, however, often went hand in hand with revulsion at the prospect of governments controlled by what North Carolina Governor Worth called "the dregs of society." Universal suffrage—government by "mere *numbers*," Worth wrote, "I regard as undermining civilization." Civilization he defined as "the possession and protection of property." It was clear that such remarks did not apply to blacks alone. In 1867, when Gen. Daniel E. Sickles declared all taxpayers eligible for jury duty, Worth protested: "To say nothing of negroes, juries drawn from the whites only under the order, would not be fit to pass on the rights of their fellow men." If North Carolina's constitution needed revision, Worth and other Democratic leaders preferred a return to the frame of government of 1776, which contained substantial property requirements for voting. Although it was most pronounced in states like North Carolina, where a considerable number of white yeomen seemed ready to join the Republicans, elitist opposition to Reconstruction could be found throughout the South. Suffrage, declared a New Orleans newspaper, was "a duty rather than a right, and [we] regret that there is so much of it among the whites." An elaborate "Remonstrance" delivered to Congress by the state's Democrats took pains to demonstrate that few convention delegates paid state taxes and castigated the idea of a school system serving the poor but funded by property owners. ("What the protest claimed as grievances," Thaddeus Stevens remarked, Republicans like himself "regarded as virtues.")[91]

With blacks all but unanimous for the constitutions, Republicans no less than Conservatives focused their attention on the white vote, although their strategy differed markedly from state to state. Florida moderates secured the gubernatorial nomination of carpetbagger Harrison

90. William M. Browne to Samuel L. M. Barlow, April 22, 1868, Samuel L. M. Barlow Papers, HL; Conway, *Reconstruction of Georgia*, 154; manuscript account of Democratic Executive Committee meeting, January 15, 1868, Robert McKee Papers, ASDAH; "Anti-Constitution," *To the White Men of North Carolina*, broadside, 1868.

91. Hamilton, ed., *Worth Correspondence*, 2:1155–56, 1201, 1217–18; Elizabeth G. McPherson, ed., "Letters from North Carolina to Andrew Johnson," *NCHR*, 29 (January 1952), 110–11; T. Harry Williams, "An Analysis of Some Reconstruction Attitudes," *JSH*, 12 (November 1946), 478; KKK Hearings, South Carolina, 1241–44.

Reed, a former supporter of President Johnson who made a tacit agreement with Conservatives to run the state on principles of white supremacy and economic development. In Georgia, whose constitutional convention delegates nominated Rufus Bullock for governor to avoid holding a state Republican convention that might be dominated by the Union Leagues, the party adopted a "two-faced" strategy, running blacks for local positions in plantation counties, while devoting exclusive attention to the relief issue in the upcountry and assuring whites there that the constitution did not allow blacks to hold office. Elsewhere, Republicans presented themselves as the "poor man's party," castigating the "landed aristocracy" that had kept "the mass of white people in poverty and ignorance." Gubernatorial candidate William H. Smith pledged to bring to Alabama factories, mines, and other enterprises previously blocked by a political regime that condemned the upcountry to "underdevelopment and its people to poverty." And Republicans reminded Southern loyalists of how the same "aristocrats and secession oligarchs" who now opposed Reconstruction had plunged the region into rebellion.[92]

Except in Georgia and Florida, Republicans did not abandon the principle of black political equality to appeal to white voters. Yet a combination of circumstances reduced the influence of blacks and white Radicals in nearly every Southern state. General Schofield, alarmed by the Virginia constitution's disenfranchisement provisions and convinced that "worthless radicals, white and black" would soon control the state, refused to authorize a ratification election. And fear of alienating white voters kept blacks off every state ticket except in South Carolina and Louisiana. Already, blacks confronted a political dilemma that would plague them throughout Reconstruction—their very unanimity as Republicans meant their ballots could be taken for granted by party leaders seeking the white vote.[93]

In some ways the most unpropitious events of all took place in Louisiana, where divisions among black Republicans enabled moderate carpetbaggers to seize control of the party. At the state convention, Henry C. Warmoth narrowly won the gubernatorial nomination over Francis E. Dumas, a wealthy French-born octoroon backed by the New Orleans *Tribune* and its circle. Perhaps because he had been one of the state's largest slaveholders, Dumas failed to win the support of freedmen (al-

92. Jerrell H. Shofner, "Florida: A Failure of Moderate Republicanism," in Otto H. Olsen, ed., *Reconstruction and Redemption in the South* (Baton Rouge, 1980), 15–21; Atlanta *Daily New Era*, March 11, 1868; Nathans, *Losing the Peace*, 88–92; Harris, *Day of the Carpetbagger*, 187; Harry K. Benson, "The Public Career of Adelbert Ames, 1861–1876" (unpub. diss., University of Virginia, 1975), 129; Mobile *Nationalist*, January 30, 1868; Raleigh *Daily Standard*, March 16, 1868.

93. John M. Schofield to Ulysses S. Grant, April 18, 1868 (copy), John M. Schofield Papers, LC; Jack P. Maddex, Jr., *The Virginia Conservatives 1867–1879* (Chapel Hill, 1970), 60–63.

though by all accounts he had treated his slaves well and during the Civil War, when he served as a Union officer under General Butler, Dumas urged his slaves to enlist as well). Warmoth also gained the backing of free blacks like P. B. S. Pinchback and James H. Ingraham, whose origins lay outside the French-speaking free elite. When the *Tribune* refused to go along with Warmoth's nomination, it lost its state and federal printing contracts and was forced to suspend publication. Its editor, Jean-Charles Houzeau, departed to settle on a Jamaica banana plantation, eventually returning to his native Belgium to direct the Royal Observatory. The *Tribune*'s demise deprived the South of an eloquent advocate of Radicalism just as Republican rule, which it had done so much to bring about, commenced.[94]

The elections of winter and spring 1868 to ratify the constitutions and elect state officials brought mixed results for Southern Republicans. The black vote again held firm, but the party's attempt to lure white voters did not always bear fruit. It required a conservative Republican ticket (in Florida), a strong tradition of upcountry Unionism (in Arkansas and northern Alabama), or one of these combined with direct economic appeals to white yeomen (in North Carolina and Georgia) to produce a significant white Republican turnout. The party's most remarkable success occurred in North Carolina, where the cry of racial unity failed to deter over 20,000 white voters, about a quarter of the white electorate, from supporting the new constitution. William W. Holden finally won election as governor, and Republicans swept to a commanding majority in the state legislature. "In North Carolina at least," wrote one observer, "it cannot be attributed to foreign influence that the Constitution was adopted." Bullock became Georgia's governor, defeating Confederate Gen. John B. Gordon by 7,000 votes, and the constitution won ratification by more than twice that margin. Buoyed by the relief issue and Joseph Brown's popularity among small farmers, Republicans carried twelve upcountry counties and won significant support in many more. Few whites voted Republican in South Carolina, but none were needed, as the party rolled up a majority of better than two to one.

Elsewhere, the results proved more ominous. Warmoth won a resounding victory, but carried only a tiny number of Louisiana's white votes. Even Unionist Winn Parish went Democratic, although its 113 white Republicans represented the largest concentration of scalawags in the state. Moreover, by taking advantage of divisions among carpetbaggers, freedmen, and free blacks, Democrats were able to carry some plantation parishes. And Alabama and Mississippi defeated the new constitutions. In the former, a white boycott prevented approval by a majority of registered voters (a requirement Congress quickly changed to a majority of those actually casting ballots). In Mississippi, whose disenfranchisement clause alienated nearly all whites, Democrats even carried Choctaw, the "banner county" of wartime Unionism.[95]

Thus, three years after the death of the Confederacy, Republicans came to power in most of the South. "These constitutions and governments," a Democratic newspaper vowed, "will last just as long as the bayonets which ushered them into being, shall keep them in existence, and not one day longer." The fate of Reconstruction still depended on the national political contest of 1868.[96]

Impeachment and the Election of Grant

To the many dramatic innovations Reconstruction brought to American politics, the spring of 1868 added yet another: the unprecedented spectacle of the President's trial before the Senate for "high crimes and misdemeanors." The roots of the impeachment of Andrew Johnson lay not only in the increasingly hostile relations between himself and Congress, but in a peculiar feature of Republican Reconstruction policy itself. For Congress had enjoined the army to carry out a policy its commander-in-chief resolutely opposed. Even before 1867, a number of Radicals had called for Johnson's removal, fearing that Reconstruction could never be successful so long as he remained in office. Ohio Congressman James M. Ashley became obsessed with the issue, attempting to prove that, like William Henry Harrison and Zachary Taylor (who, he contended, had been poisoned), Lincoln had been murdered to place his Vice President in the White House. Instead of following Ashley down the road to impeachment, however, Congress preferred to shield its policy, and the Republican party, against Presidential interference. In 1867, it required that all orders to subordinate army commanders pass through General Grant and, in the Tenure of Office Act, authorized officials appointed with the Senate's consent to remain in office until a successor had been approved. Intended primarily to protect lower level patronage functionaries, the law also barred the removal, without Senate approval, of Cabinet members during the term of the President who had appointed them. It remained uncertain, however, whether this applied to Secretary of War Stanton, who had been named to his post by Lincoln.

95. *Tribune Almanac*, 1869, 75–84; Jesse Wheeler to Benjamin S. Hedrick, May 27, 1868, Hedrick Papers; Donald W. Davis, "Ratification of the Constitution of 1868—Record of Votes," *LaH*, 6 (Summer 1965), 301–305; William T. Blain, "Banner Unionism in Mississippi, Choctaw County 1861–1869," *Mississippi Quarterly*, 29 (September 1976), 218–19. In addition to Virginia, no election was held in Texas, where the convention did not complete its work until 1869.

96. Charleston *Mercury*, February 5, 1868.

Some supporters urged the President to remove Stanton and declare the Reconstruction Act null and void. But while determined to obstruct the implementation of Congressional policy, Johnson did not covet the role of sacrificial lamb. Instead, he waited until midsummer to take advantage of a provision of the Tenure of Office Act allowing him to suspend Stanton while Congress was not in session, pending a vote on his permanent removal after the Senate reconvened. So long as Johnson remained within the law, impeachment got nowhere. Early in December, Ashley's motion from the previous winter finally came to the House floor, only to suffer an overwhelming defeat, with every Democrat and a majority of Republicans in opposition. Coming hard on the heels of the fall elections, the House vote convinced the President that the Northern public had at last rallied to his policies. He now set about actively encouraging Southern opponents of Reconstruction and ousted several military commanders in favor of more conservative replacements. And when the Senate refused to concur in Stanton's removal, Johnson, on February 21, 1868, ousted him from his office.[97]

Having failed to "play the part of Moses for the colored people," commented The Nation, Johnson had succeeded in doing so "for the impeachers." A combination of motives and calculations shaped his course—certainty of popular support, a desire to reassert the powers of the Presidency, even the hope of commending himself to the Democrats as their 1868 nominee. Yet, as so often in the past, Johnson had miscalculated. General Sherman, who mostly shared the President's views, found his actions indefensible: "He attempts to govern after he has lost the means to govern. He is like a General fighting without an army." With unanimous Republican support, the House voted to impeach the President. Although the decision represented, in part, a cathartic gesture, a determination to be rid once and for all of this irritating Chief Executive, Republicans also had practical reasons for desiring Johnson out of office, especially the growing conviction that his actions threatened the success of Reconstruction. Radicals, in particular, were daily receiving letters from the South reporting that opponents of Congressional policy monopolized federal patronage and declaring Johnson's removal essential to the survival of Southern Republicanism.[98]

Yet from the outset, the case against the President was beset with weaknesses. Of the eleven articles of impeachment, nine hinged on either

97. Robert F. Horowitz, The Great Impeacher: A Political Biography of James M. Ashley (Brooklyn, 1979), 123–42; Hans L. Trefousse, Impeachment of a President: Andrew Johnson, the Blacks, and Reconstruction (Knoxville, 1975), 44–50, 96–112; Howard K. Beale, ed., Diary of Gideon Welles (New York, 1960), 3:160–61; Benedict, Compromise of Principle, 276–94.

98. Nation, February 27, 1868; James E. Sefton, Andrew Johnson and the Uses of Constitutional Power (Boston, 1980), 157–59; M. A. De Wolfe Howe, ed., Home Letters of General Sherman (New York, 1909), 373; Eric L. McKitrick, Andrew Johnson and Reconstruction (Chicago, 1960), 490, 507.

the removal of Stanton or an alleged attempt to induce Gen. Lorenzo Thomas to accept orders not channeled through Grant. Two others, drafted by Butler and Stevens, charged the President with denying the authority of Congress and attempting to bring it "into disgrace." Nowhere were the real reasons Republicans wished to dispose of Johnson mentioned—his political outlook, the way he had administered the Reconstruction Acts, and his sheer incompetence. In a Parliamentary system, Johnson would long since have departed, for nearly all Republicans by now agreed with Supreme Court Justice David Davis, who described the President as "obstinate, self-willed, combative," and totally unfit for his office. But these, apparently, were not impeachable offenses. Despite the changes made by Butler and Stevens, the articles as a whole implicitly accepted what would become the central premise of Johnson's defense: that only a clear violation of the law warranted a President's removal.[99]

Other factors enhanced Johnson's chances for retaining office. Since the Vice Presidency remained vacant, his successor would be Ben Wade, president pro tem of the Senate and a man disliked by moderates for his radical stance on Reconstruction, and by many businessmen and laissez-faire ideologues for his high tariff, soft-money, prolabor views. "All the great Northern capitalists," a Radical editor observed, feared impeachment would shatter public confidence in the government and its securities. ("Five-twenties [federal bonds] at forty cents on the dollar," he added, "are strong reasons for preserving the Constitutional Powers of the President.") Many who would have rejoiced to "have Johnson out of the way" hesitated to elevate a man "who thinks our greenbacks are the best currency in the world, with the single defect that there are not enough of them."[100]

When Johnson's trial before the Senate opened, Chief Justice Chase, who like the defendant thought of himself as Democratic Presidential timber, steered the proceedings in a narrowly legalistic direction. Unable to raise broad issues of policy or get at the President's real "crimes," the House managers were forced to concentrate exclusively on Stanton's removal. The defense case, however, had its own weaknesses, notably the fact that its arguments seemed patently contradictory. On the one hand, Johnson's attorneys argued that since the Tenure of Office Act did not apply to Stanton, his removal was perfectly legal, although in that case

99. Richardson, ed., *Messages and Papers*, 6:709–18; Thaddeus Stevens to Benjamin F. Butler, February 28, 1868, Butler Papers; Willard L. King, *Lincoln's Manager: David Davis* (Cambridge, Mass., 1960), 260; Michael L. Benedict, *The Impeachment and Trial of Andrew Johnson* (New York, 1973), 26–34.

100. Hans L. Trefousse, *Benjamin Franklin Wade: Radical Republican from Ohio* (New York, 1963), 295–309; T. W. Egan to Andrew Johnson, October 7, 1867, Andrew Johnson Papers, LC; Horace White to Elihu B. Washburne, May 1, 1868, Washburne Papers; *Nation*, February 27, April 30, 1868.

it was hard to explain why in 1867 the President had followed the letter of the law by suspending the Secretary and informing the Senate of his reasons. On the other hand, they contended that Johnson had violated the statute in order to allow the Supreme Court to rule on its constitutionality, an argument that would have allowed the President to determine which laws he was required to obey.[101]

Whatever the merits of the case, it soon became apparent that an influential group of Senate Republicans preferred, as Johnson's attorney William M. Evarts remarked, "the present situation to the change proposed." Trumbull, Fessenden, and Grimes, who shared moderate views on Reconstruction and held free trade convictions, feared both the damage to the separation of powers that would result from conviction and the political and economic policies that might characterize a Wade Presidency. Moreover, the crisis atmosphere of February receded. Republicans triumphed in a string of Southern elections, and the Supreme Court acceded to a law rushed through Congress stripping it of jurisdiction in habeas corpus cases, thus rendering moot the appeal of an imprisoned Mississippi editor that might have raised the question of the constitutionality of Reconstruction. Meanwhile, Evarts quietly passed along assurances that Johnson, if acquitted, would cease his efforts to obstruct Republicans' Southern policy.[102]

By April, Washington bookmakers' odds had shifted in favor of acquittal, and when the decision finally came in mid-May, only thirty-five Senators voted for conviction, one short of the required two thirds. The votes of seven Republicans supplied Johnson's narrow margin of victory, although a number of others stood ready to support the President if necessary. Wade failed to disqualify himself, voting for a verdict that would have placed him in the White House. (Kansas Republican Sen. Edmund G. Ross, who voted for acquittal, also invited charges of impropriety, for he quickly cashed in on the President's gratitude by securing lucrative patronage posts for several friends.) Contrary to later myth, Republicans did not read the "seven martyrs" out of the party, and all campaigned for Grant that fall. It would be more accurate to suggest that the impeachment affair formed an important link in a chain of events that left the seven, and some who had voted for conviction, increasingly disillusioned with Reconstruction. All four who survived to 1872 would join the Liberal Republicans.[103]

101. Clemenceau, *American Reconstruction*, 174, 182; Benedict, *Impeachment*, 144–54.

102. William M. Evarts to Edwards Pierrepont, April 16, 1868, Edwards Pierrepont Papers, Yale University; James W. Grimes to Charles H. Ray, May 7, 1868, Charles H. Ray Papers, HL; Theodore C. Smith, *The Life and Letters of James A. Garfield* (New Haven, 1925), 1:425; Charles Fairman, *Reconstruction and Reunion 1864–1888: Part One* (New York, 1971), 415–67; Mantell, *Johnson, Grant, and Reconstruction*, 90–96.

103. Patrick J. Furlong and Gerald E. Hartdagen, "Schuyler Colfax: A Reputation Tarnished," in Ralph D. Gary, ed., *Gentlemen from Indiana: National Party Candidates 1836–1940*

Johnson's acquittal further weakened the Radicals' position within the party and made the nomination of Ulysses S. Grant all but inevitable. A career army officer except for an unhappy civilian interlude in the 1850s, Grant before the war had displayed little interest in politics, although what views he had inclined him toward the Democrats. But if he lacked political experience, Grant did not want for political intuition. During the war, he rose to prominence not only by virtue of his courage and military genius, but by cooperating fully with Lincoln and Congress in implementing such policies as emancipation and the raising of black troops. He emerged from the conflict as the preeminent Union military hero, and as early as 1866 was well aware that influential Republicans were predicting his nomination. Although Grant at first sought to shield himself, and the army, from the political conflict over Reconstruction, by 1868 he had committed himself fully to Congressional policy. Nonetheless, as one Republican put it, his candidacy had a "conservative odor," both because of his apparent lack of ideological convictions, and because his earliest promoters had been the "great conservative and commercial interests in New York" who had once supported Johnson. The city's business elite feared a Democratic victory would reopen the now settled Reconstruction question, while a Grant Presidency promised moderation, fiscal responsibility, and stable conditions for Southern investment. They were joined by moderates who aligned themselves with Grant as a way of weakening Radical influence within state parties. Other Republicans, lamented an Illinois Radical, supported Grant merely because "they think he can be elected. It is a bad sign when we take men instead of principles."[104]

Many Radicals had hoped against hope that, if Wade assumed the Presidency, he might secure the nomination. When the Grant tide became irresistible, they retreated to the demand that he run on "a frank and outspoken radical platform." But this too was denied them by the convention that assembled in Chicago shortly after Johnson's acquittal.

(Indianapolis, 1977), 72; Charles A. Jellison, *Fessenden of Maine* (Syracuse, 1982), 244–45; Mark A. Plummer, "Profile in Courage? Edmund G. Ross and the Impeachment Trial," *Midwest Quarterly*, 27 (Autumn 1985), 39–46; Ralph J. Roske, "The Seven Martyrs?" *AHR*, 66 (January 1959), 323–29. The seven were Fessenden, Grimes, Trumbull, Ross, Joseph Fowler of Tennessee, John B. Henderson of Missouri, and Peter Van Winkle of West Virginia. Fessenden died in 1869, Grimes and Van Winkle in 1872. Henderson joined the Liberals in 1870 but returned to the Republican fold in 1872.

104. William S. McFeely, *Grant: A Biography* (New York, 1981); William H. Trescot to James L. Orr, March 4, 1866, South Carolina Governor's Papers; Harold M. Hyman, "Johnson, Stanton, and Grant: A Reconsideration of the Army's Role in the Events Leading to Impeachment," *AHR*, 66 (October 1960), 96–98; Thompson Campbell to Cornelius Cole, February 11, 1868, Cornelius Cole Papers, University of California, Los Angeles; William B. Hesseltine, *Ulysses S. Grant, Politician* (New York, 1935), 81–102; Cyrus H. McCormick to Samuel L. M. Barlow, December 20, 1867, Barlow Papers; J. M. Edmunds to William Sprague, July 26, 1867, William Sprague Papers, Columbia University; Eva I. Wakefield, ed., *The Letters of Robert G. Ingersoll* (New York, 1951), 151.

Since the defeats of 1867, a Radical Congressman noted, moderates had adopted "the girlish practice of standing timidly on the defensive and discussing [black suffrage] as though of doubtful propriety." This hesitancy was reflected in the party's platform, which declared black voting essential in the South for reasons "of public safety, of gratitude, and of justice," while consigning the issue in the North to the voters of each state. Even this, reported a Massachusetts delegate, was better than the "simply atrocious" original draft, which all but apologized for having forced black voting upon the South. The fiscal plank, drawn up by Chicago *Tribune* editor Horace White, denounced "all forms of repudiation" as a "national crime." All in all, said the New York *Journal of Commerce*, the platform was "far more conservative than anticipated." Yet equality before the law and the right of freedmen to participate in Southern government remained intact as cardinal principles of the Republican party.[105]

Once Grant had been nominated, Congress moved to consolidate the party's position for the fall campaign, readmitting seven Southern states to the Union. Some Radicals, aware of the fragility of Southern Republicanism, criticized what they considered this undue haste. At their behest, Congress prohibited these states from ever limiting the suffrage on racial grounds. One state constitution did attract adverse attention: Georgia's, whose clause abrogating pre-1865 debts, an official of a Southern collection agency pointed out, did not "seem consistent" with the Chicago platform. For Republicans to campaign as "the party of good faith," declared Sen. Oliver P. Morton, they must revoke Georgia's own act of "repudiation." And so Congress did, despite warnings that this would "prove disastrous" to the Georgia party. By the time Congress adjourned, the Republican campaign image had been established—Reconstruction for the South, respectability for the nation, all overseen by the man whose slogan was "Let Us Have Peace."[106]

The Democrats, still exultant over their 1867 triumphs, produced no shortage of men aspiring to oppose Grant. Johnson believed logic dictated his renomination. "Why should they not take me up?" he remarked as the Democratic convention gathered in July. "They profess to accept my measures; they say I have stood by the Constitution and made a noble

105. Schuyler Colfax to Theodore Tilton, January 4, 1868, Schuyler Colfax Papers, NYPL; *CG*, 40th Congress, 2d Session, Appendix; 300; Kirk H. Porter and Donald B. Johnson, eds., *National Party Platforms 1840–1956* (Urbana, Ill., 1956), 39; Francis W. Bird to Charles Sumner, May 25, 1868, Sumner Papers; Adam S. Hill, "The Chicago Convention," *North American Review*, 107 (July 1868), 176–78; New York *Journal of Commerce*, May 23, 1868. House Speaker Schuyler Colfax received the Vice Presidential nomination, a rare instance of a teetotaler defeating a candidate (Ben Wade) who distributed liquor freely to the delegates.
106. *CG*, 40th Congress, 2d Session, 1868, 2464, 2602–3, 2742–44, 2970, 2999–3002; Jonathan McK. Gunn to Thaddeus Stevens, May 26, 1868, Stevens Papers; Joseph E. Brown to Schuyler Colfax, June 9, 1868, Brown Letterbook, Hargrett Collection.

struggle." The Blair family promoted Gen. Francis P. Blair, Jr., whose nomination would safeguard the party against charges of disloyalty. James R. Doolittle also harbored secret hopes. But as the convention approached, Ohio's George H. Pendleton, whose plan to pay the 5-20 bonds in greenbacks enjoyed considerable support in the West, emerged as the frontrunner. So crucial did Pendleton deem "the greenback issue" that he even expressed willingness to run on a platform accepting black suffrage. Many Democrats who disapproved of Pendleton's ideas thought him the most attractive choice. The point, insisted a Chicago editor, was "to get votes in the election," not worry about "sound political economy."[107]

Pendleton's popularity thoroughly alarmed Bourbon Democrats like New York's Samuel L. M. Barlow, Samuel J. Tilden, and August Belmont, all intimately tied to financial circles that viewed the Ohio Plan as a form of repudiation. But they found it difficult to come up with a viable alternative. The most obvious choice, former New York Gov. Horatio Seymour, adamantly refused to be a candidate (although skeptics noted he had made a similar disavowal in every campaign since 1850, when he first ran for governor). For a time, anti-Pendleton Democrats flirted with Chief Justice Chase, a long-time advocate of abolition and black suffrage, who some Radicals, ironically, had earlier seen as a possible Republican nominee. Although the ambitious Chief Justice modified his position on black voting as the spring progressed, finally agreeing to leave the matter to the states, his candidacy would have signaled Democrats' willingness to bury the past, accept the results of the Civil War, and abandon the race issue. For precisely this reason, he generated little enthusiasm in the North and outright hostility in the South, where, as one party leader noted, " 'a white man's government' " remained "the most popular rallying cry we have." Ironically, Eastern leaders used the race issue against Pendleton much as they were accustomed to employing it against Republicans. Financial matters, Belmont insisted, paled in comparison with black suffrage (an odd position for the American representative of the Rothschilds to take).[108]

In the end, after the announced candidates fought to a draw for twenty-one ballots, the convention drafted Horatio Seymour, exactly as those

107. New York *World*, January 10, 1868; Colonel William G. Moore Diary (transcript), July 3, 1868, Johnson Papers; Edward L. Gambill, *Conservative Ordeal: Northern Democrats and Reconstruction, 1865–1868* (Ames, Iowa, 1981), 105, 123–29; James R. Doolittle to Mary Doolittle, July 13, 1868, James R. Doolittle Papers, SHSW; James C. Olsen, *J. Sterling Morton* (Lincoln, 1942), 148–49; W. F. Strong to Manton Marble, February 12, 1868, Marble Papers.

108. Irving Katz, *August Belmont: A Political Biography* (New York, 1968), 164–70; Jerome Mushkat, *The Reconstruction of the New York Democracy, 1861–1874* (Rutherford, N. J., 1981), 129–133; J. W. Schuckers, *The Life and Public Services of Salmon Portland Chase* (New York, 1874), 575–88; L. C. Washington to Howell Cobb, June 21, 1868, Cobb-Erwin-Lamar Papers, UGa.

with a suspicious cast of mind believed the New Yorkers had planned in the first place. (Johnson, who had been second on the first two ballots, believed himself defeated by "duplicity, deceit, cunning management and sharp scheming.") While rejecting Pendleton, however, the delegates adopted his financial views, in a platform calling for the taxation of federal bonds and their repayment in "lawful money"—planks the nominee subsequently repudiated. The colorless Seymour seemed a most unsatisfactory choice; Democrats, commented Secretary of State Seward, "could have nominated no candidate who would have taken away fewer Republican votes." Seymour's conduct during the war (when he addressed New York draft rioters as "my friends") surrendered the loyalty issue to Republicans, and his close ties with New York financiers neutralized any hope of appealing to Western economic resentments.[109]

Thus, Democrats were forced back to a single issue, opposition to Reconstruction, a strategy rendered even more inevitable by the conduct of Frank Blair, Seymour's running mate. In a widely publicized letter written on the eve of the convention, Blair rejected the idea that Reconstruction was a fait accompli: A Democratic President could restore "white people" to power in the South by declaring the new governments "null and void," and using the army to disperse them. The letter injured Democratic prospects by raising the specter of a second civil war, a possibility avidly discussed within the Blair family, but relished by few outside it. Moreover, against Seymour's wishes, Blair embarked on a speaking campaign as disastrous in its own way as Johnson's swing around the circle had been. In blatantly racist language, he excoriated Republicans for placing the South under the rule of "a semi-barbarous race of blacks who are worshippers of fetishes and poligamists," and longed to "subject the white women to their unbridled lust." Offensive as they were, Blair's speeches said little not repeated in the family's private correspondence. Reading Darwin's *The Origin of Species* had reinforced Frank's long-standing fear of racial intermixing, which, he now asserted, would reverse evolution, produce a less-advanced species incapable of reproducing itself, and destroy "the accumulated improvement of the centuries." These were the convictions of the Democratic candidate for Vice President of the United States.[110]

An influential Democratic Congressman would blame Seymour's de-

109. Mantell, *Johnson, Grant, and Reconstruction*, 121–27; Alexander C. Flick, *Samuel Jones Tilden: A Study in Political Sagacity* (New York, 1939), 176–79; Porter and Johnson, eds., *Platforms*, 37–39; Beale, ed., *Welles Diary*, 3:457; Hesseltine, *Grant*, 123–24.

110. McPherson, *Political History*, 380–81; John D. Van Buren to Horatio Seymour, September 5, 1868, Horatio Seymour Papers, New-York Historical Society; Montgomery Blair to Samuel L. M. Barlow, January 29, 1868, Barlow Papers; Forrest G. Wood, *Black Scare: The Racist Response to Emancipation and Reconstruction* (Berkeley, 1968), 126–28; Francis P. Blair, Jr., to Francis P. Blair, January 24, March 1, April 14, August 2, 1868, Blair-Lee Papers, Princeton University.

feat on Blair's "stupid and indefensible" behavior. But Blair set the tone of the Democratic campaign, the last Presidential contest to center on white supremacy. The party's prospects, according to one tactician, depended on whipping up "the aversion with which the masses contemplate the equality of the negro." For their part, Republicans urged veterans to "fall in" behind their old commander and warned of the turmoil that would follow a Democratic victory. "Seymour," said one campaign slogan, "was opposed to the late war, and Blair is in favor of the next one." Many Northern conservatives who had supported Johnson now endorsed Grant, including three members of the Cabinet. Once assured that the bonds he had so successfully marketed would not be redeemed in greenbacks, Jay Cooke donated more than $20,000 to the Republican campaign, and other business leaders followed suit, deeming their contributions "a business investment, the best one they could make." Not that Democrats lacked for funds. Agricultural machinery manufacturer Cyrus McCormick and financiers linked to Belmont donated money, but it is significant that the largest contribution came from a patent medicine advertiser, H. T. Humbolt. For, more than in any previous election, Northern capitalists had united behind the Republican party.[111]

In the South, the prospect that a Seymour victory would undo Reconstruction dominated the Democratic campaign. A war of the races threatened, declared Oran Roberts, chief justice of Confederate Texas, and "nothing short of the disfranchisement of the negro race can stop it." Increasingly, events took on a sinister tone, with widespread threats of economic reprisal against black Republicans. One Democratic address to black voters put the issue with remarkable candor:

> We have the capital and give employment. We own the lands and require labor to make them productive. . . . You desire to be employed. . . . We know we can do without you. We think you will find it very difficult to do without us. . . . We have the wealth.

In case freedmen failed to get the message, Democrats unsheathed their "employing power," with merchants cutting off credit to blacks attending Republican meetings, and landlords threatening to evict from plantations "any negro who will not swear never again to vote the Radical ticket."[112]

111. Michael C. Kerr to Manton Marble, November 8, 1868, Marble Papers; Lawrence Grossman, *The Democratic Party and the Negro: Northern and National Politics 1868–92* (Urbana, Ill., 1976), 10–12; Joseph L. Brent to Samuel L. M. Barlow, n.d. [1868], Samuel Ward to Barlow, July 22, 1868, Barlow Papers; Mary R. Dearing, *Veterans in Politics: The Story of the G. A. R.* (Baton Rouge, 1952), 165; Benedict, *A Compromise of Principle*, 324; Leon B. Richardson, *William E. Chandler: Republican* (New York, 1940), 96–97; Flick, *Tilden*, 182.

112. Olsen, *Tourgée*, 168; Oran M. Roberts to James M. Burroughs, June 20, 1868 (typescript), Oran M. Roberts Papers, UTx; *Address of the Democratic White Voters of Charleston to the Colored Voters of Charleston, the Seaboard, and of the State Generally* (Charleston, 1868), 3–4; M. M. Cooke to Robert McKee, July 13, 1868, McKee Papers; Athens *Southern Watchman*, May

Violence, an intrinsic part of the process of social change since 1865, now directly entered electoral politics. Founded in 1866 as a Tennessee social club, the Ku Klux Klan now spread into nearly every Southern state, launching a "reign of terror" against Republican leaders black and white. Those assassinated during the campaign included Arkansas Congressman James M. Hinds, three members of the South Carolina legislature, and several men who had served in constitutional conventions. In northern Alabama, the Klan spread "a *nameless terror* among negroes, poor whites," and other Republicans. "Tongues can not express the time here," a black veteran wrote Alabama Gov. William H. Smith. "Our foreparents was broth from Africa and here we are in the way without a resting place to stand on in the God's . . . world. . . . Save us if you can." But military authorities did nothing, since Johnson had filled the ranks of commanding officers with men opposed to Reconstruction, and the new Republican governments proved unable to stop the violence.[113]

Although largely confined in 1868 to Piedmont counties where whites outnumbered blacks, violence in Georgia and Louisiana spread into the heart of the black belt. In the southwest Georgia village of Camilla, 400 armed whites, led by the local sheriff, opened fire on a black election parade, and then scoured the countryside for those who had fled, eventually killing and wounding more than a score of blacks. Blacks had no doubt who was behind the violence. "We don't call them democrats," a local leader commented of the assailants, "we call them southern murderers." Similar events occurred in Louisiana, where even moderate ex-Governor Hahn by October complained that "murder and intimidation are the order of the day in this state." White gangs roamed New Orleans, intimidating blacks and breaking up Republican meetings. In St. Landry Parish, a mob destroyed a local Republican newspaper, drove young teacher and editor Emerson Bentley from the area, and then invaded the plantations, killing as many as 200 blacks. Commanding Gen. Lovell Rousseau, a friend and supporter of the President, refused to take action, urging blacks to stay away from the polls for self-protection and exulting that the "ascendance of the negro in this state is approaching its end." Unable to hold meetings and fearful that attempts to bring out their vote would only result in further massacres, Georgia and Louisiana Republicans abandoned the Presidential campaign.[114]

6, 1868; Jonathan J. Knox to M. Frank Gallagher, May 6, 1868, Letters Received, B243 (1868), Third Military District, RG 393, NA.

113. Allen W. Trelease, *White Terror: The Ku Klux Klan Conspiracy and Southern Reconstruction* (New York, 1971), 3–154; Amos T. Akerman to Foster Blodgett, April 23, 1868, Amos T. Akerman Papers, GDAH; William B. Figures to William H. Smith, October 20, 1868, W. G. Crittenden to Smith, October 10, 1868, Alabama Governor's Papers. For a full discussion of the Ku Klux Klan, see Chapter 9.

114. Trelease, *White Terror*, 117–19, 127–32; Theodore B. Fitzsimmons, "The Camilla Riot," *GaHQ*, 35 (March 1951), 116–25; William Mills to R. C. Drum, September 29, 1868,

In some parts of the South, armed whites blocked blacks from going to vote or prevented polls from opening on election day. But in areas free of violence, blacks cast their ballots confident, as Robert G. Fitzgerald noted in his diary, that "the Great Epoch in the history of our race has at last arrived." As expected, Grant emerged victorious, although by a margin whose closeness James G. Blaine found "unaccountable." Nationally, he won every state except eight, but received less than 53 percent of the vote. It is more than likely that Seymour carried a majority of the nation's white electorate. For the first time, two Northern states—Iowa and Minnesota—approved constitutional amendments enfranchising blacks.[115]

In the South, Seymour won Georgia and Louisiana, where violence had decimated the Republican organization and made it impossible for blacks to vote. Eleven Georgia counties with black majorities recorded no votes at all for the Republican ticket. The Klan's effectiveness was also demonstrated in states Grant did carry, for the Republican vote fell off sharply in middle Tennessee, northern Alabama, and upcountry South Carolina. There were other grounds for Republican concern. In Georgia, many white yeomen who in the spring had embraced the party for economic reasons now returned to the Democratic fold. "If the white papol had . . . bin as well awaken as the colord," wrote a party organizer in one Piedmont North Carolina county, "we wold of given a large majority in favor of Republican." White Republicanism remained firmest in the strongholds of mountain Unionism. Taking the Southern mountains as a whole, Grant won over 60 percent of the vote.[116]

In one respect, 1868 marked a startling reversal of the political traditions of the Civil War era. Republicans, for a generation the party of change, campaigned on a platform of order and stability, while Democrats, who had appealed to continuity with the past, cast themselves as

---

Philip Joiner to John E. Bryant, September 5, 1868, John E. Bryant Papers, DU; Michael Hahn to Elihu B. Washburne, October 26, 1868, Washburne Papers; Melinda M. Hennessey, "Race and Violence in Reconstruction New Orleans: The 1868 Riot," *LaH*, 20 (Winter, 1979), 77–91; Carolyn E. De Latte, "The St. Landry Riot: A Forgotten Incident of Reconstruction Violence," *LaH*, 17 (Winter 1976), 41–49; Dawson, *Army Generals and Reconstruction*, 86–90.

115. D. Knight to William Stone, November 6, 1868, South Carolina Governor's Papers; Robert G. Fitzgerald Diary, November 3, 1868, Fitzgerald Papers; James G. Blaine, *Twenty Years of Congress* (Norwich, Conn., 1884–86), 2:408; Charles H. Coleman, *The Election of 1868* (New York, 1933), 363–65; Robert K. Dykstra, "The Issue Squarely Met: Toward an Explanation of Iowans' Racial Attitudes, 1865–1868," *Annals of Iowa*, 3 Ser., 47 (Summer 1984), 430–50.

116. KKK Hearings, Georgia, 454–59; F. Wayne Binning, "The Tennessee Republicans in Decline, 1869–1876," *THQ*, 29 (Winter 1980), 473; Kibler, *Perry*, 475; Steven Hahn, *The Roots of Southern Populism: Yeoman Farmers and the Transformation of the Georgia Upcountry, 1850–1890* (New York, 1983), 207–209, 214–16; Benjamin S. Field to William W. Holden, November 5, 1868, North Carolina Governor's Papers; Gordon B. McKinney, *Southern Mountain Republicans 1865–1900* (Chapel Hill, 1978), 32.

virtual revolutionaries. And if Grant's election guaranteed that Reconstruction would continue, it also confirmed a change in the Republican leadership that would preside over its future. Gone from the scene was Thaddeus Stevens, who died in August, his body attracting a throng of mourners to the Capitol second only to Lincoln's. For one last time, Stevens challenged his countrymen to rise above their prejudices, for he was buried in an integrated Pennsylvania cemetery in order, according to the epitaph he had composed, "to illustrate in my death the principles which I advocated through a long life, Equality of Man before his Creator." The Radical generation was passing, eclipsed by politicos like Blaine and Henry L. Dawes (who succeeded Stevens as chairman of the House Appropriations Committee). To such men, Stevens' death seemed "an emancipation for the Republican party." The "struggle over the negro," the party's rising leaders believed, must give way to economic concerns. Significantly, the first statute enacted after Grant's inauguration was the Public Credit Act, pledging to pay the national debt in gold. "I look to Grant's administration," wrote one federal official, "as the beginning of a real and true conservative era."[117]

Moderating tendencies were also apparent within the South. In cities, official ward committees and party conventions with duly elected delegates increasingly replaced the militant mass meetings of 1867. In many rural areas, the Union League was in disarray, its infrastructure destroyed by violence. "It is all broke up," said one black member from Graham, North Carolina, an area of rampant Klan activity. Even where the leagues survived, Republican leaders moved to assimilate them into a more disciplined party apparatus, evoking strong protests from black Radicals. To be sure, the tradition of local political organization embodied in the leagues persisted, in different guises, throughout Reconstruction. In Abbeville County, South Carolina, the Union League was succeeded by The Brotherhood, the United Brethren, and finally in 1875, the Laboring Union, for as one freedman explained, "they was all laboring men." But the league's decline eliminated a radical caucus within the Southern Republican party, just as more moderate men were coming to power on the state level.[118]

As Congressional Reconstruction began, a Georgia attorney voiced the fears of many white Southerners: "My chief alarm is for the formation of

117. Michael Perman, *The Road to Redemption: Southern Politics, 1869–1879* (Chapel Hill, 1984), 3–5; W. R. Brock, *An American Crisis* (London, 1963), 282; George F. Hoar, *Autobiography of Seventy Years* (New York, 1903), 1:239; James A. Putnam to Elihu B. Washburne, June 29, 1868, Washburne Papers; Joseph Logsdon, *Horace White: Nineteenth Century Liberal* (Westport, Conn., 1971), 156; J. T. Barnett to Samuel L. M. Barlow, November 6, 1868, Barlow Papers.

118. Rachleff, *Black Labor*, 50; KKK Hearings, Alabama, 8; Georgia, 28, 48; Mississippi, 77; North Carolina, 145; Fitzgerald, "Union League," Chapter 6; *Trial of Holden*, 2:1201; Harris, *Day of the Carpetbagger*, 189; Moneyhon, *Republicanism in Reconstruction Texas*, 155–57; 44th Congress, 2d Session, House Miscellaneous Document 31, 1:221, 235.

a party South which will be more radical than the Radical party north." By 1868, national events, together with the nature of its own coalition, had eliminated some of the more radical policy alternatives envisioned the year before. And yet, among those who voted in November, the dreams of 1867 survived. As one resident of upcountry South Carolina wrote the state's Republican governor four days after Grant's victory:

> I am . . . a native borned S. C. a poor man never owned a Negro in my life. . . . I am hated and despised for nothing else but my loyalty to the mother government. . . . But I rejoice to think that God almighty has given to the poor of S. C. a Gov. to hear to feel to protect the humble poor without distinction to race or color.[119]

The new Reconstruction governments now turned to the task of fulfilling the aspirations of their humble constituents for a new and more just South.

119. R. J. Moses to Alexander H. Stephens, April 2, 1867, Stephens Papers, LC; W. O. B. Houth to Robert K. Scott, November 7, 1868, South Carolina Governor's Papers.

CHAPTER 8

Reconstruction: Political and Economic

Party and Government in the Reconstruction South

UNPRECEDENTED challenges confronted the Southern Republicans who came to power between 1868 and 1870. Bequeathed few accomplishments and nearly empty treasuries by their predecessors, they faced the mammoth problems of a society devastated by warfare, new public responsibilities entailed by emancipation, and the task of consolidating an infant political organization. Most of all, both the party and its governments faced a crisis of legitimacy. Ordinarily, political parties take for granted the authority of the government and the integrity of their foes. Reconstruction's opponents, however, viewed the new regimes as alien impositions, and their black constituency as not entitled to a permanent role in the body politic. For Southern Republicans, as a result, the give-and-take of "normal politics" was superseded by a desperate struggle for political survival.[1]

Like beauty, political legitimacy resides in the eyes of the beholder. To blacks, these governments possessed a greater claim to authority than any in the South's history. But with the region's traditional economic and political leadership arrayed almost unanimously against them and Northern Republicans urging Southern counterparts to expand their political base among whites, the new governors set out to court their political foes. Their course fueled factional conflicts that weakened an already vulnerable organization, and quickly led blacks to demand a greater voice in party affairs.

Southern Republicanism's "very existence," one member declared in 1868, depended upon winning over white voters. But even as Radical Reconstruction began, events in Georgia underscored the dangers of forsaking the black constituency in search of white support. Still tantal-

1. Lawrence N. Powell, "Southern Republicanism During Reconstruction: The Contradictions of State and Party Formation" (unpub. paper, annual meeting of Organization of American Historians, 1984) develops the idea of a legitimacy crisis.

ized by the prospect of confecting a Republican majority from a program of debtor relief, economic development, and white supremacy, party leaders advised freedmen to abandon politics (except for casting Republican ballots) and concentrate on "earn[ing] an honest subsistence." In September 1868, Democrats moved to expel black members of the state legislature, on the grounds that Georgia's new constitution failed to guarantee their right to hold office. Despite the opposition of Governor Bullock, enough white Republicans voted in favor or abstained for the motion to carry. "Without the aid of the Radical members," a Savannah newspaper noted, "the Democrats could never have unseated the negroes." The inevitable consequence of white leaders' year-long campaign to relieve their party of the stigma of racial equality, the expulsion shattered the party's black-white alliance and revealed a lack of principle and indifference to black concerns that made the early collapse of Georgia Reconstruction all but inevitable.[2]

Outside Georgia, efforts to establish the new governments' legitimacy followed a less extreme path. While not repudiating black political equality, Republicans wooed white support by removing voting restrictions. By 1871 only Arkansas among the reconstructed states still retained suffrage restrictions based on Civil War loyalties. At the same time, Republican governors used their extensive appointment powers to award patronage to established local leaders, thereby attempting to create an image of moderation and defuse fears of black or carpetbagger domination. Louisiana Governor Warmoth followed the recommendations of "leading conservative citizens" in naming judges and parish officials. Mississippi's Alcorn appointed nearly as many Democrats (especially former Whigs like himself) as Republicans. Even in South Carolina, Robert K. Scott, who initially appointed a significant number of black trial justices, soon replaced many with white Democrats.[3]

Such policies won the party a few notable converts, including Presidential Reconstruction governors James L. Orr and Lewis Parsons. But they failed to produce large-scale defections from the Democracy. Even in Mississippi, Alcorn attracted at most a few thousand Old Line Whigs. And the cost far exceeded these modest gains, for placing Democrats in office produced deep disaffection among the Republican rank and file. Few blacks had confidence in appointees who "will not give us justis and

2. William B. Rodman to William W. Holden, May 5, 1868, William W. Holden Papers, NCDAH; Joseph E. Brown to Henry Wilson, December 19, 1868, Rufus Bullock Collection, HL; Edmund L. Drago, *Black Politicians and Reconstruction in Georgia* (Baton Rouge, 1982), 48–52; Savannah *Daily News and Herald*, September 15, 1868.

3. Michael Perman, *The Road to Redemption: Southern Politics, 1869–1879* (Chapel Hill, 1984), 26; Henry C. Warmoth, *War, Politics and Reconstruction* (New York, 1930), 81, 88; Lillian A. Pereyra, *James Lusk Alcorn: Persistent Whig* (Baton Rouge, 1966), 129–33; Jackson *Daily Mississippi Pilot*, August 23, 1871; 42d Congress, 2d Session, House Report 22, South Carolina, 11–12 (hereafter cited as KKK Hearings).

. . . who has opposed us all through the conflict and did all they cold against the gover and against us lik wise." Nor could scalawags understand why "*full* Democrats" and "strait out Rebels" enjoyed "the spoils of the victory," while wartime loyalists were "left out in the cold." After one shift in Louisiana patronage, five Republicans protested: "It is a shame to erase a Radical Republican off the School Board to take a dam rebel."[4]

Thus, the conciliatory patronage policy contributed to the factionalism that plagued Republicans in every Southern state. But it was hardly the only cause, for rivalries also fed upon policy differences (which will be discussed shortly), long-standing intrastate regional conflicts, and tensions between native and Northern Republicans, and blacks and whites. "We must keep together, scallawags, carpetbaggers and niggers," declared a North Carolina Republican. But party leaders failed to develop a political culture that made a virtue of solidarity. Political factionalism, of course, was not confined to the South in these years, but given Southern Republicans' other weaknesses, it was a luxury the party could scarcely afford. Nonetheless, Republicans denounced one another in language vitriolic even by nineteenth-century standards. (Georgia editor J. Clarke Swayze at least exhibited racial impartiality, describing a white party rival as "an unprincipled hermaphrodite bastard" and a black one as "the Reverend blackguard, whoremaster, forger and passer of counterfeit money.") Meanwhile, walkouts and fistfights disrupted party conventions, members of some factions connived with Democrats to defeat their rivals, Republican legislatures impeached Republican governors, and Florida's lieutenant governor even seized the state seal and claimed the right to rule. Such displays weakened the party's coherence and its image in the North, and undermined its claim of bringing to the South a new era of responsible government.[5]

Since conflict usually centered on the spoils of office, one Republican faction in each state was often made up of state and local officeholders,

4. Roger P. Leemhuis, *James L. Orr and the Sectional Conflict* (Washington, D.C., 1979), 154–55; Sarah W. Wiggins, *The Scalawag in Alabama Politics, 1865–1881* (University, Ala., 1977), 50–53; William C. Harris, "Mississippi: Republican Factionalism and Mismanagement," in Otto H. Olsen, ed., *Reconstruction and Redemption in the South* (Baton Rouge, 1980), 81; Salvy Benjamin to William W. Holden, March 5, 1869, North Carolina Governor's Papers, NCDAH; W. H. Mounce to Robert K. Scott, July 4, 1869, South Carolina Governor's Papers, SCDA; T. W. Greene to "my friend," December 16, 1868, Alabama Governor's Papers, ASDAH; James McBride et al. to Thomas W. Conway, August 28, 1870, State Board of Education Correspondence, Louisiana State Archives.

5. Otto H. Olsen, *Carpetbagger's Crusade: The Life of Albion Winegar Tourgée* (Baltimore, 1965), 170–71; John W. Morris to William Claflin, September 14, 1868, William E. Chandler Papers, LC; Macon *American Union*, April 30, August 20, 1869; Peggy Lamson, *The Glorious Failure: Black Congressman Robert Brown Elliott and the Reconstruction in South Carolina* (New York, 1973), 158–62; James H. Atkinson, "The Arkansas Gubernatorial Campaign and Election of 1872," *ArkHQ*, 1 (December 1942), 311; Jerrell H. Shofner, *Nor Is It Over Yet: Florida in the Era of Reconstruction, 1863–1877* (Gainesville, Fla., 1974), 204–206, 220–22.

who coalesced around the governor, while another consisted of holders of federal patronage posts, whose appointments Senators and Congressmen controlled. In Alabama, for example, Gov. William H. Smith and Sen. George Spencer headed rival wings. The governor's support came from native whites, who made exaggerated claims that "newcomers" monopolized patronage. (In fact, scalawags received the lion's share of federal and state positions.) In Mississippi, Republicans disaffected from the administration of James L. Alcorn rallied around Sen. Adelbert Ames. By 1871, Mississippi Republicanism resembled "an army of recruits" lacking organization and discipline, and "fighting each for himself."[6]

Probably the most byzantine factional struggles occurred in Louisiana. "Old Union citizens" charged, with some justification, that a small group of carpetbaggers had seized control of the party. Upon this rivalry was soon superimposed a feud between Governor Warmoth and the Custom House ring headed by federal marshal Stephen B. Packard, a Maine carpetbagger, and Collector of the Port James F. Casey, President Grant's brother-in-law. From this conflict flowed a series of tragicomic episodes: rival Republican state conventions in 1871; several legislators going into hiding on a champagne-stocked federal revenue cutter to prevent a quorum; the arrest of the governor and seventeen lawmakers in 1872 by deputy marshals dispatched by the Custom House group; and Warmoth's use of the state militia to regain control of the statehouse.[7]

"Our party," observed a Republican newspaper in the waning days of Reconstruction, "[has been] disorganized, disrupted, and demoralized, . . . rent and torn by internal feuds." Behind this debilitating factionalism lay the reality of a party whose leaders lacked a secure place in the South's business and professional world, and depended upon political position for their very livelihood. For carpetbaggers who went into politics in 1867 after failing at cotton planting, and lawyers, merchants, and artisans boycotted by Democratic patrons, office became more than a matter of prestige or an interlude in a successful business career; it was often the

6. Perman, *Road to Reunion*, 42–52; Wiggins, *Scalawag in Alabama*, 39, 46–48, 57, 136–53; H. C. Blackman to Adelbert Ames, February 13, 1871, A. R. Howe to Ames, January 7, 1871, Ames Family Papers, SC; G. Wiley Wells to O. O. Howard, September 1, 1871, O. O. Howard Papers, Bowdoin College. The 113 white Republicans to sit in Congress during Reconstruction included sixty carpetbaggers and fifty-three scalawags. Although carpetbaggers at one time or another were elected governors of South Carolina, Florida, Mississippi, Louisiana, Georgia, and Arkansas, their power within the party was greatest in states with large black populations and few native white Republicans (such as Louisiana and South Carolina), and least in North Carolina, Alabama, and Texas, with their large concentrations of scalawags. (This did not prevent historians from inventing a myth of carpetbagger control here as well. W. C. Nunn could entitle a book *Texas Under the Carpetbaggers* [Austin, 1962] despite the fact that virtually every important leader of Texas Reconstruction was Southern-born.)

7. Michael Hahn to Elihu B. Washburne, October 26, 1868, Elihu B. Washburne Papers, LC; Joe G. Taylor, *Louisiana Reconstructed, 1863–1877* (Baton Rouge, 1974), 210–26; Althea D. Pitrie, "The Collapse of the Warmoth Regime, 1870–72," *LaH*, 6 (Spring 1965), 161–88.

only way for a Republican to earn a living. As Georgia carpetbagger John E. Bryant explained in applying for a federal appointment:

Republican leaders are not situated in the South as they are in the North nor . . . as party leaders were elsewhere in the South before the war. At the North . . . it does not injure the business of the lawyer, the merchant, the manufacturer or the mechanic to take an active part in promoting the interests of the Republican party.[8]

Thus, Republican legislators raised substantially the compensation of positions from governor down to justice of the peace, and party members pursued offices aggressively and clung to them tenaciously. Southern Democrats could leave office and return to careers as merchants, planters, and lawyers, but for Republicans losing an office often meant economic disaster. "I do not know what I shall do," lamented a Louisiana scalawag removed as New Orleans inspector of weights and measures. "My own relatives have turned their backs and . . . it will be impossible for me to get any employment." So too, Republican newspapers, essential to party organization, lived a precarious existence. Condemned to a tiny readership by the poverty and illiteracy of their constituents (the average weekly circulation of South Carolina's ten Republican newspapers in 1873 barely exceeded 500), and denied advertising by the vast majority of businesses, they depended entirely on government printing contracts and "political favoritism." Even well-established journals faced certain demise if they backed the wrong side in intraparty disputes.[9]

Among blacks, party infighting produced growing alarm. Republicans, declared a Virginia freedman in 1868, "ought to combine all of their influence, and not fight one against the other," and political leaders like Mississippi's James Lynch and North Carolina's James W. Hood attempted to "harmonize" competing factions. Others, inevitably, found themselves drawn into the fray. Although a few black politicos, like Louisiana lieutenant governors Oscar J. Dunn and P. B. S. Pinchback, emerged as independent power brokers within the party, most coalesced around competing factions of white politicians. In every state, each major grouping had its own black wing, but since most black leaders opposed the white-oriented patronage policies of the early Reconstruction governors (which often went hand in hand with a reluctance to support civil

8. Lawrence N. Powell, "The Politics of Livelihood: Carpetbaggers in the Deep South," in J. Morgan Kousser and James M. McPherson, eds., *Region, Race, and Reconstruction: Essays in Honor of C. Vann Woodward*, (New York, 1982), 315–48; John E. Bryant to Amos T. Akerman, May 15, 1871, John E. Bryant Papers, DU.

9. Horace W. Raper, *William W. Holden: North Carolina's Political Enigma* (Chapel Hill, 1985), 108–109; E. W. Fostrick to William P. Kellogg, January 9, 1875, William P. Kellogg Papers, LSU; Henry L. Suggs, ed., *The Black Press in the South, 1865–1979* (Westport, Conn., 1983), 293–95; Frank B. Williams, "John Eaton, Jr., Editor, Politician, and School Administrator, 1865–1870," *THQ*, 10 (December 1951), 292–95; Charleston *Daily Republican*, September 8, 1871.

rights measures that might alienate potential white support), the majority sided with factions headed by federal officeholders, generally carpetbaggers. Their opposition eventually helped to overturn the policy of minimizing their role within the party, and to push aside the governors associated with it.[10]

Initially, however, blacks stood aside when the political "loaves and fishes" were divided up—not only because white Republicans so desperately coveted office, but because many black leaders did not wish to embarrass their party, heighten internal contention, or lend credence to Democratic charges of "black supremacy." Georgia blacks "went from door to door in the 'negro belt' " seeking white Republican candidates, and in North Carolina, James H. Harris, fearing his selection "would damage the party at the North," declined a Congressional nomination. Others deemed themselves unqualified for office ("I refused to run because I knew nothing about what was needed to be done," Georgia schoolteacher Houston H. Holloway later recalled of his decision to refuse a legislative nomination). Not all blacks, of course, proved so retiring. Some Louisiana free blacks "modestly yielded place to men, often their inferior in intellect and culture," but others from the outset pressed for a fair share of offices. In South Carolina, the Northern-born in particular insisted that blacks must be "admitted to a full participation in the control of affairs." Over white opposition, they placed one of their own, Benjamin F. Randolph, at the head of the state Republican committee in 1868. Even here, however, most blacks feared doing "anything that would injure the party." Francis L. Cardozo (boosted for the lieutenant governorship) and Martin R. Delany (promoted for a Congressional seat) declined to run in 1868, citing the need for "the greatest possible discretion and prudence," and the legislature's lower house, which contained a black majority, chose white scalawag Franklin J. Moses, Jr., as Speaker over black carpetbagger Robert B. Elliott.[11]

From the top of the political order to the bottom, blacks initially received a lower share of offices than their proportion of the party's elector-

10. *The Debates and Proceedings of the Constitutional Convention of the State of Virginia* (Richmond, 1868), 492; James Lynch to Adelbert Ames, December 12, 1871, Ames Family Papers; Raleigh *Tri-Weekly Standard*, September 7, 1867; Atkinson, "Arkansas Gubernatorial Campaign," 312; W. B. Raymond to Adelbert Ames, December 17, 1871, Ames Family Papers.
11. Dale A. Somers, "James P. Newcomb: The Making of a Radical," *SWHQ*, 72 (April 1969), 466; Edwin S. Redkey, ed., *Respect Black: The Writings and Speeches of Henry McNeal Turner* (New York, 1971), 17; Raleigh *Sentinel*, February 29, 1868; Houston H. Holloway Autobiography, Miscellaneous Manuscript Collections, LC; New Orleans *Louisianian*, February 19, 1871; Charleston *Mercury*, July 8, 1868; John W. Morris to William Claflin, September 14, 1868, William E. Chandler Papers; William B. Nash to Charles Sumner, August 22, 1868, Charles Sumner Papers, HU; William C. Hine, "Dr. Benjamin A. Boseman, Jr.: Charleston's Black Physician-Politician," in Howard N. Rabinowitz, ed., *Southern Black Leaders of the Reconstruction Era* (Urbana, Ill., 1982), 340; unidentified newspaper clipping (Martin Delany letter), February 1876, SCHS; *New York Times*, July 7, 1868.

ate warranted. Sixteen blacks sat in Congress during Reconstruction, but of these only three served in the Forty-First Congress (which met from 1869 to 1871). Hiram Revels of Mississippi, a North Carolina-born minister and educator, in February 1870 became the first black to serve in the United States Senate. In the House sat freeborn South Carolina barber Joseph H. Rainey and Jefferson Long, a Georgia freedman elected to the two-month short term of 1871 in which "no one was particularly interested." In the Forty-Second Congress, the black component rose to five, two of whom, however, spent much of their time fending off charges of election irregularities and were eventually unseated.[12]

Nor were blacks well-represented at the highest levels of state government. In five states—Texas, North Carolina, Alabama, Georgia, and Virginia—none held a major office during Reconstruction. Philadelphia-born missionary and educator Jonathan C. Gibbs was the only black to win a major post in Florida, serving as secretary of state from 1868 to 1872 and then as superintendent of education. Blacks came to exercise far greater power in Mississippi and South Carolina, but here, too, whites initially all but monopolized statewide positions. In Mississippi, Secretary of State James Lynch was, at first, the only black state official, and until 1870, Secretary of State Francis L. Cardozo was his lone South Carolina counterpart. Only in Louisiana did blacks hold more than one major position from the beginning of Republican rule. In 1868, Oscar J. Dunn became lieutenant governor and wealthy free sugar planter Antoine Dubuclet state treasurer, a post he retained until 1877.[13]

It did not take long for black leaders to become dissatisfied with the role of junior partners in the Republican coalition. Especially in states with large black populations, they increasingly demanded "a fair proportion of the offices" and began to assume a larger role in party affairs. The results were most dramatic in South Carolina, where in 1870 black leaders, as the result of a concerted campaign for greater power, received half the eight executive offices, elected three Congressmen, and placed Jonathan J. Wright on the state supreme court, the only black in any state to hold this position during Reconstruction. In Mississippi, they mobilized

12. Samuel D. Smith, *The Negro in Congress 1870–1901* (Chapel Hill, 1940), 5–6; KKK Hearings, Georgia, 1037. Along with Rainey, black Congressmen in the 42d Congress were Benjamin S. Turner, a self-educated freedman and prosperous Selma, Alabama, merchant, Josiah T. Walls, a former slave who had fought in the Union Army and represented Florida, and from South Carolina, free Charleston tailor Robert G. DeLarge and black carpetbagger Robert B. Elliott (described by one observer as "the ablest negro, intellectually, in the South"). Chicago *Tribune*, November 2, 1872. Walls and Delarge were the two unseated. Of the sixteen blacks to serve in Congress during Reconstruction, nine had been born slave and seven free.

13. Joe M. Richardson, "Jonathan C. Gibbs: Florida's Only Black Cabinet Member," *FHQ* 42 (April 1964), 363–66; Charles Vincent, "Aspects of the Family and Public Life of Antoine Dubuclet: Louisiana's Black State Treasurer, 1868–1878," *JNH*, 66 (Spring 1981), 29–36.

effectively against Governor Alcorn. "The complexion of political affairs in our State perceptibly *darkens*," a white Mississippi Republican remarked in 1872; the following year, blacks engineered the replacement of Alcorn's handpicked successor, Ridgely Powers, by Adelbert Ames and won half the statewide offices for themselves. Arkansas in 1872 elected its first black state officials: Superintendent of Education Joseph C. Corbin, a college-educated editor from Ohio, and Commissioner of Public Works James T. White, a minister from Indiana.[14]

By the end of Reconstruction, eighteen blacks had served as lieutenant governor, treasurer, superintendent of education, or secretary of state. (See Table.)

TABLE

Major black state officials during Reconstruction

GOVERNOR:
    Louisiana: P.B.S. Pinchback, December 9, 1872—January 13, 1873
LIEUTENANT GOVERNOR:
    Louisiana: Oscar J. Dunn, 1868–71; P.B.S. Pinchback, 1871–72; Caesar C. Antoine, 1873–77
    Mississippi: Alexander K. Davis, 1874–76
    South Carolina: Alonzo J. Ransier, 1871–73; Richard H. Gleaves, 1873–77
TREASURER:
    Louisiana: Antoine Dubuclet, 1868–77
    South Carolina: Francis L. Cardozo, 1873–77
SUPERINTENDENT OF EDUCATION:
    Arkansas: Joseph C. Corbin, 1873–74
    Florida: Jonathan C. Gibbs, 1873–75
    Louisiana: William G. Brown, 1873–77
    Mississippi: Thomas W. Cardozo, 1874–76
SECRETARY OF STATE:
    Florida: Jonathan C. Gibbs, 1868–1873
    Mississippi: James Lynch, 1869–72; Hiram Revels, 1872–73;
        Hannibal C. Carter, 1873; M. M. McLeod, 1874; James Hill, 1874–78
    South Carolina: Francis L. Cardozo, 1868–73; Henry E. Hayne, 1873–77

Apart from former slave James Hill, and Davis and McLeod, whose origins I have been unable to determine, all of the above are known to have been born free.

14. W. A. Collett to Tod R. Caldwell, June 9, 1872, Tod R. Caldwell Papers, DU; Thomas Holt, *Black over White: Negro Political Leadership in South Carolina During Reconstruction* (Urbana, Ill., 1977), 105–108; Perman, *Road to Redemption*, 40; William J. Davis to Adelbert Ames, May 1, 1872, Ames Family Papers; Elizabeth Rothrock, "Joseph Carter Corbin and Negro Education in the University of Arkansas," *ArkHQ,* 30 (Winter 1971), 277–314; Walter Nunn, "The Constitutional Convention of 1874," *ArkHQ,* 27 (Autumn 1968), 186–88.

It is difficult to gauge precisely how much power these men enjoyed. During Reconstruction more blacks served in the essentially ceremonial office of secretary of state than any other post, and by and large, the most important political decisions in every state were made by whites. On the other hand, the four black superintendents of education exerted a real influence on the new school systems, and black lieutenant governors presided over state senates and exercised gubernatorial powers when their chief executives were ill or out of the state. In December 1872, P. B. S. Pinchback became the only black governor in American history when he succeeded Henry C. Warmoth, who had been suspended as a result of impeachment proceedings.

A similar pattern of initial underrepresentation, followed in some states by growing black influence, appeared in state legislatures. Only fourteen blacks sat in Texas's first Republican legislature, which included just over 100 members, and twenty-one in North Carolina's even larger General Assembly, proportions that remained steady as Reconstruction progressed. In Georgia and Florida, with far larger black populations in percentage terms, the system of apportionment minimized their legislative influence (of 216 Georgia lawmakers elected in 1868, all but thirty-two were white). Blacks constituted a larger portion of the first Reconstruction legislatures of Mississippi, Alabama, and Louisiana, and here, moreover, not only did their numbers increase as time went on, but as white Republicanism waned, blacks formed a larger and larger portion of the party's representation. But in these states as well, whites controlled nearly all the important committees, and most bills introduced independently by black lawmakers failed to pass. Only in South Carolina did blacks come to dominate the legislative process. Throughout Reconstruction, they comprised a majority of the House of Representatives, controlled its key committees, and, beginning in 1872, elected black speakers. "Sambo . . ." reported Northern journalist James S. Pike in 1873, "is already his own leader in the Legislature. . . . The Speaker is black, the Clerk is black, the doorkeepers are black, the little pages are black." The next year, blacks gained a majority in the state senate as well (elsewhere, nearly all black lawmakers sat in the lower house).[15]

15. Alwyn Barr, "Black Legislators of Reconstruction Texas," *CWH*, 32 (December 1986), 340–52; Elizabeth Balanoff, "Negro Leaders in the North Carolina General Assembly, July, 1868–February, 1872," *NCHR*, 49 (Winter 1972), 55; Drago, *Black Politicians*, 38; Joe M. Richardson, *The Negro in the Reconstruction of Florida, 1865–1877* (Tallahassee, 1965), 187; Buford Satcher, *Blacks in Mississippi Politics 1865–1900* (Washington, 1978), 203–207; Wiggins, *Scalawag in Alabama*, 147–51; Charles Vincent, *Black Legislators in Louisiana During Reconstruction* (Baton Rouge, 1976), 71–83, 201–204; Holt, *Black over White*, 108; James S. Pike, *The Prostrate State* (New York, 1874), 15. The following blacks served as Speaker of the House: Mississippi, John R. Lynch 1872–73, I. D. Shadd 1874–76; South Carolina, Samuel J. Lee 1872–74, Robert B. Elliott 1874–76. As can best be determined, the following is the number of black Reconstruction legislators, counted to the date of full Democratic Redemp-

Despite the overall pattern of white political control, the fact that well over 600 blacks served as legislators—the large majority, except in Louisiana and Virginia, former slaves—represented a stunning departure in American politics. Moreover, because of the black population's geographical concentration and the reluctance of many scalawags to vote for black candidates, nearly all these lawmakers hailed from plantation counties, home of the wealthiest and, before the war, most powerful Southerners. The spectacle of former slaves representing the lowcountry rice kingdom or the domain of Natchez cotton nabobs epitomized the political revolution wrought by Reconstruction.

An equally remarkable transformation occurred at the local level, where the decisions of public officials directly affected daily life and the distribution of power. Although the structure of government varied from state to state, justices of the peace generally ruled on minor criminal offenses as well as a majority of civil cases, while county commissioners established tax rates, controlled local appropriations, and administered poor relief, and sheriffs enforced the law, selected trial jurors, and carried out foreclosures and public sales of land. Such officials, in the words of an Alabama lawyer, dealt with "the practical rights of the people, . . . our 'business and lives'." In the antebellum South, these positions had been monopolized by local elites, and the prospect of Republicans, whether former slaves or whites of modest wealth, occupying them alarmed the old establishment even more than their loss of statewide control. Howell Cobb said he could tolerate freedmen sitting in the Georgia legislature, "but when it comes to the home municipal government—all the blacks who vote against my ticket shall walk the plank."[16]

Although the largest number served in South Carolina, Louisiana, and Mississippi, and the fewest in Florida, Georgia, and Alabama, no state lacked its black local officials. A handful held the office of mayor, including Pierre Landry of Donaldson, Louisiana, a slave pastry chef educated before the war by his father/master, and Robert H. Wood of Natchez, member of one of that city's most respected free families. A far larger number served on city and town councils in communities from Richmond to Houston. The nation's capital itself elected two black aldermen in 1868. Some of the South's most important cities, and many towns of lesser note, became centers of black political power during Reconstruction. Republicans controlled the major rail and industrial center of

tion in each state (some of the figures are approximate): Alabama, 69; Arkansas, 8; Florida, 30; Georgia, 41; Louisiana, 87; Mississippi, 112; North Carolina, 30; South Carolina, 190; Texas, 19; Virginia, 46.

16. Eric Anderson, "James O'Hara of North Carolina: Black Leadership and Local Government," in Rabinowitz, ed., *Southern Black Leaders*, 108; James W. Garner, *Reconstruction in Mississippi* (New York, 1901), 284; Powhaten Lockett to Joseph Wheeler, January 1, 1876, Joseph Wheeler Papers, ASDAH; Paul D. Escott, *Many Excellent People: Power and Privilege in North Carolina, 1850–1900* (Chapel Hill, 1985), 144; Drago, *Black Politicians*, 79.

Petersburg from 1870 to 1874, and blacks held posts ranging from councilman to deputy customs collector and overseer of the poor. Nashville's council was about one third black, and Little Rock's at times had a black majority. Black aldermen were even elected in predominantly white upcountry Republican communities like Rutherfordton, North Carolina, and Knoxville and Maryville in East Tennessee.[17]

In virtually every county with a sizable black population, blacks served in at least some local office during Reconstruction. Atop the pyramid of local power stood the sheriff, described by one Mississippi politico as "the best paying office in the state," with annual salary and fees often amounting to thousands of dollars. In most Republican counties the post remained in white hands throughout Reconstruction, partly because black aspirants found it nearly impossible to post the bond required to assume office. But as time went on, black sheriffs appeared in many plantation counties. Eventually, nineteen held the office in Louisiana, and fifteen in Mississippi (where over one third of the black population lived in counties that elected black sheriffs). Blacks in increasing numbers also assumed such powerful offices as county supervisor and tax collector, especially in states where these posts were elective. By 1871, former slaves had taken control of the boards of supervisors throughout the Mississippi black belt. Most local black officials served in lesser posts like school board member, election commissioner, and justice of the peace. Yet even these positions, as former Alabama Governor Patton remarked, were "of considerable importance to the people."[18]

As in the North, control of offices from sheriff to justice of the peace, supplemented by state and federal patronage appointments and occasional legislative redistricting, served as the foundation for local party machines. Louisiana's legislature in 1871 created Red River Parish, which quickly became the site of a powerful organization headed by Vermont carpetbagger Marshall H. Twitchell. After achieving success as a cotton

17. "Dunn-Landry Papers," *Amistad Log*, 2 (August 1984), 1–3; Vernon L. Wharton, *The Negro in Mississippi 1865–1890* (Chapel Hill, 1947), 167; Michael B. Chesson, "Richmond's Black Councilmen, 1871–96," in Rabinowitz, ed., *Southern Black Leaders*, 198–99; Barry A. Crouch, "Self-Determination and Local Black Leaders in Texas," *Phylon*, 39 (December 1978), 347; Constance M. Green, *The Secret City: A History of Race Relations in the Nation's Capital* (Princeton, 1967), 89–91; William D. Henderson, *The Unredeemed City: Reconstruction in Petersburg, Virginia: 1865–1874* (Washington, D.C., 1977); Monroe M. Work, "Some Negro Members of Reconstruction Conventions and Legislatures and of Congress," *JNH*, 5 (January 1920), 115; Bobby L. Lovett, "Some 1871 Accounts for the Little Rock, Arkansas Freedman's Savings and Trust Company," *JNH*, 66 (Winter 1981–82), 328; Clarence W. Griffen, *History of Old Tryon and Rutherford Counties, North Carolina 1730–1936* (Asheville, 1937), 321; Alrutheus A. Taylor, *The Negro in Tennessee, 1865–1880* (Washington, D.C., 1941), 248; Maryville Union League Minute Book, McClung Collection, LML.

18. *New National Era*, October 26, 1871; J. Mason Brewer, *Negro Legislators of Texas* (Dallas, 1935), 50–52; John H. Moore, ed., *The Juhl Letters to the "Charleston Courier"* (Athens, Ga., 1974), 247; Vincent, *Black Legislators*, 219–21; Satcher, *Blacks in Mississippi Politics*, 38–39, 53–54; Robert M. Patton to J. Hayden, March 6, 1868, Letters Received, Third Military District, RG 393, NA.

planter, Twitchell brought his entire family to the South. He filled local offices with his brother and brothers-in-law, obtained a teaching position in a black school for his wife, retained four important posts for himself, and won state printing contracts for the Sparta *Times*, which he published. Meanwhile, Twitchell earned the freedmen's confidence by appointing blacks to lesser offices and actively promoting public education.[19]

Generally, whites dominated Republican organizations and the political rewards they dispensed. Throughout the South, blacks could be found in minor patronage posts ranging from postmaster to land office clerk, and a few held more lucrative offices like internal revenue collector. But major patronage plums eluded them, with rare exceptions like James H. Ingraham, who in 1872 became surveyor of the port of New Orleans, at an annual salary of $6,000. In Memphis, for example, three carpetbaggers controlled the Republican machine: John Eaton (Grant's wartime supervisor of freedmen's affairs and Tennessee superintendent of education), his brother Lucien Bonaparte Eaton (editor of the Memphis *Post* and federal marshal for West Tennessee), and Barbour Lewis (chairman of the city's Republican committee, who insisted local blacks "ought to be willing to wait awhile" for patronage).[20]

As Reconstruction progressed, dissatisfaction with being used as "stepping stones to office" led blacks to claim a larger role in many local Republican organizations. This was the case in Edgefield County, located in South Carolina's Piedmont cotton belt. Although blacks constituted 60 percent of its population, white Republicans, native and carpetbagger, initially dominated local and county offices. But by the 1870s, blacks had assumed such positions as sheriff, magistrate, school commissioner, and officer of the state militia. A similar situation developed in Beaufort, an antebellum center of the lowcountry aristocracy. "Here the revolution penetrated to the quick," commented reporter Edward King in 1873; the mayor, police force, and magistrates were all black, and the celebrated former slave Robert Smalls dominated local politics. Other enclaves of black political power emerged in the Mississippi and Louisiana plantation belts. Blanche K. Bruce created a powerful machine in Bolivar County, Mississippi, simultaneously holding the offices of sheriff, tax collector, and superintendent of education. This local organization became the springboard from which Bruce in 1875 reached the United States Senate. Sheriff Henry Demas, a former slave, controlled the politics of St. John the Baptist Parish, Louisiana, into the 1890s.[21]

19. Ted Tunnell, *Crucible of Reconstruction: War, Radicalism, and Race in Louisiana 1862–1877* (Baton Rouge, 1984), 173–88.

20. Holt, *Black over White*, Appendix; KKK Hearings, Florida, 105–107; Donald B. Sanger and Thomas R. Hay, *James Longstreet* (Baton Rouge, 1952), 358; Walter J. Fraser, Jr., "Black Reconstructionists in Tennessee," *THQ*, 34 (Winter 1975), 367–68.

21. Orville V. Burton, *In My Father's House Are Many Mansions: Family and Community in Edgefield, South Carolina* (Chapel Hill, 1985), 299; Edward King, *The Southern States of North*

Less frequently, blacks rose to similar positions in other states. Georgia's only real enclave of black political power lay in coastal McIntosh County. Here, Tunis G. Campbell, a New Jersey-born participant in the prewar abolition movement who had come South during the war to take part in the Sea Island experiment, served as state senator and justice of the peace. Campbell insisted that trial juries include equal numbers of blacks and whites, and used his power to defend the economic interests of local freedmen. In one controversial case, he ordered the arrest of a sea captain who had abused black sailors and failed to pay their wages. One overseer considered himself "powerless" to enforce work rules, for in the event of a labor dispute, "I should only get myself into trouble, and have the negro sheriff sent over by Campbell to arrest me." Local whites viewed Campbell as a "constant annoyance"; the freedmen saw him as "the champion of their rights and the bearer of their burden."[22]

Among black local officials were many whose prominence predated emancipation—like South Carolina magistrate Edward Waddill, who had officiated at slave marriages "long before the advent of freedom," and Hamilton Gibson, a "conjurer," elected a Louisiana justice of the peace in 1868. Others were veterans of the postwar black conventions or of the Union Leagues. But Reconstruction also witnessed the emergence of a new group of leaders, who gradually displaced the organizers of 1865–67. As Republican rule progressed, former slaves increasingly came to the fore, a process already noticeable in 1867. In St. Landry Parish, Louisiana, for example, freedmen in 1873 for the first time outnumbered the freeborn on the Republican executive committee. In addition, although ministers continued to play a significant role in politics, rural Baptist preachers began to outnumber the urban-based AME prelates prominent in the early days of freedom. Many of the black community's new political leaders were remarkable for their precocity. Most of Edgefield County's black politicians had not reached twenty-five when they assumed public office, John Gair served in Louisiana's legislature at twenty-one, and John R. Lynch was only twenty-four when he became speaker of Mississippi's House. But whatever their age, and whether free or freed, a web of connections bound this first generation of black officials to the network of religious, fraternal, and educational institutions created after emancipation. In cities like Memphis and New Orleans, a distinction was already evident between political leaders, who tended to be prosperous free

America (London, 1875), 426–28; William C. Harris, "Blanche K. Bruce of Mississippi: Conservative Assimilationist," in Rabinowitz, ed., Southern Black Leaders, 7–8; Clara L. Campbell, "The Political Life of Louisiana Negroes, 1865–1900" (unpub. diss., Tulane University, 1971), 135.

22. Russell Duncan, Freedom's Shore: Tunis Campbell and the Georgia Freedmen (Athens, Ga., 1986), 14–21, 42–80; Frances Butler Leigh, Ten Years on a Georgia Plantation Since the War (London, 1883), 132–36; W. Gignilliat to Rufus Bullock, January 15, 1872, Testimony of C. H. Hopkins, W. R. Gignilliat, W. J. Dunwoody, Tunis G. Campbell Papers, GDAH; KKK Hearings, Georgia, 846–58.

mulattoes, and men prominent in religious and benevolent associations, generally unskilled former slaves. But in the rural black belt, a far greater overlap existed between political and social leadership. Black schools, fire companies, churches, and fraternal orders were intertwined with local party organizations and served as training grounds for black officehold- ers, especially the former slaves who assumed increasing prominence as Reconstruction progressed.[23]

Not surprisingly, a considerable number of black officials had not en- joyed the advantages of an education. Many were entirely illiterate and had to rely on other blacks or white Republicans to conduct official business. "I have a son I sent to school when he was small," said Georgia legislator Abram Colby, an illiterate barber and minister emancipated by his master in 1851. "I make him read all my letters and do all my writing. I keep him with me all the time." But on both the local and state levels, large numbers of black politicians, especially but not exclusively those born free, had managed to acquire an education. Atlanta City Council- man William Finch had been taught to read and write by the family of his owner, Georgia Chief Justice Joseph Lumpkin. Other black officials had attended Bureau schools after the Civil War or studied in the North after being freed by the Union Army. Virtually all the black carpetbaggers and Louisiana and South Carolina free blacks who held state and local office were literate, and some had completed college educations. Mifflin Gibbs (brother of Florida official Jonathan C. Gibbs), had been a Vancouver city councilman and had studied law at Oberlin before holding a Little Rock judgeship. By the 1870s, moreover, products of newly established black universities swelled the ranks of black politicians. Among those who studied law at Howard before assuming office in the South were James E. O'Hara, elected in 1874 to the Halifax County, North Carolina, Board of Supervisors, and Matthew M. Lewey, who served as a Florida postmas- ter and justice of the peace toward the end of Reconstruction. Such men, at home in both the larger world and the local black community, func- tioned as "political middlemen," helping to interpret national and local political cultures and social experiences to one another.[24]

23. R. Turner et al. to Robert K. Scott, February 3, 1872, South Carolina Governor's Papers; George P. Deweese to Henry C. Warmoth, April 25, 1868, Henry C. Warmoth Papers, UNC; Geraldine McTigue, "Forms of Racial Interaction in Louisiana, 1860–1880" (unpub. diss., Yale University, 1975), 306–308; *Christian Recorder*, October 28, 1875; Emma L. Thornbrough, ed., *Black Reconstructionists* (Englewood Cliffs, N.J., 1972), 11; Armstead L. Robinson, "Plans Dat Comed from God: Institution Building and the Emergence of Black Leadership in Reconstruction Memphis," in Orville V. Burton and Robert C. McMath, Jr., eds., *Toward a New South* (Westport, Conn., 1982), 89–92; John W. Blassingame, *Black New Orleans 1860–1880* (Chicago, 1973), 157–58; Vernon Burton, "Race and Reconstruction: Edgefield County, South Carolina," *JSocH*, 12 (Fall 1978), 32; Alrutheus A. Taylor, *The Negro in the Reconstruction of Virginia* (Washington, D.C., 1926), 55.

24. KKK Hearings, Georgia, 702; Florida, 105–107; James M. Russell and Jerry Thorn- bery, "William Finch of Atlanta: The Black Politician as Civic Leader," in Rabinowitz, ed., *Southern Black Leaders*, 309; David C. Rankin, "The Origins of Black Leadership in New Orleans During Reconstruction," *JSH*, 40 (August 1974), 432–33; Holt, *Black over White,*

For ambitious individuals within the black community, politics offered a rare opportunity for dignified employment and personal advancement. The rewards of major posts far outstripped what blacks could ordinarily earn, and even the two or three dollars a day for jury duty seemed dazzlingly high to rural freedmen. Black lawmakers consistently opposed efforts to shorten legislative sessions or reduce official salaries—after all, one explained, "all our troubles have arisen from not paying people for their services." In occupation, local officeholders were generally artisans, shopkeepers, and small landowners, or professionals like ministers and teachers. But while better off economically than the mass of the freedmen, most were men of modest circumstances, whose personal experiences and economic interests did not diverge sharply from those of their constituents. Many had known the horrors of slavery, like William H. Councill, an Alabama legislator, editor, and lawyer who had seen two of his brothers sold in 1857 and never heard from them again. State legislator William H. Harrison, the favored slave of a prominent Georgia family, owned thirteen acres of land after the war, yet thought of himself as a man who lived by "hard work"—he "split rails and picked cotton, pulled fodder and worked on the Western and Atlantic Railroad." In neither life-style nor values did Harrison differ substantially from the bulk of the freedmen, and he entered politics because the Reconstruction Act recognized "me and my people as men, while before we had been chattels."[25]

The black political leadership did include a few men of substantial wealth. Among the Charleston and New Orleans free elite were planters and businessmen worth tens of thousands of dollars before the Civil War, some of whom assumed Reconstruction offices. Others, like white counterparts North and South, translated office into financial gain. Florida Congressman Josiah T. Walls purchased a large estate formerly owned by Confederate Gen. James H. Harrison, and about one third of Virginia's black legislators used their salaries to purchase land. Rural artisan-politicians often prospered from local government construction contracts. Some black leaders enjoyed an aristocratic life-style during Reconstruction, like Louisiana Lieut. Gov. C. C. Antoine, owner of an expensive race horse, and Blanche K. Bruce, who acquired both a fortune in real estate and "the manners of a Chesterfield." Black carpetbagger William Whipper, a rice planter in the South Carolina lowcountry, was

52–55; Mifflin W. Gibbs, *Shadow and Light: An Autobiography* (Washington, D.C., 1902); Anderson, "James O'Hara," 101–104; Suggs, ed., *Black Press*, 97; Marc J. Swartz, ed., *Local-Level Politics* (Chicago, 1968), 200–202.

25. Garner, *Reconstruction in Mississippi*, 324–25; *Virginia Convention Debates*, 35; William C. Hine, "Black Politicians in Reconstruction Charleston, South Carolina: A Collective Study," *JSH*, 49 (November 1983), 561; Edward Magdol, *A Right to the Land: Essays on the Freedmen's Community* (Westport, Conn., 1977), 113–36; Charles A. Brown, "William Hooper Councill: Alabama Legislator, Editor, and Lawyer," *Negro History Bulletin*, 26 (February 1963), 171; KKK Hearings, Georgia, 923–31.

said to have lost $75,000 in a single night of poker (nearly half on four aces defeated by a straight flush held by another black legislator).[26]

Although some black politicians enjoyed bourgeois status and others aspired to it, few proved able to translate political power into a share of the economic growth of their states. Even in South Carolina, prominent black officeholders who formed a Charleston streetcar company chartered by the state found it impossible to raise the capital needed to build the line, and a phosphate company formed by black legislators, unable to afford dredges and barges for deep-water mining, was reduced to working marginal deposits in shallow water. Nor did the commission house of Pinchback and Antoine, or a steamboat company the two organized with other black Louisiana politicians, meet with success. The fact is that black politicians' wealth, while impressive when compared to most freedmen, paled before that of Conservatives and white carpetbaggers. Even prominent leaders like Hiram Revels, Robert B. Elliott, and Prince Rivers sometimes had to request small loans from white politicians to meet day-to-day expenses. Most really wealthy blacks avoided politics, either because their business commitments took precedence, or so as not to jeopardize the personal connections with wealthy whites on which their economic standing depended. Indeed, the better-off black political leaders, largely skilled craftsmen, had less in common with the white elite than with other radical nineteenth-century artisans on both sides of the Atlantic—men committed to recasting society in accordance with the principles of republican equality and open opportunity.[27]

For many blacks, political involvement proved less a vehicle for social mobility than a cause of devastating loss. Former slave Henry Johnson, a South Carolina Union League organizer and state legislator, was a bricklayer and plasterer by trade. "I always had plenty of work before I went into politics," he remarked, "but I have never got a job since. I suppose they do it merely because they think they will break me down and keep me from interfering with politics." Congressman Jefferson Long, a tailor, commanded "much of the fine custom" of Macon before embarking upon a political career, but "his stand in politics ruined his business

26. Peter D. Klingman, *Josiah Walls* (Gainesville, 1976), 52; Luther P. Jackson, *Negro Office-Holders in Virginia 1865–1895* (Norfolk, Va., 1945), 49; New Orleans *Louisianian*, March 26, 1871; Samuel Shapiro, "A Black Senator from Mississippi: Blanche K. Bruce (1841–1898)," *Review of Politics*, 44 (January 1982), 85–88; Frank A. Montgomery, *Reminiscences of a Mississippian in Peace and War* (Cincinnati, 1901), 279; "Campaign of 1876," manuscript, Robert Means Davis Papers, USC.

27. Hine, "Boseman," 346–47; Tom W. Schick and Don H. Doyle, "The South Carolina Phosphate Boom and the Stillbirth of the New South, 1867–1920," *SCHM*, 86 (January 1985), 9–10; Blassingame, *Black New Orleans*, 72; Hiram Revels to Adelbert Ames, April 4, 1870, Ames Family Papers; Robert B. Elliott to Franklin J. Moses, Jr., June 9, 1874, Prince Rivers to Robert K. Scott, December 29, 1871, South Carolina Governor's Papers; Chesson, "Richmond's Black Councilmen," 196; Michael P. Johnson and James L. Roark, *Black Masters: A Free Family of Color in the Old South* (New York, 1984), 203–205; Hine, "Black Politicians," 563–64.

with the whites who had been his patrons chiefly." Thus, if some blacks reaped the rewards of leadership, others paid a heavy cost, and not a few eventually retired from the fray. "I was a big politics man then," an Arkansas freedman recalled decades later. "Lost all I had and quit politics." Those who persisted increasingly came to depend on office for economic survival. Indeed, blacks' clamor for a larger share of elective and patronage posts may reflect less an upsurge of nationalism among black voters (although many resented white attempts to monopolize these positions) than the same economic pressures that produced the "politics of livelihood" among white Republicans. Throughout the South, turnover among black lawmakers from one session to the next often exceeded 50 percent, reflecting the departure of some individuals in the face of economic coercion or violence, and the intense competition for office as ambitious new leaders emerged at the local level.[28]

The presence of sympathetic Republican officials, whether black or white, made a real difference in the freedmen's day-to-day lives. Many took an active interest in improving blacks' neighborhoods and securing them a fair share of jobs on municipal construction projects. "They look upon me as a protector," wrote the white mayor of Salisbury, North Carolina, "and not in vain. . . . The colored men placed me here and how could I do otherwise than to befriend them." In Louisiana, the state employed blacks, whites, and Chinese to work repairing the levees, and, in a departure from traditional practice, all received the same wages. As the chief engineer reported, "our 'Cadian friends were a little disgusted at not being allowed double (colored) wages, and the Chinamen were astonished at being allowed as much and the American citizens of African descent were delighted at being 'par'."[29]

To those accustomed to experiencing the law as little more than an instrument of oppression, moreover, it seemed particularly important that the machinery of Southern law enforcement now fell into Republican hands. Tallahassee and Little Rock chose black chiefs of police, and New Orleans and Vicksburg had black captains empowered to give orders to whites on the force. By 1870, hundreds of blacks were serving as city policemen and rural constables; they comprised half the police force in Montgomery and Vicksburg, and more than a quarter in New Orleans, Mobile, and Petersburg. In the courts, defendants now frequently confronted black magistrates and justices of the peace, and racially inte-

28. KKK Hearings, South Carolina, 325; Theophilus G. Steward, *Fifty Years in the Gospel Ministry* (Philadelphia, 1921), 129; George P. Rawick, ed., *The American Slave: A Composite Autobiography* (Westport, Conn., 1972–79), 8, pt. 2:201–202; Vincent, *Black Legislators*, 114; *Christian Recorder*, January 28, 1875; Balanoff, "Negro Legislators," 24.

29. Russell and Thornbery, "William Finch," 318–19; Howard N. Rabinowitz, *Race Relations in the Urban South 1865–1890* (New York, 1978), 265; Thomas B. Sony to William W. Holden, February 12, 1869, North Carolina Governor's Papers; Jeff Thompson to Henry C. Warmoth, May 10, 1871, Warmoth Papers.

grated juries. One white lawyer observed that being compelled to address blacks as "gentlemen of the jury" was "the severest blow I have felt."[30]

Throughout Reconstruction, planters complained it was impossible to obtain convictions in cases of theft and that in contract disputes, "justice is generally administered solely in the interest of the laborer." Nor could vagrancy laws be used, as they had been during Presidential Reconstruction, to coerce freedmen into signing labor contracts. "There is a vagrant law on our statute books . . . ," observed an Alabama newspaper in 1870, "but it is a dead letter because those who are charged with its enforcement are indebted to the vagrant vote for their offices." Black criminals, in fact, did not commonly walk away from Reconstruction courts scot-free. Indeed, as frequent victims of violence, blacks had a vested interest in effective law enforcement; they merely demanded that officials, in the words of one petition, "rise above existing prejudices and administer justice with . . . an even hand." Yet this basic notion of equal justice challenged deeply rooted traditions of Southern jurisprudence. Republican jurors and magistrates now treated black testimony with respect, the state attempted to punish offenses by whites against blacks, and minor transgressions did not receive the harsh penalties of Presidential Reconstruction.[31]

For these reasons and more, Republican officeholders found themselves facing claims from their constituents far more sweeping than in normal political times. John R. Lynch later recalled how, when he served as justice of the peace, freedmen "magnified" his office "far beyond its importance," bringing him cases ranging from disputes with their employers to family squabbles. Greene County, Alabama, freedmen expected local officials "to tell them what to do and what not to do as to the selling of their corn." Most of all, ordinary Republicans of both races deluged the new governors with letters detailing their grievances and aspirations, requesting financial assistance, and seeking advice about all kinds of public and private matters. As one family, complaining of being evicted for voting Republican, explained to Governor Holden, "we consider our selves under your protection [and] care." In part, such letters reflected the political inexperience of so much of the Republican constituency. But they also flowed from an understanding that the utopian

30. Susan B. Eppes, *Through Some Eventful Years* (Macon, 1926), 351–52; Rawick, ed., *American Slave*, 9, pt. 4:42; "Copy of Police Roll, City of Vicksburg," manuscript, 1871, Mississippi Governor's Papers, MDAH; Dennis C. Rousey, "Black Policemen in New Orleans During Reconstruction," *Historian*, 49 (February 1987), 223–36; KKK Hearings, North Carolina, 234; Katherine W. Springs, *The Squires of Springfield* (Charlotte, 1965), 251.

31. F. W. Loring and C. F. Atkinson, *Cotton Culture and the South, Considered with Reference to Emigration* (Boston, 1869), 29; Selma *Southern Argus*, February 3, 1870; Donald G. Nieman, "Black Political Power and Criminal Justice: Washington County, Texas as a Case Study, 1868–1885" (unpub. paper, annual meeting of Organization of American Historians, 1985); Petition, 109 names, November 22, 1869, South Carolina Governor's Papers.

hopes invested in Reconstruction now depended upon the policies undertaken by the new Southern governments.[32]

Southern Republicans in Power

As in so many other aspects of life, the combined effects of war, emancipation, and Reconstruction fundamentally altered the nature of Southern government. The task confronting Republicans, wrote North Carolina carpetbagger Albion W. Tourgée, was nothing less than to "make a state," and the Reconstruction state differed profoundly from anything the antebellum South had known. Slavery had sharply curtailed the scope of public authority, for, as James L. Alcorn pointed out, it produced a society of "patriarchal groupings," with blacks subject to the authority of their owners instead of coming fully "under the cognizance of the government." With planters enjoying a disproportionate share of political power, taxes and social welfare expenditures remained low and Southern education, as one Democrat admitted after the war, was a "disgrace." Some states did expand public responsibility during the 1850s, building schools and asylums for the white insane, deaf, dumb, and blind. But when Henry C. Warmoth became Louisiana's governor in 1868, the state lacked a hard-surfaced public road, and New Orleans, which contained only two hospitals, had a primitive water system that contributed· to regular outbreaks of yellow fever and malaria.[33]

Serving an expanded citizenry and embracing a new definition of public responsibility, Republican government affected virtually every facet of Southern life. Not only the scope of its activity, but the interests it aspired to serve distinguished the Reconstruction state from its predecessors and successors. Public schools, hospitals, penitentiaries, and asylums for orphans and the insane were established for the first time or received increased funding. South Carolina funded medical care for poor citizens, and Alabama provided free legal counsel for indigent defendants. The law altered relations within the family, widening the grounds for divorce, expanding the property rights of married women, protecting minors from parental abuse, and holding white fathers responsible for the support of mulatto children. A parallel enhancement of government's scope occurred in many Republican-controlled localities. Under carpetbagger

32. John R. Lynch, *Reminiscences of an Active Life: The Autobiography of John Roy Lynch*, edited by John Hope Franklin (Chicago, 1970), 60–64; 43d Congress, 2d Session, House Report 262, 706; James P. Burton to William W. Holden, August 28, 1868, John Shepperd et al. to Holden, February 4, 1869, North Carolina Governor's Papers.

33. *National Anti-Slavery Standard*, October 19, 1867; James L. Alcorn, Inaugural Address, galley proof, James L. Alcorn Papers, MDAH; Raleigh A. Suarez, "Chronicle of a Failure: Public Education in Antebellum Louisiana," *LaH*, 12 (Spring 1971), 120–22; Henry Ewbank to Robert K. Scott, December 15, 1870, South Carolina Governor's Papers; Warmoth, *War, Politics, Reconstruction*, 79–80.

Mayor Augustus E. Alden, Nashville expanded its medical facilities and provided bread, soup, and firewood to the poor. Petersburg created a thriving school system, regulated hack rates, repaved the streets, and established a Board of Health that provided free medical care in the smallpox epidemic of 1873. Washington itself embarked on a public works program, including the laying of much-needed sewer lines. All these activities inevitably entailed a dramatic growth in the cost of government. Charged with doubling the state budget between 1860 and 1873, South Carolina Republicans pointed out that much of the increase arose from support of the lunatic asylum, orphan house, state penitentiary, and public schools, none of which had existed before the war.[34]

As will be related, the rising tax burden fueled opposition to Reconstruction among both planters and yeomen. But blacks embraced the activist, reforming state as a counterbalance to the forces of wealth and tradition arrayed against them. "They look to legislation," commented an Alabama newspaper, "because in the very nature of things, they can look nowhere else." Black lawmakers not only supported appropriations for schools, asylums, and social welfare, but unsuccessfully advanced proposals to expand public responsibility even further, including regulation of private markets and insurance companies, compulsory school attendance, restrictions on the sale of liquor, and even, in Louisiana, the outlawing of fairs, gambling, and horse racing on Sunday.[35]

Four interrelated areas reveal the extent, and limits, of Republican efforts to reshape Southern society and establish their own legitimacy: education, race relations, the labor system, and economic development. All inspired intraparty disputes between those seeking to appeal to the broadest possible spectrum of voters, and others more concerned with addressing the needs of the party's most loyal constituents. All witnessed dramatic departures from the traditions of the prewar South but in all, Republicans' achievements, while substantial, failed to live up fully to the lofty goals with which Reconstruction began.

Veterans of the Freedmen's Bureau and freedmen's aid movement, the new superintendents of education viewed schooling as the foundation of

34. Robert H. Bremner, *The Public Good: Philanthropy and Welfare in the Civil War Era* (New York, 1980), 166; Hine, "Boseman," 344; *Alabama Acts 1868*, 490; Balanoff, "Negro Legislators," 43–44; Suzanne Lebsock, "Radical Reconstruction and the Property Rights of Southern Women," *JSH*, 43 (May 1977), 195–216; Dorothy Sterling, ed., *We Are Your Sisters: Black Women in the Nineteenth Century* (New York, 1984), 341; Gary L. Kornell, "Reconstruction in Nashville, 1867–1869," *THQ*, 30 (Fall 1971), 277–78; Henderson, *Unredeemed City*, 183–86, 235; Green, *Secret City*, 91–92; *New National Era*, April 16, 1874.

35. Philip S. Foner and Ronald L. Lewis, eds., *The Black Worker: A Documentary History from Colonial Times to the Present* (Philadelphia, 1978–84), 2:149; Vincent, *Black Legislators*, 105, 180, 199; Howard N. Rabinowitz, "Holland Thompson and Black Political Participation in Montgomery, Alabama," in Rabinowitz, ed., *Southern Black Leaders*, 257–58; *AC*, 1874, 17; Charleston *Daily Republican*, January 21, 1870; Charles Vincent, "Louisiana's Black Legislators and their Effort to Pass a Blue Law During Reconstruction," *Journal of Black Studies*, 7 (September 1976), 47–56.

a new, egalitarian social order. But their goal of creating modern, centralized systems, modeled on the most advanced educational thinking in the
North, proved unattainable. No state could meet entirely the enormous
cost of building a school system virtually from scratch, and traditions of
local autonomy and low taxation valued by native whites, Republican as
well as Democrat, limited the authority of state education officials. North
Carolina's scalawag lawmakers insisted that local boards of education
oversee the schools, and initially relied mainly on the state poll tax and
local levies for financing. (Because many counties refused to appropriate
the necessary funds, the legislature subsequently imposed a statewide
property tax for educational expenses.) Several states, moreover, di-
verted school funds to such purposes as paying interest on the state debt,
and teachers often became demoralized when not paid for months on
end, or when, as often happened, they received their salaries in depreciated state warrants.[36]

Despite these obstacles, a public school system gradually took shape in
the Reconstruction South. A Northern correspondent in 1873 found
adults as well as children crowding Vicksburg schools and reported that
"female negro-servants make it a condition before accepting a situation,
that they should have permission to attend the night-schools." Whites,
too, increasingly took advantage of the new educational opportunities.
Texas had 1,500 schools by 1872, with a majority of the state's children
attending classes. In Mississippi, Florida, and South Carolina, enrollment
grew steadily until by 1875 it accounted for about half the children of
both races. In many ways, educational progress must have appeared
painfully slow; schooling continued to be far more available in towns and
cities than in rural areas, and in 1880, 70 percent of the black population
remained illiterate. Nonetheless, Republicans had established, for the
first time in Southern history, the principle of state responsibility for
public education.[37]

Building schools was one thing, making them the cornerstone of an
egalitarian society quite another. Racially mixed schools were far from
common in the North, and the Peabody Fund, a Northern philanthropy

36. Robert C. Morris, *Reading, 'Riting, and Reconstruction: The Education of Freedmen in the
South 1861–1870* (Chicago, 1981), 235–36; William C. Harris, "The Creed of the Carpet-
baggers: The Case of Mississippi," *JSH*, 40 (May 1974), 209–12; Daniel J. Whitener, "Public
Education in North Carolina During Reconstruction, 1865–1876," in Fletcher M. Green,
ed., *Essays in Southern History* (Chapel Hill, 1949), 79–82; James Norton to Justus K. Jillson,
March 11, 1871, State Superintendent of Education Papers, SCDA; J. L. McDowell to John
Eaton, January 27, 1868, John Eaton Papers, TSLA; Orval T. Driggs, Jr., "The Issues of
the Powell Clayton Regime, 1868–1871," *ArkHQ*, 8 (Spring 1949), 40–44.

37. *The Jewish Times* (New York), February 7, 1873; John P. Carrier, "The Political History
of Texas During the Reconstruction" (unpub. diss., Vanderbilt University, 1971), 455–56;
William C. Harris, *The Day of the Carpetbagger: Republican Reconstruction in Mississippi* (Baton
Rouge, 1979), 328; Shofner, *Nor Is It Over*, 151; Joel Williamson, *After Slavery: The Negro in
South Carolina During Reconstruction* (Chapel Hill, 1965), 224–37; U. S. Bureau of the Census,
Historical Statistics of the United States, Colonial Times to 1970 (Washington, 1975), 382.

whose aid supplemented Southern school budgets, vigorously opposed integration. Whatever their political affiliation, moreover, white parents proved unwilling to have their children sit alongside blacks in the classroom. Any legislator advocating mixed schools, warned the staunchly Republican Rutherford *Star*, would find himself politically "dead, so far as the people of Western North Carolina are concerned." In 1868, Republicans had assured white voters that their children would not be forced to attend classes with blacks, and they generally kept their word. A Texas board of education that attempted to establish an integrated school was removed by the Republican state superintendent. Only Louisiana, where the free black leadership had long demanded integration and the education superintendent, former state Freedmen's Bureau head Thomas W. Conway, believed schools should be open to children of both races, attempted to create an integrated system. Although schools in most rural parishes remained segregated, New Orleans witnessed an extraordinary experiment in interracial education. In its first year, white enrollment plummeted and new segregated private and parochial schools sprang into existence. But many participants in this "white flight" soon returned, and by 1874 several thousand were attending integrated classes.[38]

As for blacks, many surely agreed with Edward Shaw, the militant Memphis political leader, that racial segregation attached a "stigma of inferiority to the black child." Few, however, seem to have believed integration practicable. Generally, black politicians acquiesced when officials established separate schools, and even in Mississippi and South Carolina, where blacks came to control many local boards of education, segregated facilities remained the rule. (South Carolina's School for the Deaf and Blind, initially integrated, established separate classes after the white faculty resigned in protest; thus, "color was distinguished where no color was seen.") Most blacks seem to have judged the new school systems less in terms of an abstract ideal of integration than in relation to the experience of having been denied access to education altogether. Even Frederick Douglass' *New National Era* deemed separate schools "infinitely superior" to no schools at all. Black parents appeared mainly concerned with ensuring an equitable division of school funds, and many believed all-black schools more open to parental control and more likely to hire black teachers than those enrolling whites. Indeed, as graduates of black colleges swelled the ranks of teachers in the 1870s, many found

38. William P. Vaughan, *Schools for All: The Blacks and Public Education in the South, 1865–1877* (Lexington, Ky., 1974), 55; F. Bruce Rosen, "The Influence of the Peabody Fund on Education in Reconstruction Florida," *FlHQ*, 55 (January 1977), 310–20; Rutherford *Star*, February 27, 1869; C. T. Garland to Charles Sumner, December 9, 1872, Sumner Papers; Samuel J. Powell to James McCleery, June 25, 1870, W. O. Davis to T. W. Conway, February 15, 1871, State Board of Education Correspondence; Louis R. Harlan, "Desegregation in New Orleans Public Schools During Reconstruction," *AHR*, 67 (April 1962), 663–75.

employment in the new public schools. The number of black educators in South Carolina rose from fifty in 1869 to over 1,000 six years later.[39]

With one notable exception, segregation also predominated at public institutions of higher learning. Partly to head off black demands for admission to the state university, Mississippi established Alcorn University with a black board of trustees and former Sen. Hiram Revels as president. Several black students gained admission to the University of Arkansas when it opened in 1872, but they were taught separately by the president after regular school hours to avoid embarrassing the faculty. Only the University of South Carolina, before the war the domain of "sons of the aristocracy," attempted to integrate and democratize higher education. Henry E. Hayne, South Carolina's secretary of state, became the university's first black student when he enrolled in the Medical School in 1873, whereupon a majority of white students withdrew, along with much of the faculty. In response, the legislature brought in professors from the North, abolished tuition charges, and established preparatory courses for those unable to meet admission requirements. Although mostly black, the Radical University managed to attract white students as well, and the two races, according to one undergraduate, "study together, visit each other's rooms, [and] play ball together."[40]

If its assumption of responsibility for education brought the Reconstruction state in line with an earlier expansion of public authority in the North, its effort to guarantee blacks equal treatment in transportation and places of public accommodation launched it into a realm all but unknown in American jurisprudence. But, as in the case of education, establishing a legal doctrine of equal citizenship proved easier than putting the principle into effect. In these days before the rise of a comprehensive system of legalized segregation, racial discrimination took a variety of forms. Many institutions, public and private, excluded blacks altogether; others provided separate and ostensibly equal facilities; still others offered blacks markedly inferior services. Railroads and steamboats generally refused to allow blacks, regardless of their ability to pay, access to first-class accommodations, relegating them instead to "smoking cars" or lower decks along with lower-class whites.

Clearly, some forms of discrimination, such as their exclusion from

39. Lester C. Lamon, *Blacks in Tennessee 1791–1970* (Knoxville, 1981), 38; Balanoff, "Negro Legislators," 35; Vaughan, *Schools for All*, 57–70; Williamson, *After Slavery*, 222; *New National Era*, June 16, 1870; 46th Congress, 2d Session, Senate Report 693, pt. 3:143; Jacob Broadnax to John Eaton, January 2, 1868, Eaton Papers; Howard N. Rabinowitz, "Half a Loaf: The Shift from White to Black Teachers in the Negro Schools of the Urban South, 1865–1890," *JSH*, 40 (November 1974), 566, 578; *South Carolina Reports and Resolutions 1869–70*, 406–407, *1875–76*, 409.

40. Harris, *Day of the Carpetbagger*, 347–51; Guerdon D. Nichols, "Breaking the Color Barrier at the University of Arkansas," *ArkHQ*, 27 (Spring 1968), 3; *Charleston Republican*, April 13, 1872; Daniel W. Hollis, *University of South Carolina* (Columbia, S.C., 1951–56), 2:44–79; *New National Era*, July 9, 1874.

public facilities like schools and benefits like poor relief, were a matter of concern to the entire black population. The actions of private businesses, however, in many cases affected only a small minority. The New Orleans *Tribune*, which saw legislation guaranteeing blacks' access to public accommodations as the "foremost task" facing the Republican government, assured whites that since most freedmen remained "too poor to frequent them," restaurants, theaters, and hotels would find themselves serving only "a small number, and most of these, possessed of property and educated." Well-off, cultured blacks deeply resented the affronts they encountered when attempting to obtain a meal, lodging, or transportation, and demanded treatment appropriate to their station. When the New Orleans Opera House excluded free black V. E. Macarty, "a respectable gentleman of polished manner," the *Tribune* commented, "there was perhaps not a man in the whole audience who was more fit than Mr. Macarty to frequent a refined society." Mississippi Sheriff John M. Brown expressed vividly the resentment felt by prominent blacks over discriminatory treatment: "Education amounts to nothing, good behavior counts for nothing, even money cannot buy for a colored man or woman decent treatment and the comforts that white people claim and can obtain." It would be wrong, however, to view these grievances either as purely selfish or as largely irrelevant to ordinary freedmen. Many of the latter participated in the campaign to integrate urban streetcars, which now spread to cities like Louisville and Savannah. Even those who could not afford a first-class rail ticket may well have viewed the treatment accorded their elected representatives as an insult to the black electorate as a whole.[41]

More than any other issue, demands by blacks, supported by many carpetbaggers, for the outlawing of racial discrimination exposed and sharpened the Republican party's internal divisions. Should such a law be adopted, declared a white South Carolina legislator, "I am no longer with the party. I am willing to give the Negro political and civil rights, but social equality, never." In states like Alabama and North Carolina, dominated by native white Republicans, bills introduced by black legislators protecting equal rights on public conveyances failed to pass. (A black legislator chastized north Alabama scalawags for their opposition to one civil rights proposal. During the war, he reminded them, upcountry families had been destitute "and I have seen colored people to steal corn from their own masters to keep your women and children from starving.") Two states enacted laws that appeared to sanction segregation: Georgia, which

41. New Orleans *Tribune*, January 1, 10, 21, 1869; 46th Congress, 2d Session, Senate Report 693, pt. 2:361–63; Marjorie M. Norris, "An Early Instance of Nonviolence: The Louisville Demonstrations of 1870–1871," *JSH*, 32 (November 1966), 496–503; August Meier and Elliott M. Rudwick, "A Strange Chapter in the Career of 'Jim Crow'," in August Meier and Elliott M. Rudwick, eds., *The Making of Black America* (New York, 1969), 2:15–16.

ordered common carriers to furnish all passengers with "like and equal accommodations," and Arkansas, which required "seats, equal to first class seats," for black travelers.[42]

Even where blacks enjoyed greater influence within the party, Republican governors initially employed their influence to defeat civil rights bills or vetoed them when passed, fearing that such measures threatened the attempt to establish their administrations' legitimacy by wooing white support. "No legislation . . . for the establishment of 'Social Equality' between the races has been enacted," boasted the Republican state committee of Mississippi after Governor Alcorn vetoed a bill barring railroads chartered by the state from discriminating against blacks. In 1872, a measure imposing criminal penalties upon hotel owners and operators of railroads and steamboats guilty of discrimination passed the legislature (with virtually no scalawag support) but was mysteriously lost or stolen on its way to the executive mansion and thus failed to become law. Florida Gov. Harrison Reed in 1868 vetoed a bill guaranteeing equal treatment on public conveyances, and Warmoth twice rejected civil rights measures passed by the Louisiana legislature, actions that won applause from native whites of both parties, but alienated the bulk of the black leadership.[43]

As blacks' influence and assertiveness grew and the cautious policies of the first governors fell into disrepute, laws guaranteeing equal access to transportation and public accommodations were enacted throughout the Deep South. Texas in 1871 replaced its Presidential Reconstruction law requiring segregation by the state's railroads with one barring them from "making any distinction in the carrying of passengers." Two years later, Mississippi, Louisiana, and Florida mandated fines and imprisonment (up to $1,000 and three years in jail in Mississippi) for railroads, steamboats, inns, hotels, and theaters denying "full and equal rights" to any citizen. Generally, the burden of enforcement rested with the injured party, but some states authorized their attorneys general to file damage suits and threaten common carriers and chartered businesses with the loss of their license to operate. The most sweeping legislation of all was enacted in South Carolina. Here, black lawmakers' initial attempt to outlaw discrimination died in the white-controlled Senate, but beginning in 1869, a series of laws required equal treatment by all places of public

42. Columbia *Daily Phoenix*, September 18, 1868; Richard Bailey, "Black Legislators During the Reconstruction of Alabama, 1867–1878" (unpub. diss, Kansas State University, 1984), 171; Balanoff, "Negro Legislators," 41; Allen W. Trelease, "Radical Reconstruction in North Carolina: A Roll-Call Analysis of the State House of Representatives, 1868–1870," *JSH*, 42 (August 1976), 331; *Georgia Acts 1870*, 398, 427–28; *Arkansas Acts 1868*, 39–40.

43. *Address of the State Central Committee of the Republican Party*, broadside, December 8, 1870, MDAH; Albert T. Morgan to Adelbert Ames, March 24, April 6, 1872, Ames Family Papers; Shofner, *Nor Is It Over*, 202; Roger A. Fischer, *The Segregation Struggle in Louisiana 1862–77* (Urbana, Ill., 1974), 61–73.

accommodation and any business licensed by municipal, state, or federal authority. The maximum penalty for individual violators ranging from theater owners to railroad conductors was a $1,000 fine and five years in prison, and offending corporations would forfeit their charters.[44]

A few localities experienced genuine progress toward racial integration during Reconstruction. In New Orleans, with its distinctive French heritage, Mardi Gras celebrations involved all citizens, many resorts near the city served blacks and whites, and equality reigned on the streetcars. Austin, "a cosmopolitan city, albeit on a small scale," witnessed "free intermingling of colors" in restaurants, shops, and saloons. A white boycott of Charleston's integrated streetcars faded as many became "legweary and tired of the expense of paying for private conveyances." A Confederate cavalry officer who visited Jackson in 1870 found blacks and whites walking "arm-in-arm or side by side, and the sight was stranger to me than the transformation of King George's portrait into that of General Washington" in Irving's "Rip Van Winkle."[45]

At the lower reaches of Southern urban society, casual contact between the races, not uncommon before the war, seems to have increased during Reconstruction. Grogshops, brothels, and gambling dens served integrated clienteles, sports teams were generally segregated but sometimes played interracial contests, and cockfights attracted crowds in which "kid gloved gentlemen . . . jostled against squalid negroes," as a Democratic journal complained. (The victory of South Carolina's champion gamecocks over challengers from Georgia inspired the integrated onlookers to loud cries of "South Carolina forever." "What effect will this have on reconstruction?" a Republican newspaper wondered.) Occasionally, aggrieved blacks launched successful suits under the new laws. Georgia legislator James Simms recovered nearly $2,000 from a Virginia railroad after being denied access to the first-class car, and Pinchback won an out-of-court settlement after being refused a Pullman berth.[46]

By and large, however, civil rights laws remained unenforced. State facilities like orphan asylums dealt separately with white and black inmates, railroads continued to direct "colored ladies and gentlemen" to

44. *Texas General Laws, Second Session 1871*, 16; *Mississippi Laws 1873*, 66–69; *Louisiana Acts 1873*, 156–57; *Florida Acts and Resolutions 1873*, 25–26; *Revised Statutes South Carolina 1873*, 338, 739–40, 897–98.

45. Dale A. Somers, "Black and White in New Orleans: A Study in Urban Race Relations, 1865–1900," *JSH*, 40 (February 1974), 32–34; Henry C. Warmoth to H. S. McComb, July 11, 1871 (copy), Warmoth Papers; A. C. Green, "The Durable Society: Austin in the Reconstruction," *SWHQ*, 72 (April 1969), 501–502; *CG*, 41st Congress, 3d Session, 1058; Montgomery, *Reminiscences*, 275.

46. Ira Berlin, *Slaves Without Masters: The Free Negro in the Antebellum South* (New York, 1974), 260–65; Michael B. Chesson, *Richmond After the War, 1865–1890* (Richmond, 1981), 101–102; Dale A. Somers, *The Rise of Sport in New Orleans 1850–1900* (Baton Rouge, 1972), 120, 204–205; *Charleston Daily Republican*, April 9, 1870; Taylor, *Negro in Virginia*, 53; New Orleans *Louisianian*, July 9, 1871.

the smoking car where, according to one complaint, "smoking, drinking, and obscene conversation were carried on continually by low whites," and only a few blacks thought it worth the effort to insist upon equal privileges in theaters, hotels, and restaurants. In Mississippi, despite the law forbidding segregated transport, blacks "inevitably go into those [cars] provided for 'Jim Crow'." Indeed, with the rise of independent black churches and fraternal societies, the establishment of separate schools, and the rapid expansion of black facilities from skating rinks to barrooms, separation, not integration, characterized Reconstruction social relations. Yet while most blacks valued these autonomous institutions and did not object to voluntary racial separation, they insisted the state must remain color-blind. And in fact politics and government were the most integrated institutions in Southern life. Blacks and whites sat together on juries, school boards, and city councils, and the Republican party provided a rare meeting ground for likeminded men of both races. Thus, if Reconstruction did not create an integrated society, it did establish a standard of equal citizenship and a recognition of blacks' right to a share of state services that differed sharply from the heritage of slavery and Presidential Reconstruction, and from the state-imposed segregation that lay in the future.[47]

Like education and race relations, economic legislation reflected the expanded scope and altered purposes of the Reconstruction state. The most dramatic transformation concerned labor relations. For the first time in Southern history, planters found themselves unable to use public authority to bolster their control of the black labor force, a change equally visible in the repeal of existing laws, the refusal to enact others, and new legislative departures. Southern lawmakers swept away the remnants of the Black Codes, with their limits on the freedmen's physical mobility and economic options, required parental consent for apprenticeships, and rewrote vagrancy laws so as to narrow the crime's definition and prohibit the hiring out of offenders who could not pay their fines. Republicans humanized the harsh penal codes of Presidential Reconstruction, outlawing corporal punishment and sharply reducing the number of capital offenses and the penalties proscribed for theft.[48]

Equally significant was the regularity with which lawmakers turned down proposals to reinforce labor discipline. Throughout Reconstruction, planters pressed unsuccessfully for measures securing "the observance and performance of labor contracts," preventing an employer from "enticing" away another's workers, and barring the sale of agricultural

47. Fischer, *Segregation Struggle*, 83–85; Taylor, *Negro in Tennessee*, 78–79; *New National Era*, January 18, 1872; Somers, "Black and White," 25–30; Williamson, *After Slavery*, 274–77.
48. Daniel A. Novack, *The Wheel of Servitude: Black Forced Labor After Slavery* (Lexington, Ky., 1978), 22–26; Vincent, *Black Legislators*, 108–109; *Texas General Laws, 1871*, 90–91; *Georgia Acts 1868*, 16–17; *AC*, 1869, 633; J. G. de Roulhac Hamilton, *Reconstruction in North Carolina* (New York, 1914), 355.

produce at night so as to combat theft by laborers and restrict black farmers' economic options. (Even conservative Florida rejected such "sunset" bills, and when one came before Alabama's lawmakers, blacks attacked it as "class legislation" and it went down to defeat.) Nor did planters and agricultural reformers succeed in their campaign to revise existing fence laws. An 1870 proposal to require Georgians to fence in their livestock aroused "animated opposition" from both upcountry and black legislators, and when Alabama made owners of animals liable for damages to crops in one black belt county, the law caused much "dissatisfaction among the negroes" and was quickly repealed. Efforts to secure laws restricting hunting on private property proved equally unsuccessful. "Now that the white man is so poorly represented in the Legislature," lamented a South Carolina planter, "the poacher wanders unreproved."[49]

Positive legislation also reflected the new aims of labor policy. During Presidential Reconstruction, anyone who advanced supplies to a planter or farmer was allowed to hold a lien on the crop, sometimes leaving nothing with which to pay a share worker whose employer fell into debt. Now, building upon precedents established by the Freedmen's Bureau, the law awarded workers a lien superior to all other claims upon an employer's property, prohibited the discharge of laborers for political reasons, and required that evicted tenants be compensated for time worked before their dismissal. Between 1865 and 1868, claimed Florida black legislator Robert Meacham, blacks were regularly dismissed for trifling causes just before the harvest, thereby forfeiting their share; "now . . . there is a law that allows a man to get what he works for." South Carolina went even further, securing laborers' wages from attachment for debt, barring the crop's removal before hands were paid, and authorizing laborers to demand that a disinterested party oversee its division. Mississippi's legislature instructed justices of the peace to construe the law "in the most liberal manner for the protection and encouragement of labor." No wonder one planter complained in 1872, "under the laws of most of the Southern States, ample protection is afforded to tenants and very little to landlords."[50]

Republican governments also shaped economic legislation to establish their legitimacy in the white upcountry, where debt still rested "like

49. Montgomery *Alabama State Journal,* February 10, 1871; New Orleans *Picayune,* January 22, 1871; *Southern Field and Factory,* 1 (January 1871), 15; Shofner, *Nor Is It Over,* 130; Mobile *Register,* April 16, 1873; Atlanta *Constitution,* August 30, 1870; *Alabama Acts 1868,* 473, *1870–71,* 93; "A. L." to C. W. Dustan, July 14, 1869, C. W. Dustan Papers, ASDAH; King, *Southern States,* 434.

50. *Revised Statutes South Carolina 1873,* 490–91, 550, 557–58, 728; *North Carolina Public Laws 1869–70,* 253–55; Novack, *Wheel of Servitude,* 24; *Florida Acts and Resolutions 1870,* 29–31, 39, *1872,* 54; KKK Hearings, Florida, 101–106; *Mississippi Laws 1872,* 135; *Rural Carolinian,* 3 (March 1872), 335.

an incubus upon the people." When North Carolina's Supreme Court in 1869 voided a law staying debt collections, there was an outcry among the yeomanry. "It will be the poore men of the county that will suffer . . ." complained one mountain Republican. "Poore men will have to sell the hat off thare head to pay thare debts." In fact, as state after state exempted specified amounts of land, personal property, mechanics' tools, and agricultural implements from seizure for debt (except claims for wages), the majority of small property holders, and some large ones as well, found themselves protected against losing their property to creditors. In the upcountry, debtor relief for a time remained "a strong card in the party of Reconstruction." And blacks, few of whom benefited directly, supported the policy to strengthen the party. When Alonzo J. Ransier, campaigning for lieutenant governor in 1870, reminded an audience in South Carolina's upcountry that yeomen no longer need fear "being driven out of doors" for debt, he did not fail to point out that the measure owed its passage to black votes—"colored men and legislation by colored men did it."[51]

Blacks, of course, had their own economic agenda. The coming to power of Republican governments rekindled the persistent dream of landownership. One carpetbagger, describing his party as representing "the poorer and laboring classes," evoked "deafening cheers" and cries of "Amen, thank God" when he told an Edgefield County, South Carolina, audience in 1869 that he expected within a few years "to see nearly every voter, white and black, on farms of their own." Nor did this goal appear entirely utopian. The postwar South was a sparsely settled region, containing vast tracts of uncultivated land (less than 10 percent of Louisiana's 30 million acres were being tilled at the end of Reconstruction). With many landowners desperately in need of cash, moreover, and land values having plummeted, the cost of acquiring property did not appear prohibitive. "I bought everything a man had lost one day for $85," a South Carolinian reported after a sheriff's sale in 1868, "place, household and kitchen furniture, stock and all." Few freedmen, however, possessed even the meager resources needed to acquire land on the open market, even at these bargain prices. Thus, blacks looked to the state. As Florida leader Emanuel Fortune put it: "We have to purchase land from the Government . . . otherwise we cannot get it." Their demands were echoed by at least some white Republicans. "I have the opinion that our party will go down unless there is some way provided for the landless population to have homes," a resident of the North Carolina mountains

51. "One of the People" to Tod R. Caldwell, February 26, 1869, Lucius Dockworth to Caldwell, November 22, 1868, Caldwell Papers; Kenneth E. St. Clair, "Debtor Relief in North Carolina During Reconstruction," *NCHR*, 18 (July 1941), 223; J. H. Thomas, "Homestead and Exemption Laws of the Southern States," *American Law Register*, 19 (1871), 137–50; John M. Cain to Rufus Bullock, May 10, 1870, Reconstruction File, GDAH; Charleston *Daily Republican*, October 8, 1870.

warned Governor Holden. "The rebels hold the most of the land property in our county and a large majority of the poor union men is destitute of land."[52]

Given the importance of the land question, however, it is remarkable how few concrete actions the Republican governments took. As on so many other issues South Carolina proved the exception, adopting a path-breaking program of land distribution overseen by a land commission with the power to purchase real estate and resell it on long-term credit. Greeted with enthusiasm by the state's freedmen, the commission was initially plagued by mismanagement and corruption. It accomplished little until Secretary of State Francis L. Cardozo reorganized its affairs in 1872. But by 1876 some 14,000 black families (about one seventh of the state's black population), plus a handful of whites, had been settled on homesteads. One Abbeville County tract, purchased from the commission by a group of black families, survives to this day as the community of Promised Land. All this represented a remarkable achievement, but one no other state came close to matching.[53]

Rather than emulate South Carolina, Republicans in other states chose to employ taxation as an indirect means of weakening the plantation and promoting black ownership. Not only did rising state expenses and the fall in Southern property values necessitate an increase in rates, but the new tax system dramatically altered the relative burden borne by different groups of Southerners. In contrast to the prewar policy of low land taxes supplemented by levies on slaves and an array of licenses and fees, and the exorbitant poll taxes of Presidential Reconstruction, revenue now derived primarily from ad valorem taxation of landed and personal property. Moreover, while planters had previously been allowed to assess the value of their property for tax purposes, Republican officials, sometimes their own former slaves, now performed that task. Thus, planters and white farmers, many for the first time, paid a significant portion of their income as taxes, while the law, complained an Alabama Democrat, " 'lets out,' as the phrase is, every negro," since a small amount of personal property, tools, and livestock was exempted from the levies. "The new system," a Republican newspaper announced, "aims to tax property. . . . A man who has nothing should pay no tax save, perhaps, a poll tax."[54]

52. Charleston *Daily Republican*, August 31, 1869; William I. Hair, *Bourbonism and Agrarian Protest: Louisiana Politics 1877–1900* (Baton Rouge, 1969), 46; John J. Ragin to John H. King, April 20, 1868, John H. King Papers, USC; KKK Hearings, Florida, 95; William W. Buchanan to William W. Holden, November 30, 1868, North Carolina Governor's Papers.
53. Carol R. Bleser, *The Promised Land: The History of the South Carolina Land Commission, 1869–1890* (Columbia, S.C., 1969); Elizabeth Bethel, *Promiseland: A Century of Life in a Negro Community* (Philadelphia, 1981), 20–29. Florida in 1872 offered homesteaders up to 160 acres of unsold state swamp lands. *Florida Acts and Resolutions 1872*, 16–18.
54. J. Mills Thornton III, "Fiscal Policy and the Failure of Radical Reconstruction in the Lower South," in Kousser and McPherson, eds., *Region, Race, and Reconstruction*, 349–94; J. M. Hollander, ed., *Studies in State Taxation* (Baltimore, 1900), 85–92; KKK Hearings, Missis-

Rising taxes quickly emerged as a rallying cry for Reconstruction's opponents, but many blacks hoped high levies would make extensive holdings of uncultivated land unprofitable, and force real estate onto the market. "I want to see the man who owns one or two thousand acres of land, taxed a dollar on the acre," declared Abraham H. Galloway, "and if they can't pay the taxes, sell their property to the highest bidder . . . and then we negroes shall become land holders." During Reconstruction, immense tracts fell into the hands of state governments for nonpayment of taxes—in Mississippi alone over 6 million acres, one fifth of the entire area of the state, was forfeited in this way. Stephen Duncan, the antebellum South's largest cotton producer, saw seven of his Louisiana plantations seized and sold for back taxes in 1874. And nearly every state provided that such property should be divided into small lots when thrown on the market. Yet relatively little in fact changed hands, for often the threat of sale led owners to satisfy their tax liabilities, and neighbors frequently conspired to prevent bids on lands placed at auction. And, while tax sales aroused enormous interest among the freedmen, and some—their number as yet undetermined—acquired land in this manner, most lacked the resources to compete with the wealthy investors and speculators (including Republican politicos like Louisiana's Marshall Twitchell) who engrossed the bulk of the acreage placed on the market. Nor were titles to tax lands always secure, for most states allowed owners to redeem forfeited holdings by paying the tax bill even after a sale had been completed. In Mississippi, 95 percent of the forfeited acreage eventually found its way back to the owner. Judged by the standard of basic equity, the Reconstruction tax system appeared to Republicans preferable to its predecessor, which had "borne so unjustly and unequally upon the people." But as a means of land distribution, it proved singularly ineffective.[55]

Outside South Carolina, in fact, the same political and ideological barriers that had prevented the constitutional conventions from grap-

---

sippi, 289, 369–73, 384, 414; Georgia, 304; South Carolina, 117, 775; Alabama, 963–64; Jackson *Weekly Mississippi Pilot*, August 29, 1870. Georgia had introduced ad valorem taxation in 1852 but without increasing the burdens of planters and farmers, for most revenues derived from the state-owned Western & Atlantic Railroad. Peter Wallenstein, *From Slave South to New South: Public Policy in Nineteenth-Century Georgia* (Chapel Hill, 1987), 54–57.

55. Raleigh *Tri-Weekly Standard*, September 7, 1867; KKK Hearings, South Carolina, 238–39, 1434–36; Alabama, 1416–17; Mississippi, 502–503; E. N. Farrar to Stephen Duncan, May 25, 1875, Stephen Duncan Papers, Natchez Trace Collection, UTx; *Texas General Laws 1873*, 191–92; *North Carolina Public Laws 1868–69*, 174–75; *Florida Acts and Resolutions 1871*, 331; John W. Graves, "Town and Country: Race Relations and Urban Development in Arkansas 1865–1905" (unpub. diss., University of Virginia, 1978), 151–52; Jimmy G. Shoalmire, "Carpetbagger Extraordinary: Marshall Harvey Twitchell 1840–1905" (unpub. diss., Mississippi State University, 1969), 105; Michael Wayne, *The Reshaping of Plantation Society: The Natchez District, 1860–1880* (Baton Rouge, 1983), 95–97; Edward McPherson, *The Political History of the United States of America During the Period of Reconstruction* (Washington, D.C., 1875), 481.

pling successfully with the land issue frustrated the freedmen's hope that Reconstruction would place land in their hands. For one thing, Republican politics never escaped a "colonial" pattern. Unable to establish their legitimacy in the eyes of powerful opponents, the new governments' survival ultimately rested on federal support. Thus, Washington determined the boundaries of change; as Albert T. Morgan observed, "Three fourths of the republican members of the Legislature regard themselves as still under the control and dominion of *Congress.* . . . They can hardly settle our per diem without the feeling of *subserviency to Congress.*" And with Radicalism waning in the North, the national party seemed in no mood to countenance "agrarian" departures in economic policy. Then too, the desire to attract the political support of progressive planters often stood at cross-purposes to the aim of assisting the freedmen. And the free labor ideology itself condemned the idea of direct state intervention on the land issue. Most white Republicans, and many freeborn blacks, while perfectly willing to guarantee the freedmen their rights as free laborers and equal citizens, opposed using the power of the state to redistribute property. Reconstruction, declared one Southern Republican newspaper, meant "protection and fair play," not "free gifts of land or money."[56]

Black dissatisfaction with the failure of Reconstruction governments to do more for plantation laborers, as well as the difficulty even among black leaders of transcending free labor precepts, were evident at the statewide "labor conventions" held in Georgia and South Carolina at the end of 1869, and in Alabama early in 1871. Although artisans and agricultural laborers predominated among the delegates, black politicians rather than men "actually engaged in farming or in some mechanical occupation" controlled the proceedings. Some delegates hoped to apply the expansive Reconstruction definition of public responsibility directly to economic relations, advocating that the state government regulate rents and "fix a price for the labor of plantation hands." Such ideas aroused strenuous opposition from political leaders like Jefferson Long, the "ruling spirit" of the Georgia gathering, and State Senator Lucius Wimbush, who told South Carolina's convention that wages and prices could not be regulated by law.[57]

The South Carolina gathering did, however, propose a series of laws to improve the bargaining power of black labor. An official would be appointed in each county to supervise contracts and provide "advisory counsel" to workers, trial juries would adequately represent "the labor-

56. Albert T. Morgan to Adelbert Ames, April 16, 1870, Ames Family Papers; Charleston *Daily Republican*, April 19, 1869.
57. Foner and Lewis, eds., *The Black Worker*, 2:4–15; Montgomery *Alabama State Journal*, January 4, 1871; Charleston *Daily Republican*, November 27–30, 1869; Charleston *Daily News*, November 29, 1869.

ing classes" and nine hours would constitute a legal day's work for manufacturing and skilled labor. But when a bill embodying some of these ideas came before the legislature early in 1870, it encountered a hail of criticism. "The law of supply and demand," declared upcountry scalawag John Feriter, "must regulate the matter. . . . It is wrong for the General Assembly to convey the idea that it is able to raise wages one cent." Black carpetbagger William J. Whipper warned against any "invasion of private rights." With freedmen providing far more support than white Republicans and free blacks, Northern or Southern, a watered-down bill passed the House. It was further weakened in the white-controlled Senate by the removal of a clause for public funding of laborers' lawsuits. "If a plaintiff, however poor and humble," commented the Charleston *Republican*, "has a good cause of action, he will have no difficulty in securing the assistance of an attorney to prosecute his case." The following winter, South Carolina freedman Aaron Logan introduced what the New York *World* called "one of the most extraordinary [bills] in the annals of legislation"—a measure to regulate the profits of retail merchants and empower plantation laborers to "meet and fix by ballot the rate of wages which their employer shall pay them." The bill never came to a vote, but it underscored the continuing desire among at least some former slaves to use the Reconstruction state to address their economic plight.[58]

In Georgia and Alabama, the labor conventions also established statewide unions to press for higher agricultural wages ($30 was mentioned in Georgia as an acceptable monthly rate) and raise money to "give homes to our poor peoples." The Alabama Labor Union, which involved many of the state's Union League organizers, survived into the mid-1870s. Although Reconstruction witnessed scattered plantation strikes, the unionization of agricultural workers was an idea whose time had not yet come. Yet the attempt illustrated how Reconstruction created possibilities for rural organization inconceivable during slavery and all but impossible under the Redeemers. And that prominent black officeholders countenanced such a step reflected their growing awareness of the economic questions Reconstruction had failed to solve.[59]

As the New Orleans *Tribune* repeatedly pointed out, the realities confronting the freedmen rendered free labor assumptions utterly unrealistic: lacking land and encountering not social harmony but the persistent

58. Foner and Lewis, eds., *The Black Worker*, 2:22–23; Holt, *Black over White*, 154–62; Charleston *Daily Republican*, December 3, 21, 1869, January 12, 17, February 2, 3, 8, 9, 25, December 17, 24, 1870; New York *World*, February 27, 1871.

59. Foner and Lewis, eds., *The Black Worker*, 2:7; Charles R. Edwards to John E. Bryant, December 27, 1869, Bryant Papers; Michael W. Fitzgerald, "The Union League Movement in Alabama and Mississippi: Politics and Agricultural Change in the Deep South during Reconstruction" (unpub. diss., University of California, Los Angeles, 1986), Chapter 5; William W. Rogers, *The One-Gallused Rebellion: Agrarianism in Alabama 1865–1896* (Baton Rouge, 1970), 12.

hostility of planters and merchants, they "cannot rise, except in extraordinary cases. . . . They must be servants to others, with no hope of bettering their condition." But most Republican leaders, while convinced that the plantation system and the existence of "a great landless class" threatened regional progress and effective government, preferred to await the freedmen's slow acquisition of property through the natural workings of the market. They invested their hopes for change not in a policy of economic redistribution, but in a program of regional development, with railroad construction as its centerpiece.[60]

The Gospel of Prosperity

More than any other, the issue of economic development preoccupied Republican leaders in the first years of Reconstruction. With the aid of the state, they believed, the backward South could be transformed into a society of booming factories, bustling towns, a diversified agriculture freed from the plantation's dominance, and abundant employment opportunities for black and white alike. "A free and living Republic," declared a Tennessee scalawag, would "spring up in the track of the railroad as inevitably, as surely as grass and flowers follow in the spring." Here was a vision with appeal to a broad array of Southerners—entrepreneurs who hoped to exploit the region's untapped mineral resources and harness its rivers to power cotton factories, town artisans and merchants long resentful of planters' monopoly of the South's economic resources, and blacks seeking new opportunities for work and advancement. Two thousand freedmen in his county, wrote Mississippi black leader Merrimon Howard, anxiously awaited jobs in railroad construction: "The day we commens to work on a Rail Road . . . it would make this whole South flourish." Most of all, the "gospel of prosperity" captivated the moderate leaders of early Republican rule, who hoped it would recast politics along nonracial lines and win legitimacy for the Reconstruction state. "The party that first completes the Internal Improvement System . . ." wrote a follower of Gov. William W. Holden, "will hold the reins of power here for years to come."[61]

In some respects, little in the program of railroad aid was new, for the promotion of entrepreneurial activity had long been a major concern of American government. Before the Civil War, state and local authorities throughout the country had granted special privileges to railroad compa-

---

60.New Orleans Tribune, January 8, 1869; Samuel S. Ashley to Charles L. Woodworth, January 1, 1869 (misdated 1868), AMA Archives, Amistad Research Center, Tulane University.
61. Mark W. Summers, Railroads, Reconstruction, and the Gospel of Prosperity: Aid Under the Radical Republicans, 1865–1877 (Princeton, 1984); CG, 40th Congress, 2d Session, 976; Merrimon Howard to O. O. Howard, January 16, 1868, H-33 1868, Letters Received, Washington Headquarters, RG 105, NA; unknown to William W. Holden, August 20, 1868, North Carolina Governor's Papers.

nies and other corporations, and lent their credit to help finance all kinds of economic enterprises. (Their experience ought to have sounded a cautionary note, for more than one state, finding itself hopelessly in debt, had been forced into temporary bankruptcy.) Presidential Reconstruction, too, had witnessed numerous efforts to assist railroad building and industrial development, but with few tangible results. The scope and purposes of Republican aid, however, differed substantially from those of their predecessors. In the prewar South, railroads had generally been conceived as adjuncts of the plantation system, a way to facilitate the marketing of staple crops. And while the policies of 1865–67 envisioned a more diversified economy, they failed to make any provision for blacks except as dependent plantation laborers. Republicans, however, saw railroads as catalysts of a peaceful revolution that would dislodge the plantation from its economic throne. Even Republican promoters who saw the freedmen mainly as a source of labor were more consistent modernizers than their predecessors, for they understood that efforts to tie blacks to the plantations would leave them unavailable for work in the mines and factories of a revitalized South.[62]

In the first years of Republican rule, every Southern state extended munificent aid to railroad corporations, either in enactments providing direct payments to a particular company, or in the form of general laws authorizing the state's endorsement of a railroad's bonds once a specified number of track miles had been laid (thus guaranteeing potential investors that the state would assume responsibility if the company proved unable to meet interest payments). County and local governments subscribed directly to railroad stock as well, from Mobile, which spent over $1 million, to tiny Spartanburg, South Carolina, which appropriated $50,000. Republican legislatures also chartered scores of banks and manufacturing companies, appropriated money to repair Mississippi levees and reclaim swamp lands, and in several cases continued their predecessors' policy of encouraging European and Northern immigration.[63]

No matter how extensive, state aid alone could not bring capitalist development to the South. For this, private investment was indispensable, and the frantic effort to attract outside capital led Republican governments down paths that in many ways contradicted their professed image as spokesmen for the black and white poor. To tap the mineral

62. Carter Goodrich, "State In, State Out—A Pattern of Development Policy," *Journal of Economic Issues*, 2 (December 1968), 365–67; Carl Moneyhon, *Republicanism in Reconstruction Texas* (Austin, 1980), 144; Summers, *Railroads, Reconstruction, and the Gospel of Prosperity*, 158; Jackson *Daily Mississippi Pilot*, January 13, 1871.

63. Summers, *Railroads, Reconstruction, and the Gospel of Prosperity*, 32–46, 63–84; Carter Goodrich, "Public Aid to Railroads in the Reconstruction South," *Political Science Quarterly*, 71 (September 1956), 407–42; Rowland T. Berthoff, "Southern Attitudes Toward Immigration, 1865–1914," *JSH*, 17 (August 1951), 336–37.

RECONSTRUCTION: POLITICAL AND ECONOMIC

resources of western North Carolina, one supporter advised Governor Holden, "it will be necessary to enlist the aid of northern capitalists. Unless some protective enactment is made in regard to taxation, I doubt exceedingly if we could interest a single [one]." Such logic had led several states during Presidential Reconstruction to exempt railroads and industrial enterprises from taxes for varying periods of time; Republicans, while having to fund vastly increased public responsibilities, continued and expanded the policy. In Mississippi, for example, railroad companies, factories, and banks virtually escaped taxation altogether. Some states also authorized the leasing of convicts, a system that supplied cheap labor to entrepreneurial allies of Republican governors like Bullock, Reed, and Warmoth, although it involved far smaller numbers than would become common under the Redeemers.[64]

The Southern law, moreover, was redesigned so as to encourage the free flow of capital and enhance the property rights of corporations. The repeal of usury laws in several states left money "free [to] regulate its own price"; the government employed its power of eminent domain to assist railroad and factory construction; and general incorporation laws made their appearance on the Southern statute books. Such measures, noted one aspiring South Carolina entrepreneur, "may be new to us, but they are common in all the manufacturing states of the Union"—and, indeed, these changes recapitulated the way the Northern law had earlier been transformed to facilitate capitalist development.[65]

To some extent, the "gospel of prosperity" succeeded in bringing the two parties closer together. Democrats sat on the boards of railroads that received public aid, and supported subsidy bills in many legislatures. In Alabama, the Republican-controlled Alabama & Chattanooga Railroad and the Louisville & Nashville, allied with the Democrats, vied for public assistance and access to the state's mineral resources. The election of Democratic Gov. Robert Lindsay in 1870 altered the direction of state aid, but not the overall policy. On the local level, leaders of the two parties often cooperated in railroad promotion. Prominent Edgefield County Democrats temporarily put aside their white supremacist convictions to join black officials in an effort to attract a railroad, and Charleston's Board of Trade did not hesitate to lobby among black legislators for state aid to the Blue Ridge Railroad, hoping that a link to Western

64. N. L. Stith to William W. Holden, February 10, 1869, North Carolina Governor's Papers; Harris, *Day of the Carpetbagger*, 336, 572; A. Elizabeth Taylor, "The Origins and Development of the Convict Lease System in Georgia," *GaHQ*, 26 (June 1942), 113–15; J. Thorsten Sellin, *Slavery and the Penal System* (New York, 1976), 145–62.

65. Knoxville *Whig*, January 15, 1868; *Mississippi Laws 1873*, 81–82; *Arkansas Acts 1868*, 32; Jonathan M. Wiener, *Social Origins of the New South: Alabama 1860–1885* (Baton Rouge, 1978), 151; Benjamin Evans to Robert K. Scott, July 13, 1868, South Carolina Governor's Papers; Morton J. Horwitz, *The Transformation of American Law, 1790–1860* (Cambridge, Mass., 1977).

markets would reverse the city's economic decline. Arkansas Democratic leader David Walker stubbornly denied the legal authority of the Republican state government, but invoked its aid law in an effort to interest Northern capitalists in building a railroad to Fayetteville.[66]

Nor were Republicans always united in support of favors to corporations. Although initially enjoying support among all factions, the railroad aid program was most closely identified with party moderates. From the outset, some Republicans insisted the funding of public schools and other social programs should take priority over economic modernization, and others opposed aiding railroads until they conceded equal treatment to black passengers. When Mississippi in 1871 granted the Mobile & Northwestern Railroad nearly 2 million acres of land, twenty-five Republican legislators denounced the measure, since these lands had previously been earmarked to be sold for the public school fund. Texas presented the unusual spectacle of a Republican governor—Radical scalawag Edmund J. Davis—unsuccessfully opposing a massive program of railroad aid pushed through the legislature by a coalition of party moderates and Democrats. Many blacks did not share the modernizing mentality. When Jay Gould sought aid for a railroad project from Harrison County, in the Texas black belt, a freedman later recalled, "They told him they didn't need his railroad."[67]

Nonetheless, in most cases, railroad aid bills were drafted by Republican committees, passed by Republican legislatures, and implemented by Republican governors. Black lawmakers, like white, pressed for aid to enterprises that promised to enrich their communities. And Republicans throughout the South identified their party's fortunes with the "gospel of prosperity."[68] But the program of state-sponsored capitalist development, inaugurated with grandiose hopes, proved in many ways the party's undoing. Vastly increasing the financial claims upon government, it produced ever-rising state debts and taxes and drained resources from schools and other programs. Far from enhancing Republicans' respectability, it opened the door to widespread corruption. The state aid program alienated voters in the white upcountry, where many yeomen and artisans feared the railroad would subordinate their self-sufficient society

66. Summers, *Railroads, Reconstruction, and the Gospel of Prosperity*, 68–84; Horace Mann Bond, *Negro Education in Alabama: A Study in Cotton and Steel* (Washington, D.C., 1939), 41–51; Burton, "Race and Reconstruction," 39; Jamie W. Moore, "The Lowcountry in Economic Transition: Charleston since 1865," *SCHM*, 80 (April 1979), 156–58; W. L. Trenholm to Robert K. Scott, July 21, 1868, South Carolina Governor's Papers; W. J. Lemke, ed., *Judge David Walker: His Life and Letters* (Fayetteville, N.C., 1957), 78–79; George H. Thompson, *Arkansas and Reconstruction* (Port Washington, N.Y., 1976), 86.

67. Summers, *Railroads, Reconstruction, and the Gospel of Prosperity*, 83–84; Charleston *Daily Republican*, March 19, June 10, 1870; Harris, *Day of the Carpetbagger*, 533; Moneyhon, *Republicanism in Texas*, 137–50; Rawick, ed., *American Slave*, Supplement, Ser. 2, 8:3096.

68. Summers, *Railroads, Reconstruction, and the Gospel of Prosperity*, 83–84; Klingman, *Walls*, 29; Drago, *Black Politicians*, 70–72; Holt, *Black over White*, 141.

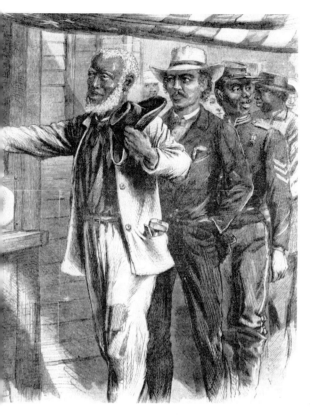

"The First Vote." Note that the voters represent key sources of the black political leadership—the artisan carrying his tools, the well-dressed urbanite, and the soldier. (*Harper's Weekly*, November 16, 1867)

ctioneering at the South." Women as well as men attended these early political gather-. (*Harper's Weekly*, July 25, 1868)

(Above left) Benjamin S. Turner, Congressman from Alabama. (Library of Congress)

(Above right) P. B. S. Pinchback, Lieutenant Governor and Governor of Louisiana. (Library of Congress)

(Left) Blanche K. Bruce, Senator from Mississippi. (Library of Congress)

(Below left) Robert Smalls, Civil War hero and Congressman from South Carolina. (Library of Congress)

(Below right) Robert B. Elliott, Congressman from South Carolina. (Library of Congress)

Above left) James L. Alcorn, Governor of Mississippi. Library of Congress)

Above right) William G. Brownlow, Governor of Tennessee. (Library of Congress)

Right) Henry C. Warmoth, Governor of Louisiana. Library of Congress)

Below left) Albion W. Tourgée, North Carolina Jurist. Library of Congress)

Below right) Adelbert Ames, Governor of Mississippi. Library of Congress)

OPPONENTS OF RECONSTRUCTION

(*Above left*) Democratic Campaign Badge, 1868. (New York Public Library, Schomburg Center for Research in Black Culture)

(*Above right*) Zebulon Vance, Redeemer Governor of North Carolina. (Library of Congress)

(*Left*) Wade Hampton, Redeemer Governor of South Carolina. (Library of Congress)

(*Below left*) John B. Gordon, Democratic candidate for governor and head of the Ku Klux Klan in Georgia (Library of Congress)

(*Below right*) William H. Trescot, South Carolina Planter (Library of Congress)

Klan Warning: a graphic prediction of the fate of Ohio carpet-bagger A. S. Lakin and scalawag Noah B. Cloud in the event of a Democratic victory in the 1868 presidential election. (Tuscaloosa *Independent Monitor,* September 1, 1868)

"Two Members of the Ku-Klux Klan in Their Disguises." *Harper's Weekly,* December 19, 1868)

Klansmen Firing into a Home. (Dorothy Sterling)

FACES OF THE GRANT ERA

(*Top left*) Victoria Woodhull, Feminist (New-York Historical Society)

(*Top right*) Jay Gould, Financier (Library of Congress)

(*Left*) Roscoe Conkling, Senator from New York (Library of Congress)

(*Below left*) Crazy Horse, Conqueror of General George A. Custer at the Little Big Horn (Chicago Historical Society)

(*Below right*) Susan B. Anthony, Feminist (Metropolitan Museum of Art)

Railroad Workers. (American Social History Project)

"Capital and Labor": from a cotton textile depicting scenes of labor and industry, ca. 1870. The harmony of interests of capital and labor was a central tenet of the free labor ideology. (Metropolitan Museum of Art)

The Corliss Engine, symbol of the new industrial age at the Philadelphia Centennial Exposition. (*Harper's Weekly*, May 27, 1876)

Ruins of Pittsburgh Round House, the "Great Strike," July 1877. (*Scribner's Magazine*, July 1895)

"This is a White Man's Government."
(*Harper's Weekly*, September 5, 1868)

"And Not This Man?" (*Harper's Weekly*,
August 5, 1865)

"Colored Rule in a Reconstructed (?) State."
(*Harper's Weekly*, March 14, 1874)

Changes in graphic artist Thomas Nast's depiction of blacks mirrored the evolution of
Republican sentiment in the North. Upper left, the black soldier as upstanding citizen
deserving of the franchise; upper right, the former slave as innocent victim of Irish immigrants, Confederate veterans, and Wall Street financiers (ostensibly, the three pillars of the
Democratic party); bottom, black legislators as travesties of democratic government.

to the tyranny of the marketplace. And the failure to produce tangible benefits among the freedmen produced growing disillusionment with the leaders they had helped to place in power. What had begun in the hope of bolstering Republican government soon threatened to undermine the entire Reconstruction experiment.

Rising taxes, steadily increasing debts, and the declining value of state bonds offered the most visible evidence of the new governments' financial problems. To a large extent, the fiscal crisis of the Reconstruction state flowed inexorably from the war and emancipation, and the circumstances under which Republicans assumed power. In the absence of massive federal aid, the rebuilding of damaged facilities and the expansion of public responsibilities to serve the needs of a now doubled citizenry required a dramatic rise in state expenditures, while property taxes had to rise sharply simply to compensate for the decline in Southern land values. The first Republican governors, moreover, inherited from their predecessors substantial public debts and empty treasuries. Upon taking office in Florida in 1868, Harrison Reed discovered that no account existed of how state monies had been spent between 1848 and 1860 or during Presidential Reconstruction. Reed had to dip into his own pocket to meet the government's initial expenses. When the new legislatures issued bonds in 1868, they found it impossible to market them except at huge discounts. With the national Democratic party threatening to oust the fledgling regimes and local opponents of Reconstruction promising to repudiate the "bayonet bonds," Northern investors proved unwilling to touch Southern securities until "political uncertainty" subsided.[69]

Grant's victory assured Reconstruction's survival, and Southern bonds quickly rose to 80 or 90 percent of par, still below the prices commanded by Northern securities, but a substantial improvement nonetheless. But as the program of railroad aid took hold, bonds issued or endorsed by Republican governments flooded the market. By 1872, North Carolina, whose debt had doubled because of an extravagant program of railroad aid, saw its credit virtually disappear, and South Carolina, whose bonds were selling at 25 cents on the dollar, was unable to meet either current expenses or outstanding liabilities. Meanwhile, property taxes rose steadily. In Mississippi, where the rate had been 1 mill (one tenth of 1 percent) before Republican rule, it reached 9 mills in 1871, and over 12 two years later. To this were added rising local and county levies. Since the value of property had declined so sharply, these figures exaggerate the actual increase in taxes paid, but the rise was real enough. When tax revenues and bond sales failed to provide enough income to cover government

69. B. U. Ratchford, *American State Debts* (Durham, 1941), 182–83; Thornton, "Fiscal Policy," 352; Shofner, *Nor Is It Over*, 200–201, 218–19; Ella Lonn, *Reconstruction in Louisiana After 1868* (New York, 1918), 18–19; Charleston *Daily News*, July 17, 1868; H. H. Kimpton to Robert K. Scott, October 1, 1868, South Carolina Governor's Papers.

needs, some states took to issuing scrip in small denominations (Mississippians dubbed such certificates of indebtedness "Alcorn money"), but these rapidly depreciated in value, to the chagrin of officials forced to accept them as tax payments and public employees who received them for wages.[70]

Many factors contributed to the growing crisis, including renewed threats by Democrats and disaffected Republicans that Reconstruction state debts would be repudiated, but the main cause was fear that Reconstruction governments lacked the resources to meet their burgeoning obligations. It would be fiscal suicide, a New Yorker advised Governor Holden, to increase the state debt "so long as the debtor is not in condition to pay the interest." Yet such warnings were ignored in the rush to shower aid upon the railroads. Ironically, the very effort to attract outside capital ended up destroying "the confidence of capitalists" in Southern securities.[71]

If the collapse of their credit reduced Southern governments' ability to finance the ambitious programs they had undertaken, widespread corruption undermined the effort to establish their political legitimacy. There was nothing unique about the corruption that cast a shadow over the conduct of public affairs in nearly every Reconstruction state. Ever since Capt. Samuel Argall, a deputy governor of early colonial Virginia, took advantage of the opportunity to "make hay whilst the sunne doth shine, however it may fare with the generality," bribery, fraud, and influence peddling have been endemic to American politics. In the Jacksonian era, the collector of New York Port fled to England, having embezzled $1 million. Such incidents were not unknown in the prewar South, and during Presidential Reconstruction Atlanta's treasurer "retired from the place taking with him the city funds." Nor did government in the Reconstruction North—the era of the Tweed and Whiskey Rings—offer a model of probity. Throughout the country, public honor was among the casualties of the Civil War.[72]

Corruption may be ubiquitous in American history, but it thrived in the Reconstruction South because of the specific circumstances of Republi-

70. CG, 41st Congress, 2d Session, 2811; Summers, Railroads, Reconstruction, and the Gospel of Prosperity, 137–38; Raper, Holden, 127; Francis B. Simkins and Robert H. Woody, South Carolina During Reconstruction (Chapel Hill, 1932), 147–54; 44th Congress, 1st Session, Senate Report 527, Documentary Evidence, 149; Cecil E. McNair, "Reconstruction in Bullock County," AlHQ, 15 (Spring 1953), 106; Driggs, "Clayton Regime," 51–53; Harris, Day of the Carpetbagger, 291–94.

71. New National Era, August 8, 1872; C. B. Curtis to William W. Holden, February 18, 1869, North Carolina Governor's Papers; J. G. Neagle to Robert K. Scott, October 4, 1871, South Carolina Governor's Papers.

72. Edwin G. Burrows, "Corruption in Government," in Jack P. Greene, ed., Encyclopedia of American Political History (New York, 1984), 1:417–18; Edward Pessen, "Corruption and the Politics of Pragmatism: Reflections on the Jacksonian Era," in A. S. Eisenstadt et al., eds. Before Watergate: Problems of Corruption in American Society, eds. (Brooklyn, 1979), 79–80; Hartford Courant, February 6, 1867.

can rule. The expansion of public responsibilities and the rapid growth of capitalist enterprise linked to the state dramatically increased both the size of budgets and the demands placed upon them. Officials regularly handled unprecedented sums of money, corporations vied for the benefits of state aid, and communities competed for routes that would supposedly guarantee their future prosperity—conditions that offered numerous opportunities for bribery and plunder. The prevailing spirit of economic promotion fostered a get-rich-quick mentality, and many officials saw nothing wrong with taking a piece of the expanding economic pie for themselves. "Ten years from now," wrote an ally of one entrepreneur, "the people . . . will see so clearly the vast benefit of the present proposed Railroads that they will care very little how or by what means they were built." The same social marginality that impelled Republican officials, black and white, to cling to office for a livelihood rendered corruption all but inevitable. Owning relatively little property and facing uncertain political and personal futures, many Republicans determined to make the most of their tenure in office. "I hoped to make money—dreamed of thousands," Daniel H. Chamberlain later recalled of his entrance into South Carolina politics. Finally, the very newness of the Reconstruction political system meant the channels through which business normally influences government (contributing money to party organizations, for example, rather than bribing officials) were far less developed than in other states.[73]

Corruption took many forms in the Reconstruction South. Bribery, either distributed voluntarily by railroads and other enterprises seeking state aid or demanded of them by state officials, became widespread in several states. Henry C. Warmoth established a system of "exacting tribute" from Louisiana's railroads and left the governorship a wealthy man. (Louisiana had a long history of corrupt politics—its prewar lawmakers had not even bothered to make bribery a crime.) Although railroad aid, which involved the largest state expenditures, generated the most corruption, other areas of public life were not immune. Georgia carpetbagger John E. Bryant received a $30,000 loan from New York publishing house Harper and Brothers, in exchange for a promise to have their textbooks adopted by the new state school system.[74]

These examples do not exhaust the ways in which the pursuit of private

73. Summers, *Railroads, Reconstruction, and the Gospel of Prosperity*, 98–117; Henry E. Colton to George W. Swepson, November 16, 1868, George W. Swepson Papers, NCDAH; Daniel R. Chamberlain to Francis W. Dawson, June 9, 1875, Francis W. Dawson Papers, DU; James C. Scott, "The Analysis of Corruption in Developing Nations," *Comparative Studies in Society and History*, 15 (June 1969), 321–25.
74. Summers, *Railroads, Reconstruction, and the Gospel of Prosperity*, 104–105; William P. Kellogg to J. J. Creswell, January 29, 1874, Kellogg Papers; Joe G. Taylor, *Louisiana Reconstructed, 1863–1877* (Baton Rouge, 1974), 194–200, 250; John E. Bryant to Harper and Brothers, April 1, June 15, 1868, Bryant Papers.

advantage sometimes overwhelmed public responsibilities. Railroad laws usually contained strict clauses to safeguard the state's interest: requirements that aid be issued only after a specified track mileage had been laid and that the state assume a first lien upon the company's assets to protect against the consequences of default. But bonds were not infrequently overissued, partly because railroads misrepresented the extent and quality of their construction, but also when governors like Warmoth, Bullock, and Scott allowed friendly companies to evade legal requirements. Bullock illicitly signed over bonds to railroads controlled by his close ally, Hannibal I. Kimball, despite the fact that they had not met construction requirements. (Kimball also received the contract to build a new Atlanta fairgrounds, purchased the city's opera house and then resold it to the state for ten times it� cost for use as the capitol, and employed the proceeds to build Kimball House, reputedly the South's finest hotel.)[75]

South Carolina Gov. Robert K. Scott and state fiscal agent H. H. Kimpton (the college roommate of party leader Daniel H. Chamberlain) speculated in state bonds, while at the same time issuing and attempting to market far more securities than authorized by the legislature. Scott also joined the "ring" of officials who fraudulently acquired the state's shares in the Greenville & Columbia Railroad at a fraction of their true value. And he was deeply involved in land-commission corruption, which involved the purchase at inflated prices of useless swamp lands owned by well-connected individuals. These frauds injured blacks' prospects of acquiring land but not the governor's, for Scott managed to invest $100,000 in real estate in his hometown of Napoleon, Ohio. Reconstruction also witnessed numerous conflicts of interest that, while neither uncommon in the contemporary North nor against the law, created ample opportunities for political and economic favoritism. Many legislators held directorships or stock in railroads receiving aid, as did officials like Louisiana's chief justice and Georgia's attorney general, who were called upon to rule whether companies with which they were connected had fulfilled their legal obligations.[76]

While political leaders often orchestrated corruption for their own gain, the largest schemes originated with economic buccaneers who looked to state aid to finance grandiose dreams of railroad empire. Most notorious was the "ring" headed by Milton J. Littlefield, a former Union

75. Summers, *Railroads, Reconstruction, and the Gospel of Prosperity,* 42–55; W. C. Morrill to Simon Cameron, November 22, 1871, Simon Cameron Papers, LC; Alice E. Reagan, *H. I. Kimball: Entrepreneur* (Atlanta, 1983), 15–23; Willard Range, "Hannibal I. Kimball," *GaHQ,* 29 (June 1945), 47–70.

76. H. H. Kimpton to Robert K. Scott, September 25, 28, 1868, South Carolina Governor's Papers; Kimpton to Scott, July 9, 1869, September 10, 1870, Robert K. Scott Papers, Ohio Historical Society; Williamson, *After Slavery,* 383–86; Bleser, *Promised Land,* 78–81; E. Dale Odom, "The Vicksburg, Shreveport & Texas: The Fortunes of a Scalawag Railroad," *Southwestern Social Science Quarterly,* 44 (December 1963), 280–84; Henry P. Farrow to J. B. Hargroves, September 25, 1871, Henry P. Farrow Papers, UGa.

general who had allegedly mustered in fictitious black soldiers and pocketed their bounty money, and George W. Swepson, a North Carolina entrepreneur who hoped to create a vast Southern transportation network. Swepson, an adviser to North Carolina Governor Holden, possessed the vision of Northern railroad magnates like Cornelius Vanderbilt and Thomas A. Scott, but lacked their access to capital, so for financing he turned to the legislature. Disbursing $200,000 in bribes, loans, and lavish entertainments for the lawmakers, the ring succeeded in obtaining millions in appropriations for roads under its control. Instead of expending the money on construction, however, they employed it to purchase the stock of other railroads, speculate in state bonds, reward political friends with extravagant legal fees and bogus building contracts, and finance a European tour. In one instance, Swepson used funds embezzled from a state-aided North Carolina railroad, plus a bad personal check for half a million dollars, to purchase three bankrupt Florida railroads, which he then persuaded that state's legislature to aid. Holden does not seem to have profited from these manipulations, but while aware of the ring's *"damnable rascality"* he took no action, from either personal loyalty or incompetence.[77]

Although the charge of corruption did much to discredit Reconstruction, bribery, conflict of interest, and the misuse of public position for private gain transcended party lines. "The briber, my moral philosophy teaches, is just as bad as the bribed," commented a black Congressman, and Democratic lobbyists and railroad directors dispensed money as eagerly as Republican. "You are mistaken," a Louisiana Democrat wrote a Northern party leader, "if you suppose that all the evils . . . result from the carpetbaggers and negroes—the democrats are leagued with them when anything is proposed which promises to pay." In the scramble for personal gain, men "of high repute" lent their names to "rotten corporations, lotteries, and other enterprises" seeking state aid. Those benefiting financially from the Swepson-Littlefield frauds included prominent Democrats Augustus S. Merrimon, who drafted much of North Carolina's railroad legislation, and David S. Walker, a former governor of Florida. In Georgia, leaders of the opposing parties joined forces to lease the state-owned Western & Atlantic Railroad under a remarkable arrangement by which "none need advance one dollar, and from which all will receive large dividends." And leading South Carolina Democrats Matthew C. Butler and Martin W. Gary, both former Confederate generals and "first class gentlemen," arranged to have a Taxpayers' Convention

77. Jonathan Daniels, *Prince of Carpetbaggers* (New York, 1958); Charles L. Price, "The Railroad Schemes of George W. Swepson," *East Carolina Publications in History*, 1 (1964), 32–50; Paul E. Fenlon, "The Notorious Swepson-Littlefield Fraud: Railroad Financing in Florida (1868–1871)," *FlHQ*, 32 (April 1954), 231–61; Tod R. Caldwell to William W. Holden, August 30, 1868, John Pool to Holden, December 13, 1868, North Carolina Governor's Papers; Holden to George W. Swepson, November 1, 1869, Swepson Papers.

endorse the state's bonds, after contracting with a group of New York bankers for a share of the profits from the rise in market value they expected would follow. Although Democrats in several states launched investigations of corruption after overturning Republican rule, they proved reluctant to seek either criminal indictments or "any return of the loot," probably for fear of drawing into the net members of their own party.[78]

Black Republicans were hardly immune to the lure of illicit gain. Pinchback used his position on the New Orleans Park Commission to orchestrate the purchase of land, at an inflated price, of which he was part-owner. He also made a handsome profit speculating in state bonds. Inside information, he frankly admitted, allowed his operations to succeed: "I belonged to the General Assembly, and knew about what it would do, etc. My investments were made accordingly." Thomas W. Cardozo appears to have embezzled state funds while serving as Mississippi's superintendent of education. And black lawmakers supported aid to corporations in which they held directorships or stock, and took money for other votes. The bribing of black legislators helped Texas railroads overcome Governor Davis' opposition to state aid. "Our leading men . . ." wrote one disgusted freedman, "sold themselves for gold." Social marginality and the "politics of livelihood" weighed even more heavily upon black officials than white. As a South Carolina legislator explained, after selling his vote in an election for U. S. Senator, "I was pretty hard up, and I did not care who the candidate was if I got two hundred dollars." Overall, however, blacks' share of the take paled before that of governors like Warmoth and Scott, for few occupied political positions that allowed them access to plunder. And in the bribery of lawmakers, the going rate for whites usually exceeded that for blacks.[79]

Even where whites monopolized Reconstruction graft, as in North Carolina and Georgia, few blacks raised voices against it, either from fear of weakening the Republican party or because notorious corruptionists shrewdly took steps to enhance their standing among the black electorate. Littlefield donated funds to AME educational activities, and Hannibal

78. *CG*, 45th Congress, 3d Session, Appendix, 267; R. Hutcheson to Alexander Long, May 27, 1871, Alexander Long Papers, Cincinnati Historical Society; [Belton O'Neall Townsend] "South Carolina Morals," *Atlantic Monthly*, 39 (April 1877), 469; Benjamin H. Hill to Columbus Delano, January 27, 1871, Cameron Papers; Price, "Swepson," 34–35; New York *Sun*, February 1, 1878; Agreement, April 17, 1871 (copy), Martin W. Gary Papers, USC; Memorandum, December 1875, Dawson Papers; W. E. B. Du Bois, *Black Reconstruction in America* (New York, 1935), 622.

79. James Haskins, *Pinckney Benton Stewart Pinchback* (New York, 1973), 85–86; New Orleans *Times* in New Orleans *Louisianian*, March 14, 1872; Euline W. Brock, "Thomas W. Cardozo: Fallible Black Reconstruction Leader," *JSH*, 47 (May 1981), 203–204; Okon E. Uya, *From Slavery to Public Service: Robert Smalls 1839–1915* (New York, 1971), 63; *New National Era*, November 3, 1870; Williamson, *After Slavery*, 394; Drago, *Black Politicians*, 67; W. W. Howe to Henry C. Warmoth, July 21, 1870, Warmoth Papers.

I. Kimball employed black laborers in building his hotel and other projects. (A song popular among Atlanta blacks went: "H. I. Kimball's on de floor/'Taint gwine ter rain no more.") Increasingly, however, black leaders realized that widespread corruption placed them at a disadvantage within the party. A brilliant Republican organizer, Robert B. Elliott "knew the political condition of every nook and corner" of South Carolina. But when he sought a U. S. Senate seat in 1872, he discovered that access to money far outweighed talent or service to the party. Elliott turned down an offer of $15,000 to withdraw from the race but, lacking funds, proved no match for carpetbagger "Honest John" Patterson, who proceeded to spend three times that amount bribing legislators. "The potent influence of the 'almighty dollar'," in Elliott's words, determined the outcome. Corruption also drained money from state programs of special concern to blacks. The theft of school funds for a time left several Louisiana parishes unable to provide for education, and black teacher Robert G. Fitzgerald was told by North Carolina education officials he "could do better in the field," since school monies "had been squandered by 'Littlefield,' the railroad bills, and those high in authority." South Carolina's land-commission frauds appeared particularly outrageous. "These people," wrote one black, "are not only thieves . . . but robbers; having robbed the freedman of his chances of obtaining a homestead."[80]

Most important, corruption threatened to undermine the integrity of Reconstruction itself, not simply in the eyes of Southern opponents but in the court of Northern public opinion. It was true, but beside the point, that corruption flourished outside the South, and that charges of malfeasance generally arose from hostile sources with a vested interest in exaggerating its scope. In normal times, corruption seemed an unfortunate but unavoidable aspect of political life, but Southern Republicans would have benefited enormously from a reputation for honesty. "We cannot afford," one party newspaper correctly remarked, "to bear the odium of profligates in office."[81]

Along with rising taxes and swollen state debts, corruption not only handed the Democrats a potent issue, but contributed to the demise of the entire "gospel of prosperity." The high tide of state aid came between 1868 and 1871. Then, disillusionment set in as a result of the program's high cost, corrupt practices, and meager results. The reaction coincided with growing opposition to the party's initial conciliatory policies, and

80. Balanoff, "Negro Legislators," 39; Drago, *Black Politicians*, 67–68; Lamson, *Elliott*, 164–67; Sterling, ed., *Trouble They Seen*, 205; Robert B. Elliott to Franklin J. Moses, Jr., November 23, 1872, South Carolina Governor's Papers; R. K. Diossy to Thomas W. Conway, May 16, 1871, State Board of Education Correspondence; Robert G. Fitzgerald Diary, January 26, 1870, Robert G. Fitzgerald Papers, Schomburg Center for Research in Black Culture; "Kush," *The Political Battle Axe for the Use of the Colored Men of the State of South Carolina* (Charleston, 1872), 6.
81. Harris, "Creed of the Carpetbaggers," 216.

with blacks' increasing political assertiveness, for railroad aid epitomized the close working relationship the first Republican governors tried to forge with Democrats and their assumption that the party's future lay in appeasing its opponents rather than vigorously promoting the interests of its own constituents. The black labor conventions of 1869–71 reflected growing disenchantment with the economic priorities of Reconstruction government and the assumptions underlying them. South Carolina's urged the legislature to withhold further aid from railroads, using the funds instead to secure land for the freedmen. "What reason have we to suppose," legislator George Cox asked Alabama's gathering, "that [new mines and factories] would benefit us?" Manufacturers' exemption from taxation now aroused charges of "class legislation . . . and discrimination in favor of capital," and increasing numbers of Republicans demanded that corporations bear their fair share of the tax burden. Black legislators launched attacks on the convict lease system, restricting its scope in Alabama and Georgia, and in South Carolina abolishing it altogether. The South Carolina House even voted in 1871 to reestablish usury laws, although the Senate killed the measure.[82]

By the early 1870s, the program of railroad aid ground to a halt. North Carolina's legislature in 1870 repealed its generous aid laws, ordered railroads to return unsold bonds to the state treasury, and established a committee to investigate the Swepson-Littlefield Ring. The inquiry accomplished nothing, for Swepson warned that he would not bear "all the odium of these matters" without revealing "what members of the General Assembly have been paid." But the railroad program was effectively dead. Similar reactions occurred in other states. Railroads did not lose all their friends in party circles, but they found it increasingly difficult to obtain aid.[83]

Perhaps the reaction would have been less severe had the program of state-sponsored economic development come close to achieving its ambitious goals. Compared with the unsuccessful programs of Presidential Reconstruction, the Republicans could, in fact, claim real accomplishments. Between 1868 and 1872, Southern railroads had been rebuilt and some 3,300 miles added, an increase of nearly 40 percent. But this progress was confined to Georgia, Alabama, Arkansas, and Texas; other states had little to show for their generosity except enormous debts. And throughout the South, railroads faced increasing economic difficulties as

82. Summers, *Railroads, Reconstruction, and the Gospel of Prosperity,* 243–49; Charleston *Daily Republican,* November 30, 1869; Foner and Lewis, eds., *The Black Worker,* 2:122; Aiken *Tribune,* February 15, 1873; Jackson *Daily Mississippi Pilot,* June 3, 1871; *Alabama Acts 1872–73,* 133; *Georgia Acts and Resolutions 1870* 421–22; Charleston *Daily Republican,* February 15, 17, 20, 25, 1871.

83. Raper, *Holden,* 140–41; George Swepson to A. J. Jones, February 12, 1870, Swepson Papers; Carrier, "Political History of Texas," 499; Summers, *Railroads, Reconstruction, and the Gospel of Prosperity,* 250–67.

state aid proved insufficient to meet the enormous costs of construction and operation. By 1872, many lived under the shadow of bankruptcy, and the economic empires of Kimball and Swepson had collapsed. Nor did immigration or the anticipated industrial revolution materialize. "I can't imagine why any one would turn from this rich west and go South," a prominent Georgia scalawag exclaimed on a visit to St. Paul. Tennessee industry made "remarkable advances" between 1865 and 1870, and in the South as a whole the number of workers employed in manufacturing showed a modest increase, but mostly in small concerns processing agricultural goods. The real expansion of Southern textile factories did not occur until the 1880s, nor was Alabama's industrial potential awakened until then.[84]

Part of the problem was that despite the inducements offered by new laws, Northern and European capital failed to arrive in the amounts necessary to spark an economic revolution. Northerners proved perfectly willing to finance cotton production from a distance, and to assist in reviving prewar establishments like Richmond's Tredegar Iron Works and the Shelby Iron Company in Alabama, but they shied from investing in new enterprises. Exceptions did exist, such as the expanding Texas cattle kingdom, South Carolina phosphate mining, and Florida's tourist industry, already a mecca for Northern vacationers each winter. But most industrial enterprises founded in these years had to locate funding at home. Capital, after all, generally shuns political uncertainty, and despite legislative efforts to attract investment, the prevailing "social discord," as New York diarist George Templeton Strong commented, made the South "the last region on earth in which . . . a Northern or European capitalist [would] invest a dollar." Indeed, one reason the South fell prey to buccaneers like Swepson was that more established entrepreneurs (with the exception of New Yorkers with long-standing Southern connections) avoided the region. The postwar years witnessed massive capital movements within the country, and large-scale inflows from Europe, but most investors preferred the lure of the North and West. The flood of capital anticipated by Republicans (bringing with it outside control of Southern railroads and many industrial enterprises) would not come until after the end of Reconstruction.[85]

84. Goodrich, "Public Aid to Railroads," 436; Raper, Holden, 128–32; Range, "Kimball," 57–61; Henry P. Farrow to "My dear Carrie," May 27, 1868, Farrow Papers; Constantine G. Bellissary, "The Rise of Industry and the Industrial Spirit in Tennessee, 1865–1885," JSH, 19 (May 1953), 197–99; Dwight B. Billings, Jr., Planters and the Making of a "New South": Class, Politics, and Development in North Carolina, 1865–1900 (Chapel Hill, 1979), 42; Ethel Armes, The Story of Coal and Iron in Alabama (Birmingham, 1910), 253.

85. Charles B. Dew, Ironmaker to the Confederacy: Joseph R. Anderson and the Tredegar Iron Works (New Haven, 1966), 305–11; Robert H. McKenzie, "Reconstruction Historiography: The View from Shelby," Historian, 36 (February 1974), 213–19; Lewis Atherton, The Cattle Kings (Bloomington, Ind., 1961), 1–5; Frederic C. Jaher, The Urban Establishment (Urbana, Ill., 1982), 403; George W. Smith, "Carpetbag Imperialism in Florida 1862–1868," FlHQ, 27

Thus, the gospel of prosperity failed in both its aims, for it produced neither a stable Republican majority nor a modernizing economy. Within Southern society, nonetheless, profound changes were underway, affecting the internal structure of the black and white communities and their relations with one another. Begun by the Civil War and emancipation, the transformation of Southern life was in some ways accelerated, in others redirected, by the years of Republican rule.

Patterns of Economic Change

Slowly, a new social order took shape in the Reconstruction South. While far from the modernizing economy envisioned by Republican policymakers, it differed in crucial respects from its antebellum predecessor. The demise of slavery and the rapid spread of market relations in the predominantly white upcountry produced new systems of labor and new class structures among both black and white Southerners. The burden of history weighed heavily upon the South's economic transformation, which took place in a war-torn, capital-scarce region that lacked the institutional base for sustained economic growth, faced a slowing world demand for its major export, and was excluded from a significant share of national political power. In retrospect, it appears all but inevitable that the postwar South would descend into a classic pattern of underdevelopment, its rate of economic growth and per capita income lagging far behind the rest of the nation. Yet the advent of Republican rule subtly affected the process of change, accelerating the commercialization of the Southern upcountry while increasing the freedmen's bargaining power in the ongoing struggle over labor on the plantations.[86]

"One thing is perfectly certain," Kentucky Democrat Garrett Davis told the Senate in 1867: "[Cotton] is to be made by negro labor." No aphorism could have been more misleading. The absorption of white farmers into the cotton economy took place at different speeds in different parts of the South. But where railroads penetrated already settled upcountry counties, as in South Carolina, Georgia, and Alabama, or opened new areas to cultivation in Texas, Arkansas, and Louisiana, the shift from subsistence to commercial farming proceeded apace. The increased avail-

(January 1949), 298–99; Billings, Planters and the Making of a "New South", 60, 94–95; Allan Nevins and Milton H. Thomas, eds., The Diary of George Templeton Strong (New York, 1952), 4:158; Allan G. Bogue, Money at Interest: The Farm Mortgage on the Middle Border (Ithaca, N.Y., 1955), 7; John F. Stover, The Railroads of the South 1865–1900 (Chapel Hill, 1955), 55.

86. Richard A. Easterlin, "Regional Income Trends, 1840–1950," in Seymour E. Harris, ed., American Economic History (New York, 1961), 528; Barbara J. Fields, "The Nineteenth-Century American South: History and Theory," Plantation Society in the Americas, 2 (April 1983), 8–9, 22–25; Peter Temin, "The Post-Bellum Recovery of the South and the Cost of the Civil War," JEcH, 36 (December 1976), 898–907; Gavin Wright, "Cotton Competition and the Post-Bellum Recovery of the American South," JEcH, 34 (September 1974), 610–35.

ability of commercial fertilizers also encouraged the spread of cotton cultivation to poorer upcountry soils. Although the full impact of commercialization did not occur until after the end of Republican rule, the trend was already evident during Reconstruction. A Georgia newspaper in 1872 reported a "considerable revolution" in upcountry agriculture: "Cotton, formerly cultivated on a very limited extent, has increased rapidly in the last few years." A year later, Northern journalist Edward King reported that the railroads and the "universal use of the new fertilizers" had introduced cotton production to counties where it had been all but unknown before the war.[87]

By the mid-1870s, both the geographical and racial locus of cotton production had been transformed. Output now nearly equaled the level of slavery days—a situation unique among the South's major staples, and exceedingly rare in any postemancipation society. But white labor, as much as black, was responsible. Nearly 40 percent of the crop was raised west of the Mississippi River (where the majority of cotton farmers were white), and in the older states production had increasingly shifted to the upcountry. Black laborers, who cultivated nine tenths of the South's cotton crop in 1860, grew only 60 percent in 1876.[88]

At the close of the Civil War, Edward Atkinson predicted that the demise of slavery would cause a cotton boom among enterprising white farmers who, like their Northern counterparts, would prosper from an increasing involvement with the market. But things did not work out this way. Some yeomen found economic salvation in commercial agriculture but for others, burdened by Civil War debts, postwar crop failures, rising Reconstruction taxes, and, during the 1870s, a precipitous decline in agricultural prices, cotton proved a recipe for disaster. Wherever possible, yeomen clung to their farms. As one planter commented, it was impossible to hire poor whites to "step into the shoes of the negro" on the plantations: "They most have small plots of land and prefer tending them, poor as may be the return, to lowering themselves, as they think it, by hiring to another." Independence, as we have seen, was the credo of upcountry yeomen, and independence meant, among other things, freedom from debt. "By the goodness of God," wrote a farm woman in 1869, "I do not owe ten dollars in the world. I'd rather be poor than in

87. CG, 39th Congress, 2d Session, 1931; David F. Weiman, "The Economic Emancipation of the Non-Slaveholding Class: Upcountry Farmers in the Georgia Cotton Economy," JEcH, 45 (March 1985), 71–94; William W. Rogers, ed., "From Planter to Farmer: A Georgia Man in Reconstruction Texas," SWHQ, 72 (April 1969), 527–28; Peter Temin, "Patterns of Cotton Agriculture in Post-Bellum Georgia," JEcH, 43 (September 1983), 661–74, Steven Hahn, The Roots of Southern Populism: Yeoman Farmers and the Transformation of the Georgia Upcountry, 1850–1890 (New York, 1983), 141–51; King, Southern States, 515.
88. Report of the Commissioner of Agriculture for the Year 1876 (Washington, D.C., 1877), 119, 136; Lawrence D. Rice, The Negro in Texas 1874–1900 (Baton Rouge, 1971), 159; Walter L. Fleming, Civil War and Reconstruction in Alabama (New York 1905), 726; Robert P. Brooks, The Agrarian Revolution in Georgia, 1865–1912 (Madison, Wis., 1914), 79–80.

debt." But as Reconstruction progressed, yeomen found their autonomy more and more seriously undermined.[89]

Increasingly, upcountry farmers fell into new forms of dependency, their situation exacerbated by the credit system that took hold in the postwar South. Because of the region's chronic capital shortage and scarcity of banking institutions, local merchants generally represented the only source of available credit. With land values having plummeted, merchants would usually advance loans only in exchange for a lien on the year's cotton crop, rather than take a mortgage on real estate, as was conventional in the North. The emergence of the crop lien as the South's major form of agricultural credit forced indebted farmers to concentrate on cotton, further expanding production and depressing prices. Too often, the result was the loss of cherished independence. Many yeomen, a report from Alabama lamented in 1874, had become "but the tenants of the merchants." Others were forced to seek employment as agricultural laborers. The depression of the mid-1870s greatly accelerated the process of dispossession. By 1880, one third of the white farmers in the cotton states were tenants renting either for cash or a share of the crop, and a region self-sufficient before the Civil War was no longer able to feed itself.[90]

The growth of tenancy, one journalist reported from North Carolina in 1871, "does not seem to be very popular with the poor whites." And many yeomen attributed its spread to the Republican program of railroad promotion, which brought market relations into the upcountry, and to rising Reconstruction taxes, which increased small farmers' need for ready cash. Indeed, these grievances proved almost as great a liability for Southern Republicans among white Piedmont voters as the party's identification with black political equality. It is significant that the white Republican vote remained most secure in the Southern mountains, for here not only had wartime Unionism sunk deepest roots, but during Reconstruction the railroad had not yet penetrated, most farmers still tilled their own land, and the transition to cotton made little headway. Among the results of the "gospel of prosperity," which promised to modernize the South's economy and attract white voters to the Republican party, none proved more ironic than that state-sponsored economic growth

89. Edward Atkinson, *On Cotton* (Boston, 1865), 40–46; Hahn, *Roots of Southern Populism*, 186–93; Augustine T. Smythe to "My dear brother," December 5, 1867, Augustine T. Smythe Letters, USC; Mrs. N. F. Norton to "My dear brother," January 30, 1869, Joseph J. Norton Papers, USC.

90. *Rural Carolinian*, 5 (July 1874), 545–46; Gavin Wright, *The Political Economy of the Cotton South* (New York, 1978), 164–74; Forrest McDonald and Grady McWhiney, "The South from Self-Sufficiency to Peonage: An Interpretation," *AHR*, 85 (December, 1980), 1113–16. Recent research suggests that tenancy was more common among antebellum white farmers than once believed, but there is no question that it expanded rapidly after the war. Frederick A. Bode and Donald E. Ginter, *Farm Tenancy and the Census in Antebellum Georgia* (Athens, Ga., 1986).

helped King Cotton extend its sway into the upcountry, or that the party's white support proved most resilient where capitalist development made the least impact during Reconstruction.[91]

The rise of upcountry cotton farming represented only one part of a fundamental reorientation of Southern trading patterns, and a wholesale shift in regional economic power. As railroad and telegraph lines worked their way into the interior, merchants in rapidly developing market towns for the first time found it possible to trade directly with the North, bypassing the coastal cities that had traditionally monopolized Southern commerce. Atlanta, whose rise was stimulated by its selection as state capital and the opening of rail connections to the North, was the quintessential upcountry boom city, serving as a gathering point for cotton grown in Georgia's Piedmont and a distribution center for Northern goods. Here a visitor in 1870 found "more of the life and stir of business than in all the other southern cities, which I have seen, put together." The railroad also transformed smaller towns like Selma and Macon into commercial entrepôts and called still others into existence seemingly from nowhere. The number of South Carolina communities described as towns by the census doubled between 1860 and 1870 and tripled in the following decade. Tiny Sumter, a newspaper reported in 1868, "had become quite a cotton market, and by consequence a heavy distributor of all kinds of merchandise." Meanwhile, older port cities languished. By 1880, Charleston was a minor seaport of little commercial significance. New Orleans found itself unable to compete with St. Louis for access to the expanding cotton trade of East Texas, and Richmond, Savannah, and Mobile, bypassed by the railroad, also stagnated economically.[92]

The "economic reconstruction of the upcountry" laid the foundation for the rise of a bourgeoisie composed of merchants, railroad promoters, and bankers resident in interior towns. Nationally, this new class wielded little economic power, for it depended for credit and supplies on financiers and wholesale merchants in the North. But within the South, it reaped the benefit of the spread of cotton agriculture. Although some of these business leaders emerged from the old planter class, the economic

91. New York Herald, June 13, 1871; Hahn, Roots of Southern Populism, 9n., 144; Brooks, Agrarian Revolution, 72–74; U.S. Census Office, Tenth Census, 1880, 3:236–37, 240–41; Fleming, Alabama, 788–805; William Barney, "Patterns of Crisis, Alabama Families, and Social Change," Sociology and Social Research, 63 (April 1979), 529.

92. Roger L. Ransom and Richard Sutch, One Kind of Freedom: The Economic Consequences of Emancipation (New York, 1977), 116; David R. Goldfield, "The Urban South: A Regional Framework," AHR, 86 (December 1981), 1015–17; Cleveland Leader, February 11, 1870; Jerry W. DeVine, "Town Development in Wiregrass Georgia, 1870–1900," Journal of Southwest Georgia History, 1 (Fall 1983), 1–6; King, Southern States, 265; Moore, ed., Juhl Letters, 257; John P. Radford, "Culture, Economy and Urban Structure in Charleston, South Carolina, 1860–1880" (unpub. diss., Clark University, 1974), 221; L. Tuffly Ellis, "The Revolutionizing of the Texas Cotton Trade, 1865–1885," SWHQ, 73 (April 1970), 478–89; Chesson, Richmond After the War, xvii, 436–45; Justin Fuller, "Alabama Business Leaders, 1865–1900," Alabama Review, 16 (October 1963), 280–81.

importance of the new upcountry elite rested more on access to credit and ties with the North than connections with the prewar aristocracy. The full impact of these changes would only come after the chaotic years of Republican rule, but by the end of Reconstruction an area dominated before the war by self-sufficient yeomen was well on the way to becoming a commercial economy peopled by merchants, tenants, farm laborers, and commercially oriented yeomen—groups of far lesser importance before the war.[93]

Towns and cities also exerted a growing influence on black life, although migration from the countryside slowed considerably after 1870. In urban centers were concentrated the schools, churches, newspapers, and fraternal societies that produced many of the articulate leaders of Reconstruction politics. And the variety of employment opportunities far outstripped those available in the countryside. The twelve blacks giving court testimony in 1871 about the Meridian, Mississippi, riot included four barbers and a shoemaker, hotel porter, depot man, merchant, laborer, drayman, carpenter, and a man "working streets." But the black urban social structure that began to emerge during Reconstruction was, compared with that of whites, extremely truncated and heavily weighted toward the bottom. Nearly all urban blacks lived by manual labor, the vast majority as servants, porters, and unskilled day laborers. They received subsistence wages, faced unemployment rates far exceeding those of whites, and enjoyed virtually no opportunities for the accumulation of property or for upward mobility.[94]

The black urban community contained no well-established elite of wealthy bankers and merchants, and found most white collar positions closed to it. Nor did more than a tiny minority achieve professional status during Reconstruction, although the number of lawyers and doctors began to grow in the 1870s thanks to the new black universities. Ministers, the largest group of black professionals, often had to rely on voluntary contributions from poor parishioners for a living, and supplemented their religious calling by other kinds of labor. Artisans, perhaps a quarter of employed blacks in most Southern towns and cities, constituted the

93. Lacy K. Ford, "Rednecks and Merchants: Economic Development and Social Tensions in the South Carolina Upcountry, 1865–1900," *JAH*, 71 (September 1984), 294–318; Harold D. Woodman, *King Cotton and His Retainers: Financing and Marketing the Cotton Crop of the South, 1800–1925* (Lexington, Ky., 1968), 348–55; David L. Carlton, *Mill and Town in South Carolina, 1880–1920* (Baton Rouge, 1982), 9–10; Hahn, *Roots of Southern Populism,* 201–202.

94. Orville V. Burton, *In My Father's House,* 295–96; Wharton, *Negro in Mississippi,* 106–107; KKK Hearings, Mississippi, 127–64; Frank J. Huffman, "Town and Country in the South, 1850–1880: A Comparison of Urban and Rural Social Structures," *SAQ,* 76 (Summer 1977), 375–76; Green, *Secret City,* 82; Richard J. Hopkins, "Occupational and Geographic Mobility in Atlanta, 1870–1896," *JSH,* 34 (May 1968), 200–213; Herbert A. Thomas, Jr., "Victims of Circumstance: Negroes in a Southern Town, 1865–1880," *Register of the Kentucky Historical Society,* 71 (July 1973), 262.

largest group above the ranks of the unskilled. A significant number managed to acquire property, although generally in very modest amounts. (In Louisville in 1870, only 11 percent of black families owned property worth more than $100.) Instead of prospering, artisans' economic position grew increasingly precarious. Denied access to credit, threatened by the growing availability of manufactured goods from the North, and driven from many skilled crafts by white employers and competitors, black artisans found themselves mostly confined to trades that required little capital, like carpenter, blacksmith, brickmason, and shoemaker, or to occupations like barber, traditionally avoided by whites. As time went on, artisans constituted a shrinking proportion of the black community. By 1880, their average age considerably exceeded that of their white counterparts, an indication that the younger generation found it almost impossible to move into skilled work. The crisis of black artisanship helps explain why many were forced to rely on government employment for a livelihood.[95]

At the apex of the urban community stood those few blacks able to escape manual labor for entrepreneurial occupations. In cities like Charleston, New Orleans, and Natchez, where wealthy free mulattoes brought skills and capital into the postwar years, color and class position overlapped well into the twentieth century. Even in inland towns mulattoes, often able to draw upon white kin for credit, constituted a far higher percentage of artisans and businessmen than they did of the black community as a whole. Yet in many ways, Reconstruction was a time of crisis for the old free elite. Challenged for positions of political leadership by former slaves, they also faced severe economic difficulties because of the decline of the old port cities. Especially in growing inland cities like Atlanta and Montgomery, where newcomers did not have to contend with a preexisting black elite, a new business class arose, composed of enterprising freedmen who served a black clientele.[96]

In a few cities, the black upper class was large enough to constitute a distinct segment of society, with its own life-style and entertainments. Washington's mostly light-skinned "elite of colored fashion and high life"

95. David Sowell, "Racial Patterns of Labor in Postbellum Florida," *FlHQ*, 63 (April 1985), 438–39; William Harris, "Work and the Family in Black Atlanta," *JSocH*, 9 (Spring 1976), 320; Drago, *Black Politicians*, 39; John W. Blassingame, "Before the Ghetto: The Making of the Black Community in Savannah, Georgia, 1865–1880," *JSocH*, 6 (Summer 1973), 465–67; Ransom and Sutch, *One Kind of Freedom*, 32–35; Herbert G. Gutman, *The Black Family in Slavery and Freedom, 1750–1925* (New York, 1976), 479–82, 626; Huffman, "Town and Country," 375–76; Thomas, "Victims of Circumstance," 264.

96. Arnold H. Taylor, *Travail and Triumph: Black Life and Culture in the South Since the Civil War* (Westport, Conn., 1976), 185–90; Thomas, "Victims of Circumstance," 267–70; Peter Kolchin, *First Freedom: The Responses of Alabama's Blacks to Emancipation and Reconstruction* (Westport, Conn., 1972), 145; Russell and Thornbery, "William Finch," 312; Rabinowitz, *Race Relations*, 84.

included barbers, caterers, professors at Howard University, political appointees, and a few black lawyers and doctors. The New Orleans elite, comprising wealthy members of the old free group supplemented by newcomers like Pinchback, was renowned for its lavish entertainments and "sumptuous repasts." In Columbia, the Rollin sisters (Catherine de Medici, Charlotte Corday, and Louisa Muhlbach) presided over upper-crust black social life. Mulatto daughters of a prominent slaveholder, they had been educated in Boston before the war, and their home now served as "a kind of Republican headquarters" as well as a gathering place for "the elite of our colored society."[97]

These examples, however, should not obscure the essential facts about the black upper class—its tiny size and negligible economic importance. Only in life-style and aspirations did this elite constitute a "black bourgeoisie," for it lacked capital and economic autonomy, and did not own the banks, stores, and mills that could provide employment for other blacks. Black business was small business: grocery stores, restaurants, funeral parlors, and boarding houses—individually owned, and devoid of economic significance. Black proprietors formed no part of the national or regional bourgeoisie, and their businesses faced bleak prospects for long-term survival (a majority disappeared from city directories after only a year or two). When black businessmen did acquire wealth, they tended to invest it in real estate rather than in economic enterprises. During Reconstruction, the black upper class remained too small and too weak to enjoy either political or cultural hegemony within its own community. Yet the seeds of its future prominence were already being sown as the steady growth of separate black neighborhoods created the economic foundation for the consolidation of a black business class, while the declining position of black artisans and the port city free elite forecast their inevitable eclipse as leaders of black politics.[98]

The black community, a white North Carolina Republican observed in 1879, was only beginning to be divided into "classes or ranks of society." The distillation of a new system of social stratification to supersede pre-war divisions between free and slave, mulatto and black, had only begun during Reconstruction, a fact that helps explain the black community's remarkable political unity. Outside the largest cities, black social life had not yet separated along class lines; community activities like dances, picnics, concerts, and sporting events catered to large crowds rather than small, class-defined groups, and many black businessmen had roots in slavery and family members in the laboring class. Yet as in the cities and

97. *New National Era*, May 26, 1870; Green, *Secret City*, 94; Blassingame, *Black New Orleans*, 159–60; New Orleans *Louisianian*, December 20, 1870; New York *Sun*, March 29, 1871.

98. Taylor, *Negro in Tennessee*, 154–56; Blassingame, "Before the Ghetto," 466–68; E. Franklin Frazier, *Black Bourgeoisie* (New York, 1957), 43, 53, 153, 168; Rice, *Negro in Texas*, 97, 189; Rabinowitz, *Race Relations*, 97, 189; John Kellogg, "The Evolution of Black Residential Areas in Lexington, Kentucky, 1865–1887," *JSH*, 48 (February 1982), 21–52.

white upcountry, a new class structure in the plantation belt slowly replaced the shattered world of master and slave.[99] The planter still stood atop this new social pyramid. Slavery was gone, but in the absence of large-scale land distribution, the plantation system endured. The degree of "planter persistence" varied considerably among the South's crop regions. Louisiana sugar planters, unable to raise the enormous sums required to repair wartime destruction, saw themselves replaced by newcomers; by 1870, about half the estates had fallen into the hands of Northern investors. In the lowcountry rice kingdom, with its equally large capital requirements, continuing labor turmoil and the opening of new lands to rice cultivation west of the Mississippi discouraged outside investment. Many planters, unable to resume production, were reduced to penury. "I do not believe that the ruin of the French nobility at the first Revolution," commented a Northern reporter, "was more complete than . . . that of the proud, rich, and cultivated aristocracy of the low country of South Carolina." Elsewhere, however, the plantation South seemed at first glance to offer a remarkable example of social and economic continuity. No "revolution in land-titles" swept the tobacco and cotton belts. Studies of rural counties from Virginia to Texas have demonstrated that the majority of planter families managed to retain control of their land. Yet despite their uncanny capacity for survival, war, emancipation, and Reconstruction fundamentally altered the planters' world, and their own role within it.[100]

"The sum of the planters' experience since the war," observed the owner of a Louisiana plantation in 1874, "has the balance largely on the side of disappointed hopes and expectations." In the antebellum years, planters had dominated regional politics and enjoyed a significant share of national authority. Now, they exerted little influence in Washington and in several states found themselves without power on the state and local levels. At the same time, emancipation and the decline of Southern land values meant planters' wealth was only a tiny fraction of its prewar total. Once alone at the apex of Southern society, they now saw other groups rising in economic importance. Planters constituted a large ma-

99. 46th Congress, 2d Session, Senate Report 693, 403; Barbara A. Richardson, "A History of Blacks in Jacksonville, Florida, 1860–1895: A Socio-Economic and Political Study" (unpub. diss., Carnegie-Mellon University, 1975), 130–31; Taylor, *Travail and Triumph*, 194.

100. J. Carlye Sitterson, *Sugar Country* (Lexington, Ky. 1953), 312; Eric Foner, *Nothing But Freedom: Emancipation and Its Legacy* (Baton Rouge, 1983), 81–84; Wiener, *Social Origins of the New South*, 5–33; Crandall A. Shifflett, *Patronage and Poverty in the Tobacco South: Lousia County, Virginia, 1860–1900* (Knoxville, 1982), 3–4; Kenneth G. Greenberg, "The Civil War and the Redistribution of Land: Adams County, Mississippi, 1860–1870," *AgH*, 25 (April 1978), 292–307; Randolph B. Campbell, "Population Persistence and Social Change in Nineteenth-Century Texas: Harrison County, 1850–1880," *JSH*, 48 (May 1982), 185–204; A. Jane Townes, "The Effect of Emancipation on Large Landholdings, Nelson and Goochland Counties, Virginia," *JSH*, 45 (August 1979), 403–12.

jority of the wealthiest Texans in 1860, but only 17 percent ten years later. Unlike the situation in the Old South, possession of a plantation no longer guaranteed wealth. "Although the owner of two of the best and largest cotton plantations in . . . South Carolina," lamented former Gov. Milledge Luke Bonham in 1874, "my life has been absorbed in trying to keep my head above water. The effect has been crushing." Alexander H. Stephens and Gen. John B. Gordon were reduced to earning money by writing testimonials for Darby's Prophylactic Fluid, a medicine manufactured in New York. ("No family should be without it," declared the Confederacy's Vice President.) Many who had once seen planting as the only honorable profession now urged their sons to take up careers in business and the professions. "The day of the wealthy and independent planter," commented a South Carolinian, "is past and gone. For anyone to pursue agriculture, he must have other pecuniary resources besides, or else descend to the plow handle."[101]

Most of all, control of land no longer translated automatically into control of labor, a situation evident since the end of slavery but in many ways exacerbated by the advent of Reconstruction. "You know how intolerable a life on the plantation is now to your father," a planter's wife wrote her son in 1870, "and I see no prospect of any better while we have . . . the utter impossibility of commanding labor." In 1869, two Boston cotton brokers sent questionnaires to dozens of planters, inquiring about agricultural conditions. With virtually one voice, the respondents complained of a continuing labor problem. "Once we had reliable labor, controlled at will," commented a Georgia planter. "Now . . . it is both uncertain and unreliable; and our contracts must often be made at great disadvantage." "Labor," agreed a Louisianan, "is more scare, harder to procure at the present time than at any time since the close of the rebellion." Many black women continued to shun work in the fields, black children were attending school, and in areas where new economic enterprises competed for their services, freedmen seemed eager to take advantage of the opportunity to escape the plantations. Railroad construction drew "thousands of hands out of the fields." When the work was completed, laborers returned, bringing with them a new sense of independence; planters found them "unruly, insolent, and disobedient," their "value for systematic work . . . destroyed." Phosphate mining companies,

101. Albert A. Batchelor to "Dear Ned," October 16, 1874, Albert A. Batchelor Papers, LSU; Jayne Morris-Crowther, "An Economic Study of the Substantial Slaveholders of Orangeburg County, 1860–1880," *SCHM*, 86 (October 1985), 300–305, 313; Ralph A. Wooster, "Wealthy Texans, 1870," *SWHQ*, 74 (July 1970), 24–35; Wayne, *Reshaping of Plantation Society*, 75–83, 99–109; Milledge L. Bonham to Joseph E. Johnston, January 19, 1874 (draft), Milledge L. Bonham Papers, USC; Atlanta *Constitution*, January 10, 1869; James L. Roark, *Masters Without Slaves: Southern Planters in the Civil War and Reconstruction* (New York, 1977), 150–58; John W. Kirk to Emily K. Moore, June 28, 1868, Kirk Family Papers, Private Collection.

which advertised for large numbers of laborers in the early 1870s, enabled lowcountry freedmen to earn "cash money" without working in the rice fields, while the rapidly expanding cattle industry employed thousands of Texas freedmen as cooks and cowboys. And many blacks squatted on abandoned land, or attempted to subsist by hunting and fishing.[102]

"We are the capital," declared a Mississippi planter in 1871, "then let us dictate terms." Instead, planters continued to compete with one another for labor, offering better employment conditions than their neighbors and "enticing" away blacks already under contract. To retain the services of a freedman, one Florida planter told him, "Whenever he wanted to ride, say nothing to nobody, go and get a mule, and travel." In addition, it seemed impossible to establish effective plantation discipline. "They lose many days, they go to work late, remain longer at meals, and refuse to work on Saturdays after noon," reported a South Carolina planter. Even rising wages often translated into less, rather than more, available labor. "Instead of availing himself of the high rates and advantages offered," complained a Mississippi planter, the freedman "prefers to make use of his power to reduce his labor, rather than increase his compensation."[103]

Republican rule subtly altered the balance of power in the rural South. Planters, as we have seen, complained that Republican officials did nothing to discourage vagrancy and theft and interfered in plantation disputes. The end of the state's efforts to bolster labor discipline and the coming to power of local officials sympathetic to the freedmen produced during Reconstruction a kind of stalemate on the plantations. "Capital is powerless and labor demoralized," wrote South Carolina agricultural reformer D. Wyatt Aiken in 1871. Labor was "scarce" not merely because fewer blacks were willing to work on plantations, but because those who did so were unmanageable: "Though abundant, this labor is virtually scarce because not available, and almost wholly unreliable." "The power to control [black labor]," the Selma *Southern Argus* agreed, "is gone."[104]

As late as 1869, the *Annual Cyclopedia* commented on the "transitory state of the labor system in the Southern States." By then, however, new

102. Mrs. Basil G. Kiger to Basil G. Kiger, Jr., November 27, 1870, Basil G. Kiger Papers, Natchez Trace Collection, UTx; Loring and Atkinson, *Cotton Culture*, 13–16, 22–24; Charleston *Daily Republican*, July 28, 1870; Uya, *Robert Smalls*, 73; Kenneth W. Porter, "Negro Labor in the Western Cattle Industry," *Labor History*, 10 (Summer 1969), 346–75; Ronald L. F. Davis, "Labor Dependency Among Freedmen, 1865–1880," in Walter J. Fraser, Jr., and Winfred Moore, Jr., eds., *From Old South to New: Essays on the Transitional South* (Westport, Conn., 1981), 162.

103. *Southern Field and Factory*, 1 (March 1871), 116; J. M. Perry to Grant Perry, February 19, 1869, J. M. Perry Family Papers, Atlanta Historical Society; Ulrich B. Phillips and James D. Glunt, eds., *Florida Plantation Records from the Papers of George Noble James* (St. Louis, 1927), 182–85; Loring and Atkinson, *Cotton Culture*, 8, 110.

104. *Rural Carolinian*, 2 (January 1871), 195, 3 (January 1872), 173; Selma *Southern Argus*, February 17, 1870.

patterns of social relations had already appeared in the different crop regions of the plantation South, their form prefigured in the early days of freedom but affected as well by the course of Reconstruction. Nowhere was the dialectic of continuity and change more evident than in the sugar districts of Louisiana, where the centralized plantation employing gang labor survived the end of slavery. An influx of outside capital rescued the industry, and, indeed, produced a concentration of land ownership in the sugar parishes. Although sugar requires extremely arduous field labor and round-the-clock work in processing mills during the harvest season, it has been grown successfully elsewhere by sharecroppers, who bring their crop to central mills at the end of the season. In Louisiana, however, each plantation possessed its own costly steam-powered sugarhouse, thus requiring a large harvest and a labor force available throughout the year, and reducing the economic viability of sharecropping and small owner-ship. Many estates, as well, had extensive irrigation and levee systems, whose maintenance demanded coordination of the labor force.

Along with the liquidation of the old planter class and the new owners' access to cash, this situation helped produce the most rapid transition to capitalist labor relations in any part of the South. Blacks quickly became a wage-earning labor force, receiving daily or monthly wages considera-bly exceeding those elsewhere in the South, and enjoying, as well, the traditional right to garden plots on which to raise vegetables and keep poultry and livestock. Yet the system did not end the conflict over labor discipline. Successful sugar production required a "thorough control of ample and continuous labor," yet planters throughout Reconstruction complained of a shortage of workers, especially at harvest time, and of blacks' continuing demands for higher pay. Indeed, the advent of Repub-lican rule coincided with a sharp increase in sugar wages, which lasted until the economic collapse of 1873. Moreover, planters charged, blacks spent more time on their garden plots than in the sugar fields. During Reconstruction, many sugar estates lay idle, and production did not regain the level of 1861 until the 1890s.[105]

An even more complex economic transition took place in the rice kingdom, where planter Ralph I. Middleton described labor relations as "a continuous struggle where the planter is all the time at a great disad-vantage." As in the early days of freedom, a variety of labor systems coexisted during Reconstruction. Despairing of resuming production themselves, some planters turned their land over to freedmen to cultivate

105. *AC*, 1869, 207; Sitterson, *Sugar Country*, 221, 291–301, 312; Joseph P. Reidy, "Sugar and Freedom: Emancipation in Louisiana's Sugar Parishes" (unpub. paper, annual meeting of American Historical Association, 1980); Robert Somers, *The Southern States Since the War 1870–71* (London, 1871), 222–25; Charles Nordhoff, *The Cotton States in the Spring and Summer of 1875* (New York, 1876), 70; *Rural Carolinian*, 2 (August 1871), 11; New York *Journal of Commerce*, September 1, 1871; Walter Prichard, "The Effects of the Civil War on the Louisiana Sugar Industry," *JSH*, 5 (August 1939), 322–31.

as they pleased. Others, staring bankruptcy in the face, sold their holdings to blacks or to the state land commission. In Colleton County, South Carolina, several large plantations in the early 1870s were operating under what a newspaper called "a sort of communism," with black laborers forming societies, electing officers, and purchasing the estates collectively. Some freedmen squatted on abandoned lands, working as plantation day laborers at harvest time; others labored on the "two-day" system. Local political power, moreover, quickly fell into black hands. "The negro magistrate or majesty as they call him," observed Middleton, "tells them that no rice is to be shipped until it is all got out and divided 'according to law'." The planter did not even enjoy control over the disposition of his own crop.

Lowcountry rice output never regained its prewar levels. In 1874, Edward King found the region dotted with abandoned manorial houses, standing "like sorrowful ghosts lamenting the past." In the end, the great plantations fell to pieces, their lands rented or sold to blacks—a process well underway by the end of Reconstruction. In their place emerged a pattern of small-scale farming, with black families growing their own food and supplementing their income by marketing farm produce or seeking day labor in Charleston and Savannah. Here in the lowcountry a unique combination of circumstances—initial federal policy, the inability to attract outside investment, a prolonged period of local black political power, and the cohesion and militancy of the black community—promoted black landownership. As a result, more than in any other region of the South, the freedmen succeeded in shaping labor relations in accordance with their own aspirations.[106]

Entirely different was the economic transition in the far larger plantation regions of the tobacco and cotton South. Here, where farming required less capital and a less coordinated labor force, the planter class by and large retained control of its land and resumed production. At the same time, Reconstruction reinforced what a Southern newspaper called blacks' "growing disposition to throw off all restraint and to assert the dignity of freedom by 'setting up' for themselves." Few blacks managed to fulfill this aspiration by acquiring homesteads of their own, for even those who possessed the necessary resources found most whites adamantly opposed to black landownership. In Mississippi, lands for which blacks had offered $10 an acre if divided into small lots were sold to whites in large units for

106. Foner, *Nothing But Freedom*, 85–90, 107–10; Ralph I. Middleton to Henry A. Middleton, August 24, 1869, April 16, 1870, Middleton Papers, Langdon Cheves Collection, SCHS; *New National Era*, September 25, 1873; John S. Strickland, "Traditional Culture and Moral Economy: Social and Economic Change in the South Carolina Low Country, 1865–1910," in Steven Hahn and Jonathan Prude, eds., *The Countryside in the Age of Capitalist Transformation* (Chapel Hill, 1985), 154–63; King, *Southern States*, 427–28, 451; James M. Clifton, "Twilight Comes to the Rice Kingdom: Postbellum Rice Culture on the South Atlantic Coast," *GaHQ*, 62 (Summer 1978), 146–54.

half that price—thus seemingly defying the laws of the marketplace and demonstrating, a prominent Northern economist commented, how "little efficacy has *interest* when opposed to passion or sentiment." In Florida, where many freedmen took advantage of the Southern Homestead Act, and South Carolina, with its land commission, 10 percent or more of black farmers acquired land by the end of Reconstruction. But in most of the cotton states, the commissioner of agriculture reported in 1876, only about one black family in twenty had managed to do so.[107]

Like urban businessmen, landowners stood atop the emerging black class structure, but possessed little real significance in the economy at large. Compared to whites, they owned smaller farms, less machinery and livestock, and used fertilizer less frequently. Typically, one or two whites in a county owned real estate worth more than the entire value of black-owned farms. Many black farmers, their tiny plots unable to provide a family with subsistence, found it necessary to engage in occasional plantation labor. Below landowners, although closest to them in the degree of independence from white control, stood those who rented for a fixed payment in cash or cotton. Renting seems to have involved as many as 20 percent of black farmers by the end of Reconstruction. At the bottom of the black social order were wage-earning farm laborers, whose numbers remained high throughout Reconstruction, especially in the Upper South, where a continuing shift from tobacco to grain and truck farming lessened the need for resident year-round workers. But by the early 1870s, especially in the cotton belt, sharecropping had become the dominant form of black labor.[108]

Politics and economics combined to promote the system's rapid spread. The Union Leagues, as we have seen, encouraged blacks' quest for economic autonomy. "Their leaders," complained a South Carolina planter, "counsel them *not* to work for wages at all, but to insist upon setting up for themselves." The 1869 cotton crop—the best since the end of the war—proved so lucrative for blacks working for part of the crop that others vowed to "do nothing but lease or go on shares" the following year. Thus, both large gangs and the squads of a dozen or so workers that had superseded them on many plantations dissolved into family units. The transition was hardly complete by the end of the decade, and sometimes cash wages and various forms of tenancy coexisted on the same

107. Selma *Southern Argus*, February 17, 1870; KKK Hearings, Georgia, 213–14; E. Pershine Smith to Henry C. Carey, November 30, 1868, Edward Carey Gardiner Collection, HSPa; Claude F. Oubre, *Forty Acres and a Mule: The Freedmen's Bureau and Black Landownership* (Baton Rouge, 1978), 194–95; *Report of Commissioner of Agriculture 1876*, 137.

108. Shifflett, *Patronage and Poverty*, 18–23, 39–40, 58–61; Edward Bonekemper III, "Negro Ownership of Real Property in Hampton and Elizabeth City County, Virginia, 1860–1870," *JNH*, 55 (July 1970), 175–78; Ransom and Sutch, *One Kind of Freedom*, 183; Brooks, *Agrarian Revolution*, 24–25, 35, 44, 53; Barbara J. Fields, *Slavery and Freedom on the Middle Ground: Maryland During the Nineteenth Century* (New Haven, 1985), 4–5, 170–71; Nan Netherton et al., *Fairfax County, Virginia: A History* (Fairfax, Va., 1978), 382, 448.

plantation. But by 1870 a traveling British journalist reported that share-cropping prevailed "so generally that any other form of contract is but the exception."[109]

Sharecropping, according to an 1870 report of the Department of Agriculture, developed not as "a voluntary association from similarity of interests, but an unwilling concession to the freedman's desire to become a proprietor." Many planters continued to resist the system because, under the conditions of Reconstruction, it failed to allow for adequate supervision of the labor force even when, as was sometimes the case, share contracts contained detailed instructions as to what corps were to be raised and how. "Wages," wrote one planter, "are the only successful system of controlling hands." But efforts to introduce a wage system were stymied by "the inveterate prejudices of the freedmen, who desire to be masters of their own time." Agricultural reformer D. Wyatt Aiken railed against the inefficiency of sharecropping, but found himself forced to adopt it on his own plantation: "I had to yield, or lose my labor." His situation was hardly unique. When Georgia officials surveyed a group of "intelligent and experienced" landowners as to which form of labor they preferred, two thirds chose wages, but four out of five actually employed either renting or sharecropping. The growing predominance of share-cropping had profound implications for the structure of rural black society. With the demise of the communal slave quarter, seen by many freed-men as a "relic of their former subjection," black families, living on isolated tenancies scattered across the length and breadth of the planta-tion, emerged as the basic units of social organization. Thus, institutions like the church, fraternal orders, and the political party became increas-ingly important elements of cohesion within the black community.[110]

To some extent, sharecropping "solved" the plantation labor shortage. Since each family had a vested interest in larger output, black women and children returned in large numbers to field work (although the extent and allocation of their labor now derived from a family decision, instead of being dictated by whites). In other ways, however, the system merely shifted the focus of labor conflict. Planters strongly resented the sense of

109. Fitzgerald, "Union League," Chapter 5; Gerald D. Jaynes, Branches Without Roots: Genesis of the Black Working-Class In the American South, 1862–1882 (New York, 1986), 280–301; Mary C. Oliphant et al., eds., The Letters of William Gilmore Simms (Columbia, S.C., 1952–56), 5:276; David C. Barrow, Jr., "A Georgia Plantation," Scribner's Monthly, 21 (March 1881), 830–36; Ronald F. Davis, Good and Faithful Labor: From Slavery to Sharecropping in the Natchez District, 1860–1890 (Westport, Conn., 1982), 102–104; Southern Field and Factory, 1 (May 1871), 237–38; Somers, Southern States, 128.
110. Rural Carolinian, 1 (June 1870), 9; Joseph D. Reid, Jr., "Sharecropping in History and Theory," AgH, 49 (April 1975), 427–28; Tenth Census, 1880, 5:84, 154, 6:156; Loring and Atkinson, Cotton Culture, 13, 32; Southern Fertilizer Company, The Cotton Question (Rich-mond, 1876), 24–25; Rural Carolinian, 2 (March 1871), 323–24; Ransom and Sutch, One Kind of Freedom, 103; [Belton O'Neall Townsend] "South Carolina Society," Atlantic Monthly, 39 (June 1877), 678.

"quasi-proprietorship" blacks derived from the arrangement—the notion that sharecropping made the tenant "part owner of the crop" and therefore entitled to determine his family's own pace of work. Republican laws awarding laborers a first lien on the crop reinforced blacks' contention that they owned their share and did not just receive it as a wage from the planter. Indeed, property rights in the growing crop remained indeterminate during Reconstruction, a fact illustrated by the Mississippi planter who in a single letter referred to paying freedmen "half the cotton" as wages, and to making a living from his portion of "their crop." While sharecropping did not fulfill blacks' desire for full economic autonomy, the end of planters' coercive authority over the day-to-day lives of their tenants represented a fundamental shift in the balance of power in rural society, and afforded blacks a degree of control over their time, labor, and family arrangements inconceivable under slavery.[111]

For many blacks, however, the credit system that grew up alongside sharecropping quickly undermined its promise of autonomy. Before the war, individual planters had purchased goods like clothing in bulk for the entire work force; now, blacks formed an independent market for supplies, and their level of consumption rose well above the meager levels of slavery. "I have sold Jack Peters' negroes more goods this year and last year," reported an Alabama merchant, "than ever I sold Peters, and he owned 450 negroes." In addition, direct trade with the North became increasingly available to black belt counties. As a result, the coastal factors who had financed and marketed prewar crops by dealing exclusively with planters found themselves supplanted by local merchants, who furnished supplies to planter and tenant alike in exchange for a lien on the growing crop, and shipped the cotton north at the end of the season. Although the black belt experienced little urban growth, tiny towns and crossroads stores spread across the region, offering a wide array of merchandise and serving as centers of political and social life. "We have stores at almost every crossroad," reported a South Carolina journalist, "and at the railway stations and villages they have multiplied beyond precedent." By 1880, the cotton South counted over 8,000 rural stores. Many landlords established stores on their own plantations, sometimes finding the business of supplying tenants "as lucrative, if not more so, than planting or renting."[112]

111. Burton, *In My Father's House*, 282; Loring and Atkinson, *Cotton Culture*, 29–32, 131; W. B. Jones–D. H. Smith Plantation Journal, January 1872, MDAH; *Rural Carolinian*, 1 (February 1870), 317; S. Z. Williamson to Richard Sadler, March 24, 1870, David Hope Sadler Papers, Winthrop College; J. William Harris, "Plantations and Power: Emancipation on the David Barrow Plantations," in Burton and McMath, eds., *Toward a New South?* 256–57; Davis, *Good and Faithful Labor*, 100–101.

112. Woodman, *King Cotton*, 260–308; *New York Times*, March 12, 1866; Thomas D. Clark, *Pills, Petticoats and Plows: The Southern Country Store* (New York, 1974); Moore, ed., *Juhl Letters*, 163; Loring and Atkinson, *Cotton Culture*, 75.

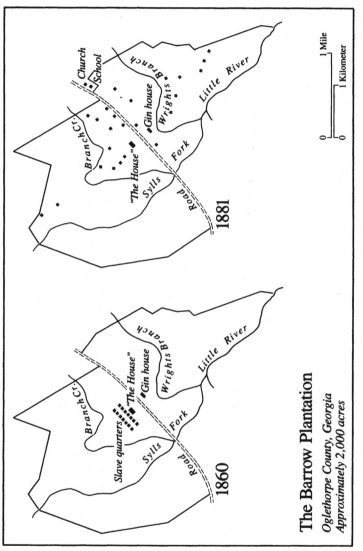

The Barrow Plantation

Oglethorpe County, Georgia
Approximately 2,000 acres

The Barrow Plantation in 1860 and 1881. Based on drawings in *Scribner's Monthly*, April 1881, these maps dramatically illustrate some of the changes in black life wrought by emancipation. In 1860, the plantation's entire black population lived in the communal slave quarters. Two decades later, the families of sharecroppers and cash renters were scattered on individual plots of land, and a church and school, built by the blacks after 1865, were in operation. Over half the black families in 1881 contained at least one member who had been a slave on the plantation before the Civil War.

Between black belt planters and the expanding merchant class, whose origins not infrequently lay in the North or Europe, tensions often ran high. Planters charged that the proliferation of country stores undermined good order by encouraging tenants to steal and sell growing cotton before the crop had been divided. (Some planters, however, were reported to be doing the same thing.) They also objected to lien laws that failed to give a landowner's claim for rent priority over debts owed by the tenant to local merchants. Yet as planters and their children went into merchandising and storekeepers acquired land, the two groups tended to coalesce. Reconstruction witnessed the origin of a new landlord-merchant class that by the end of the century would dominate Southern rural life.[113]

As for black tenants, the ability to seek supplies off the plantation enhanced their independence, since merchants rarely demanded the right to supervise day-to-day labor. On the other hand, interest rates for goods purchased on credit rose to exorbitant levels (often exceeding 50 percent), reflecting both the South's capital shortage and the fact that many rural merchants faced no local competition. In many cases, too, they inflicted outright fraud upon illiterate tenants. "A man that didn't know how to count would always lose," an Arkansas freedman later remarked, and in some rural areas, blacks banded together to find men able to check merchants' calculations "to keep them from taking all our labor away from us." As cotton prices declined during the 1870s, many tenants, finding themselves without sufficient income to settle their accounts at the end of the year, carried indebtedness over to the new season. To obtain additional credit, they were compelled to concentrate ever more heavily on cotton.[114]

For the region as a whole, for farmers white and black, the crop lien produced a growing overreliance on cotton and neglect of food, a pattern already clear by the 1870s. "The credit system," reported a resident of Mississippi, "has been pushed to such an extent that crops have been mortgaged for supplies before they have been planted. The culture of everything but cotton has either been abandoned or greatly curtailed." Blacks found the accelerating trend away from self-sufficiency especially dismaying. In 1873, the Alabama Labor Union urged the state government to exempt from taxation all farms with corn, peas, and potatoes as

113. Wiener, *Social Origins of the New South*, 78–81, 110–20; Wayne, *Reshaping of Plantation Society*, 165–67; M. H. Wetson to William W. Holden, October 16, 1868, North Carolina Governor's Papers; E. W. Rogers and Co. to Edward S. Canby, December 26, 1867, Canby to James L. Orr, January 7, 1868, South Carolina Governor's Papers; Ransom and Sutch, *One Kind of Freedom*, 146–47; C. Vann Woodward, *Origins of the New South* (Baton Rouge, 1951), 184.

114. Ransom and Sutch, *One Kind of Freedom*, 120–31; John A. James, "Financial Underdevelopment in the Postbellum South," *Journal of Interdisciplinary History*, 11 (Winter 1981), 444–45; Graves, "Town and Country," 166; 44th Congress, 1st Session, Senate Report 527, 1585–86.

the principal crops, with lost revenue to be recovered from increased levies on the holdings of "monopolists and speculators." Increasingly, sharecropping seemed to offer only permanent impoverishment. As a black Georgian wrote, "No man can work another man's land . . . even for half and board and clothe himself and family and make any money. The consequence will be the freedmen will become poorer and poorer every year." Thus, the same credit system that helped reduce many upcountry whites from yeomen to tenants made it impossible for blacks to use sharecropping as a springboard for the accumulation of money and the acquisition of land.[115]

In retrospect, sharecropping appears as a transitional arrangement, a way station between slavery and wage labor that would eventually become synonymous with economic oppression and, in some cases, debt peonage. Even during Reconstruction, large numbers of freedmen remained vulnerable to myriad forms of economic injustice, a fact underscored at the statewide labor conventions of 1869 and 1871, where reports on local conditions painted a melancholy picture: Landlords combined to deny blacks access to land, merchants charged exorbitant interest rates with the result that sharecroppers "generally come out in debt," laborers were often defrauded of their earnings, and, in conservative Georgia, planters employed "whipping same as in slavery" without interference from civil authorities. "We are today where 1866 left us," lamented George W. Cox, a former Union League organizer and member of Alabama's legislature.[116]

Like all the Reconstruction South's labor systems, sharecropping was not born fully formed, and its contours were continually reshaped by the ongoing day-to-day conflict over control of plantation labor. Only after the depression of the 1870s and the end of Reconstruction combined to limit severely the bargaining power of black laborers, would the exploitative implications of sharecropping become fully clear. Indeed, the early 1870s witnessed a partial economic recovery, stimulated by rising cotton production at a time when prices remained high by historic standards. Although still unable to obtain land, many blacks shared in the relative prosperity. Reports from throughout the region spoke of increasing wages for agricultural laborers, rising deposits in the Freedman's Savings Bank, and a growing number of sharecroppers rising to the status of renter. "The colored people . . ." claimed a Mississippi Republican newspaper in 1870, "were never so well off as at present. They have money to spend in the stores."[117]

115. E. Fisher to Joseph Holt, February 10, 1874, Joseph Holt Papers, LC; *AC*, 1873, 19; Harrison Berry to John M. Broomall, November 17, 1867, John M. Broomall Papers, HSPa.

116. T. J. Byres, "Historical Perspectives on Sharecropping," *Journal of Peasant Studies*, 11 (Winter 1983), 7–41; Macon *American Union*, October 29, 1869; Foner and Lewis, eds., *Black Worker*, 2:122–29.

117. Eugene Lerner, "Southern Output and Agricultural Income, 1860–1880," *AgH*, 33 (July 1959), 124; Williamson, *After Slavery*, 168; Taylor, *Louisiana Reconstructed*, 350–51, 376,

* * *

In 1872, newspapers recorded the passing of Maurice Jasper, the last surviving slave of George Washington. Born of parents brought from Africa over a century before, Jasper had lived, the *New National Era* commented, to see the "second revolution."[118] He died at a time of stocktaking, North and South, as the first period of Republican Reconstruction drew to a close. In political, economic, and social life, these years presented a complex pattern of achievement and disappointment. Compared with their predecessors, the accomplishments of the new governments were indeed impressive. Biracial democratic government, a thing unknown in American history, was functioning effectively in many parts of the South. Men only recently released from bondage cast ballots and sat on juries, and, in the Deep South, enjoyed an increasing share of authority at the state level, while the conservative oligarchy that had dominated Southern government from colonial times to 1867 found itself largely excluded from power. Public facilities had been rebuilt and expanded, school systems established, and tax codes modernized. Occurring at a crucial juncture in the transition from slavery to freedom, Reconstruction had nipped in the bud the attempt to substitute a legalized system of labor discipline for the coercion of slavery, and enhanced blacks' bargaining power on the plantations. All in all, declared a white South Carolina lawyer in 1871, "we have gone through one of the most remarkable changes in our relations to each other, that has been known, perhaps, in the history of the world."

Most striking of all, perhaps, was the impact upon the freedmen. For Reconstruction transformed their lives and aspirations in ways unmeasurable by statistics and in realms far beyond the reach of the law. A Northern correspondent, reporting from Vicksburg early in 1873, captured something of the change in blacks' conception of themselves: "One hardly realizes the fact that the many negroes one sees here . . . have been slaves a few short years ago, at least as far as their demeanor goes as individuals newly invested with all the rights and privileges of an American citizen. They appreciate their new condition thoroughly, and flaunt their independence. . . ." In many ways, of course, the "second revolution" was woefully incomplete. Republicans had undertaken "to promote political equality in a society characterized by equality in nothing else."[119] Blacks still suffered from dire poverty, and the old ruling class remained

385; Harris, *Day of the Carpetbagger,* 274–79; Charleston *News and Courier,* July 8, 1873; Carl R. Osthaus, *Freedmen, Philanthropy, and Fraud: A History of the Freedman's Savings Bank* (Urbana, Ill., 1976), 125–30; Jackson *Weekly Mississippi Pilot,* April 2, 1870.

118. *New National Era,* May 30, 1872. Jasper may, in fact, have been born free shortly after his parents were emancipated under the terms of George Washington's will. See Donald Sweig, ed., *Registrations of Free Negroes Commencing September Court 1822 . . .* (Fairfax, Va., 1977), 19, 71.

119. KKK Hearings, South Carolina, 796; *The Jewish Times* (New York), February 7, 1873; W. McKee Evans, *Ballots and Fence Rails: Reconstruction on the Lower Cape Fear* (Chapel Hill, 1967), 251.

largely intact, implacably hostile to the new order of things. But as long as Reconstruction survived, so too did the possibility of further change. "Nothing renders society more restless," Carl Schurz had written in 1865, "than a . . . revolution but half accomplished."[120] If Reconstruction's failures betrayed the utopian aspirations with which Republican rule began, its successes persuaded those accustomed to controlling the South's destiny that the entire experiment must be overturned. While Republican leaders struggled to consolidate a party and government plagued by factionalism, racism, and corruption, develop a program capable of capturing white support while serving the needs of the freedmen, and establish themselves as a permanent fixture on the South's political landscape, their opponents were organizing to bring Reconstruction to a violent, irrevocable end.

120. Frederic Bancroft, ed., *Speeches, Correspondence and Political Papers of Carl Schurz* (New York, 1913), 1:354.

CHAPTER 9

The Challenge of Enforcement

The New Departure and the First Redemption

I F Southern Republicans suffered from factional, ideological, and racial strife, their opponents encountered difficulties of their own. In the aftermath of Grant's victory, with Reconstruction seemingly a fait accompli, Southern Democrats confronted their own legitimacy crisis: the need to convince the North that their party stood for something other than simply a return to the old regime. Much of the Democratic constituency still refused to accept black suffrage as a permanent feature of the Southern political landscape. But a growing number of Democratic leaders saw little point in denying the reality that blacks were voting and holding office. These advocates of a New Departure argued that their party could only return to power by putting the issues of Civil War and Reconstruction behind them. Thus began a period in which Democrats, like Republicans, proclaimed their realism and moderation and promised to ease racial tensions. But if, in political rhetoric, "convergence"[1] reigned, in practice the New Departure only underscored the chasm separating the parties on fundamental issues, and the limits of Democrats' willingness to come to terms with the changes in Southern life intrinsic to Reconstruction.

Southern Democrats made their first attempts to seize the political center in 1869. Instead of running its own candidates for state office, the party threw support to disaffected Republicans, and focused its campaigns on the restoration of voting rights to former Confederates rather than on opposition to black suffrage. In Virginia and Tennessee, the strategy paid immediate dividends. Virginia, it will be recalled, remained under military control into 1869, since Gen. John M. Schofield had refused to authorize a referendum on the document drafted by the state's Radical-dominated constitutional convention. Soon after assuming

1. Michael Perman, *The Road to Redemption: Southern Politics, 1869–1879* (Chapel Hill, 1984), 21.

office, President Grant called for a popular vote on the constitution (with separate balloting on the stringent disenfranchisement clause) and the simultaneous election of state officials. Republicans nominated a ticket headed by Gov. Henry H. Wells, a former Union general who, after being appointed to direct state affairs by Schofield in 1868, had moved to the Radical wing of the party. Most prominent among the governor's critics was powerful railroad promoter William Mahone, whose plans to consolidate a group of Virginia railroads into a network that rivaled the Baltimore & Ohio Wells opposed.

A man whose diminutive stature (he weighed under 100 pounds) belied enormous talent and ambition, Mahone was a Democrat who believed his party should "gracefully acquiesce in accomplished facts" by "acknowledging [blacks'] legal status" and appealing to them "as Virginians" to oppose control of the local economy by out-of-state corporations like the B&O. He quickly emerged as the chief financier of a "new movement" designed to bring together Democrats and moderate Republicans on a program of accepting the new constitution, defeating its disenfranchisement section, and supporting Mahone's railroad schemes. The coalition's gubernatorial candidate was Gilbert C. Walker, a Northern-born Republican banker, manufacturer, and railroad man who had moved to the state in 1864 and thus had no association with secession. Walker's backers secured the withdrawal of a ticket headed by Col. Robert E. Withers, a Democrat of the old school who "did not propose to ask [for] . . . the vote of any negro." Meanwhile, Republicans found themselves "discouraged and demoralized." Blacks complained of exclusion from juries and local offices under Wells's administration, potential white voters were alienated by the choice of a black candidate for lieutenant governor, and the Grant administration, viewing Walker's candidacy as an opportunity to reshape Virginia politics along nonracial lines, extended little aid. With the Republican vote largely confined to the plantation counties of eastern and southern Virginia, Walker was elected governor, his supporters swept the legislature, and Virginia became the only Confederate state to avoid a period of Radical Reconstruction.[2]

Similar events transpired in Tennessee, although here the Republican governor himself initiated the political realignment. Assuming office in February 1869 when "Parson" Brownlow departed for the U. S. Senate, DeWitt Senter set out to win election in his own right by conciliating the state's Democrats. His policy split his already factionalized party, whose hold on statewide power still rested on widespread disenfranchisement.

2. Jack P. Maddex, Jr., *The Virginia Conservatives 1867–1879* (Chapel Hill, 1970), 66–82; Nelson M. Blake, *William Mahone of Virginia: Soldier and Political Insurgent* (Richmond, 1935), 83–144; Robert E. Withers, *Autobiography of an Octogenarian* (Roanoke, 1907), 246–49; Allen H. Curtis to Charles Sumner, April 2, 1869, Charles Sumner Papers, HU; Richard Lowe, "Another Look at Reconstruction in Virginia," *CWH*, 32 (March 1986), 71–75.

At the Republican state convention "blows were given, pistols and dirks were drawn," and the delegates adjourned without making a nomination. Challenged for reelection by Congressman William B. Stokes, a Union Army veteran and opponent of conciliation, the governor set aside the suffrage law and allowed thousands of former Confederates to register, whereupon the Democrats endorsed his candidacy. The result was an overwhelming victory for Senter, who carried the state by better than two to one and even edged ahead in East Tennessee.[3]

In Mississippi and Texas, the "new movements" of 1869 proved less successful. Mississippi Republicans themselves seized the moderate high ground by nominating Whig planter James L. Alcorn for governor on a platform repudiating the proscriptive features of the constitution the state's voters had rejected in 1868. Although Democrats endorsed a hastily organized National Union Republican ticket headed by Louis Dent (the President's brother-in-law) and pledged to accept black suffrage, the administration, alarmed by the Tennessee result, gave its full backing to Alcorn. The Dent forces made every effort to appeal to blacks, even nominating a freedman for secretary of state, but their efforts appeared clumsy at best. "You cannot blame us for your slavery," declared one Dent supporter to a black audience. "Most of you were well treated as slaves." On election day, nearly every freedman voted for Alcorn, large numbers of Democratic voters remained at home, and the result was the ratification of the constitution, the defeat of its disenfranchisement clause, and an overwhelming Republican victory. In Texas, Radical scalawag Edmund J. Davis defeated Democrat-endorsed moderate Republican Andrew J. Hamilton, although the margin of victory—fewer than 800 votes of nearly 80,000 cast—did not augur well for future Republican success.[4]

Although none of these "new movements" attracted significant black support, the New Departure gathered strength in 1870. As in Virginia and Tennessee, Missouri's Democrats formed a victorious coalition with self-styled Liberal Republicans, adopting a platform promising "universal amnesty and universal suffrage." Only the small-farming areas of western Missouri, ravaged by guerrilla bands during the Civil War, remained steadfastly Republican in the state's gubernatorial election, and the enrollment of 20,000 blacks, casting ballots for the first time thanks to the Fifteenth Amendment, could not offset the split within the party or the Democrats' majority among the thousands of immigrants who had entered the state since 1865. South Carolina Democrats created a Union

3. Thomas B. Alexander, *Political Reconstruction in Tennessee* (Nashville, 1950), 198–222; *CG*, 41st Congress, 2d Session, 137.

4. William C. Harris, *The Day of the Carpetbagger: Republican Reconstruction in Mississippi* (Baton Rouge, 1979), 219–37; *Speech of Hon. Amors R. Johnston, at Sardis, Mississippi* (n.p., 1869), 7; Carl H. Moneyhon, *Republicanism in Reconstruction Texas* (Austin, 1980), 104–24.

Reform party that nominated for governor disaffected Republican judge Richard B. Carpenter and actively courted the black vote, although to little avail as Gov. Robert K. Scott swept to reelection.[5]

In other states, Democrats accepted Reconstruction "as a finality," but retained their party identity rather than merge into new organizations or endorse dissident Republicans. Alabama's successful Democratic gubernatorial candidate, Robert Lindsay, insisted that his party had abandoned racial issues for economic ones, and openly sought black votes. Benjamin H. Hill of Georgia, an uncompromising opponent of black suffrage in 1867, now announced his willingness to recognize blacks' right to the "free, full, and unrestricted enjoyment" of the ballot. Louisiana Democrats, too, abandoned the rhetoric of race—"our platform," announced the party's state convention, "is retrenchment and reform." Some New Departure advocates sincerely expected to attract black votes; others believed this unlikely but anticipated winning the support of "moderate and conservative *white* republicans" once they became convinced that Democrats honestly accepted "the *accomplished facts of the war.*" Whatever the electoral outcome, Democrats hoped the New Departure would increase their legitimacy in Northern eyes. Since Grant's election, commented Mississippi Congressman Lucius Q. C. Lamar in 1873, Democratic nominations and platforms had been designed primarily with the aim of "secur[ing] the sanction and favor" of the national administration.[6]

In place of racial issues, Democratic leaders now devoted their energies to financial criticisms of Republican rule. In several states, they organized Taxpayers' Conventions, whose platforms denounced Reconstruction government for corruption and extravagance and demanded a reduction in taxes and state expenditures. "Grumbling" about high taxes, Republicans replied, was "confined to the large landowners" who resented their increased liability and the loss of their prewar right to assess their own holdings for tax purposes. And it is true that self-described men of "wealth, virtue and intelligence" took the lead in organizing the Taxpayers' Conventions—South Carolina's included bankers, merchants, and planters, among them four ex-governors, fifty-nine former members of the legislature, and eleven Confederate generals. Yet complaints about rising taxes cut across class lines, and became an effective rallying cry for opponents of Reconstruction. In largely self-sufficient areas like western

5. Thomas S. Barclay, *The Liberal Republican Movement in Missouri 1865–1871* (Columbia, Mo., 1926), 154–85, 241–70; William E. Parrish, *Missouri Under Radical Rule 1865–1870* (Columbia, Mo., 1965), 269–310; Francis B. Simkins and Robert H. Woody, *South Carolina During Reconstruction* (Chapel Hill, 1932), 448–53.

6. Perman, *Road to Redemption*, 16–21; Joshua B. Moore Diary (typed copy), March 7, 1870, ASDAH; Benjamin H. Hill, Jr., *Senator Benjamin H. Hill of Georgia: His Life, Speeches and Writings* (Atlanta, 1893), 58; *AC*, 1871, 457, 36; James H. Stone, "L. Q. C. Lamar's Letters to Edward Donaldson Clark, 1868–1885, Part I: 1868–1873," *JMH*, 35 (February 1973), 70.

North Carolina, where "the family do not see as much as $20 in money all the year," even a few dollars tax could seem onerous. "You cannot have an idea how destitute of money the country is," a resident of up-country South Carolina reported to Governor Scott in 1871. "The taxes now are the cause of the greatest anxiety and to meet them, people are selling every egg and chicken they can get." In retrospect, the antebellum years appeared to small farmers as a fiscal golden age. Asked if his tax of four dollars on 100 acres of land seemed excessive, one replied: "It appears so, sir, to what it was formerly, . . . next to nothing."[7]

Despite the broad appeal of calls for tax reduction, the growing prominence of the issue proved less of a transformation in Reconstruction politics than meets the eye. Indeed, while accepting the "finality" of Reconstruction and the principle of civil and political equality, the Taxpayers' Conventions simultaneously exposed the limits of political "convergence." For what they offered blacks with one hand, Democratic leaders simultaneously withdrew with the other. Convinced that Reconstruction denied men of property their rightful control of public affairs, the conventions and their supporters attributed the increasing tax burden to the fact that "nine-tenths of the members of the Legislature own no property and pay no taxes." Most Democrats objected not only to the amount of state expenditures (which they persistently exaggerated), but to such new purposes of public spending as tax-supported common schools. An amalgam of racism, elitism, and even traditional American republicanism (with its stress on economic autonomy as the only basis of independent political judgment), Democratic calls for a return to rule by "intelligent property-holders" not only would have excluded many whites from government, but implicitly denied blacks any role in the South's public affairs except to cast ballots for their social betters.[8]

Even at its height, the New Departure failed to win unanimous support among Southern Democrats. A large portion of the party's constituency still refused to acquiesce in the legitimacy of Reconstruction or the permanence of black suffrage. In 1870, for example, the year Democrats "redeemed" Missouri by allying with Liberal Republicans, their counter-

7. *Proceedings of the Tax-Payers' Convention of South Carolina* (Charleston, 1871); Ernest W. Winkler, *Platforms of Political Parties in Texas* (Austin, 1916), 140; W. C. Venning to Robert K. Scott, November 9, 1869, Robert K. Scott Papers, Ohio Historical Society; J. Mills Thornton III, "Fiscal Policy and the Failure of Radical Reconstruction in the Lower South," in J. Morgan Kousser and James M. McPherson, eds., *Region, Race, and Reconstruction: Essays in Honor of C. Vann Woodward* (New York, 1982), 367–71; J. G. de Roulhac Hamilton, ed., *The Papers of Randolph Abbott Shotwell* (Raleigh, 1929–39), 2:284; "Anonymous" to Robert K. Scott, March 13, 1871, South Carolina Governor's Papers, SCDA; 42d Congress, 2d Session, House Report 22, South Carolina, 892 (hereafter cited as KKK Hearings).

8. Eric Foner, *Politics and Ideology in the Age of the Civil War* (New York, 1980), 116; *Proceedings of the Tax Payers' Convention of South Carolina* (Charleston, 1874), 12–14; *South Carolina Before the National Government. Address of L. Cass Carpenter, Esq.* (Washington, D.C., 1874), 3.

parts in West Virginia chose to conduct an independent, avowedly anti-black, campaign. Despite the misgivings of the state's Republicans, the Fifteenth Amendment had recently enfranchised a few thousand black voters, and Democrats seized the occasion to raise again the cry of "white man's government." This was a principle that hardly seemed in peril in a state more than 95 percent white, but it proved extremely effective politically. Although laws barring "rebels" from voting remained on the books, Democrats won the governorship and legislature. "The spirit of the late rebellion is in the ascendant. . . ." a Republican leader concluded. "Hostility to negro suffrage was the prime element of our defeat." The policy of conciliation gained its greatest support in Deep South states with black voting majorities. But even where their state platforms appealed for black support, many local Democratic organizations continued to campaign "on a strictly white man basis."[9]

Even among its advocates, the New Departure smacked less of a genuine accommodation to the democratic revolution embodied in Reconstruction, than a tactic for reassuring the North about their party's intentions. Indeed, there was always something grudging about Democrats' embrace of black civil and political rights. "We are led to this course," declared a Mississippi newspaper in 1871, "not through choice, but by necessity—by the stern logic of events." Publicly, Democratic leaders spoke of a new era in Southern politics; privately, many hoped to undo the "evil" of black suffrage "as early . . . as possible." One prominent Alabama Democrat urged his party to "rise above . . . prejudice and contempt for the negro," but added that fifteen years hence, whites might be able to settle "the suffrage question" without Northern interference. And even centrist Democrats could not countenance independent black political organization. South Carolina's Taxpayers' Convention, for example, called for the dissolution of the Union Leagues. Alabama Gov. Robert Lindsay and the South Carolina Union Reform candidate for lieutenant governor, Matthew C. Butler, courted the black vote in 1870, but the following year, when asked to define the body politic of their states, each replied: "I mean the white people." Such statements helped convince blacks that, as one put it, Democrats "are the same today as they were, when Judge Taney gave his famous Dred Scott decision."[10]

Democratic economic policy presents a similar picture of "conver-

9. Milton Gerofsky, "Reconstruction in West Virginia, Part I," *WVaH*, 6 (July 1945), 329–54; Waitman T. Willey Diary, October 26, 1870, University of West Virginia; Perman, *Road to Redemption*, 66; W. M. Byrd to John W. A. Sanford, November 9, 1870, John W. A. Sanford Papers, ASDAH.

10. Thomas B. Alexander, "Persistent Whiggery in Mississippi Reconstruction: The Hinds County Gazette," *JMH*, 23 (April 1961), 84; John P. Carrier, "A Political History of Texas During the Reconstruction, 1865–1874" (unpub. diss., Vanderbilt University, 1971), 355; Albert Elmore to John W. A. Sanford, June 8, 13, 1870, Sanford Papers; KKK Hearings, Alabama, 206; South Carolina, 1190; Matthew T. Newsome to Adelbert Ames, November 1, 1872; Ames Family Papers, SC.

gence" more apparent than real. Advocates of a racial New Departure also spoke the language of an economic New South and often lent support to Republican programs of state-sponsored economic development. Acknowledging "the superiority of Yankee civilization," Benjamin H. Hill called upon Southerners to support railroad and factory construction, urban growth, and the education of labor (black as well as white) as the route to regional prosperity. A host of magazines associated politically with the Democratic party—*The Land We Love, Rural Carolinian, Southern Cultivator, Southern Field and Factory,* and *DeBow's Review*—preached the gospel of agricultural modernization (even proposing the introduction of exotic new crops like nutmeg and bananas) and scornfully rebuked planters for overdependence on cotton, recklessness in incurring debts, and backward farming techniques. Even George Fitzhugh, renowned before the war for his relentless criticism of Northern class relations, now emerged as a "Chamber of Commerce booster" of capitalist enterprise. In its most radical form, among a rising generation of urban newspaper editors and entrepreneurs, this latest incarnation of New South ideology called for replacing plantation agriculture with diversified, scientific small farms. One contributor to *DeBow's Review* even proposed that the state take the lead in breaking up the large plantations.[11]

"To say that the South ought to be and can be self-sustaining," noted one journalist, "is to reiterate a declaration which was by no means new even fifty years ago." And like its predecessors, including Presidential Reconstruction programs of which New Departure economics was a lineal descendant, this new attempt to supersede the plantation's dominance foundered on a combination of natural, economic, and political obstacles. Although mixed farming could and did make headway in the Upper South, the soil and climate of the cotton states doomed agricultural diversification without massive applications of fertilizer, an expense only the wealthiest planters could afford. In a credit system dominated by the crop lien, moreover, loans were readily available to finance cotton cultivation, but not to promote agricultural modernization.[12]

Instead of offering a successful blueprint for economic change, this

11. Hill, *Hill,* 334–49; E. Merton Coulter, "The New South: Benjamin H. Hill's Speech Before the Alumni of the University of Georgia, 1871," *GaHQ,* 57 (Summer 1973), 179–85; William E. Highsmith, "Louisiana Landholding During War and Reconstruction," *LaHQ,* 37 (January 1955), 46; Harris, *Day of the Carpetbagger,* 376; *Southern Field and Factory,* 1 (January 1871), 2; *Rural Carolinian,* 1 (October 1869), 11–12, (November 1869), 73; Jonathan M. Wiener, "Coming to Terms with Capitalism: The Postwar Thought of George Fitzhugh," *VaMHB,* 87 (October 1979), 438–47; Paul M. Gaston, *The New South Creed* (New York, 1970), 26–30, 65–66; Jonathan M. Wiener, *Social Origins of the New South: Alabama 1860–1885* (Baton Rouge, 1978), 186–95.

12. John H. Moore, ed., *The Juhl Letters to the "Charleston Courier"* (Athens, Ga., 1974), 266–67; Julius Rubin, "The Limits of Agricultural Progress in the Nineteenth-Century South," *AgH,* 69 (April 1975), 364–70; Gilbert C. Fite, "Southern Agriculture Since the Civil War: An Overview," *AgH,* 53 (January 1979), 9–10.

new New South program merely underscored planters' continued influence within the Democratic party, and reformers' continued inability to think of blacks as anything but dependent agricultural laborers. Advocates of industrialization generally expected poor whites and immigrants to supply the labor force. When they spoke of a new yeomanry, reformers had in mind farmers drawn from among landless Southern whites and immigrants from the North or Europe, not the freedmen. The triumph of diversified agriculture, contended Matthew C. Butler, would induce blacks to emigrate from South Carolina, leaving in their wake "lands divided into small farms, with an intelligent, thrifty, white population." Indeed, the entire agricultural reform program would have proved extremely detrimental to blacks' interests. New fence laws, necessary if planters were to shift from cotton production to raising cattle, would have prevented landless freedmen from owning stock. Because "progressive" farming, with its scientific fertilizers and carefully chosen mix of crops, required close supervision of the labor force, reformers denounced the sharecropping system and advocated a return to wage labor. "A farmer should have full control of the management of his crops, to manage them successfully," wrote one reform-minded planter, "and this you can hardly do, when your laborers are partners."[13]

Not surprisingly, most planters took only what they wanted from the New South ideology. The prospect of increased labor discipline *on* the plantation, rather than a new economic order off it, attracted many "disgusted with negro labor" to reform ideas. While some New South advocates spoke of attracting white immigrants as a prelude to the subdivision of the large estates, planters remained infatuated with the idea of reinforcing their own position by bringing in a new class of dependent laborers from abroad. The argument for immigration, commented an Alabama editor, "when stripped of its verbosity, is about as follows: 'We have lands but can no longer control the niggers; . . . hence we want Northern laborers, Irish laborers, German laborers, to come down and take their places, to work our lands for ten dollars a month'." Even more attractive were indentured laborers from China, whose "natural" docility would bolster plantation discipline and whose arrival, by flooding the labor market, would reduce the wages of blacks. With the coming of the Chinese, said one newspaper, "the tune . . . will not be 'forty acres and a mule,' but . . . 'work nigger or starve'." Such talk sometimes took on a grandiose flavor. "Give us five million of Chinese laborers in the valley of the Mississippi," wrote a planter's wife, "and we can furnish the world

13. *Letter from Chas. A. Fonde*, broadside, October 1868, South Carolina Governor's Papers; Atlanta *Constitution*, November 17, 1868; KKK Hearings, South Carolina, 1216; *Southern Field and Factory*, 1 (September 1871), 306; *Rural Carolinian*, 2 (April 1871), 382–83, 3 (February 1872), 229; Edmund L. Drago, *Black Politicians and Reconstruction in Georgia* (Baton Rouge, 1982), 128.

with cotton and teach the negro his proper place." Similarly, planters saw in cattle raising, grain production, and labor-saving machinery not only a route to regional self-sufficiency, but a way of combating the labor shortage by eliminating the need for some black labor and rendering the rest "less irksome and more profitable." Agricultural machinery was particularly enticing; always "under perfect control," it possessed the added advantage that it "cannot vote."[14]

Cognizant of their need for planter support, and sharing more assumptions with tradition-bound planters than they cared to recognize, many reformers encouraged these attitudes. New South advocates combined attacks upon the plantation system with "bitter denunciations of the Negroes," insisting that blacks' incorrigible laziness, exacerbated by the lax supervision prevalent under sharecropping, helped explain the South's economic backwardness. Instead of accepting blacks' right to compete in the marketplace, reformers seemed to fear their economic success, sometimes promoting scientific agriculture precisely as a weapon against black independence. "If the negroes all go to farming on their own hook," wrote a Mississippi advocate of modern farming methods in 1869, "we must drive them out by our superior system of culture." In advanced agriculture, declared the Selma Southern Argus, a tireless advocate of economic modernization, "the planters of the South have within their reach a solution of the labor question [and] . . . the proper subordination of the inferior race among us." Labor, reformers insisted, was entitled to no voice in managing a plantation and should be bought and sold in the marketplace like any other item, but, almost in the same breath, they urged planters to combine among themselves to fix the price of this "commodity." Thus, just as the political New Departure failed to accommodate itself fully to the democratic revolution at the heart of Reconstruction, proponents of an economic New South clung to hopes for labor control and racial subordination that belied their rhetorical commitment to far-reaching social change.[15]

Nor did official conduct in communities that remained under Democratic control inspire confidence that a real shift in policy or ideology had

14. William Hones to John and Joseph Le Conte, January 14, 1868, Le Conte Family Papers, Bancroft Library, University of California, Berkeley; Robert A. Gilmour, "The Other Emancipation: Studies in the Society and Economy of Alabama Whites During Reconstruction" (unpub. diss., Johns Hopkins University, 1972), 41; Gunther Barth, Bitter Strength: A History of the Chinese in the United States 1850–1870 (Cambridge, Mass., 1964), 188; Southern Field and Factory, 1 (February 1871), 55; Mrs. Basil G. Kiger to Basil G. Kiger, Jr., November 4, 1869, Basil G. Kiger Papers, Natchez Trace Collection, UTx; Rural Carolinian, 1 (November 1869), 71; J.C.A. Stagg, "The Problem of Klan Violence: the South Carolina Up-Country, 1868–1871," Journal of American Studies, 8 (December 1974), 312.

15. Highsmith, "Louisiana Landholding," 48; Sam Nostlethwaite to James A. Gillespie, March 14, 1869, James A. Gillespie Papers, LSU; Selma Southern Argus, March 3, 1870; Claude H. Nolen, The Negro's Image in the South (Lexington, Ky., 1967), 156–57; Southern Field and Factory, 1 (January 1871), 27.

occurred. Here, blacks complained of exclusion from juries, severe punishment for trifling crimes, the continued apprenticeship of their children against parental wishes, and a general inability to obtain justice. "I do not see none of my color in office. . . ." a former Tennessee slave complained in 1869. "When a white man kills a black man by having black men on the juree bench we then could defend our rights before the laws." In one Democratic Alabama county in 1870, a black woman brutally beaten by a group of whites was ordered to raise $16.45 for court costs before her complaint was heard. After she did, the judge released the offenders and instructed the injured woman to drop the matter or face a jail term. A group of prisoners in the Charlotte jail reported that the keeper routinely whipped Republican inmates, while the "one democrat in this god forsaken place . . . fares like a lord." Some Democratic judges, refusing to recognize the legality of homestead exemptions, ordered the sale of entire farms for debt. "I thought the Constitution that we radicals ratified was to let the poor man have a home. . . ." wrote a North Carolina scalawag. "The rebels say we have no constitution."[16]

Equally revealing were the statewide policies Democrats adopted in the border states and Upper South: Kentucky and Delaware, which Republicans never controlled; Maryland and West Virginia, "redeemed" by the Democrats, respectively, in 1867 and 1870; and Virginia, Tennessee, and Missouri, captured by "new movements" in 1869 and 1870. In the latter three states, Democrats controlled the legislature from the outset, and quickly shouldered aside the moderate Republican governors their votes had helped to elect. An indication of where power lay in these coalitions came when Missouri Democrats in 1871 ignored Carl Schurz's plea to send a man "at least semi-Republican" to the Senate, and chose Francis P. Blair, Jr., still an outspoken opponent of black suffrage.[17]

Having led the way in wartime Reconstruction, the border and Upper South now blazed the trail of Redemption. The threat of federal intervention restrained the most extreme proposals, and the heterogeneous nature of the Democratic coalition ensured that specific policies varied from state to state, but it remained perfectly clear that the party was still devoted to white supremacy and labor control. As late as 1872, Kentucky still barred blacks from testifying in court. These states also pioneered in legal segregation, Delaware authorizing hotels, theaters, and common carriers to refuse admission to persons "offensive" to other customers,

16. New York *Herald*, August 30, 1871; J. W. Bailey to DeWitt Senter, May 15, 1869, Tennessee Governor's Papers, TSLA; Joseph Eddens to William H. Smith, January 18, 1870, Jacob Fisher to Smith, June 17, 1870, Alabama Governor's Papers, ASDAH; Ben Smith et al. to William W. Holden, February 3, 1869, A. N. Barker to Holden, January 29, 1869, North Carolina Governor's Papers, NCDAH.

17. Maddex, *Virginia Conservatives*, 100–103; Parrish, *Missouri*, 312–17; Carl Schurz to William Grosvenor, January 7, 1871, William Grosvenor Papers, Columbia University; *CG*, 42d Congress, 1st Session, 701.

while Tennessee Democrats repealed the Republican law penalizing railroads for discriminating against blacks and drafted a new constitution requiring racial separation in the public schools. This last provision appeared somewhat redundant, for the superintendent of public instruction elected in 1869 believed "it was not necessary to educate the farmer, mechanic, or laborer" at all, and the legislature repealed the state education law, leaving schooling a voluntary decision of each county and destroying public education for blacks except in Memphis and Nashville. Delaware, Kentucky, and Maryland Democrats initially made no provision at all for black education, then ordered that these schools be financed by taxes on black parents. Outside Virginia, whose education superintendent, William H. Ruffner, successfully badgered lawmakers for funds, schooling was a major casualty of Democratic rule.[18]

Despite the ratification in 1870 of the Fifteenth Amendment prohibiting disenfranchisement because of race, border Democrats developed ingenious methods of limiting black voting power. Delaware, whose Democratic party insisted that the state was not "morally bound" by any of the postwar Constitutional Amendments (a position that, taken literally, appeared to envision a return to slavery) in 1873 made payment of a poll tax a requirement for voting, effectively disenfranchising the bulk of the black population and ensuring over twenty years of uninterrupted Democratic ascendancy in the state. Tennessee's 1870 constitution also included a provision requiring payment of a poll tax to vote, and Maryland in the same year enacted a property qualification. Virginia's Redeemers gerrymandered cities so as to ensure Democratic control, reduced the number of polling places in black precincts, empowered the legislature to appoint local governments, and barred from voting all those who had failed to pay a poll tax or been convicted of petty larceny. Maryland's Constitution of 1867 reoriented representation toward the plantation counties at the expense of Baltimore and the small farming regions to its north and west. "This they call a white man's Government," remarked one critic. "That is the right of a few white men, by counting the disfranchised blacks, to govern a great many white men. This is progress backwards. . . ."[19]

18. Victor B. Howard, "The Black Testimony Controversy in Kentucky, 1866–1872," *JNH*, 58 (April 1973), 140–65; Harold B. Hancock, "The Status of the Negro in Delaware After the Civil War, 1865–1875," *Delaware History*, 13 (April 1968), 64–65; Lester C. Lamon, *Blacks in Tennessee 1791–1970* (Knoxville, 1981), 47; Leroy P. Graf, ed., "Education in East Tennessee, 1867–1869: Selections from the John Eaton, Jr. Papers," *ETHSP*, 23 (1951), 111; Henry L. Swint, ed., "Reports from Educational Agents of the Freedmen's Bureau in Tennessee, 1865–1870," *THQ*, 1 (June 1942), 164–66; *CG*, 41st Congress, 3d Session, 1039; *AC*, 1870, 467; Walter J. Fraser, Jr., "William Henry Ruffner and the Establishment of Virginia's Public School System, 1870–1874," *VaMHB*, 79 (July 1971), 272–73.

19. *AC*, 1872, 235; Amy H. Hiller, "The Disfranchisement of Delaware Negroes in the late Nineteenth Century," *Delaware History*, 13 (October 1968), 124–54; Joshua W. Caldwell, *Studies in the Constitutional History of Tennessee* (Cincinnati, 1895), 150–51; William Gillette,

Thus, even before the end of Reconstruction, Democratic governments had begun to undermine the legal and political gains blacks had achieved. No similar unity of purpose informed the early Redeemers' economic policies. Tennessee dismantled the Republican railroad aid program, but Virginia's legislature granted "enormous concessions" to railroad companies. West Virginia's so-called Lawyer's Constitution of 1872 barred railroad officials from seats in the legislature but also included a section on land titles—described as "a contrivance gotten up to make litigation the principal business" of the state—that facilitated the transfer of real estate and mineral rights from small proprietors to mining and lumber companies. Whatever their other priorities, however, these early Redeemers did not hesitate to pass measures to bolster labor control. Tennessee restricted hunting and fishing, repealed the Republicans' mechanics' lien law and a statute expediting workers' claims for wages, and allowed property like wagons, teams, and furniture to be seized and sold to pay the state poll tax. Virginia nullified its constitution's homestead exemption and reestablished whipping as a punishment for petty theft—a penalty, noted black legislator George Teamoh, likely to affect only the poor: "The rich man never . . . steals on a small scale. . . . No danger of his being stretched upon that whipping post."[20]

The most comprehensive effort to undo Reconstruction, however, occurred in Georgia, whose legislature fell into Democratic hands in 1870, followed by the governorship a year later. A poll tax, coupled with new residency and registration requirements, sharply reduced the number of black voters, while a shift from ward to citywide elections eliminated Republicans from Atlanta's city council. To demolish the enclave of black political power in McIntosh County, the legislature ousted Tunis G. Campbell from his seat in favor of a white Democrat and appointed a board of commissioners to replace the elected local government. A state court subsequently sentenced Campbell (at age sixty-three) to a year of hard labor on the flimsy pretext that as justice of the peace he had improperly arrested a white man. Redeemer Gov. James M. Smith ad-

Retreat from Reconstruction 1869–1879 (Baton Rouge, 1979), 40–41; James T. Moore, "Black Militancy in Readjuster Virginia, 1879–1883," *JSH*, 41 (May 1975), 168–69; *CG*, 41st Congress, 3d Session, 282; Charles E. Wynes, *Race Relations in Virginia 1870–1902* (Charlottesville, 1961), 13; Barbara J. Fields, *Slavery and Freedom on the Middle Ground: Maryland During the Nineteenth Century* (New Haven, 1985), 134.

20. Caldwell, *Constitutional History*, 153–63; Allen W. Moger, "Railroad Practices and Policies in Virginia after the Civil War," *VaMHB*, 69 (October 1951), 427–39; Milton Gerofsky, "Reconstruction in West Virginia, Part II," *WVaH*, 7 (October 1945), 39; John A. Williams, "The New Dominion and the Old: Ante-Bellum and Statehood Politics as the Background of West Virginia's 'Bourbon Democracy'," *WVaH*, 33 (July 1972), 366–67; *Fur, Fin, and Feather: A Compilation of the Game Laws of the Principal States and Provinces of the United States and Canada*, 5th ed. (New York, 1872), 58–59; *Tennessee Acts 1869–70*, 102, 112; Maddex, *Virginia Conservatives*, 167–69; *New National Era*, January 26, March 16, 1871.

vised blacks to abandon politics and "get down to honest hard work," an injunction Democratic judges and legislators set about to imbue with the force of law. Measures impossible to enact during Reconstruction now appeared on the statute book, prohibiting the sale of farm products at night without the landlord's permission, making it a criminal offense to hire a laborer already under contract, making a laborer's lien on the crop inferior to that of the planter, restricting hunting and fishing, and facilitating changes in fence laws to the detriment of landless laborers. Some affected the poor of both races, but many applied only in black belt counties. All in all, Georgia's Redeemers demonstrated the truth of Governor Smith's remark that the state could "hold inviolate every law of the United States and still so legislate upon our labor system as to retain our old plantation system."[21]

Were the Democratic party to regain power throughout the South, a group of North Carolina Republicans "both white and colored" warned in 1869, "they can nullify the republican form of government and place the colored race and laboring class of white people in the same position only worse as they were before." If the First Redemption did not go this far, it certainly belied the idea that Southern Democrats acquiesced in the democratic and free labor advances of Reconstruction. The result served as a rallying cry for Republicans in other states, a warning of what to expect should they, too, fall into Democratic hands. "I ask you to turn to Tennessee," Robert B. Elliott told a South Carolina audience in 1870. "As soon as they secured their seats in the Legislature . . . they dropped from the statute book the school law which guaranteed to colored men the chance to educate their children."[22]

For their part, whether "white line" or New Departure, adherents of the traditional agrarian order or a New South, all Democrats viewed the overthrow of Republican rule as a prerequisite for pursuing their political and economic goals. Those interested in lower taxes and tighter state budgets had little confidence that Republicans could accomplish these ends, since "the *negro vote* will always be unaffected by regard for *economy* in government." Willingness to work with Republicans in promoting

21. Horace C. Wingo, "Race Relations in Georgia 1872–1890" (unpub. diss., UGa, 1969), 17–21; James M. Russell and Jerry Thornbery, "William Finch of Atlanta: The Black Politician as Civic Leader," in Howard N. Rabinowitz, ed., *Southern Black Leaders of the Reconstruction Era* (Urbana, Ill., 1982), 323; Russell Duncan, *Freedom's Shore: Tunis G. Campbell and the Georgia Freedmen* (Athens, Ga., 1986), 83–87; *Sufferings of the Rev. T. G. Campbell and His Family, in Georgia* (Washington, D. C., 1877); Atlanta *Herald* in Savannah *Morning News*, December 8, 1874; Charles L. Flynn, Jr., *White Land, Black Labor: Caste and Class in Late Nineteenth-Century Georgia* (Baton Rouge, 1983), 87–95, 124–30; Savannah *Advertiser and Republican*, August 17, 1873.

22. Silas L. Curtis et al. to William W. Holden, August 11, 1869, North Carolina Governor's Papers; *Speech of Col. R. B. Elliott, at Anderson Court House, South Carolina* (Charleston, 1870), 9.

railroads and industrial development did not alter New South Democrats' conviction that capital could never be attracted to the South until "Negro government" had been overthrown and "the statesmen and leading intelligent men take the guidance of affairs." Nor, Democrats of all persuasions believed, could plantation discipline be reinforced so long as Republicans remained in power. Whatever their blueprints for the future, all Democrats shared the conviction of New Departure spokesman Benjamin H. Hill: "The first thing to be done is to secure Home Government for Home Affairs."[23] In pursuit of "home rule," the opponents of Reconstruction launched a campaign of violence that made a mockery of talk of political convergence, and confronted Reconstruction governments with the most basic of challenges—a threat to their very physical survival.

The Ku Klux Klan

To Europeans, America has always seemed a violent land—a consequence, perhaps, of its individualist values, frontier traditions, and weak governmental authority. In its pervasive impact and multiplicity of purposes, however, the wave of counterrevolutionary terror that swept over large parts of the South between 1868 and 1871 lacks a counterpart either in the American experience or in that of the other Western Hemisphere societies that abolished slavery in the nineteenth century. It is a measure of how far change had progressed that the reaction against Reconstruction proved so extreme.

Violence, as has been related, had been endemic in large parts of the South since 1865. But the advent of Radical Reconstruction stimulated its further expansion. By 1870, the Ku Klux Klan and kindred organizations like the Knights of the White Camelia and the White Brotherhood had become deeply entrenched in nearly every Southern state. One should not think of the Klan, even in its heyday, as possessing a well-organized structure or clearly defined regional leadership. Acts of violence were generally committed by local groups on their own initiative. But the unity of purpose and common tactics of these local organizations makes it possible to generalize about their goals and impact, and the challenge they posed to the survival of Reconstruction. In effect, the Klan was a military force serving the interests of the Democratic party, the planter class, and all those who desired the restoration of white supremacy. Its purposes were political, but political in the broadest sense, for it sought to affect power relations, both public and private, throughout

23. Linton Stephens to Alexander H. Stephens, April 25, 1872, Alexander H. Stephens Papers, Manhattanville College; Daniel Pratt to Robert M. Patton, August 20, 1867, Alabama Governor's Papers; *Southern Field and Factory*, 1 (August 1871), 243; Gaston, *New South Creed*, 37; Eric Foner, *Nothing But Freedom: Emancipation and Its Legacy* (Baton Rouge, 1983), 48; Savannah *Morning News*, August 19, 1873.

Southern society. It aimed to reverse the interlocking changes sweeping over the South during Reconstruction: to destroy the Republican party's infrastructure, undermine the Reconstruction state, reestablish control of the black labor force, and restore racial subordination in every aspect of Southern life.[24]

In parts of the South, blacks holding public office daily faced the threat of violence. Congressman Richard H. Cain and his family lived "in constant fear," their home guarded day and night by armed men. More typically, however, violence was directed at local leaders. As Emanuel Fortune, driven from Jackson County, Florida, by the Klan, explained: "The object of it is to kill out the leading men of the republican party . . . men who have taken a prominent stand." At least one tenth of the black members of the 1867–68 constitutional conventions became victims of violence during Reconstruction, including seven actually murdered. Greene County, Georgia, Klansmen in 1869 forced Abram Colby into the woods "and there stripped and beat him in the most cruel manner for nearly three hours." His assailants, reported the local agent of the American Missionary Association, had "as they said, many old scores against him, as a leader of his people in the county," for Colby had organized "one of the largest and most enthusiastic branches" of Georgia's Equal Rights Association in 1866, and won election to the legislature two years later. (Colby's mother, wife, and daughter witnessed the assault. "My little daughter came out and begged them not to carry me away. . . . She never got over it" and died soon theralfter.) Legislator Richard Burke, murdered in 1870, was considered "obnoxious" by Sumter County, Alabama, whites as "a leader in the Loyal League" who had "acquired a great influence over people of his color." His killing was part of a wave of violence that resulted in the deaths of several blacks "who went through the county doing the best they could, keeping the party up." Jack Dupree, victim of a particularly brutal murder in Monroe County, Mississippi—assailants cut his throat and disemboweled him, all within sight of his wife, who had just given birth to twins—was "president of a republican club," and known as a man who "would speak his mind."[25]

Countless other local leaders were forced to flee their homes after brutal whippings. Andrew Flowers was whipped in 1870 after defeating

24. Ray Granada, "Violence: An Instrument of Policy in Reconstruction Alabama," *AlHQ*, 30 (Fall–Winter 1968), 182–83; Ryland Randolph to Walter L. Fleming, October 15, 1901, Walter L. Fleming Papers, NYPL; Allen W. Trelease, *White Terror: The Ku Klux Klan Conspiracy and Southern Reconstruction* (New York, 1971), xlvi.
25. George P. Rawick, ed., *The American Slave: A Composite Autobiography* (Westport, Conn., 1972–79), Supplement 2, 4:1272; KKK Hearings, Florida, 94–95; Richard L. Hume, "Negro Delegates to the State Constitutional Conventions of 1867–69," in Rabinowitz, ed., *Southern Black Leaders*, 146; R. H. Gladding to Rufus B. Bullock, November 29, 1869, Georgia Governor's Papers; KKK Hearings, Georgia, 696–97; Alabama, 334–37, 998–99; 46th Congress, 2d Session, Senate Report 693, pt. 1:399; KKK Hearings, Mississippi, 270, 360, 435, 809.

a white candidate in a Chattanooga contest for justice of the peace. "They said they had nothing particular against me," he related, "that they didn't dispute I was a very good fellow . . . but they did not intend any nigger to hold office in the United States." And many blacks became targets merely for exercising their rights as citizens. Alabama freedman George Moore reported how, in 1869, Klansmen came to his home, administered a beating, "ravished a young girl who was visiting my wife," and wounded a neighbor. "The cause of this treatment, they said, was that we voted the radical ticket."[26]

Nor did white Republicans escape the violence. Klansmen murdered three scalawag members of the Georgia legislature and drove ten others from their homes. North Carolina State Sen. John W. Stephens, a harnessmaker who allegedly avoided the Confederate draft, "displayed a particular hostility toward the ex-slaveholding gentry," and rose to prominence in Caswell County Republican affairs, was assassinated in 1870. Warned repeatedly that his life was in danger, Stephens had been "accustomed to say that 3,000 poor ignorant colored Republican voters had stood by him . . . at the risk of persecution and starvation, and that he had no idea of abandoning them to the Ku-Klux." The Klan in western North Carolina settled old scores with wartime Unionists, burned the offices of the Rutherford *Star,* and brutally whipped Aaron Biggerstaff, a Hero of America and Republican organizer. Indeed, the situation of scalawags became increasingly desperate as Reconstruction progressed. "There are but few white males in our county belonging to the Republican party," one wrote from Calhoun County, Alabama, "and those few are ostracized and cursed and . . . in some instances their houses have been surrounded by disguised bands in the night time, threatening their lives if they do not desist in their political course."[27]

On occasion, violence escalated from the victimization of individuals to wholesale assaults upon the Republican party and its leadership. In October 1870, a group of armed whites broke up a Republican campaign rally at Eutaw, the county seat of Greene County, Alabama, killing four blacks and wounding fifty-four. In the same month, on the day after Republicans carried Laurens County, in South Carolina's Piedmont cotton belt, a racial altercation at Laurensville degenerated into a "negro chase" in which bands of whites scoured the countryside, driving 150 freedmen from their homes and committing thirteen murders. The victims included

26. *New National Era,* January 12, 1871; KKK Hearings, Florida, 41–43; South Carolina, 292; Alabama, 1188.

27. Drago, *Black Politicians,* 144; Jonathan W. O'Neil et al. to Senate of the United States, January 1, 1869, Georgia Governor's Papers; Otto H. Olsen, *Carpetbagger's Crusade: The Life of Albion Winegar Tourgée* (Baltimore, 1965), 162–64; New York *Tribune,* August 3, 1870; KKK Hearings, North Carolina, 20–21, 29, 106, 111–13, 127–30; Hamilton, ed., *Shotwell Papers,* 2:400–35; John A. Detman et al. to William H. Smith, August 7, 1869, Alabama Governor's Papers.

the newly elected white probate judge, a black legislator, and others "known and prominent as connected with politics." In Meridian, a small Mississippi town to which many blacks had fled from centers of Klan activity in nearby western Alabama, three black leaders were arrested in March 1871 on charges of delivering "incendiary" speeches. Firing broke out at their court hearing, the Republican judge and two defendants were killed, and a day of rioting followed, which saw perhaps thirty blacks murdered in cold blood, including "all the leading colored men of the town with one or two exceptions."[28]

The Klan's purposes, however, extended far beyond party politics. Indeed, the organization flourished in Kentucky well before blacks obtained the right to vote there in 1870. Institutions like churches and schools, embodiments of black autonomy, frequently became targets. "Nearly every colored church and school-house" in the Tuskegee area was burned in the fall of 1870. Female teachers were attacked as well as male, and white educators victimized along with black. North Carolina Klansmen assaulted Alonzo B. Corliss, a crippled Northern-born Quaker, and drove him from the state, for "teaching niggers and making them like white men." In Georgia, "a quiet Christian citizen" was "most cruelly whipped" for renting a room to a Northern teacher. William Luke, an Irish-born teacher in a black school, suffered verbal abuse, saw shots fired into his home, and finally, in 1870, was lynched at Cross Plains, Alabama, along with four black men. Those blacks who managed to acquire an education were often singled out for attack. The Georgia Klan murdered freedman Washington Eager because, according to his brother, he was "too big a man . . . he can write and read and put it down himself." It also destroyed a teacher's library, declaring "they would just dare any other nigger to have a book in his house."[29]

Equally important among the aims of violence was the restoration of labor discipline on white-owned farms and plantations. In a sense, the Klan sought to take the place of both the departed personal authority of the master and the labor control function the Reconstruction state had abandoned. A scalawag judge reported that the Klan in his section of Alabama originated among white employers no longer able to "control the labor . . . through the courts" but determined to "compel them to

28. Melinda M. Hennessey, "Political Terrorism in the Black Belt: The Eutaw Riot," *Alabama Review*, 33 (January 1980), 35–48; W. W. Ball, *A Boy's Recollections of the Red Shirt Campaign of 1876 in South Carolina* (Columbia, S.C., 1911), 3–4; John Hubbard to Robert K. Scott, February 14, 1871, South Carolina Governor's Papers; KKK Hearings, South Carolina, 1306–7, 1333; Mississippi, 7–204; Jackson *Daily Mississippi Pilot*, March 8, 11, 1871; John R. Lynch to Adelbert Ames, March 10, 1871, Ames Family Papers.

29. Trelease, *White Terror*, 89, 124, 294; KKK Hearings, Alabama, 747, 1024–25; Mrs. H. J. Moore to William W. Holden, January 16, 1869, North Carolina Governor's Papers; KKK Hearings, North Carolina, 144–50; D. A. Newsom to Rufus B. Bullock, September 13, 1869, Georgia Governor's Papers; Gene L. Howard, *Death at Cross Plains: An Alabama Reconstruction Tragedy* (University, Ala., 1984); KKK Hearings, Georgia, 668, 402.

do by fear what they were unable to make them do by law." Blacks who disputed the portion of the crop allotted them at year's end were frequently whipped, and as in 1865 and 1866, violent bands drove freedmen off plantations after the harvest, to deprive them of their share. In one Georgia county, Klansmen in 1869 lynched a freedman and his wife accused of "resenting a blow from their employer"; in another they murdered a blacksmith who refused to work for a white employer until paid for previous labor. Blacks working on a South Carolina railroad construction gang were whipped and told to go "back to the farms to labor." Victims of the Mississippi Klan included a black woman whipped for "laziness" and a freedman beaten for taking a white debtor to court; "darkeys were through with suing white men," his assailants declared. Blacks who sought to change employers were also victimized. "If we got out looking for some other place to go," a Texan later recalled, "them KKK they would tend to Mister negro good and plenty."[30]

But the most "offensive" blacks of all seemed to be those who achieved a modicum of economic success, for, as a white Mississippi farmer commented, the Klan "do not like to see the negro go ahead." Night riders in Florence, South Carolina, killed a freedman on one plantation "because it is rented by colored men, and their desire is that such thing ought not to be." The Noxubee County, Mississippi, Klan, according to a former Confederate Army officer who knew the organization from the inside, aimed "to keep them from renting land, so that the majority of the white citizens may control labor." Klansmen also objected to blacks owning animals of their own, sometimes killing the freedmen's livestock in order to make them more dependent upon their employers. Florida, with more available land and a higher proportion of black landowners than other Southern states, saw the Klan direct much of its activity against economically independent freedmen. "Didn't I know they didn't allow damned niggers to live on land of their own," one black homesteader reported being told after he and his wife and children were beaten and ordered to go to work for a white employer. Whites charged with encouraging black economic autonomy, including merchants who purchased cotton from black tenants, also drew the Klan's wrath. One elderly North Carolina farmer who distributed land to his former slaves suffered a whipping. Even James L. Alcorn, shortly before his election as Mississippi's governor, saw several plantation buildings burned as a punishment for renting land to freedmen.[31]

30. KKK Hearings, Alabama, 1758; A. S. Bradley to John Eaton, December 13, 1868, John Eaton Papers, TSLA; Morgan L. Ogden to J. H. Taylor, August 5, 1869, Georgia Governor's Papers; KKK Hearings, Georgia, 8, 12; South Carolina, 27–28; Mississippi, 672, 277; Rawick, ed., *American Slave*, Supplement 2, 2:101.

31. P. C. Cudd to Robert K. Scott, May 17, 1869, South Carolina Governor's Papers; KKK Hearings, Mississippi, 233–35, 355, 376; W. R. Jones to William H. Smith, August 9, 1870,

Ultimately, as a former Confederate officer put it, the Klan aimed to regulate blacks' "status in society." Sometimes, in vigilante fashion, it punished those accused of crimes. A Tennessee freedman and his family were whipped in 1869 for allegedly poisoning a white neighbor's horse, a charge the victims vehemently denied. More commonly, violence was directed not at alleged criminals, but against "impudent negroes"—those who no longer adhered to patterns of behavior demanded under slavery. A North Carolina freedman related how, after he was whipped, Klan assailants "told me the law—their law, that whenever I met a white person, no matter who he was, whether he was poor or rich, I was to take off my hat." Other victims included blacks accused of speaking disrespectfully to whites, those who did not yield the sidewalk to white passersby, and women who "dress up . . . and fix up, like they thought anything of themselves." Those most certain to suffer abuse were interracial couples in which the male was black (a breach of traditional mores that appears more frequently in records of the time than might be imagined). Black Georgian Marion Bellups in 1870 described in pathetic detail the "persecutions" he and his pregnant white wife endured. Klansmen drove them from their home and brutally whipped Mrs. Bellups; when she "went to her own color for protection they driven her off." Local officials then jailed the woman and her child from a previous marriage. "I have worked hard and made something and made her look like some body," Bellups wrote Gov. Rufus Bullock, "[and] they are now trying to break us up. . . . You know that it is getting very cold [in jail] for her and her little child."[32]

By and large, Klan activity was concentrated in Piedmont counties where blacks comprised a minority or small majority of the population and the two parties were evenly divided. But no simple formula can explain the pattern of terror that engulfed parts of the South while leaving others relatively unscathed. Georgia's Klan was most active in a cluster of black belt and Piedmont cotton counties east and southeast of Atlanta, and in a group of white-majority counties in the northwestern part of the state. Unknown in the overwhelmingly black South Carolina and Georgia lowcountry, the organization flourished in the western Alabama plantation belt. Scattered across the South lay counties particularly notorious for rampant brutality. Carpetbagger Judge Albion W. Tourgée counted twelve murders, nine rapes, fourteen cases of arson, and over

Alabama Governor's Papers; KKK Hearings, Florida, 279–80; Trelease, *White Terror*, 336; Lillian A. Pereyra, *James Lusk Alcorn: Persistent Whig* (Baton Rouge, 1966), 79.

32. KKK Hearings, Georgia, 37; North Carolina, 54; J. B. to DeWitt Senter, May 18, 1869, Tennessee Governor's Papers; *Trial of William W. Holden* (Raleigh, 1871), 2:1214; KKK Hearings, Georgia, 74–78, 402; Marion Bellups to Rufus B. Bullock, August 10, 1870, Marion Bellups Papers, GDAH.

and over 700 beatings (including the whipping of a woman 103 years of age) in his judicial district in North Carolina's central Piedmont. An even more extensive "reign of terror" engulfed Jackson, a plantation county in Florida's panhandle. "That is where Satan has his seat," remarked a black clergyman; all told over 150 persons were killed, among them black leaders and Jewish merchant Samuel Fleischman, resented for his Republican views and reputation for dealing fairly with black customers.[33]

Nowhere did the Klan become more deeply entrenched than in a group of Piedmont South Carolina counties where medium-sized farms predominated and the races were about equal in number. An outbreak of terror followed the October 1870 elections, in which Republicans retained a tenuous hold on power in the region. Possibly the most massive Klan action anywhere in the South was the January 1871 assault on the Union county jail by 500 masked men, which resulted in the lynching of eight black prisoners. Hundreds of Republicans were whipped and saw their farm property destroyed in Spartanburg, a largely white county with a Democratic majority. Here, the victims included a considerable number of scalawags and wartime Unionists, among them Dr. John Winsmith, a member of "the old land aristocracy of the place" wounded by Klansmen in March 1871. In York County, nearly the entire white male population joined the Klan, and committed at least eleven murders and hundreds of whippings; by February 1871 thousands of blacks had taken to the woods each night to avoid assault. The victims included a black militia leader, found hanging from a tree in March with a note pinned to his breast, "Jim Williams on his big muster," and Elias Hill, a self-educated black teacher, minister, and "leader amongst his people." Even by the standards of the postwar South, the whipping of Hill was barbaric: A dwarflike cripple with limbs "drawn up and withered away with pain," he had mistakenly believed "my pitiful condition would save me." Hill had already been organizing local blacks to leave the region in search of the "peaceful living, free schools, and rich land" denied them in York County. Not long after his beating, together with some sixty black families, he set sail for Liberia.[34]

Contemporary Democrats, echoed by subsequent scholars, often attributed the Klan's sadistic campaign of terror to the fears and prejudices of poorer whites. (More elevated Southerners, one historian contends,

33. Charles L. Flynn, "The Ancient Pedigree of Violent Repression: Georgia's Klan as a Folk Movement," in Walter J. Fraser, Jr., and Winfred B. Moore, Jr., eds., *The Southern Enigma: Essays on Race, Class, and Folk Culture* (Westport, Conn., 1983), 190–91; KKK Hearings, Georgia, 1007–1009; South Carolina, 87; Trelease, *White Terror*, 246; New York *Tribune*, August 3, 1870; KKK Hearings, Florida, 174, 222–23; Ralph L. Peek, "Lawlessness in Florida, 1868–1871," *FIHQ,* 40 (October 1961), 172–80.

34. KKK Hearings, South Carolina, 29, 299, 367–69, 624, 897–98, 919–22, 943, 1406–12, 1465–85; Trelease, *White Terror*, 356–58, 362–72; *Proceedings in the Ku Klux Klan Trials at South Carolina* (Columbia, S.C., 1872), 221–24; R. M. Camden to Robert K. Scott, January 23, 1871, South Carolina Governor's Papers; *New National Era*, August 24, 1871.

could never have committed these "horrible crimes.") The evidence, however, will not sustain such an interpretation. It is true that in some upcountry counties, the Klan drove blacks from land desired by impoverished white farmers and occasionally attacked planters who employed freedmen instead of white tenants. Sometimes, violence exacerbated local labor shortages by causing freedmen to flee the area, leading planters to seek an end to Klan activities. Usually, however, the Klan crossed class lines. If ordinary farmers and laborers constituted the bulk of the membership, and energetic "young bloods" were more likely to conduct midnight raids than middle-aged planters and lawyers, "respectable citizens" chose the targets and often participated in the brutality.[35]

Klansmen generally wore disguises—a typical costume consisted of a long, flowing white robe and hood, capped by horns—and sometimes claimed to be ghosts of Confederate soldiers so, as they claimed, to frighten superstitious blacks. Few freedmen took such nonsense seriously. "Old man, we are just from hell and on our way back," a group of Klansmen told one ex-slave. "If I had been there," he replied, "I would not want to go back." Victims, moreover, frequently recognized their assailants. "Dick Hinds had on a disguise," remarked an Alabama freedman who saw his son brutally "cut to pieces with a knife." "I knew him. Me and him was raised together." And often, unmasked men committed the violence. The group that attacked the home of Mississippi scalawag Robert Flournoy, whose newspaper had denounced the Klan as "a body of midnight prowlers, robbers, and assassins," included both poor men and property holders, "as respectable as anybody we had there." Among his sixty-five Klan assailants, Abram Colby identified men "not worth the bread they eat," but also some of the "first-class men in our town," including a lawyer and a physician.[36]

Personal experience led blacks to blame the South's "aristocratic classes" for violence and with good reason, for the Klan's leadership included planters, merchants, lawyers, and even ministers. "The most respectable citizens are engaged in it," reported a Georgia Freedmen's Bureau agent, "if there can be any respectability about such people." Editors Josiah Turner of the Raleigh *Sentinel,* Ryland Randolph of the Tuscaloosa *Monitor* (who years later recalled administering whippings "in the regular *ante bellum style*"), and Isaac W. Avery of the Atlanta *Constitu-*

35. Francis B. Simkins, "The Ku Klux Klan in South Carolina, 1868–1871," *JNH*, 12 (October 1927), 618, 622; Granada, "Violence," 196; KKK Hearings, Georgia, 420; South Carolina, 47; N. J. Reynolds et al. to Rufus B. Bullock, June 3, 1869, Georgia Governor's Papers; Trelease, *White Terror*, 52; Paul D. Escott, *Many Excellent People: Power and Privilege in North Carolina, 1850–1900* (Chapel Hill, 1985), 154–58; Annie Young to Robert K. Scott, July 8, 1869, South Carolina Governor's Papers.
36. KKK Hearings, North Carolina, 2–3; Mississippi, 662–63, 667; Alabama, 674–76; Simkins, "Ku Klux Klan," 618; Nannie Lacey, "Reconstruction in Leake County," *PMHS*, 11 (1910), 275–76; KKK Hearings, Mississippi, 93; Georgia, 696–97.

tion were prominent Klansmen, along with John B. Gordon, Georgia's Democratic candidate for governor in 1868.[37] When the Knights of the White Camelia initiated Samuel Chester in Arkansas, the pastor of his church administered the oath and the participants included Presbyterian deacons and elders "and every important member of the community." In Jackson County, Florida, the "general ring-leader of badness . . . the generalissimo of Ku-Klux" was a wealthy merchant; elsewhere in the black belt, planters seem to have controlled the organization. Even in the upcountry, "the very best citizens" directed the violence. "Young men of the respectable farming class" composed the Klan's rank and file in western North Carolina, but its leaders were more substantial—former legislator Plato Durham, attorney Leroy McAfee (whose nephew, Thomas Dixon, later garbed the violence in romantic mythology in his novel *The Clansman*), and editor Randolph A. Shotwell. As the Rutherford *Star* remarked, the Klan was "not a gang of *poor trash,* as the leading Democrats would have us believe, but men of property . . . respectable citizens."[38]

Many "respectable citizens," of course, had no connection with the violence, and a few spoke out manfully against it. When the son of former North Carolina Chief Justice Thomas Ruffin joined the Klan, his father dashed off a stinging rebuke: "I am satisfied that such associations [are] . . . dangerous to the community, and highly immoral. . . . It is wrong—all wrong, my son, and I beg you to have nothing to do with it." Yet even Ruffin said nothing in public. Indeed, the silence of the most prominent white Southerners spoke volumes of what Maj. Lewis Merrill, who investigated the Klan in York County, South Carolina, called "the demoralization of public opinion." In many communities, a few individuals still shaped white political sentiment. Orange County, where the North Carolina Klan originated, was said to follow the lead of former Gov. William A. Graham "as a good Catholic does . . . the Pope." "One short letter" by Graham, black legislator James H. Harris believed, "could put down all these outrages." Yet like so many others, Graham remained silent, in effect offering tacit approval to Klan activities.[39]

37. Montgomery *Alabama State Journal,* January 6, 1871; Stagg, "Klan Violence," 305; W. C. Morrill to "Dear General," November 2, 1868, Georgia Governor's Papers; Trelease, *White Terror,* 191; Ryland Randolph to Walter L. Fleming, August 21, 1901, Fleming Papers; I. W. Avery, *The History of the State of Georgia from 1850 to 1881* (New York, 1881), 382; Allen P. Tankersley, *John B. Gordon: A Study in Gallantry* (Atlanta, 1955), 249–56.

38. Samuel H. Chester, *Pioneer Days in Arkansas* (Richmond, 1927), 63; KKK Hearings, Florida, 151; Drago, *Black Politicians,* 151–52; J. R. Davis, "Reconstruction in Cleveland County," *Trinity College Historical Society Historical Papers,* 10 (1914), 27; Hamilton, ed., *Shotwell Papers,* 2:256, 345–49; Trelease, *White Terror,* 338–40; Rutherford *Star,* December 10, 1870.

39. J. G. de Roulhac Hamilton, ed., *The Papers of Thomas Ruffin* (Raleigh, 1918-20), 4:225–27; KKK Hearings, South Carolina, 1482; Olsen, *Tourgée,* 158; Zebulon Vance to

Republicans, white and black, heaped scorn upon "respectables" who did not participate directly in the violence but "could not stop their sons from murdering their inoffensive neighbors in broad daylight." Yet their complicity went beyond silence in the face of unspeakable crimes. Through their constant vilification of blacks, carpetbaggers, scalawags, and Reconstruction, the "old political leaders" fostered a climate that condoned violence as a legitimate weapon in the struggle for Redemption. Democratic leaders, moreover, proved less than candid about their party's connection with the Klan. Even Alabama Gov. Robert Lindsay, who sought to discourage violence after his election in 1870, could declare, "I do not think there ever was a political motive in any outrage committed on a colored man." Both "white line" and New Departure party leaders put up bail for arrested Klansmen, spoke of the "good" the organization accomplished despite its "excesses," and strongly opposed federal intervention. Rather than dissociate themselves from the campaign of terror, prominent Democrats either minimized the Klan's activities or offered thinly disguised rationalizations for them. Some denied the organization's existence altogether, dismissing reports of violence as electoral propaganda, products of a Republican "slander mill." Others sought to discredit the victims, portraying them as thieves, adulterers, or men of "bad character" who more or less deserved their fate. Still others fell into a web of contradictions by attributing violence to personal grudges and family feuds rather than politics, and then justifying the Klan as a response to the growth of the Union League.[40]

In ordinary times, commented Major Merrill, "every man's hand will be against criminals." In a democratic society, law enforcement ultimately rests on public willingness to cooperate with the police, a condition that did not exist in the Reconstruction South where, as William H. Trescot put it, a majority of the white population "do not believe that the present state government is legitimate." Much Klan activity took place in those Democratic counties where local officials either belonged to the organization or refused to take action against it. Even in Republican areas, however, the law was paralyzed. When sheriffs overcame fear of violence and arrested suspects, witnesses proved reluctant to testify, Klansmen perjured themselves to provide one another with alibis, and, as a Florida Republican leader observed, if "any one of these men is on

Cornelia Spencer, April 27, 1866, Transcripts, William A. Graham Papers, UNC; *Speech of Hon. James Harris on the Militia Bill* (Raleigh, 1870), 21–22.
    40. W. J. Purman to "Dear Captain," October 17, 1869, Black History Collection, LC; KKK Hearings, Mississippi, 75; Alabama, 200; Trelease, *White Terror*, 332; George J. Leftwich, "Reconstruction in Monroe County," *PMHS*, 9 (1906), 69; David Schenck Diary, December 18, 1869, UNC; Charleston *Courier*, September 28, 1871; Atlanta *Constitution*, April 23, 1870; KKK Hearings, South Carolina, 94–99; Mississippi, 377; Georgia, 184; Alabama, 349–50, 872, 907. Michael Perman's contention that "the New Departurists did, by mid-1871, find the will and the means to help bring the Klan to heel" cannot be sustained by the evidence. Perman, *Road to Redemption*, 63–64.

the jury . . . you cannot convict." Only rarely would a member give evidence against his compatriots. (The organization's sense of solidarity even extended to an Alabama Klansman rescued from jail despite having raped the twelve-year-old daughter of a fellow member. This led the outraged father to reveal to a grand jury details of the murder of teacher William Luke, but still no indictment resulted.) Community support extended far beyond the Klan's actual membership, embracing the numerous Southern women who sewed costumes and disguises for night riders, and manufacturers who sought to increase sales by adopting "Ku-Klux" as a name for their products. Whites unconnected with the Klan, moreover, still seemed to view violence against blacks as something less than a crime. "If a white man kills a colored man in any of the counties of this State," lamented a Florida sheriff, "you cannot convict him."[41]

No one can doubt the political commitment or personal courage of victims who risked their lives to testify before grand juries and Congressional investigators or affirmed their convictions in the face of certain reprisal. After being beaten by the Klan, Abram Colby was asked whether he would again vote Republican. "If there was an election tomorrow, I would vote the radical ticket," he replied, whereupon "they set in and whipped me a thousand licks more." George Roper, an incautious black veteran, was shot and beaten after rushing into the street and shouting, "Hurrah for Grant and Colfax" as Klansmen rode through the streets of Huntsville. Some Republicans armed themselves and repulsed attacks on their homes. A white Georgian who joined the party because of a revulsion against violence ("if this is what they call Democracy in the South," he declared, "I am done with it"), singlehandedly fought off twenty attackers, killing four. Occasionally, organized groups successfully confronted the Klan. White Union Army veterans in mountainous Blount County, Alabama, organized "the anti-Ku Klux," which put an end to violence by threatening Klansmen with reprisal unless they stopped whipping Unionists and burning black churches and schools. Armed blacks patrolled the streets of Bennettsville, South Carolina, to prevent Klan assaults. And groups of freedmen sometimes threatened to take their own retribution on known criminals if violence persisted. At one near-confrontation in Alabama's plantation belt, "the blacks . . . rather invited a contest, saying they were willing to go out into an open field and 'fight it out'."[42]

The scale of violence, however, dwarfed these efforts at extralegal

41. KKK Hearings, South Carolina, 1603; Wilbur E. Miller, "Reconstruction as a Police Problem" (unpub. paper, annual meeting of Organization of American Historians, 1978); William H. Trescot to Robert K. Scott, October 24, 1868, South Carolina Governor's Papers; Trelease, *White Terror*, 376–84; KKK Hearings, Georgia, 960; South Carolina, 1371–72; Florida, 153; Howard, *Death at Cross Plains*, 115; Kathryn R. Schuler, "Women in Public Affairs in Louisiana During Reconstruction," *LaHQ*, 19 (July 1936), 675; Atlanta *Constitution*, June 8, 1872; KKK Hearings, Florida, 125.

42. KKK Hearings, Georgia, 414–20, 696; Alabama, 668–71, 687–89, 777–79; John P. Green, *Fact Stranger than Fiction* (Cleveland, 1920), 143–44; Pierce Burton to William H. Smith, August 4, 1870, Alabama Governor's Papers.

reprisal. Indeed, many Northerners wondered aloud, in an accusatory tone, how Republican communities allowed themselves to be terrorized by bands of night riders. "The capacity of a people for self-government," a Northern friend lectured Judge Tourgée, "is proved, first of all, by its inclination and capacity for self-protection. . . . If people are killed by the Ku-Klux, why do they not kill the Ku-Klux?" Tourgée's reply seemed almost to blame the victims for their own plight: "Our entire party consists of poor men . . . without the self-respect so natural to the Yankee." Others sought an answer in the legacy of slavery. "They are not like men who have been always free," commented a white Republican sheriff, "not by a great deal. . . . They do not know how to resist white men." Nor were whites alone in falling back on such explanations. "The colored men, as a general thing, . . ." declared Congressman Jeremiah Haralson, who had known bondage until 1865, "are afraid of the white men. He has been raised to be afraid of them."[43]

It is indeed true that slavery, which gave rise to numerous forms of black resistance, did not produce a broad tradition of violent retaliation against abuse. And after emancipation, freedmen proved far less prone than whites to commit violent acts, either within their own community or against outsiders. Of the large number of blacks incarcerated in state penitentiaries, nearly all had been convicted of crimes against property rather than assaults upon persons. But the failure of nerve, if such it was, extended up and down the Republican hierarchy, and was not confined to one race. James M. Justice, a Republican leader in western North Carolina, confessed to being "ashamed" that his scalawag constituents "have not resented and resisted our wrongs; we ought to have done it, but we have not done it." Local Republican officials, both black and white, discouraged talk of violence and urged angry bands of armed blacks to disperse. Some black political leaders, indeed, seemed entirely at a loss as to how to counsel fearful constituents. Oscar J. Dunn later admitted that he avoided his own people during Louisiana's violent 1868 Presidential campaign, for "I had no advice to give them."[44]

Perhaps the problem was that Republicans, black and white, took democratic processes more seriously than their opponents. A mob of Arkansas freedmen in 1871 lynched three whites accused of murdering a black lawyer, but this was indeed a rare occurrence. Blacks sometimes intimidated members of their community who voted Democratic, ridiculing them at church and, on a few occasions, threatening them with violence, but no Republicans rode at night to murder their political foes, nor did

43. Albion W. Tourgée, *A Fool's Errand* (Cambridge, Mass., 1961 [orig. pub. 1879]), 233–34; Catherine Silverman, " 'Of Wealth, Virtue, and Intelligence': The Redeemers and Their Triumph in Virginia and North Carolina, 1865–1877" (unpub. diss., Graduate Center, City University of New York), 184; KKK Hearings, Mississippi, 259; 44th Congress, 2d Session, Senate Report 704, 191.

44. KKK Hearings, North Carolina, 169; Mississippi, 670; Florida, 134–35; Georgia, 1035; 41st Congress, 2d Session, House Miscellaneous Document 154, pt. 1:178.

armed bands seek to drive Democrats from the polls. "We could burn their churches and schoolhouses but we don't want to break the law or harm anybody," wrote one black from a violence-torn part of Georgia. "All we want is to live under the law."[45]

Whatever the "legacy of slavery" or the strength of blacks' commitment to legal processes, the practical obstacles to armed resistance were immense. Many rural freedmen owned firearms, but these were generally shotguns, much inferior to the "first-class weapons" like Winchester rifles and six-shooters in the hands of the Klan. Although many had served in the Union Army, blacks with military experience were far outnumbered in a region where virtually every white male had been trained to bear arms. The fate of individuals who successfully repelled Klan assaults did not inspire confidence in extralegal resistance, for many were forced to flee their homes for fear of subsequent attack, and saw family members victimized in retaliation. As for organized self-defense, the specter of armed blacks taking the law into their own hands was certain to enrage the white community and produce a further escalation of violence. "It would be annihilation to the negroes if they should undertake such a thing," commented a white Republican official in Alabama, an appraisal borne out in Louisiana in 1873. The election of 1872 produced rival claimants for the governorship, a situation paralleled in localities throughout the state. In Grant Parish, freedmen who feared Democrats would seize the government cordoned off the county seat of Colfax and began drilling and digging trenches under the command of black veterans and militia officers. They held the tiny town for three weeks; on Easter Sunday, whites armed with rifles and a small cannon overpowered the defenders and an indiscriminate slaughter followed, including the massacre of some fifty blacks who lay down their arms under a white flag of surrender. Two whites also died.

The bloodiest single instance of racial carnage in the Reconstruction era, the Colfax massacre taught many lessons, including the lengths to which some opponents of Reconstruction would go to regain their accustomed authority. Among blacks, the incident was long remembered as proof that in any large confrontation, they stood at a fatal disadvantage. "The organization against them is too strong. . . ." Louisiana black teacher and legislator John G. Lewis later remarked. "They attempted [armed self-defense] in Colfax. The result was that on Easter Sunday of 1873, when the sun went down that night, it went down on the corpses of two hundred and eighty negroes."[46]

45. *AC*, 1871, 35; 42d Congress, 1st Session, Senate Report 1, 147; *New National Era*, February 17, 1870.

46. KKK Hearings, Mississippi, 244; South Carolina, 15; 46th Congress, 2d Session, Senate Report 693, pt. 2:357, 373, 409, 433; Paul D. Escott, *Slavery Remembered* (Chapel Hill, 1979), 157–58; KKK Hearings, Georgia, 209–10; Alabama, 63; Joe G. Taylor, *Louisiana Reconstructed, 1863–1877* (Baton Rouge, 1974), 268–70; Ted Tunnell, *Crucible of Reconstruction: War, Radicalism, and Race in Louisiana 1862–1877* (Baton Rouge, 1984), 189–93.

Ultimately, of course, the responsibility for suppressing crime rests not with the victim but with the state. And an avalanche of heart-rending pleas for protection descended upon the South's Republican governors. "If Chief Justice Taney was wrong in saying that a colored man had no rights a white man was bound to respect, they are not respected. . . ." a South Carolina Republican wrote Governor Scott. "I feel assured, that if you really knew or saw, the demoniacal acts of these lawless people, you would at once urge protection, to the innocent laborers." Local officials, declaring themselves unable to enforce the law, called for the rapid organization of state militias or the dispatch of federal troops, demands echoed by rank and file Republicans. "We are more slave today in the hand of the wicked than before. . . ." declared six Alabama Union League members. "Only a standing army in this place can give us our right and life." On occasion, local Republicans (usually scalawags) asked governors to dispatch arms, "and then let us fight it out." More frequently, the letters testified to a pervasive sense of helplessness in the face of Klan terror. "The ku kluks klan is shooting our familys and beating them notoriously," one letter informed North Carolina Governor Holden. "We do not know what to do."[47]

"Put on your iron gloves," one Northerner advised Southern Republicans. And on paper, the new governments did take decisive steps, outlawing going about in disguise, raising the penalties for assault, murder, and conspiracy, authorizing ordinary citizens to arrest Klan members, sometimes even requiring counties to pay damages to citizens whose rights were abridged or property destroyed by a mob. Yet when it came to enforcing these laws, Republican leaders vacillated. Nowhere in the country in these years was the militia an effective fighting force, and in states with few white Republicans, governors proved extremely reluctant to employ an organization composed almost entirely of freedmen. Willingness to fight alongside blacks had proved essential to winning the Civil War, but Deep South governors had little confidence in the freedmen's prospects when confronting well-trained Confederate veterans, and feared the arming of a black militia would inaugurate all-out racial warfare. Such a step, moreover, was certain to destroy efforts to attract white support and demonstrate the Republican party's moderation. Thus, the same governors who opposed strong civil rights legislation and appointed Democrats to local office failed to respond effectively to calls for protection.[48]

47. John T. Henderson to Robert K. Scott, November 30, 1868, South Carolina Governor's Papers; A. A. Smith to William H. Smith, April 7, 1870, James Martin et al. to Smith, May 25, 1869, Alabama Governor's Papers; W. A. Patterson to William W. Holden, July 1, 1868, M. Hester et al. to Holden, October 9, 1869, North Carolina Governor's Papers.
48. A. W. Spies to Robert K. Scott, March 8, 1871, South Carolina Governor's Papers; *South Carolina Acts and Joint Resolutions 1870–71*, 559–62; James W. Patton, *Unionism and Reconstruction in Tennessee 1860–1869* (Chapel Hill, 1934), 195; *Alabama Acts 1868*, 444–46;

Instead of uniting Republicans in a struggle for survival, the question of violence further exacerbated the party's internal discord. Urged by local Republicans to "throw aside your conservatism," Alabama Gov. William H. Smith declared reports of violence exaggerated and insisted that local officials bore the primary responsibility for law enforcement. Smith did issue a proclamation ordering armed bands to disperse, but on the day of its publication 300 Klansmen invaded Huntsville, threatening freedmen and driving them from their churches—their answer, one scalawag observed acidly, to the policy of "talking without actions." The governor's caution undermined his credibility among many rank-and-file Republicans and strengthened black support for his rival, Sen. George Spencer. Mississippi Gov. James L. Alcorn proved as reluctant as Smith to use a black militia, proposing instead to create an elite, presumably white, cavalry (an idea defeated in the legislature), and establishing a seven-man Secret Service—hardly a force capable of suppressing violence. Georgia's Republican administration declined even to organize a militia, and although Florida Gov. Harrison Reed did recruit a mostly black force, he never sent it into action against the Klan.[49]

South Carolina Gov. Robert K. Scott did enroll thousands of blacks in the state militia in 1870, creating a patronage machine for political allies who filled the officer corps, and helping ensure a heavy Republican turnout in the fall balloting. Having won reelection, however, Scott's policy changed, and when violence swept across the Piedmont, the governor disarmed many units to appease his opponents. And although Louisiana created a state-controlled Metropolitan Police Force commanded by former Confederate Gen. James Longstreet, to patrol New Orleans, the Warmoth regime's main response to violence was to establish a returning board, empowered to alter or throw out election returns from areas where violence and intimidation prevailed.[50]

If Deep South governors sought stability through conciliation, those able to draw upon large populations of white Republicans took decisive action. Gov. William G. Brownlow recruited a militia, manned largely by

---

Jerry M. Cooper, *The Army and Civil Disorder* (Westport, Conn., 1980), 11–12; George W. Williams to Robert K. Scott, April 3, 1869, South Carolina Governor's Papers; Trelease, *White Terror*, 116; Otto H. Olsen, "Reconsidering the Scalawags," *CWH*, 12 (December 1966), 318–19; Perman, *Road to Redemption*, 34–35.

49. W. B. Jones to William H. Smith, April 3, 1870, John H. Wager to Smith, July 3, 1869, Alabama Governor's Papers; John Z. Sloan, "The Ku Klux Klan and the Alabama Election of 1872," *Alabama Review*, 18 (April 1965), 116–20; Trelease, *White Terror*, 261–64, 299–301; William T. Blain, "Challenging the Lawless: The Mississippi Secret Service, 1870–1871," *JMH*, 40 (May 1978), 119–31; Pereyra, *Alcorn*, 134–38; Otis A. Singletary, *Negro Militia and Reconstruction* (Austin, 1957), 11.

50. Joel Williamson, *After Slavery: The Negro in South Carolina During Reconstruction* (Chapel Hill, 1965), 261–62; Herbert Shapiro, "The Ku Klux Klan During Reconstruction: The South Carolina Episode," *JNH*, 49 (January 1964), 44–45; Taylor, *Louisiana Reconstructed*, 174–80.

East Tennessee Unionists, and early in 1869 declared martial law in nine violence-plagued counties, a step which led to a drastic curtailment of Klan activities. But to garner Democratic support, his successor, DeWitt Senter, who assumed office at the end of February, restored civil authority throughout the region and disbanded the militia. Governors Powell Clayton in Arkansas and Edmund J. Davis in Texas achieved more enduring results. Clayton placed ten counties under martial law at the end of 1868 and dispatched a state militia composed of blacks and scalawags (usually in segregated units) and commanded by former Gen. Robert F. Catterson. Scores of suspected Klansmen were arrested; three were executed after trials by military courts, and numerous others fled the state. By early 1869, order had been restored and the Klan destroyed. Davis proved equally decisive, organizing a crack two-hundred-member State Police, 40 percent of whose members were black. Between 1870 and 1872, the police made over 6,000 arrests, effectively suppressing the Klan and providing freedmen with a real measure of protection in a state notorious for widespread violence.[51]

As Clayton and Davis demonstrated, a government willing to suspend normal legal processes and employ armed force could mount an effective response to the Klan. But as many a modern government has discovered, the suspension of constitutional rights in the interest of law enforcement carries its own risks, especially the possibility of transforming perpetrators of violence from criminals into victims in the eyes of citizens who sympathize with the motives, if not the methods, of those arrested. Nowhere was this dilemma more apparent than in North Carolina, where Gov. William W. Holden's use of the militia provoked a reaction that helped bring down his administration. In the face of mounting violence in 1869 and early 1870, Holden had vacillated, doing little more than calling upon local sheriffs to enforce the law and sending sympathetic notes to victims of assault. But the murder of John W. Stephens, and Clayton's success in Arkansas, led the governor to dispatch white militia units raised in the western North Carolina mountains to Caswell and Alamance counties, under the command of former Union Army officer George W. Kirk. About 100 men were arrested, and although the state constitution did not authorize the governor to declare martial law, Holden suspended the local courts (which the Klan controlled), ordered the prisoners tried before a military commission, and refused to honor a writ of habeas corpus issued by the state's chief justice. Ironically, Democrats then appealed to the federal courts under the Habeas Corpus

51. Trelease, White Terror, 43–44, 155–79; CG, 42d Congress, 1st Session, Appendix, 199; Orval T. Driggs, Jr., "The Issues of the Powell Clayton Regime, 1868–1871," ArkHQ, 8 (Spring 1949), 15–27; Carrier, "Political History of Texas," 428–30, 443–50; Ann P. Baenziger, "The Texas State Police During Reconstruction: A Reexamination," SWHQ, 72 (April 1969), 475–76.

Act of 1867, originally enacted to protect blacks and white Unionists. Holden was forced to release the captives, and the campaign against the Klan collapsed.[52]

Although some prisoners suffered rough treatment, no blood was shed in the Kirk-Holden War. Yet the affair provided a field day for Holden's opponents, and in the legislative elections of 1870, which occurred in the midst of the furor over habeas corpus, Democrats swept to victory. In some parts of the state, the Klan had crippled the Republican organization (ten of the fifteen counties that moved from their column to the Democratic had experienced significant violence). In others, Holden's use of troops solidified the opposition and alarmed yeomen who resented the intrusion of an outside force into a local community. Democrats even carried the western North Carolina mountains. In the aftermath of their victory, Holden's opponents expelled enough Republican legislators to obtain a two-thirds majority, and proceeded to impeach the governor for "subvert[ing] personal liberty" in the state. (An article linking Holden to railroad frauds mysteriously disappeared, probably because leading Democrats had much to fear from an exposure of corruption.) Voluminous evidence of Klan atrocities introduced by his attorneys had no effect on the preordained outcome, and Holden became the first governor in American history to be removed from office by impeachment. All in all, it was an inglorious end to his long, erratic political career, and to Reconstruction in a state where its prospects had once appeared so bright.[53]

North Carolina was not the only state where Republican control unraveled in 1870, for Alabama and Georgia also fell into Democratic hands. A host of factors produced these results, some of the Republicans' own doing. The feud between Governor Smith and Senator Spencer and the nomination of black leader James Rapier for secretary of state contributed to the Democrats' narrow conquest of Alabama's governorship. Georgia Republicans had been demoralized ever since 1868 by the conservatism and racism of the Bullock administration; the Democrats' sweep of the legislative elections of 1870 only confirmed Reconstruction's failure in a state where it had never really gotten off the ground. Throughout the South, the black vote, where it could be cast without fear of violence, remained solidly Republican. Among the white electorate, however, Democrats continued to benefit from superior economic resources, a near-monopoly of the press, and the potent appeal of white supremacy (whether couched in the extreme language of "white line"

52. William W. Holden to Mrs. H. J. Moore, January 4, 1869, Letterbook, North Carolina Governor's Papers; Trelease, *White Terror*, 209–22.

53. Horace W. Raper, *William W. Holden: North Carolina's Political Enigma* (Chapel Hill, 1985), 158; Olsen, *Tourgée*, 165–66; Escott, *Many Excellent People*, 156, 163; Gordon B. McKinney, *Southern Mountain Republicans 1865–1900* (Chapel Hill, 1978), 47; Cortez A. M. Ewing, "Two Reconstruction Impeachments," *NCHR*, 15 (July 1938), 204–21.

leaders or the more genteel tones of the New Departure). Republicans, moreover, had weakened themselves by factionalism and corruption, and the growing black demand for office seems to have led at least some scalawags to leave the party. The party remained powerful in traditional centers of upcountry Unionism, among voters whose political loyalties had been established during the Civil War, but despite its ambitious economic programs and conciliatory policy toward former Confederates, it was failing to attract new white support, either from younger men now coming onto the voting rolls or Democratic defections.

The role of violence in producing the results of 1869 and 1870 varied considerably from state to state. In Missouri, West Virginia, Virginia, and Tennessee, where Republican majorities based on disenfranchisement had evaporated and party splits opened the door to Democrats' resumption of power, the Klan played little or no part in Democratic victories. In the Deep South, its impact was far more significant. Violence had made it virtually impossible for Republicans to campaign or vote in large parts of Georgia. The Alabama returns revealed that Democrats could not have carried the state in a peaceful election, for the decline in the Republican vote in Greene County alone (scene of the Eutaw riot and numerous other acts of terror), exceeded Governor Lindsay's entire statewide majority. In Florida, meanwhile, violence wiped out Republican majorities in all but the most heavily black counties, and helped a Democrat win election as lieutenant governor.[54]

In other ways as well, violence had a profound effect on Reconstruction politics. For the Klan devastated the Republican organization in many local communities. By 1871, the party in numerous locales was "scattered and beaten and run out." "They have no leaders up there—no leaders," a freedman lamented of Union County, South Carolina. No party, North or South, commented Adelbert Ames, could see hundreds of its "best and most reliable workers" murdered and still "retain its vigor." Indeed, the black community was more vulnerable to the destruction of its political infrastructure by violence than the white. Local leaders played such a variety of roles in schools, churches, and fraternal organizations that the killing or exiling of one man affected many institutions at once. And for a largely illiterate constituency, in which political information circulated orally rather than through newspapers or pamphlets, local leaders were bridges to the larger world of politics, indispensable sources of political intelligence and guidance. Republican officials, black and white, epitomized the revolution that seemed to have put the bottom rail on top.

54. Loren Schweninger, "Black Citizenship and the Republican Party in Reconstruction Alabama," *Alabama Review*, 29 (April 1976), 90–91; Elizabeth S. Nathans, *Losing the Peace: Georgia Republicans and Reconstruction, 1865–1871* (Baton Rouge, 1968), 204; Hennessey, "Eutaw Riot," 48; Peek, "Lawlessness," 185. Election returns may be found in *Tribune Almanac*, 1871.

Their murder or exile inevitably had a demoralizing impact upon their communities.[55]

The violence of 1869–71 etched the Klan permanently in the folk memory of the black community. "What cullud person dat can't 'membahs dem, if he lived dat day?" an elderly Texas freedman asked six decades later. The issue of protection transcended all divisions within the black community, uniting rich and poor, free and freed, in calls for drastic governmental action to restore order. To blacks, indeed, the violence seemed an irrefutable denial of the white South's much-trumpeted claims to superior morality and higher civilization. "Pray tell me," asked Robert B. Elliott, "who is the barbarian here?"[56]

More immediately, violence underscored yet again the "abnormal" quality of Reconstruction politics. Before the war, Democrats and Whigs had combated fiercely throughout the South, but neither party, as Virginia Radical James Hunnicutt pointed out, advised its supporters "to drive out, to starve and to perish" its political opponents. Corrupt election procedures, political chicanery, and even extralegal attempts to oust the opposition party from office were hardly unknown in the North, but not pervasive political violence. "I never knew such things in Maine," commented an Alabama carpetbagger. "Republicans and Democrats were tolerated there." Democracy, it has been said, functions best when politics does not directly mirror deep social divisions, and each side can accept the victory of the other because both share many values and defeat does not imply "a fatal surrender of . . . vital interests." This was the situation in the North, where, an Alabama Republican observed, "it matters not who is elected." But too much was at stake in Reconstruction for "normal politics" to prevail. As one scalawag pointed out, while Northern political contests focused on "finances, individual capacity, and the like, our contest here is for life, for the right to earn our bread . . . for a decent and respectful consideration as human beings and members of society."[57]

Most of all, violence raised in its starkest form the question of legitimacy that haunted the Reconstruction state. Reconstruction, concluded Klan victim Dr. John Winsmith, ought to begin over again: "I consider a government which does not protect its citizens an utter failure." Indeed, as a former Confederate officer shrewdly observed, it was precisely the

55. KKK Hearings, South Carolina, 1161; *CG*, 42d Congress, 1st Session, 197; Vernon Burton, "Race and Reconstruction: Edgefield County, South Carolina," *JSocH*, 12 (Fall 1978), 32.

56. Rawick, ed., *American Slave*, Supplement 2, 9:3635; *CG*, 42d Congress, 1st Session, 391–92.

57. *The Debates and Proceedings of the Constitutional Convention of the State of Virginia* (Richmond, 1868), 72; George L. Woods Recollections, Bancroft Library, University of California, Berkeley; Carl Becker, *New Liberties for Old* (New Haven, 1941), 106–107; John H. Henry to William E. Chandler, July 15, 1872, B. F. Saffold to Edwin D. Morgan, October 5, 1872, William E. Chandler Papers, LC.

Klan's objective "to defy the reconstructed State Governments, to treat them with contempt, and show that they have no real existence." The effective exercise of power, of course, can command respect if not spontaneous loyalty. But only in a few instances had Republican governments found the will to exert this kind of force. Only through "decided action," wrote an Alabama scalawag, could "the state . . . protect its citizens and vindicate its own authority and *right to be.*" Yet while their opponents acted as if conducting a revolution, Republicans typically sought stability through conciliation.[58]

Soon after winning control of North Carolina's legislature, Democrats elected former Confederate Gov. Zebulon Vance to the U. S. Senate. "It seems that we are drifting," wrote one black Republican, "drifting back under the leadership of the slaveholders. Our former masters are fast taking the reins of government." Yet the disappointing results of 1870 did not signal Reconstruction's immediate demise. The Redemption of Georgia proved irreversible, but in Alabama and North Carolina, Republican control was succeeded by stalemate—a legislature divided between the two parties in the former, continued Republican control of the governorship in the latter. In both states, Republicans could claim that in peaceful elections, they remained the majority party. Indeed when North Carolina Democrats in 1871 proposed a constitutional convention to "restore the old order of things," voters decisively defeated the idea, fearing the abrogation of the homestead exemption and the resurrection of the oligarchic prewar structure of local government. Such resilience, however, required an end to violence. As Klan activity escalated after the 1870 elections, Southern Republicans once again turned to Washington for salvation.[59]

"Power from Without"

The President, forced to cope with Southern violence, had been elected on the slogan "Let Us Have Peace." While Ulysses S. Grant had clearly identified himself with Republican Reconstruction policies, no one could be certain what attitude toward the South would characterize his Administration. His inaugural address, "a string of platitudes that deserved praise only for its brevity," offered few clues. And when the new Cabinet was announced, friend and foe alike expressed astonishment. Unlike Lincoln, who had surrounded himself with the most powerful figures in

58. KKK Hearings, South Carolina, 625–28; J. W. Williams to William H. Smith, February 6, 1869, William B. Figures to Smith, July 26, 1869, Alabama Governor's Papers.

59. George M. Arnold to Charles Sumner, December 2, 1870, Sumner Papers; Sarah W. Wiggins, *The Scalawag in Alabama Politics, 1865–1881* (University, Ala., 1977), 66–67; Olsen, *Tourgée,* 129; Aubrey L. Brooks and Hugh T. Lefler, eds., *The Papers of Walter Clark* (Chapel Hill, 1948-50), 1:179; David Schenck Diary, August 4, 1871; Escott, *Many Excellent People,* 164.

his party, Grant, coming from a military background, looked upon Cabinet members as "staff officers," whose main qualification was that they enjoyed his confidence or had done him personal favors. Composed largely of men with little political influence and "abilities below mediocrity," Grant's Cabinet seemed oddly detached from the debates over Reconstruction. Initially, former supporters of Andrew Johnson outnumbered those identified with Congressional policy, and representatives of Southern Republicanism were excluded altogether. The new Secretary of the Navy, retired Philadelphia merchant Adolph Borie, was apparently chosen because he had entertained Grant at his estate on the Delaware River. One Pennsylvania newspaper congratulated the President for having discovered a man of whom no one in the state had ever heard.[60]

An even more unusual selection was Secretary of the Treasury Alexander T. Stewart, for as the nation's largest importer, he did more business with the department he had been chosen to head than any other citizen. Stewart had built the era's most famous department store in New York's Astor Place, where he pioneered modern retailing techniques and employed a small army of spies to apprehend shoplifters. Not until after his confirmation by the Senate did anyone notice a statute dating from the 1780s that barred persons engaged in trade from heading the Treasury. Despite the President's request, Congress declined to repeal this provision, a decision to which Stewart contributed by announcing his intention to appoint a complete new slate of officials at the New York Custom House. Fundamentally, however, the rebuff reflected Congressional displeasure at Grant's evident desire to stand above partisanship. "The Republican party was built up by its leaders," commented Tammany Hall's Peter B. Sweeney, who understood the nature of political obligation, "and they should have been allowed to administer the estate."[61]

Grant quickly learned the rules of party politics. He came increasingly to rely on leading members of Congress for advice and guidance, and brought Radical George S. Boutwell into the Cabinet as Stewart's replacement. Yet the fact that nearly all his initial appointees held moderate or conservative views on Reconstruction revealed a grasp of political realities. For Grant's election both confirmed the "finality" of Southern Reconstruction, and suggested that the issues arising from the slavery controversy had at last been settled. Even as he assumed office, what one Republican called "the vexed question of suffrage" appeared to have

60. Gillette, *Retreat from Reconstruction*, 20; Adam Badeau, *Grant in Peace* (Hartford, Conn., 1887), 163; William B. Hesseltine, *Ulysses S. Grant, Politician* (New York, 1935), 133–39; Henry C. Baird to Charles Sumner, March 8, 1869, Sumner Papers; Allan Nevins, *Hamilton Fish: The Inner History of the Grant Administration* (New York, 1936), 1:100, 108–109; Erwin S. Bradley, *The Triumph of Militant Republicanism* (Philadelphia, 1964), 303.

61. William Leach, *True Love and Perfect Union: The Feminist Reform of Sex and Society* (New York, 1980), 222–28; Henry E. Resseguie, "Federal Conflict of Interest: The A. T. Stewart Case," *New York History*, 47 (July 1966), 271–78; New York *Herald*, November 26, 1869.

been laid to rest. In February 1869, Congress approved the Fifteenth Amendment, prohibiting the federal and state governments from depriving any citizen of the vote on racial grounds. Republicans, declared Boutwell, who introduced the proposal in the House, could no longer "escape this issue as a Congress and as a party." A little over a year later, it became part of the Constitution.[62]

To Democrats, the Fifteenth Amendment seemed "the most revolutionary measure" ever to receive Congressional sanction, the "crowning" act of a Radical conspiracy to promote black equality and transform America from a confederation of states into a centralized nation. Yet while clothing black suffrage with constitutional sanction, the Amendment said nothing about the right to hold office and failed to make voting requirements "uniform throughout the land," as many Radicals desired. Henry Wilson, indeed, decried its "lame and halting" language, for the Amendment did not forbid literacy, property, and educational tests that, while nonracial, might effectively exclude the majority of blacks from the polls. (Under the Fourteenth Amendment, to be sure, states that adopted such restrictions would sacrifice part of their Congressional representation.) Democrats, however, were not entirely wrong to contend that the measure singled out "the colored race as its special wards and favorites." For unlike the Fourteenth Amendment, with its universalist language, the Fifteenth failed to expand the definition of citizenship for all Americans. Republicans' concern was to enfranchise blacks in the border states and prevent a retreat from Reconstruction in the South. (Some, in addition, expected a marginal benefit from the enfranchisement of Northern blacks.) Congress had rejected a far more sweeping proposal barring discrimination in suffrage and officeholding based on "race, color, nativity, property, education, or religious beliefs." Nor did the Amendment break decisively with the notion that the vote was a "privilege" that states could regulate as they saw fit. "The whole question of suffrage," declared one Senator, "subject to the restriction that there shall be no discrimination on account of race, is left as it now is."[63]

Thus, remarked Henry Adams, the Fifteenth Amendment was "more remarkable for what it does not than for what it does contain." The failure to guarantee blacks' right to hold office arose from fear that such a provision would jeopardize the prospects of ratification in the North. More significant, Congress rejected suffrage provisions that "covered the

62. Morton Keller, *Affairs of State: Public Life in Late Nineteenth Century America* (Cambridge, Mass., 1977), 260; Eugene H. Berwanger, *The West and Reconstruction* (Urbana, Ill., 1981), 173; *CG*, 40th Congress, 3d Session, 555.

63. Charles C. Tansill, *The Congressional Career of Thomas Francis Bayard 1869–1885* (Washington, D. C., 1946), 36, *AC*, 1869, 665; 1870, 335; *CG*, 40th Congress, 3d Session, 728, 979, 1625–26; William Gillette, *The Right to Vote* (Baltimore, 1969 ed.), 52–71; Michael L. Benedict, *A Compromise of Principle: Congressional Republicans and Reconstruction 1863–1869* (New York, 1974), 331–34.

white man" as well as the black. Southern Republicans, joined by many Northern Radicals, feared that a blanket guarantee of the right to vote would void the disenfranchisement of "rebels." Equally important, Northern states wished to retain their own suffrage qualifications. In the West, the Chinese could not vote; if the Fifteenth Amendment altered this situation, warned California's Republican Sen. Cornelius Cole, it would "kill our party as dead as a stone." Pennsylvania demanded the payment of state taxes to vote; Rhode Island required foreign-born citizens to own $134 worth of real estate; Massachusetts and Connecticut insisted upon literacy. Nativist prejudice and partisan advantage explained the survival of these restrictions, for the poor, illiterate, and foreign-born generally supported the Democracy. But whatever the reason, the Northern states during Reconstruction actually abridged the right to vote more extensively than the Southern. "Equality of the law," said John A. Bingham, who favored an amendment sweeping away all these limitations, "is the very rock of American institutions." Yet this principle Republicans were unwilling to accept at home.[64]

In a reversal of long-established political traditions, support for black voting rights now seemed less controversial than efforts to combat other forms of inequality. Thus, it was not a limited commitment to blacks' rights, but the desire to retain other inequalities, affecting whites, that produced a Fifteenth Amendment that opened the door to poll taxes, literacy tests, and property qualifications in the South. It is interesting that blacks who commented on the Amendment preferred language explicitly guaranteeing all male citizens the right to vote. Not for the first time in the nation's history, their commitment to the ideal of equal citizenship exceeded that of other Americans.[65]

And, of course, proponents of both a "strong" and "weak" Fifteenth Amendment ignored the claims of women. To feminists like Elizabeth Cady Stanton and Susan B. Anthony, the Amendment seemed another in a long train of "humiliations" Republicans had inflicted on their cause. Rejecting the idea that the Constitution should prohibit racial discrimination in voting while countenancing disclusions based on sex, they opposed ratification, thus dealing a final blow to the old abolitionist-feminist alliance, already mortally wounded by the dispute over the Fourteenth Amendment and the Kansas campaign of 1867. As her regard

64. Henry Adams, "The Session," *North American Review*, 108 (April 1869), 613; *CG*, 40th Congress, 3d Session, 725, 1427; Cornelius Cole to Olive Cole, June 25, 1870, Cornelius Cole Papers, University of California, Los Angeles; Simeon E. Baldwin, "Recent Changes in Our State Constitutions," *Journal of Social Science*, 10 (1879), 146–47; Patrick H. Conlan to Benjamin F. Butler, December 17, 1874, Benjamin F. Butler Papers, LC; Richard H. Abbott, *Cobbler in Congress: The Life of Henry Wilson, 1812–1875* (Lexington, Ky., 1972), 205; Charles Fairman, *Reconstruction and Reunion 1864–1888: Part One* (New York, 1971), 1266–67.

65. William D. Forten to Charles Sumner, February 1, 1869, Sumner Papers.

for her erstwhile allies waned, Stanton increasingly voiced racist and elitist arguments for rejecting the enfranchisement of black males while women of culture and wealth remained excluded. "Think of Patrick and Sambo and Hans and Ung Tung," she wrote, "who do not know the difference between a Monarchy and a Republic, who never read the Declaration of Independence . . . making laws for Lydia Maria Child, Lucretia Mott, or Fanny Kemble." In May 1869, the annual meeting of the Equal Rights Association, an organization devoted to both black and female suffrage, dissolved in acrimony. "The white women," commented black feminist Frances Harper, "all go for sex, letting race occupy a minor position." Frederick Douglass pleaded with the delegates to recognize the special character and immediate urgency of the problems confronting blacks, but his resolution endorsing the Amendment as a welcome, although partial, step toward universal suffrage went down to defeat. Out of the wreckage emerged rival national organizations: Stanton and Anthony's National Woman Suffrage Association, an embodiment of independent feminism, and the American Woman Suffrage Association, still linked to older reform traditions. Not until the 1890s would the two groups be reconciled.[66]

Most reformers, nonetheless, hailed the Fifteenth Amendment as a triumphant conclusion to four decades of agitation on behalf of the slave. Abolitionists and Radicals could be forgiven for savoring the triumph. A few years earlier, a Constitutional Amendment guaranteeing black suffrage would have been utterly inconceivable. As late as 1868, only eight Northern states allowed blacks to vote, and Republicans had shied away from the issue at their national convention. "Nothing in all history," exulted William Lloyd Garrison, equaled "this wonderful, quiet, sudden transformation of four millions of human beings from . . . the auction-block to the ballot-box." In March 1870 the American Anti-Slavery Society disbanded, its work, members believed, now complete. Yet amid the blaze of celebration, a few voices of caution could be heard. The Radical ideology of civil and political equality overseen by the national state had triumphed. But was this enough to give full meaning to black freedom? Wendell Phillips, while supporting the official dissolution of abolitionism, warned that the "long crusade" had not really ended, for as victims of "cruel prejudice" and "accumulated wrongs," the freedmen would continue to deserve the nation's "special sympathy." Even among reformers, however, this view came under attack. For their own benefit, said Thomas Wentworth Higginson, who had assisted John Brown and com-

66. Susan B. Anthony to Charles Sumner, February 8, 1870, Sumner Papers; Ellen C. DuBois, *Feminism and Suffrage: The Emergence of an Independent Women's Movement in America, 1848–1869* (Ithaca, N. Y., 1978), 163–97; Dorothy Sterling, ed., *We Are Your Sisters: Black Women in the Nineteenth Century* (New York, 1984), 414–15; Philip S. Foner, ed., *The Life and Writings of Frederick Douglass* (New York, 1950–55), 4:43.

manded black troops, the freedmen "should not continue to be kept wards of the nation."[67]

Such opinions acquired increasing prominence in Republican circles. Even as they inscribed black suffrage in the Constitution, many party spokesmen believed the troublesome "Negro question" had at last been removed from national politics. With their civil and political equality assured, blacks no longer possessed a claim upon the federal government; the competitive rules of the free market would determine their station in society. "The Fifteenth Amendment," declared influential Congressman James A. Garfield, "confers upon the African race the care of its own destiny. It places their fortunes in their own hands." "The negro is now a voter and a citizen," echoed an Illinois newspaper. "Let him hereafter take his chances in the battle of life."[68]

Like all great social and political transformations, the Second American Revolution had arrived at a period of consolidation. "Our *day*," Phillips warned Radicals and abolitionists, "is fast slipping away. Once let public thought float off from the great issue of the war, and it will take perhaps more than a generation to bring it back again." Increasingly, however, Northern public opinion turned to other questions. Letters received by Republican Congressmen, concerned mainly with Southern affairs until 1867, now concentrated on economic issues like currency, taxation, and internal improvements. Early in 1870, during debates on the readmission of Virginia, a Boston newspaper commented that while the issue caused "considerable excitement in Congress," it had produced "none whatsoever among the people." Even the Radical Boston *Commonwealth* announced that a new political era had dawned: "A party cannot be maintained on past traditions. It must move on to new conquests."[69]

Nor did the party's Southern wing inspire much fraternal concern in the North. Lacking strong support from the local business community, Southern Republicans were forced to rely on outside aid to finance newspapers and conduct campaigns. They found little forthcoming. Generally, the national party ignored Southern state organizations and local campaigns, and even when the Presidency was at stake, few experienced

67. Harold M. Hyman, ed., *The Radical Republicans and Reconstruction 1861–1870* (Indianapolis, 1967), 493; James M. McPherson, *The Struggle for Equality: Abolitionists and the Negro in the Civil War and Reconstruction* (Princeton, 1964), 428; *National Anti-Slavery Standard,* February 5, 1870.

68. James A. Garfield to Robert Folger, April 16, 1870, Letterbook, James A. Garfield Papers, LC; Roger D. Bridges, "Equality Deferred: Civil Rights for Illinois Blacks, 1865–1885," *JISHS,* 74 (Spring 1981), 92–95.

69. *National Anti-Slavery Standard,* November 13, 1869; Berwanger, *West and Reconstruction,* 241–47; Hans L. Trefousse, *The Radical Republicans: Lincoln's Vanguard for Racial Justice* (New York, 1969), 437–39, 464–65; Charles Blank, "The Waning of Radicalism: Massachusetts Republicans and Reconstruction Issues in the Early 1870's" (unpub. diss., Brandeis University, 1972), 15, 18; John L. Larson, *Bonds of Enterprise: John Murray Forbes and Western Development in America's Railway Age* (Cambridge, Mass., 1984), 112.

speakers ventured south and little money was dispatched by the Republican National Committee. In 1868, the committee "bankrupted ourselves to aid Pennsylvania and Indiana," leaving little for Southern needs. "I never knew the national committee to do anything sensible for the South," a prominent Georgia scalawag would later comment as Reconstruction drew to a close, "and would not expect anything so unlikely now."[70]

In Washington, Southern Republicans found themselves treated less as objects of special concern than as poor relations, their very presence slightly embarrassing. Northerners had grown "tired of this word reconstruction," one Senator told a South Carolina Congressman. Southerners received few important committee assignments and often found it difficult to obtain the floor to deliver speeches. Because Northerners controlled the key legislative posts and continued to view the South as a land of "rebels" undeserving of federal largesse, Southern Republicans could not obtain a fair share of spending. Of funds allocated for internal improvements by the Forty-First Congress, the entire South received only 15 percent, mostly, complained Mississippi Congressman George McKee, aid to railroads controlled by "northern capitalists." (McKee moved that his bill for improvements in the Yazoo River "be considered as lying within the boundaries of Wisconsin and Michigan," the only way, he contended, to ensure approval.) The handful of black Congressmen seemed particularly ineffective. Although they managed to obtain patronage appointments like postmaster, customs inspector, and internal revenue agent for black constituents, most of their proposals to assist corporations and institutions in their districts disappeared without a trace in Congressional committees. Hiram Revels, the first black Senator, secured the employment of black mechanics at the Navy Yard in Washington, but introduced only one successful bill—a measure relieving the political disabilities of former Confederate Brig. Gen. Arthur E. Reynolds.[71]

With national economic policies severely disadvantageous to their region, Southern Republicans were torn between loyalty to the party and the interests of their constituents. While repealing many wartime levies, Congress retained the excise tax on liquor (often the "chief marketable commodity" in isolated upcountry communities)—a cruel reward for the

70. Terry L. Seip, *The South Returns to Congress: Men, Economic Measures, and Intersectional Relationships, 1868-1879* (Baton Rouge, 1983), 87–92; Richard H. Abbott, *The Republican Party and the South, 1855-1877: The First Southern Strategy* (Chapel Hill, 1986), 194–95, 219–21; William E. Chandler to William Sprague, October 16, 1868, William Sprague Papers, Columbia University; Leon B. Richardson, *William E. Chandler, Republican* (New York, 1940), 113; James Atkins to James A. Garfield, February 1, 1877, Garfield Papers.

71. *CG*, 41st Congress, 1st Session, 437; 3d Session, 200; 42d Congress, 2d Session, 795; 3d Session, 220; Seip, *South Returns to Congress*, 115–19, 219–26; *CR*, 43d Congress, 1st Session, 88, 1443; John Hosmer and Joseph Fineman, "Black Congressmen in Reconstruction Historiography," *Phylon*, 39 (Summer 1978), 97–107; Julius E. Thompson, "Hiram R. Revels, 1827–1901: A Biography" (unpub. diss., Princeton University, 1973), 92, 98.

wartime loyalty of residents of the Southern mountains. The government's deflationary fiscal policy and a system of allocating national bank notes that reserved 70 percent to New England and the Middle States aggravated the economic problems of Southern farmers and entrepreneurs. Efforts to refund the tax on cotton collected immediately after the war, supported by Southerners of both parties, attracted few Northern votes. All in all, commented North Carolina Sen. John Pool, Southern Republicans had been severely injured by "the want of friendly legislation" in Washington.[72]

A handful of party leaders seemed almost ready to accept Reconstruction's eventual demise as inevitable. "This southern business must have its run," wrote William E. Chandler, chairman of the Republican National Committee. "We are bound to be overwhelmed by the new rebel combinations in every southern state." Few, however, as yet shared Chandler's pessimism. Indeed, in a party internally divided on economic questions, the issues of war, emancipation, and Reconstruction provided a "cohesive power," a raison d'être for Republicans' continued rule. Nearly all Republicans still believed Reconstruction must be defended. But it became increasingly clear that calls for further efforts to promote social change in the South stood little chance of success. A bill establishing a national land commission, introduced by black Congressman Benjamin S. Turner with a moving speech about the plight of former slaves whose labor had enriched the nation but who "have consumed less of [its] substance . . . than any other class of people," never even came to a vote. George W. Julian's efforts to prevent the renewal of prewar land grants to Southern railroads in the hope that such land, reassigned to the public domain, would be settled by freedmen, proved equally unsuccessful.[73]

So too, while hardly abandoning the broad conception of national power engendered by the Civil War, Republicans proved reluctant to promote the state's expansion into new realms. Measured by the magnitude of the federal budget, the size of the bureaucracy, and the number of bills brought before Congress, the scope of national authority far exceeded antebellum levels. "The system is outgrown. . . ." wrote Henry Adams in 1870. "New persons, new duties, new responsibilities, new burdens of every sort, are increasingly crowding upon the government." Yet even among Republicans, doubts about the activist state persisted, and of the numerous initiatives envisioning a continued expansion of federal authority, only a handful were enacted into law. One idea that did reach fruition was the Weather Bureau, established in 1870. But propos-

72. Horace Kephart, *Our Southern Highlanders* (New York, 1922 ed.), 450; Seip, *South Returns to Congress*, 5–6, 174–87; *CG*, 42d Congress, 2d Session, 2723, Appendix, 411; 3d Session, 891.

73. William E. Chandler to Benjamin F. Butler, August 10, 1869, Butler Papers; Willard H. Smith, *Schuyler Colfax: The Changing Fortunes of a Political Idol* (Indianapolis, 1952), 325; *CG*, 42d Congress, 2d Session, Appendix, 541; 41st Congress, 2d Session, 1762–63.

als for a national Bureau of Health, a federal railroad commission, and the nationalization of the telegraph industry died in Congress.[74]

Nor did Congress assume an active role in promoting public education, an idea strongly supported by Northern Radicals and by Southern Republicans alarmed at efforts of the first Redeemers to dismantle Reconstruction school systems. (James Rapier said he would like to see the words "United States" emblazoned on every schoolhouse in the land, with a "national series of textbooks" outlining the duties of citizens.) If a state failed to provide common schooling, George F. Hoar of Massachusetts proposed, the federal government should do so itself. Democrats, as usual, cried "centralization," one Congressman painting a horrifying portrait of national officials requiring all schools to adopt textbooks "published near Cambridge or Boston, with the Puritan confession of faith attached to the multiplication table." But many Republicans also viewed education as a state responsibility. Hoar's bill generated little support and never came to a vote. A weaker measure introduced by Mississippi carpetbagger Winfield Perce, offering to help finance education in states desiring federal assistance, passed the House early in 1872, but died in the Senate.[75]

While unwilling to push further with Reconstruction, however, party leaders remained committed to what had been accomplished. Alarmed by the overthrow of Republican rule in the border states and the apparent fragility of Reconstruction regimes farther south, and aware that the normal ebb and flow of the political tide would eventually bring Democrats to power, at least for a time, in most Southern states, Congress searched for additional ways to ensure that states could not in the future erode blacks' civil and political rights. When Virginia, Mississippi, and Texas finally approved their Reconstruction constitutions, Congress early in 1870 added new requirements to the process of restoration, barring all future amendments that abridged the right to vote and hold office or denied citizens access to an education.

The imposition of "fundamental conditions" that went beyond the Reconstruction Acts of 1867 seemed a triumph for Radicalism. Charles Sumner, who pressed vigorously for the new requirements, found himself

74. Margaret S. Thompson, The "Spider Web": Congress and Lobbying in the Age of Grant (Ithaca, N. Y., 1985), 45–48; Alan Trachtenberg, The Incorporation of America: Culture and Society in the Gilded Age (New York, 1982), 164–65; Keller, Affairs of State, 101, 107; Harold M. Hyman, A More Perfect Union: The Impact of the Civil War and Reconstruction on the Constitution (New York, 1973), 382, 404.

75. Philip S. Foner and Ronald L. Lewis, The Black Worker: A Documentary History From Colonial Times to the Present (Philadelphia, 1978–84), 2:136; CG, 41st Congress, 3d Session, 809, 1244, 1371, Appendix, 101; 42d Congress, 2d Session, 862; CR, 44th Congress, 1st Session, Appendix, 318; Richard E. Welch, Jr., George Frisbie Hoar and the Half-Breed Republicans (Cambridge, Mass., 1971), 21–25; Kenneth R. Johnson, "Legrand Winfield Perce: A Mississippi Carpetbagger and the Fight for Federal Aid to Education," JMH, 34 (November 1972), 340–53.

in the unaccustomed position of enjoying majority support within his party. Yet, as Missouri's Radical Sen. Charles Drake observed, the debate marked an "extraordinary" change in Congressional politics. In the final votes on previous Reconstruction measures, each party had arrayed itself "in solid column." Now, Republican ranks were sundered—nineteen Senators and thirty-seven Representatives opposed the conditions appended to the readmission of Virginia, among them such rising powers as Roscoe Conkling, John Sherman, and James A. Garfield. Joining them in dissent was Lyman Trumbull, who insisted that Congress could no more dictate the contents of Southern state constitutions than interfere in the affairs of the North. Not content with presenting this argument, Trumbull unleashed a bitter attack on Sumner, charging him with having always obstructed "practical efficient measures" of Reconstruction in favor of his own "idiosyncratic, . . . impractical measures." Nevada's Sen. William M. Stewart joined the assault. Commented Drake, "This is one of the days of the Senate's degradation before the country."[76]

Behind this acrimony lay not only political and ideological divisions, but a sense of frustration that increasingly permeated Republican ranks. The debates, commented Sen. Willard Warner of Alabama, revealed "a want of confidence, a lack of faith in the reconstruction of these States, in the hold that Republican ideas have gained upon them." Many Republicans shared a sense of foreboding about Reconstruction, but were unable to see what more could be done. The imposition of "fundamental conditions" (which many Congressmen believed could never be enforced) offered an alternative to more far-reaching attempts to redefine Reconstruction. The bill admitting Virginia, wrote a New York Representative, "was very unsatisfactory to many members but it was passed by a strong vote as the only available mode of bringing the southern states to a sense of their real dependence upon Congress."[77]

Even as they sent this message to the South, however, Congress rejected the idea of intervening in state affairs to rescue Reconstruction regimes from unfavorable election results. The power to require states to adopt a republican form of government, commented Senator Stewart, conveyed "no right to legislate to make them belong to the Republican party." After the fall of the Brownlow government, Tennessee Republicans clamored for a second period of military rule in their state. They found Congress "heartily sick of this reconstruction legislation" and "much averse to interfering in any way." When Georgia's legislature in 1868 expelled its black members, Congress, after waiting more than a

76. David Donald, *Charles Sumner and the Rights of Man* (New York, 1970), 421–26; Edward McPherson, *The Political History of the United States of America During the Period of Reconstruction* (Washington, D. C., 1875), 573–78; *CG*, 41st Congress, 2d Session, 421–22, 541, 1174, 1183, 1361.

77. *CG*, 41st Congress, 2d Session, 1359; Hyman, *More Perfect Union*, 520–21; Noah Davis to Isaac Sherman, January 25, 1870, Isaac Sherman Papers, Private Collection.

year, ordered the seating of the ousted lawmakers. But in a series of extremely close votes, it rejected Governor Bullock's request to prolong the legislature's term beyond the fall 1870 elections. Trumbull was particularly critical of this effort to "usurp unauthorized power." Such a move, he warned, would hang "like a millstone around the neck of the party" in the eyes of Northern voters. The result, as a petition by black legislators had predicted, was a Democratic sweep, "deliver[ing] us, bound hand and foot, into the hands of our most bitter and relentless enemies."[78]

Thus, Congress in 1869 and 1870 stood poised between retreating from Reconstruction and pressing further with its Southern policy. Visiting Washington soon after Grant's inauguration, Georgia scalawag Amos T. Akerman observed that while the postwar amendments had made the government "more national in theory," he had observed "even among Republicans, a hesitation to exercise the powers to redress wrongs in the states." Akerman was alarmed, for he believed that "unless the people become used to the exercise of these powers now, while the national spirit is still warm with the glow of the late war, . . . the 'state rights' spirit may grow troublesome again." Yet just as the intransigence of Andrew Johnson and his Southern governments had helped to radicalize Congress in 1866, the Ku Klux Klan's campaign of terror overcame Republicans' growing reluctance to intervene in Southern affairs. "If that is the only alternative," declared Sherman, "I am willing to . . . again appeal to the power of the nation to crush, as we once before have done, this organized civil war."[79]

Adopted in 1870 and 1871, a series of Enforcement Acts embodied the Congressional response to violence. The first, "a criminal code upon the subject of elections," forbade state officials to discriminate among voters on the basis of race and authorized the President to appoint election supervisors with the power to bring to federal court cases of election fraud, the bribery or intimidation of voters, and conspiracies to prevent citizens from exercising their constitutional rights. A second act, designed with Democratic practices in Northern cities more fully in mind than conditions in the South, strengthened enforcement powers in large cities. But as violence persisted, Congress enacted a far more sweeping measure—the Ku Klux Klan Act of April 1871. This for the first time designated certain crimes committed by individuals as offenses punishable under federal law. Conspiracies to deprive citizens of the right to vote,

78. A. J. Ricks to Oliver P. Temple, January 15, 1870, Oliver P. Temple Papers, University of Tennessee; Nathans, *Losing the Peace*, 164–71; Donald, *Sumner*, 449–50; *CG*, 41st Congress, 2d Session, 326, 1856, 2677, 2820–21, 4797, 5378, 5621, Appendix, 291–93; Mark M. Krug, *Lyman Trumbull: Conservative Radical* (New York, 1965), 280–94; Lyman Trumbull to Henry P. Farrow, August 26, 1870, Henry P. Farrow Papers, UGa.

79. Amos T. Akerman to Charles Sumner, April 2, 1869, Sumner Papers; *CG*, 42d Congress, 1st Session, 820.

hold office, serve on juries, and enjoy the equal protection of the laws, could now, if states failed to act effectively against them, be prosecuted by federal district attorneys, and even lead to military intervention and the suspension of the writ of habeas corpus.[80]

The Ku Klux Klan Act pushed Republicans to the outer limits of constitutional change. The Civil Rights Act and postwar amendments, designed largely to protect freedmen against hostile state actions, had left private criminal acts within the purview of local law enforcement officials. Now, in making violence infringing civil and political rights a federal crime and calling upon the full authority of the national state to suppress it, Congress "moved tentatively into modern times." "These are momentous changes. . . ." observed *The Nation.* "They not only increase the power of the central government, but they arm it with jurisdiction over a class of cases of which it has never hitherto had, and never pretended to have, any jurisdiction whatever." Bewailing this "crowning act of centralization and consolidation," Democrats warned that if the national government could punish crime within the states, local self-government would perish. "The Radical laws to enforce the 15th or 14th Amendment," declared California Sen. Eugene Casserly, "are unconstitutional clearly so far as they deal with *individuals* and not with *states.* . . . This is the rock which is to wreck these scoundrel bills." Democrats never accepted the legitimacy of the Enforcement Acts. Over twenty years later, when the party for the first time since the 1850s controlled both the White House and Congress, it repealed nearly all of their provisions.[81]

In defending the Enforcement Acts, Republicans again appealed to the broad conception of national authority spawned by the Civil War and embodied in the postwar amendments. "The Constitution is not now what it was. . . ." insisted Job Stevenson of Ohio. "The old, vexed question whether this was really a national Union or merely a disjoined confederation . . . has been settled forever." "If the Federal Government," asked Benjamin F. Butler, "cannot pass laws to protect the rights, liberty, and lives of citizens of the United States in the States, why were guarantees of those fundamental rights put in the Constitution at all?" Understandably, the most sweeping defense of federal power came from the black Congressmen, whose constituents had deluged Congress with pleas for forceful action to "enable us to exercise the rights of citizens." Democrats stigmatized the new laws as Force Acts, dire threats to individual freedom. But one person's force may spell another's liberty. With

80. *CG*, 41st Congress, 2d Session, 3656; Hyman, *More Perfect Union,* 526–30; Trelease, *White Terror,* 385–91.

81. Herman Belz, *Emancipation and Equal Rights* (New York, 1978), 127–28; Hyman, *More Perfect Union,* 530; *Nation,* March 23, 1871; *CG*, 41st Congress, 3d Session, 1271; 42d Congress, 1st Session, Appendix, 223; Eugene Casserly to Manton Marble, March 12, 1871, Manton Marble Papers, LC. (This "great point," the immodest Senator went on to say, "is *my* discovery.")

terror rampant in large parts of the South, black Congressmen expressed impatience with "legal technicalities," and evinced little interest in abstract debates about the Constitution. "I desire that so broad and liberal a construction be placed upon its provisions," declared Joseph Rainey, "as will insure protection to the humblest citizen. Tell me nothing of a constitution which fails to shelter beneath its rightful power the people of a country."[82]

"It was not often that the line of party was so strictly drawn," commented Blaine on the acrimonious partisan division provoked by the Enforcement Acts. Yet a small but articulate group of Republicans recoiled from this latest expansion of federal authority. Most outspoken was Lyman Trumbull, who echoed Democratic charges that the Ku Klux Klan Act would revolutionize the federal system. The states, he insisted, remained "the depositories of the rights of the individual"—if Congress could enact a "general criminal code" and punish offenses like assault and murder, "what is the need of the State governments?" Trumbull's views were seconded by Carl Schurz, who considered the Ku Klux Klan Act unwarranted by the Constitution; privately, he characterized it as "insane."[83]

These criticisms portended a major breach over Southern policy, which would culminate in the Liberal Republican movement of 1872. For the moment, however, the reality of Southern violence reinvigorated public concern for the fate of Reconstruction and isolated those, like Trumbull and Schurz, who advocated a laissez-faire approach to the South. "All the freedom you have has been through the action of the Federal Government," declared Henry Wilson, and most Republicans agreed that citizens' rights remained in greater danger of abridgment from state officials and criminals than from vigorous exercise of national authority. Many, however, were troubled by the provision authorizing the suspension of habeas corpus—that is, allowing suspects to be arrested and held without charge. The clause survived only because of the nearly unanimous support of Southern Republicans, and was subsequently amended so as to lapse the following year. The party's mainstream opinion was probably voiced by Massachusetts Congressman Henry L. Dawes, who admitted the "medicine" was "extreme," but asked whether any alternative existed: "Am I to abandon the attempt to secure to the American citizen these rights, given to him by the Constitution?"[84]

82. CG, 42d Congress, 1st Session, 394–95, Appendix, 299; 2d Session, 448, 808–10, 1987; Herbert Aptheker, ed., A Documentary History of the Negro People in the United States (New York, 1969 ed.), 594–95.

83. James G. Blaine, Twenty Years of Congress (Norwich, Conn., 1884), 2:466; CG, 42d Congress, 1st Session, 575–79; Carl Schurz to E. L. Godkin, March 31, 1871, E. L. Godkin Papers, HU.

84. Joseph Logsdon, Horace White: Nineteenth Century Liberal (Westport, Conn., 1971), 192–94; CG, 42d Congress, 1st Session, 579, 519, 477.

Although, under the Enforcement Acts, aggrieved individuals could file suit against their assailants, the major burden of suppressing violence now fell to the federal government. Overseeing the process were two representatives of Southern Republicanism: Amos T. Akerman, a New Hampshire-born lawyer long resident in Georgia, who had assumed the Attorney Generalship in mid-1870, and Solicitor General Benjamin H. Bristow, a Union Army veteran from Kentucky. Both were committed to vigorous enforcement of the new laws. Although a moderate on many Reconstruction issues (he had opposed the debtor relief provisions of Georgia's 1868 constitution and favored property and educational qualifications for voting), Akerman had become convinced that suppression of the Klan required "extraordinary means." "These combinations," he wrote in 1871, "amount to war, and cannot be effectually crushed on any other theory." As United States attorney for Kentucky, Bristow had distinguished himself for energetic efforts to secure blacks' civil rights in a state that still barred them from testifying in court. Under Akerman and Bristow stood the recently established Department of Justice and an array of federal marshals and district attorneys. Then there was the army, although its limited numbers (excluding Texas, with its continuing Indian wars, fewer than 6,000 troops still garrisoned the South) and the reluctance of military commanders to take on the role of law enforcement made its employment desirable only as a last resort.[85]

Despite a limited budget, difficulties in securing evidence, the reluctance of some victims to testify, and the fact that defendants—assisted by legal defense funds, including one headed by Wade Hampton—employed talented and experienced lawyers to oppose the overworked district attorneys, the prosecution of accused Klansmen began in earnest in 1871. Hundreds of men were indicted in North Carolina, where federal troops helped apprehend suspects. Many ended up in prison, including Rutherford County Klan leader Randolph Shotwell, who served two years in an Albany, New York, penitentiary. United States attorney G. Wiley Wells secured nearly 700 indictments in Mississippi, although most escaped with suspended sentences and the threat of sterner punishment if violence resumed. Only in South Carolina did the military provisions of the Enforcement Acts come into play. At Akerman's urging, Grant in October 1871 proclaimed a "condition of lawlessness" in nine upcountry counties and suspended the writ of habeas corpus. Federal troops occu-

85. William S. McFeely, "Amos T. Akerman: The Lawyer and Racial Justice," in Kousser and McPherson, eds., *Region, Race, and Reconstruction*, 396–404; *Journal of the Proceedings of the Constitutional Convention of the People of Georgia* (Augusta, 1868), 132–35, 266; Amos T. Akerman to Foster Blodgett, November 8, 1871, Akerman to B. Silliman, November 9, 1871, Amos T. Akerman Papers, University of Virginia; Ross A. Webb, "Benjamin H. Bristow: Civil Rights Champion, 1866–1872," *CWH*, 15 (March 1969), 39–53; Joseph G. Dawson III, *Army Generals and Reconstruction: Louisiana, 1862–1877* (Baton Rouge, 1982), 150–53.

pied the region, making hundreds of arrests, and perhaps 2,000 Klansmen fled the state. Personally directing the government's legal strategy, the Attorney General allowed those who confessed and identified the organization's leaders to escape without punishment, while bringing a few dozen of the worst offenders to trial before predominantly black juries. Most of those indicted eventually pleaded guilty and received prison sentences.[86]

The legal offensive of 1871, culminating in the use of troops to root out the South Carolina Klan, represented a dramatic departure for the Grant Administration, which in its first two years had launched few initiatives in Southern policy. Much of the credit belongs to Akerman. And the revelations that unfolded in the course of the trials deeply affected the Attorney General. "Though rejoiced at the suppression of KuKluxery even in one neighborhood," he wrote, "I feel greatly saddened by this business. It has revealed a perversion of moral sentiment among the Southern whites which bodes ill to that part of the country for this generation." Akerman embarked upon a personal crusade to make known the full reality of Southern violence, lecturing in the North and providing detailed accounts of Klan atrocities at Cabinet meetings. Not all of Grant's advisers shared his preoccupation. The Attorney General, complained Secretary of State Fish, had the Klan "on the brain. . . . It has got to be a bore to listen twice a week to this thing." In December 1871, in the midst of the trials, Akerman was suddenly dismissed, apparently because his rulings concerning certain railroad land grants had displeased influential Republicans connected to Collis P. Huntington and Jay Gould. But his successor, former Oregon Sen. George H. Williams, continued the prosecutions.[87]

Judged by the percentage of Klansmen actually indicted and convicted, the fruits of "enforcement" seem small indeed, a few hundred men among thousands guilty of heinous crimes. But in terms of its larger purposes—restoring order, reinvigorating the morale of Southern Republicans, and enabling blacks to exercise their rights as citizens—the policy proved a success. "The law on the side of freedom," Frederick Douglass would later remark, "is of great advantage only where there is power to make that law respected." By 1872, the federal government's evident willingness to bring its legal and coercive authority to bear had

86. Robert J. Kaczorowski, *The Politics of Judicial Interpretation: The Federal Courts, Department of Justice and Civil Rights, 1866–1876* (New York, 1985), 57–61, 90; Kermit L. Hall, "Political Power and Constitutional Legitimacy: The South Carolina Ku Klux Klan Trials, 1871–1872," *Emory Law Journal,* 33 (Fall 1984), 936, 938, 941; Clarence W. Griffin, *History of Old Tryon and Rutherford Counties, North Carolina 1730–1936* (Asheville, 1937), 326–32; Trelease, *White Terror,* 345–48, 399–409; Harris, *Day of the Carpetbagger,* 399–400; McFeely, "Akerman," 407–408; Amos T. Akerman to B. Silliman, November 9, 1871, Akerman Papers.

87. Gillette, *Retreat from Reconstruction,* 103, 166–68; Amos T. Akerman to Alfred H. Terry, November 18, 1871, Akerman to Lewis W. Merrill, November 9, 1871, Akerman Papers; Nevins, *Fish,* 2:591; McFeely, "Akerman," 409–11.

broken the Klan's back and produced a dramatic decline in violence throughout the South.[88]

So ended the Reconstruction career of the Ku Klux Klan, certainly one of the most ignoble chapters in all of American history. National power had achieved what most Southern governments had been unable, and Southern white public opinion unwilling, to accomplish: acquiescence in the rule of law. Yet the need for outside intervention was a humiliating confession of weakness for the Reconstruction regimes. "The Enforcement Act," wrote a Mississippi Republican, "has a potency derived alone from its source; no such law could be enforced by state authority, the local power being too weak." The outcome further reinforced Southern Republicans' tendency to look to Washington for protection. Only *"steady, unswerving power from without,"* believed carpetbagger Albert T. Morgan, could ensure the permanence of Reconstruction.[89] Whether in the future such power would be forthcoming depended not only on events in the South, but on how the North responded to its own experience of Reconstruction.

88. Gillette, *Retreat from Reconstruction,* 42–45; Kaczorowski, *Politics of Judicial Interpretation,* 79; James W. Garner, *Reconstruction in Mississippi* (New York, 1901), 343–44; Trelease, *White Terror,* 287, 348, 414–15; [Frederick Douglass] *Life and Times of Frederick Douglass* (New York, 1962 ed.), 377.

89. Caleb Lindsay to Adelbert Ames, May 13, 1872, Ames Family Papers; A. T. Morgan, *Yazoo: or, On the Picket Line of Freedom in the South* (Washington, D.C., 1884), 323.

CHAPTER 10

The Reconstruction of the North

L IKE the South, the victorious North experienced a social transformation after the Civil War. And if the North's reconstruction proved less
revolutionary than the South's, the process of change catalyzed by the
war continued to accelerate in peacetime. Throughout the western world,
the classic Age of Capital, a period of unprecedented economic expansion presided over by a triumphant industrial bourgeoisie, had entered
its heady final years. In the United States, evidence abounded of the
consolidation of the capitalist economy: in the South the end of slavery,
the expansion of market relations among white farmers, and the rise of
upcountry commercial centers; in the North a manufacturing boom, the
spread of new forms of industrial organization, the completion of the
railroad network, and the opening of the Trans-Mississippi West to mining, lumbering, ranching, and commercial farming. The North's social
structure, like the South's, was altered in these years. An increasingly
powerful class of industrialists and railroad entrepreneurs took its place
alongside the older commercial elite; the number of professionals and
white-collar workers grew dramatically, and the wage-earner irrevocably
supplanted the independent craftsman as the typical member of the working class. Returning from England in 1868, Charles Francis Adams and
his family were stunned by the triumph of the new economy of coal, iron,
and steam over a world centered on agriculture and artisanship: "Had
they been Tyrian traders of the year B.C. 1000, landing from a galley fresh
from Gibraltar, they could hardly have been stranger on the shore of a
world, so changed from what it had been ten years before."[1] As in the
South, moreover, economic and social change produced new demands
on the state and altered the terms of political debate and modes of party
organization. From labor relations to party politics to attitudes toward
the postwar South, no aspect of life remained unaffected by the reconstruction of the North.

1. Eric Hobsbawm, *The Age of Capital 1848–1875* (London, 1975); [Henry Adams] *The
Education of Henry Adams* (Boston, 1907), 237–38.

The North and the Age of Capital

The war-inspired boom in industrial profits and investment slackened only momentarily with the coming of peace, and manufacturing output quickly resumed its relentless upward course. By 1873, the nation's industrial production stood 75 percent above its 1865 level, a figure all the more remarkable in view of the South's economic stagnation. In the same eight years, 3 million immigrants entered the country, nearly all destined for the North and West, their labor fueling the rapid growth of metropolitan centers like New York and smaller industrial cities from Paterson to Milwaukee. Outside New England, the typical manufacturing establishment remained the small workshop owned by an individual or family rather than the large-scale factory organized on a corporate basis. But the 1860s did witness a notable growth of the industrial labor force and a shift within it toward iron and steel, machine tools, and other highly mechanized branches of heavy industry. By 1873, with the United States second only to Britain in manufacturing production and the number of farmers outstripped by nonagricultural workers, the North had irrevocably entered the industrial age.[2]

If the cotton mill symbolized the early industrial revolution, the railroad epitomized the maturing capitalist order. Between 1865 and 1873, 35,000 miles of track were laid, a figure that exceeded the entire rail network of 1860. Railroad construction helped pull the economy out of the downturn of 1865, and inspired a boom in coal and pig iron production (both of which more than doubled in the postwar decade), and the rapid spread of the new Bessemer process for making steel. Railroads opened vast new areas to commercial farming and helped cities like Chicago and Kansas City extend their economic sway over agricultural hinterlands. Their voracious appetite for funds absorbed much of the nation's investment capital (leaving little available for the credit-hungry South), contributed to the development of banking, and facilitated the further concentration of the nation's capital market in Wall Street.[3]

As significant as the completion of the nation's transportation system,

2. Allan Nevins, *The Emergence of Modern America 1865-1878* (New York, 1927), 31–33; Ralph Andreano, ed., *The Economic Impact of the American Civil War*, 2d ed. (Cambridge, Mass., 1967), 226; U. S. Bureau of the Census, *Historical Statistics of the United States, Colonial Times to 1870* (Washington, D. C., 1975), 106; David Montgomery, *Beyond Equality: Labor and the Radical Republicans 1862-1872* (New York, 1967), 4–13; David T. Gilchrist and W. David Lewis, eds., *Economic Change in the Civil War Era* (Greenville, Del., 1965), 160; Clarence H. Danhof, *Change in Agriculture: The Northern United States, 1820-1870* (Cambridge, Mass., 1969), 10.

3. U. S. Bureau of the Census, *Historical Statistics*, 731; Rendigs Fels, *American Business Cycles 1865-1897* (Chapel Hill, 1959), 92–98; Stephen Salisbury, "The Effect of the Civil War on American Industrial Development," in Andreano, ed., *Economic Impact of the Civil War*, 180–87; Alfred D. Chandler, *The Visible Hand: The Managerial Revolution in American Business* (Cambridge, Mass., 1977), 258–67; Dolores Greenberg, *Financiers and Railroads: A Study of Morton, Bliss and Company* (Newark, Del., 1980), 13–45.

however, was its coordination and consolidation, as large trunk lines, mostly controlled by Eastern capitalists, increasingly absorbed smaller companies. Under the aggressive leadership of Thomas A. Scott, the Pennsylvania Railroad, the nation's largest corporation, forged an economic empire that stretched across the continent and included coal mines and oceangoing steamships. Only two nations in the world, Britain and France, possessed more track than the Pennsylvania's 6,000 miles. The "first modern business enterprises," the trunk railroads far outstripped the largest manufacturing concerns in capitalization, operating expenses, and number of employees, and, with an army of professional managers overseeing their far-flung activities, pioneered new forms of labor control and bureaucratic management. For the entrepreneurs who came to symbolize economic enterprise in the postwar years—Collis P. Huntington, James J. Hill, Jay Gould, and the like—railroads created opportunities for unheard-of wealth (even though many roads never paid a dividend and eventually went bankrupt). For other Americans, the railroad, like the war itself, acted as a nationalizing force, sharply reducing transportation costs and establishing a vast national market that laid the foundation for the explosive growth that, within a quarter century, would make the United States the world's preeminent industrial nation.[4]

Nowhere did capitalism penetrate more rapidly or dramatically than in the Trans-Mississippi West, whose "vast, trackless spaces" (as Walt Whitman called them) were now absorbed into the expanding economy. At the close of the Civil War, the frontier of settlement did not extend far beyond the Mississippi River. To the west lay millions of acres of fertile and mineral-rich land roamed by immense buffalo herds that provided food, clothing, and shelter for a population of perhaps a quarter of a million Indians, many of them members of Eastern tribes forced inland two centuries before from the East Coast, and moved again earlier in the nineteenth century to open the Old Northwest and Southwest to white farmers and planters. Although Indian policy provoked much controversy during the Grant years, nearly all military and civilian officials shared a common assumption: that the federal government should persuade or coerce the Plains Indians to exchange their religion, communal form of property, and "nomadic" way of life for Christian worship and settled agriculture on federally supervised reservations. In a word, they should surrender most of their land and cease to be Indians. "If the Indians had tried to make the whites live like them," said Big Eagle of the Santee Sioux, "the whites would have resisted, and it was the same way with many Indians." And despite Grant's much-publicized Peace Policy,

4. Nevins, *Emergence*, 63–64; Chandler, *Visible Hand*, 79–80, 87–88, 95–115, 123–24, 151–55, 289–90; Thomas K. McCraw, *Prophets of Regulation* (Cambridge, Mass., 1984), 4; Thomas C. Cochran and William Miller, *The Age of Enterprise: A Social History of Industrial America* (New York, 1942), 131–34.

conflict continued to rage between the army and various tribes, with Civil War generals employing methods (like the destruction of the buffalo and, with it, the infrastructure of the Indian economy) not unlike those that had produced victory over the Confederacy.[5]

Like efforts to "uplift" the freedmen, the attempt to "civilize" the Indians by teaching them Christianity and the ways of free labor did not lack for noble motives. In practice, however, whenever their rights conflicted with the land hunger of settlers, ranchers, mining companies, and railroads, the Indians came out second best. In 1871, Congress abrogated the treaty system that dealt with Indians as independent nations—a step strongly supported by railroad corporations, which found tribal sovereignty an obstacle to construction, and by Radical Republicans, to whom the traditional system seemed a form of local autonomy incompatible with the uniform nationality born of the Civil War. By the time Grant left office, railroads traversed the Great Plains, farmers and cattlemen had replaced the buffalo, most Indians had been concentrated on reservations, and although warfare did not end until the massacre of the Sioux at Wounded Knee in 1890, the world of the Plains Indians had come to an end.[6]

The Indians' subjugation formed an essential prelude to the economic exploitation of the West. Even as the Indian wars continued, settlers poured into the Trans-Mississippi region; more land came into cultivation in the thirty years after the Civil War than in the previous two and a half centuries of American history. While the South struggled with the problems of recovery, a new agricultural empire arose on the Middle Border (Minnesota, the Dakotas, Nebraska, and Kansas), whose population grew from 300,000 in 1860 to well over 2 million twenty years later. Inexorably, the railroad pulled the West, the fabled home of the independent yeoman, into the capitalist orbit. Already, Californians noted a marked trend toward the concentration of landownership into large tracts tilled by indentured Chinese and migrant Mexicans—a troubling tendency in a free labor society. "California is not a country of farms, but . . . of plantations and estates," wrote the young journalist Henry George in 1871, urging that the government employ taxation to combat "land monopoly" and "give all men an equal chance" to achieve republican

5. Alan Trachtenberg, *The Incorporation of America: Culture and Society in the Gilded Age* (New York, 1982), 19; Nevins, *Emergence*, 101–14; Richard R. Levine, "Indian Fighters and Indian Reformers: Grant's Indian Peace Policy and the Conservative Consensus," *CWH*, 31 (December 1985), 329–50; Francis P. Prucha, *The Great Father: The United States Government and the American Indians* (Lincoln, Neb., 1984), 1:481–520, 534–35; Dee Brown, *Bury My Heart at Wounded Knee* (New York, 1970), 38; Robert G. Athearn, *William Tecumseh Sherman and the Settlement of the West* (Norman, Okla., 1956), 230–31.

6. Richard Slotkin, *The Fatal Environment: The Myth of the Frontier in the Age of Industrialization 1800–1890* (New York, 1985), 311–19; Prucha, *Great Father*, 1:529–31; H. Craig Miner, *The Corporation and the Indian* (Columbia, S.C., 1976), 32, 76; *CG*, 41st Congress, 2d Session, 1258; Edward E. Dale, *The Range Cattle Industry* (Norman, Okla., 1930), 36–46.

independence. But elsewhere in the West, although a few "bonanza farms" made their appearance, these years marked the last great heyday of the family farmer. And while the census reported numerous tenants, most, unlike their Southern counterparts, still entertained realistic prospects of working their way up the agricultural ladder to farm ownership. Nonetheless, in both the Old and New Wests, self-sufficient agriculture waned as farmers became increasingly subject to the vagaries of the world market and dependent on credit for mortgages, fertilizer, and machinery.[7]

The West, however, was more than an agrarian empire. Around the Great Lakes and in the Ohio Valley arose new mining and industrial complexes geared to processing the farmer's expanding output and meeting the railroad's enormous demand for machinery, coal, and iron products. The most startling growth occurred in Chicago, which, thanks to the railroad, dominated the region's grain, meat, and lumber trades and emerged as a major industrial center, the home of iron and steel mills, agricultural machinery factories, and meat-packing plants. Between 1860 and 1870, the city's population nearly tripled, and its expansion propelled Illinois, fifteenth among manufacturing states in 1850, to sixth place two decades later. Throughout the West, highly capitalized corporate enterprises appeared with remarkable rapidity. Although Chicago lagged far behind Philadelphia and New York, the nation's leading manufacturing centers, in investment and output, a larger proportion of its labor force worked for firms with fifty or more employees. The Wisconsin and Michigan lumber industry, dominated by small-scale producers in 1860, came under the control of Eastern-financed corporations that engrossed immense tracts of forest, constructed sawmills whose output far exceeded anything known before the war, and employed armies of loggers. Flour milling, centered in Milwaukee and Minneapolis, experienced a similar trend toward concentrated ownership and large-scale production. Western mining, whether Michigan iron ore and copper or gold and silver in California, Nevada, and Colorado, increasingly fell under the sway of corporations that mobilized Eastern and European capital to introduce the most advanced technology and replace the independent prospector, working a surface mine with his pick and shovel, with wage-earning deep-shaft miners.[8]

7. Nevins, *Emergence*, 154; U. S. Bureau of the Census, *Historical Statistics*, 457; Paul W. Gates, "Public Land Disposal in California," *AgH*, 49 (January 1975), 158–78; Henry George, Jr., ed., *The Complete Works of Henry George*, (New York, 1911), 8:68–69, 86–88; Paul W. Gates, "Frontier Landlords and Pioneer Tenants," *JISHS*, 38 (June 1945), 159–60; Donald L. Winters, *Farmers Without Farms: Agricultural Tenancy in Nineteenth-Century Iowa* (Westport, Conn., 1978).

8. Frederick C. Jaher, *The Urban Establishment* (Urbana, Ill., 1982), 472–86; John H. Keiser, *Building for the Centuries: Illinois, 1865 to 1898* (Urbana, Ill., 1977), xiii, 182; John B. Jentz and Richard Schneirov, *The Origins of Chicago's Industrial Working Class* (forthcoming), Chapter 3; Lawrence Costello, "The New York City Labor Movement, 1861–1873" (unpub. diss.,

"The locomotive," observed *The Nation* in 1873, "is coming in contact with the framework of our institutions." And as in the South, the North's political structure was ill-equipped to deal with the demands upon the resources of the state, and the new opportunities for corruption, created by the capitalist economy's rapid development. "The galleries and lobbies of every Legislature. . . ." observed an Illinois Republican leader, "are thronged with men seeking to procure an advantage" for one corporation or another. Railroad lobbyists in 1867 pushed through the Massachusetts legislature a $3 million grant to the Boston, Hartford, & Erie Railroad, a scheme, charged the Springfield *Republican*, "to make money by . . . getting it voted out of the Massachusetts treasury into [the directors'] pockets." Despite the existence of general incorporation laws, legislatures throughout the North awarded thousands of special charters to railroad, manufacturing, and mining companies, conveying privileges regarding debt limits and director liability. In Pennsylvania's legislature, reputedly the nation's most corrupt public body (one former member said he never admitted having held a seat "unless closely pressed upon the subject"), the "third house" of railroad lobbyists exerted as much influence as the elected chambers. The lawmakers allowed the state's railroads to endow employees with police power (spurring the growth of private detective agencies like that headed by former Union spy Allan Pinkerton), and enacted countless bills in aid of the state's corporations.[9]

But it was in the West that a "politics of development" akin to that of the Reconstruction South emerged most dramatically, with leading officials merging public and private interests, and railroad, lumber, and mining companies exerting inordinate influence on government policy. Western state constitutions contained antebellum provisions barring direct state aid to corporations, but counties and municipalities throughout the region vied with each other to lavish assistance upon railroads, and legislatures often turned a blind eye to the legal prohibitions. Under Gov. Samuel J. Crawford, a lawyer closely associated with the Union Pacific, the Kansas legislature divided hundreds of thousands of acres of public land among the state's railroads. Illinois Gov. John M. Palmer, a former Democrat with a Jacksonian suspicion of corporations, vetoed seventy-two special charters and railroad aid bills in 1869 alone, but hundreds

Columbia University, 1967), 20–22; Frederick Merk, *Economic History of Wisconsin During the Civil War Decade* (Madison, Wis., 1916), 71–75; Richard N. Current, *The Civil War Era, 1848–1873* (Madison, Wis., 1976), 478; Rodman Paul, *Mining Frontiers of the Far West 1848–1880* (New York, 1963), 90–91, 193; Mark Wyman, *Hard Rock Epic* (Berkeley, 1979), 6–18.

9. *Nation*, April 10, 1873; *CG*, 40th Congress, 3d Session, 652; Samuel Shapiro, *Richard Henry Dana, Jr., 1815–1882* (East Lansing, Mich., 1961), 134; John Cadman, Jr., *The Corporation in New Jersey: Business and Politics 1791–1875* (Cambridge, Mass., 1949), 160–72; William A. Russ, Jr., "The Origin of the Ban on Special Legislation in the Constitution of 1873," *PaH*, 11 (October 1944), 261; Frank Morn, *"The Eye That Never Sleeps": A History of the Pinkerton National Detective Agency* (Bloomington, Ind., 1982), 43–52, 94–95.

more became law. In Wisconsin, many legislators held stock or directorships in lumber companies and railroads that benefited from state aid.
Lobbyists thronged the statehouse; "the worst traits of the human character are brought out by them," observed Gov. Cadwallader C. Washburn,
"and to see the majority of our legislators struggling for the grab . . . is
humiliating." Under "Boss" Elisha W. Keyes, the Republican organization regularly performed favors for major corporations. In 1873, for
example, Keyes used "all the resources at my command" to secure a land
grant and tax exemption for the West Wisconsin Railroad.[10]

As in the South, the relationship of politics and business transcended
partisan lines. Railroad corporations prudently made donations to both
parties and included representatives of both among their directors. Democratic legislators and localities in many parts of the North supported
railroad aid, and, in the words of a Pennsylvania newspaper, "corruption
belongs to no one party but has invaded all." In Wisconsin, as Keyes
noted, the "business interests" of some railroads were "interminably
connected with *our* political fortunes," but the Milwaukee and St. Paul
employed "all the mighty engines at their command" to support the
Democracy. Yet because the war had bound the fortunes of so many
entrepreneurs to the national state, and since Republicans exuded an air
of economic respectability and appeared more attuned to the promo-
tional role of government, they developed the closest ties with the increasingly powerful industrialists and railroad men.[11]

The growing connection of Republican leaders with business corporations seemed even more apparent at Washington. Even officials of unimpeachable integrity felt comfortable with arrangements that today seem
egregious conflicts of interest. Lyman Trumbull, for example, received
an annual retainer from the Illinois Central Railroad while serving in the
Senate. Less scrupulous lawmakers devoted much of their time to enrich-
ing themselves and serving the interests of companies with which they
and their party had a connection. Pennsylvania Sen. Simon Cameron, an
investor in coal, oil, and railroad enterprises, worked closely with the
Pennsylvania Railroad to elect unknown corporate attorney John Scott to

10. Howard R. Lamar, "Carpetbaggers Full of Dreams: A Functional View of the Arizona
Pioneer Politician," *Arizona and the West*, 7 (Autumn 1965), 187–206; Carter Goodrich,
Government Promotion of American Canals and Railroads 1800–1890 (New York, 1960), 230–42,
275; Mark A. Plummer, *Frontier Governor: Samuel J. Crawford of Kansas* (Lawrence, Kans.,
1971), 92; John M. Palmer, *Personal Recollections of John M. Palmer* (Cincinnati, 1901), 290–93,
312–14; Helen J. Williams and T. Harry Williams, "Wisconsin Republicans and Reconstruction, 1865–70," *WMH*, 23 (September 1939), 18–21; Cadwallader C. Washburn to Lucius
Fairchild, February 24, 1873, Lucius Fairchild Papers, SHSW; Elisha W. Keyes to D. A.
Baldwin, March 13, 1873, Letterbooks, Elisha W. Keyes Papers, SHSW.

11. Lee Benson, *Merchants, Farmers and Railroads: Railroad Regulation and New York Politics
1850–1887* (Cambridge, Mass., 1955), 62; Robert D. Marcus, *Grand Old Party: Political
Structure in the Gilded Age 1880–1896* (New York, 1971), 246–47; Frank B. Evans, *Pennsylvania
Politics, 1872–1877: A Study in Political Leadership* (Harrisburg, Pa., 1966), 43–44; Elisha W.
Keyes to James H. Howe, October 6, 1872, Letterbooks, Keyes Papers.

the Senate in 1869. Iowa Congressman Grenville Dodge looked after the interests of the Union Pacific Railroad, which he served as chief civil engineer. Numerous Congressmen received favors from railroad companies, ranging from free passes to land and stock. The Central Pacific rewarded Sen. William M. Stewart of Nevada with 50,000 acres of land for his services on the Committee on the Pacific Railroad. Banker Jay Cooke, the "financier of the Civil War" and leading individual contributor to Grant's Presidential campaigns, not only had the Republican party in his debt, but a remarkable number of its leading officials as well. Cooke took a mortgage on Speaker of the House James G. Blaine's Washington home, sold a valuable piece of Duluth land to Ohio Gov. Rutherford B. Hayes at "a great bargain," and employed as lobbyists such out-of-office politicos as Ben Wade and Ignatius Donnelly. Corporations having business with the government eagerly sought the influence of William E. Chandler, chairman of the Republican National Committee. At one point, Chandler found himself simultaneously on the payroll of four railroads, including the Union Pacific.[12]

Like the states, the federal government proved remarkably solicitous of the interests of the railroads and other corporations. The National Mineral Act of 1866 dispensed millions of acres of mineral-rich public land to mining companies free of charge. The Supreme Court repeatedly prevented municipalities from repudiating railroad-aid bonds even when evidence came to light that bribery accounted for their being issued. Between 1862 and 1872, the government awarded over 100 million acres of land and millions of dollars in direct aid to support railroad construction, mostly to help finance the transcontinental lines chartered during and after the Civil War. Blacks could not help noting the contrast between such largesse and the failure to provide the freedmen with land. Why, asked Texas freedman Anthony Wayne, "whilst Congress appropriated land by the million acres to pet railroad schemes . . . did they not aid poor Anthony and his people starving and in rags?" No one ever offered an answer.[13]

As in the Reconstruction South, much of the Grant era's corruption

12. Mark M. Krug, *Lyman Trumbull: Conservative Radical* (New York, 1965), 277; Erwin S. Bradley, *The Triumph of Militant Republicanism* (Philadelphia, 1964), 316–19; Stanley P. Hirshson, *Grenville M. Dodge* (Bloomington, Ind., 1967), 128–61; Russell R. Elliott, *Servant of Power: A Political Biography of Senator William M. Stewart* (Reno, 1983), 47–74; David S. Muzzey, *James G. Blaine: A Political Idol of Other Days* (New York, 1935), 64; Charles R. Williams, ed., *Diary and Letters of Rutherford Birchard Hayes* (Columbus, Ohio, 1922–26), 3:89–91; Hans L. Trefousse, *Benjamin Franklin Wade: Radical Republican From Ohio* (New York, 1963), 315; Martin Ridge, *Ignatius Donnelly: The Portrait of a Politician* (Chicago, 1962), 129–30; Margaret S. Thompson, *The "Spider Web": Congress and Lobbying in the Age of Grant* (Ithaca, N. Y., 1985), 169.

13. Elliott, *Stewart*, 55; Charles Fairman, *Reconstruction and Reunion 1864–88: Part One* (New York, 1971), 918–1116; Goodrich, *Government Promotion*, 193–201; Anthony Wayne to Charles Sumner, September 2, 1872, Charles Sumner Papers, HU.

stemmed from the government's policy of promoting railroad development. Possibly the most notorious example involved Crédit Mobilier, a
dummy corporation formed by an inner ring of Union Pacific stockhold-
ers to oversee the line's government-assisted construction. Essentially a
means by which the participants contracted with themselves, at an exorbitant profit, to build their own line, the arrangement was protected by the
distribution of Crédit Mobilier shares to influential Congressmen. After
a newspaper broke the story, a Congressional investigation resulted in
the expulsion of two members and damaged the reputations of numerous
other public officials, including House Speaker Blaine, and Grant's Vice
Presidents Schuyler Colfax and Henry Wilson.[14]

The reckless awarding of land, money, and special favors by the federal
and state governments also encouraged promoters and speculators to vie
for control of railroad companies, through which they could reap enormous profits by watering stock and diverting government aid for private
gain. The resulting distortion of both politics and economic enterprise
was vividly illustrated in the battle for control of the Erie Railroad between Cornelius Vanderbilt, the steamship and railroad magnate whose
fortune exceeded $100 million, and Erie managers Daniel Drew, Jay
Gould, and Jim Fisk, men expert at "amassing wealth without labor."
Among other things, this titanic struggle involved the printing of new
Erie stock as fast as Vanderbilt bought up existing issues, thus preventing
him from obtaining a majority of the shares, and contradictory injunc-
tions and contempt citations issued by judges beholden to the competing
parties. Meanwhile, the 1868 New York legislature, the "worst assemblage of official thieves" in the state's history, attempted to make the most
of a situation in which both sides offered bribes for favorable legislation.
But, as Charles Francis Adams, Jr., concluded in a brilliant account of the
affair that shocked and titillated the middle-class readers of the *North
American Review*, the Erie battle seemed most of all to demonstrate that
"our great corporations are fast emancipating themselves from the State,
or rather subjecting the State to their own control."[15]

Adams, however, exaggerated, for the political hegemony of railroad
men and industrialists was far from complete. It would be wrong to
accuse every lawmaker of malfeasance, or to assume that corporate lobbying and political corruption fully explained the remarkable governmental
generosity of what Mark Twain called America's Gilded Age. Ultimately
more significant was the widespread belief in the social benefits and
political advantages of capitalist development. Despite their growing

14. Willard H. Smith, *Schuyler Colfax: The Changing Fortunes of a Political Idol* (Indianapolis, 1952), 369–97; Allan Peskin, *Garfield* (Kent, Ohio, 1978), 355–62.
15. Charles Francis Adams, Jr., and Henry Adams, *Chapters of Erie and Other Essays* (Boston, 1871), 1–96; Wheaton J. Lane, *Commodore Vanderbilt: An Epic of the Steam Age* (New York, 1942), 223–60.

power, moreover, the new corporations were only one among many groups struggling to use the state to advance their own interests. Indeed, as in the South, the process of social change profoundly affected politics and government in the postwar North. Here, too, the state emerged as a battleground for competing claims upon its authority—not only entrepreneurs seeking economic advantage, but Radicals, blacks, and women hoping to extend northward the process of government-promoted social and racial reconstruction, farmers, laborers, and others bent on redressing inequities caused by capitalism's rapid expansion, and a newly self-conscious intelligensia determined to redefine the meaning of "reform."

The Transformation of Politics

In some respects, changes in the scope and purposes of Northern public authority paralleled those in the Reconstruction South. State governments greatly expanded their responsibility for public health, welfare, and common schooling, and cities invested heavily in public works such as park construction and improved water and gas services. Since government activism invited a political backlash, it flourished in cities like Philadelphia, ruled by impregnable political machines, and states like Michigan, Massachusetts, and New York, where Republicans enjoyed secure majorities or Radicals commanded the greatest power. It proved weakest in places like Connecticut and California, where the party's position was most precarious, and Ohio, whose multiplicity of Republican factions rendered new policy initiatives all but impossible. Massachusetts, where a tight-knit cadre of Radicals known as the Bird Club (after veteran reformer Francis W. Bird), controlled Republican politics throughout the 1860s, established state boards of public charities and public health and a state constabulary. Michigan Republicans expanded state facilities for homeless children and the insane, deaf, and blind; they established a board of health, made school attendance compulsory for the first time, outlawed cruelty to animals, and created a metropolitan police board for Democratic Detroit. Coupled with aid to corporations, such initiatives produced, as in the South, a rapid increase in state and city budgets, debts, and tax rates. Between 1860 and 1870, the tax burden tripled in five Northern states, quintupled in Michigan, and rose by six times in New Jersey, while in the six years preceding the Panic of 1873, the bonded debt of Boston, New York, and Chicago tripled. (Of course, unlike the South, the North's ability to pay taxes—its aggregate wealth and per capita income—were also rising in these years.)[16]

16. Morton Keller, *Affairs of State: Public Life in Late Nineteenth Century America* (Cambridge, Mass., 1977), 117, 124, 134; John C. Teaford, *The Unheralded Triumph: City Government in America, 1870–1900* (Baltimore, 1984), 285–94; Howard F. Gillette, "Corrupt and Contented: Philadelphia's Political Machine, 1865–1887" (unpub. diss., Yale University, 1970),

Under Radical Gov. Reuben Fenton, New York enacted the North's most ambitious reform program. Between 1865 and 1867, the legislature established eight new teacher-training colleges, created the state's first Board of Charities, eliminated fees for common schooling, and set minimum housing standards for New York City. A professional fire department replaced the city's notoriously inefficient volunteer companies, and the legislature created a new Board of Health, one of the few administrative agencies of the period with the power not simply to collect and publicize data, but to enforce its own regulations (including the awarding of licenses to sell liquor). Clearly, this program embodied a variety of purposes. Sabbatarians had long campaigned to impose their distaste for Sunday drinking upon the city's German and Irish immigrants; the new licensing procedures, wrote one critic, would make "the poor man's day of rest and recreation . . . a day of gloom and privation." The new laws also aimed to weaken the Democratic organization that controlled city politics, for the volunteer fire companies, apart from having been implicated in the outbreak of the 1863 draft riot, formed important cogs in the Tammany machine. Inescapable urban realities, however, inspired both the regulation of housing conditions in the nation's most crowded city and the establishment of the Board of Health to combat New York's high death rate and notoriously unhealthy sanitation facilities (attributed by some to the mayor's practice of awarding streetcleaning contracts to political cronies).[17]

Throughout the North, moreover, Republicans took steps to improve the lot of the region's blacks. The party could hardly expect to reap political benefits from efforts to combat racial discrimination, for white voters alienated by such a policy were likely to outnumber the North's tiny black population. Indeed, along with German defections because of the Sunday law, New York Republicans' attempt to eliminate the state's property qualification for black voters was widely blamed for Democrats' capture of the governorship in 1868 and legislature in 1869, abruptly ending the era of state-sponsored reform. Rather, the drive for civil and political equality arose from the ideological commitment of Radicals, the course of Reconstruction politics, and the agitation of blacks themselves, whose churches, newspapers, and state conventions pressed for the re-

---

125; John Niven, "Connecticut: 'Poor Progress' in the Land of Steady Habits," Felice A. Bonadio, "A 'Perfect Contempt of All Unity'," Richard H. Abbott, "Maintaining Hegemony," George M. Blackburn, "Quickening Government in a Developing State," in James C. Mohr, ed., *Radical Republicans in the North: State Politics During Reconstruction* (Baltimore, 1976), 3–8, 26–49, 82–103, 130–38; C. K. Yearley, *The Money Machines: The Breakdown and Reform of Governmental and Party Finance in the North, 1860–1920* (Albany, 1970), 10.

17. James C. Mohr, *The Radical Republicans and Reform in New York During Reconstruction* (Ithaca, N. Y., 1973); Matthew P. Breen, *Thirty Years of New York Politics* (New York, 1899), 109–12; Jerome Mushkat, *The Reconstruction of the New York Democracy, 1861–1874* (Rutherford, N. J., 1981), 92–93.

peal of discriminatory laws, access to public facilities like schools and streetcars, and an extension of Reconstruction's democratic revolution to the North.[18]

Partly because of Congressional measures that applied throughout the country, and partly due to actions at the state and local level, the decade following the Civil War witnessed astonishing advances in the political, civil, and social rights of Northern blacks. The Civil Rights Act of 1866 and postwar amendments voided laws barring blacks from entering Northern states, testifying in court, and voting, and were successfully employed by individuals pressing damage claims against railroads and streetcars that excluded them altogether or barred them from first-class compartments. Although state courts generally held that segregated facilities, if truly equal, did not violate the Fourteenth Amendment, discrimination in transportation faded in many parts of the North. Pennsylvania's legislature prohibited streetcar segregation in 1867 and New York Republicans six years later enacted a pioneering civil rights law that outlawed discrimination in public accommodations. Blacks also gained access to public schools in states that had previously made no provision for their education. Some cities with sizable black populations, like New York and Cincinnati, maintained separate schools, but others, like Chicago, Cleveland, and Milwaukee, not only operated integrated systems but occasionally employed a black teacher. In a few states, integrated education now became the norm. Michigan's legislature outlawed school segregation in 1867 (although Detroit's school board refused to comply for four years), and the state university admitted its first black students in 1868. And Iowa's Supreme Court ruled separate schooling a violation of the Fourteenth Amendment.[19]

Despite the rapid toppling of traditional racial barriers, the North's racial Reconstruction proved in many respects less far-reaching than the South's. Prejudice, lamented Cincinnati black leader Peter Clark, remained ubiquitous: "It hampers me in every relation of life, in business,

18. Mohr, *Radical Republicans in New York*, 264–65; Emma L. Thornbrough, *The Negro in Indiana Before 1900* (Indianapolis, 1957), 232, 238; *Proceedings of the Iowa State Colored Convention* (Muscatine, Iowa, 1868); [Lewis H. Putnam] *Review of the Revolutionary Elements of the Rebellion . . . By a Colored Man* (Brooklyn, 1868).

19. Robert J. Chandler, "Friends in Time of Need: Republicans and Black Civil Rights in California During the Civil War Era," *Arizona and the West*, 22 (Winter 1982), 319–31; David A. Gerber, *Black Ohio and the Color Line 1860–1915* (Urbana, Ill., 1976), 47, 53–57; Harry C. Silcox, "Nineteenth Century Philadelphia Black Militant: Octavius V. Catto (1839–1871)," *PaH*, 44 (January 1977), 53–76; Jonathan Lurie, "The Fourteenth Amendment: Use and Application in Selected State Court Civil Liberties Cases, 1870–1890," *American Journal of Legal History*, 28 (October 1984), 304–13; Leslie H. Fishel, Jr., "Repercussions of Reconstruction: The Northern Negro, 1870–1883," *CWH*, 14 (December 1968), 340–41; David M. Katzman, *Before the Ghetto: Black Detroit in the Nineteenth Century* (Urbana, Ill., 1973), 50, 84–88; Elizabeth G. Brown, "The Initial Admission of Negro Students to the University of Michigan," *Michigan Quarterly Review*, 2 (October 1963), 233–36; Robert R. Dykstra, "Iowa: 'Bright Radical Star'," in Mohr, ed., *Radical Republicans*, 186.

in politics, in religion, as a father or as a husband." The bulk of the North's black population remained trapped in urban poverty and confined to inferior housing and menial and unskilled jobs (and even here their foothold, challenged by the continuing influx of European immigrants and discrimination by employers and unions alike, became increasingly precarious). A survey of New York City's black community in 1871 found some 400 waiters and 500 longshoremen, but only two physicians and a handful of skilled craftsmen. Black politicians in the North lacked the militancy of their Southern counterparts, and failed to develop a viable strategy for addressing the economic plight of their community. Perhaps this was inevitable for a group mostly derived from the tiny black business class and representing a politically marginal constituency (blacks still comprised less than 2 percent of the North's population). Although black politicos won seats in the Massachusetts and Illinois legislatures, most, with no realistic prospect of elective office, found themselves beholden for position to patronage from white Republicans.[20]

Nonetheless, blacks now found the North's public life open to them in ways inconceivable before the war. A recognition of their claim to equal civil and political rights had become so much a part of what it meant to be a Republican that no fewer than 90 percent of the party's voters in New York State supported equal suffrage in an unsuccessful 1869 referendum.[21] Blacks, however, were not the only group demanding an extension of Reconstruction into the North. More divisive were other claims that tested the limits of the activist state and Republicans' commitment to equality before the law.

With blacks' enfranchisement, women remained "the only class of citizens wholly unrepresented in the government." And despite the disintegration of the abolitionist-feminist alliance in disputes over the postwar amendments, a broadly based, independent movement continued to demand an end to the restrictions on women's social and legal rights. Feminist leaders like Elizabeth Cady Stanton and Susan B. Anthony envisioned changes in male-female relations going far beyond the suffrage, claiming for women the same self-determination and opportunity for individual advancement so prized by American men. "To build a true republic," Stanton wrote, "the church and the home must undergo the

20. Lawrence Grossman, "In His Veins Coursed No Bootlicking Blood: The Career of Peter H. Clark," *Ohio History*, 86 (Spring 1977), 94; H. D. Bloch, "The New York City Negro and Occupational Eviction, 1860–1910," *International Review of Social History*, 5 (1960), 26–31; New York *Tribune*, August 25, 1871; Fishel, "Repercussions of Reconstruction," 325–30, 344; Kenneth L. Kusmer, *A Ghetto Takes Shape: Black Cleveland, 1870–1930* (Urbana, Ill., 1976), 98–100, 116; Katzman, *Before the Ghetto*, 177–79; CG 39th Congress, 2d Session, 104; Bridges, "Equality Deferred," 98.

21. Phyllis F. Field, *The Politics of Race in New York: The Struggle for Black Suffrage in the Civil War Era* (Ithaca, N. Y., 1982), 205.

same upheaving we now see in the state." The principle of free contract, central to Reconstruction ideology, should, she insisted, be extended to the family itself, with marriage recognized as a voluntary agreement dissolvable at will, and married women enjoying an independent claim to their earnings. Stanton even embraced the heretical doctrine of "free love" and championed a woman's right to birth control. The vote, however, remained central to feminist thinking, for it promised to transform women's lives by recognizing their equal standing in public life, overturning the ideology that defined the home as women's sphere, and making it possible for women to undertake collective political action to address their other grievances.[22]

"If I were to give free vent to all my pent-up wrath concerning the subordination of women," Lydia Maria Child wrote Charles Sumner in 1872, "I might frighten *you*. . . . Suffice it, therefore, to say, either the theory of our government is *false*, or women have a right to vote." Child's bitterness was not uncommon among nineteenth-century women, and her logic was irrefutable. Radicals like Sumner remained more sympathetic to women's suffrage than other politicians, but even in Massachusetts the issue split the party. Under Bird Club leadership, the state enlarged the property and contractual rights of married women and expanded the grounds for divorce to include extreme cruelty, habitual intoxication, and a husband's failure to provide for his family. But although Gov. William Claflin supported suffrage, the 1871 party convention sidestepped the question—noting only that it "deserves the most careful and respectful attention"—and a constitutional amendment failed in the legislature. The same fate befell proposals in Iowa and Maine, and when staunchly Republican Michigan held a referendum in 1874, women's suffrage suffered a decisive defeat. Two territories—Wyoming in December 1869, followed by Utah two months later—did enfranchise women, but neither acted for feminist reasons. With males comprising six sevenths of its adult population, Wyoming hoped suffrage would call forth "an immigration of ladies" from the East, while Utah Mormons expected it to counterbalance the votes of a rising "Gentile" population (mostly unmarried miners), and enhance the political power of men who headed polygamous households.[23]

22. *Official Proceedings of the National Democratic Convention Held at New York* (Boston, 1868), 29; Elizabeth Cady Stanton to Charles Sumner, April 3, 1866, Charles Sumner Papers, HU; Amy Dru Stanley, "Status or Free Contract: Marriage in the Age of Reconstruction" (unpub. paper, annual meeting of American Historical Association, 1985); Ellen C. DuBois, ed., *Elizabeth Cady Stanton, Susan B. Anthony: Correspondence, Writings, Speeches* (New York, 1981), 92–100; Steven M. Buechler, *The Transformation of the Woman Suffrage Movement: The Case of Illinois, 1850–1920* (New Brunswick, N. J., 1986), 26–27, 38–43, 79–81, 91–99.

23. Lydia Maria Child to Charles Sumner, July 9, 1872, Sumner Papers; Dale Baum, "Woman Suffrage and the 'Chinese Question': The Limits of Radical Republicanism in Massachusetts, 1865–1876," *New England Quarterly*, 56 (March 1983), 63–69; Abbott, "Maintaining Hegemony," 6–7; Edward McPherson, *A Handbook of Politics for 1872* (Wash-

If women's agitation for the vote exposed Republicans' reluctance to carry the principle of equal rights to its logical conclusion, the political demands of economically aggrieved Northerners confronted the party with a different set of problems. Especially in the West, the railroad network's expansion called forth a growing chorus of protest. Commercial farmers complained of exorbitant freight rates, discrimination in favor of large shippers, and the high price of manufactured goods. River and lake ports resented being bypassed, and even Chicago merchants contended that other localities enjoyed preferential freight charges and protested high fees by railroad-controlled warehouses. To all these groups, the railroad increasingly appeared less an embodiment of progress than an alien, intrusive force that disrupted traditional economic arrangements and channels of commerce and threatened the independence of individuals and local communities. One response was the rapid spread in the Upper Mississippi Valley of the Patrons of Husbandry, or Grange, which moved to establish cooperatives so as to circumvent exploitative middlemen and "*compel* the carriers to take our produce at a fair price." Although the Grange itself remained nonpartisan, its members quickly turned to politics, demanding that the activist state that had lavished aid upon corporations now redress the imbalances caused by the capitalist economy's rapid growth.[24]

In increasing numbers, the railroad's critics called upon state governments to regulate freight rates and warehouse charges. Railroad objections that such action would violate the rights of private property struck proponents of regulation as the height of hypocrisy. "On the plea of public interests," one Granger pointed out, railroads clamored for government aid and the right to condemn property in the path of construction. "To do this they are public corporations, acting for the public good. The charter and right of way once granted . . . and railroad companies are private institutions not amenable to Legislatures or Courts." The demand for regulation transcended party lines, but since Republicans controlled nearly all state governments, and commanded substantial majorities in areas of mature commercial agriculture, like northern Illinois, where antirailroad sentiment flourished most strongly, they faced the task of fashioning a response to the demand for regulation.[25]

---

ington, D. C., 1872), 120–21; Harriette M. Dilla, *The Politics of Michigan 1865–1878* (New York, 1912), 172; T. A. Larson, "Woman Suffrage in Western America," *Utah Historical Quarterly*, 38 (Winter 1970), 9–17; Alan P. Grimes, *The Puritan Ethic and Woman Suffrage* (New York, 1967), 28–34, 53–58.

24. George H. Miller, *Railroads and the Granger Laws* (Madison, Wis., 1971), 15–16, 19–21; Irwin Unger, *The Greenback Era: A Social and Political History of American Finance 1865–1879* (Princeton, 1964), 200–204; North Star Grange, Minutes of Meetings, February 5, 1870, Minnesota Historical Society.

25. Rasmus S. Saby, "Railroad Legislation in Minnesota, 1849–1875," *Minnesota Historical Society Collections*, 15 (1915), 84; Miller, *Granger Laws*, 165–66; Philip D. Swenson, "Illinois: Disillusionment with State Activism," in Mohr, ed., *Radical Republicans*, 108–109.

Although this early "antimonopoly" movement pales before the agrarian revolt of subsequent years, it did lead several states to attempt to control freight rates and warehouse charges. The best-known efforts—the Massachusetts and Illinois railroad commissions—illustrated Republicans' sharp divisions over state interference with private economic interests. Established in 1869, the Massachusetts body was inspired by Charles Francis Adams, Jr., who knew all too well the dangers posed by the railroads' power, but feared "coercive legislation" would do more harm than good by interfering with management's legitimate prerogatives. Far better, Adams thought, to create an agency staffed by nonpartisan experts, whose authority rested on publicity and persuasion. Following this blueprint, the legislature created a commission that lacked coercive power and quickly became a profoundly conservative instrument, shielding the state's railroads from further legislative action and opposing the unionization of their workers. (Adams, while a member, speculated in railroad bonds and lands near proposed lines; his behavior did not appear to differ markedly from that he had condemned in "Chapters of Erie.") The Illinois Board of Railroad and Warehouse Commissioners, by contrast, embodied a far broader definition of the state's regulatory powers, for the legislature clothed it with the authority to eliminate rate discrimination, establish maximum charges, and take legal action to enforce its decisions. A legal challenge by railroads and warehouse operators produced the landmark 1877 Supreme Court decision in *Munn v. Illinois*, vindicating the state's right to regulate businesses of a quasi-public character.[26]

Even more thorny political and ideological questions arose from the North's revived labor movement. Organized labor's dramatic growth came as a surprise to many contemporaries, for a labor shortage caused by the postwar economic boom and a decline in prices as wartime inflation waned resulted in a 40 percent increase in average real wages between 1865 and 1873. Nor had capitalism's complex, uneven development created anything resembling a homogenous working class, for large-scale factories coexisted with thousands of small enterprises, and traditional artisans with wage-earning operatives and seasonally employed unskilled "outdoor workers" (railroad freight handlers, coal heavers, longshoremen, and the like). The average earnings of unskilled Massachusetts factory operatives amounted to some $600 per year, a sum on which a family could not make ends meet, and in cities like New York, thousands of seamstresses and tailors labored for even less in squalid sweatshops. But Pittsburgh remained a "craftsmen's empire," whose

26. McCraw, *Prophets of Regulation*, 2–20; Edward C. Kirkland, *Charles Francis Adams, Jr., 1835–1915: The Patrician at Bay* (Cambridge, Mass., 1984), 41–53, 67–77; Keller, *Affairs of State*, 178–79; Miller, *Granger Laws*, 76–96. Wisconsin, Iowa, and Minnesota also enacted railroad regulation laws in these years.

"Sunshine and Shadow in New York": frontispiece of Matthew Smith's 1868 best-seller.

labor aristocracy of skilled iron and steel workers controlled the labor process and commanded high wages. Nativity, religion, and party politics all produced further divisions in the house of labor.[27]

Fostering a sense of unity despite the diversity of working-class experience, however, were aspects of economic life that nearly all laborers (and many other Americans) found alarming. Despite widespread prosperity, the unprecedented fortunes accumulated by the nation's captains of commerce and industry helped create one of the highest levels of income inequality in all of American history. Matthew Smith's 1868 best-seller *Sunshine and Shadow in New York* opened with an engraving that contrasted department store magnate Alexander T. Stewart's $2 million mansion with the city's slum tenements; even Rochester railroad president Isaac Butts identified the concentration of wealth "in fewer and fewer hands" as the nation's most serious problem. The inexorable tendency toward the mechanization of industry and a larger scale of production transformed skilled craftsmen into "tenders of machinery" and threatened to entrench within American life what *The Nation* called "the great curse of the Old World—the division of society into classes." Massachusetts industry, Edward Atkinson had noted during the Civil War, stood "in a transition state, changing from transient help, farmer's daughters, etc., working for a few years and living in boarding houses, to a permanent factory population." By 1870, the same tendency was evident farther west. In Cincinnati, for example, a few large factories employed as many workers as the city's thousands of small shops. All this raised troubling questions about the continued validity of free labor axioms: that liberty rested on ownership of productive property, and that working for wages was merely a temporary resting place on the road to economic autonomy.[28]

Drawing upon traditional ideals of artisan independence and republican equality, the postwar labor movement mobilized the skilled and unskilled, the native and foreign-born. To the reigning economic orthodoxy that preached the virtues of acquisitive individualism and the iron law of supply and demand, labor counterposed an ethic of mutuality and what

27. Philip R. P. Coelho and James F. Shepherd, "Regional Differences in Real Wages: The United States, 1851–1880," *Explorations in Economic History*, 13 (April 1976), 212–13; Steven J. Ross, *Workers on the Edge: Work, Leisure, and Politics in Industrializing Cincinnati, 1788–1890* (New York, 1985), 67–68, 81–83; Richard Schneirov, "Chicago's Great Upheaval of 1877," *Chicago History*, 9 (Spring 1980), 3; *Report of Bureau of Statistics of Labor* (Boston, 1872), 340–41; Francis G. Couvares, *The Remaking of Pittsburgh: Class and Culture in an Industrializing City, 1877–1919* (Albany, 1984), 9–30; Montgomery, *Beyond Equality*, 40–42, 75.

28. Jeffrey G. Williamson and Peter H. Lindert, *American Inequality: A Macroeconomic History* (New York, 1980), 75; Matthew H. Smith, *Sunshine and Shadow in New York* (Hartford, Conn., 1868), frontispiece; Isaac Butts to Isaac Sherman, February 6, 1872, Isaac Sherman Papers, Private Collection; *Report of Bureau of Statistics of Labor*, 342; *Nation*, June 27, 1867; Edward Atkinson to Charles Eliot Norton, March 7, 1864, Charles Eliot Norton Papers, HU; Ross, *Workers on the Edge*, 80.

one mechanic called a *"moral* economy" that insisted economic activity, like other endeavors, must be judged by ethical standards. By the early 1870s, the three national unions that had survived the Civil War had grown to twenty-one; local unions and municipal labor councils existed in cities throughout the North, and strikes had become a regular feature of industrial life. While many conflicts centered on the level of wages, broader concerns galvanized the labor movement, especially a long-standing aversion to "the wages system." Reconstruction, claimed labor leader Ira Seward, was a national problem, for "something of slavery" remained within the North. "The masses," declared one union, "will never be completely free . . . until they have thrown off the system of working for hire."[29]

Reflecting the transitional state of the economy itself, the labor movement stood poised between old and new ideas. Despite the reality of increasing class conflict, most union leaders could not liberate themselves completely from the influence of free labor precepts (a situation all but inevitable given the broad hegemony of market values in the Age of Capital, and paralleled in the contemporaneous British labor movement). Instead of conceiving of themselves as spokesmen for a wage-earning class with interests inherently antagonistic to those of their employers, labor reformers viewed "cooperation between capital and labor" as a natural and desirable state of affairs and insisted that America must avoid the emergence of permanent class divisions. Men like Ira Steward sometimes spoke of an "irrepressible conflict" between employer and employee, but at other times declared that the problem could be resolved by allowing "labor to obtain capital" through the establishment of worker-owned cooperatives (a plausible strategy at a time when the scale of enterprise in most industries remained small).[30]

If cooperation combated the spread of "wage slavery" through workers' own efforts, other planks in labor's platform envisioned action by the government. Greenbackism—the elimination of bank notes in favor of abundant national paper money—promised to lower interest rates, thereby facilitating the establishment of cooperatives, promoting economic growth, and transferring control of the currency from a parasitical "money power" to the democratic state. Even more popular was the

29. David Montgomery, "Working-Class Radicalism in America 1860–1920" (unpub. paper, Milan [Italy] Conference on American Radicalism, 1979); *Nation*, November 16, 1865; Samuel Bernstein, "American Labor in the Long Depression, 1873–1878," *Science and Society*, 20 (Winter 1956), 66; New York *Tribune*, May 13, 14, 16, 1867; Cincinnati *Commercial*, April 15, 1867; David Roediger, "Ira Steward and the Anti-Slavery Origins of American Eight-Hour Theory," *Labor History*, 27 (Summer 1986), 424; *The Proceedings of the Annual Meeting of the Knights of St. Crispin for 1869* (Boston, 1870), 33–34.

30. Montgomery, *Beyond Equality*, 4, 178–80, 446; Hobsbawm, *Age of Capital*, 114, 249; John R. Commons et al., eds., *A Documentary History of American Industrial Society* (Cleveland, 1909–11), 9:151; Robert P. Sharkey, *Money, Class, and Party: An Economic Study of Civil War and Reconstruction* (Baltimore, 1967 ed.), 206; Roediger, "Steward," 413–23.

demand for legislation reducing the daily hours of labor, an issue that mobilized craftsmen, factory operatives, and day laborers. Instead of viewing the eight-hour day as a narrow "economic" question, labor invested it with almost utopian meaning, seeing shorter hours as a "new gospel" that would enable workers to devote time to education, self-improvement, and participation as citizens, thus establishing "the independence of the working class" and "the success of our republican institutions."[31]

"All political Republics," declared a labor newspaper in 1869, "have heretofore perished because not based on a social Republic." But if organized labor envisioned a broad extension of the Reconstruction principle of equal rights, its own conception of equality remained in many respects thoroughly conventional. Composed mostly of artisans and skilled industrial workers, unions proved unwilling to expand their membership beyond the ranks of white men. Despite the growing number of women in the labor force and the emergence of a Working Women's Association that sought cooperation with male labor organizations, most unions viewed women as a threat to the wage levels, skills, and social prerogatives of men, including their ability to command a "family wage" that enabled their wives and daughters to remain at home. Nor did organized labor give evidence of racial inclusiveness. In California, where indentured Chinese immigrants by 1870 constituted a quarter of the wage labor force, the agitation for their exclusion, more than any other issue, shaped the labor movement's development.[32]

Throughout the country, moreover, nearly all unions barred blacks from membership. The postwar National Labor Congresses either advocated the formation of segregated black locals, or concluded that the whole question of black labor involved "so much mystery, and upon it so wide diversity of opinion among our members," as to defy resolution. These assemblies also ignored Reconstruction issues, aside from calling for the "speedy restoration" of the South to the Union and noting how much Northern employment depended, directly or indirectly, on the revival of cotton production. Many labor leaders sympathized with Andrew Johnson—"the old tailor," a Cincinnati worker called him. Others feared that to endorse blacks' political and economic aspirations meant associating with the Republican party, a step that would torpedo independent labor politics and offend workers loyal to the Democracy. Even those who advocated the organization of black labor expressed little interest in

31. Montgomery, *Beyond Equality*, 260, 422–44; Chester McA. Destler, *American Radicalism 1865–1901* (New London, 1946), 7–8; Commons et al., eds., *Documentary History*, 9:145, 177–79, 206, 216–18, 234–35; *Workingman's Advocate* (Chicago), May 11, 1866.

32. *Workingman's Advocate* (Chicago), September 25, 1869; Ellen C. DuBois, *Feminism and Suffrage: The Emergence of an Independent Women's Movement in America, 1848–1869* (Ithaca, N. Y., 1978), 111–33; Alexander Saxton, *The Indispensable Enemy: Labor and the Anti-Chinese Movement in California* (Berkeley, 1971).

blacks' own concerns. William Sylvis, president of the Iron Moulder's Union, toured the South early in 1869 recruiting members of both races, but simultaneously called the Freedmen's Bureau a "huge swindle upon the honest workingmen of the country" and blamed carpetbaggers for the South's woes. Thus, despite the parallels between blacks' quest for economic autonomy and its own hostility to "wage slavery," the Northern labor movement failed to identify its aspirations and interests with those of the former slaves.[33]

In addition to the inescapable reality that white labor did not wish to cooperate with black, the differences in the experiences and preoccupations of workingmen of the two races prevented joint action between their labor organizations. One delegate on his way to the Colored National Labor Convention that met in 1869 (composed mostly of politicians, religious leaders, and professionals, rather than sons of toil) was evicted from a first-class railroad car, a situation no white labor reformer would ever encounter. The convention's discussions focused on equal access to employment regardless of race, equality before the law, and ways of assisting the freedmen in acquiring land—issues ignored at gatherings of white labor. And the resolutions laid far greater stress than at white conventions on the futility of strikes and the harmony of interests between capital and labor, reflecting the experience of many black artisans. Indeed, the convention's president, Baltimore ship caulker Isaac Myers, had seen a "great strike against the colored mechanics and longshoremen" result in the discharge of 1,000 black shipyard workers in 1865. White employers had opposed the strike, and a white merchant subsequently aided Myers and a group of black businessmen in leasing their own shipyard. Unlike white labor gatherings, in addition, those of blacks remained firmly attached to the Republican party, rejecting talk of independent labor politics, as well as ideas like the payment of the national debt in greenbacks that, although popular among white workers, diverged from Grant Administration policy.[34]

When the labor question entered the political arena, it produced alignments quite unlike those surrounding Reconstruction. In Congress, George W. Julian's bill to establish an eight-hour day for federal employees won its most consistent support from an unusual coalition of

33. Philip S. Foner, *Organized Labor and the Black Worker 1619–1973* (New York, 1974), 18–23; Sumner E. Matison, "The Labor Movement and the Negro During Reconstruction," *JNH*, 33 (October 1948), 445–53; Commons et al., eds., *Documentary History*, 9:139, 185, 190–91, 227, 237; James M. Morris, "The Road to Trade Unionism: Organized Labor in Cincinnati to 1893" (unpub. diss., University of Cincinnati, 1969), 216; *Workingman's Advocate* (Chicago), February 13, March 6, 27, 1869.

34. Philip S. Foner and Ronald L. Lewis, eds., *The Black Worker: A Documentary History from Colonial Times to the Present* (Philadelphia, 1978–84), 2:36–47, 53–54, 58, 63–78, 85–86, 90, 95, 113; Martin E. Dann, ed., *The Black Press 1827–1890* (New York, 1971), 231–33; Bettye C. Thomas, "A Nineteenth Century Black Operated Shipyard, 1866–1884: Reflections Upon Its Inception and Ownership," *JNH*, 59 (January 1974), 1–12.

Northern Democrats, many representing urban working-class constituencies, and Radical Republicans accustomed to using the democratic state for reform purposes. Although Thaddeus Stevens and Charles Sumner, concerned almost exclusively with Reconstruction, remained indifferent or hostile to the eight-hour cause, most of the criticism came from moderate and conservative Republicans. "I am utterly opposed to the idea of regulating hours of labor by law," intoned Fessenden. "Let the matter be regulated by that great regulator, supply and demand." The eight-hour bill passed the House in March 1867 (with nearly all the Radicals and Democrats voting in favor), only to die in a Senate committee. The bill became law in 1868, after Radicals and Democrats combined to defeat John Sherman's proposal that when the federal workday fell from ten hours to eight, wages should be reduced proportionately. Nonetheless, the "old tailor's" Attorney General ordered a cut in the daily wages of Navy Yard workers, an edict withdrawn after Grant assumed office.[35]

One Republican newspaper considered Congressional Radicals' support for the eight-hour bill perfectly natural, since labor and the Radicals shared a commitment to "the essential dignity of man" and "universal equality." Yet at the state level, the issue proved far more complex, for while the federal law applied only to government-employed workers, state legislation would affect employees in private industry. Some Radical newspapers, like the Chicago *Republican,* edited by former utopian socialist Charles A. Dana, supported the eight-hour movement, but others warned against legislative interference with the "natural laws" of the economy, and reiterated their conviction that "the interests of the employer and the employed are identical." With the labor movement too weak to impose its will on Northern governments, both parties sought political mileage from ineffectual measures that paid little more than lip service to the eight-hour idea. Seven Northern legislatures declared eight hours a legal day's work, but lacking enforcement provisions and incorporating "liberty of contract" clauses that allowed longer hours by mutual consent, these remained dead letters. Not a single employer, one newspaper reported in 1867, had heeded New York's eight-hour law. The only workers who achieved reduced hours did so through their own efforts. Employers' refusal to abide by the Illinois law triggered an unsuccessful general strike of Chicago workers in 1867, but five years later walkouts involving some 100,000 workers brought the eight-hour day, at least temporarily, to New York City.[36]

35. Ira Steward to George W. Julian, March 19, 1866, George W. Julian Papers, Indiana State Library; *CG,* 39th Congress, 1st Session, 1969; 40th Congress, 1st Session, 413–14, 425, 2d Session, 3424–29; Montgomery, *Beyond Equality,* 312–18.

36. Washington *Weekly Chronicle,* July 4, 1868; Jentz and Schneirov, *Chicago's Working Class,* Chapter 5; Montgomery, *Beyond Equality,* 301–5, 311; New York *Journal of Commerce,* October 3, 1867; Costello, "New York City Labor Movement," 369–75. Connecticut, New York, Pennsylvania, Illinois, Missouri, Wisconsin, and California enacted eight-hour laws. Most,

The divisive potential of the labor question was strikingly illustrated in Massachusetts, the nation's most thoroughly industrialized state. Here, the labor movement and Radical Republicans shared considerable common ground, for many union leaders had been influenced by abolitionism before the war, and the Boston *Daily Evening Voice*, a labor journal established after the city's publishers locked out unionized printers in 1864, supported Congressional Reconstruction and urged unions to stop excluding blacks. For their part, many Massachusetts Radicals, including House Speaker James M. Stone, endorsed legislation to limit the working day, prohibit child labor, and insure factory safety. Others, however, including Radical kingpin Francis W. Bird (a successful paper manufacturer), insisted labor's demands would violate the law of supply and demand and contradict the principle of free contract Republicans were attempting to introduce in the South. Dominated by Radicals, a succession of government commissions investigated the need for labor legislation, but all opposed state regulation of working hours as "subverting the right of individual property." (The law, one report concluded, could assist workers in other ways, such as establishing "parks, menageries, and botanic gardens" for their recreation.) In 1866, the legislature prohibited the employment of children under ten in factories, and eight years later established a maximum work week of sixty hours for women and children, but it refused to enact a general eight-hour law. And when the Knights of St. Crispin, the shoemakers union, applied for a state charter to enable it to establish cooperative enterprises, the lawmakers refused.[37]

The fate of the Crispin charter demonstrated, according to one shoemaker, that in Massachusetts, "capital holds labor in a system of slavery." The shoeworkers and other unions responded by launching a Labor Reform party on a platform calling for reduced hours, the taxation of government bonds, and the payment of the national debt in greenbacks. Its support concentrated in shoe towns like Lynn and Haverhill (where it received over a third of the vote), the new party polled 10 percent statewide in 1869, and elected twenty-three members of the legislature. The following year, labor reformers staked their claim to be the inheritors of Massachusetts Radicalism. Adopting the slogan "Equal Rights for All," the new party nominated Wendell Phillips for governor (the only time the

---

but not all, of these legislatures were controlled by Republicans, but the pattern of voting did not adhere to party lines, or to divisions on Reconstruction. In New Jersey, for example, Republican support exceeded Democratic, while in New York the reverse was true. Kent A. Peterson, "New Jersey Politics and National Policy-Making, 1865–1868" (unpub. diss., Princeton University, 1970), 202–203; Mohr, *Radical Republicans in New York*, 134.

37. Philip S. Foner, "A Labor Voice for Black Equality: The *Boston Daily Evening Voice*, 1864–1867," *Science and Society*, 38 (Fall 1974), 304–325; Baum, "Woman Suffrage," 70–71; Montgomery, *Beyond Equality*, 120–26, 266–68; James Leiby, *Carroll Wright and Labor Reform: The Origins of Labor Statistics* (Cambridge, Mass., 1960), 47–48, 51; *Seventh Annual Report of the Bureau of Statistics of Labor* (Boston, 1876), 274–77, 293.

great orator sought political office) and black legislator George L. Ruffin for attorney general. Phillips appealed to immigrant voters by embracing Irish nationalism, and condemned the "wages system" for preventing a "fair division of the joint profits of labor and capital." When a North Adams shoe factory brought in seventy-five Chinese workers to break a Crispins strike during the campaign, Phillips refused to be swept up in the ensuing wave of Oriental exclusion sentiment, condemning the use of indentured strikebreakers but insisting the state should welcome voluntary immigrants of all races. Lambasted by his erstwhile Radical and abolitionist allies, Phillips captured 16 percent of the vote, doing especially well in industrial cities and among Irish-Americans (despite the fact that he had also been endorsed by the Prohibition party). And Ruffin ran ahead of the ticket.[38]

The election of 1870 marked the high tide of independent labor politics in Massachusetts, but the movement had already wrung one concession from the legislature: the nation's first Bureau of Labor Statistics. Headed by an enlightened factory manager and an active unionist, its first report painted a melancholy picture of long hours, low wages, and widespread child labor in the state's factories, and offered a sensational exposé of living conditions in Boston's tenements, complete with the owners' names and earnings. The report attracted enough attention that Massachusetts Congressman George F. Hoar proposed the establishment of a federal labor commission, an idea that produced an acrimonious debate in 1871 and 1872 over the propriety of government intervention in the labor market. Although Hoar had in mind a body confined to gathering and disseminating information, other Republicans called for more drastic action. "The laboring people of this nation," declared Indiana Congressman John P. Shanks, "think today that they are subjected unjustly to capital. . . . If they are correct, it is the duty of Congress to redeem them from that thralldom." Here, exulted one of Hoar's admirers, was "an entirely new programme for the legislation of the Nation." All this proved too much for both Democratic exponents of limited government and Republicans retreating from the idea of an activist state. "Where shall we stop?" one Congressman asked. "What is there in private life that shall be sacred from the intrusion of this Government?" Hoar's proposal passed the House, but died in the Senate after being first transformed by amendment into an investigation of the tariff.[39]

38. *Report of the Bureau of Statistics of Labor 1870*, 281–82; *AC*, 1869, 416; Dale Baum, *The Civil War Party System: The Case of Massachusetts, 1848–1876* (Chapel Hill, 1984), 147–53; Irving H. Bartlett, *Wendell Phillips: Brahmin Radical* (Boston, 1961), 351–55; Frederick Rudolph, "Chinamen in Yankeedom: Anti-Unionism in Massachusetts in 1870," *AHR*, 53 (October 1947), 1–29; *AC*, 1870, 474; S. P. Cummings to Charles Sumner, December 14, 1871, Sumner Papers.

39. William R. Brock, *Investigation and Responsibility: Public Responsibility in the United States, 1865–1900* (New York, 1984), 151; Leiby, *Carroll Wright*, 57–58; S. G. F. Spackman, "Na-

Henry Wilson, who supported the Hoar bill, found the debate's language deeply disturbing: "I never heard the term 'laboring class' here without the same sort of sensation which I used to have on hearing the word 'slave'," he told the Senate, adding that the law should never acknowledge the existence of "classes in this land of equality." Wilson's discomfiture illustrates how the labor reform program implicitly challenged the free labor ideology and the adequacy of the Radical vision of a beneficent state guaranteeing its citizens legal and political equality. The aim of Reconstruction, a Republican newspaper had announced in 1865, was to resolve the Southern "labor question" with "as little injustice to either side as . . . here in the North." By questioning the equity of Northern labor relations, the labor movement raised fundamental questions for the free labor ideology. Like freedmen seeking land and Western farmers advocating regulation of the railroads, labor reformers called upon the Republican party to move beyond its commitment to equality before the law to consider the realities of unequal economic power and widespread economic dependence, and the state's responsibility for combating them. By itself, class conflict was not "the submerged shoal on which Radical dreams foundered"; Radical Republicanism, as we have seen, ran aground on the all too visible politics of race and Reconstruction. Yet what *The Nation* called the emerging "politics of class feeling" did weaken Radicalism by fostering a new kind of political leadership: nonideological power brokers in a contentious, pluralistic society.[40]

The very diversity of the claims upon postwar Northern state governments facilitated a shift within the Republican party from an ideological to an organizational mode of politics. The party still contained characters like abolitionist editor Morgan Bates, Michigan's lieutenant governor between 1868 and 1872, who had begun his career as a journeyman printer on Greeley's *Tribune* and now lent money to poor homesteaders. But especially after Grant's assumption of the Presidency defused the ideological crisis of the Johnson years and weakened Radical influence in Washington, state parties fell under the control of powerful Senators who saw government less as an instrument of reform than as a means of obtaining office and mediating the rival claims of the diverse economic and ethnic groups that made up Northern society. The power of these leaders rested on control of federal patronage, which oiled political machines that ranged from local newspaper editors, postmasters, and revenue agents, up through city, county, and statewide party committees. Although many of these "Stalwarts" had cut their political eyeteeth in the

tional Authority in the United States: A Study of Concepts and Controversy in Congress 1870–1875" (unpub. diss., Cambridge University, 1970), 214–26; *CG*, 42nd Congress, 2d Session, 102–104, 221–22, 4016, 4042–44.

40. *CG*, 42d Congress, 2d Session, 4018, 4040; *New York Times*, July 22, 1865; Montgomery, *Beyond Equality*, ix–x; *Nation*, June 27, 1867.

antislavery crusade, most had grown impatient with the ideological mode of politics that had shaped the party at its birth and been further strengthened by the crises of war and Reconstruction. The organization itself, not the issues that had once created it, commanded their highest loyalty. And, on their own terms, they succeeded, for parties remained objects of intense devotion for the mass of Northerners, voter turnout reached incredibly high levels (in some states, 90 percent of those eligible regularly cast ballots), and Republicans continued to reign as the North's "natural" majority.[41]

The transition from the ideological politics of the Civil War era to the "professionally managed politics" of the Gilded Age, while not complete until the end of Reconstruction, was well underway during Grant's first term. The emerging politicos included both a rising second generation of Republican leaders and party founders able to adjust to the organizational mode. Probably the quintessential Stalwart was New York Senator Roscoe Conkling, a man of shrewdness, wealth, and good looks, whose obsession with physical fitness has a modern aspect (he had a passion for riding, boxing, and exercise, and neither smoked nor drank). After winning control in 1870 of the New York Custom House—whose army of employees spent far more time on party affairs than their ostensible jobs—Conkling's power quickly eclipsed that of former Governor Fenton and his Radical supporters. Similarly, wartime Gov. Andrew Curtin saw himself supplanted as head of Pennsylvania's party by Sen. Simon Cameron, backed by the Pennsylvania Railroad and powerful urban machines in Pittsburgh and Philadelphia. (The latter's election laws enabled local officials, aided by Nick English, the legendary "lightning calculator," to produce whatever majority the party needed to win each state election.) And John A. Logan eclipsed his Senate colleague Lyman Trumbull in Illinois. Veterans were the backbone of ex-General Logan's machine: In one county, Republicans "elected a *one-legged soldier* for County Treasurer, a *one-armed* soldier for circuit clerk, and a good soldier for County judge." Other powerful bosses included Senators William B. Allison, Zachariah Chandler, and Oliver P. Morton (the last of whom, it was said, opposed slavery everywhere except within his own organization, where he demanded "absolute servitude").[42]

41. Lewis M. Miller, "Reminiscences of the Michigan Legislature of 1871," *Michigan Pioneer and Historical Society Historical Collections*, 32 (1903), 425–46; Montgomery, *Beyond Equality*, 359–60; David Donald, *Charles Sumner and the Rights of Man* (New York, 1970), 349–50; Morton Keller, "The Politicos Reconsidered," *Perspectives in American History*, 1 (1967), 406–407; James D. Norris and Arthur H. Shaffer, eds., *Politics and Patronage in the Gilded Age: The Correspondence of James A. Garfield and Charles E. Henry* (Madison, Wis., 1970), xx–xxvii; Richard L. McCormick, "The Party Period and Public Policy: An Exploratory Hypothesis," *JAH*, 66 (September 1979), 281–83; Melvyn Hammarberg, *The Indiana Voter: The Historical Dynamics of Party Allegiance During the 1870s* (Chicago, 1977), 27.

42. Baum, *Civil War Party System*, 209; David M. Jordan, *Roscoe Conkling of New York* (Ithaca, N. Y., 1971), 133–53; Erwin S. Bradley, *Simon Cameron: Lincoln's Secretary of War*

In a sense, the growing ascendancy of organization-minded politicos represented a throwback to the nonideological mass politics that had flourished in the Age of Jackson. But the greatly expanded functions of government and the increased patronage made it possible to build political machines far larger and more powerful than before the war. Staffed by thousands of men who earned all or part of their livelihood from politics, the party organizations were among the largest institutions of the day. They were also among the most expensive, with state machines employing thousands of individuals, and their increasingly elaborate campaigns—replete with millions of pamphlets, broadsides, buttons, banners, and other paraphernalia—costing hundreds of thousands of dollars. Usually, candidates and officeholders were required to finance their own party, through "voluntary" assessments that could amount to a significant portion of their annual salaries. (Levies in Pennsylvania ranged from fifteen dollars for letter carriers to $4,000 for county nominees.)[43]

Coupled with the vast sums now handled by officials and the new functions assumed by the state, the increasing cost of political organization helped produce the political scandals of the Gilded Age. Much of the corruption of the Grant era involved payments to public officials by businesses seeking state aid, and as in the Reconstruction South it often proved difficult to tell where bribery left off and extortion began. The exceedingly complex import laws allowed customs officials to fine merchants for minor, technical violations, with the proceeds sometimes going directly into the party's coffers. Republican leaders expected public works contractors to provide kickbacks and businessmen who benefited from legislative favors to contribute to campaigns. And the notorious Whiskey Rings of St. Louis, Milwaukee, and other cities united Republican officials, internal revenue inspectors, and distillers in a massive scheme that defrauded the federal government of millions of dollars in excise taxes, while funneling money into both the pockets of the participants and campaign war chests. These scandals illustrate the complexity of the business-government relationship in the Grant era. The party's new bosses served business, but also preyed upon it, and many entrepreneurs resented a system that regularly required them to pay bribes to

(Philadelphia, 1966), 285–324; Robert Harrison, "The Structure of Pennsylvania Politics, 1876–1880" (unpub. diss., Cambridge University, 1970), 67–68; James P. Jones, *John A. Logan: Stalwart Republican from Illinois* (Tallahassee, 1982), 17–28; James D. Kilpatrick to Richard Yates, February 12, 1869, Richard Yates Papers, Illinois State Historical Society; Keller, *Affairs of State*, 255–56; David Turpie, *Sketches of My Own Times* (Indianapolis, 1903), 221–22.

43. Yearley, *Money Machines*, 104–107; Keller, *Affairs of State*, 241–44, "List of Officeholders in Wisconsin Who Have Contributed to the Funds of the U[nion] R[epublican] Cong. Ex. Comm., Washington," n.d. [October 1872], Keyes Papers; Harrison, "Pennsylvania Politics," 183–84.

individuals and tribute to the party. Nobody "owned" the Republican party, or, to put it another way, its first loyalty was always to itself.[44]

On both the state and national levels, the rise of organizational politics helped to eclipse the highly ideological issues associated with Reconstruction. Some rising Republican politicos evinced no interest whatever in such questions. Congressman Philetus Sawyer, a key member of the Wisconsin machine, for example, devoted himself entirely to obtaining internal improvements funds for the state; his sole contribution to the impeachment debate was a reference to a move to cut off debate: "Oh, give them ten minutes." Slowly, Reconstruction gave way to issues arising from the economic legacy of the war and the impact of capitalism's rapid expansion—questions defined by shifting alliances along East-West, urban-rural, and occupational lines rather than Radical-moderate or even Democrat-Republican divisions. This development placed a premium on the political arts of brokerage and compromise. Rhetorically, for example, Republicans remained committed to "sound money," but in practice the Grant Administration resisted a precipitous resumption of specie payments, and Congress in 1870 moved to quiet Western and Southern demands for easier credit by increasing the amount of national bank currency in circulation and redistributing it slightly away from the Northeast. So too, the tariff question, as James A. Garfield put it, required "the most careful and prudent management" lest it threaten party unity. And while the Republican party often found ethno-cultural appeals—temperance, Sabbatarianism, the use of the King James Bible in public schools— an effective means of mobilizing its voters at the local level, such issues also alienated key voting blocs, like Protestant Germans, whose loss the party could not afford.[45]

Nevertheless, the fact that Republicans remained internally divided on virtually every other question heightened the importance of the Civil War and Reconstruction as touchstones that transcended local differences and served as a continuing definition of the party's identity. For some politicians, this meant cynically "waving the bloody shirt" before each election; for others, a genuine concern for the fate of Southern blacks endured.[46]

44. Edwin G. Burrows, "Corruption in Government," in *Encyclopedia of American Political History*, ed. Jack P. Greene (New York, 1984), 1:429; Leonard D. White, *The Republican Era: 1869–1901* (New York, 1958), 118–26; Amy Mittelman, "The Politics of Alcohol Production: The Liquor Industry and the Federal Government 1862–1900" (unpub. diss., Columbia University, 1986), 76–101; Matthew Josephson, *The Politicos 1865–1896* (New York, 1938), 101–104; Yearley, *Money Machines*, 122.

45. James C. Mohr, "New York: The Depoliticization of Reform," in Mohr, ed., *Radical Republicans*, 76–77; Richard N. Current, *Pine Logs and Politics: A Life of Philetus Sawyer 1816–1900* (Madison, 1950), 55, 98–99; Norris and Shaffer, eds., *Politics and Patronage*, xv–xvi, 15; Unger, *Greenback Era*, 66–67, 164–65; Harold M. Helfman, "The Cincinnati 'Bible War,' 1869–1870," *Ohio Archaeological and Historical Quarterly*, 60 (October 1951), 369–86; Keller, *Affairs of State*, 129–30.

46. David J. Rothman, *Politics and Power: The United States Senate 1869–1901* (Cambridge, Mass., 1966), 87; Marcus, *Grand Old Party*, 6–11; Swenson, "Disillusionment," 109–15. The

Thus, if the Southern question waned as an inspiration for far-reaching reform or a source of new policy initiatives, a commitment to Reconstruction remained a powerful point of unity for party leaders in the North. Ironically, the rise of the Stalwarts did less to undermine Republican Southern policy than the emergence of an influential group of party reformers whose revolt against the new politics of the Grant era came to include the demand for an end to Reconstruction.

The Rise of Liberalism

By the end of Grant's first term, one disgruntled party leader complained, the "grand Republican organization" had "degenerated into thirty-seven Senatorial Cabals or 'Rings'," each monopolizing its state's share of federal patronage, and each having earned a reputation for corruption. And the new politics of the Gilded Age produced a growing chorus of demands for "reform," a refrain that won the support of Northerners from businessmen tired of being despoiled by party machines to politicos on the losing side of patronage battles. But at reform's cutting edge stood a collection of intellectuals, publicists, and professionals whose strategic location and literary talents more than made up for their meager numbers. Among the group's leading members were editors E. L. Godkin, Horace White, and Samuel Bowles, academic economists Francis Amasa Walker and David A. Wells, intellectually inclined businessmen Edward Atkinson and Isaac Sherman, and reform-minded politicians Carl Schurz and James A. Garfield. Mostly college graduates who resided in the urban Northeast or Western cities like Cincinnati and Chicago, these self-styled "best men" articulated a shared ideology and developed a sense of collective identity through organizations like the American Social Science Association (founded in 1869 with a membership that included Godkin, Atkinson, Wells, and the presidents of Harvard and Yale), and a network of influential journals, among them *The Nation, North American Review,* Springfield *Republican,* and Chicago *Tribune.* Their growing prominence reflected the coming of age of an American intelligentsia determined to make its mark on the politics of the Gilded Age.[47]

Central to reform ideology was the conviction that the political world, no less than the natural, was regulated by ascertainable laws that provided the basis for "scientific legislation." Classical liberalism supplied

term "waving the bloody shirt" refers to the practice of literally displaying the bloodstained uniforms of Civil War soldiers as a Republican campaign device. Used figuratively, it describes the party's practice of using the memory of the war to solidify its electoral support.

47. Hermann K. Platt, ed., *Charles Perrin Smith: New Jersey Political Reminiscences 1828–1882* (New Brunswick, N.J., 1965), 181; John G. Sproat, *"The Best Men": Liberal Reformers in the Gilded Age* (New York, 1968); Michael G. McGerr, *The Decline of Popular Politics: The American North, 1865–1928* (New York, 1986), 43–45; *Journal of Social Science,* 1 (1869), 195–99; Stephen Skowronek, *Building a New American State* (New York, 1982), 42–46.

the axioms of reformers' "financial science"—free trade, the law of supply and demand, and the gold standard. These doctrines had assumed the character of a stolid orthodoxy in the nation's colleges and were elevated to a moral dogma by the reformers, who tirelessly advocated a dismantling of the war's financial legacy. As special commissioner of the revenue, David A. Wells called for sweeping tariff reductions, a demand echoed by the American Free Trade League, founded in 1865 by "a few young men of wealth and college training, who . . . were enthusiastic disciples of Adam Smith." (Considered "a set of monomaniacs" by Wendell Phillips, league members saw themselves as "new abolitionists," and the protective tariff as a form of "commercial slavery.")[48] Equally insistently, reformers advocated a return to specie payments and deprecated calls for inflation of the nation's paper currency and the redemption of government bonds in greenbacks. Like free trade, the "honest dollar" of gold embodied natural law, while inflation, wrote *North American Review* editor Charles Eliot Norton, "means dishonesty, corruption, repudiation." Limited government formed another pillar of reform faith, for state intervention in economic affairs fostered all the distasteful aspects of Gilded Age politics: bribery, high taxes, and public extravagance. "The Government," declared Godkin, "must get out of the 'protective' business and the 'subsidy' business and the 'improvement' and 'development' business. . . . It cannot touch them without breeding corruption."[49]

Measures like the tariff, greenback inflation, and the income tax, which used the state to benefit one "class interest" at the expense of others, exemplified the reformers' bête noire, "class legislation." In their own eyes, liberal reformers stood above social divisions as disinterested spokesmen for the common good. Yet at the same time, the ideology of reform helped to crystallize a distinctive and increasingly conservative middle-class consciousness. In a reversal of the traditional glorification of productive labor, a writer for the American Social Science Association identified the "commercial class" as the backbone of society, whose role in public affairs had been eclipsed by new classes spawned by the industrial age. "An ignorant proletariat and a half-taught plutocracy," wrote historian Francis Parkman, had "risen like spirits of darkness on our

48. Sidney Fine, *Laissez-Faire and the General-Welfare State* (Ann Arbor, 1956), 47–50; Bernard Newton, *The Economics of Francis Amasa Walker: American Economics in Transition* (New York, 1968), 2–3, 153; Edwin L. Godkin, "Legislation and Social Science," *Journal of Social Science*, 3 (1871), 122; James A. Garfield to Edward Atkinson, August 11, 1868, Edward Atkinson Papers, MHS; Roeliff Brinkerhoff, *Recollections of a Lifetime* (Cincinnati, 1904), 194; Joanne H. Reitano, "Free Trade in the Gilded Age" (unpub. diss., New York University, 1974), 82–83, 104; Walter M. Merrill and Louis Ruchames, eds., *The Letters of William Lloyd Garrison* (Cambridge, Mass., 1971–81), 6:105.

49. Sproat, *"Best Men,"* 4, 145–46, 184–85, 246; Charles Eliot Norton to Charles Sumner, February 9, 1868, Sumner Papers; Richard L. McCormick, "The Discovery That 'Business Corrupts Politics': A Reappraisal of the Origins of Progressivism," *AHR*, 86 (April 1981), 255.

social and political horizon." Yet while the reformers railed against the railroad men and "iron and coal rings" whose actions distorted both marketplace and government, and lamented the eclipse of "the smaller workshop under the personal control of the owner" by the "larger corporate manufactory owned by absentee stock-holders," they seemed far more alarmed by a growing political danger from below.[50]

Granger laws providing for government regulation of the railroads appeared to reformers as egregious examples of "class legislation." (Instead of rate regulation, they much preferred the kind of railroad commission headed by Charles Francis Adams, Jr., in Massachusetts, which published detailed evaluations of business practices prepared by disinterested "experts," thus, ostensibly, enabling the market to function more efficiently while avoiding any hint of governmental coercion.) And while hardly unaware of labor's problems—*The Nation* took note of "the almost incredible fact that the condition of the working-classes throughout Massachusetts is a declining one"—the "best men" responded to strikes and demands for eight-hour legislation with a violent counterattack. These "schemes for interference with property" violated the "eternal laws of political economy" by undermining the principle of supply and demand and employing government's coercive power in the interests of a single class of citizens. *The Nation* even praised Calvin T. Sampson, the shoe manufacturer who brought Chinese strikebreakers to North Adams, Massachusetts, for his entrepreneurial initiative. The violence associated with the 1871 Paris Commune, coupled with large-scale rioting the same year between Catholic and Protestant Irish New Yorkers, rekindled fears aroused by the wartime draft riots about the North's "dangerous classes." Joel T. Headley's influential history of New York riots concluded that the city must protect itself by organizing a crack force of specially trained police, since the army could not always be relied upon "to shoot down their friends and acquaintances."[51]

Much of the degradation of Gilded Age politics, reformers believed, arose from the success of demagogues and spoilsmen who rose to power by playing upon the prejudices of working-class voters. New York's liberal reformers joined with a far larger number of businessmen, lawyers, financiers, and opposition politicians to overturn the Democratic ma-

50. Boston *Journal*, April 21, 1869, supplement; Michael L. Benedict, "Laissez-Faire and Liberty: A Re-Evaluation," *Law and History Review*, 1985, 293–331; Hamilton A. Hill, "Relations of Business Men to National Legislation," *Journal of Social Science*, 3 (1871), 149–50; Francis Parkman, "The Failure of Universal Suffrage," *North American Review*, 127 (July–August 1878), 14; George S. Merriam, *The Life and Times of Samuel Bowles* (New York, 1885), 2:196; Edward Atkinson, "An Easy Lesson in Money and Banking," *Atlantic Monthly*, 34 (August 1874), 204–205.

51. Sproat, *"Best Men"*, 164, 208–11; *Nation*, June 16, 1866, June 27, 1867, June 8, 1871; Joseph Logsdon, *Horace White: Nineteenth Century Liberal* (Westport, Conn., 1971), 142–43; Rudolph, "Chinamen in Yankeedom," 18; Joel T. Headley, *The Great Riots of New York, 1712–1873* (New York, 1873), 20–22.

chine of "Boss" William M. Tweed, which had plundered the city of tens of millions of dollars (the scale of its depredations dwarfed anything in the Reconstruction South). To the "best men," the Tweed Ring epitomized the symbiotic relationship between corruption, organizational politics, the political power of both railroad men and the urban working class, and the misuses of the state. For the Boss controlled the city's patronage, forged an alliance with Jim Fisk and Jay Gould during the Erie Wars, established close ties with labor unions, and fashioned a kind of welfare system that used municipal funds to aid Catholic schools and provide food and fuel for the poor. To the reformers' annoyance, moreover, the exposure of Tweed's activities failed to diminish his popularity among lower-class voters. Popular Irish-American ballads of the 1870s recalled him as an urban Robin Hood and friend of labor, "poverty's best screen. . . . no matter who may you condemn." "The word 'reform'," noted New York Democratic chieftain Horatio Seymour, "is not popular with the workingmen. To them it means less money spent and less work."[52]

Even more outrageous than Tweed, because he was a Republican, was Massachusetts Congressman Benjamin F. Butler, who flamboyantly identified himself with the causes that appalled reformers: the eight-hour day, inflation, and payment of the national debt in greenbacks. He even shocked respectable opinion by embracing women's suffrage and Irish nationalism and praising the Paris Commune. "Butlerism" became a shorthand for a new kind of mass politics that had infused American public life with "the spirit of the European mob." The juxtaposition of Butler and Karl Marx as advocates of the "spoilation" of property occurred more frequently in the reform press than the modern reader might imagine. In 1868, determined to "stamp on Butler and bury him," the wealthy Harvard-educated lawyer and author Richard Henry Dana, Jr., challenged Butler for his Congressional seat. Despite the enthusiastic backing of "all the best men of the party all over the country," as well as funding from Boston bondholders and bankers alarmed by the incumbent's financial heresies, Dana's campaign was a fiasco. Pilloried as "an aristocrat, of the snobbiest sort," handicapped by his opposition to the eight-hour movement and distaste for mingling "with the common people," he ended up receiving only 9 percent of the vote. Reformers were more successful three years later, when they helped block Butler's unorthodox bid for the gubernatorial nomination (he canvassed the state denouncing capitalists who ex-

52. Alexander B. Callow, Jr., *The Tweed Ring* (New York, 1966); John W. Pratt, "Boss Tweed's Public Welfare Program," *New York Historical Society Quarterly*, 45 (October 1961), 396–411; Michael A. Gordon, "Studies in Irish and Irish-American Thought and Behavior in Gilded Age New York City" (unpub. diss., University of Rochester, 1977), 387; Alexander C. Flick, *Samuel Jones Tilden: A Study in Political Sagacity* (New York, 1939), 307.

ploited honest workingmen, and urged the party to adopt labor reform as its rallying cry).[53]

Clearly, liberal reform was at one and the same time a moral creed, part of an emerging science of society, and the outcry of a middle-class intelligentsia alarmed by class conflict, the ascendancy of machine politics, and its own exclusion from power. Its emergence reflected a splintering of the Radical impulse that had flourished during the Civil War and early Reconstruction. Nearly all the liberal reformers had been early advocates of emancipation and black suffrage. (Isaac Sherman had even helped pay for John Brown's New York farm and contributed to his "military chest.") Yet if all Radicals agreed the state should establish the principle of civil and political equality, liberals increasingly insisted it should do little else. As labor, farmers, and other groups of Northerners demanded a new round of government-sponsored social and economic change, the same liberal reformers who had once exalted the power of the activist state retreated into what one Democratic Congressman recognized as a "genuine conservatism." Fearing that strong government could be used for the wrong purposes, they attacked "the fallacy of attempts to benefit humanity by legislation," and insisted public authority was "by nature wasteful, corrupt, and dangerous."[54]

Inspired by this critique of governmental activism, reformers searched for ways to insulate politics from the influence of the North's urban working class, and limit the scope and expense of government. If their economic outlook reflected the teachings of British political economy, they increasingly took England as a political model as well. To call the reformers Anglophiles, however, would not be strictly accurate, for their loyalty was not to the real Britain, a society of mature industrialism and expanding political participation, but to an idealized English past marked by hierarchy, order, and social harmony. Many reformers retreated from democratic principles altogether, advocating educational and property qualifications for voting, especially in the nation's large cities, and an increase in the number of officials appointed rather than elected. As early as 1866, Godkin proposed that taxpayers should exercise a veto on city expenditures "made by the representatives of mere numbers." Such views became increasingly fashionable in intellectual circles as time went on. "The tendency among thoughtful men who desire honesty, economy

53. Hans L. Trefousse, *Ben Butler: The South Called Him BEAST!* (New York, 1957), 219–24; William D. Mallam, "Butlerism in Massachusetts," *New England Quarterly*, 33 (June 1960), 186–206; *Nation*, November 12, 19, 1868; Edward Atkinson to Henry L. Dawes, March 17, 1874, Henry L. Dawes Papers, LC; Samuel Shapiro, " 'Aristocracy, Mud and Vituperation': The Butler-Dana Campaign in Essex County in 1868," *New England Quarterly*, 31 (September 1958), 340–60; Baum, *Civil War Party System*, 156–61.

54. F. B. Sanborn to Isaac Sherman, December 1, 1875, Sherman Papers; Michael C. Kerr to Manton Marble, August 26, 1872, Manton Marble Papers, LC; Charles Moran to Edward Atkinson, June 30, 1871, Atkinson Papers; Harold M. Hyman, *A More Perfect Union: The Impact of the Civil War and Reconstruction on the Constitution* (New York, 1973), 533.

and a good deal of intelligence in legislation," reported a New York newspaper in 1869, "is towards a restriction of the right of suffrage considerably inside its present limits."[55]

As one reformer recognized, proposals for sweeping restrictions on the ballot stood little chance of approval, since "men will not vote to disfranchise themselves." Other reforms offered greater prospect of success. Together with "active business men and merchants" who believed they bore the brunt of corruption and rising tax rates, reformers pressed for changes in the structure of city government so as to oust the spoilsmen, reduce expenditures, and bring "business principles" to municipal affairs. The destruction of the Tweed Ring was not their only success. A Citizen's Ticket organized by the Chicago *Tribune* temporarily swept the city's corrupt Republican machine out of office in 1869. And a group of Philadelphia reformers and businessmen persuaded the state government to call a constitutional convention, hoping thereby to undermine their city's "Ring." Composed of "one hundred lawyers and thirty-three honest men," the convention heeded both the outcry against corruption and the growing desire to replace the system of special legislative favors, so unpredictable and fraught with opportunities for extortion, with general laws and charters. The new constitution abolished the office of city alderman, a cog in the party machine generally occupied by men of modest circumstances, prohibited the granting of special charters, set a limit on Philadelphia's debt, and doubled the number of legislators, all in the hope of discouraging bribery. Several other states instituted similar reforms.[56]

These changes, a retreat from the period of state activism that followed the Civil War, failed to end patronage-based politics or achieve efficient government. Nor did civil service reform, the liberals' favored means of breaking the power of party machines and opening positions of responsibility to men like themselves. Beginning in 1865, Rhode Island Congressman Thomas Jenckes introduced proposals, modeled on British precedent, for a system of competitive examinations and permanent tenure in office for federal employees. Here, Jenckes maintained, was a way to elevate the tone of public life by attracting men who had achieved "reputable positions in the learned professions or in business." Predictably, party leaders remained unenthusiastic about a proposal that threatened the lifeblood of their organization. For their own reasons, they pointed

55. Donald, *Sumner*, 234; Horace White, "An American's Impressions of England," *Fortnightly Review*, September 1, 1875, 291–305; Sproat, *"Best Men,"* 250–53; *Nation*, October 18, 1866; New York *Journal of Commerce*, February 2, 1869.

56. Dorman B. Eaton, "Municipal Government," *Journal of Social Science*, 5 (1873), 7; Boston *Journal*, April 21, 1869, Supplement; Charles Nordhoff, "The Misgovernment of New York—a Remedy Proposed," *North American Review*, 113 (October 1871), 321–43; Hill, "Relations of Business Men," 155–64; Logsdon, *White*, 175–76; Gillette, "Corrupt and Contented," 43–76; Evans, *Pennsylvania Politics*, 81–84; Keller, *Affairs of State*, 112, 140.

out the antidemocratic assumptions that underlay the plan. Life tenure, critics charged, would "establish an aristocracy of office-holders" insulated from the will of the people. The examination system, at a time when only a tiny percentage of Americans enjoyed access to a college education, would limit office to wealthy "dunce[s]" who had "crammed up to a diploma at Yale" but knew nothing of practical affairs. The outcry against corruption induced Congress in 1871 to authorize the President (apparently against his better judgment) to appoint a commission to prescribe rules for examining applicants for the civil service. But its guidelines, promulgated the following year, were largely ignored.[57]

The lack of progress in reforming the civil service was only one among several reasons for reformers' growing estrangement from the Grant Administration. Although initially entertaining high hopes for the new President, who had never been known as a party man, reformers soon chided Grant for surrounding himself with cronies and spoilsmen. They saw the only members of the Administration close to the "best men"— Secretary of the Interior Jacob Cox, Attorney General Ebenezer R. Hoar, and Special Commissioner of the Revenue David A. Wells—forced out of office in 1870. Nor did the idea of free trade make headway. But nothing revealed more starkly the Presidents' lack of concern for reformers' sensibilities than his effort to annex the Dominican Republic.[58]

A number of factors came together to produce this disreputable scheme. The navy desired a Caribbean base, American businessmen who owned property on the island pressed for annexation, and the President believed that Southern blacks might wish to seek refuge in the Dominican Republic and that the success of free labor there would compel nearby Cuba and Puerto Rico to abolish slavery. In addition, the ruling Dominican political faction hoped to save itself from overthrow in a continuing civil war, and Grant's private secretary Orville Babcock, who, without informing Secretary of State Hamilton Fish, negotiated the agreement, owned land on the island that would appreciate in value under American rule. Nobody considered the wishes of the people who lived there, although a plebiscite, held with four days' advance notice and coupled with a warning that opponents of the treaty would be exiled or shot, produced a resounding vote in favor of annexation.[59]

The debate over the treaty, which came before the Senate during the

57. Ari Hoogenboom, *Outlawing the Spoils: A History of the Civil Service Reform Movement 1865–1883* (Urbana, Ill., 1961), 16–19, 87–95, 132–33; *CG*, 39th Congress, 2d Session, 838; 41st Congress, 2d Session, 3257; 42d Congress, 2d Session, 458; Yearley, *Money Machines*, 99–101; W. C. Dodge to Horace Austin, July 17, 1873, Horace Austin Papers, Minnesota Historical Society.

58. Sproat, *"Best Men,"* 74–75; Peskin, *Garfield*, 339; Logsdon, *White*, 179–86.

59. Allan Nevins, *Hamilton Fish: The Inner History of the Grant Administration* (New York, 1936), 1:250–77, 315; *AC*, 1868, 687; Daniel Ammen, *The Old Navy and the New* (Philadelphia, 1891), 509; James D. Richardson, ed., *A Compilation of the Messages and Papers of the Presidents 1789–1897* (Washington, D.C., 1896–99), 7:62, 99–100.

first half of 1870, underscored once again the buoyant nationalism the Civil War had inspired in the Republican party. Already during the Johnson years Secretary of State Seward had engineered the purchase of Alaska and unsuccessfully attempted to acquire the Danish West Indies; now, some supporters pictured annexation as one step in a process by which Canada, Mexico, and "all the West Indian islands" would be absorbed into the American republic. Just as its own commitment to emancipation provided a moral justification for Britain's late nineteenth-century imperialism, Republicans who before the war had opposed "manifest destiny" because it would "rivet the bonds of slavery" upon new territories now maintained that with the peculiar institution abolished, America's expansion would enable "our neighbors to join with us in the blessings of our free institutions." Southern Republicans, including such black leaders as Sen. Hiram Revels and Congressman Joseph Rainey, shared in this expansionist spirit. Frederick Douglass, who served as secretary of a commission that visited the Dominican Republic in 1871 in an attempt to drum up Congressional support for the treaty, returned appalled by the poverty and backwardness of the inhabitants, convinced annexation would uplift the island nation and "transplant within her tropical borders the glorious institutions" of the United States.[60]

The treaty, however, quickly encountered the formidable opposition of Charles Sumner, chairman of the Senate Foreign Relations Committee. Despite the President's personal plea for his support, the Foreign Relations Committee in March reported the treaty adversely to the Senate. In a lengthy speech, Sumner insisted annexation would complicate relations with European powers, involve the United States in the Dominican civil war, and—the argument closest to his heart—threaten the independence of neighboring Haiti, the hemisphere's only black republic. (In reply, Oliver P. Morton of Indiana waxed eloquent about the island's natural resources, bringing a large block of Dominican salt onto the Senate floor, which the solons took turns licking.) With a majority of party leaders supporting the President, Sumner was joined in opposition by Democrats, New England allies, and Republicans like Lyman Trumbull and Carl Schurz attuned to the ideology of liberal reform and convinced that the treaty was another result of corrupt influences upon government. In June 1871, the President suffered a humiliating defeat when the treaty, requiring a two-thirds majority for approval, mustered only a 28-28 tie. Although the Administration made one further attempt to revive the idea at the end of the year, annexation was effectively dead.[61]

60. Glyndon G. Van Deusen, *William Henry Seward* (New York, 1967), 526–43; New York *Evening Post*, May 13, 1870; David B. Davis, *Slavery and Human Progress* (New York, 1984), 298–312; *CG*, 41st Congress, 3d Session, 226, 416, 427–28; Nevins, *Fish*, 2:497–99; Nathan I. Huggins, *Slave and Citizen: The Life of Frederick Douglass* (Boston, 1980), 125–26; *New National Era*, January 12, 1871.
61. Donald, *Sumner*, 434–52.

The "furious contest" over the Dominican treaty had profound consequences for Republican politics, solidifying Grant's alliance with powerful Stalwarts like Conkling, Morton, and Butler and thoroughly alienating the reformers. Grant never abandoned his commitment to annexation, even including in his second inaugural a curious passage linking it with a divine plan to unite the world as "one nation, speaking one language, [with] armies and navies . . . no longer required." Convinced he had been betrayed, the President removed Sumner's close friend John Lathrop Motley as ambassador to Great Britain. For his part, Sumner's attacks on the President became increasingly personal and vitriolic. One philippic equated the treaty with the Kansas-Nebraska Act and Grant with Pierce, Buchanan, and Andrew Johnson. The breach reached its conclusion at the opening of the Forty-Second Congress in December 1871, when the Republican caucus, at Grant's behest, removed Sumner as head of Foreign Relations, ostensibly because the Senator was no longer on speaking terms with either the President or Secretary of State Fish. Simon Cameron succeeded to the chairmanship—"swapping an eagle for a toad," as the New York *World* described it. Nothing could have demonstrated the ascendancy of organizational politics more emphatically than Sumner's humiliation.[62]

Among the most striking features of the whole affair was the ease with which the treaty's critics fell back upon racism to bolster their position. Sumner, concerned for the right of nonwhite peoples to pursue an independent destiny, found himself in an uncomfortable alliance with those who deemed Dominicans unfit for the rights and responsibilities of American citizenship. It was not surprising to hear Democratic Congressmen deny that the country could safely absorb "degenerate races." But Carl Schurz also sought to demonstrate that the incorporation of tropical peoples would destroy the foundations of the republic. The very lushness of the tropics, Schurz insisted, enabled its inhabitants to live without labor; they therefore descended into "shiftlessness" and became incapable of the self-discipline necessary for productive free labor and active citizenship. Because tropical peoples could not rule themselves democratically, annexation would introduce a "poison" into the nation's public life, leading inexorably down the road to despotic government. Schurz claimed his elaborate circular argument rested on environment, not race, but other reformers proved less fastidious. In *The Nation*, Godkin condemned the idea that "ignorant Catholic spanish negroes" could become American citizens, while Charles Nordhoff of the New York *Evening Post* spoke of the Dominican Republic's "barbarous population." Their cri-

62. Jonathan B. Brownlow to Leonidas C. Houk, December 23, 1870, Leonidas C. Houk Papers, McClung Collection, LML; Donald, *Sumner*, 445–500; Richardson, ed., *Messages and Papers*, 7:222; *CG*, 41st Congress, 3d Session, 226–31; 42d Congress, 1st Session, 35–52; New York *World*, March 14, 1871.

tique of annexation laid the groundwork for the racial anti-imperialism that flourished at the turn of the century and involved many of the same individuals, including Godkin and Schurz.[63]

Ironically, as Illinois Sen. Richard Yates pointed out, opponents of expansionism employed arguments extremely reminiscent of proslavery ideology, while its supporters upheld the principle that nonwhites could be successfully incorporated into the body politic. (No people, quipped Nevada Sen. James W. Nye, were "too degraded" for citizenship: "We have New Jersey, and all things considered, it has proven a success.") All this held profound implications for Reconstruction. Sumner, commented a Massachusetts Republican, "has shown that he can oppose annexation, without losing one spark of the old fire." But for others, opposition to the treaty helped crystallize a growing disillusionment with both egalitarian ideology and the entire Reconstruction experiment. Schurz made the connection explicit. The Deep South—"the semi-tropical portion of this Republic"—could never come up to the standard of stable republican government. Further federal intervention toward this end was, it seemed, pointless.[64]

The reformers' accelerating disenchantment with Reconstruction, however, derived from a variety of sources. Many saw the Southern question as an annoying distraction that enabled party spoilsmen to retain the allegiance of voters by waving the bloody shirt, while preventing tariff reduction, civil service reform, and good government from taking the center stage of politics. Although they had supported, indeed helped to formulate, the Reconstruction Acts and postwar amendments, reformers insisted more strenuously than other Republicans that with the principle of equal rights secured, the party should move on to the "living issues" of the Gilded Age. Increasingly, Democratic criticisms of Southern government found a receptive audience among liberal reformers of the North. Like the Tweed Ring and "Butlerism," Reconstruction underscored the dangers of unbridled democracy and the political incapacity of the lower orders. "Universal suffrage," wrote Charles Francis Adams, Jr., in 1869, "can only mean in plain English the government of ignorance and vice:—it means a European, and especially Celtic, proletariat on the Atlantic coast, an African proletariat on the shores of the Gulf; and a Chinese proletariat on the Pacific." To reformers like Adams, egalitarian ideas seemed an anachronism, a throwback to the unscientific sentimentalism of an earlier era. Class and racial prejudices reinforced one an-

63. CG, 41st Congress, 3d Session, 798–99, Appendix, 26–34; Charles Nordhoff to Charles Sumner, December 21, 1870, Sumner Papers; Robert L. Beisner, *Twelve Against Empire* (New York, 1968), 561–77.

64. CG, 41st Congress, 3d Session, 429; Jud Samon, "Sagebrush Falstaff: A Biography of James Warren Nye" (unpub. diss., University of Maryland, 1979), 393; Thomas Russell to J. A. Smith, April 22, 1871, Sumner Papers; CG, 41st Congress, 3d Session, Appendix, 29.

other, as the reformers' concern with distancing themselves from the lower orders at home went hand in hand with a growing insensitivity to the egalitarian aspirations of the former slaves.[65]

In addition to obvious parallels between Southern and Northern corruption, the economic and social policies of Reconstruction governments struck reformers as unacceptable examples of "class legislation." The complaints of taxpayers' conventions about extravagant expenditures by propertyless lawmakers won "the hearty sympathy of the best Northerners," because of increasing taxes and swollen public budgets in the North, and the fact that in both regions, the general property tax allowed the poor to escape taxation altogether. The "best men" even organized their own Tax-Payers' Union, complete with a journal, People's Pictorial Taxpayer, to lobby for revenue reform. Nor did Southern Republican Congressmen win reformers' approbation, for most opposed tariff reduction, favored currency inflation, displayed little sympathy for the idea of civil service reform (an understandable position for men who relied upon the spoils of office even more heavily than their Northern counterparts), and even supported the removal of Sumner from his chairmanship.[66]

To reformers, Reconstruction increasingly seemed to exemplify all the deleterious consequences of state activism. Even at the height of nationalist exuberance in 1865 and 1866, men like Schurz had feared that an effective exercise of federal power to protect the freedmen, however necessary in the short run, might produce a "habit of overriding State rights" and a "fatal familiarity with arbitrary processes." Freedom, reformers insisted, meant not economic autonomy or the right to call upon the aid of the activist state, but the ability to compete in the marketplace and enjoy protection against an overbearing government. More insistently than other Republicans, reformers argued that the nation had done all it could for blacks; it was up to the freedmen to make their own way in the world. "The removal of white prejudice against the negro," declared The Nation in 1867, "depends almost entirely on the negro himself." Legislation could never counteract the "great burden" weighing upon the race: its "want of all the ordinary claims to social respectability." Reformers' opposition to continuing federal intervention, as we have seen, came to a head in debates over the Ku Klux Klan Act of 1871 (managed in the House by the hated Butler). Many now described the

65. Nation, October 17, 1867; Harris L. Dante, "The Chicago Tribune's 'Lost' Years, 1865–1874," JISHS, 58 (Summer 1965), 150–51; Logdson, White, 178; Roger A. Cohen, "The Lost Jubilee: New York Republicans and the Politics of Reconstruction and Reform, 1867–1878" (unpub. diss., Columbia University, 1975), 96–102; McGerr, Decline of Popular Politics, 46.

66. Nation, July 15, 1869, July 6, 1871; Yearley, Money Machines, xii–xv, 10, 37–45; William M. Grosvenor to Edward Atkinson, February 1, 1872, Atkinson Papers; Reitano, "Free Trade," 120–21; Terry L. Seip, The South Returns to Congress: Men, Economic Measures, and Intersectional Relationships, 1868–1879 (Baton Rouge, 1983), 162–64, 178–87; CG, 42d Congress, 1st Session, 52, 2d Session, 3159, 2d Session, Appendix, 128–29; Logdson, White, 168.

very goal of protecting blacks' equal rights as quixotic. "There are many social disorders," declared Schurz in the Senate, "which it is very difficult to cure by laws." And if local authorities proved unwilling or unable to put down violence? Blacks, said Godkin, should move to states where their rights were still respected.[67]

Fundamentally, reformers believed, Southern violence arose from the same cause as political corruption: the exclusion from office of men of "intelligence and culture." If in the North, civil service reform offered a solution, in the South, reformers advocated the removal of political disabilities that barred prominent Confederates—the region's "natural leaders"—from office. "We have got to go back to the doctrine of Gov. Andrew's valedictory . . ." wrote Jacob Cox, "and recognize the fact that the South can only be governed through the part of the community that embodies the intelligence and the capital." Edward Atkinson came to the same conclusion after a brief visit to the South in 1870. Putting aside "the question of color," he insisted, Southern misgovernment stemmed inevitably from the misguided attempt to place "the control of affairs in the hands of the more ignorant classes" while "a large portion of the most active and intelligent people [remain] . . . under political disability." Thus, a remarkable reversal of sympathies took place, with Southern whites increasingly portrayed as the victims of injustice, while blacks were deemed unfit to exercise suffrage and carpetbaggers denounced as unprincipled thieves. Originating as a critique of social and political changes in the North, liberal reform had come to view Reconstruction as an expression of all the real and imagined evils of the Gilded Age. "Reconstruction," declared The Nation, "seems to be morally a more disastrous process than rebellion." As an experiment in government, it had "totally failed."[68]

The Election of 1872

As Grant's first term drew to a close, the categories moderate and Radical—defined by policies regarding blacks, the South, and Reconstruction—no longer adequately described Republican factions. With the decline of Radicalism, the ascendancy of organizational politics, and the emergence of liberal reform, party alignments now centered on attitudes toward Grant himself and the new politics of the Gilded Age.

Conscious of their weakness within Republican ranks and Grant's con-

67. Carl Schurz, "The True Problem," Atlantic Monthly, 19 (March 1867), 371–78; William M. Armstrong, ed., The Gilded Age Letters of E. L. Godkin (Albany, 1974), 75; Nation, August 1, 1867; Charles Nordhoff to Charles Sumner, March 15, 1871, Sumner Papers; CG, 42d Congress, 1st Session, 686–87; Nation, March 23, 1871.
68. Nation, December 2, 1869, June 16, 1870; Springfield Republican, April 7, 1871; Jacob D. Cox to James A. Garfield, March 27, 1871, Edward Atkinson to Garfield, May 13, 1871, James A. Garfield Papers, LC; CG, 42d Congress, 2d Session, 699; Nation, March 23, July 6, 1871, March 28, 1872.

tinuing popularity despite his Administration's blunders, reformers increasingly turned to thoughts of a new party, which would "enact on the *national stage*" the stunning victories achieved by alliances of reform Republicans and Democrats in Virginia, Tennessee, and Missouri. In a September 1871 speech in Nashville, Schurz launched Liberal Republicanism as a national movement, outlining a platform that included civil service reform, tariff reduction, lower taxes, the resumption of specie payments, and an end to land grants to railroads. As befitted its location, however, the speech focused on the South. While insisting that the postwar constitutional amendments must be held inviolate, Schurz advocated political amnesty, an end to federal intervention, and a return to "local self-government" by men of "property and enterprise." Schurz sincerely believed blacks' rights would be more secure under such governments than under the Reconstruction regimes. But whether he quite appreciated it or not, his program had no other meaning than a return to white supremacy. As Oliver P. Morton would remark a few years later, "when certain men talk about local self-government by the people, they mean the white people." In January 1872, Missouri Liberals called for a national nominating convention to meet in Cincinnati in early May. The gathering, wrote Godkin, would herald "the break-up of old parties": "Reconstruction and slavery we have done with; for administrative and revenue reform we are eager."[69]

Nothing underscored more dramatically how the Republican organization had changed than the large number of party founders who now rallied to the Liberal cause. The list included numerous former Democrats, like Lyman Trumbull, who believed "old Whiggery dominated the [Republican] party." There were also antislavery pioneers such as George W. Julian, convinced that politics had degenerated into little more than "a struggle between the ins and outs for the loaves and fishes" and that Liberalism promised a reaffirmation of the politics of principle over the claims of organizational discipline. But what united nearly all these disaffected politicians was the experience of having been pushed aside by the party's new leadership. A few Liberal leaders, like Illinois Gov. John M. Palmer and Francis W. Bird of Massachusetts, retained considerable power (although the Bird Club resented the treatment of Sumner and feared the rise of Ben Butler). More typical was the experience of three Republican Civil War governors: New York's Reuben Fenton, his influence now eclipsed by Senator Conkling; Pennsylvania's Andrew Curtin, whose supporters had been ousted from power by the Cameron Ring; and Austin Blair, his ambition for a Michigan Senate seat frustrated by the Zachariah Chandler machine. A group of disgruntled

69. James R. Doolittle to Manton Marble, January 5, 1872, Marble Papers; Frederic Bancroft, ed., *Speeches, Correspondence and Political Papers of Carl Schurz* (New York, 1913), 2:258–85; *CR*, 43d Congress, 2d Session, 371; *Nation*, March 21, 1872.

Radical Congressmen shared similar grievances. Julian had seen the Indiana legislature, at the behest of Senator Morton, gerrymander him out of office in 1870. Ignatius Donnelly blamed Minnesota party boss Alexander Ramsey for his own defeat in 1868. And Ohio's James M. Ashley believed Sen. John Sherman responsible for his dismissal as governor of Montana, a position to which he had been appointed after losing his seat in Congress. Illinois produced the most impressive Liberal roster, for here, along with Trumbull and Palmer, the movement's leaders included such intimates of Lincoln as Gustav Koerner, David Davis, Jesse K. DuBois, and Orville H. Browning—all bypassed by John A. Logan's machine.[70]

Given this long list of experienced if mostly out-of-office politicos, it is not surprising that no dearth of candidates presented themselves for the Cincinnati Convention's nomination. Illinois alone harbored three aspirants: Governor Palmer (whose states rights opposition to federal aid at the time of the Chicago Fire won him considerable Southern support); Supreme Court Justice Davis (already nominated by a Labor Reform Convention); and Senator Trumbull (who followed in Lincoln's 1860 footsteps by delivering a speech at New York's Cooper Union that amounted to a declaration of availability). Many Easterners supported Charles Francis Adams, who, however, departed for a European vacation after cautiously throwing his hat into the ring. New York *Tribune* editor Horace Greeley, a close ally of Fenton whose own appetite for office had not been sated by a single term, twenty years earlier, in Congress, also coveted the nomination, as did Missouri Gov. B. Gratz Brown. (Foreign birth barred Schurz, the movement's preeminent spokesman, from consideration.)[71]

A heterogeneous collection of men alienated from the Grant regime assembled in Cincinnati: reformers, free traders, antislavery veterans (including a large number of "old Free Soilers of 1848 and 1852"), and

70. Michael E. McGerr, "The Meaning of Liberal Republicanism: The Case of Ohio," *CWH*, 28 (December 1982), 311–12; Brinkerhoff, *Recollections*, 191–93; Evarts B. Greene, "Some Aspects of Politics in the Middle West, 1860–72," *Proceedings, SHSW*, 59 (1911), 72–75; Abbott, "Maintaining Hegemony," 21; Bradley, *Triumph of Militant Republicanism*, 416–19; Robert C. Harris, "Austin Blair of Michigan: A Political Biography" (unpub. diss., Michigan State University, 1969), 207–13; George W. Julian, *Political Recollections, 1840 to 1872* (Chicago, 1884), 303; George W. Julian Journal, April 25, December 21, 1870, Julian Papers; Ridge, *Donnelly*, 108–18; Robert F. Horowitz, *The Great Impeacher: A Political Biography of James M. Ashley* (Brooklyn, 1979), 162–65; Stephen L. Hansen, "Principles, Politics, and Personalities: Voter and Party Identification in Illinois, 1850–1876" (unpub. diss., University of Illinois, Chicago, 1978), 278–83.

71. George H. Work to John M. Palmer, February 21, 1872, John M. Palmer Papers, Illinois State Historical Library; Willard L. King, *Lincoln's Manager: David Davis* (Cambridge, Mass., 1960), 278–80; *Speech of Hon. Lyman Trumbull* (Washington, D.C., 1872); Earle D. Ross, *The Liberal Republican Movement* (New York, 1919), 82; Charles Francis Adams Diary, April 1–24, 1872, Adams Family Papers, MHS; Norma L. Peterson, *Freedom and Franchise: The Political Career of B. Gratz Brown* (Columbia, Mo., 1965), 212–15.

a considerable body of men who "had been turned out of office or expected to get in." For a gathering dedicated to reforming politics, the convention witnessed a remarkable amount of behind-the-scenes maneuvering. Even before it assembled, Schurz had engaged in secret negotiations with Democratic National Chairman August Belmont, with the aim of uniting the two parties behind an Adams-Trumbull ticket. Meanwhile, four prominent reform editors agreed to publish simultaneous attacks upon Davis, thereby creating a "wondrous consensus of public opinion" against him. But, as Henry Watterson later related, all these schemes were "knocked into a cocked hat" when, to the astonishment of everybody except *Tribune* managing editor Whitelaw Reid, Horace Greeley emerged as the nominee, with Brown as his running mate.[72]

The weakness of the other leading candidates helped produce this unexpected result. The genteel Adams, the intellectuals' choice, was undercut by his old antagonists in the Bird Club, and at any rate, as "the greatest iceberg in the northern hemisphere," hardly inspired confidence as an effective campaigner. Nor did the bookish Trumbull possess much popular appeal. Davis was suspect for not declining the nomination of the labor convention, whose platform included the eight-hour day, greenbackism, and the taxation of government bonds. The state's excess of favorite sons prevented Illinois from uniting behind any candidate. Greeley, to be sure, had a history of erratic judgment, including his support of peaceable secession in the winter of 1860–61, and a private peace effort he launched in 1864. But he was genuinely popular, his long editorial career having endeared him to the many delegates with antislavery backgrounds. The platform echoed the principles laid down by Schurz in 1871 (equality before the law, amnesty, local self-government, civil service reform) with one large exception. Because Greeley's name was synonymous with the protective tariff, the delegates sidestepped the issue of free trade.[73]

Many Liberals considered Greeley a formidable candidate. "He is popular from Maine to California," wrote one New Yorker, and had far more support among "manufacturers and working-men" than anyone committed to tariff reduction. To reformers, however, the nomination of a man opposed to free trade and specie currency, indifferent to civil service reform, and with a career of supporting prohibition and other efforts by

72. George W. Julian Journal, May 9, 1872, Julian Papers; *CG*, 42d Congress, 2d Session, 4015; Irving Katz, *August Belmont: A Political Biography* (New York, 1968), 198; Henry Watterson, "The Humor and Tragedy of the Greeley Campaign," *Century*, 85 (November 1912), 32–39; James G. Smart, "Whitelaw Reid and the Nomination of Horace Greeley," *Mid-America*, 49 (October 1967), 227–43.

73. Matthew T. Downey, "Horace Greeley and the Politicians: The Liberal Republican Convention in 1872," *JAH*, 53 (March 1967), 727–50; Smart, "Whitelaw Reid," 231–32; Samuel Bowles to Charles Sumner, May 21, 1872, Samuel Bowles Papers, Yale University; Logsdon, *White*, 228–39; Martin Duberman, *Charles Francis Adams 1807–1886* (Stanford, 1960), 356–66.

"quacks, charlatans, ignoramuses, and sentimentalists" to use the state for their own purposes, came as a stunning blow. Schurz, convinced that the outcome had deprived the movement of "its higher moral character," sent a long, barely civil letter to the nominee, inviting him to withdraw. When Greeley refused to do so, some free traders vainly tried to entice another candidate into the field. In the end, most reformers deemed Greeley preferable to Grant, although a significant number, including Godkin and Atkinson, ended up supporting the President. ("That Grant is an Ass," wrote one Ohio Liberal, "no man can deny, but better an Ass than a mischievous Idiot.") The outcome discredited the movement for tariff reduction (the Free Trade League "went to Cincinnati but never came back") and, wrote Godkin, made it "impossible for anyone to speak of 'reform' . . . without causing shouts of laughter." Equally important, it ensured that the Greeley campaign would focus on the one issue capable of holding the heterogeneous Liberal coalition together: a new policy for the South.[74]

On this question, Greeley possessed impeccable reform credentials. From the moment the war ended, he had opposed confiscation and treason trials, and called upon "gentlemen" of both sections to unite in support of a magnanimous Reconstruction based on "Universal Amnesty and Impartial Suffrage." In 1867, he had provided part of the bond that freed Jefferson Davis from prison (resulting, for a time, in a sharp drop in the *Tribune*'s circulation). And while denouncing the Klan and supporting federal enforcement efforts, Greeley became convinced that high taxes and the exclusion of the "best men" from office had prevented the Northern migration and investment that, he felt, held the key to regional development and national reconciliation. Meanwhile, this longtime defender of black rights increasingly echoed Democratic complaints about the freedmen. "They are an easy, worthless race, taking no thought for the morrow, and . . . your course aggravates their weaknesses," he wrote in 1870, criticizing the relief efforts of Josephine Griffing, a "one-woman welfare agency" in the nation's capital. Blacks, Greeley insisted, must fend for themselves; his harsh injunction was "Root, Hog, or Die!" In early 1872, ambition for the Presidency led Greeley to increase the ferocity of his attacks upon Reconstruction governments; regimes founded upon "ignorance and degradation," the *Tribune* now called them. Having won the Cincinnati nomination in part because of strong support among Southern delegates, Greeley penned a letter of acceptance that stressed the issue of "local self-government" and called upon Americans to "clasp

74. Walter M. Dwyer to Lyman Trumbull, April 29, 1872, Lyman Trumbull Papers, LC; Horace Greeley, *Essays Designed to Elucidate the Science of Political Economy* (Philadelphia, 1869), 57, 74–78; *Nation*, May 23, 1872; Bancroft ed., *Carl Schurz*, 2:361–69, 377–78, 384; Brinkerhoff, *Recollections*, 216–20; Reitano, "Free Trade," 176–78; S. Lester Taylor to Edward Atkinson, July 27, 1872, Atkinson Papers; Armstrong, ed., *Letters of Godkin*, 186.

hands across the bloody chasm" by putting the war and Reconstruction behind them.[75]

If Greeley's candidacy "were as weak as it is ridiculous," remarked James A. Garfield (who, as a reformer enjoying considerable power within the Republican party, had remained aloof from the Cincinnati Convention), "it might be laughed at and passed by, but unfortunately, it has much strength." Faced with this unexpected challenge, Republicans responded as American parties usually do in such circumstances— they moved to steal their opponents' thunder. Within a month of Greeley's nomination, Congress cut tariff duties by 10 percent and enacted an amnesty law restoring the right to hold office to nearly all former Confederates still excluded under the Fourteenth Amendment. Similar legislation had come before Congress in 1870 and 1871, only to fail to obtain the required two-thirds majority. Since the bill affected only a small number of individuals, the amnesty debates had more symbolic than practical importance. The bill aroused the opposition of a majority of the black Congressmen, who blamed prominent "rebels" for Ku Klux Klan violence and feared amnesty presaged a complete abandonment of Reconstruction.[76]

Before embracing the Amnesty Act, Republican leaders had to dispose of another measure that had bounced around Congress since 1870, Charles Sumner's Civil Rights Bill. Guaranteeing all citizens equal access to public accommodations, common carriers, public schools, churches, cemeteries, and jury service, the bill possessed weaknesses and strengths characteristic of its author. The church clause appeared to violate the First Amendment, and Sumner showed little interest in offering a constitutional rationale for the bill as a whole, preferring instead to quote the Declaration of Independence and Sermon on the Mount. (Other Republicans, however, insisted it was fully justified under the Fourteenth Amendment.) The measure's enforcement machinery consisted mainly of empowering aggrieved parties to bring suit in federal court. Nonetheless, as a broad statement of principle, the bill challenged the nation to live up to what Sumner called the principle of "equal rights promised by a just citizenship." He explicitly, moreover, repudiated the legitimacy of separate but equal facilities; "equivalent" was not the same thing as

75. New York *Tribune*, April 6, 10, May 27, September 12, 1865, May 30, 1866, January 9, 1868; Glyndon G. Van Deusen, *Horace Greeley: Nineteenth Century Crusader* (Philadelphia, 1953), 317–25, 343–56, 381–86; *Mr. Greeley's Letters from Texas and the Lower Mississippi* (New York, 1871), 48–53; James M. McPherson, *The Struggle for Equality: Abolitionists and the Negro in the Civil War and Reconstruction* (Princeton, 1964), 392; Cohen, "The Lost Jubilee," 51–56, 205–12, 267–71, 307–308; Samuel Bowles to Frederick Law Olmsted, May 11, 1872, Bowles Papers.

76. Peskin, *Garfield*, 350; F. W. Taussig, *The Tariff History of the United States*, 8th ed., (New York, 1931), 184–85; James A. Rawley, "The General Amnesty Act of 1872: A Note," *MVHR*, 47 (December 1960), 480–84; *CG*, 42d Congress, 2d Session, 398–99, 524, 1908–12, 3382–83, 3660.

"equality." Letters from blacks pleaded for the bill's passage. "We ask this," read one petition, "that our Government may be in fact a true Republic."[77]

Viewed by Congressional Republicans as a political liability, the Civil Rights Bill languished in committee in 1870 and 1871. But when Sumner, in December 1871, moved it as an amendment to a general amnesty bill, party leaders suddenly seized upon his measure as a means of ensuring amnesty's defeat and blaming their opponents to boot. By Vice President Colfax's tie-breaking vote, the Senate in February approved Sumner's motion, whereupon the combined bill, with Democrats solidly opposed, failed to receive the necessary two-thirds majority. The same sequence of events was repeated in early May. But toward the end of the month, Republican leaders concerned by Greeley's nomination negotiated a deal with Democratic Senators. With the ailing Sumner absent, the Senate at an all-night session passed a weakened Civil Rights Bill, shorn of clauses relating to schools and juries and leaving enforcement to state courts. Arriving on the floor just after the vote, Sumner moved his own far stronger measure, only to see it fail by an overwhelming margin. Next, the Amnesty Bill sailed through by a nearly unanimous vote. And, as expected, the "emasculated civil rights bill" soon died in the House. "The juggles and tricks of this session," remarked Conkling, no stranger to parliamentary maneuvering, "exceeded all the kind I have ever seen before."[78]

If Greeley's nomination sent Republicans scurrying to solidify a moderate image in relation to Reconstruction, it confronted Democrats with a different sort of problem. The Liberal movement was a godsend to those Democrats who sought to demonstrate that their party had, at last, come to terms with the results of the Civil War. But the Cincinnati Convention's choice of candidate fell upon them "like a wet blanket." Could Democrats adopt as their candidate a man who had devoted his career to lambasting the party and everything it stood for? Greeley's nomination, declared Democratic National Chairman August Belmont, was "one of those stupendous mistakes which it is difficult even to comprehend." But, he added, the party had no alternative if it hoped to defeat Grant. Democrats' decision was made easier by the fact that after a furious effort to defeat the Fifteenth Amendment, the party in the North had recently embarked on its own New Departure, deemphasizing traditional

77. Donald, *Sumner*, 529–34; Bertram Wyatt-Brown, "The Civil Rights Act of 1875," *Western Political Quarterly*, 18 (December 1965), 767; Richard T. Greener to Charles Sumner, January 15, 1872, George W. Richardson to Sumner, January 27, 1872, James White to Sumner, January 27, 1872, Landon Kurdle to Sumner, February 3, 1872, Sumner Papers; *CG*, 42d Congress, 2d Session, 381, 432, 842–84.
78. Donald, *Sumner*, 535–39, 544–46; *CG*, 42d Congress, 2d Session, 919, 3268–70, 3730–39, 3932; James M. McPherson, "Abolitionists and the Civil Rights Act of 1875," *JAH*, 52 (December 1965), 502.

racial appeals and concentrating on issues like economy in government, lower taxes, and political reform.[79]

As with its Southern counterpart, there was perhaps less to the Northern New Departure than met the eye, for while acquiescing in "the *Constitution* as it stands *now,* " the party continued to oppose all federal enforcement efforts, a position that would effectively emasculate the postwar amendments. Nonetheless, a striking change came over the Democracy's rhetoric and platform, as veterans of racial politics abandoned the term "nigger" in favor of "negro" or "colored person." Peter B. Sweeney of New York's Tammany Hall called on Democrats to drop "the negro agitation," which enabled Republicans to "carry away many voters" by identifying themselves with equality and progress. Even Ohio's Clement Vallandigham, the notorious wartime Copperhead, endorsed the new policy. Reform, not race, declared the New York *World,* should be the party's rallying cry. (Moreover, it pointed out, "the negroes are of little account, if we can control the judgment of the white voters.") In July, the Democratic National Convention endorsed the Liberal candidate and platform as its own. Whether Greeley won or lost, declared the *World,* was irrelevant; his nomination had "cut the party loose from the dead issues of an effete past."[80]

As in the North, the Southern Liberal movement attracted the support of prominent Republicans who had lost out in intraparty power struggles. Former Alabama Sen. Willard Warner joined the Liberals, as did Louisiana Gov. Henry C. Warmoth, whose policies had been repudiated by the bulk of his party, and disaffected Texan Andrew J. Hamilton, who had been pushed aside by Gov. Edmund J. Davis. Numerous Old Line Whigs alarmed by blacks' increasing influence in the Mississippi party also supported Greeley. Initially, too, the Greeley campaign expected to make significant inroads into the black vote, given their candidate's long record of support for equal rights and his endorsement by Charles Sumner. After three months of badgering by his friends in the Bird Club, Sumner at the end of July addressed a public letter to the nation's black voters. Grant's policies, he insisted, had been too weak to suppress Southern violence, but strong enough to antagonize the region's white population; the President had worked assiduously for the Dominican treaty, but failed to force the Civil Rights Bill through Congress. The freedmen's rights, Sumner concluded, would be safer under a Greeley Administration than a continuation of the Grant Presidency. The let-

79. John Tapley to James R. Doolittle, May 4, 1872, James R. Doolittle Papers, SHSW; Katz, *Belmont,* 200–202; New York *World,* June 3, 1871.

80. August Belmont to G. W. McCook, June 5, 1871, Marble Papers; Lawrence Grossman, *The Democratic Party and the Negro: Northern and National Politics 1868–92* (Urbana, Ill., 1976), 24–27, 35–36; Springfield *Weekly Republican,* December 16, 1870; New York *Herald,* November 26, 1869; George T. McJimsey, *Genteel Partisan: Manton Marble, 1834–1917* (Ames, Iowa, 1971), 154–56; New York *World,* July 11, 17, 1872.

ter's "tortured reasoning" revealed the inner turmoil of a man who detested Grant, but understood in his heart of hearts that a Greeley victory spelled the end of Reconstruction.[81]

Most of Sumner's old abolitionist allies repudiated his arguments. "If the Devil himself were at the helm of the ship of state," wrote Lydia Maria Child, "my conscience would not allow me to aid in removing him to make room for the Democratic party." But the most poignant protests came from those for whom Sumner, to quote Frederick Douglass, had been "mind and voice for a quarter of a century." Some blacks dismissed the letter as a hoax, convinced their champion "would never go back on them." Others pointed out what Sumner had failed to mention—that Democrats, along with Liberal leaders Schurz and Trumbull, had consistently opposed the Ku Klux Klan and Civil Rights Bills in Congress. And despite Greeley's personal record, blacks had little confidence in Democratic professions of respect for equal rights. "If I go, sir, into a hotel here in the South," one correspondent informed Sumner, "where they have hung up large glaringly colored maps or charts with portraits of Horace Greeley and Gratz Brown . . . I am [told]: 'Go out of here, this is no place for niggers'!"[82]

Nor did "reform" hold much appeal for blacks. To them, retrenchment and lower taxes meant fewer government services, and behind reformers' clamor against "class legislation" and admonitions to the freedmen to "work out their own destiny," they discerned a refusal to acknowledge blacks' unique historical experience, and a cruel indifference to their fate. "Does individual or national duty end," asked the New National Era, "where ceasing to do evil begins?" Or, as William Whipper reminded Liberals, "the white race have had the benefit of class legislation ever since the foundation of our government." Blacks recognized, moreover, that civil service reform would effectively bar "the whole colored population" from office. As a Mississippi freedman eloquently explained after taking a federal examination for lighthouse keeper:

> It was not my good fortune to receive an education. I was in bondage from youth to manhood. . . . But I learned all that was practicable for one in my situation. . . . I can navigate a vessel on our sound. Yet I cannot stand an

81. Michael Perman, The Road to Redemption: Southern Politics, 1869–1879 (Chapel Hill, 1984), 112–14; Joe G. Taylor, Louisiana Reconstructed, 1863–1877 (Baton Rouge, 1974), 228–36; Carl H. Moneyhon, Republicanism in Reconstruction Texas (Austin, 1980), 176; Richard A. McLemore, ed., A History of Mississippi (Hattiesburg, 1973), 1:581; Francis W. Bird to Charles Sumner, April 15, 1872, Edward Atkinson to Sumner, April 8, 11, 1872, Sumner Papers; The Works of Charles Sumner (Boston, 1870–83), 15:175–95; James M. McPherson, "Grant or Greeley? The Abolitionist Dilemma in the Election of 1872," AHR, 71 (October 1965), 56.

82. McPherson, "Grant or Greeley," 43–61; Lydia Maria Child to Charles Sumner, June 28, 1872, Frederick Douglass to Sumner, July 19, 1872, H. Boemler to Sumner, July 15, 1872, Edwin Belcher to Sumner, August 5, 1872, Sambo Estelle to Sumner, August 5, 1872, Sumner Papers.

examination in geometry, in scientific engineering, in mathematics, or even in artistic penmanship. . . . I think, without vanity, that I am qualified to keep a Light House. . . . If this test be the rule, you will exclude every colored man on this seaboard from position. You tell us that all the honors of the nation are open to us, yet you exclude us by ordeals that none of us can pass.[83]

As the results made plain, blacks continued to regard the Republican party, for all its faults, as the only institution capable of securing the South's "new order of freedom and civilization." Except for Georgia, whose Democratic administration did nothing to deter election day violence, 1872 witnessed the most peaceful election of the entire Reconstruction period. Blacks overwhelmingly cast Republican ballots (further reinforcing reformers' conviction that they were "incapable of comprehending the issues of a political campaign"). At Hurricane plantation, Davis Bend, the tally stood Grant 442, Greeley 1. Greeley won only three ex-Confederate states (Georgia, Tennessee, and Texas), along with Kentucky, Maryland, and Missouri. And Republicans made remarkable comebacks in states lost in 1869 and 1870, carrying West Virginia, winning a majority of the Congressmen in Tennessee and Virginia, and electing governors in North Carolina and Alabama. Although its legislature remained divided between the parties, Alabama became the only instance during Reconstruction in which Republicans regained control of a state administration previously "redeemed" by the Democrats. The outcome reflected the loyalty of the black electorate, a stabilization of scalawag support, and the fact that a number of Democrats, unable to stomach Greeley, remained at home. The narrowness of their victory in some states (fewer than 2,000 votes in North Carolina), did not prevent Republicans from claiming that in a peaceful election, they constituted the South's natural voting majority. Redemption, it appeared, was neither inevitable nor irreversible, and Republicans, newly elected black Congressman John R. Lynch later recalled, seemed to be "the party of the future."[84]

In the North, the Greeley campaign found itself beset with difficulties from the start, beginning with the candidate's own history. "I am curious to see the biography that his friends will put out," remarked one Democrat, and Greeley's supporters were forced to spend much of their time explaining past references to Democrats as "murderers, adulterers, drunkards, cowards, liars, thieves." ("I never said all Democrats were

83. *New National Era*, April 14, 1870, December 12, 1872, April 10, 1873; Pinckney Ross to John R. Lynch, March 7, 1873, Ames Family Papers, Smith College. (This letter is in the handwriting of J.F.C. Claiborne but signed by Ross.)
84. *New National Era*, April 4, 1872; William L. Scruggs to William E. Chandler, October 5, 1872: Logsdon, *White*, 247–48; Dorothy Sterling, ed., *We Are Your Sisters: Black Women in the Nineteenth Century* (New York, 1984), 469; *Tribune Almanac*, 1873, 59–63, 72–83; Allen W. Trelease, "Who Were the Scalawags?" *JSH*, 29 (November 1963), 452–58; John R. Lynch, "The Tragic Era," *JNH*, 16 (January 1931), 110.

saloon keepers," the candidate declared at one point. "What I said was that all saloonkeepers were Democrats.") And although Republican defectors (including gubernatorial candidates Francis W. Bird in Massachusetts and Gustav Koerner in Illinois) sometimes headed fusion state tickets, Democrats, to the Liberals' chagrin, refused to disband their separate party organization. "The great difficulty" facing the Greeley campaign, wrote one Liberal, "is the persistence of the Democratic party in not dying." Indeed, Cyrus McCormick, traditionally a major source of Democratic campaign funds, refused to open his checkbook until he was allowed to replace Governor Palmer at the head of the Illinois Greeley committee.[85]

For both parties, internally divided on other issues, Southern policy became the keynote of the fall campaign. The speeches of Greeley and his supporters focused almost exclusively on the evils of Reconstruction and the need to restore "local self-government." "How is it," asked *The Nation*, "that we hear constantly of the condition of the South in Mr. Greeley's canvass, and not at all, or very rarely, of any of the other reforms which brought the Cincinnati Convention together?" Thus, the campaign culminated the process by which opposition to Reconstruction became inextricably linked with the broader crusade for reform and good government. For their part, Republicans fell back upon the technique of waving the bloody shirt, reinforced by references to "Ku Klux outrages and midnight murders and maraudings." "Go vote to burn school houses, desecrate churches and violate women," Ben Butler told one audience, "or vote for Horace Greeley, which means the same thing." But beyond such rhetoric, the party appealed to the commitment to blacks' rights to equal citizenship and protection against violent assault, concepts now deeply ingrained in the Republican electorate. "The cry that the liberties of the blacks are still in danger . . ." concluded Liberal editor Horace White, "has been the most potent weapon of the campaign."[86]

In the end, Grant carried every state north of the Mason-Dixon line. Only in New England did the Liberal revolt translate into sizable defections among the electorate; about 10 percent of Massachusetts Republicans appear to have voted for Greeley. In the Midwest more Democrats switched to Grant (especially Germans alarmed by Greeley's prohibitionist leanings) than Republicans abandoned him. Nationally, the Republi-

85. Edward Mayes, *Lucius Q. C. Lamar: His Life, Times, and Speeches* (Nashville, 1896), 170; Samuel J. Bayard to J. B. Guthrie, September 8, 1872, Samuel J. Bayard Papers, Princeton University; Horace S. Merrill, *Bourbon Democracy of the Middle West 1865–1896* (Seattle, 1953), 73; Baum, *Civil War Party System*, 170; Thomas J. McCormack, ed., *Memoirs of Gustave Koerner 1809–1896* (Cedar Rapids, Iowa, 1909), 2:562; W. C. Flagg to Lyman Trumbull, July 15, 1872, Trumbull Papers; William T. Hutchinson, *Cyrus Hall McCormick* (New York, 1930–35), 2:319–24.

86. Ross, *Liberal Republican Movement*, 175; *Nation*, August 1, 1872; *Behold! The Contrast*, broadside, September 1872, Keyes Papers; James R. Doolittle to Augustus Schell, December 4, 1872, Doolittle Papers; Logsdon, *White*, 252.

can total of over 55 percent of the vote represented the largest majority in any Presidential election between 1836 and 1892. "I was the worst beaten man that ever ran for that high office," Greeley lamented as the results became known. "And I have been assailed so bitterly that I hardly know whether I was running for President or the penitentiary." Debilitated by the rigors of the campaign, depressed by the vilification he had received and by the death of his wife shortly before election day, Greeley died on November 29.[87]

The election of 1872 confirmed and reinforced the reign of organizational politics. For one thing, the fact that a significant number of former Radicals supported Greeley demonstrated once for all the death of Radicalism as both a political movement and a coherent ideology. The bolt of a group of prominent party leaders, moreover, did not affect the Republican structure, except to solidify the Stalwarts' hold on power. Individuals, it seemed, were dispensable, the party permanent. Some bolters, including Schurz, returned to the Republican fold, but for many others, especially those of Democratic antecedents, the Liberal movement served as a way station on a path leading back to their old allegiance. In 1876, Trumbull, Palmer, Julian, Austin Blair, and Charles Francis Adams, Jr., to name only a few leading Liberals, campaigned for the Democrats. Indeed, in a majority of Ohio's gubernatorial campaigns between 1877 and 1910, a former Liberal Republican headed the Democratic ticket. As for reform journalists and professionals, most after 1872 adopted a stance of nonpartisan independence, preferring, like prewar Garrisonian abolitionists, to advance their views by arousing public opinion and pressuring politicians, rather than by forming a party of their own.[88]

Ironically, the 1872 election both demonstrated the Republican North's commitment to Reconstruction and strengthened tendencies within both parties that would soon set Southern policy on a course Liberals had pioneered. It was "scarcely possible," wrote one Republican newspaper, "to doubt that this will be the death of the Democratic party." In fact, the Democracy demonstrated once again its uncanny talent for survival. Northern Democrats, indeed, emerged from defeat strategically positioned to profit from the coming economic depression, for by abandoning explicit appeals to the race issue and absorbing a highly visible

87. Baum, *Civil War Party System,* 172–73; W. Dean Burnham, *Presidential Ballots 1836–1892* (Baltimore, 1955), 109–10; Hutchinson, *McCormick,* 2:326; Horace Greeley to Mrs. Jennie Mason, November 8, 1872 (copy), Horace Greeley Papers, NYPL.

88. Patrick W. Riddleberger, "The Break in the Radical Ranks: Liberals vs. Stalwarts in the Election of 1872," *JNH,* 44 (April 1959), 136–57; Hansen, "Principles, Politics, and Personalities," 300–305; Krug, *Trumbull,* 338; Patrick W. Riddleberger, *George Washington Julian: Radical Republican* (Indianapolis, 1966), 279–86; Harris, "Blair," 253; Duberman, *Adams,* 393–94; McGerr, "Liberal Republicanism," 319–22; Skowronek, *Building American State,* 42.

group of Liberal Republicans, they partially dissolved their old association with racism and disloyalty, while in no way altering their intention of undoing federal protection for blacks' rights. In the South, meanwhile, the election returns discredited advocates of fusion, convergence, and the New Departure, ensuring that the party would henceforth fight its battles on the "straight-out" basis that seemed best able to mobilize the white electorate. As for Republicans, the Democratic-Liberal demand for "local self-government," critique of Reconstruction corruption, and call for sectional reconciliation found a more receptive audience than the returns appeared to indicate. Despite their espousal of black voting rights, few Northern Republicans actually defended the Reconstruction governments. More commonly, they seemed to accept the characterization of these regimes as inept and venal, adopting a tone of apology or embarrassment and falling back upon the hope that Grant's reelection would bring the South stability and good government. Henceforth, despite the President's sweeping victory, Reconstruction would be on the defensive in the North as well as the South.[89]

89. Dilla, *Politics of Michigan*, 147; Dale Baum, "The 'Irish Vote' and Party Politics in Massachusetts, 1860–1876," *CWH*, 26 (June 1980), 138; Mushkat, *Reconstruction of New York Democracy*, 226; Perman, *Road to Redemption*, 121–31; Cohen, "The Lost Jubilee," 380–85.

CHAPTER 11

The Politics of Depression

The Depression and Its Consequences

THE intoxicating economic expansion of the Age of Capital came to a wrenching halt in 1873. In September, Jay Cooke and Company, a pillar of the nation's banking establishment, collapsed after being unable to market millions of dollars in bonds of the Northern Pacific Railroad. Within days, a financial panic engulfed the credit system. Banks and brokerage houses failed, the stock market temporarily suspended operation, and factories began laying off workers. Throughout the western world, the Panic of 1873 ushered in what until the 1930s was known as the Great Depression, a downturn that lasted, with intermittent periods of recovery, nearly to the end of the century. This first great crisis of industrial capitalism permanently altered the nature of economic enterprise, and had profound political and ideological consequences. For by shattering the mid-Victorian era's faith in the inevitability of progress and exacerbating class conflict, the depression propelled the "labor question" to the forefront of social thought, undermined assumptions at the core of the free labor ideology, and reshaped the nation's political agenda and the balance of power between the parties.[1]

In a way, it was fitting that the Northern Pacific's financial problems triggered the Panic, for if the railroad boom nourished postwar growth, the network's overexpansion, paid for by an outpouring of speculative credit, created a financial house of cards whose eventual collapse was only a matter of time. By 1876, over half the nation's railroads had defaulted on their bonds and were in the hands of receivers. And as track construction halted, the industries that had prospered from the railroad's growth suffered disastrous reverses. By the end of 1874, nearly half the nation's iron furnaces had suspended operation. Not until 1878, a year that saw more than 10,000 businesses fail, did the depression reach bottom. Growth resumed early the following year, but the sixty-five months fol-

1. Rendigs Fels, *American Business Cycles 1865–1897* (Chapel Hill, 1959), 99–107; Eric Hobsbawm, *The Age of Capital 1848–1875* (London, 1975), 4–5, 46.

lowing the Panic of 1873 remains the longest perod of uninterrupted economic contraction in American history.[2]

Compared with the inflationary profit boom of the 1860s, the depression ushered in an entirely new business environment, one of cutthroat competition and a relentless downward price spiral. By 1879, wholesale prices stood 30 percent below their level six years earlier. Ruinous competition for traffic drove down railroad freight rates and spurred the formation of "pools" through which the major lines attempted to stabilize charges and apportion traffic among themselves. Although most such agreements quickly fell apart, they reflected a quest for economic order that increasingly preoccupied the business world. Leaders in industry after industry established nationwide trade associations, aiming, mostly unsuccessfully, to stabilize prices and set production quotas. Meanwhile, a new generation of entrepreneurs seized the opportunity to reorganize their concerns so as to increase productivity and lower costs. Depressions often provide the occasion for structural changes in capitalist production, and that of the 1870s was no exception. Convinced the "new era" required new business practices, John D. Rockefeller seized control of the oil industry, and Andrew Carnegie laid the foundations of his steel empire. Most industries, to be sure, remained plagued by excessive competition rather than monopoly. But while overall output dipped and the total of establishments remained constant, the number of workers employed in manufacturing continued to rise, reflecting a growing concentration of capital and the triumph of large-scale mechanized production. The 1880 census reported four fifths of manufacturing workers laboring "under the factory system."[3]

For workers, the depression was nothing short of a disaster, whose victims included surviving cooperatives, the majority of the nation's unions, and, where it had been achieved, the eight-hour day. Real wages declined only slightly, as the steady fall in prices compensated for wage cuts by employers struggling to remain competitive, but widespread unemployment appeared in the major urban centers. According to one estimate, a quarter of New York City's labor force could not find jobs in 1874. Arriving in San Francisco the following year, Irish immigrant Frank

2. Allan Nevins, *The Emergence of Modern America* (New York, 1927), 293–303; U. S. Bureau of the Census, *Historical Statistics of the United States, Colonial Times to 1970* (Washington, 1975), 732; Samuel Reznck, "Distress, Relief, and Discontent in the United States During the Depression of 1873–78," *Journal of Political Economy*, 58 (December 1950), 495–97; Irwin Unger, *The Greenback Era: A Social and Political History of American Finance 1865–1879* (Princeton, 1964), 213–24, 265–66; Fels, *Business Cycles*, 83, 107–11.

3. U. S. Bureau of the Census, *Historical Statistics*, 139, 201, 666–67; Gabriel Kolko, *Railroads and Regulation 1877–1916* (Princeton, 1965), 7–10; Alfred D. Chandler, Jr., *The Visible Hand: The Managerial Revolution in American Business* (Cambridge, Mass., 1977), 135–41, 246, 316–17; David Montgomery, "Radical Republicanism in Pennsylvania, 1866–1873," *PaMHB*, 85 (October 1961), 455–57; David Hawke, *John D.: The Founding Father* (New York, 1980), 94–108, 154–55; Nevins, *Emergence*, 33–43; Steven J. Ross, *Workers on the Edge: Work, Leisure, and Politics in Industrializing Cincinnati, 1788–1890* (New York, 1984), 219–32; U. S. Census Office, *Tenth Census, 1880*, 2:16.

Roney found factories closed, life a constant struggle for survival, and hope absent from working-class life. So many men took to the roads in search of work that by mid-decade the "tramp" had become "a fixed institution" on the social landscape. Never in the country's history, commented Pennsylvania's Bureau of Labor Statistics, had "so many of the working classes, skilled and unskilled, . . . been moving from place to place seeking employment that was not to be had."[4]

The depression had a profound impact on the labor movement, shifting its focus from the issues of the 1860s—greenbackism, cooperation, and the eight-hour day—to demands for public relief, the desperate struggle to maintain predepression wage levels, and, for a few workers, socialism. In the winter of 1873–74, cities from Boston to Chicago witnessed massive demonstrations demanding that authorities ease the economic crisis by inaugurating such projects as street and park improvements and new rapid transit systems—a remarkable expansion of labor's conception of government's role and responsibilities. The movement for "Work or Bread" reached its climax in New York, where on January 13, 1874, the city police violently dispersed a crowd of 7,000 demonstrators who had assembled at Tompkins Square, arrested scores of workers, and inaugurated a period of "extreme repression" against subsequent labor gatherings.[5]

Although sporadic demonstrations of jobless workers continued throughout the depression, the Tompkins Square "riot" marked the effective end of the movement for public employment. In its wake, the labor reform impulse splintered. Some workers, especially German and Bohemian immigrants, now moved in an explicitly socialist direction. Viewing traditional labor ideologies based on republican and free labor precepts as hopelessly outmoded, they created tiny organizations like the Workingmen's party of Illinois, devoted to public ownership of railroads and factories. Many native and English-born labor leaders (including the young Samuel Gompers), drew from Tompkins Square a different lesson: that labor must not allow itself to be associated in the public mind with violence and "communism." Workers, they believed, should concentrate on maintaining wage levels, work rules, and unions—a difficult task at

4. Philip R. P. Coelho and James F. Shepherd, "Regional Differences in Real Wages: the United States, 1851–1880," *Explorations in Economic History,* 13 (April 1976), 212–13; Neil L. Shumsky, ed., "Frank Roney's San Francisco—His Diary: April 1875–March 1876," *Labor History,* 17 (Spring 1976), 245–47; Paul T. Ringenbach, *Tramps and Reformers 1873–1916: The Discovery of Unemployment in New York* (Westport, Conn., 1973), 3; Samuel Bernstein, "American Labor in the Long Depression, 1873–1878," *Science and Society,* 20 (Winter, 1956), 60–82.

5. Herbert G. Gutman, "The Failure of the Movement by the Unemployed for Public Works in 1873," *Political Science Quarterly,* 80 (June 1965), 254–76; Herbert G. Gutman, "The Tompkins Square 'Riot' in New York City on January 13, 1874: A Re-examination of Its Causes and Its Aftermath," *Labor History,* 6 (Winter 1965), 44–70; Samuel Gompers, *Seventy Years of Life and Labor* (New York, 1925), 1:92–97.

best, since employers found it easy to call upon the unemployed to take the jobs of strikers. Seeking to shield themselves against the depression's impact, unions of the labor aristocracy retreated into a defensive stance, eschewing both confrontations with capital and cooperation with the unskilled. Many mine and factory workers, on the other hand, resisted wage cuts in militant, violent strikes. The year 1874 witnessed bitter labor disputes on the railroads and in midwestern mines, and the following year 15,000 textile workers stayed out for two months in unsuccessful opposition to wage reductions. Also in 1875, the "long strike" in Pennsylvania's anthracite coal fields ended with the defeat of the powerful Workingmen's Benevolent Association and produced the celebrated Molly Maguire trials, which culminated in the hanging of twenty militant miners.[6]

To one Boston newspaper, these strikes denoted "a transition period" in the nation's history. The depression, it seemed, had brought European-style class conflict to America and, as a Pennsylvania official commented, "fearfully" strengthened "the antagonism between rich and poor." Whether inclined toward socialism, bread-and-butter unionism, or militant confrontations with capital, labor increasingly abandoned older free labor shibboleths in favor of a more forthright recognition of the permanence of the wage system and the reality of conflict between employer and employee. The ideals of the independent producer and the language of "equal rights" and free labor survived (to be reinvigorated by the Knights of Labor in the 1880s), but they increasingly served as a "protest ideal," a critique of the emerging capitalist order, rather than an expression of faith in individual mobility and the harmony of all interests in society.[7]

For farmers, too, the depression brought economic dislocation and galvanized new forms of protest. Farm population and output continued to increase throughout the decade, but as agricultural prices and land values tumbled, postwar prosperity gave way to hard times. Throughout the world, the depression devastated small farmers producing for the international market. In the United States, Western farmers in particular found themselves falling further and further into debt, while the number

6. John B. Jentz and Richard Schneirov, *The Origins of Chicago's Industrial Working Class* (forthcoming); Gompers, *Seventy Years*, 1:97; Leonard S. Wallock, "Chapel, Custom, Craft: The Transformation of the Struggle to Control the Labor Process Among the Journeyman Printers of Philadelphia, 1850–1886" (unpub. diss., Columbia University, 1983), 386–89, 416, 437–38; Herbert G. Gutman, "Trouble on the Railroads in 1873–74: Prelude to the 1877 Crisis?" *Labor History*, 2 (Spring 1961), 215–35; Herbert G. Gutman, "Reconstruction in Ohio: Negroes in the Hocking Valley Coal Mines in 1873 and 1874," *Labor History*, 3 (Fall 1962), 243–64; Eric Foner, *Politics and Ideology in the Age of the Civil War* (New York, 1980), 170–75.

7. Wayne G. Broehl, Jr., *The Molly Maguires* (Cambridge, Mass., 1964), 205; Bernstein, "American Labor," 73; *Report of the Committee of the Senate upon the Relations Between Labor and Capital, and Testimony Taken by the Committee* (Washington, D.C., 1885), 1:49, 358; Jentz and Schneirov, *Chicago's Working Class*. The concept of a "protest ideal" is developed in Franco Venturi, *Utopia and Reform in the Enlightenment* (Cambridge, Mass., 1971), especially Chapter 3.

of tenants unable to acquire land rose dramatically and the wages of agricultural laborers plummeted. By 1878, the Midwest even witnessed episodes of machine-breaking by groups of farm workers unable to obtain jobs. While large-scale "gentleman farmers" weathered the storm by diversifying production, small farmers flooded into the Grange; some of its leaders now appealed to "brother workers" to join in an attack upon railroads and land speculators, even proposing that state legislatures repeal railroad charters "and remand their property into the hands of Commissioners, who shall run them for the people." More moderate Grangers called for increased railroad regulation and currency inflation, to raise agricultural prices and enable indebted farmers to meet their mortgage payments.[8]

In a number of Western states, the depression shattered the mold of two-party politics, spawning insurgent movements that achieved remarkable, if temporary, successes in 1873. "Antimonopoly" coalitions uniting Democrats with disgruntled Republican farmers and merchants elected two state officials in Minnesota and captured one house of the Iowa legislature. In Wisconsin, Reform candidate William Taylor ousted Gov. Cadwallader C. Washburn, a stunning upset produced by an unusual combination of circumstances and allies. The onset of the depression fostered a mood of "outright rebellion" among farmers and Milwaukee merchants against high freight rates, German voters resented a recently enacted temperance law (although Taylor himself belonged to the antiliquor Good Templars), and railroads that had failed to receive aid from the Washburn administration secretly supported the reformers. (The outgoing governor described the victorious alliance as representing "the combined powers of darkness, whiskey, beer, railroads, and a sprinkling of Grangers." Once the legislature met, this unlikely coalition quickly fell apart, enabling Republicans to "outreform the Reformers" by enacting the Potter law creating a three-member commission empowered to reduce railroad rates.) As falling freight charges defused the movement for regulation, and Democrats moved to capitalize on the depression themselves rather than merge into new political alignments, the Antimonopoly parties of 1873 faded from the scene. But their brief success illustrated how the crisis had catapulted economic issues to the center stage of politics.[9]

8. U.S. Bureau of the Census, *Historical Statistics*, 201; Margaret B. Bogue, *Patterns from the Sod* (Springfield, Mass., 1959), 139–76; Peter H. Argensinger and JoAnn E. Argensinger, "The Machine Breakers: Farm Workers and Social Change in the Rural Midwest of the 1870s," *AgH*, 58 (July 1984), 395–406; Gerald Prescott, "Gentleman Farmers in the Gilded Age," *WMH*, 55 (Spring 1972), 197–205; Robert McCluggage, "Joseph H. Osborn, Grange Leader," *WMH*, 35 (Spring 1952), 183–84; A. Gaylord Spalding to Joseph H. Osborn, December 23, 1874, Joseph H. Osborn Papers, Oshkosh Public Museum; Henry S. Magoon to Elisha W. Keyes, June 9, 1874, Elisha W. Keyes Papers, SHSW.

9. Solon J. Buck, *The Granger Movement* (Cambridge, Mass., 1913), 88; *Tribune Almanac*, 1874, 65–69; Dale E. Treleven, "Railroads, Elevators, and Grain Dealers: The Genesis of Antimonopolism in Milwaukee," *WMH*, 52 (Spring 1969), 205–15; W. W. Fields to Lucius Fairchild, November 20, 1873, Lucius Fairchild Papers, SHSW; Horace Rublee to Cadwal-

Rudely disrupting visions of social harmony, the depression of the 1870s marked a major turning point in the North's ideological development. As widespread tension between labor and capital emerged as the principal economic and political problem of the day, public discourse fractured along class lines. In small industrial centers throughout the North, striking workers continued to enjoy support among local officials and small-town businessmen, many of whom shared labor's resentment against the disruptive impact of large corporations controlled from outside the community. But in the nation's large cities, and at the upper echelons of both major parties, older notions of equal rights and the dignity of labor gave way before a sense of the irreducible barriers separating the classes and a preoccupation with the defense of property, "political economy," and the economic status quo.[10]

As the depression deepened and class conflict intensified, the critique of labor and farmers' movements and of the activist democratic state pioneered by liberal reformers found a widening audience among the North's urban middle and upper classes. To some, Wisconsin's Potter law epitomized the various threats to private property that dotted the political landscape—"the most ignorant, arbitrary and wholly unjustifiable law to be found in the history of railroad legislation," was how Charles Francis Adams, Jr., described it. The movement for "Work or Bread" also helped propel the urban bourgeoisie to the right, as newspapers of both parties joined in denouncing the idea of public employment, raised anew the specter of the Paris Commune, and praised New York's police for effectively defending law and order. Castigating labor leaders as "enemies of society" who believed "the world owes them a living," the urban press attributed poverty to laziness and extravagance, and insisted the laws of political economy dictated only one way out of the depression: "Things must regulate themselves."[11]

These beliefs quickly broadened to include a critique of both nongovernmental charity and workers' own collective efforts. A special report on "pauperism" for the American Social Science Association not only rejected the idea of public assistance outside workhouses, but blamed "indiscriminate" and "over-generous" private relief for exacerbating labor

lader C. Washburn, December 4, 1873, Cadwallader C. Washburn Papers, SHSW; Herman J. Deutsch, "Disintegrating Forces in Wisconsin Politics of the Early Seventies," *WMH*, 15 (March 1932), 296; Graham A. Cosmas, "The Democracy in Search of Issues: The Wisconsin Reform Party, 1873–77," *WMH*, 46 (Winter 1962–63), 97–105; George H. Miller, *Railroads and the Granger Laws* (Madison, Wis., 1971), 153–60.

10. Herbert G. Gutman, "The Workers' Search for Power," in H. Wayne Morgan, ed., *The Gilded Age* (Syracuse, 1970 ed.), 37–52; Morton Keller, *Affairs of State: Public Life in Late Nineteenth Century America* (Cambridge, Mass., 1977), 162; Walter T. K. Nugent, *Money and American Society 1865–1880* (New York, 1968), 177, 205–207.

11. Miller, *Railroads and the Granger Laws*, 140; Gutman, "Failure of the Unemployed," 254–60, 270–75; Samuel Bernstein, "The Impact of the Paris Commune in the United States," *Massachusetts Review*, 12 (Summer 1971), 444–45; New York *World*, December 21, 1873, January 10, 1874; Gutman, "Tompkins Square," 47, 56–57, 68–70.

unrest by encouraging the unemployed to turn down jobs paying reduced wages. Meanwhile, antiunion sentiment continued to harden. Whitelaw Reid, who succeeded Horace Greeley as editor of the New York *Tribune*, adopted a particularly ruthless attitude toward unions, reflected both in his editorials and in his success at reducing printers' wages and bringing in Italian replacements when builders constructing the paper's new offices struck in 1874. Indeed, many urban newspapers viewed the depression as "not an unmixed evil," since it promised to lower wages, discipline labor, and curb the power of unions.[12]

Such attitudes toward strikes and poverty had always comprised one strand of free labor thought. Now, cut loose from the more egalitarian aspects of that ideology, they formed part of a self-conscious bourgeois outlook whose emergence and consolidation liberal reformers directly encouraged. "All industries in our day stand or fall together," wrote Godkin, calling upon businessmen to put aside parochial concerns and recognize their common class interests. As the depression deepened, consciousness of being members of a separate capitalist class (rather than a broader grouping of "producers") spread within the business community. Industrialists largely abandoned their earlier advocacy of easy money, and many urban merchants previously critical of the railroads pulled back from the idea of regulation. The proliferation of trade associations during the 1870s also reflected the growth of bourgeois class consciousness, as did the formation of new political organizations among businessmen, such as the Citizens' Association in Chicago. Thus, in the face of agrarian unrest and working-class militancy, metropolitan capitalists united as never before in defense of fiscal conservatism and the inviolability of property rights.[13]

The depression also pushed reformers' elitist hostility to political democracy and government activism (except in the defense of law and order) to almost hysterical heights. Horace White of the Chicago *Tribune*, who had earlier welcomed the rise of the Grange, since the farmers at least agreed politics should no longer revolve around "the dead corpse of slavery," now veered sharply to the right. He condemned agrarian and labor organizations for initiating "a communistic war upon vested rights and property," and insisted that universal suffrage had "cheapened the ballot" by throwing political power into the hands of those influenced by the "harangues of demagogues." Reid conjured up images of "ignorant

12. Robert H. Bremner, *The Public Good: Philanthropy and Welfare in the Civil War Era* (New York, 1980), 200–207; "Pauperism in the City of New York," *Journal of Social Science*, 6 (1874), 74–83; Bingham Duncan, *Whitelaw Reid: Journalist, Politician, Diplomat* (Athens, Ga., 1975), 61; Gutman, "Worker's Search for Power," 36.

13. *Nation*, November 6, 1873; Walter T. K. Nugent, *The Money Question During Reconstruction* (New York, 1967), 62–63; Lee Benson, *Merchants, Farmers, and Railroads: Railroad Regulation and New York Politics 1850–1887* (Cambridge, Mass., 1955), 67–72; Richard Schneirov, "Class Conflict, Municipal Politics, and Governmental Reform in Gilded Age Chicago," in Hartmut Keil and John B. Jentz, eds., *German Workers in Industrial Chicago, 1850–1910: A Comparative Perspective*, (DeKalb, Ill., 1983), 194–95.

voters" forming "a party by themselves as dangerous to the interests of society as the communists of France," while *The Nation* linked the Northern poor and Southern freedmen as members of a dangerous new "proletariat" as different "from the population by which the Republic was founded, as if they belonged to a foreign nation." Godkin's invocation of the founding fathers was revealing, for liberal reformers were increasingly obsessed by the same dilemma with which men like Madison had wrestled a century earlier—how to reconcile private property with political democracy.[14]

Although reformers failed to achieve their cherished goal of establishing property qualifications for urban voters, the economic crisis greatly strengthened the movement to insulate municipal and state government from the vagaries of the popular will. New state constitutions of the 1870s, noted one commentator, represented "a wide departure from the theories of government so long and unquestioningly accepted among us," for they extended the terms of office of governors and judges, established limits on the duration of legislative sessions, and in a few cases even curtailed the jury system. With the depression exacerbating the fiscal problems of state and city governments and threatening to push taxes to even higher levels, retrenchment became the order of the day. Several states prohibited public aid to railroads, placed limits on budgets and tax rates, and armed governors with the item veto. State authorities, moreover, became increasingly willing to use the courts and militia on the side of capital, as Republican Gov. John Hartranft of Pennsylvania did in the miners' "long strike" of 1875, to the applause of newspapers of both parties. Perhaps even more indicative of new tendencies in Northern public life was the proliferation of vagrancy laws designed to combat the menace thought to be posed by tramps. Although intended more to clear towns of beggars than to establish a system of quasi-free labor, by making unemployment a crime the laws bore more than a passing resemblance to the Southern Black Codes of 1865–66. Indiana even directed that those who refused employment be put to work on city streets, and leased out convicts to a manufacturer of railroad cars. Such policies illustrated how free labor principles seemed to have crumbled in the depression's wake.[15]

14. Joseph Logsdon, *Horace White: Nineteenth Century Liberal* (Westport, Conn., 1971), 263–67; Roger A. Cohen, "The Lost Jubilee: New York Republicans and the Politics of Reconstruction and Reform, 1867–1878" (unpub. diss., Columbia University, 1975), 201–202; *Nation*, April 9, 1874.

15. Simeon E. Baldwin, "Recent Changes in Our State Constitutions," *Journal of Social Science*, 10 (1879), 138–39; C. K. Yearley, *The Money Machines: The Breakdown and Reform of Governmental and Party Finance in the North, 1860–1920* (Albany, 1970), 5–10; Jon C. Teaford, *The Unheralded Triumph: City Government in America, 1870–1900* (Baltimore, 1984), 105, 285–94; Carter Goodrich, "The Revulsion Against Internal Improvements," *JEcH*, 10 (November 1950), 152; Keller, *Affairs of State*, 110–14; Frank B. Evans, *Pennsylvania Politics, 1872–1877: A Study in Political Leadership* (Harrisburg, 1966), 233–34; Bernstein, "American Labor," 76–77; Emma L. Thornbrough, *Indiana in the Civil War Era 1850–1880* (Indianapolis, 1965), 316–17, 588.

Changes in the women's rights movement also reflected the growing influence of notions of respectability. A crusade against drink—with bands of women kneeling in prayer before saloons and destroying liquor in the presence of "excited crowds"—swept across Pennsylvania and the Middle West, and the Women's Christian Temperance Union, founded in 1874, quickly became the nation's largest female organization, mobilizing far more women than the quest for the ballot ever had. By bringing thousands of women into the public sphere, the WCTU pointed the way toward an expansion of female political activism. Eventually, it would wed its condemnation of the saloon to the demand for the vote. But in the 1870s, under the banner of "Home Protection," the organization disavowed women's suffrage. The rapid rise of the temperance movement and the temporary waning of interest in the vote symbolized the eclipse of the broad feminist tradition that stressed the common humanity of men and women by a narrower movement emphasizing women's unique moral purity and their responsibility for bringing feminine virtues to bear upon a dissolute male world.[16]

As the ideological center of gravity shifted at feminism's grass roots, an analogous transformation occurred among its national leadership. Here, the catalyst was the Beecher-Tilton scandal, a sensational affair in which the Rev. Henry Ward Beecher was accused by reform editor Theodore Tilton of maintaining an illicit relationship with Tilton's wife. Although Beecher's guilt seemed beyond dispute, respectable opinion rallied to his defense, and an 1875 trial—"the greatest national spectacle" of the decade—ended with a hung jury. The unfortunate Tilton fled to Paris, and bourgeois outrage descended upon Victoria Woodhull, who had broken the scandal in her weekly newspaper.

Apart from being an editor, Woodhull was the first woman to open a brokerage house (thanks to the aid of Cornelius Vanderbilt, who had become infatuated with her vivacious sister Tennessee Claflin), and the first to testify before a Congressional committee on behalf of female suffrage. This remarkable woman was also a controversial member of the International Workingmen's Association (the First International), where her intense individualism set her at odds with European socialists like Karl Marx. She was best known, however, as a outspoken critic of the sexual double standard. Indeed, Woodhull's complaint against Beecher involved not immorality but hypocrisy, for the good reverend, as she pointed out, condemned free love from his pulpit while practicing it in

16. T. A. Goodwin, *Seventy-Six Years' Tussle With the Traffic* (Indianapolis, 1883), 26–27; Ernest L. Bogart and Charles M. Thompson, *The Industrial State 1870–1893* (Springfield, Ill. 1920), 48; Mari Jo Buhle, *Women and American Socialism 1870–1920* (Urbana, Ill., 1981), 53–56, 60–65; Steven M. Buechler, *The Transformation of the Woman Suffrage Movement: The Case of Illinois, 1850–1920* (New Brunswick, N.J., 1986), 102–107, 118, 138; Barbara Epstein, *The Politics of Domesticity: Women, Evangelicism, and Temperance in Nineteenth-Century America* (Middletown, Conn., 1981), 125–27.

private. Woodhull's views had strongly influenced demands for sexual equality that circulated among feminist leaders. But the reaction to the scandal, coupled with the threat of legal action under the Comstock Law, which prohibited the circulation of "obscene" materials through the mails, led leading feminists to retreat from public discussion of sexual matters and to take great pains to dissociate their position on marriage and divorce reform from that of the "free lovers." (Woodhull herself departed for England, where she married a London banker and lived into the 1920s, helping create, among other groups, a Ladies Automobile Club and a Women's Aerial League.)[17]

The clash between demands for government intervention to ease the depression's impact and the growing clamor for law, order, and respectability quickly spilled over into national politics. The Forty-Third Congress, which assembled in December 1873, became embroiled in controversy over Iowa Republican George W. McCrary's proposal for a national commission to establish "reasonable" railroad rates. Although a handful of Democrats supported the plan as a legitimate response to "the great labor question," most denounced it as yet another example of "centralization." Meanwhile, Republicans saw their party sunder along regional lines, with Westerners and Southerners (including all seven black Congressmen) favoring the bill and most Easterners opposed. To its supporters, federal railroad regulation seemed a legitimate exercise of the expanded sovereignty of the national state. "This question," commented Iowa's James Wilson, "has been beyond the reach of States. . . . The influence is national, the interest is national, the benefits are national, and the wrongs are national." But Eastern Republicans not only denounced the bill as a form of "communism," but appealed to the principle of states rights—a doctrine, observed a Tennessee Congressman, "which now emanates from the North so suddenly and unexpectedly." By a five-vote margin, the bill passed the House, only to expire in the Senate.[18]

The currency issue, on which both parties contained a full complement of positions from "wild" inflationists to advocates of the immediate resumption of specie payments, proved even more disruptive. A flood of proposals, reflecting a "diversity of sentiment and opinion almost infinite," descended upon Congress. Early in 1874, Sherman presented a plan, "the result of great labor, long consideration, and the consequence of compromise," to stabilize the currency. But a series of amendments, pressed by Westerners and Southerners of both parties, transformed the measure into an "Inflation Bill" that proposed to add $64 million to the

17. Altina Waller, *Reverend Beecher and Mrs. Tilton* (Amherst, Mass., 1982), 1–11, 146–47; Madeline B. Stern, ed., *The Victoria Woodhull Reader* (Weston, Mass., 1979), 1–10; William Leach, *True Love and Perfect Union: The Feminist Reform of Sex and Society* (New York, 1980), 61–63; Buechler, *Transformation of Woman Suffrage*, 8.
18. CR, 43d Congress, 1st Session, 490–91, 1946–47, 2046, 2180, 2240, 2421, 2493.

circulation of greenbacks and national bank notes. Although the magnitude of the increase was hardly extravagant, the bill embodied the idea that the government could regulate the money supply to meet the economy's fluctuating needs. To the metropolitan bourgeoisie, it epitomized all the heretical impulses and dangerous social tendencies unleashed by the depression. "Nearly every man of intelligence and experience in commercial and financial matters," claimed *The Nation,* opposed the measure, and the reform press rallied respectable opinion against it. Secretary of State Hamilton Fish, who spoke for the New York business community in the Cabinet, argued strenuously for a veto, as did the state's elderly Gov. John A. Dix, a hard-money fanatic who carried a few gold coins in his pocket, "occasionally refreshing himself with a look at them." Although the President initially leaned toward signing the measure, these pressures had their impact. At the end of April he vetoed the bill, and Congress failed to muster the two-thirds vote needed to make it law.[19]

Although highly unpopular among Western farmers and small-town entrepreneurs desperate for an easing of credit, Grant's veto of the Inflation Bill won the approbation, even in the West, of "the men of money, the stable business men, the solid substantial interests." Among Eastern "men of note," only the iconoclastic Benjamin F. Butler voiced dissent. Reformers found themselves lavishing unaccustomed praise upon the President. The veto, declared *Harper's Weekly,* rescued "the national honor" and constituted "the most important event of his administration." What Edward Atkinson called the "civil war in the Republican ranks" over the money question did not suddenly cease. But the veto marked a milestone in the process by which "the slow, conservative sentiment," as Henry L. Dawes called it, gained ascendancy in Republican circles and economic respectability replaced equality of rights for black citizens as the essence of the party's self-image.[20]

Whatever its appeal in metropolitan centers, the Republican party's economic stance, commented Wisconsin Democratic leader George H. Paul, "ought to revolutionize the North-west." But fiscal conservatism was only one of many burdens borne by Northern Republicans as the

19. Unger, *Greenback Era,* 216, 233–40, 410; Roscoe Conkling to Isaac Sherman, December 26, 1873, Isaac Sherman Papers, Private Collection; Nugent, *Money and American Society,* 221–24; John Sherman, *Recollections of Forty Years in the House, Senate, and Cabinet* (Chicago, 1895), 1:495–96, 504; Cohen, "The Lost Jubilee," 435–43; *Nation,* March 26, 1874; Allan Nevins, *Hamilton Fish: The Inner History of the Grant Administration* (New York, 1936), 2:704–13; Morgan Dix, *Memoirs of John Adams Dix* (New York, 1883), 2:190; William S. McFeely, *Grant: A Biography* (New York, 1981), 395–97.

20. H. M. Jones to Richard Ogelsby, April 25, 1874, N. C. Thompson to Ogelsby, April 25, 1874, B. F. Parks to Ogelsby, April 24, 1874, Richard Ogelsby Papers, Illinois State Historical Library; New York *Commercial Advertiser,* undated clipping, E. S. Keitt Scrapbook, SCHS; Thomas J. McCormick, ed., *Memoirs of Gustave Koerner 1809–1896* (Cedar Rapids, Iowa, 1909), 2:583; *Harper's Weekly,* May 9, 1874; Edward Atkinson to Henry L. Dawes, February 21, 1874, Henry L. Dawes Diary, March 14, 1876, Henry L. Dawes Papers, LC.

elections of 1874 approached. The 1873 Salary Grab, by which Congress doubled the President's compensation and awarded itself a 40 percent retroactive pay increase, proved highly unpopular. The following year, Secretary of the Treasury William A. Richardson was forced to resign after it came to light that he had allowed a prominent Massachusetts politico to take over the investigation of tax evaders and pocket fees totaling $200,000. All this provided a field day for Democrats; "the strumpet of corruption . . ." exclaimed one St. Paul newspaper, "strides in naked horror through the land." Meanwhile, the women's temperance crusade, along with laws enacted by certain Western legislatures restricting the sale of liquor, wreaked havoc on voting alignments by alienating German Republicans.[21]

Whatever voters' local grievances, however, only the depression can explain the electoral tidal wave that swept over the North in 1874. As they would in 1896 and 1932, voters reacted to hard times by turning against the party in power. In the greatest reversal of partisan alignments in the entire nineteenth century, they erased the massive Congressional majority Republicans had enjoyed since the South's secession, transforming the party's 110-vote margin in the House into a Democratic majority of sixty seats. "The election is not merely a victory but a revolution," declared a newspaper in New York, where Samuel J. Tilden rolled up a 50,000-vote triumph over Governor Dix. Democrats won the governorship of Massachusetts for the first time, and swept to victory in New Hampshire, New Jersey, Pennsylvania, Illinois, Indiana, and Ohio. Republicans retained control of the U.S. Senate, but losses in state legislatures meant that seven of their Senators would not return to Washington.[22]

"What does it mean?" one Democrat wondered as the results became known. "Is it permanent or only the result of a combination of circumstances amongst the most prominent of which is the widespread prostration of industries?" Time would reveal that 1874 inaugurated a new era in national politics, although one of stalemate rather than Democratic ascendancy. Not until 1896 would Republicans reestablish their electoral dominance; until then, the same party would control both Houses only three times, and only twice the White House and Congress. More than once, Congress would find itself all but paralyzed as important bills

21. George H. Paul to Joseph H. Osborn, August 20, 1874, Osborn Papers; Hans L. Trefousse, *Ben Butler: The South Called Him BEAST!* (New York, 1957), 225–26; Ross A. Webb, *Benjamin Helm Bristow: Border State Politician* (Lexington, Ky., 1969), 134; Henry A. Castle, "Reminiscences of Minnesota Politics," *Minnesota Historical Society Collections*, 15 (1915), 568; Thornbrough, *Indiana*, 262–64.

22. Dale Baum, *The Civil War Party System: The Case of Massachusetts, 1848–1876* (Chapel Hill, 1984), 183; William Gillette, *Retreat from Reconstruction 1869–1879* (Baton Rouge, 1979), 246; Jerome Mushkat, *The Reconstruction of the New York Democracy 1861–1879* (Rutherford, N. J., 1981), 242; *Tribune Almanac*, 1875, 49–94.

shuttled back and forth between House and Senate, committees of conference failed, and special sessions became necessary. Given the political stalemate, new departures in national policy became unthinkable. Americans would have to weather the economic crisis without leadership from Washington.[23]

Although the depression far outweighed Reconstruction as a cause of Republican defeat, the implications for the party's Southern wing were indeed ominous. When Democrats assumed control of the House in 1875, the South would receive half the committee chairmanships. Southern Republicans feared the results portended the abandonment of national efforts to bolster their party and protect their rights. Blacks were particularly grieved by the defeat of Butler, one of the Republican Congressmen to fall before the Democratic landslide. "I must say," wrote a black resident of Baltimore, "that by your defeat . . . the colored people have lost one of their best men. You have been with us ever since the war commenced in regard to our liberty and equal rights." All in all, exulted the New York *Herald,* the election results suggested that white Southerners would be welcomed back as "our brothers and our fellow-citizens." If, as the *Herald* suggested, the North had finally put the Civil War behind it, could Reconstruction long survive?[24]

Retreat from Reconstruction

The election of 1874 offered only one indication of a pronounced shift in Northern attitudes toward the South during Grant's second term. As evidence multiplied of a growing spirit of sectional reconciliation, Reconstruction's defenders found themselves on the losing side in what one Southern Democrat called "the war of words which has followed the battles of the rebellion." One index of the new mood was the response to Mississippi Democratic Congressman Lucius Q. C. Lamar's much-publicized 1874 eulogy on the death of Charles Sumner. Singling out for special praise his subject's "devotion to the great principle that liberty is the birthright of all humanity," Lamar affirmed that however difficult the North might find it to believe, "Mississippi regrets the death of Charles Sumner and sincerely unites in paying honor to his memory." Lamar had not abandoned his bitter opposition to Reconstruction; as he explained privately, Southern whites would never be restored "to the control of

23. Gustavus Fox to Gideon Welles, November 11, 1874, Gideon Welles Papers, HL; Leonard D. White, *The Republican Era: 1869–1901* (New York, 1958), 59; Baum, *Civil War Party System,* 17–18.

24. Margaret S. Thompson, *The "Spider Web": Congress and Lobbying in the Age of Grant* (Ithaca, N.Y., 1985), 195, 207; John R. Lynch, *Reminiscences of an Active Life: The Autobiography of John Roy Lynch,* edited by John Hope Franklin (Chicago, 1970), 145; George W. Gilham to Benjamin F. Butler, December 31, 1874, Benjamin F. Butler Papers, LC; Mushkat, *Reconstruction of New York Democracy,* 244.

their own affairs" until the Northern public became convinced "that the results of the war were fixed beyond the power of reaction." His speech surprised and gratified Northerners, and won encomiums from newspapers of both parties. The following year, a Confederate military unit received a hearty public welcome on visits to Boston and New York.[25]

The shift of Republican opinion in a conservative direction during the depression strongly affected prevailing attitudes toward Reconstruction. In 1875, former Freedmen's Bureau agent and Georgia legislator John E. Bryant learned just how profoundly Northern Republican thought had changed when he wrote for the New York Times a discussion of the political situation in his adopted state. Bryant's analysis rested upon classic free labor premises. At the heart of Reconstruction, he insisted, lay the same issue that had brought on the Civil War, the struggle between "two systems of labor." Northerners felt "that the laboring man should be as independent as the capitalist"; Southern whites still, in their heart of hearts, believed workers "ought to be slaves." Although overstated, Bryant's analysis underscored the centrality of labor relations to Reconstruction politics. But on one point, he appeared out of date. Was it still an article of faith in the Republican North that the laborer should be "as independent as the capitalist?" As one of Bryant's Northern friends informed him, "there was reason to believe" that the attitude toward "the labor question," and the "general view of society and government" held by the South's "old ruling class" were now "substantially shared by a large class in the North."[26]

Only a few years earlier, Republicans had united in the determination to remake Southern society in accordance with the principles of free labor and legal and political equality. Now, the erosion of the free labor ideology made possible a resurgence of overt racism that undermined support for Reconstruction. Perhaps the most influential exemplar of this development was James S. Pike, a veteran antislavery journalist and the nation's wartime ambassador to the Netherlands, whom the New York Tribune dispatched to South Carolina early in 1873. Appearing first as a series of articles and then as the book, The Prostrate State, Pike's reports depicted a state engulfed by political corruption and governmental extravagance, and wholly under the control of "a mass of black barbarism . . . the most ignorant democracy that mankind every saw." Pike was hardly a model of objectivity—he had long held racist views, had taken an active part in the Greeley campaign of 1872 (when his brother campaigned for Congress as a Liberal Republican), and had incorporated essentially the same

25. CR, 44d Congress, 1st Session, 2105; 43d Congress, 1st Session, 3410–11; Mattie Russell, ed., "Why Lamar Eulogized Sumner," JSH, 21 (August 1955), 374–78; James B. Murphy, L. Q. C. Lamar: Pragmatic Patriot (Baton Rouge, 1973), 113–18; Willie Lee Rose, Rehearsal for Reconstruction: The Port Royal Experiment (Indianapolis, 1964), 400.
26. New York Times, April 26, 1875; Robbins Little to John E. Bryant, July 10, 1875, John E. Bryant Papers, DU.

critique in articles written before ever visiting the state. Moreover, he acquired much of his information from interviews with white Democratic leaders and seems to have spoken with only one black Carolinian. Nor did Pike place Southern corruption in a national context. (When a Democrat began reading from *The Prostrate State* on the House floor, black Congressman Robert Smalls inquired, "Have you the book there of the city of New York?")

Although little in Pike's polemic was new, his book represented a significant shift in anti-Reconstruction propaganda. Unlike Democratic critics of Southern government, liberal reformers had previously eschewed overt racism. Pike, however, focused primarily on the innate incapacity of blacks. As *The Nation*'s response to *The Prostrate State* revealed, his account struck a responsive chord. To his familiar equation of Reconstruction, the Tweed Ring, and Butlerism as forms of "Socialism," a critique that had more to do with class relations than race, Godkin now added an antiblack animus new less in underlying assumptions than in explicitness and virulence. The "blackest" legislators, he claimed, were the worst offenders, and South Carolina freedmen enjoyed an overall "average of intelligence but slightly above the level of animals."[27]

Occasionally, a dissenting voice could be heard in the Northern press. The New York *Herald* published a remarkably balanced report from South Carolina in 1874, pointing to evidence of educational and economic progress and noting that Pike's observations "derived from less than cosmopolitan study." But despite its many inadequacies, *The Prostrate State* not only helped make South Carolina a byword for corrupt misrule but reinforced the idea that the cause lay in "negro government." In its wake, even newspapers that had long supported Reconstruction joined in the condemnation of the state's black legislators, and a spate of articles appeared in middle-class journals like *Scribner's*, *Harper's*, and *The Atlantic Monthly*, echoing the conclusion that good government and regional prosperity would return to the South only when Reconstruction came to an end. After *The Prostrate State*, Charles Nordhoff's *The Cotton States*, a collection of articles written for the *Herald* in 1875, probably had the widest impact. Like Pike, Nordhoff had been a vocal supporter of the Greeley campaign, and like his predecessor, he found in the South precisely what he had come to see: leading whites prepared to safeguard blacks' rights if restored to local power, and freedmen too ignorant to take a responsible role in politics and too lacking in enterprise to lift themselves out of poverty. Meanwhile, these same journals expressed

27. James S. Pike, *The Prostrate State* (New York, 1874), 12, 67; Robert S. Durden, *James Shepherd Pike: Republicanism and the American Negro, 1850–1882* (Durham, 1957); Okon E. Uya, *From Slavery to Public Service: Robert Smalls 1839–1915* (New York, 1971), 95; *Nation*, April 16, 1874. Reprimanded by Thomas Wentworth Higginson, Godkin retreated a bit, confining his remarks to "the negroes of the low-country," who, he maintained, were "just about as degraded as Mr. Pike describes them to be." *Nation*, April 30, 1874.

increasingly retrograde racial attitudes, reflected visually in a shift from engravings depicting the freedmen as upstanding citizens harrassed by violent opponents, to vicious caricatures presenting them as little more than unbridled animals. Ironically, therefore, even as racism waned as an explicit component of the Northern Democratic appeal, it gained a hold on respectable Republican opinion, offering a convenient explanation for Reconstruction's "failure."[28]

A "Counter-Revolution," Vice President Henry Wilson lamented to William Lloyd Garrison, was overtaking Reconstruction: "Men are beginning to hint at changing the condition of the negro . . . our Anti-slavery veterans must again speak out." But by the mid-1870s, with "reform" now suggesting rule by the "best men" rather than the desire to purge American life of racial inequality, survivors of the Radical generation seemed relics of a bygone era. Visiting Ohio's Western Reserve, George W. Julian found that the "abolition element" had "almost died out in that old stronghold of radicalism, . . . and the few antislavery pioneers who remain seem to feel lonely and lost." Many antislavery veterans, indeed, shared in the mood of national reconciliation. Partly because the depression devastated their sources of income, the missionary zeal of Northern aid societies waned considerably during the 1870s. The New England Freedmen's Aid Society disbanded in 1874, while the American Missionary Association, displaying a growing concern with winning the goodwill of Southern whites, pronounced black suffrage a failure and the freedmen ungrateful for the organization's many efforts on their behalf.[29]

By mid-decade, commented the *New York Times* in noting the passing of abolitionist Gerrit Smith, the "era of moral politics" had come to a definitive end. When a large body of "grey-headed men and women" gathered in Chicago for an 1874 abolitionist reunion, the meeting failed to speak out as forcefully as some desired in defense of black civil rights. Only a few discordant notes disturbed the sense that the mission to the slave had ended. From Virginia, the daughter of early abolitionist Myron Holley related how a mob had recently vandalized her school for black children. "Surrounded as we are by such evidences that the spirit of American slavery still lives," she added, "I am not yet ready to celebrate the completion of the anti-slavery work." Another abolitionist wrote that

28. New York *Herald*, October 19, 1874; Columbia *Daily Union*, February 17, 1874; Richard H. Abbott, *The Republican Party and the South, 1855–1877: The First Southern Strategy* (Chapel Hill, 1986), 229; Vincent P. DeSantis, *Republicans Face the Southern Question* (Baltimore, 1959), 41–42; Charles Nordhoff, *The Cotton States in the Spring and Summer of 1875* (New York, 1876), 10–21, 36–39, 97; Gillette, *Retreat from Reconstruction*, 368–69.

29. James M. McPherson, "Coercion or Conciliation? Abolitionists Debate President Hayes' Southern Policy," *New England Quarterly*, 39 (December 1966), 476; Gillette, *Retreat from Reconstruction*, 238; Rose, *Rehearsal*, 389; Richard B. Drake, "Freedmen's Aid Societies and Sectional Compromise," *JSH*, 29 (May 1963), 175, 182–84; Joe M. Richardson, *Christian Reconstruction: The American Missionary Association and Southern Blacks, 1861–1890* (Athens, Ga., 1986), 252–53.

after a trip South, he remained convinced of the need for national action on the freedmen's behalf; if they failed to act now, "anti-slavery men had better have done nothing."[30]

Buffeted by the shifting tides of public opinion, preoccupied first with the economic depression and later with yet another wave of political scandals, the second Grant Administration found it impossible to devise a coherent policy toward the South. "*Radicalism* is dissolving—going to pieces," wrote a Southern Democratic Senator, "but what is to take its place, does not clearly appear."[31] Events lurched from crisis to crisis, with alternating moments of conciliation and firmness but no overriding sense of purpose. In retrospect, however, it is clear that, partly due to events outside his own control, Grant in his second term presided over a broad retreat from the policies of Reconstruction.

Having won reelection, Grant quickly moved to avoid further national intervention. Federal patronage flowed freely to "respectable" Southern Democrats—"in the strife . . . to placate the rebels and get their votes," complained one Northern politico, "the government is being filled with them in every Dept." Simultaneously, the Justice Department severely curtailed prosecutions under the Enforcement Acts, and many convicted Klansmen received hasty pardons. And when a minor civil conflict erupted in Arkansas in 1874, the President refused to assist local Republicans. Even by Reconstruction standards, the state's political alignments were chaotic. Republican Elisha Baxter, a loyalist slaveholder and Union Army veteran, had won a disputed gubernatorial victory in 1872 over Joseph Brooks, a carpetbagger backed by Liberals, Democrats, and native Unionists who believed Northerners had shouldered them aside in party affairs. Once in office, however, Baxter alienated his supporters by pushing through a constitutional amendment enfranchising ex-Confederates and refusing to release additional state bonds to Arkansas railroads. Whereupon Republicans disavowed their own governor and took up Brooks's cause. There followed the "Brooks-Baxter War," a melodramatic title for skirmishes between supporters of the rival claimants, each backed by his own legislature and militia. In May 1874, after a period of indecision, Grant recognized Baxter as governor and ordered the forces supporting Brooks to disperse, a decision that sealed the doom of Arkansas Reconstruction. In the fall, Democrat Augustus Garland, who had masterminded Baxter's strategy, overwhelmingly won election as his successor.[32]

30. *New York Times*, December 30, 1874; George W. Julian Journal, June 28, 1874, George W. Julian Papers, Indiana State Library; Larry Gara, "A Glorious Time: The 1874 Abolitionist Reunion in Chicago," *JISHS*, 64 (Autumn 1972), 280–92; Sallie Holley to Zebina Eastman, June 8, 1874, Zebina Eastman Papers, Chicago Historical Society; Lawrence J. Friedman, *Gregarious Saints: Self and Community in American Abolitionism, 1830–1870* (New York, 1982), 278–80.

31. Gillette, *Retreat from Reconstruction*, 180–82; Augustus S. Merrimon to D. F. Caldwell, April 29, 1874, D. F. Caldwell Papers, UNC.

32. W. C. Dodge to Horace Austin, July 17, 1873, Horace Austin Papers, Minnesota Historical Society; Robert J. Kaczorowski, *The Politics of Judicial Interpretation: The Federal*

Even had the will for an interventionist Southern policy survived in the White House, a series of Supreme Court decisions during Grant's second term undercut the legal rationale for such action. Previously, the Court had proved reluctant to become involved in Reconstruction controversies. But during the 1870s, responding to the shifting currents of Northern public opinion, it retreated from an expansive definition of federal power, and moved a long way toward emasculating the postwar amendments—a crucial development in view of the fact that Congress had placed so much of the burden for enforcing blacks' civil and political rights on the federal judiciary.

The first pivotal decision, in the *Slaughterhouse Cases*, was announced in 1873. Four years earlier, Louisiana had chartered a corporation to monopolize butchering in New Orleans, ostensibly to protect public health but actually to encourage the construction of modern meat-packing facilities that would enable the city to compete for control of the Texas cattle trade. Butchers suddenly deprived of employment sued in federal court, contending that the monopoly violated their right to pursue a livelihood, guaranteed, they insisted, by the Fourteenth Amendment. In effect, they asked the Court to decide whether the Amendment had expanded the definition of national citizenship for all Americans, or simply accorded blacks certain rights already enjoyed by whites. Speaking for the five-man majority, Justice Samuel F. Miller rejected the butchers' plea, insisting Congress had intended primarily to enlarge the rights of the former slaves. Blacks might have been forgiven for thinking this construction would bolster federal action on their behalf, but Miller went on to distinguish sharply between national and state citizenship, and to insist that the Amendment only protected those rights that owed their existence to the federal government. What were these federal rights? Miller mentioned access to ports and navigable waterways, and the ability to run for federal office, travel to the seat of government, and be protected on the high seas and abroad. Clearly, few of these rights were of any great concern to the majority of freedmen. The Fourteenth Amendment, Miller declared, had not fundamentally altered traditional federalism; most of the rights of citizens remained under state control, and with these the Amendment had "nothing to do." As Justice Stephen J. Field pointed out in a stinging dissent, if this were the Amendment's meaning, "it was a vain and idle enactment, which accomplished nothing and most unnecessarily excited Congress and the people on its passage."[33]

Courts, Department of Justice, and Civil Rights, 1866–1876 (New York, 1985), 105–13; George H. Thompson, *Arkansas and Reconstruction* (Port Washington, N.Y., 1976), 88–166; Earl F. Woodward, "The Brooks and Baxter War in Arkansas, 1872–1874," *ArkHQ,* 30 (Winter, 1971), 315–36.

33. Charles Fairman, *Reconstruction and Reunion 1864–1888: Part One* (New York, 1971), 1321–59; Kaczorowski, *Politics of Judicial Interpretation,* 143–59; Harold M. Hyman and William M. Wiecek, *Equal Justice Under Law: Constitutional Development 1835–1875* (New York, 1982), 475–81.

Ironies abounded in what *The Nation* called this "curious case," and not only because the parties seeking relief in federal court were Southern whites, not aggrieved freedmen. The butchers' attorney, John A. Campbell, a Democrat, prewar Supreme Court Justice, and the Confederacy's Assistant Secretary of War, invoked free labor principles to urge the Court to protect a citizen's right to choose a livelihood. Justice Field, in his dissent, agreed that the "right of free labor" was "a distinguishing feature of our republican institutions." (He did not, apparently, believe women entitled to this right, for in a simultaneous case, he joined the majority in rejecting the suit of Myra Bradwell, who sought to overturn an Illinois court ruling barring females from practicing law.) Field's broad interpretation of the Fourteenth Amendment, however, had little to do with Reconstruction, which, as a Democrat, he opposed, or with blacks' rights, in which he had almost no interest. Rather, he had become convinced by the Grange, Paris Commune, and other "class" movements, that the federal government must exercise some restraint on unwise actions by the states. His argument in *Slaughterhouse* blazed a trail toward the judicial conservatism of the 1880s and 1890s, when the federal courts became a refuge for those seeking to protect property rights against local restrictions on economic enterprise. Justice Miller, for his part, opposed the butchers because he did not wish to see the Court become "a permanent censor upon all the legislation of the states." A founder of the Iowa Republican party, he remained committed to the freedmen's enjoyment of equal rights before the law. But his judgment that primary authority over citizens' rights rested with the states led lower federal courts to limit national jurisdiction over the administration of justice, further weakening civil rights enforcement.[34]

Slaughterhouse at least affirmed the indisputable fact that the postwar amendments had been designed to protect black rights (although the Court's denial of their applicability to whites, and its studied distinction between the privileges deriving from state and national citizenship, should have been seriously doubted by anyone who read the Congressional debates of the 1860s). Even more devastating was the 1876 decision in *U.S.* v. *Cruikshank.* This case arose from the Colfax massacre, the bloodiest single act of carnage in all of Reconstruction. Indictments were brought under the Enforcement Act of 1870, alleging a conspiracy to deprive the victims of their civil rights. On the grounds that the wording

34. *Nation,* December 1, 1870; Michael L. Benedict, "Preserving Federalism: Reconstruction and the Waite Court," *Supreme Court Review,* 1978, 55–57; Fairman, *Reconstruction and Reunion,* 1364–66; Carl B. Swisher, *Stephen J. Field, Craftsman of the Law* (Washington, D.C., 1930), 376–83, 418–20, 429; William E. Forbath, "The Ambiguities of Free Labor: Labor and the Law in the Gilded Age," *Wisconsin Law Review,* 1985, 772–99; Robert G. McCloskey, *American Conservatism in the Age of Enterprise 1865–1910* (Cambridge, Mass., 1951), 73–84; Charles Fairman, *Mr. Justice Miller and the Supreme Court, 1862–1890* (Cambridge, Mass., 1939), 138, 193–94; Kaczorowski, *Politics of Judicial Interpretation,* 173–93.

failed to specify race as the rioters' motivation, the Supreme Court overturned the only three convictions the government had managed to obtain. More, however, was at stake than faulty language, for the Court went on to argue that the postwar amendments only empowered the federal government to prohibit violations of black rights by *states;* the responsibility for punishing crimes by individuals rested where it always had—with local and state authorities. The decision did uphold Washington's authority to protect the "attributes of national citizenship," but these had been defined so narrowly in *Slaughterhouse* as to render them all but meaningless to blacks. In the name of federalism, the decision rendered national prosecution of crimes committed against blacks virtually impossible, and gave a green light to acts of terror where local officials either could not or would not enforce the law.[35]

Additional evidence of the nation's waning commitment to the freedmen abounded during the 1870s. Partly because his own son was among the offenders, the President did nothing to force the United States Military Academy to take action against classmates who ostracized and harassed its first black cadets. James W. Smith, the product of a South Carolina Freedmen's Bureau school whom Connecticut philanthropist David Clark took into his home and educated, in 1870 broke West Point's color line. After enduring three years of persecution, Smith was dismissed after failing a test administered privately (in defiance of custom) by his philosophy professor. His successors endured similar ordeals. Not until 1877 did the stoical Henry O. Flipper, the son of a Georgia slave artisan, graduate and become the first black to receive a commission in the regular army.[36]

Of far more import to most blacks was the fate of the Freedman's Savings and Trust Company, one of the many financial institutions to succumb to the depression. Chartered in 1865, the bank had actively sought deposits from the freedmen, while at the same time instructing them in the importance of thrift. To inspire confidence in its activities, it employed local black leaders as cashiers and members of branch advisory boards. Blacks by the thousands came to the bank with tiny deposits—the majority of accounts were under fifty dollars and some amounted only to a few pennies—and used it to handle their financial affairs and remit funds to distant relatives. And organizations from churches to benevolent societies entrusted their modest treasuries to its vaults. The freedmen's money, declared Henry Wilson in 1867, was "just as safe there as if it were in the Treasury of the United States." Unfortunately, the bank's directors soon lost sight of both their reforming zeal and prudent busi-

35. Kaczorowski, *Politics of Judicial Interpretation,* 205–16; Benedict, "Preserving Federalism," 69–73; Fairman, *Reconstruction and Reunion,* 1371–78.
36. McFeely, *Grant,* 375–77; David Clark to Charles Sumner, December 24, 1870, Charles Sumner Papers, HU; Henry O. Flipper, *The Colored Cadet at West Point* (New York, 1878).

ness practices and became caught up in the speculative fever of the early 1870s, investing heavily in Washington real estate and making large, unsecured loans to railroads and other companies. When the Panic came, they made frantic attempts to maintain depositors' confidence, even appointing Frederick Douglass as the institution's president and inducing him to write a personal check for $10,000, ostensibly to cover funds delayed in the mail. Douglass never saw the money again and in June 1874, with only $31,000 on hand to cover obligations to its 61,000 depositors, the Freedman's Savings Bank suspended operations.[37]

Although legally a private corporation rather than an arm of the government, the bank often shared offices with the Freedmen's Bureau, used army officers to solicit customers, and through newspaper advertisements and circulars festooned with images of Lincoln, encouraged blacks to think the federal government stood behind its activities. Even a Kentucky Democrat hardly known as a friend of the freedmen told the House that the government was "as morally bound to see to it that not a dollar is lost . . . as it is possible to establish a moral obligation." But although a succession of Presidents and comptrollers of the currency urged that the freedmen be repaid with federal funds, Congress did little more than assist the bank in winding up its affairs. Eventually, half the depositors received compensation—an average of $18.51 per person, or about three fifths the value of their accounts. The remainder, who died, lost hope, or failed, as required, to send their bankbooks to Washington, received nothing. The collapse had a "paralyzing effect" upon black organizations and destroyed confidence in banks among individual freedmen for years to come. Well into the twentieth century, pathetic appeals arrived in Washington from depositors seeking the balance of their funds. "Mr. President," one wrote in 1921, "I pray you to consider us old People . . . our best life spent in slavery. . . . Just asking for what we worked for." Their letters now gather dust in the National Archives.[38]

And then there was the Civil Rights Bill, still mired in Congressional debate in 1874 despite the fact that Grant had, in his second inaugural address, publicly endorsed it for the first time. This measure, it will be recalled, made it illegal for places of public accommodation and entertainment to make any distinction between black and white patrons, and outlawed racial discrimination in public schools, jury selection, churches, cemeteries, and transportation. That the bill survived at all was due to

37. Carl R. Osthaus, *Freedmen, Philanthropy, and Fraud: A History of the Freedman's Savings Bank* (Urbana, Ill., 1976); Savannah *Freemen's Standard*, February 15, 1868; Bobby L. Lovett, "Some 1871 Accounts for the Little Rock, Arkansas Freedman's Savings and Trust Company," *JNH*, 66 (Winter 1981–82), 322–28; Robert Somers, *The Southern States Since the War 1870–71* (London, 1871), 54; *CG*, 40th Congress, 1st Session, 79; Robert E. Withers, *Autobiography of an Octogenarian* (Roanoke, Va., 1907), 334.

38. Mobile *Nationalist*, May 23, 1867; *CR*, 43d Congress, 1st Session, Appendix, 477; John W. Blassingame, *Black New Orleans 1860–1880* (Chicago, 1973), 67; Dorothy Sterling, ed., *The Trouble They Seen* (Garden City, N.Y., 1976), 260–61; *Report of Committee on Labor and Capital*, 4:420–21; Osthaus, *Freedman's Savings Bank*, 211–19.

the tireless advocacy of Charles Sumner, and as he lay dying in March 1874, Sumner whispered to a visitor: "You must take care of the civil-rights bill, . . . don't let it fail." Two months later, his colleagues adopted Sumner's bill, shorn only of its church provision—a tribute to the man whose twenty-three-year career in the Senate had been devoted to the principle of racial equality, even though many who voted for the bill believed it stood no chance of passage in the House. There, the bill's manager was Benjamin Butler, who had introduced it earlier in the session with an eloquent account of how the wartime conduct of black soldiers had tempered his own racial prejudices. Over 500 blacks, Butler recalled, had died in one engagement on the James River: "As I looked on their bronzed faces upturned in the shining sun to heaven as if in mute appeal against the wrong of the country for which they had given their lives . . . feeling I had wronged them in the past . . . I swore to myself a solemn oath . . . to defend the rights of these men who have given their blood for me and my country."[39]

The *Slaughterhouse* decision added a new dimension to the civil rights debate. Democrats seized upon the Court's restrictive interpretation of the Fourteenth Amendment to insist the rights protected by Sumner's bill lay outside federal jurisdiction. Like a ghost of some earlier time, Confederate Vice President Alexander H. Stephens, now representing Georgia in the House, delivered the major speech in opposition, arguing that the measure would transform the republic "into a centralized empire." To illustrate that the bill exceeded the federal government's constitutional authority, Kentucky Democrat James F. Beck read aloud the first ten amendments, to enumerate the privileges the Fourteenth Amendment required states not to abridge. They contained no reference, he noted, to schools or public accommodations. (As Beck's speech indicated, the doctrine of "incorporation"—that the states were now required not to violate the Bill of Rights—had by 1874 become a virtually noncontroversial *minimum* Congressional interpretation of the Amendment's purposes.) The bill's defenders construed the Amendment far more broadly. Sumner's bill, they insisted, was a legitimate expression of the nationalization of individual rights brought about by the postwar amendments.[40]

Seven blacks sat in the Forty-Third Congress, and all spoke, with vigor and eloquence, on the Civil Rights Bill. Before galleries crowded with black spectators, their speeches invoked both personal experience and the black political ideology that had matured during Reconstruction. Several related their own "outrages and indignities." Joseph Rainey had

39. James D. Richardson, ed., *A Compilation of the Messages and Papers of the Presidents 1789–1897* (Washington, D.C., 1896–99), 7:221; David Donald, *Charles Sumner and the Rights of Man* (New York, 1970), 586; Gillette, *Retreat from Reconstruction*, 200–206; *CR*, 43d Congress, 1st Session, 457.
40. *CR*, 43d Congress, 1st Session, 312, 342, 412, 427; S. G. F. Spackman, "American Federalism and the Civil Rights Act of 1875," *Journal of American Studies*, 10 (December 1976), 317–23.

been thrown from a Virginia streetcar, John R. Lynch forced to occupy a railroad smoking car with gamblers and drunkards, Richard H. Cain and Robert B. Elliott excluded from a North Carolina restaurant, James T. Rapier denied service by inns at every stopping point between Montgomery and Washington. To Rapier, such "anti-republican" discrimination recalled the class and religious inequalities of other lands: In Europe, "they have princes, dukes, lords;" in India, "brahmans or priests, who rank above the sudras or laborers;" in America, "our distinction is color." Cain reminded the House that "the black man's labor" had for generations enriched the nation, and "hurl[ed] back with contempt" a North Carolina Democrat's description of his people as barbarians to whom slavery had brought civilization and Christianity. His colleague, Cain suggested, had probably never heard of Hannibal, Hamilcar, and other accomplished blacks of antiquity, "for that kind of literature does not come to North Carolina." Alonzo J. Ransier drew on free labor principles to contend that blacks could not enjoy "an equal chance in the race of life" while constantly subject to "humiliating discriminations." In a speech that won national attention, Elliott recalled the sacrifices of black soldiers and carefully analyzed the *Slaughterhouse* decision in order to demonstrate that it cast no doubt on Congressional authority to enact the bill.[41]

Especially in "redeemed" states, where "separate but equal" education was little more than a mockery, Southern blacks considered the bill's schools clause its most important feature. The state's segregated education, declared a Tennessee black convention in April 1874, was "defective" and "anti-republican," teaching whites "the spirit of caste and hate" and blacks "their inferiority." But Butler came under tremendous pressure to bury the bill. Many Southern Republicans warned that its passage would injure their party and threaten the "immediate destruction" of public education in their region. In Congress, only Boutwell among prominent Republicans forthrightly defended school integration as a way of eroding racial prejudices. Already facing more than enough political liabilities, House Republicans put off consideration of the Civil Rights Bill until after the fall 1874 elections. Along with the shifting positions of the President and federal courts, Congress' handling of the bill suggested that the South's defenders of Reconstruction would face alone both the consequences of the depression and opponents again turning to violence.[42]

41. Alfred H. Kelly, "The Congressional Controversy over School Segregation, 1867–1875," *AHR*, 64 (April 1959), 552; *CR*, 43d Congress, 1st Session, 344, 382, 407–409, 565, 901–902, 4782, 4785, 2d Session, 945; Peggy Lamson, *The Glorious Failure: Black Congressman Robert Brown Elliott and the Reconstruction in South Carolina* (New York, 1973), 171.

42. Nashville *Union and American*, April 30, 1874; W. G. Eliot to Benjamin F. Butler, May 28, 1874, Eugene B. Drake to Butler, June 2, 1874, D. H. Graves to Butler, June 12, 1874, Butler Papers; *CR*, 43d Congress, 1st Session, 4116, 4592.

The Waning of Southern Republicanism

The depression of the 1870s dealt the South an even more severe blow than the rest of the nation. Between 1872 and 1877, the price of cotton fell by nearly 50 percent, until it more or less equaled the cost of production, and tobacco, rice, and sugar also suffered precipitous declines. The effects rippled through the entire economy, plunging farmers into poverty and drying up the region's already inadequate sources of credit. The depression disrupted the commerce of port and interior cities, bankrupted merchants, seriously undermined the economic prospects of artisans (especially blacks, who generally had fewer resources to fall back on than their white counterparts), and all but eliminated prospects for social mobility among unskilled laborers of both races. It rudely shattered whatever hope still existed for the early emergence of a modernized, prosperous Southern economy, and forced into bankruptcy even such long-established bulwarks of Southern industry as Richmond's Tredegar Iron Works. In 1880, with a per capita income only one third that of the rest of the nation, the South lagged farther behind the North in the total value of its agricultural and industrial output than when the decade began. The economic disaster dealt yet another blow to the credibility of surviving Reconstruction governments, helping to propel the Deep South down the road to Redemption. "Had the volume of business remained unbroken . . . " a South Carolina journalist would write in 1877, "the negro and the carpet-bagger would have retained power indefinitely."[43]

As in the North, far-reaching structural changes accompanied the crisis, permanently altering the balance of economic power. Ironically, the depression facilitated the penetration of Northern capital that Reconstruction governments had previously been unable to attract. Railroad construction came to an even more abrupt halt than in the North (the South's share of national track mileage fell by 20 percent during the 1870s) and a majority of the region's lines fell into the hands of receivers. Outside investors like Collis P. Huntington took advantage of the opportunity to purchase bankrupt Southern railroads at bargain prices, greatly accelerating the consolidation of the Southern rail system into fewer and larger lines and its integration into national traffic patterns. Not one of the Southern roads that fell into receivership escaped Northern financial

43. U.S. Bureau of the Census, *Historical Statistics of the United States*, 208; Eugene Lerner, "Southern Output and Agricultural Income, 1860–1880," *AgH*, 30 (July 1959), 124; Joe G. Taylor, *Louisiana Reconstructed, 1863–1877* (Baton Rouge, 1974), 360–61; Richard J. Hopkins, "Occupational and Geographic Mobility in Atlanta, 1870–1896," *JSH*, 34 (May 1968), 200–13; Charles B. Dew, *Ironmaster to the Confederacy: Joseph R. Anderson and the Tredegar Iron Works* (New Haven, 1966), 318; Richard A. Easterlin, "Regional Income Trends, 1840–1950," in Seymour E. Harris, ed., *American Economic History* (New York, 1961), 528; [George W. Bagby] *Selections from the Miscellaneous Writings of Dr. George W. Bagby* (Richmond, 1884–85), 2:163.

influence, thus fixing the regional economy even more firmly in a colonial mold.[44]

But it was in agriculture that the depression's impact proved most severe. As upcountry yeomen found themselves once again engulfed by poverty and indebtedness, the trend toward cotton production and reliance on merchants for food and supplies, evident since the war, rapidly accelerated. Cleveland County, in western North Carolina, produced only 520 bales of cotton in 1870 but grew more than ten times that amount a decade later. Similar increases took place in upcountry Alabama and Georgia. And along with cotton came the inexorable growth of tenancy. In one upcountry Georgia county, two thirds of the white farmers were renting for cash or shares by 1880; in another, where the Census Bureau found only forty white families renting their farms in 1870, it reported nearly 250 a decade later. The rapid growth of a white population unable to make ends meet on their own land or uprooted from it during the depression created an abundant supply of cheap labor that helped make possible the rapid expansion of the cotton textile industry in the 1880s.[45]

In the black belt, many planters who had weathered the uncertainties of the postwar years saw declining land prices sharply reduce the value of their holdings, while falling agricultural earnings made it impossible for them to discharge their debts to local merchants at the end of the year. In the rich cotton region surrounding Natchez, over 150 planters had forfeited all or part of their land by 1875 for debt or nonpayment of taxes. The depression dealt the final blow to the lowcountry rice aristocracy of South Carolina and Georgia, hastening the breakup of their plantations into tiny plots owned by black farmers. "Many families who have managed thus far to hold out against the immediate distress which has increased upon us from year to year," wrote rice magnate Louis Manigault in 1876, "have finally succumbed . . . and I hear only of poverty and misery amongst those who were the richest and oldest families prior to the war." The trend toward outside control of Louisiana's sugar plantations continued apace, as many small planters saw their holdings pass

44. John F. Stover, *The Railroads of the South 1865–1900* (Chapel Hill, 1955), xv–xviii, 66–67, 125–54, 282; Robert L. Brandfon, *Cotton Kingdom of the New South* (Cambridge, Mass., 1967), 70; Maury Klein, "The Strategy of Southern Railroads," *AHR*, 73 (April 1968), 1060–61.

45. J. R. Davis, "Reconstruction in Cleveland County," *Trinity College Historical Society Historical Papers*, 10 (1914), 14; Walter L. Fleming, *Civil War and Reconstruction in Alabama* (New York, 1905), 804–805; Steven Hahn, *The Roots of Southern Populism: Yeoman Farmers and the Transformation of the Georgia Upcountry, 1850–1890* (New York, 1983), 146–47; Frank J. Huffman, "Town and Country in the South, 1850–1880: A Comparison of Urban and Rural Social Structures," *SAQ*, 76 (Summer 1977), 376–77; Frank J. Huffman, "Old South, New South: Continuity and Change in a Georgia County, 1850–1880" (unpub. diss., Yale University, 1974), 222; Gavin Wright, "Cheap Labor and Southern Textiles Before 1880," *JEcH*, 39 (September 1979), 655–57, 678–80.

into the hands of owners of larger estates. With fewer and fewer planters able to pay money wages, the depression also gave the final impetus to the spread of sharecropping as the nearly universal labor system in cotton agriculture. When the Census Bureau made an intensive study of cotton production at the end of the decade, it found the share system prevalent in nearly every county surveyed, including those where wage labor had prevailed ten years earlier.[46]

If the spread of sharecropping fulfilled, in part, the freedmen's aspiration for day-to-day autonomy, in other ways the depression proved a disaster for blacks, severely limiting their power to influence working conditions. Where wage labor persisted, as in the Louisiana sugar fields and on diversified Upper South farms, monthly payments plummeted and agricultural workers faced chronic bouts of unemployment. Except in the rice kingdom and some parts of the Upper South, hard times arrested the modest progress of the late 1860s and early 1870s toward the growth of a class of independent black farmers, and drove many owners and cash renters back into the ranks of sharecroppers and laborers. Many blacks who did acquire land confronted dire circumstances. "I thought I had seen poverty in the great cities of the country . . . " declared one federal official, commenting on Liberty County in lowcountry Georgia, "but I never saw anything to compare with the poverty of those negroes there." The value of the property they had owned as slaves, he thought, exceeded their holdings at the end of the 1870s.[47]

By a cruel irony, the depression hit just as blacks were acquiring increased political influence in states where Reconstruction still survived. This development reflected the waning of white Republicanism, a sense among grass-roots black voters that although the party's "fountain of power" they had too often seen their interests neglected by white Republicans, and a growing resentment on the part of black political leaders at serving as their party's "hewers of wood and drawers of water." Throughout the Republican South, the number of black officials rose significantly

46. Jayne Morris-Crowther, "An Economic Study of the Substantial Slaveholders of Orangeburg County, 1860–1880," SCHM, 86 (October 1985), 305, 313; Michael Wayne, The Reshaping of Plantation Society: The Natchez District, 1860–1880 (Baton Rouge, 1983), 84; "Season of 1876," in "Statement of Sales, Gowrie Plantation, Savannah River," Manigault Family Papers, UNC; Walter Prichard, "The Effects of the Civil War on the Louisiana Sugar Industry," JSH, 5 (August 1939), 330; Clifton Paisley, From Cotton to Quail: An Agricultural Chronicle of Leon County, Florida 1860–1967 (Gainesville, Fla., 1968), 34; Tenth Census, 1880, 5:83, 104–105, 154, 161, 6:61, 155, 172.

47. Ralph Shlomowitz, " 'Bound' or 'Free'? Black Labor in Cotton and Sugarcane Farming, 1865–1880," JSH, 50 (November 1984), 578; Barbara J. Fields, Slavery and Freedom on the Middle Ground: Maryland During the Nineteenth Century (New Haven, 1985), 188–89; Foner, Politics and Ideology, 120; Roger Ransom and Richard Sutch, One Kind of Freedom: The Economic Consequences of Emancipation (New York, 1977), 83, 181; 46th Congress, 2d Session, Senate Report 693, pt. 2:261.

in the early 1870s. Black representation in Congress grew from five to seven in 1873 and reached a Reconstruction peak of eight (representing six different states) in 1875. (It was indicative of the changing origins of black political leadership that of the nine men who served for the first time after 1872, only three had been born free.) In 1872, Florida and Arkansas chose their first black state officials, and Louisiana and South Carolina each elected three. The number of black legislators rose dramatically in several states, and even where it did not, more blacks acquired committee chairmanships and seats in state senates. At the local level, too, black officeholding continued to increase. Traveling across the South in 1873 and 1874, reporter Edward King encountered black city councilmen in Petersburg, Houston, and Little Rock, parish jury members in Louisiana, sheriffs scattered across the black belt. After Republicans regained control of Charleston in 1873, blacks made up half the police and board of aldermen, and nearly the entire legislative delegation.[48]

This rising presence in office did not always translate into augmented power at the highest echelons of state government. Sixteen prominent black politicians in 1874 publicly complained of being excluded from "any knowledge of the confidential workings of the party and government" in Louisiana, and "not infrequently humiliated in our intercourse with those whom we have exalted to power." In Mississippi, on the other hand, black political leaders, alienated by the conservative policies of Governors James L. Alcorn and Ridgely Powers, were instrumental in engineering the 1873 gubernatorial nomination of Sen. Adelbert Ames, along with a six-man state ticket that included three black candidates.

48. Petition, April 1876, Tunica, Louisiana, William P. Kellogg Papers, LSU; Nashville *Union and American*, April 30, 1874; Michael Perman, *The Road to Redemption: Southern Politics, 1869–1879* (Chapel Hill, 1984), 137–39; Charles Vincent, *Black Legislators in Louisiana During Reconstruction* (Baton Rouge, 1976), 143–55, 192–94; Edward King, *The Southern States of North America* (London, 1875), 113, 281, 293, 448, 581–82. For a complete list of black state officials, see page 353. In the 43d Congress, the black Congressmen were Richard H. Cain, Robert B. Elliott, Joseph H. Rainey, and Alonzo J. Ransier of South Carolina, Josiah Walls of Florida, James T. Rapier of Alabama, and John R. Lynch of Mississippi. In the 44th Congress, they were Rainey and Robert Smalls of South Carolina, Lynch and Senator Blanche K. Bruce of Mississippi, Charles E. Nash of Louisiana, John Hyman of North Carolina, Jeremiah Haralson of Alabama, and Walls. Of the new members, Bruce was a Virginia planter's privileged slave educated at Oberlin after the war, who had established a political power base in Bolivar County, Mississippi, Cain a freeborn AME minister, Haralson a self-educated former slave fieldhand who spoke "with the brogue of the cornfield" (John H. Henry to William E. Chandler, July 15, 1872, William E. Chandler Papers, LC). Hyman was a farmer, storekeeper, and former slave who had been "bought and sold as a brute" (John A. Hyman to Charles Sumner, January 24, 1872, Sumner Papers). Lynch was a slave freed by the Union Army and educated at a Natchez freedmen's school, who worked as a photographer. Nash was a freedman and bricklayer who had lost a leg serving in the Union Army, Ransier a freeborn Charleston shipping clerk, Rapier a member of a prosperous free Alabama family who acquired a plantation during Reconstruction, and Smalls a celebrated former slave, political boss of Beaufort, and postwar merchant. For the first black Congressmen, see page 352.

With the solid backing of black voters, Ames handily defeated Alcorn, who ran with the support of the Democrats. Even Alcorn's plantation hands voted against him. At the same time, blacks substantially increased their representation in the legislature, and consolidated their hold on local offices throughout the black belt. When the lawmakers assembled, they chose a black speaker of the house, and elected Blanche K. Bruce to the U. S. Senate.[49]

In a sense, blacks' growing assertiveness represented not only a departure from their earlier willingness to step aside in favor of white candidates, but a waning of the very ideal of a polity in which color was irrelevant. The hard realities of Southern political life had taught the lesson that black constituents needed to be represented by black officials. (In 1874, an Alabama black convention even resolved that a black party to a lawsuit had a right to demand a jury "composed of not less than one-half of his own race.") Accused by white Republicans of drawing a political "color line," black leaders responded that race, historically the "cause of exclusion," must now become a "ground of recognition until the scales are once more balanced."[50]

Under other circumstances, blacks might have used their enhanced political power to press for bold new political and economic programs. As 1873 began, however, Republicans enjoyed undisputed control of state government only in Arkansas, Louisiana, Mississippi, and South Carolina. Tennessee, Georgia, and Virginia had been "redeemed," while in Alabama, Florida, North Carolina, and Texas, Republican governors confronted hostile or divided legislatures. Blacks' political role waxed, moreover, precisely as the Northern outcry against corruption and "extravagance" reached its height. The depression, which devastated the Southern states' credit, tax rolls, and budgets, added further urgency to demands for retrenchment. Together with the persistent need to attract white voters, these developments significantly narrowed Southern Republicans' policy options. The shifting balance of racial power within the party does help account for the passage of civil rights laws in Florida, Louisiana, and Mississippi in 1873. And Florida's black lawmakers, responding to strikes among timber workers and longshoremen, pushed a number of prolabor measures through the legislature in 1874 and 1875. One gave workers a first lien on the timber they had produced to guarantee payment of wages, another required six-months' prior residence for licensed dock workers (to protect black stevedores against migrant Cana-

49. New Orleans *Louisianian*, October 3, 1874; John R. Lynch to Adelbert Ames, January 31, 1873, J. M. P. Williams to Ames, February 19, 1873, Ames Family Papers, SC; William C. Harris, *Day of the Carpetbagger: Republican Reconstruction in Mississippi* (Baton Rouge, 1979), 468–79; Lillian A. Pereyra, *James Lusk Alcorn: Persistent Whig* (Baton Rouge, 1966), 160; James W. Garner, *Reconstruction in Mississippi* (New York, 1901), 307–308.

50. 46th Congress, 2d Session, Senate Report 693, pt. 2:395; Garner, *Mississippi*, 293; *New National Era*, August 28, 1873.

MEMBERS OF THE LEGISLATURE
STATE OF MISSISSIPPI, 1874-'75.
Photographed by E. von SEUTTER, Jackson, Miss.

SENATE

1. Lt. Gov. A.K. Davis, Pres.
2. W.C. White, Secretary.
3. Little, Finis H.
4. Warner, Alex.
5. Campbell, M.
6. McClure, H.B.
7. Carter, J.P.
8. Thornton, P.R.
9. Mendenhall, J.L.
10. Sessions, J.F.
11. Metts, M.A.
12. Tuttle, M.H.
13. Taylor, R.H.
14. Furlong, C.E.
15. Graham, T.B.
16. Price, W.M.
17. Stone, J.M.
18. Allen, R.H.
19. Everett, J.E.
20. Bridges, N.B.
21. McNeil, J.A.
22. Cullens, C.
23. Bennett, Jos.
24. Steel, S.A.D.
25. Gillmer, J.P.
26. Henderson.
27. Gray, Will.
28. Barrow, P.B.
29. White, G.W.
30. Smith, G.C.
31. Stuart, Isham.
32. Gleed, Rob.
33. Williams, J.M.P.
34. Caldwell, Chas.
35. Albright, G.W.
36. Miss Adies Ball, Post M.

Mississippi's Last Reconstruction Senate (Library of Congress)

dian laborers who in winter competed for their jobs.) Louisiana (temporarily, alas) abolished the convict lease system. In general, however, blacks found themselves helping to preside over a period of moderation and consolidation, rather than striking out in radical new directions.[51]

Ironically, Reconstruction governments now took up their opponents' cry of retrenchment and reform. Although support for railroad aid had already waned, the crisis dealt the final blow to the "gospel of prosperity." Republican administrations imposed strict limits on state debts and replaced outstanding bonds with new ones bearing a uniform interest rate and representing smaller amounts of principal (inspiring cries of "repudiation" on Wall Street). The process of retrenchment progressed farthest in Florida, which did away with special legislation entirely by passing a general incorporation law, and amended its constitution to prohibit lending the state's credit to any corporation. By 1876, official salaries had been slashed, the legislature met only every other year, and the cost of state government barely exceeded that of Presidential Reconstruction. Louisiana Republicans also reduced the state's debt, expenses, and tax rates, and reined in the "flamboyant thievery" of the Warmoth era. As Mississippi's Ames discovered, such policies sometimes exacerbated the Republican party's internal discord. Although he owed his nomination and election to blacks, Ames entered office in 1874 convinced that the depression and the growing clamor for tax reduction compelled his administration to observe "rigid economy and a strict accountability" in fiscal affairs. Some of Ames's proposals were adopted, including a reduction in the property tax and an end to the practice of accepting depreciated state warrants for tax payments, but the legislature refused to cut expenses significantly. Most white Republicans supported his program, but blacks united to defeat efforts to lower the pay of lawmakers, reduce the school tax, and establish biennial legislative sessions.[52]

Black Republicans remained considerably less enthralled than white by the lure of reform. The depression had enhanced many black leaders' dependence on their official salaries for a livelihood, and both officeholders and ordinary freedmen feared retrenchment inevitably meant cuts in state programs like education, of special concern to their community.

51. Perman, Road to Redemption, 139, 213; Jerrell H. Shofner, "Militant Negro Laborers in Reconstruction Florida," JSH, 39 (August 1973), 402–408; Florida Acts and Resolutions 1874, 57–58; Taylor, Louisiana Reconstructed, 259.

52. Perman, Road to Redemption, 143–45; Mark W. Summers, Railroads, Reconstruction, and the Gospel of Prosperity: Aid Under the Radical Republicans, 1865–1877 (Princeton, 1984), 268–86; J. Mills Thornton III, "Fiscal Policy and the Failure of Radical Reconstruction in the Lower South," in J. Morgan Kousser and James M. McPherson, eds., Region, Race, and Reconstruction: Essays in Honor of C. Vann Woodward (New York, 1982), 383–85; Cohen, "The Lost Jubilee," 451; Carter Goodrich, "Public Aid to Railroads in the Reconstruction South," Political Science Quarterly, 71 (September 1956), 423–37; Joe M. Richardson, The Negro in the Reconstruction of Florida, 1865–1877 (Tallahassee, 1965), 205–11; Taylor, Louisiana Reconstructed, 260–65; Harry K. Benson, "The Public Career of Adelbert Ames, 1861–1876" (unpub. diss., University of Virginia, 1975), 226–33, 270–71; Harris, Day of the Carpetbagger, 430–36, 603–10, 633.

Many blacks, moreover, continued to look to the activist state to ease their economic plight. "We will never be any good only to serve all the days of our lives if we don't get help from our government," wrote one freedman, proposing that the states provide credit for those seeking to acquire land. "We have been supplied by our merchants here . . . but we cannot make a living at it for when our crops is shipped is never nothing left behind," complained another letter, which asked the Kellogg administration to advance supplies to tenant farmers. But such ideas lay far beyond the straitened resources of Southern governments and the conception of the state's proper role held by national party leaders.[53]

Ames's inaugural address, which condemned the crop lien and the system of plantation agriculture for impoverishing the state's farmers and making it impossible for blacks to acquire land, did hint at an alternative approach. Although he offered no specific program to promote land distribution, the legislature quickly moved to open property held by the state to settlement, mostly lands forfeited for nonpayment of taxes. Although a number of timber companies and railroads took advantage of the law to acquire holdings in the Yazoo-Mississippi delta, few blacks were as fortunate, partly because the state government also made it easier for landowners to redeem delinquent tax lands. And when the lawmakers stayed the collection of debts and repealed the state's lien law, Ames reversed himself and vetoed both bills, fearing they would completely dry up supplies and credit for landowners and tenants alike. Thus ended Ames's brief flirtation with the idea of promoting economic change. When a black constituent at the end of 1874 appealed to the governor for "Republican protection," insisting "we cannot . . . get the complete benefit of our labor," Ames could only reply, "time works many changes and patience is a virtue we must all cultivate." On the back of the freedman's letter, the governor's private secretary wrote his own one-word reaction to its contents: "annoyance."[54]

Nowhere was the need for change greater, or the conservative implications of "reform" more evident, than in South Carolina. Here, the first Republican governor, Robert K. Scott, who served from 1868 to 1872, had compiled an unenviable record of malfeasance in office, and his successor, scalawag Franklin J. Moses, Jr., proved to be "entirely devoid of moral sense." During the Moses administration, South Carolina did join in the nationwide reaction against railroad aid and expensive govern-

53. Lawrence N. Powell, "The Politics of Livelihood: Carpetbaggers in the Deep South," in Kousser and McPherson, eds., *Region, Race, and Reconstruction*, 324; Henry Pearce to Benjamin F. Butler, June 13, 1874, Butler Papers; D. P. Carter to William P. Kellogg, January 19, 1875, Kellogg Papers.

54. *Inaugural Address of Gov. Adelbert Ames* (Jackson, Miss., 1874), 7; Harris, *Day of the Carpetbagger*, 507, 612; John C. Hudson, "The Yazoo-Mississippi Delta and Plantation Country," *Proceedings, Tall Timbers Ecology and Management Conference*, 16 (1979), 67–74; Perman, *Road to Redemption*, 147–48; J. D. Penn to Adelbert Ames, November 5, 1874, Ames to Penn, November 14, 1874, Letterbook B, Mississippi Governor's Papers, MDAH.

ment, reducing its public debt and state taxes. But the governor's extravagant life-style in the mansion he somehow managed to acquire in Columbia alienated much of the party. In 1874, Moses was denied renomination in favor of carpetbagger Daniel H. Chamberlain, who in the fall defeated a coalition of Democrats and reform-minded Republicans. (John T. Green, the reform candidate, died shortly after the election; had he won, his black running mate Martin R. Delany would have become South Carolina's governor.)[55]

An ardent Massachusetts abolitionist and wartime officer of a black regiment, Chamberlain had come to South Carolina in 1866 hoping to plant Sea Island cotton. Like other aspiring Northern planters, he failed economically and turned to politics for a living. In some ways, Chamberlain seemed a poor choice to root out corruption, for as the state's attorney general he had been party to several of the Scott administration's frauds. Yet Chamberlain also personified the North's elitist liberal intelligensia. A Harvard graduate, he delighted in reading the classics, and his dignified, aloof bearing reminded black leader Beverly Nash of the "old silk stocking Whigs." For Chamberlain, reform offered a vehicle for strengthening white control of the Republican party and wooing respectable members of the Democracy. Earlier in the decade, he had served as vice president of a Taxpayers' Convention and had once offered to run for governor "to keep the party from going over to *negroism*."[56]

Chamberlain entered office promising sweeping changes to ensure "economy and honesty in the administration of the government," and to a large extent, he succeeded. His administration reorganized state finances so as to consolidate and properly fund the debt, reduced taxes and equalized assessments, slashed public printing costs, and launched investigations into previous frauds. All these measures won the support of South Carolina's legislature, both houses of which blacks now controlled. But other "reform" proposals aroused widespread black opposition. To further curtail expenses and win the backing of white taxpayers, Chamberlain reduced the size of the state militia and removed many black trial justices and local school officials, frequently replacing them with white Democrats. He also proposed to reinstitute the leasing of convicts and reduce appropriations for the insane asylum, state university, and public schools, but failed to gain legislative approval. By mid-1875, the governor had become the toast of upper-crust Democratic society, mak-

55. Robert H. Woody, "Franklin J. Moses, Jr., Scalawag Governor of South Carolina, 1872–1874," *NCHR*, 10 (April 1953), 111–23; Interview with Francis L. Cardozo, Southern Notebook E, Frederic Bancroft Papers, Columbia University; Goodrich, "Public Aid," 432; Francis B. Simkins and Robert H. Woody, *South Carolina During Reconstruction* (Chapel Hill, 1932), 464–73.

56. Lamson, *Elliott*, 154; Daniel H. Chamberlain to Francis W. Dawson, April 20, 1876, Francis W. Dawson Papers, DU; Interview with Beverly Nash, Southern Notebook C, Bancroft Papers; Thomas Holt, *Black over White: Negro Political Leadership in South Carolina During Reconstruction* (Urbana, Ill., 1977), 176–77; Charleston *News and Courier*, September 16, 1876.

ing the rounds of Charleston literary associations and forging an open alliance with the Charleston *News and Courier*'s influential editor, Francis W. Dawson. "If Governor Chamberlain continues to pursue the course he has done for the last twelve months," ex-governor Benjamin F. Perry commented in November 1875, "I think it would be exceedingly unwise and ungrateful for the Democratic party to oppose his reelection."[57]

Black annoyance over Chamberlain's course came to a head at the end of 1875, when the governor refused to sign commissions allowing William J. Whipper and Franklin J. Moses, Jr., to assume Charleston judgeships to which the legislature had elected them. Moses scarcely possessed the character required of a judge, but Chamberlain's indignation centered upon Whipper, a flamboyant Northern-born black lawyer and planter who admitted to having been a gambler in "the degenerate days of the past," but claimed to have reformed after a religious awakening. Whipper's election, Chamberlain exclaimed, was "a horrible disaster" that imperiled "the civilization of the Puritan and the Cavalier." To Dawson, it suggested a heinous plot to "Africanize South Carolina, to make it a Black Republic." Chamberlain's "coup d'état" successfully blocked the appointments, but embittered most of South Carolina's black Republicans. It met, however, with rapturous applause from the state's "most intelligent and influential citizens," and won considerable praise as well from Northern Republicans. "If South Carolina," wrote the Cincinnati *Gazette*, "cannot be saved to the Republican party by the course of Governor Chamberlain, it is not worth saving."[58]

The Whipper affair and the reaction to it offered incontrovertible evidence that blacks remained junior partners at the highest echelons of Southern politics, and that Northern support for Reconstruction was on the wane. Together with frustration at Reconstruction's failure to bring about a radical improvement in black living conditions and a growing sense of foreboding as one state after another fell to the Redeemers, these realities inspired some black leaders to seek new strategies for community advancement. But they met with little success. Independent black politics surfaced from time to time during the 1870s, only to be

57. *Inaugural Address of Gov. D. H. Chamberlain . . .* (Columbia, S.C., 1874), 3; *The Campaign in South Carolina . . . Letter of Gov. Chamberlain to the Chairman of the State Democratic Executive Committee* (Columbia, S.C., 1876), 4; Alrutheus A. Taylor, *The Negro in South Carolina During the Reconstruction* (Washington, D.C., 1924), 213–26; Holt, *Black over White*, 179–82; Joseph Middleton et al. to Daniel H. Chamberlain, January 6, 1875, John B. Dennis to Chamberlain, January 15, 1875, A. W. Folger to Chamberlain, May 8, 1875, South Carolina Governor's Papers, SCDA; Francis W. Dawson to Mrs. Dawson, September 12, 13, 1874, Daniel H. Chamberlain to Dawson, February 18, 1875, Dawson Papers; Lillian A. Kibler, *Benjamin F. Perry: South Carolina Unionist* (Durham, 1946), 483.

58. Walter Allen, *Governor Chamberlain's Administration in South Carolina* (New York, 1888), 192–201, 237–40; Charleston *News and Courier*, December 18, 20, 22, 1876; F. J. Donaldson to Daniel H. Chamberlain, January 3, 1876, South Carolina Governor's Papers.

doomed by ordinary freedmen's unwavering loyalty to the Republican party. When Edward Shaw, a prominent Memphis black leader, ran for Congress against the white Republican incumbent, he received only 165 votes. Despite his personal popularity among Georgia's lowcountry freedmen, Aaron A. Bradley's successive independent campaigns for office also came to naught. Indeed, ordinary blacks reacted with extreme hostility to any political strategy that threatened to weaken the party of emancipation. When Alabama legislator Charles S. Smith proposed that blacks declare their political independence, he won a respectful although generally unfavorable hearing from the 1876 national black convention, but "colored men on street corners" spoke of cutting his throat. Even a measured Senate speech criticizing Northern indifference to the freedmen's plight earned Blanche K. Bruce a grass-roots reputation as a "conservative negro." Thus, black political leaders had few options outside the party, since gestures toward political autonomy not only threatened their access to patronage, but undermined their credibility in their own community.[59]

With black votes taken for granted and political independence all but impossible, black politicos had little choice other than to act as "field hands" for the Republican party (a term Frederick Douglass used to describe his own political role during Reconstruction, and whose implications he does not seem to have fully appreciated). A number of black spokesmen, however, urged their constituents to rely less on the state and more on individual self-help as a strategy for advancement. Such ideas, never entirely absent from black thought, gained increasing currency as the 1870s progressed. As early as 1871, James Lynch, observing that Reconstruction had not arrested the spread of new forms of economic dependence, urged his people to deemphasize the "enjoyment of political honors" as a "means of our elevation," and look to their own resources. Henry O. Flipper, who believed that despite prejudice black cadets could "make our life at West Point what we will," exemplified the new attitude. He even suggested that James W. Smith, his predecessor at the military academy, had been responsible for his own difficulties because of his outspoken insistence on equal treatment. Louisiana legislator and editor David Young agreed there was "no use in keeping up the same old whine" about the denial of blacks' rights. In 1876, one delegate told the national black convention that the race would only gain recognition of its rights by proving itself worthy of acceptance—"the educated col-

59. Walter J. Fraser, Jr., "Black Reconstructionists in Tennessee," *THQ,* 34 (Winter 1975), 373–74; Joseph P. Reidy, "Aaron A. Bradley: Voice of Black Labor in the Georgia Lowcountry," in Howard N. Rabinowitz, ed., *Southern Black Leaders of the Reconstruction Era* (Urbana, Ill., 1982), 297; Alexander A. Lawrence, ed., "Some Letters from Henry C. Wayne to Hamilton Fish," *GaHQ,* 43 (December 1959), 400; Nashville *Daily American,* April 1876, undated clipping, William Pledger Scrapbook, John E. Bryant Papers, DU; Maurice Bauman to Blanche K. Bruce, February 17, 1876, Blanche K. Bruce Papers, Howard University.

ored man must be seen in the foundry, in the machine shop, in the carpenter shop."[60]

Such talk of an individual route to advancement that eschewed political action in favor of economic self-help anticipated the fully developed conservative ideology associated with Booker T. Washington that would emerge in the post-Redemption South. Like Washington, conservative blacks during Reconstruction urged their constituents to seek out political alliances with "the Independent-Conservative element of the South." And like him, they viewed an aggressive demand for equal access to public facilities as unrealistic and counterproductive. When New Orleans integrated its public schools, some blacks warned that the only result would be "to make the entire white community turn against us, and without the substantial white people of the South we cannot get along."[61]

So long as Reconstruction survived, the new conservatism proved of little attraction to ordinary freedmen and remained a secondary theme in black thought. But its very appearance suggested both the political implications of the growing class differentiation within the black community, and the continuing importance in certain states of the antebellum distinction between free blacks and slaves. Conservative ideas found their greatest support among the emerging class of black businessmen—the same group that would later supply the principal base for Washington's ideology. Robert Gleed, a black merchant reportedly worth $15,000 in the early 1870s, served on the central committee of Mississippi's 1871 Planters' Convention. Other black landowners and entrepreneurs echoed the shibboleth that government should be "carried on by men of refinement," and, especially with the advent of the depression, shared Democratic resentments about high taxes and state expenditures. After establishing a land and brokerage business in 1871, Martin R. Delany lectured blacks repeatedly on the harmony of interests between capital and labor and spoke out against carpetbaggers—representatives of "the lowest grade of northern society." Delany's increasingly conservative outlook blended personal economic interests, disappointed political ambition, and pessimism about the future. Utopian hopes for a permanent change in Southern life, he counseled, should be abandoned; property would eventually rule in South Carolina as it did elsewhere, and blacks should

60. [Frederick Douglass] *Life and Times of Frederick Douglass* (New York, 1962 ed.), 416; Jackson *Daily Mississippi Pilot*, February 26, 1871; William C. Harris, "James Lynch: Black Leader in Southern Reconstruction," *Historian*, 34 (November 1971), 55–57; Flipper, *Colored Cadet*, 134–38, 160–65; New Orleans *Louisianian*, August 14, 1875; Nashville *Daily American*, April 1876, undated clipping, Pledger Scrapbook, Bryant Papers.

61. Nashville *Daily American*, April 1876, undated clipping, Pledger Scrapbook, Bryant Papers; "Colored men" to Thomas W. Conway, May 19, 1870, State Board of Education Papers, Louisiana State Archives.

strike a deal with leading whites while they still retained significant bargaining power.[62]

But it was among the well-to-do free blacks of South Carolina and Louisiana that the notion of replacing existing Republican governments with interracial conservative coalitions won the most enthusiastic hearing. In both states, free blacks had taken the early lead in political organizing, only to see their influence weakened by carpetbagger control of state offices and the freedmen's growing political assertiveness. In South Carolina, Chamberlain's conservative policies won the support of prosperous mulatto legislators. In Louisiana, many New Orleans free blacks were attracted to the Unification Movement of 1873, a political alignment independent of the two existing parties that promised to restore racial harmony, economic prosperity, and social peace to the state. Organized by some of the city's most prominent businessmen and professionals— "the flower of [its] wealth and culture"—the movement made far greater concessions to blacks' aspirations than "reform" coalitions in other states, not only pledging to guarantee their civil and political rights, but accepting the integration of schools and public accommodations and an equal division of offices between the races. It even promised to encourage the acquisition of land by the freedmen. Its black support came largely from former free men of color who resented the carpetbaggers and believed "unworthy" blacks had obtained office at the expense of "more wealthy, intelligent and refined colored men." Such pioneers of wartime black politics as Louis C. Roudanez, the founding editor of the New Orleans *Tribune*, endorsed the Unification movement, along with representatives of the property-owning free elite like wealthy philanthropist Aristide Mary (who had failed in an 1872 bid for the Republican gubernatorial nomination).[63]

Despite a fanfare of publicity, the Unification movement did not survive for long, since most freedmen distrusted the motives of its white organizers, while its genuine concessions to blacks alienated the bulk of the white electorate. Indeed, even as the Unifiers searched for a new political middle ground, Democrats throughout the South were abandoning the centrist rhetoric of the New Departure in favor of a return to the open racism of early Reconstruction. Various considerations facilitated the triumph of

62. Jackson *Daily Mississippi Pilot*, January 27, 1871; *New National Era*, February 22, 1872; 43d Congress, 2d Session, House Report 261, pt. 3:460–62; Victor Ullmann, *Martin Delany: The Beginnings of Black Nationalism* (Boston, 1971), 420–23, 440–50; Martin Delany, "A Warning Voice," unidentified newspaper clipping, SCHS; Robert H. Woody, *Republican Newspapers of South Carolina* (Charlottesville, Va., 1936), 17.

63. Holt, *Black over White*, 190–92; T. Harry Williams, "The Louisiana Unification Movement of 1873," *JSH*, 11 (August 1945), 349–60; *New York Herald*, October 9, 1874; David C. Rankin, "The Impact of the Civil War on the Free Colored Community of New Orleans," *Perspectives in American History*, 11 (1977–78), 403–404.

"white line" politics, including the Greeley debacle, the failure of the New Departure to attract black votes, and the candid judgment that the cry of white supremacy offered the best prospects for mobilizing the Democratic electorate and winning over remaining scalawags. And the reassertion of the political color line had profound implications for the Democracy's political and economic strategies. White-line politics went hand in hand with the adoption of new modes of political organization—with intensive local canvassing replacing traditional "barbecues and mass meetings"—to maximize white turnout. Especially in the Deep South, where Democratic victory was impossible without the neutralization of part of the black electorate, it also implied a revival of political violence. And, a noticeable shift away from support for state-sponsored modernization (the economic corollary of the discredited New Departure) accompanied the reemergence of white supremacist rhetoric. The depression heightened the attractiveness of retrenchment and tax reduction to white voters who associated expensive government with new state programs that primarily benefited corporations and blacks, and who feared that high taxes threatened both planters and yeomen with the loss of their land. And with Republicans proposing as an economic program little more than a milder version of "reform," they had little to offer white voters to counteract Democrats' racist appeals.[64]

The "white line" strategy meant the Democratic quest for votes would now focus on counties outside the plantation belt. But planters retained a decisive influence in the movement for Redemption, a role reinforced by their leadership of the Grange, which expanded rapidly in the South after the Panic of 1873. While ostensibly nonpartisan, the organization, as North Carolina diarist David Schenck noted, was "really a political society," which excluded blacks from membership and took an active part in mid-decade Redemption campaigns. Its leading spirits included D. Wyatt Aiken (who had reportedly masterminded the assassination of South Carolina black political leader Benjamin F. Randolph in 1868), Mississippi Democratic editor Ethelbert Barksdale, and William D. Bloxham, one of the Florida's most prominent Democratic politicians. So many members ran as Democratic candidates in the Alabama campaign of 1874, that the subsequent General Assembly became known as the "Grange legislature."[65]

64. Taylor, *Louisiana Reconstructed*, 278–79; Williams, "Louisiana Unification Movement," 362–67; Perman, *Road to Redemption*, 127–28, 150–55, 173–77.
65. David Schenck Diary, July 19, 1873, UNC; Buck, *Granger Movement*, 59, 74n.; 44th Congress, 2d Session, Senate Miscellaneous Document 45, 224; J. S. Corthian to D. Wyatt Aiken, January 15, 1869, D. Wyatt Aiken Papers, USC; Willie D. Halsell, "The Bourbon Period in Mississippi Politics 1875–1890," *JSH*, 11 (November 1945), 529; Ruby L. Carson, "William Dunnington Bloxham: The Years to the Governorship," *FlHQ*, 27 (January 1949), 227; William W. Rogers, *The One-Gallused Rebellion: Agrarianism in Alabama 1865–1896* (Baton Rouge, 1970), 76.

"Co-operation among farmers," wrote Aiken, could "redeem the entire South from political thralldom, because it is our employees and they alone, who impede the return [to Democratic rule]." As Aiken's words made clear, while Grangers claimed to represent an undifferentiated "agrarian interest," they spoke particularly for employing farmers, not subsistence-oriented yeomen, tenants, or laborers. Indeed, the Southern Grange drew its leadership almost exclusively from the planter class, even in upcountry counties where most of its members were small farmers. "We must have less freedom and more protection to property," declared a speaker at a Mississippi Grange meeting, and the organization spent much of its time agitating for laws to curtail the economic options of black laborers, such as restricting suffrage in fence law referenda to property owners and barring merchants from dealing in small quantities of agricultural produce. Southern Grangers' animosity to merchants reflected not only antimiddleman sentiment, but planters' determination to reassert exclusive control over their tenants' economic relations. Some local branches conspired to set the wages of black laborers, and Mississippi's Grange, while sponsoring educational programs for white farmers, called for the abolition of the school system established by the Republicans. One speaker at a regional Grange convention even reported that to keep them "under perfect control," he had resumed "whipping some of his negroes."[66]

The interconnected issues of white supremacy, low taxes, and control of the black labor force set the tone for the Democratic campaigns of the mid-1870s. And the strategy's potency became evident in 1873 and 1874 as Democrats solidified their hold on states already under their control and "redeemed" new ones. Texas Democrat Richard Coke defeated Gov. Edmund J. Davis in 1873 by a margin of better than two to one. Demographic changes had rendered the outcome all but inevitable, for Texas, one of the nation's most rapidly growing states, had experienced a massive influx of Southern white immigrants, swelling Democratic ranks and reducing blacks, Germans, and native Unionists, the backbone of the Republican organization, to a shrinking minority of the population. Meanwhile, Virginia Democrats jettisoned the moderate Republicans with whom they had cooperated in 1869 and carried the state with a "straight-out" ticket and a platform of "race against race." The 1874 Southern elections proved as disastrous for Republicans as those in the

66. *Rural Carolinian*, 7 (November 1876), 516; Robert A. Calvert, "A. J. Rose and the Granger Concept of Reform," *AgH*, 51 (January 1977), 182–84; Hahn, *Roots of Southern Populism*, 221–22; *Southern Field and Factory*, 4 (February 1874), 667; Cecil E. McNair, "Reconstruction in Bullock County," *AlHQ*, 15 (Spring 1953), 122; James S. Ferguson, "Co-operative Activity of the Grange in Mississippi," *JMH*, 4 (January 1942), 5–9; 44th Congress, 1st Session, Senate Report 527, 1744; James S. Ferguson, "The Grange and Farmer Education in Mississippi," *JSH*, 8 (November 1942), 500; Jackson *Daily Mississippi Pilot*, January 9, 1875.

North. Democrats won over two thirds of the region's House seats, redeemed Arkansas, and gained control of Florida's legislature. Throughout the Appalachian South, moreover, there was a noticeable decline in Republican voting, attributed by contemporaries to the pending Civil Rights Bill. Horace Maynard, Republicans' unsuccessful gubernatorial candidate, lamented that thousands of East Tennessee Unionists voted for his opponent, "swearing as they did so that they would never again vote with a party which supported the coeducation of the races."[67]

In these campaigns, which mostly took place in states where blacks constituted a minority of the population, Democratic victories depended mainly on the party's success at drawing the political color line. In Louisiana and Alabama, however, the darker side of white-line politics came to the fore. Events seemed to follow their own perverse logic in Reconstruction Louisiana, where every election between 1868 and 1876 was marked by rampant violence and pervasive fraud. In the much-disputed 1872 election, which saw both candidates, for a time, claim the governorship, Illinois carpetbagger William P. Kellogg defeated John McEnery. Kellogg attempted to conciliate his opponents by offering appointments to local offices and reforming the state's finances, but proved unable to assuage their bitterness. Still claiming authority as governor, McEnery organized his own militia, which in March 1873 unsuccessfully attempted to seize control of New Orleans police stations. April witnessed the Colfax Massacre, the most dramatic example of the anarchy that reigned throughout much of rural Louisiana. With federal troops, numbering well under 2,000, unable to establish order, many white parishes refused to pay taxes or otherwise recognize the authority of the state government. The situation worsened in 1874 with the formation of the White League, openly dedicated to the violent restoration of white supremacy. It targeted local Republican officeholders for assassination, disrupted court sessions, and drove black laborers from their homes. The state Democratic platform opened with the words, "We, the white people of Louisiana," and one party newspaper pronounced "a war of races" imminent. White League violence and extensive efforts to use economic intimidation against black voters dominated the campaign. Even veterans of the New Orleans Unification movement embraced these tactics. As one explained:

67. Homer L. Kerr, "Migration into Texas, 1860–1880," *SWHQ*, 70 (October 1966), 189–92; Perman, *Road to Redemption*, 154–55; James A. Bear, Jr., ed., "Henry A. Wise and the Campaign of 1873," *VaMHB*, 62 (July 1954), 332–33; Abbott, *Republican Party and the South*, 230; William W. Davis, *The Civil War and Reconstruction in Florida* (New York, 1913), 643–44; Gordon B. McKinney, *Southern Mountain Republicans 1865–1900* (Chapel Hill, 1978), 40, 49; F. Wayne Binning, "The Tennessee Republicans in Decline, 1869–1876," *THQ*, 40 (Spring 1981), 76–77.

Last summer one hundred of us, representing fairly *all* the grades of public and social status, humbled ourselves into the dust in an effort to secure the cooperation of the colored race in a last attempt to secure good government, and failed. . . . To this complexion it has come at last. The niggers shall not rule over us.[68]

In Red River Parish, the campaign degenerated into a violent reign of terror, which culminated in August in the cold-blooded murder of six Republican officials, among them three relatives of Republican leader Marshall H. Twitchell. (A man of remarkable courage, Twitchell returned to the parish to contest the fall 1874 elections. In May 1876 a disguised gunman wounded him severely enough to require the amputation of both arms.) Emboldened by their Red River "success," the league launched a full-scale insurrection in New Orleans, hoping to install McEnery as governor. On 14 September, 3,500 leaguers, mostly Civil War veterans, overwhelmed an equal number of black militiamen and Metropolitan Police under the command of Confederate Gen. James Longstreet, and occupied the city hall, statehouse, and arsenal. They only withdrew upon the arrival of federal troops, ordered to the scene by the President.[69]

For decades, the "Battle of Liberty Place" would be dredged up in Louisiana political campaigns to rally Democratic voters and solidify party unity. In 1874, it demonstrated both the ruthless determination of Reconstruction's opponents and the impotence of Kellogg's regime. The uprising shocked the Grant Administration out of the paralysis that had marked its attitude toward Louisiana. A few reform journals defended the league—*The Nation* declared it knew of "no case of armed resistance to an established government in modern times, in which the insurgents had more plainly the right on their side." But this was armed rebellion, not some taxpayers' protest meeting, and most Republicans, including anti-Administration newspapers like the New York *Tribune,* praised Grant for his firmness.[70]

Thanks in part to the Republican returning board, which threw out results from parishes plagued by violence, the Louisiana campaign of 1874 produced not a Democratic victory, but another close, disputed

68. Taylor, *Louisiana Reconstructed,* 241–55; Charles E. Nash et al. to William P. Kellogg, July 28, 1873, Kellogg Papers; Joseph G. Dawson III, *Army Generals and Reconstruction: Louisiana, 1862–1877* (Baton Rouge, 1982), 141–48; 46th Congress, 2d Session, Senate Report 693, pt. 2:114, 171; 43d Congress, 2d Session, House Report 261, pt. 3:752–53, 791; J. Dickson Burns to Manton Marble, July 2, 1874, Manton Marble Papers, LC.

69. Ted Tunnell, *Crucible of Reconstruction: War, Radicalism, and Race in Louisiana, 1862–1877* (Baton Rouge, 1984), 196–208; Frank L. Richardson, "My Recollections of the Battle of the Fourteenth of September, 1874, in New Orleans, Louisiana," *LaHQ,* 3 (October 1920), 498–501; Dawson, *Army Generals and Reconstruction,* 156–80.

70. Joy Jackson, "Bosses and Businessmen in Gilded Age New Orleans Politics," *LaH,* 5 (February 1964), 387–88; *Nation,* September 24, 1874; William D. Foulke, *Life of Oliver P. Morton* (Indianapolis, 1899), 2:351–52; Cohen, "The Lost Jubilee," 525–26.

outcome. Simultaneously, however, Alabama became the first black belt state since Georgia to be "redeemed." Here, with the black and white populations nearly equal, several thousand upcountry Unionists held the balance of political power. To assure these voters that Democratic victory would not imply a return to Confederate rule, the party nominated for governor George S. Houston, a planter from northern Alabama who had remained "neutral" during the war. At the same time, it abandoned all pretense of seeking black support (an unlikely prospect in any event, since Houston was known to have evicted black laborers from his estate for joining the Union League). "Appealing to the interests, the pride and prejudices of the white race," wrote one party strategist, offered the surest means of overcoming the regional and class divisions among white Alabamans, and while not neglecting questions like lower taxes, the party made "nigger or no nigger" the chief issue of the campaign. Democrats harped particularly on the school integration clause of the pending Civil Rights Bill, convinced that upcountry whites "are more sensitive upon this question than all others."[71]

Alabama's Republican party entered the 1874 campaign with "its leaders embittered against each other, and the rank and file . . . torn and distracted." As if the economic depression and "the complications surrounding our railroad system" were not damaging enough, the civil rights issue threw the party on the defensive and caused deep divisions between black and white Republicans. Controlled by white party leaders, the Republican convention nominated an all-white state ticket, defeated an attempt to endorse the Civil Rights Bill, and disclaimed any desire to promote "the social equality of different races" or "mixed schools and mixed accommodation." Black delegates, James T. Rapier later related, agreed to the platform because "the republicans in the northern part of Alabama . . . could not secure the white vote" without it.[72]

Democratic prospects for success, editor Robert McKee believed, depended on "the thoroughness and efficiency of party discipline." But the party also moved to disrupt Republican campaigning and reduce black belt turnout. In August, two Sumter County Republican leaders, a black and a carpetbagger, were assassinated, and others saw white mobs destroy their homes and crops. Meanwhile, Barbour County whites, led by "the *pretended* 'best citizens'," instituted a "perfect reign of terror." Nonetheless, hundreds of blacks marched on election day to Eufaula, Barbour's chief marketing center, hoping to cast their ballots.

71. Taylor, *Louisiana Reconstructed*, 299–302; Daniel H. Bingham to William H. Smith, August 7, 1867, Wager Swayne Papers, ASDAH; Edward C. Williamson, "The Alabama Election of 1874," *Alabama Review* (July 1964), 210–18; Rogers, *One-Gallused Rebellion*, 42–43; *AC*, 1874, 15; H. C. Jones to Robert McKee, May 23, 1874, W. Brewer to McKee, May 10, 1874, R. K. Boyd to McKee, May 10, 1874, Robert McKee Papers, ASDAH.

72. Montgomery *Alabama State Journal*, October 9, 1875, June 25, 27, August 23, 1874; *AC*, 1874, 15–16; *CR*, 43d Congress, 2d Session, 1001.

To avoid any pretext for violence, they came without weapons, a decision that proved disastrous when armed whites began firing into the crowd, killing seven blacks and wounding ten times that number. That evening, a mob surrounded the polling place, killed the son of scalawag judge Elias M. Keils, and burned the ballot box. (Keils fled the state, never to return, while the mob's leader two years later won election to the state senate.) In Mobile, too, armed whites drove black voters from the polls.[73]

On the strength of a landslide among white voters, Houston won the governorship, a result aided by the depression and the fact that many scalawags were "induced to withdraw their support from the republican party by the appeal made to them to support their color against the negro." Among predominantly white counties, only staunchly Unionist Winston in the Alabama mountains produced a Republican majority. Although turnout increased in much of the black belt, violence helped swing Barbour and six other black-majority counties into the Democratic column. With Democrats in control of the state offices and both houses of the General Assembly, Reconstruction in Alabama had come to an end.[74]

The Crisis of 1875

By the time the Forty-Third Congress reassembled in December, the political landscape had been transformed. As a result of the Democratic landslide, this session would be the last time for two years (indeed, as it turned out, for over a decade), that Republicans controlled both the White House and Congress. With political violence having again erupted in many parts of the South, and their party's hegemony at Washington about to expire, Benjamin Butler and other Stalwarts devised a program to safeguard what remained of Reconstruction. Their proposals included the Civil Rights Bill, a new Enforcement Act expanding the President's power to put down conspiracies aimed at intimidating voters and including the right to suspend the writ of habeas corpus, a two-year army appropriation (to prevent the incoming House from limiting the military's role in the South), a bill further expanding the jurisdiction of the federal courts, and a subsidy for the Texas & Pacific Railroad. Taken together, the package embodied a combination of idealism, partisanship, and crass economic advantage typical of Republican politics. Civil rights would be the program's spearhead, and to make it more palatable, Butler

73. Robert McKee to A. W. Dillard, May 7, 1874, McKee Papers; 43d Congress, 2d Session, House Report 262, 16–20, 130, 345–49, 425; Elias M. Keil to Mr. Gardner, August 25, 1874, Alabama Governor's Papers, ASDAH; Melinda M. Hennessey, "Reconstruction Politics and the Military: The Eufaula Riot of 1874," *AlHQ*, 38 (Summer 1976), 112–25.
74. 44th Congress, 2d Session, Senate Report 704, 139; Sarah W. Wiggins, *The Scalawag in Alabama Politics, 1865–1881* (University, Ala., 1977), 97–99.

dropped the bill's most controversial feature, the clause requiring integrated schools.[75]

Events in Louisiana disrupted the already tenuous party unity necessary to enact such a program. Having suppressed the New Orleans insurrection of September 1874, Grant, newly determined to "protect the colored voter in his rights," ordered General Sheridan to use federal troops to sustain the Kellogg administration and put down violence. On January 4, 1875, when Democrats attempted to seize control of the state assembly by forcibly installing party members in five disputed seats, a detachment of federal troops under the command of Col. Phillippe de Trobriand entered the legislative chambers and escorted out the five claimants. The following day, Sheridan wired Secretary of War Belknap, urging that military tribunals be established to try White League leaders as "banditti."[76]

The reaction to de Trobriand's "purge" (reminiscent, in some minds, of Pride's Purge during the English Civil War) could not have differed more dramatically from the response to Grant's September intervention. If, for Reconstruction's critics, South Carolina epitomized the evils of corruption and "black rule," Louisiana now came to represent the dangers posed by excessive federal interference in local affairs. The spectacle of soldiers "marching into the Hall . . . and expelling members at the point of the bayonet" aroused more Northern opposition than any previous federal action in the South. In Boston, the nation's "cradle of liberty," a large body of "highly respectable citizens" gathered at Faneuil Hall to demand Sheridan's removal and compare the White League with the founding fathers as defenders of republican freedom. Wendell Phillips was among those present. Four decades earlier, Phillips had launched his abolitionist career at a similar assembly in this very hall, rising to reprimand a speaker who praised the murderers of antislavery editor Elijah P. Lovejoy. Then, his eloquence converted the audience. Now, as he rebuked those who would "take from the President . . . the power to protect the millions" the nation had liberated from bondage, he heard only hisses, laughter, and cries of "played out, sit down." "Wendell Phillips and William Lloyd Garrison," commented the *New York Times*, "are not exactly extinct from American politics, but they represent ideas in regard to the South which the majority of the Republican party have outgrown."[77]

75. Spackman, "American Federalism," 315–16; Kelly, "Congressional Controversy," 556–57.
76. Marshall Jewell to Lucius Fairchild, December 29, 1874, Fairchild Papers; Dawson, *Army Generals and Reconstruction,* 197–211.
77. Theodore C. Smith, *The Life and Letters of James A. Garfield* (New Haven, 1925), I, 519; Gillette, *Retreat from Reconstruction,* 124–31; Walter M. Merrill and Louis Ruchames, eds. *The Letters of William Lloyd Garrison* (Cambridge, Mass., 1971–81), 6:367; *Wendell Phillips, in Faneuil Hall, on Louisiana Difficulties* (Boston, 1875), 13–15; James B. Stewart, *Wendell Phillips: Liberty's Hero* (Baton Rouge, 1986), 308–10.

The Louisiana imbroglio divided and embarrassed the Grant Administration, still smarting from the electoral disaster of 1874. When Belknap, on his own initiative, assured Sheridan that he enjoyed the President's full confidence, Secretaries Benjamin H. Bristow and Hamilton Fish dissociated themselves from the statement and urged Grant to "wash the hands of the Administration entirely of the whole business." A Presidential message to Congress offered only the most tepid defense of the army's actions, insisting that the troops had acted without the Administration's knowledge. In February, a Congressional committee headed by New York Republican William A. Wheeler worked out a face-saving compromise—Republicans would retain their majority in the Louisiana Senate, while Democrats would control the House and in return agree to allow Kellogg to continue in office undisturbed. The Wheeler Compromise extricated the Grant Administration from an unpleasant situation, but far from attempting a defense of Reconstruction, the committee's report criticized Louisiana Republicans for corrupt government and said the premature enfranchisement of blacks had been a mistake. The uproar over Louisiana convinced Grant of the political dangers posed by a close identification with Reconstruction, and made Congressional Republicans extremely wary of further military intervention in the South.[78]

As Republican resolve weakened, the new Reconstruction bills became embroiled in a complex series of parliamentary maneuvers. When Butler attempted to place the Civil Rights Bill on the legislative calendar, Democrats paralyzed the House with a barrage of motions to adjourn. An attempt by John Cessna of Pennsylvania, Butler's legislative lieutenant, to amend the rules so as to prohibit dilatory motions failed to obtain the required two-thirds majority because fifteen Republicans refused to go along. Then, two influential House Republicans, Speaker Blaine and Garfield, devised a more modest rules change. Both favored the Civil Rights Bill but opposed further federal intervention in the South, a position hardened by their belief that Reconstruction had become a political liability. Garfield suspected that the 1874 debacle stemmed, in part, from "a general apathy among the people concerning the war and the negro," while Blaine told John R. Lynch that if Democrats overturned every Reconstruction government, "the result would be a solid North against a solid South; in which case the Republicans would have nothing to fear." The rules change made it possible for the Civil Rights Bill to become law shortly before Congress adjourned. It also allowed time for the passage of the Jurisdiction and Removal Act, which facilitated the transfer from state to federal courts of suits asserting a citizen's rights under the Constitution or national law—a climax to the tremendous expansion of the

78. Webb, *Bristow*, 158–61; Benjamin H. Bristow to E. D. Forge, January 11, 1875, Benjamin H. Bristow Papers, LC; Richardson, ed., *Messages and Papers*, 7:305–11; James T. Otten, "The Wheeler Adjustment in Louisiana: National Republicans Begin to Reappraise Their Reconstruction Policy," *LaH*, 13 (Fall 1972), 349–67.

federal judiciary's powers during Reconstruction. But the Enforcement, Texas-Pacific, and two-year army appropriation bills expired with the Forty-Third Congress.[79]

The legislative infighting of January and February 1875 illustrated how divided Republicans had become over Reconstruction. "Is it possible," asked one House member, "that you can find power in the Constitution to declare war, levy taxes . . . and pass laws upon all conceivable subjects and find means to enforce them, but can find no power to protect American citizens . . . in the enjoyment and exercise of their constitutional rights?" Yet Congressional Republicans had little stomach for further intervention in Southern affairs. Even men like Connecticut's Joseph R. Hawley, who proclaimed (with some exaggeration), "I have been a radical abolitionist from my earliest days," had resigned themselves to the conclusion that the South's "social, and educational, and moral reconstruction" could "never come from any legislative halls." Others now echoed the Democratic refrain that blacks should abandon "the habit . . . [of relying] upon external aid," and sang the praises of "local self-government." Even as Congress deliberated, the Massachusetts legislature elected Henry L. Dawes to fill Sumner's seat—a choice that demonstrated how the party had changed, for the new Senator was a nondescript politico whose chief talent had been to maintain reasonably good relations with all factions while committing himself to none.[80]

Even the Civil Rights Act reflected Republicans' divided mind. Despite having been shorn of its schools provision, the law represented an unprecedented exercise of national authority, and breached traditional federalist principles more fully than any previous Reconstruction legislation. "How a Congress which enacted this bill should have a scruple about the 'Force Bill' is surprising," noted former Attorney General Amos T. Akerman. Yet in fact, the law was more a broad assertion of principle than a blueprint for further coercive action by the federal government. It left the initiative for enforcement primarily with black litigants suing for their rights in the already overburdened federal courts. Only a handful of blacks came forward to challenge acts of discrimination by hotels, theaters, and railroads, and well before the Supreme Court declared it unconstitutional in 1883, the law had become a dead letter.[81]

79. Kelly, "Congressional Controversy," 557–62; Bertram Wyatt-Brown, "The Civil Rights Act of 1875," *Western Political Quarterly*, 18 (December 1965), 772–73; Smith, *Garfield*, 1:521; Lynch, *Reminiscences*, 160; *CR*, 43d Congress, 2d Session, 1011, 1600–1, 1870; Stanley I. Kutler, *Judicial Power and Reconstruction Politics* (Chicago, 1968), 143–44; Gillette, *Retreat from Reconstruction*, 284–91.

80. *CR*, 43d Congress, 2d Session, 1838–39, 1846, 1853, 1886.

81. Spackman, "American Federalism," 325; Amos T. Akerman Diary (typescript), March 6, 1875, Amos T. Akerman Papers, GDAH; Wyatt-Brown, "Civil Rights Act," 763–65; John Hope Franklin, "The Enforcement of the Civil Rights Act of 1875," *Prologue*, 6 (Winter 1974), 225–35; Leslie H. Fishel, Jr., "Repercussions of Reconstruction: The Northern Negro, 1870–1883," *CWH*, 14 (December 1968), 342–43.

For all their divisions and uncertainties with regard to Reconstruction, Congressional Republicans managed to find the will to reaffirm their party's image as the guardian of manufacturing and defender of fiscal responsibility. As the Forty-Third Congress drew to a close, they repealed the 10 percent tariff reduction of 1872, and mandated the resumption of specie payments within four years. The latter bill was rammed through Congress with virtually no debate, a remarkable display of unity by a party that had been fractured less than a year earlier by debates over the currency. With Northern Republicans for the first time more united on economic and fiscal policy than on Reconstruction, and Democrats preparing to assume control of the House of Representatives, national politics, observed *The Nation,* had passed "out of the region of the Civil War."[82]

Two political contests of 1875 provided an early test of the Republican party's refurbished image. One took place in Ohio, whose gubernatorial election became the most closely watched state contest since the Lincoln-Douglas campaign of 1858. Taking advantage of the growing popularity of greenbackism in a state suffering severely from the economic depression, Gov. William Allen invoked Jacksonian rhetoric to denounce the Resumption Act as the fruit of a "Money Power" that "drain[ed] the life-blood of the American people." When the bloody shirt failed to galvanize the Republican electorate, the party's candidate for governor, Rutherford B. Hayes, turned to two emblems of respectability—Protestantism and fiscal orthodoxy. Seizing upon a seemingly unexceptionable Democratic law that allowed Catholic priests to minister to inmates at state hospitals and prisons, Hayes charged that popery threatened the public school system. But even more than nativism, the currency issue became the keynote of the Republican campaign. Blithely ignoring the fact that their own party had flooded the nation with greenbacks, the Republican press equated paper money with repudiation, gambling, and the confiscation of private property, even depicting Allen as a proponent of "communist revolution."

In the largest vote ever polled in Ohio, Hayes emerged victorious and his party regained control of the legislature. Although the inflation issue aided Allen in Ohio's depressed industrial and mining regions, Hayes held his party's rural Protestant support, and swamped the governor in urban centers like Cleveland, Columbus, and Cincinnati, where the frightened middle class rallied to the support of Republican respectability. Despite his narrow margin of victory (less than 1 percent of the nearly 600,000 ballots cast), Hayes's election represented a stunning reversal of

82. F. W. Taussig, *The Tariff History of the United States,* 8th ed. (New York, 1931), 190; Webb, *Bristow,* 163–64; *CR,* 43d Congress, 2d Session, 205, 208, 317–19; Unger, *Greenback Era,* 250–63; *Nation,* March 11, 1875.

the 1874 results. The campaign, exulted the New York *Tribune*, had "thoroughly educated the Republican party on the currency question . . . converted the numerous inflationists in its ranks, or driven them over to the Democratic camp," and vindicated the party's stance as representative of "the sober, industrious, practical God-fearing people who are the hope and stay of our country." It underscored as well the declining political importance of Reconstruction. Indeed, Hayes himself assured a Southern friend: " 'The let alone policy' seems now to be the true course; at any rate nothing but good will now exists towards you."[83]

The full implications of the "let alone policy" became abundantly clear in the very different political campaign that unfolded simultaneously in Mississippi. A preview of the tactics that would achieve the state's Redemption had already been seen in Vicksburg in 1874. That summer, the city's white residents organized a People's or White Man's party. At the August municipal election, it patrolled the streets in armed gangs and succeeded in intimidating enough black voters to oust the city's Republican officeholders. Meanwhile, planters in the surrounding countryside formed White League clubs, aimed at ridding the region of "all bad and leading negroes . . . and controlling more strictly our tenants and other hands." In December, inspired by Democratic successes in the Northern elections, armed league members demanded the resignation of black sheriff Peter Crosby and his board of supervisors. Crosby fled to the state capital of Jackson, and a hastily organized posse of rural blacks marched on Vicksburg, only to be dispersed by a white force rallied by city officials. (The "battle," one participant related, "didn't require any valor," for it pitted whites armed with long-range rifles against blacks who possessed only shotguns and pistols.) In the days that followed, armed bands roamed the countryside, murdering perhaps 300 blacks. Finally spurred to action, the President in early January 1875 (and on the eve of the military's far more controversial intervention in Louisiana), dispatched a company of federal troops to the city and restored Crosby to office.[84]

The Vicksburg affair demonstrated, in the words of Democratic Congressman L. Q. C. Lamar, the "absence of all the elements of real authority" in the Ames administration. A man of unimpeachable integrity, Ames was an indecisive leader, who had no emotional commitment to his

83. Forest W. Clonts, "The Political Campaign of 1875 in Ohio," *Ohio Archaeological and Historical Publications*, 31 (1922), 38–97; Irwin Unger, "Business and Currency in the Ohio Gubernatorial Campaign of 1875," *Mid-America*, 41 (January 1959), 27–39; Cincinnati *Commercial*, July 20, 1875; Charles R. Williams, ed., *Diary and Letters of Rutherford Birchard Hayes* (Columbus, Ohio, 1922–26), 3:274, 297; New York *Tribune*, October 11, 1875; Rutherford B. Hayes to Horace Austin, August 22, 1875, Austin Papers; E. W. Winkler, ed., "The Bryan-Hayes Correspondence: V," *SWHQ*, 26 (October 1922), 156, 161.

84. 43d Congress, 2d Session, House Report 265, i–xiii, 108–109, 159, 169–70, 190–93, 400–402, 467–68; James M. Batchelor to Albert A. Batchelor, October 11, 1874, Albert A. Batchelor Papers, LSU; Harris, *Day of the Carpetbagger*, 645–48; George C. Rable, *But There Was No Peace: The Role of Violence in the Politics of Reconstruction* (Athens, Ga., 1984), 145–49.

adopted state. Redemption, he wrote his wife Blanche (the daughter of Ben Butler), who remained in Massachusetts with their children, would be a "blessing" to him personally, although "I pity the poor colored people of the South when that day shall come." Nor had Ames been able to deliver on his promise of sweeping retrenchment. Mississippi's taxes remained extremely high by historic standards, and millions of acres of land had been forfeited for nonpayment. Nature itself seemed to conspire against the Ames administration, for 1874 witnessed disastrous flooding on the Mississippi River.[85]

Nonetheless, in a state with a substantial black majority, effectively mobilized at the local level, Republicans' hold on power appeared secure. White Mississippians, however, interpreted the 1874 elections as a national repudiation of Reconstruction. As one Democratic leader later remarked, "In 1874, the tidal wave, as it is called, of the North, satisfied us that if we succeeded in winning the control of the government of Mississippi we would be permitted to enjoy it." Although the Democratic state convention adopted a platform recognizing blacks' civil and political rights, the 1875 campaign quickly degenerated into a violent crusade to destroy the Republican organization and prevent blacks from voting. Democratic rifle clubs paraded through the black belt, disrupting Republican meetings and assaulting local party leaders. Unlike crimes by the Ku Klux Klan's hooded riders, those of 1875 were committed in broad daylight by undisguised men, as if to underscore the impotence of local authorities and Democrats' lack of concern about federal intervention.[86]

Two "riots" in late summer set the campaign's tone. On September 1, a Yazoo County white "military company" broke up a Republican rally, drove carpetbagger Sheriff Albert T. Morgan and other officials from the area, and went on to murder several prominent blacks, including a state legislator. Morgan had outraged the old elite of Mississippi's wealthiest plantation county by marrying a black teacher from the North, but his real offense had been assisting some 300 black families to acquire real estate and presiding over the expansion of county schools and other public facilities, all without corruption and only a slight increase in taxes. A few

85. James H. Stone, ed., "L. Q. C. Lamar's Letters to Edward Donaldson Clark, Part II: 1874–1878," *JMH*, 37 (May 1975), 191; James W. Garner to Adelbert Ames, December 4, 1899, Ames Family Papers; Blanche B. Ames, ed., *Chronicles from the Nineteenth Century: Family Letters of Blanche Butler and Adelbert Ames* (Clinton, Mass., 1957), 1:703–705; Harris, *Day of the Carpetbagger*, 612–32; Garner, *Mississippi*, 312–14. In *Profiles in Courage* (New York, 1956), John F. Kennedy, then Senator from Massachusetts, claimed that "no state suffered more from carpetbag rule than Mississippi" under Ames (161). The ex-governor's redoubtable daughter Blanche Ames, who lived into the 1960s, hounded Kennedy for years to retract this entirely unjustified remark, even enlisting her grandson, writer George Plimpton, to lobby (unsuccessfully) at a White House dinner after Kennedy had become President. See Plimpton's account in *New York Review of Books*, December 18, 1980, 56.

86. Harris, *Day of the Carpetbagger*, 634–35, 653–56; 44th Congress, 2d Session, Senate Miscellaneous Document 45, 206; Benson, "Public Career of Ames," 278; Vernon L. Wharton, *The Negro in Mississippi 1865–1890* (Chapel Hill, 1947), 182–90.

days later, Democrats assaulted a Republican barbecue at Clinton, only fifteen miles from the state capital. A few individuals on each side were killed, and armed whites went on to scour the countryside, shooting down blacks "just the same as birds." They claimed perhaps thirty victims, among them schoolteachers, church leaders, and local Republican organizers. Further violence followed in October. In Coahoma County, former Governor Alcorn, denying he had ever been a "negro republican," staged his own local redemption, organizing a group of whites who attacked a meeting being addressed by black Sheriff John Brown. Six blacks and two or three whites were killed, and Brown fled the area. "The slaughtered dead of Coahoma," declared a Republican who witnessed the affray, "speak in thundertones against the treachery of the pretended friend and betrayer of the negro—Alcorn." Perhaps they also offered a final epitaph to the persistent myth of Whig moderation.[87]

Appeals for protection poured into the offices of Governor Ames. "They are going around the streets at night dressed in soldiers clothes and making colored people run for their lives. . . ." declared a petition by black residents of Vicksburg. "We are intimidated by the whites. . . . We will not vote at all, unless there are troops to protect us." "Dear sir," read another letter, "did not the 14th Article . . . say that no person shall be deprived of life nor property without due process of law? It said all persons have equal protection of the laws but I say we colored men don't get it at all. . . . Is that right, or is it not? No, sir, it is wrong." Convinced "the power of the U. S. alone can give the security our citizens are entitled to," Ames early in September requested Grant to send troops to the state. From his summer home on the New Jersey shore, the President dispatched contradictory instructions to Attorney General Edwards Pierrepont. One widely quoted sentence came to symbolize the North's retreat from Reconstruction: "The whole public are tired out with these annual autumnal outbreaks in the South . . . [and] are ready now to condemn any interference on the part of the Government." Yet Grant, who had earlier assured Butler his son-in-law could rely on federal assistance, went on to say that federal authorities could not "evade" the governor's plea for aid. Pierrepont, however, a conservative former Democrat, subtly altered the thrust of Grant's letter, stressing in his reply to Ames that the federal government would only act after the state had exhausted

87. E. H. Anderson, "A Memoir of Reconstruction in Yazoo City," *JMH*, 4 (October 1942), 187–95; Albert T. Morgan to Adelbert Ames, September 4, 1875, Mississippi Governor's Papers; A. T. Morgan, *Yazoo: or, On the Picket Line of Freedom in the South* (Washington, D.C., 1884), 416–44, 468–74; Herbert Aptheker, "Mississippi Reconstruction and the Negro Leader, Charles Caldwell," *Science and Society*, 11 (Fall 1947), 359–61; Sterling, ed., *Trouble They Seen*, 440–43; Pereyra, *Alcorn*, 173; David H. Donald, "The Scalawag in Mississippi Reconstruction," *JSH*, 10 (November 1944), 455; John Brown to Adelbert Ames, October 8, 1875, Mississippi Governor's Papers; E. Stafford to George S. Boutwell, June 5, 1876, U.S. Senate, Select Committee to Investigate Elections in Mississippi Papers, NYPL.

its own resources. Let Republicans, he advised, raise a militia and demonstrate "the courage and the manhood to *fight* for their rights and to destroy the bloody ruffians."[88]

Such advice displayed little awareness of realities in the state. Many Democrats actually welcomed the prospect of fighting a black militia, convinced "we will wipe them from the face of the earth." But nearly all Republican leaders, black and white alike, feared that the raising of troops would inaugurate a "war of the races," and preferred to rely instead on federal intervention. With the governor alternating between moments of resolve and a longing to rejoin his family in the North, Pierrepont dispatched an aide to the state, who in October arranged a "peace agreement" whereby the only two militia companies on active service were disbanded, and whites promised to disarm. "I believe we will have peace, order, and a fair election," Ames exulted. But Democrats, as black state senator Charles Caldwell reported, held the agreement "in utter contempt." On election eve, armed riders drove freedmen from their homes and warned that they would be killed if they appeared to cast ballots. "It was the most violent time that ever we have seen," said one black official. At Aberdeen, whites equipped with rifles and a six-pounder cannon "came armed to the polls and drove colored men away." Elsewhere, Democrats destroyed the ballot boxes or replaced Republican votes with their own. "The reports which come to me almost hourly are truly sickening. . . ." Ames reported to his wife. "The government of the U. S. does not interfere."[89]

The Mississippi campaign united the white population as never before, mobilizing thousands who had not cast ballots during Reconstruction and all but eliminating the scalawag vote, estimated at over 6,000 as late as 1873. Wherever possible, however, blacks remained steadfast; indeed in some plantation counties, the Republican vote actually increased. But where violence had devastated the party's infrastructure and blacks "feared for their lives" if they presented themselves at the polls, the returns constituted a political revolution. In Yazoo, which Ames had carried by an 1,800 vote majority in 1873, the returns stood 4,044 to 7 in favor of the Democrats. The Republican vote declined substantially in other black belt counties—by approximately 800 in Coahoma, 1,600 in

<hr>

88. "Three hundred voters" to Adelbert Ames, September 14, 1875, William Carely to Ames, October 9, 1875, Mississippi Governor's Papers; Ames, ed., *Chronicles*, 2:167; Adelbert Ames to Edwards Pierrepont, September 11, 1875, Pierrepont to Ames, September 14, 1875, Mississippi Governor's Papers; Benjamin F. Butler to Adelbert Ames, March 3, 1875, Ames Family Papers; Ulysses S. Grant to Edwards Pierrepont, September 13, 1875, Edwards Pierrepont Papers, Yale University.

89. E. B. B. to Adelbert Ames, June 1875, James W. Lee to Ames, November 2, 1875, Mississippi Governor's Papers; Garner, *Mississippi*, 382–87; Morgan, *Yazoo*, 456–57; Adelbert Ames to Edwards Pierrepont, October 16, 1875, Pierrepont Papers; 44th Congress, 1st Session, Senate Report 527, 32–33, 90–94, 132–35, 863–65, 1031, Documentary Evidence, 20–23, 29.

Holmes, 1,300 in Jefferson, 1,100 in Hinds—and Democratic totals rose dramatically, suggesting the widespread stuffing of ballot boxes. The combination of an increased vote in the white counties and an electoral shift in many black ones produced a Democratic landslide in the race for treasurer (the only state office contested), and gave the party five of the six Congressmen, and a four-to-one majority in the legislature.[90] Nor did this conclude Mississippi's "Redemption." In plantation counties where local positions remained under Republican control, violence continued after the election, with officials forced to resign under threat of assassination, and vigilante groups meting out punishment to blacks accused of theft and other violations of plantation discipline. On Christmas day, Charles Caldwell, "as brave a man as I ever knew," according to one associate, was shot in the back at Clinton after being lured to take a drink with a white "friend." When the legislature assembled, it impeached and removed from office Lieut. Gov. Alexander K. Davis (so that Ames's departure would not elevate a black to take his place) and then compelled the governor to resign and leave the state rather than face his own impeachment charges.[91]

Thus, in defiance of federal law and the national Constitution, Democrats gained control of Mississippi. As Ames had written in the midst of the election campaign, "a *revolution* has taken place—by force of arms—and a race are disfranchised—they are to be returned to a condition of serfdom—an era of second slavery." Blame rested, in the first instance, on Ames's administration, which had proved too weak to execute the laws. Yet this failure did not absolve the federal government of its own responsibility for the outcome. Grant, who had sent troops to prop up the unstable and corrupt Republican regime in Louisiana, turned a deaf ear to pleas from the far stronger and more upright government of Mississippi. Could the postwar amendments be effectively nullified and the constitutional rights of American citizens be openly violated without provoking federal intervention? If so, remarked John R. Lynch, Mississippi's only Republican Congressman to survive the landslide, "then the war was fought in vain." Yet Northern Republicans now saw Reconstruction as a political liability. Lynch himself learned a lesson in political reality when he asked the President in November why he had not sent troops. Northern party leaders, Grant replied, had pressured him not to

90. 44th Congress, 2d Session, Senate Miscellaneous Document 45, 206; Warren G. Ellem, "Who Were the Mississippi Scalawags?" *JSH*, 38 (May 1972), 222; Thomas H. Kinson to Adelbert Ames, November 3, 1875, Mississippi Governor's Papers; *Tribune Almanac*, 1876, 66–67; Harris, *Day of the Carpetbagger*, 670–87. Harris' account seriously underestimates the extent of violence in 1875 and its impact on the outcome.

91. 44th Congress, 1st Session, Senate Report 527, 435–38, 623, 1630–37; C. H. Green to Adelbert Ames, November 22, 1875, Mississippi Governor's Papers; Albert D. Thompson to Blanche K. Bruce, December 8, 1875, Bruce Papers; Aptheker, "Caldwell," 369–71; Alexander Warner to James W. Garner, May 4, 1900, James W. Garner Papers, MDAH; Harris, *Day of the Carpetbagger*, 691–96.

act—there was no sense in trying to save Mississippi if the attempt to do so would lose Ohio.[92]

But the interrelation between the Ohio and Mississippi campaigns went far deeper than immediate political exigencies. Early in Reconstruction, Schuyler Colfax had noted that a heterogeneous collection of social interests made up the Republican party, among them the North's "men of business and property," and the black "humble and defenseless millions of the south." If the Ohio campaign solidified the party's association with the former, its response to Mississippi underscored its growing detachment from the latter. The depression and its consequences—an erosion of free labor thought, growing middle-class conservatism, and resurgent racism—all contributed to this shift in the party's center of gravity. Ames commented bitterly, "I am fighting for the Negro, and to the whole country a white man is better than a 'Nigger'." But as Mississippi's blacks were well aware, the campaign for white supremacy also involved a struggle to maintain the planter's economic domination. "I suppose it is a fight between the poor people and the rich man now," commented Alexander Branch, a former slave who had risen to serve on the board of police of Wilkinson County. In such a fight, as the Ohio campaign demonstrated, the sympathies of Northern Republicans would rest with men of property.[93]

All in all, from the inability of the Forty-Third Congress in its waning days to agree on a policy toward the South, to Grant's failure to intervene in Mississippi, 1875 marked a milestone in the retreat from Reconstruction. As the nation approached another Presidential election, it seemed certain that whoever emerged victorious, Reconstruction itself was doomed.

92. Ames, ed., *Chronicles*, 2:216; *CR*, 44th Congress, 1st Session, 3783; Lynch, *Reminiscences*, 171–74; Gillette, *Retreat from Reconstruction*, 159.
93. Willard H. Smith, *Schuyler Colfax: The Changing Fortunes of a Political Idol* (Indianapolis, 1952), 337; Ames, ed., *Chronicles*, 2:200; 44th Congress, 1st Session, Senate Report 527, 1594.

CHAPTER 12

Redemption and After

The Centennial Election

TO celebrate the anniversary of their nation's independence, Americans in 1876 flocked to Philadelphia for the Centennial Exposition, a monument to the "Progress of the Age." The attendance of nearly 10 million represented over one fifth of the population of the United States, and the fair's thousands of exhibits displayed everything from Siamese ivory to an animated wax replica of Cleopatra "in extreme dishabille." New inventions, harbingers of vast changes in America's economic and social life, abounded: the telephone, typewriter, and electric light, a "new floor-cloth called linoleum," packaged yeast, an internal combustion engine. But the main focus of public interest was the mighty Corliss steam engine. Weighing 700 tons and rising forty feet into the air, it symbolized the Exposition's theme—that machines were remaking the society, ushering in an era of technical progress and material abundance in which all Americans would share.[1]

In the face of the continuing economic depression, with millions of workers unemployed and labor strife widespread, the mood of self-congratulation seemed somewhat incongruous. It was only achieved by ignoring some less than admirable features of contemporary American life. Pennsylvania's exhibit made no mention of the bitter Long Strike of 1875, nor did that of Massachusetts illustrate conditions that had sparked the great walkout of Fall River textile workers, even though twenty-seven of the city's mills exhibited their wares. One labor newspaper questioned whether workers enjoyed "independence" at all, for "capital has now the same control over us that the aristocracy of England had at the time of the Revolution." Even Edward Atkinson wondered whether the effects of mechanization were as benign as the fair suggested. Factory operatives

1. Robert C. Post, ed., *1876: A Centennial Exhibition* (Washington, D.C., 1976); Eric Hobsbawm, *The Age of Capital 1848–1875* (London, 1975), 32–33; Dee Brown, *The Year of the Century: 1876* (New York, 1966), 129–32; *AC*, 1876, 262–79.

themselves, he mused, had been reduced to "machines"; they could only find personal satisfaction outside the "dreary monotony" of their working lives. Nor were women offered a real opportunity to display their contributions to American society. A Woman's Pavilion, added as "an afterthought, as theologians claim woman herself to have been," housed a collection of needlework and exhibited the talents of female carpet and silk weavers, their power looms overseen by a "lady engineer." But women's subordinate legal status received no attention, an oversight corrected when feminists led by Elizabeth Cady Stanton and Susan B. Anthony interrupted the Exposition's July 4th celebration to read Woman's Declaration of Independence.[2]

Nor did the Centennial Exposition do justice to the nation's nonwhite population. Blacks were excluded from the construction crews that built the exhibition halls and largely absent from the displays. (One exception was the Southern Restaurant, which, according to a guidebook, featured "a band of old-time plantation 'darkies' " performing songs.) A bale of Benjamin Montgomery's cotton won an agricultural medal, defeating competition from all over the South as well as Brazil, Egypt, and the Fiji Islands, but no notice was taken of the violent overthrow of Mississippi Reconstruction, or the impending legal action that would soon drive the Montgomery family from Davis Bend. The Exposition did include representations of American Indians, mostly depicting them as a harmless, "primitive" counterpoint to white civilization. In July, however, real Indians rudely interrupted the celebration when word reached Philadelphia of the massacre of Gen. George A. Custer and his command by Sioux warriors led by Sitting Bull and Crazy Horse. The Indians were defending lands reserved from them by an 1868 treaty. Although Custer's defeat only temporarily delayed the inexorable march of white soldiers, settlers, and prospectors, it said more of the twin realities of broken government promises and the Indians' tenacity in defending their way of life than all the exhibits at Philadelphia.[3]

Like its economy, America's political health offered little cause for celebration in 1876, for even as workmen put the finishing touches on the exposition, new revelations of scandal engulfed the Grant Administration. Ambassador to Britain Robert C. Schenck returned home in disgrace, after disclosures that he had become rich by lending his name to

2. Post, *1876*, 203; Philip S. Foner, ed., *We, The Other People* (Urbana, Ill., 1976), 19; *The First Century of the Republic: A Review of American Progress* (New York, 1876), 207–208; *AC, 1876*, 272–73; Theodore Stanton and Harriot Stanton Blatch, eds., *Elizabeth Cady Stanton* (New York, 1922), 1:263–69.

3. Robert W. Rydell, *All the World's A Fair: Visions of Empire at American International Expositions, 1876–1916* (Chicago, 1984), 21–31; Janet S. Hermann, *The Pursuit of a Dream* (New York, 1981), 151; Dee Brown, *Bury My Heart at Wounded Knee* (New York, 1970), 274–300; Robert M. Utley, *Frontier Regulars: The U.S. Army and the Indian, 1866–1891* (New York, 1973), 237–61.

the fraudulent prospectus of a Utah mining concern that bilked English investors of millions. Then Secretary of War William Belknap resigned to avoid an impeachment trial when it became known that he had received kickbacks from the government-appointed Indian trader at Fort Sill. But the most bizarre turn of events concerned the Whiskey Rings, which had robbed the government of millions of dollars in internal revenue. Although an investigation launched by Secretary of the Treasury Benjamin H. Bristow revealed that the President's private secretary, Orville H. Babcock, and close friend, Gen. John McDonald, stood at the center of the frauds, Grant refused to believe the charges against his associates. The Cabinet forced him to abandon a plan to travel to St. Louis to testify in Babcock's defense, but Grant insisted on dispatching an affidavit attesting to his secretary's innocence, thereby securing an acquittal. (Forced from his White House post, Babcock received a lucrative appointment as inspector of lighthouses.)[4]

The crescendo of scandal suggested that the republic had not maintained the level of civic virtue anticipated by its founding fathers. On a more mundane level, it strongly affected the scramble for the Republican Presidential nomination. Until early in the centennial year, the choice of former House Speaker James G. Blaine, the party's most popular leader, seemed all but assured. But in February stories began to circulate about "certain curious features in Blaine's personal record . . . extremely injurious" to his reputation. When a newspaper broke the story in April, it was hard to resist the conclusion that Blaine had used his influence as Speaker to secure a land grant for an Arkansas railroad in which he owned stock, and that the Union Pacific, another government-subsidized railroad, had accepted this stock as collateral for a loan that Blaine never repaid. Coming on top of the other scandals, the revelations weakened Blaine's candidacy, for, as a member of "Boss" Keyes's Wisconsin machine noted, "public opinion among Republicans" demanded "a reformer at the head of the ticket." The same consideration injured the prospects of Stalwart aspirants Roscoe Conkling and Oliver P. Morton. Many Liberals favored Secretary Bristow, whose crusade against the Whiskey Rings, however, had incurred the formidable opposition of both party spoilsmen and the President. Moreover, as George W. Julian, who supported Bristow's candidacy, admitted, "Kentucky was not exactly the right place for the coming man to be born in."[5]

4. Clark C. Spence, "Robert C. Schenck and the Emma Mine Affair," *OHQ,* 68 (April 1969), 141–60; Edward McPherson, *A Handbook of Politics for 1876* (Washington, D.C., 1876), 157–61; Ross K. Webb, *Benjamin Helm Bristow: Border State Politician* (Lexington, Ky., 1969), 187–208.

5. Keith I. Polakoff, *The Politics of Inertia: The Election of 1876 and the End of Reconstruction* (Baton Rouge, 1973), 16–23, 44–52; E. B. Wight to William W. Clapp, February 28, 1876, William W. Clapp Papers, LC; Rock J. Flint to Elisha W. Keyes, May 22, 1876, Elisha W. Keyes Papers, SHSW; Webb, *Bristow,* 219–34; Grace J. Clarke, *George Washington Julian* (Indianapolis, 1923), 365.

Thus, for the first time in sixteen years, the Republican nomination was truly up for grabs. Robert G. Ingersoll, the era's most spellbinding orator, nearly secured Blaine's nomination with a thrilling speech depicting his favorite as a "plumed knight" who had raised his "shining lance" against personal detractors and the country's enemies. But as in 1860, the nod eventually went to a lesser-known figure from the pivotal Midwest, who, while the first choice of few delegates, was acceptable to nearly all. Here, however, similarities between the Great Emancipator and Ohio Gov. Rutherford B. Hayes began and ended. For the colorless Hayes was in Henry Adams' words, a "third-rate nonentity," whose main claim to fame was managing to remain on good terms with all factions of the party. Unlike Lincoln, he seemed to shrink rather than grow in office. In Congress in the 1860s, Hayes dutifully supported party measures, without attracting much public attention. As Ohio's three-time governor, his upper-crust education (which included a degree from Harvard Law School) and hard-money stance endeared him to reformers, even though he avoided committing himself to the Liberal movement. Vice Presidential nominee William A. Wheeler of New York, a small-town lawyer and banker who had served five terms in Congress, was equally innocuous and almost entirely unknown, partly because public speaking and "the presence of crowds" made him ill. The candidates, observed *The Nation*, were "eminently respectable men—the most respectable men, in the strict sense of the word, the Republican party has ever nominated."[6]

The platform displayed the same timidity as the standard bearers. A "tepid document," it called for the prosecution of corrupt officials, termed demands for women's suffrage worthy of "respectful consideration," and deprecated appeals to sectional feeling. The only controversy concerned a plank opposing the "immigration and importation of Mongolians"—the first time, a Massachusetts delegate pointed out, that the party had included "a discrimination of race" in its national platform. Little was said of Reconstruction. "Do you mean to make good to us the promises in your constitution?" Frederick Douglass asked the gathering. But most Republican leaders had concluded that the Northern public would no longer support federal intervention in Southern affairs. Hayes, in this regard, was typical, for during Grant's second term he had come to "doubt the ultra measures relating to the South," and his carefully worded letter of acceptance pledged to bring the region "the blessings of honest and capable local self government" (code words, he well understood, for an end to Reconstruction). "Unless I am very much mistaken," exulted Carl Schurz, "the Cincinnati Convention has nominated our man without knowing it." As if to underscore a shift in Northern priorities and

6. C. P. Farrell, ed., *The Works of Robert G. Ingersoll* (New York, 1900), 9:59–60; Harry Barnard, *Rutherford B. Hayes and His America* (Indianapolis, 1954), 237–38, 277–87; Polakoff, *Politics of Inertia*, 67, 123; Charles R. Williams, ed., *Diary and Letters of Rutherford Birchard Hayes* (Columbus, Ohio, 1922–26), 3:301; *Nation*, June 22, 1876.

the certainty of a new Southern policy whatever the election results, Congress shortly before the Republican convention repealed the Southern Homestead Act of 1866, in order to open public land in the South to exploitation by timber and mining companies.[7]

Unlike the Republican nomination, little uncertainty surrounded the Democratic, for wealthy New York Congressman Abram Hewitt had personally funded an effective "propaganda" in favor of Gov. Samuel J. Tilden. Not that Tilden needed Hewitt's financial aid, for he was one of the country's richest men, having acquired a fortune (and the nickname "the Great Forecloser") by reorganizing bankrupt railroad lines, helping float sometimes watered corporate bond issues, and acting as counsel for Jim Fisk, Jay Gould, and other Gilded Age captains of industry. The cold, aloof Tilden was not a very likable individual (among his political associates, remarked John Bigelow, none "seem to care for him personally"), but his supporters considered his Wall Street ties a distinct asset. "He is connected with the moneyed men of the country," one wrote, "and will be supported by bankers. . . . This is exactly what we want. No man can be elected who has the money of the country to fight." As a leading figure in the overthrow of the Tweed Ring, Tilden possessed impeccable reform credentials, even though he had gone on to construct his own highly centralized political machine during the 1870s. By the time the Democratic convention assembled, his nomination by the nearly 1,000 delegates (whose ranks did not include a single black) was a foregone conclusion. Youngest in ideas if not age was a third candidate in the 1876 field: eighty-five-year-old Peter Cooper, an industrialist and philanthropist nominated by the new Greenback party. Although unable to wage a vigorous campaign, Cooper made known his belief that a new "oligarchy" of wealth had replaced the Slave Power and that the government should take steps to equalize property and protect "the poor toilers and producers" from exploitation.[8]

In the North, political corruption and the depression became Tilden's watchwords; issues many Republicans feared would suffice to carry the

7. Polakoff, *Politics of Inertia*, 60–61, 104–105; *Proceedings of the Republican National Convention Held at Cincinnati, Ohio . . .* (Concord, N.H., 1876), 26–27; Barnard, *Hayes*, 265; Williams, ed., *Diary and Letters of Hayes*, 3:329; Frederic Bancroft, ed., *Speeches, Correspondence and Political Papers of Carl Schurz* (New York, 1913), 3:258–59; Paul W. Gates, "Federal Land Policy in the South, 1866–1888," *JSH*, 6 (August 1940), 310–15.

8. Allan Nevins, ed., *Selected Writings of Abram Hewitt* (New York, 1937), 159; Horace S. Merrill, *Bourbon Democrats of the Middle West 1865–1896* (Seattle, 1953), 110; Alexander C. Flick, *Samuel Jones Tilden: A Study in Political Sagacity* (New York, 1939), 102–18, 164–65, 289; John Bigelow Diary, January 4, 1877, John Bigelow Papers, NYPL; S. P. Purdy to J. H. Harmon, May 14, 1876, Samuel J. Tilden Papers, NYPL; Jerome Mushkat, *The Reconstruction of the New York Democracy, 1861–1874* (Rutherford, N. J., 1981), 176–190; Brown, *Year of the Century*, 218; Allan Nevins, *Abram S. Hewitt, With Some Account of Peter Cooper* (New York, 1935), 286–89. Cooper received 81,737 votes in November.

election. "When men are suffering for want of employment," one party leader wrote, "they are very apt to think that a change in parties may bring better times." In response, Republicans fell back on the strategy of waving the bloody shirt. But if the Civil War still dominated Republican rhetoric, the defense of black rights played a secondary role. The Liberal critique of Reconstruction had by now sunk deep roots among mainstream Republicans. As one wrote privately, "the truth is, the negroes are ignorant, many of them not more than half civilized . . . [and] no match for the whites. . . . Our Southern system is wrong." With a close election at hand and the depression making it difficult to raise large contributions, the Republican campaign committee concentrated all its efforts on the North, virtually writing off the border and South to Tilden. Why, wondered a Southern Republican, did speakers "eloquently vibrate from Maine to Indiana . . . and yet never put a foot across the Potomac?" (The national committee, one Republican complained after a visit to its headquarters, was entirely unaware that West Virginia had an election upcoming in October.)[9]

Recent events—the defeat of the 1875 Force bill, the Supreme Court's *Cruikshank* decision, and the President's refusal to send troops to Mississippi—had thoroughly demoralized many Southern Republicans. Feeling themselves "abandoned . . . to the tender mercies of the Ku Klux," some abandoned the 1876 campaign. "We are helpless and unable to organize," wrote a Mississippi scalawag, "dare not attempt to canvass, or make public speeches." But especially in the four "unredeemed" states, Republicans waged a desperate struggle for political survival. North Carolina's stirring gubernatorial contest, dubbed "the battle of the giants," pitted Republican Judge Thomas A. Settle against former Confederate Gov. Zebulon Vance, a man of striking appearance who had been chosen, after an international search, to represent the Caucasian race in a geography textbook. The two crisscrossed the state, debating no fewer than fifty-seven times. In Florida, the parties competed for the mantle of moderation: Republican leaders chose an all-white state ticket and even denied renomination to black Congressman Josiah Walls, while Democrats selected for governor George H. Drew, a wealthy New Hampshire-born lumberman who had supported the Union during the war and voted for Grant in 1868. A far different situation emerged in Louisiana, where Democrats nominated "all that was left" of Confederate Gen. Francis T. Nicholls, who had lost an arm in one Civil War battle and a leg in another.

9. John D. Defrees to Benjamin Harrison, August 17, 1876, Benjamin Harrison Papers, LC; Williams, ed., *Diary and Letters of Hayes*, 3:340–43; Matilda M. Gresham, *Life of Walter Quintin Gresham 1832–1895* (Chicago, 1919), 2:459–60; Sister Mary K. George, *Zachariah Chandler: A Political Biography* (East Lansing, Mich., 1969), 252–53; Edward F. Noyes to Rutherford B. Hayes, August 30, 1876, William E. Chandler to Hayes, October 12, 1876, Rutherford B. Hayes Papers, Hayes Memorial Library; *CR*, 44th Congress, 2d Session, 1535; John Defrees to W. H. Painter, July 27, 1876, Hayes Papers.

The platform pledged to respect blacks' rights under the postwar amendments, but "where the negroes hadn't come into line as they should," Democrats fell back on political violence. Armed bands disrupted Republican meetings, whipped freedmen, and murdered local officials; their behavior, said *Harper's Weekly,* "would have disgraced Turks in Bulgaria."[10]

More than any other Southern state, however, national attention focused on South Carolina. Here, Democrats entered 1876 divided between Charleston-centered "fusionists"—who, in the face of the state's substantial black voting majority and the conciliatory policies of Gov. Daniel H. Chamberlain, advocated conceding the gubernatorial race and concentrating on local and legislative contests—and partisans of a "straight-out" campaign for white supremacy. A contest modeled on Mississippi's, insisted upcountry planter-lawyer Martin W. Gary, could redeem South Carolina. Gary's "Plan of Campaign" called upon each Democrat to "control the vote of at least one negro by intimidation, purchase, keeping him away or as each individual may determine," always bearing in mind that *"argument* has no effect on them: They can only be influenced by their *fears."*[11]

In May, South Carolina's Democratic convention adjourned without nominating candidates for state office—a victory for the fusionists. But an event in the tiny town of Hamburg soon transformed the state's political climate. Situated just across the Savannah River from Augusta, Hamburg was one of many centers of Reconstruction black power. Its local officials included trial justice Prince Rivers, a former member of Thomas Wentworth Higginson's Civil War regiment, and militia commander Dock Adams, a skilled carpenter, Union Army veteran, and former Augusta politician who had moved to the town in 1874, complaining that blacks in Redeemed Georgia "could not exercise their political opinion as they wished, and I did not desire to be oppressed in that way." Local whites claimed to have been subjected to severe indignities when passing through Hamburg: they were prevented from drinking at the public water fountain, arrested "on the slightest provocation," and forced to give way

10. 44th Congress, 2d Session, Senate Report 704, 605; J. B. Work to James A. Garfield, December 16, 1876, James A. Garfield Papers, LC; James R. Cavett to James Redpath, August 22, 1876, U.S. Senate Select Committee on Mississippi Election Papers, NYPL; Glenn Tucker, *Zeb Vance: Champion of Personal Freedom* (Indianapolis, 1965), 456–57; Otto H. Olsen, "North Carolina: An Incongruous Presence," in Otto H. Olsen, ed., *Reconstruction and Redemption in the South* (Baton Rouge, 1980), 193; Jerrell H. Shofner, "A Note on Governor George F. Drew," *FlHQ,* 48 (April 1970), 412–14; Fanny Z. L. Bone, "Louisiana in the Disputed Election of 1876," *LaHQ,* 14 (July 1931), 439–40, (October 1931), 562–65, 15 (January 1932), 93–94, 100–103; William I. Hair, *Bourbonism and Agrarian Protest: Louisiana Politics 1877–1900* (Baton Rouge, 1969), 4–6.

11. Charleston *News and Courier,* May 8, 9, 30, June 5, 1876; Francis B. Simkins and Robert H. Woody, *South Carolina During Reconstruction* (Chapel Hill, 1932), 564–69.

on the streets for militia parades—an "insult [such] as no white people upon earth had ever to put up with before."[12]

What came to be known as the Hamburg Massacre began with the black militia's celebration of the July 4 centennial. When the son and son-in-law of a local white farmer arrived on the scene and ordered the assembled militiamen to move aside for their carriage, harsh words were exchanged, although Adams eventually opened his company's ranks and the pair proceeded on their way. On the fifth, the farmer appeared before Prince Rivers, demanding that Adams be arrested for obstructing "my road." Apparently, Adams chastized the justice for even entertaining the complaint, for he found himself charged with contempt of court and ordered to stand trial on the eighth. That day, the black militia gathered in Hamburg, as did a large number of armed whites. After Adams refused a demand by Gen. Matthew C. Butler, the area's most prominent Democratic politician, to disarm his company, fighting broke out, about forty militiamen retreated to their armory, and Butler made for Augusta, returning with a cannon and hundreds of white reinforcements. As darkness fell, the outgunned and outnumbered militiamen attempted to flee the scene. Hamburg's black marshal was mortally wounded and twenty-five men captured; of these, five more were murdered in cold blood around two in the morning. After the killings, the mob ransacked the homes and shops of the town's blacks. In August, a grand jury indicted seven men for murder and named several dozen more as accessories; all were acquitted after Redemption. One white youth also lost his life in the affray.[13]

"If you can find words to characterize [this] atrocity and barbarism . . ." wrote Chamberlain, "your power of language exceeds mine." Among the affair's most appalling features was the conduct of General Butler, who either selected the prisoners to be executed (according to black eyewitnesses), or left the scene when the crowd began "committing depredations" (his own, hardly more flattering, account). In either case, Butler's conduct testified to the utter collapse of a sense of paternalist obligation, not to mention common decency, among those who called themselves the region's "natural leaders." (A few months later, he had the temerity to tell a Congressional investigating committee that blacks

12. Charleston *News and Courier*, May 8, 1876; 44th Congress, 2d Session, Senate Miscellaneous Document 48, 1:34–35, 48, 73, 2:607; "Memoirs of Reconstruction," manuscript, 1918, Matthew C. Butler Papers, DU.

13. Joel Williamson, *After Slavery: The Negro in South Carolina During Reconstruction* (Chapel Hill, 1965), 267–69; 44th Congress, 2d Session, Senate Miscellaneous Document 48, 1:34–39, 712, 1050–55, 2:326–31, 603, 3:473–76. Contemporary sources disagree as to whether six or seven blacks perished at Hamburg. In 1916, a monument to McKie Meriweather, the only white to die, was unveiled at North Augusta, into which Hamburg had been absorbed. No mention was made of the blacks who perished. Daniel S. Henderson, *The White Man's Revolution in South Carolina* (North Augusta, S.C., 1916), 1.

possessed "little regard for human life.") Certainly, no one could again claim that the South's "respectable" elite disdained such violence, for in one of its first actions, South Carolina's Redeemer legislature in 1877 elected Butler to the U. S. Senate.[14]

During the killings, Dock Adams related, whites kept repeating: "This is the beginning of the redemption of South Carolina." And by raising political and racial tensions to fever pitch, the massacre ended any possibility of "fusion" between Governor Chamberlain's Democratic and Republican supporters. In August, South Carolina Democrats chose a ticket headed by Gen. Wade Hampton, probably the state's most popular figure (at least among the white population). Still deeply divided over Chamberlain's policies, Republicans reluctantly renominated the governor but chose two of his most articulate black critics for lieutenant governor and attorney general. Despite the white-line sentiment kindled by Hamburg, Hampton's speeches promised an administration that would strengthen the educational system, avoid "vindictive discrimination," and offer the protection against violence Chamberlain seemed unable to provide. But among prominent blacks, only Martin R. Delany campaigned on Hampton's behalf. No more than a handful of freedmen supported the Democratic ticket; the vast majority, including those offered bribes or threatened with the loss of their jobs, remained loyal to the Republicans. Edgefield County deputy marshal David Graham refused to "quit Chamberlain" despite an offer of $500 and future employment. "I heap rather farm than be in politics. . . ." Graham remarked, "but here I is; these other leading fellows can't get along without me." (To Democrats, such integrity only proved that "the American negro takes no thought of the morrow.")[15]

It was not surprising that most freedmen resisted the enticements and doubted the pledges of a party whose ticket consisted entirely of former Confederate officers and whose local organization rested on rifle clubs manned by Civil War veterans. "Dey says dem *will do* dis and dat," remarked a Sea Island black. "I ain't ax no man what him *will do*— I ax him what him *hab done.*" Indeed, a close look at Hampton's own record might have raised doubts about the depth of his statesmanship and moderation, despite his eloquent appeals for racial harmony. One of the South's wealthiest planters, Hampton had seen his fortune collapse with

14. Simkins and Woody, *South Carolina*, 487; 44th Congress, 2d Session, Senate Miscellaneous Document 48, 2:242–43, 616; 44th Congress, 2d Session, House Miscellaneous Document 31, 1:304.

15. 44th Congress, 2d Session, Senate Miscellaneous Document 48, 1:46–47; Charleston *News and Courier*, August 17, 18, September 16, 1876; Thomas Holt, *Black over White: Negro Political Leadership in South Carolina During Reconstruction* (Urbana, Ill., 1977), 201; *The Pledges of Gen. Wade Hampton, Democratic Candidate for Governor, to the Colored People of South Carolina 1865–1876* (n.p., 1876), 2–7; Victor Ullmann, *Martin Delany: The Beginnings of Black Nationalism* (Boston, 1971), 489; 44th Congress, 2d Session, Senate Miscellaneous Document 48, 1:471.

emancipation and, saddled with enormous debts, had played little role in public affairs since 1865. But his correspondence and few public statements revealed a man unable to formulate a coherent response to the crisis of Reconstruction. Along with bitter denunciations of the Freedmen's Bureau, black soldiers, and even the Emancipation Proclamation, Hampton's attitudes toward blacks vacillated between early support of "qualified" impartial suffrage, predictions of the freedmen's imminent "extermination," and advocacy, as late as 1869, of their removal from the country. Two years later he told the Ku Klux Klan Committee that blacks lacked the capacity for "forethought" and suffered from "an exaggerated opinion of their own power."[16]

Labor conflict in the lowcountry rice kingdom added yet another element to the heated political climate of 1876. In May, day laborers on a number of Combahee River plantations walked off their jobs, demanding higher wages and payment in cash rather than checks redeemable only at plantation stores. Hundreds of strikers paraded through the fields, calling laborers from their work and beating those who refused to join. In August, a resumption of the strike produced a confrontation between a Democratic rifle club and armed strikers; only the intervention of Congressman Robert Smalls prevented bloodshed. Ten strikers were arrested and brought to Beaufort, where crowds of former slaves applauded them on the streets and a black trial justice dismissed all charges. Meanwhile, Chamberlain, needing every black vote he could muster, refused planters' pleas to have federal troops restore order (an odd demand for advocates of "home rule," but one necessitated by the fact that the district's militia was "composed in great part of the strikers.") In the end, planters acceded to the laborers' demands. The episode underscored the intimate relationship between political and economic power during Reconstruction. Their inability to obtain the support of local or state authorities reinforced planters' conviction that only a change in administration could restore labor discipline to the rice region.[17]

With so much at stake, the 1876 campaign became the most tumultuous in South Carolina's history, and the one significant exception to the Reconstruction pattern that cast blacks as the victims of political violence and whites as the sole aggressors. In September, black Republicans assaulted Democrats of both races leaving a Charleston meeting; several

16. Simkins and Woody, *South Carolina*, 500; Rupert S. Holland, ed., *Letters and Diary of Laura M. Towne* (Cambridge, Mass., 1912), 253–54; Hampton M. Jarrell, *Wade Hampton and the Negro: The Road Not Taken* (Columbia, S.C., 1950), 34; Wade Hampton to Armistead Burt, March 13, 1868, Wade Hampton Papers, DU; Charles E. Cauthen, *Family Letters of the Three Wade Hamptons 1782–1910* (Columbia, S.C., 1953), 129–30, 139–40; Duane Mowry, ed., "Post-Bellum Days: Selections from the Correspondence of the Late Senator James R. Doolittle," *Magazine of History*, 17 (August–September 1913), 51; 42d Congress, 2d Session, House Report 22, South Carolina, 1236.

17. Eric Foner, *Nothing But Freedom: Emancipation and Its Legacy* (Baton Rouge, 1983), 91–106.

were wounded, and one white lost his life. A month later a group of blacks began firing at a "joint discussion" at Cainhoy, a village near the city, resulting in the deaths of five whites and one black. The presence of Martin R. Delany had enraged local freedmen and set the stage for violence; "the cry," reported an eyewitness, "was that any white man had a right to be a democrat, 'but no damned black man had'." Throughout the state, black Democrats found themselves ostracized as "deserters of their race." "Black women," were said to be "worse than the men"; one "threw her husband's clothes out . . . and locked the door on him," saying she would rather "beg her bread" than live with a "democratic nigger"; another abused a Hampton supporter as a "damned democratic son of a bitch," saying he "was voting to put her and her children back into slavery."[18]

But the campaign of intimidation launched by Hampton's supporters far overshadowed such incidents. To rally white voters, many of whom had taken no part in politics since black suffrage began, the Democratic candidate embarked on a triumphant tour of the state, accompanied by hundreds of armed and mounted supporters. Later immortalized in legend as "knightly figures on prancing steeds," most of Hampton's "Red Shirts" actually bestrode mules, but the tour nonetheless attracted immense, frenzied crowds. Meanwhile, rifle clubs disrupted Republican rallies with "violent and abusive tirades." A reign of terror reminiscent of Ku Klux Klan days swept over Edgefield, Aiken, Barnwell, and other Piedmont counties, with freedmen driven from their homes and brutally whipped, and "leading men" murdered. "The way things are going is not right. . . ." complained one black to the governor. "They have kill col'd men in every precinct." The belief that they need not fear federal intervention gave Democrats a free hand. Former slave Jerry Thornton Moore, president of an Aiken County Republican club, was told by his white landlord that opponents of Reconstruction planned to carry the election "if we have to wade in blood knee-deep." "Mind what you are doing," Moore responded, "the United States is mighty strong." Replied the landlord: "but, Thornton, . . . the northern people is on our side."[19]

South Carolina's election, a Democratic observer acknowledged, "was one of the grandest farces ever seen." Despite the campaign of intimida-

18. Melinda M. Hennessey, "Racial Violence During Reconstruction: The 1876 Riots in Charleston and Cainhoy," *SCHM*, 86 (April 1985), 100–12; 44th Congress, 2d Session, House Miscellaneous Document 31, 2:68, 215; 44th Congress, 2d Session, Senate Miscellaneous Document 48, 1:592–93, 939.

19. [Belton O'Neall Townsend] "The Political Condition of South Carolina," *Atlantic Monthly*, 39 (February 1877), 180–81; Jarrell, *Hampton*, 71; Alfred B. Williams, *Hampton and His Red Shirts* (Charleston, 1935), 151, 201–12, 256; L. Cass Carpenter to William E. Chandler, August 26, 1876, William E. Chandler Papers, LC; W. J. Mixsons to Daniel H. Chamberlain, September 27, 1876, P. Jenkins to Chamberlain, September 30, 1876, E. J. Black to Chamberlain, October 2, 1876, South Carolina Governor's Papers, SCDA; 44th Congress, 2d Session, Senate Miscellaneous Document 48, 1:4–7, 2:429–30.

tion, Chamberlain polled the largest Republican vote in the state's history. But Edgefield and Laurens County Democrats effectively carried out Gary's instructions to vote "early and often" and prevent blacks from reaching the polls, thereby producing massive majorities that enabled their party to claim a narrow statewide victory. And throughout the Deep South, black belt Democrats either barred freedmen from the polls (Yazoo County recorded only two votes for Hayes) or stuffed the ballot boxes to "make it appear the negroes voted with them." Meanwhile, scalawag strength continued to decline. In North Carolina, Settle, although defeated by Vance, pulled as many white ballots as black, but in the South as a whole, only half as many counties as four years earlier recorded a significant white Republican vote.[20]

Early returns on election night appeared to foretell a Democratic victory. Tilden carried New York, New Jersey, Connecticut, and Indiana—more than enough, together with expected victories on the West Coast and a solid South, to give him the Presidency. *New York Times* editor George F. Jones even wired Hayes announcing his defeat. But in the early hours of the morning, someone at Republican headquarters in New York noticed that if Hayes lost no more Northern states and managed to carry South Carolina, Florida, and Louisiana, where the party controlled the voting machinery, he would emerge with a one-vote Electoral College majority. Both Gen. Daniel E. Sickles and William E. Chandler later claimed to have made this discovery and to have sent telegrams, over the signature of the sleeping party chairman, Zachariah Chandler, urging Republican officials to hold their states for Hayes. Soon after he awakened, Chandler announced: "Hayes has 185 electoral votes and is elected."[21]

The Electoral Crisis and the End of Reconstruction

Thus, sixteen years after the secession crisis, Americans entered another winter of political confusion, constitutional uncertainty, and talk of civil war. Predictably, Republican election boards in Florida, South Carolina, and Louisiana invalidated enough returns from counties rife with vio-

20. [Townsend] "Political Condition of South Carolina," 186–87; Benjamin R. Tillman, *The Struggles of 76* (n.p., 1909), 28–29; 44th Congress, 2d Session, House Miscellaneous Document 31, 1:237; 44th Congress, 2d Session, Senate Miscellaneous Document 48, 3:98–99; 44th Congress, 2d Session, Senate Report 704, 606; Vernon L. Wharton, *The Negro in Mississippi 1865–1890* (Chapel Hill, 1947), 199–200; W. R. Richardson to Thomas Settle, December 4, 1876, Thomas Settle Papers, UNC; Allen W. Trelease, "Who Were the Scalawags?" *JSH*, 29 (November 1963), 460.
21. George F. Jones to Rutherford B. Hayes, November 8, 1876 (telegram); William E. Chandler to Hayes, November 9, 1876, Hayes Papers; William E. Chandler, undated memorandum, Chandler Papers; Jerome L. Sternstein, ed., "The Sickles Memorandum: Another Look at the Hayes-Tilden Election-Night Conspiracy," *JSH*, 32 (August 1966), 342–57; Nevins, *Hewitt*, 320–22.

lence to declare Hayes and the party's candidate for governor victorious. Equally predictably, Democrats challenged the results. Rival state governments assembled in Louisiana and South Carolina, and rival electoral certificates were dispatched to Washington. Florida avoided a double administration when its Supreme Court ruled that Democrat George F. Drew had won the contest for governor, but the judges let Hayes's margin stand, whereupon Drew appointed a new canvassing board, which determined that Tilden had carried the state.[22]

As on so many other questions, the Constitution was maddeningly ambiguous as to how the validity of disputed returns should be decided. The Twelfth Amendment directed the President of the Senate (in 1876 Republican Thomas W. Ferry of Connecticut) to open electoral certificates in the presence of both houses: "The votes shall then be counted." It was unclear, however, whether the presiding officer decided which returns to tally, in which case a Hayes victory seemed assured, or if either chamber could dispute his decision, in which event stalemate appeared certain since Republicans controlled the Senate and Democrats the House. Meanwhile, reports reached Washington of the "greatest excitement" at the grass roots. Blacks were said to be especially agitated, convinced that in the event of a Democratic victory, "slavery is to be reestablished." With Tilden holding an insurmountable lead in the popular vote, many Democrats vowed to see him inaugurated, if necessary by force. "Tilden or War" proclaimed more than one newspaper, and letters descended on the Democratic standard-bearer announcing that thousands of "well armed men" stood ready to march on Washington. Republicans, insisting that Hayes, in a peaceful election, would have won an easy victory, pledged to resist Tilden's attempt "to be President by fraud and intimidation."[23]

Despite this bellicose rhetoric, neither candidate relished the idea of a violent seizure of the White House or a new civil war. "Everything now depends upon your nerve and your leadership," Manton Marble wrote Tilden on December 10, but the Democratic aspirant, a man with

22. Polakoff, *Politics of Inertia*, 210–18, 226–30; Jerrell H. Shofner, *Nor Is It Over Yet: Florida in the Era of Reconstruction, 1863–1877* (Gainesville, Fla., 1974), 315–39. It is probably impossible to say who "really" won the election of 1876. Hayes undoubtedly carried South Carolina. Fewer than 100 votes of 50,000 separated the candidates in Florida. In Louisiana, the truth "is well nigh hopelessly buried" in a maze of "false testimony contradicted and refuted." (Ella Lonn, *Reconstruction in Louisiana After 1868* [New York, 1918], 452.) A peaceful election would have produced a Hayes victory in at least several Southern states. The electoral situation was further complicated by the effort of Oregon's governor to replace a Republican elector with a Democrat, on the grounds that the former was ineligible by virtue of holding a federal appointment. If allowed to stand, this would erase the one-vote margin claimed by Hayes.

23. Flick, *Tilden*, 352–55; H. C. Bruce to Blanche K. Bruce, November 14, 1876, Blanche K. Bruce Papers, Howard University; Barnard, *Hayes*, 338–41; "Corse" to Samuel J. Tilden, December 7, 1876 (telegram), C. D. W. Ries to Tilden, December 13, 1876, A. N. Robinson to Tilden, December 25, 1876, Tilden Papers; A. W. Stiles to Garfield, November 13, 1876, Garfield Papers.

an abiding fear of disorder, seems to have resigned himself to defeat almost from the moment the crisis began. Instead of putting his claim effectively before the public, Tilden retreated to his study, spending most of December drafting a lengthy examination of legal precedents concerning the counting of electoral ballots. In contrast to his rival's "annoying inactivity," Hayes at least gave tacit approval to a series of complex negotiations involving his close political associates, representatives of South Carolina and Louisiana Democrats, and a group of self-appointed maneuverers who hoped to promote their own vision of a New South. As an immediate objective, these efforts sought, through discreet assurances that the next administration would treat the South with "kind consideration," to detach enough Southern Democratic Congressmen from Tilden to insure Hayes's election. But many Northern Republicans also hoped to use the crisis to jettison a Reconstruction policy they believed had failed. The freedmen, insisted former Ohio Gov. Jacob D. Cox, must moderate their "new kindled ambition" for political influence and recognize that they lacked whites' "hereditary faculty of self government."[24]

Many Republicans who rejected Cox's crude racism shared his belief that the party's Southern wing must be recast so as to reduce the influence of carpetbaggers and blacks and attract the "better class" of local whites. The persistent idea of a vast reservoir of Southern Whigs eager to join the Republican party contained more than a little wishful thinking. But with Reconstruction having demonstrably failed to produce a Republican South, few Northerners could envision an alternative. Even William D. Kelley, for three decades a leading Radical, advised Hayes that the party must look to the "Old Whig or Union" element as the basis of Southern Republicanism. Hayes, for his part, let it be known that he held "precisely" this view. Meeting with a New Orleans editor who brought the message that Louisiana Democrats cared more about control of their own state than the White House, Hayes remarked, "I believe, and I have always believed, that the intelligence of any country ought to govern it." The actions of the Grant Administration reinforced the widespread expectation of an impending change in Southern policy, for the President refused to recognize Chamberlain and Stephen B. Packard (the Republican claimant in Louisiana) as governors. Privately, Grant told the Cabinet that the Fifteenth Amendment had been a mistake: "It had done the Negro no good, and had been a hindrance to the South, and by no means a political advantage to the North."[25]

24. Manton Marble to Tilden, December 10, 1876, Manton Marble Papers, LC; John Bigelow Diary, November 11, 1876, Bigelow Papers; Polakoff, *Politics of Inertia*, 206, 222, 234–36; Flick, *Tilden*, 331; James A. Garfield to Hayes, December 12, 1876, Jacob D. Cox to Hayes, January 31, 1877, Hayes Papers.

25. Thomas A. Osborne to Edward F. Noyes, December 18, 1876, Edward F. Noyes to Rutherford B. Hayes, December 20, 1876, William D. Kelley to Hayes, December 17, 1876, Hayes Papers; George Sinkler, "Race: Principles and Policy of Rutherford B. Hayes," *Ohio*

Meanwhile, a separate but overlapping set of negotiations was under-way. These involved, among other individuals, William H. Smith, head of the Western Associated Press and a close friend of Hayes, and Memphis editor Andrew J. Kellar, a Tennessee railroad man and self-styled "Independent Conservative" committed to his state's economic development and eager to aid in "building up a Conservative Republican party in the South." To these men, a peaceful solution to the crisis, Hayes's inauguration, and a new era in Southern politics all hinged on Republican assurances of internal improvements subsidies for the South. They were particularly eager for Hayes to pledge assistance to the Texas & Pacific Railroad, a project, headed by Thomas A. Scott of the Pennsylvania Railroad, which enjoyed extensive support among Southern political leaders, but little from Northern Democratic Congressmen. Kellar himself had ties to Scott's army of 200 Washington lobbyists and hoped to see Memphis chosen as one of the road's eastern terminals. Although Smith kept Hayes informed of the group's efforts, the candidate, eager to avoid the impression of bargaining for the Presidency, made no comment on the scheme. Nonetheless, rumors of backroom maneuvers abounded, and newspapers filled in the outlines of an embryonic deal. "There is undoubtedly danger of defection among southern Democrats," the New York *Sun*'s Washington correspondent reported in mid-December. "The friends of Hayes are certainly bidding high in that direction. . . . The subsidy for T[exas] P[acific] is part of the programme, as well as counting in of Hayes. Packard and Chamberlain are to be abandoned and a new departure in Republican party policy is to date from Hayes' inauguration."[26]

Things, however, did not work out quite this smoothly. For one thing, even after Scott and his arch-rival, Collis P. Huntington of the Southern Pacific, temporarily agreed to pool their efforts to fatten at the public trough, lack of support among Northerners of both parties prevented a railroad subsidy bill from even reaching the House floor. For another, Republicans found themselves deeply divided by the electoral crisis. Stalwarts like Oliver P. Morton and Benjamin Butler had their own plan: Ferry would be directed to count the electoral vote, Hayes inaugurated "at the point of the bayonet," and Chamberlain and Packard recognized as governors. Roscoe Conkling and James G. Blaine, however, not only

History, 77 (Winter, Spring, Summer 1968), 161; Barnard, *Hayes*, 357–58; Allan Nevins, *Hamilton Fish: The Inner History of the Grant Administration* (New York, 1936), 2:853–54.

26. C. Vann Woodward, *Reunion and Reaction: The Compromise of 1877 and the End of Reconstruction*, rev. ed. (Garden City, N.J., 1956), 11, 27–35, 50, 74–81, 85–87; David M. Abshire, *The South Rejects a Prophet: The Life of Senator D. M. Key, 1824–1900* (New York, 1967), 107–15, 122–25; Grady Tollison, "Andrew J. Kellar, Memphis Republican," *West Tennessee Historical Society Papers*, 16 (1962), 29–55; David J. Rothman, *Politics and Power: The United States Senate 1869–1901* (Cambridge, Mass., 1966), 198; William H. Smith to Rutherford B. Hayes, December 14, 1876, Hayes Papers; Williams, ed., *Diary and Letters of Hayes*, 3:393; Rutherford B. Hayes to James A. Garfield, December 16, 1876, Garfield Papers; A. M. Gibson to Charles A. Dana, December 13, 1876, Tilden Papers.

feared Hayes's "reform" inclinations but, already thinking ahead to 1880, believed their own ambitions would be better served with Tilden in the White House. As for Grant, he appeared entirely indifferent as to who emerged as the winner.[27]

With neither party enjoying vigorous leadership or unity of purpose, a plan to create an independent commission to decide the disputed returns found increasing favor in Congress. Although Tilden considered the proposal an "abandonment of the Constitution" (while characteristically failing to convey his views effectively to Washington), and Speaker of the House Samuel J. Randall of Pennsylvania compared it to "raffling off the Presidency," nearly every Democrat in Congress supported the bill, since it at least offered an alternative to having Senator Ferry count the votes himself. And although a majority of Republicans opposed exchanging a sure result for what Garfield called "the uncertain chances of what a committee . . . will do," enough followers of Conkling and Blaine voted for the Electoral Commission bill to secure Senate passage. Beyond party intrigues, however, enactment of the "plan of peace" reflected a growing desire within both parties for a settlement, the outgrowth, in part, of pressure from a "mercantile and business interest" thoroughly alarmed by extreme rhetoric among supporters of both candidates. Petitions advocating a peaceful solution flooded into Washington, signed mostly, one cynic noted, by "firms doing chiefly *southern* trade." Philadelphia's Board of Trade unanimously called for the bill's passage, and, as William H. Smith's Washington operative, Gen. Henry V. Boynton, reported, "the business interests of New York (democrats and republicans) made themselves felt . . . and demanded some action that should snuff out the violent talk of the mob element." The army, too, welcomed the prospect that it could avoid having to help settle the deadlock. With the possibility receding that troops might be needed in Washington, wrote Gen. William T. Sherman, military men could devote their full attention to "operations against the hostile Sioux."[28]

Enacted late in January, the Electoral Commission Law established a body with fifteen members: ten Congressmen, divided equally between the parties, and five Supreme Court Justices. Four of the latter, two Democrats and two Republicans, were named in the bill and given the

27. Woodward, *Reunion and Reaction*, 120–23, 138–41; Ben: Perley Poore to William W. Clapp, December 29, 1876, January 8, 1877, E. B. Wight to Clapp, December 27, 1876, Clapp Papers; John Bigelow, ed., *Letters and Literary Memorials of Samuel J. Tilden* (New York, 1908), 2:491; Polakoff, *Politics of Inertia*, 259–69.

28. John Bigelow, *The Life of Samuel J. Tilden* (New York, 1895), 2:74–79; Nevins, ed., *Hewitt Writings*, 155–56, 170–71; Samuel J. Randall to Manton Marble, January 15, 1877, Marble Papers; *CR*, 44th Congress, 2d Session, 913, 1050; James A. Garfield to Rutherford B. Hayes, January 19, 1877, Hayes Papers; Allan Peskin, *Garfield* (Kent, Ohio, 1978), 415–16; Frank B. Evans, *Pennsylvania Politics, 1872–1877: A Study in Political Leadership* (Harrisburg, 1966), 295–96; Ellwood E. Thorpe to John Sherman, January 24, 1877, John Sherman Papers, LC; Henry V. Boynton to James M. Comly, January 25, 1877, Hayes Papers; William T. Sherman to Philip H. Sheridan, January 29, 1877, William T. Sherman Papers, LC.

power to select the fifth, who everyone assumed would be Justice David Davis. Tilden's supporters expressed confidence that Davis, a Republican who had spent the past few years "intriguing in politics with the Democrats and Liberals," would decide at least one state in their favor, although unbeknownst to them the justice privately believed Hayes legally elected. But even as Democrats congratulated themselves on securing for Davis "the power to make a President," their compatriots in Illinois seized an unexpected opportunity to deny John A. Logan reelection to the Senate, by joining with a group of Greenback legislators to name Davis in his place. William H. Smith considered this unexpected turn of events a scheme to obligate Davis in favor of Tilden; others thought Illinois Republicans had outmaneuvered their "slow-witted" opponents by arranging Davis' victory behind the scenes, thereby disqualifying him from the commission. In any case, Davis resigned from the commission, Republican Justice Joseph P. Bradley took his place, and by a series of 8–7 votes, the disputed electors were awarded to Hayes. "We have been cheated, shamefully cheated," protested one outraged Democrat.[29]

Further turmoil, however, lay ahead before Hayes could enter the White House. Tilden's supporters threatened, through interminable motions to adjourn and other dilatory measures, to paralyze deliberations in the House of Representatives and obstruct a final count of the electoral vote, thus preventing an inauguration on March 4. "Very anxious about the situation," Hayes's Washington representatives initiated a new round of negotiations. A caucus of "conservative" Southerners proposed that Hayes agree to appoint to the Cabinet Tennessee Sen. David M. Key, a close associate of Andrew J. Kellar, who shared his interest in the South's economic modernization and "a reorganization of parties." There was also talk of Southern Democrats aiding the Republicans in organizing the closely divided House of Representatives in the next Congress, and Scott's railroad lobbyists resumed their effort to make federal aid to the Texas & Pacific part of any deal. On February 26, four Southern Democrats met with five Ohio Republicans at Washington's Wormley House (a hotel owned by James Wormley, the city's wealthiest black resident). Hayes's confidant Stanley Matthews announced that the new President intended to recognize Nicholls as Louisiana's governor and pursue a policy of noninterference in Southern affairs, while Nicholls' emissary, Col. Edward A. Burke, pledged to avoid reprisals against the state's Republicans and recognize the civil and political equality of blacks. Similar discussions with Hampton's representatives soon followed.[30]

29. William H. Smith to John Sherman, January 24, 1877, John Sherman Papers; *CR*, 44th Congress, 2d Session, 820; Willard L. King, *Lincoln's Manager: David Davis* (Cambridge, Mass., 1960), 290–93; William H. Smith to Rutherford B. Hayes, January 24, 1877, Hayes Papers; Flick, *Tilden*, 380–84; W. W. H. Davis to Samuel J. Randall, February 12, 1877, Samuel J. Randall Papers, University of Pennsylvania.

30. E. B. Wight to William W. Clapp, February 14, 1877, Clapp Papers; William H. Smith to Rutherford B. Hayes, February 17, 1877, Hayes Papers; Abshire, *Key*, 67–85, 145–51;

To the end, "home rule" remained the central issue in the bargaining, outweighing internal improvements and other points of discussion. "It matters little to us who rules in Washington," commented an Abbeville newspaper at the end of February, "if South Carolina is allowed to have Hampton and Home Rule." It is significant that Northern Democrats played as large a role as Southern in ending the House filibuster, for few had any interest in Southern railroad aid, but all were fiercely committed to ending Reconstruction. No one played a more critical part in resolving the crisis than Speaker Randall. Having earlier delivered a "violent speech" to the Democratic caucus warning of "bayonet rule" in the South, he suddenly, on March 1, began ruling dilatory motions out of order. The discussions at Wormley House and elsewhere, coupled with letters from Pennsylvania businessmen demanding an end to the stalemate, had persuaded Randall to help clear the way for Hayes to assume office.[31]

In a sense, as journalist Henry Watterson later wrote, "both sides were playing something of a 'bluff'," and the negotiations produced results inescapable by February 1877: Hayes's inauguration and an end to Reconstruction. Once the Electoral Commission made its decision, Democrats could only obstruct the count, not place Tilden in the White House. To secure a tranquil inauguration, and in the hope of strengthening the new Administration among white Southerners, Republicans agreed to what, in any event, was a foregone conclusion—the recognition of Nicholls and Hampton. By this time, everyone understood that Hayes would adopt a new Southern policy. "As matters look to me now," wrote the chairman of Kansas' Republican state committee on February 22, "I think the policy of the new administration will be to conciliate the white men of the South. Carpetbaggers to the rear, and niggers take care of yourselves."[32]

Garfield, who left the meeting early, believed "a compact of some kind was mediated" at Wormley House, but the terms of the "Bargain of 1877" remain impossible to determine with any precision. Certainly, the filibuster ended, Hayes became President, and Key entered the Cabinet as Postmaster General. Equally significant were the appointments of William M. Evarts, Johnson's counsel at the impeachment trial, as Secretary of State, and Carl Schurz as Secretary of the Interior, further indications that Hayes planned to put Reconstruction behind him and identify his

Woodward, *Reunion and Reaction*, 188–96; Polakoff, *Politics of Inertia*, 298–312; *New National Era*, May 11, 1871; Edward A. Burke to Francis T. Nicholls, February 27, 1877 (copy), Chandler Papers.

31. George C. Rable, "Southern Interests and the Election of 1876: A Reappraisal," *CWH*, 26 (December 1980), 346–53; Abbeville (S.C.) *Press and Banner*, February 28, 1877; Michael L. Benedict, "Southern Democrats in the Crisis of 1876–1877: A Reconsideration of *Reunion and Reaction*,"*JSH*, 46 (November 1980), 497, 512–16; Charles Foster to Rutherford B. Hayes, February 21, 1877, Hayes Papers; Evans, *Pennsylvania Politics*, 304–307.

32. Mary R. Dearing, *Veterans in Politics: The Story of the G. A. R.* (Baton Rouge, 1952), 242; Polakoff, *Politics of Inertia*, 292–97; John A. Martin to "Dear Senator," February 22, 1877, Hayes Papers.

administration with the party's "reform" wing. The Texas & Pacific never did get federal assistance, nor did a single Southern Democrat support Garfield's bid the following autumn to become Speaker of the House. No flood of former Whigs entered the Republican party, and peaceful, honest elections did not soon return to the South. But "home rule" quickly came to Louisiana and South Carolina. Within two months of taking office, Hayes ordered federal troops surrounding the South Carolina and Louisiana statehouses, where Chamberlain and Packard still claimed the office of governor, to return to their barracks. (Hayes did not, as legend has it, remove the last federal troops from the South, but his action implicitly meant that the few remaining soldiers would no longer play a role in political affairs.) Hampton and Nicholls peacefully assumed office, marking the final triumph of "Redemption." "The whole South—every state in the South," lamented black Louisianan Henry Adams, "had got into the hands of the very men that held us as slaves."[33]

"To think that Hayes could go back on us," commented a South Carolina freedman, "when we had to wade through blood to help place him where he is now." But rather than an abrupt change in Northern policy, Hayes's actions, as the New York *Herald* pointed out, only confirmed in two states "what in the course of years has been done by his predecessor or by Congress" elsewhere in the South. Indeed, the abandonment of Reconstruction was as much a cause of the crisis of 1876–77 as a consequence, for had Republicans still been willing to intervene in defense of black rights, Tilden would never have come close to carrying the entire South. Nonetheless, the "withdrawal" of troops marked a major turning point in national policy. "The long controversy over the black man," announced the Chicago *Tribune*, "seems to have reached a finality." "The negro," echoed *The Nation*, "will disappear from the field of national politics. Henceforth, the nation, as a nation, will have nothing more to do with him."[34]

Among other things, 1877 marked a decisive retreat from the idea, born during the Civil War, of a powerful national state protecting the fundamental rights of American citizens. Yet the federal government was

33. Henry J. Brown and Frederick D. Williams, eds., *The Diary of James A. Garfield* (East Lansing, Mich., 1967–73), 3:449–50; Barnard, *Hayes*, 414–17; Allan Peskin, "Was There a Compromise of 1877?" *JAH*, 60 (June 1973), 63–75; C. Vann Woodward, "Yes, There Was a Compromise of 1877," *JAH*, 60 (June 1973), 215–23; Vincent P. DeSantis, "Rutherford B. Hayes and the Removal of the Troops and the End of Reconstruction," in J. Morgan Kousser and James M. McPherson, eds., *Region, Race, and Reconstruction: Essays in Honor of C. Vann Woodward*, (New York, 1982), 417–50; Clarence C. Clendenen, "President Hayes' 'Withdrawal' of the Troops—an Enduring Myth," *SCHM*, 70 (October 1969), 240–50; 46th Congress, 2d Session, Senate Report 693, pt. 2:108.

34. William T. Rodenbach to Daniel H. Chamberlain, April 8, 1877, South Carolina Governor's Papers; William Gillette, *Retreat From Reconstruction 1869–1879* (Baton Rouge, 1979), 333, 345–47; Vincent P. DeSantis, *Republicans Face the Southern Question* (Baltimore, 1959), 52; *Nation*, April 5, 1877.

not rendered impotent in all matters—only those concerning blacks. Hayes did not hesitate to employ the national state's coercive powers for other purposes. Even as the last Reconstruction governments toppled, troops commanded by former Freedmen's Bureau Commissioner O. O. Howard relentlessly pursued the Nez Percé Indians across the Far West to enforce a federal order removing them from Oregon's Wallowa Valley. After a 1,700-mile retreat, in which their courage and tactical skill in outmaneuvering the army won an embarrassed admiration from the nation, the Nez Percé were forced to surrender, although for two decades their leader Chief Joseph would importune successive Presidents for the right to return to their beloved Oregon homeland. Among those serving under Howard were soldiers transferred in midsummer from the South, where their duties had come to an end.[35]

Nor did the federal government prove reluctant to intervene with force to protect the rights of property. Within three months of the end of Reconstruction, the Hayes administration confronted one of the bitterest explosions of class warfare in American history—the Great Strike of 1877. Beginning on July 16, when workers on the Baltimore & Ohio Railroad walked off their jobs at Martinsburg, West Virginia, to protest the second wage cut in less than a year, the strike spread westward along the great trunk lines, affecting every region of the country except New England and the Deep South, and expanding to include workmen in other industries. In Pittsburgh, traffic was halted on the Pennsylvania Railroad, and miners and steel workers organized sympathy strikes. Militiamen were brought in from Philadelphia (because local units refused to act against the strikers), and when they fired on crowds that had seized the city's railroad switches and killed twenty people, outraged citizens set fire to the Pittsburgh railroad yards. Flames engulfed over 100 locomotives and 2,000 railroad cars, a substantial portion of the Pennsylvania's rolling stock. General strikes paralyzed Chicago and St. Louis, uniting skilled and unskilled workers in demands for an eight-hour day, a return to predepression wage levels, an end to child labor, the nationalization of the railroads, and the repeal of "tramp ordinances" allowing the arrest of unemployed workers. In Pennsylvania's anthracite coal region, where "wages had been cut down until once comfortably-nourished families began to languish in misery," some 40,000 workers left their jobs.[36]

"The most extensive and deplorable workingmen's strike which ever took place in this, or indeed in any other country," as The Nation described it, the labor upheaval suggested how profoundly American class

35. Utley, Frontier Regulars, 309–15; AC, 1877, 39–40; Clendenen, "Hayes' 'Withdrawal'," 247–48.
36. Philip S. Foner, The Great Labor Uprising of 1877 (New York, 1977); Robert V. Bruce, 1877: Year of Violence (Indianapolis, 1959); Harper's Weekly, August 11, 1877; David Roediger, "America's First General Strike: The St. Louis Commune of 1877," Midwest Quarterly, 21 (Winter 1980), 196–206; AC, 1877, 423–31.

relations had been reshaped during the Civil War and Reconstruction. The strike exposed the deep hostility to railroads—symbols and creators of the new industrial order—that permeated many American communities. "Public opinion," reported the New York *Tribune*, "is almost everywhere in sympathy with the insurrection." Although in San Francisco the strike degenerated into anti-Chinese rioting, elsewhere it achieved a remarkable spirit of solidarity that transcended divisions of skill, ethnicity, and race. In St. Louis, Louisville, and other cities, rallies and marches brought together "white and colored men, . . . men of all nationalities in one supreme contest for the common rights of workingmen." At the same time, the Great Strike revealed the political power and collective consciousness of the urban middle and upper classes, which joined with municipal authorities and veterans' organizations to form "citizen militias" that did battle with strikers. St. Louis' Committee of Public Safety organized a huge private army with two Civil War generals (one from each side) at its head, and effectively suppressed the city's general strike. Louisville's city hall was "converted virtually into an arsenal," from which "influential and wealthy citizens" received arms to augment the local police. Benjamin Harrison, the unsuccessful gubernatorial candidate of Indiana Republicans in 1876 and a future President, enrolled in the militia "to enable me to arm a company in aid of the suppression of the Great Strike." Harrison marched around Indianapolis in his Civil War uniform, bearing, according to one newspaper, "a striking resemblance to Napoleon."[37]

The response to the labor upheaval underscored how closely the Civil War era had tied the new industrial bourgeoisie to the Republican party and national state. Where local authorities and middle-class citizens proved unable to restore order, federal troops stepped into the breach. Hayes had filled his Cabinet with party leaders who were also corporate attorneys and railroad directors. Thomas A. Scott, the Pennsylvania's president, had direct access to the White House during the strike. As requests for troops descended upon the Administration from frightened governors and beleaguered railroad executives, Hayes neither investigated the need for troops nor set clear guidelines for their use. Thus, when soldiers were sent to cities from Buffalo to St. Louis, they acted less as impartial defenders of order than as strikebreakers, opening railroad lines, protecting nonstriking workers, and preventing union meetings.

37. *Nation*, July 26, 1877; *AC*, 1877, 424–27; Foner, *1877*, 37–39; Alexander Saxton, *The Indispensable Enemy: Labor and the Anti-Chinese Movement in California* (Berkeley, 1971), 114; David Roediger, " 'Not Only the Ruling Class to Overcome, But Also the So-Called Mob': Class, Skill, and Community in the St. Louis General Strike of 1877," *JSocH*, 19 (Winter 1985), 212–17, 221–26; Bill L. Weaver, "Louisville's Labor Disturbance, July, 1877," *Filson Club Historical Quarterly*, 48 (April 1974), 179–83. Dearing, *Veterans in Politics*, 217–18; Commission, July 27, 1877, Harrison Papers; Harry J. Sievers, *Benjamin Harrison: Hoosier Statesman* (New York, 1959), 136–39.

With the army suddenly overextended, units were hastily transferred from the South, including some who, until recently, had guarded the Louisiana statehouse. By July 29, the Great Strike had come to an end, although Pennsylvania miners held out until October, their communities, as during the Civil War, suffering the indignity of prolonged occupation by federal soldiers. "The strikers," wrote the President in his diary, "have been put down by *force.*"[38]

Thus, if the era of Reconstruction opened in 1863 with the national promise of freedom to blacks, quickly followed by an explosion of class and racial antagonisms in the streets of New York City, the restoration of white supremacy in the South coincided with an even more powerful reminder of the conflicts that divided Northern society. For labor and capital alike, the Great Strike had immense long-term consequences. Solidifying class consciousness among both, it ushered in two decades of labor conflict the most violent the country had ever known. For labor, the unprecedented cooperation between skilled and unskilled, black and white, foreshadowed the spectacular rise of the Knights of Labor in the 1880s. Among the respectable middle class, the upheaval reinforced elitist biases already evident in the rise of Liberalism and the retreat from Reconstruction. The strike demonstrated, declared the New York *Tribune*, that the radical spirit of "communism" was abroad in the land and that only the "substantial, property-owning" classes could save "civilized society." For such observers, the strike threw into question one of the most deeply rooted articles of American faith—the dream of exceptionalism, the belief that the nation could have capitalism without class conflict, industrialization without the "dark satanic mills" of Europe. "The days are over," intoned the *New York Times*, "in which this country could rejoice in its freedom from the elements of social strife which have long abounded in the old countries. . . . We cannot too soon face the unwelcome fact that we have dangerous social elements to contend with, and that they are rendered all the more dangerous by the peculiarities of our political system."[39]

Among the upheaval's casualties was middle-class confidence in local forces of law and order, since militias had proved unwilling or unable to suppress the uprising. "The 'citizen soldier' theory" had outlived its usefulness, wrote Charles Eliot Norton, Harvard professor and former

38. Philip H. Burch, Jr., *Elites in American History* (New York, 1981), 2:75–77; George F. Howe, ed., "President Hayes's Notes of Four Cabinet Meetings," *AHR*, 37 (January 1932), 286–89; Foner, *1877*, 74–76, 193; Jerry M. Cooper, *The Army and Civil Disorder* (Westport, Conn., 1980), 45–48, 64, 75–79; Clendenen, "Hayes' 'Withdrawal'," 248–50; Williams, ed., *Diary and Letters of Hayes*, 3:440.

39. David Montgomery, "Strikes in Nineteenth-Century America," *Social Science History*, 4 (Winter 1980), 95–97; Richard Schneirov, "Chicago's Great Upheaval of 1877," *Chicago History*, 9 (Spring 1980), 13–17; Gabriel Kolko, *Railroads and Regulation 1877–1916* (Princeton, 1965), 12–14; New York *Tribune*, July 25, 1877; *New York Times*, July 25, 1877.

editor of the *North American Review,* and state militias must be "essentially remodeled" so as to provide an "efficient force for the protection of life and property and the maintenance of order." In the aftermath of 1877, cities retrained and expanded their police forces, while the militia and National Guard were professionalized and equipped with more modern weapons. In the next quarter century, the Guard would be used in industrial disputes over 100 times. Meanwhile, the federal government constructed armories not in the South to protect black citizens, but in the major cities of the North, to ensure that troops would be on hand in subsequent labor difficulties. Thus, the upheaval marked a fundamental shift in the nation's political agenda. "The overwhelming labor question has dwarfed all other questions into nothing. . . ." wrote an Ohio Republican. "We have *home* questions enough to occupy attention now." "The Southern question," a Charleston newspaper agreed, was "dead"—the railroad strike had propelled to the forefront of politics "the question of labor and capital, work and wages."[40]

Enjoying respite in Europe from the cares of office, former President Grant found the events of 1877 "a little queer." During his Administration, he wrote, the entire Democratic party and the "morbidly honest and 'reformatory' portion of the Republican" had thought it "horrible" to employ federal troops "to protect the lives of negroes. Now, however, there is no hesitation about exhausting the whole power of the government to suppress a strike on the slightest intimation that danger threatens." Grant was not the only contemporary to note the ironic juxtaposition of home rule for the South and armed intervention in the North. "I wish Sheridan was at Pittsburgh," a neighbor told William Lloyd Garrison II. "Indeed," Garrison replied, "but remember how you denounced him at New Orleans."[41]

All in all, 1877 confirmed the growing conservatism of the Republican party and portended a new role for the national state in the post-Reconstruction years. The federal courts, for example, retained the greatly expanded jurisdiction born of Reconstruction; they increasingly employed it, however, to protect corporations from local regulation. To be sure, neither the humanitarian impulse that had helped create the Republican party nor the commitment to equal citizenship that evolved during the war and Reconstruction, entirely disappeared. Southern issues, however, played a steadily diminishing part in Northern Republican politics and support for the idea of federal intervention to enforce the Fourteenth

40. Sara Norton and Mark A. De Wolfe Howe, eds., *Letters of Charles Eliot Norton* (Boston, 1913), 2:69; Sidney L. Harring, *Policing A Class Society: The Experience of American Cities, 1865–1915* (New Brunswick, N.J., 1983), 27, 108–10; Cooper, *Army and Civil Disorder,* 9–13, 44; J. M. Dalzell to John Sherman, July 29, 1877, John Sherman Papers; Charleston *Daily Courier,* July 28, 1877.

41. David Ammen, *The Old Navy and the New* (Philadelphia, 1891), 537; William Lloyd Garrison II to William Lloyd Garrison, July 24, 1877, Garrison Family Papers, SC.

and Fifteenth Amendments continued to wane. Admitting that blacks confronted intolerable inequities in every aspect of their lives, a Philadelphia newspaper in 1882 nonetheless concluded: "The time has passed when the federal government can interfere for the protection of these people from this mean tyranny." The following year, the Supreme Court declared the Civil Rights Act of 1875 unconstitutional. Joseph P. Bradley, whose vote on the Electoral Commission had made Hayes President, wrote the majority opinion, which observed that blacks must cease "to be the special favorite of the laws." The only dissenter was Kentucky's John Marshall Harlan. The United States, he warned, had entered "an era of constitutional law, when the rights of freedom and American citizenship cannot receive from the nation that efficient protection which heretofore was unhesitatingly accorded to slavery." But Harlan's was a lonely voice. The general approval that greeted the decision, observed *The Nation*, revealed "how completely the extravagant expectations" aroused by the Civil War had "died out."[42]

The Redeemers' New South

As a period when Republicans controlled Southern politics, blacks enjoyed extensive political power, and the federal government accepted responsibility for protecting the fundamental rights of black citizens, Reconstruction came to an irrevocable end with the inauguration of Hayes. Of course, the coming of "home rule" did not suddenly arrest the process of change or resolve the social conflicts unleashed by the Civil War. But after 1877 these took place in a new context, in which the South's rulers enjoyed a free hand in managing the region's domestic affairs. "Left to ourselves," John C. Calhoun, an Arkansas planter named for his celebrated grandfather, told a Senate committee in 1883, the white South could settle "all questions" involving the black population. Constrained only by the increasingly remote possibility of federal intervention, the survival of enclaves of Republican political power, and fear of provoking divisions within the now dominant Democracy, the Redeemers moved in the final decades of the nineteenth century to put in place new systems of political, class, and race relations. A new social order did not come into being immediately, nor could the achievements of Reconstruction be entirely undone. But the harsh realities of Redeemer rule

42. Harry N. Scheiber, "Federalism, the Southern Regional Economy, and Public Policy Since 1865," in David J. Bodenhamer and James W. Ely, Jr., eds., *Ambivalent Legacy: A Legal History of the South* (Jackson, Miss., 1984), 75–77; Tony A. Freyer, "The Federal Courts, Localism, and the National Economy, 1865–1900," *Business History Review*, 53 (Autumn 1979), 343–63; Philadelphia *Evening Bulletin*, January 11, 1882; John A. Scott, "Justice Bradley's Evolving Concept of the Fourteenth Amendment from the Slaughterhouse Case to the Civil Rights Cases," *Rutgers Law Review*, 25 (Summer 1971), 562–68; Milton R. Konvitz, *A Century of Civil Rights* (New York, 1961), 118–19; *Nation*, October 18, 1883.

confirmed what a North Carolina Democrat had predicted as Reconstruction began: "When the bayonets shall depart . . . then look out for the reaction. Then the bottom rail will descend from the top of the fence."[43]

No single generalization can fully describe the social origins or political purposes of the South's Redeemers, whose ranks included secessionist Democrats and Union Whigs, veterans of the Confederacy and rising younger leaders, traditional planters and advocates of a modernized New South. They shared, however, a commitment to dismantling the Reconstruction state, reducing the political power of blacks, and reshaping the South's legal system in the interests of labor control and racial subordination. In a majority of Southern states, they moved, upon assuming office, to replace Reconstruction constitutions with new documents severely restricting the scope and expense of government. "Instruments of prohibition," as one newspaper described them, Redeemer constitutions reduced the salaries of state officials, limited the length of legislative sessions, slashed state and local property taxes, curtailed the government's authority to incur financial obligations (in Georgia and Louisiana, it could borrow money only to repel an invasion or suppress an insurrection), and repudiated, wholly or in part, Reconstruction state debts. Public aid to railroads and other corporations was prohibited, and several states abolished their central boards of education.[44]

Judged in terms of election pledges to reduce the cost of government and the burden of property taxes, the Redeemers were a success. Mississippi Democrats slashed the state budget by over 50 percent in the ten years following 1875 and restored to their owners millions of acres forfeited for nonpayment of taxes. (Eliminating corruption proved more difficult. Louisiana's first Redeemer treasurer, Edward A. Burke of Wormley House fame, fled to Honduras with $1 million in state funds.) But Southerners did not benefit equally from the reduction in taxes and expenditures. As land levies declined, licenses and poll taxes rose. Tenants received no benefit from the fall in taxation of landed property, and yeomen, although paying less, saw Reconstruction laws excluding a fixed amount of property from taxes replaced by exemptions only for specific items, such as machinery and implements utilized on a plantation. Laborers, tenants, and small farmers paid taxes on virtually everything they owned—tools, mules, even furniture—while many planters had thou-

43. *Report of the Committee of the Senate upon the Relations Between Labor and Capital, and Testimony Taken by the Committee* (Washington, D.C., 1885), 2:160, 169; Raleigh *Sentinel*, July 4, 1868.

44. C. Vann Woodward, *Origins of the New South, 1877–1913* (Baton Rouge, 1951), 1–5, 20, 65, 86–92; Judson C. Ward, Jr., "The New Departure Democrats in Georgia: An Interpretation," *GaHQ*, 41 (September 1957), 228–29; James T. Moore, "Redeemers Reconsidered: Change and Continuity in the Democratic South, 1870–1900," *JSH*, 44 (August 1978), 357–64; Michael Perman, *The Road to Redemption: Southern Politics, 1869–1879* (Chapel Hill, 1984), 172–210; Malcolm C. McMillan, *Constitutional Development in Alabama: 1798–1901: A Study in Politics, the Negro, and Sectionalism* (Chapel Hill, 1955), 202–206, 233–34.

sands of dollars in property excluded. "The farmer's hoe and plow, and the mechanic's saw and plane," a Georgia Republican newspaper lamented, "must be taxed to support the Government. . . . Show me the rich man who handles a hoe or pushes a plane." Thus, the tax system became increasingly regressive, as those with the least property bore the heaviest proportional burden. Moreover, although homestead exemptions remained on the books, new laws allowed their voluntary waiver, and "shark storekeepers" often refused to advance supplies until they had been set aside, threatening tenants and small farmers with the loss of their personal and landed property in the event of a poor crop.[45]

Fiscal retrenchment went hand in hand with a retreat from the idea of an activist state meeting broad social responsibilities. "Spend nothing unless absolutely necessary," Gov. George F. Drew advised the Florida legislature in 1877, and lawmakers took his advice to heart, abolishing the penitentiary, thus saving $25,000, and abandoning a nearly completed Agricultural College, leaving the state without any institution of higher learning, public or private. Alabama's Redeemers closed public hospitals at Montgomery and Talladega and Louisiana's were "so economical that . . . state services to the people almost disappeared." Similar reductions affected provisions for the insane and blind as well as appropriations for Southern paupers, despite the lingering effects of the economic depression. South Carolina Democrats tightened collections from blacks owing mortgages to the state land commission, producing a "pell-mell rout of Negro settlers." Public education—described as a "luxury" by one Redeemer governor—was especially hard hit, as some states all but dismantled the education systems established during Reconstruction. Texas began charging fees in its schools, while Mississippi and Alabama abolished statewide school taxes, placing the entire burden of funding on local communities. Louisiana spent so little on education that it became the only state in the Union in which the percentage of native whites unable to read or write actually rose between 1880 and 1900. School enrollment in Arkansas did not regain Reconstruction levels until the 1890s. Blacks suffered the most from educational retrenchment, for the gap between expenditures for black and white pupils steadily widened.[46]

45. Richard A. McLemore, ed., *A History of Mississippi* (Hattiesburg, 1973), 1:601; Charles Vincent, "Aspects of the Family and Public Life of Antoine Dubuclet: Louisiana's Black State Treasurer 1868–1878," *JNH*, 66 (Spring 1981), 32–33; Allen J. Going, *Bourbon Democracy in Alabama 1874–1890* (University, Ala., 1951), 80–82, 97–98; 44th Congress, 2d Session, Senate Report 704, 137; *The Weekly Sun*, August 6, 1874, clipping, scrapbook, John E. Bryant Papers, DU; Crandall A. Shifflett, *Patronage and Poverty in the Tobacco South: Louisa County, Virginia, 1860–1900* (Knoxville, 1982), 75–83; Wayne K. Durrill, "Producing Poverty: Local Government and Economic Development in a New South County, 1874–1884," *JAH*, 71 (March 1985), 768–69; *Report of Committee on Labor and Capital*, 4:69.

46. Woodward, *Origins of the New South*, 60–65; Edward C. Williamson, "George F. Drew, Florida's Redemption Governor," *FlHQ*, 38 (January 1960), 207–208; *Alabama Acts 1874*, 155–56; Joe G. Taylor, *Louisiana Reconstructed 1863–1877* (Baton Rouge, 1974), 508; How-

Simultaneously, the Redeemers moved to solidify their hold on state and local government and curtail Republicans' remaining political power. The quality of political life after 1877 varied considerably from state to state. In the border and Upper South, white Republican voting persisted and the advent of "home rule" had little immediate effect on blacks' ability to exercise the franchise, with the result that the party remained competitive into the 1890s. In Arkansas and Texas, cotton states with large white majorities, blacks also voted freely after Redemption. But in the Deep South, where electoral fraud was widespread and the threat of violence hung most heavily over the black community, the Republican party crumbled after 1877. Here, long before their outright disenfranchisement around the turn of the century, blacks saw their political rights progressively eroded.[47]

The Fourteenth and Fifteenth Amendments, a Southern newspaper had declared in 1875, "may stand forever; but we intend . . . to make them dead letters on the statute-book." Even though Nicholls and Hampton appointed a number of blacks to minor local positions, Redeemer promises to respect blacks' constitutional rights proved, as one former slave noted, "like pie-crust, easily broken." With Democrats in control of the electoral machinery established during Reconstruction, ballot fraud became the order of the day in counties with black majorities. "After the polls are closed the election really begins," complained one Louisiana Republican. Throughout the South, moreover, districts were gerrymandered to reduce Republican voting strength. Mississippi Redeemers concentrated the bulk of the black population in a "shoestring" Congressional district running the length of the Mississippi River, leaving five others with white majorities. Alabama parceled out portions of its black belt into six separate districts to dilute the black vote. Cities from Richmond to Montgomery redrew ward lines to ensure Democratic control. Wilmington's black wards, containing four fifths of the city's population, elected only one third of its aldermen. Georgia severely restricted black voting by a cumulative poll tax requirement, a measure adopted at the behest

ard N. Rabinowitz, *Race Relations in the Urban South 1865–1890* (New York, 1978), 134, 170; Carol R. Bleser, *The Promised Land: The History of the South Carolina Land Commission, 1869–1890* (Columbia, S.C., 1969), 126–33; Nell I. Painter, *Exodusters* (New York, 1976), 46–52; Hair, *Louisiana*, 120–23; Powell Clayton, *Aftermath of the Civil War in Arkansas* (New York, 1915), 232.

47. Gordon B. McKinney, *Southern Mountain Republicans 1865–1900* (Chapel Hill, 1978), 9–10, 63–76; Richard O. Curry, ed., *Radicalism, Racism, and Party Realignment: The Border States During Reconstruction* (Baltimore, 1969), xxv; John W. Graves, "Town and Country: Race Relations and Urban Development in Arkansas 1865–1905" (unpub. diss., University of Virginia, 1978), 109; Lawrence D. Rice, *The Negro in Texas 1874–1900* (Baton Rouge, 1971), 120; George B. Tindall, *South Carolina Negroes 1877–1900* (Columbia, S.C., 1952), 31–36, 43–44; "My Experience in Aspiring to Be a Statesman," manuscript, Ephraim S. Stoddard Papers, Tulane University.

of Robert Toombs, who professed his willingness "to face thirty years of war to get rid of negro suffrage in the South."[48]

Meanwhile, Redeemer legislatures moved to take control of all-important local offices which, in plantation counties, could still be controlled by black voters. "It is absolutely necessary . . . " wrote one North Carolina Democrat, "that the county funds shall be placed beyond the reach of the large negro majorities." The state's Redeemers responded by transferring the power to select county commissioners and justices of the peace to the legislature, in effect restoring the oligarchic antebellum system of local government. Alabama adopted the same policy for selected black belt counties. Mississippi Democrats purged many counties of Republican officials by requiring officeholders to post new bonds and empowering the governor to replace those unable to raise the necessary funds. More violent methods of unseating Republican officials were also employed—a white mob lynched William H. Foote, Yazoo City's black tax collector, in 1883.[49]

Black officeholding did not entirely cease with Redemption. Small numbers of blacks continued to sit in Southern legislatures, and a few even won election to Congress. (The last Southern black Representative until the modern era was North Carolina's George H. White, who served from 1897 to 1901.) Blacks still held seats on city councils and minor posts in some plantation counties, and enclaves of genuine black power persisted, from the "black second" Congressional district of eastern North Carolina to South Carolina's low country and the Texas black belt. Many of these local officials represented a new breed of black political leadership. Professionals trained in law or at the new black colleges and members of the emerging black middle class, they possessed far more education and owned greater amounts of property than their Reconstruction predecessors. But the political context within which they operated had changed profoundly since the days of Republican rule. Local officials confronted hostile state governments and black lawmakers found it impossible to exert any influence in Democratic legislatures. Far more than during Reconstruction, their ability to provide services for their constitu-

48. *CR*, 43d Congress, 2d Session, 1929; 46th Congress, 2d Session, Senate Report 693, pt. 2:435–36; Tindall, *South Carolina Negroes*, 22–23; Hair, *Louisiana*, 21; J. Morgan Kousser, *The Shaping of Southern Politics* (New Haven, 1974), 153, 210–15; Sarah W. Wiggins, *The Scalawag in Alabama Politics, 1865–1881* (University, Ala., 1977), 116; Loren Schweninger, *James T. Rapier and Reconstruction* (Chicago, 1978), 151; Rabinowitz, *Race Relations*, 266–73; W. McKee Evans, *Ballots and Fence Rails: Reconstruction on the Lower Cape Fear* (Chapel Hill, 1967), 167–71; William Y. Thompson, *Robert Toombs of Georgia* (Baton Rouge, 1966), 230.

49. William Eaton to David S. Reid, September 20, 1875, David S. Reid Papers, NCDAH; Paul D. Escott, *Many Excellent People: Power and Privilege in North Carolina, 1850–1900* (Chapel Hill, 1985), 167–70; McMillan, *Constitutional Development in Alabama*, 218; G. C. Chandler to Adelbert Ames, February 27, 1876, Mississippi Governor's Papers, MDAH; Bettye J. Gardner, "William H. Foote and Yazoo County Politics 1866–1883," *SS*, 21 (Winter 1982), 404.

ents depended on connections with influential white Republicans or the goodwill of prominent local Democrats, rather than leadership of a politically mobilized black community.[50]

Until legal disenfranchisement, those blacks allowed to cast ballots valiantly strove to influence Southern public life, voting Republican where the party remained viable and demonstrating a willingness to ally with groups of whites who challenged Redeemer rule. Independent and Greenback parties, which flourished in parts of the Southern upcountry in the late 1870s and early 1880s, received substantial black support. On occasion, blacks shared in remarkable, albeit temporary victories over Redeemer rule. Virginia's Readjuster movement, which proposed to repudiate part of the state debt and thereby free funds for education and other social services, drew votes from Piedmont and mountain white farmers and blacks of the plantation counties. Led by the redoubtable railroad entrepreneur William Mahone, the Readjusters swept to power in 1879. Once in office, they poured funds into the public schools, abolished the poll tax, raised taxes on corporations while reducing levies on small farmers, and moved to reinforce blacks' civil and political rights. In a sense, Virginia, which had escaped Radical rule in the 1860s, experienced its Reconstruction during the four years of Readjuster control. A decade later a Populist-Republican alliance won control of North Carolina, bringing the state a "second Reconstruction" complete with increased funding for education, the return of control of county government to local voters, and a temporary revival of black officeholding.[51]

These victories, however, could not disguise the progressive narrowing of blacks' political and social options. In some realms, to be sure, change came slowly. Although separate schooling for black and white children, which existed de facto during Reconstruction, was quickly written into Southern constitutions and laws, and the Redeemers repealed or ignored Republican civil rights legislation, Southern race relations remained for

50. Frenise A. Logan, *The Negro in North Carolina 1876–1894* (Chapel Hill, 1964), 25–37; Edward N. Akin, "When a Minority Becomes a Majority: Blacks in Jacksonville Politics, 1887–1907," *FlHQ*, 53 (October 1974), 136–37; Michael B. Chesson, "Richmond's Black Councilmen, 1871–1896," in Howard N. Rabinowitz, ed., *Southern Black Leaders of the Reconstruction Era* (Urbana, Ill., 1982), 191–92; Tindall, *South Carolina Negroes*, 54–64; Rice, *Negro in Texas*, 86–91; Eugene J. Watts, "Black Political Progress in Atlanta: 1868–1895," *JNH*, 59 (July 1974), 286; Joseph H. Cartwright, "Black Legislators in Tennessee in the 1880's: A Case Study in Black Political Leadership," *THQ*, 32 (Fall 1973), 265–66; Carl H. Moneyhon, "Black Politics in Arkansas During the Gilded Age, 1876–1900," *ArkHQ*, 44 (Autumn 1985), 222–33.

51. Kousser, *Shaping*, 11–17, 26–28; Michael Hyman, "Response to Redeemer Rule: Hill Country Dissent in the Post-Reconstruction South" (unpub. diss., City University of New York, 1986); Akin, "When a Minority," 126–27; Lawrence C. Goodwyn, "Populist Dreams and Negro Rights: East Texas as a Case Study," *AHR*, 76 (December 1971), 1435–56; James T. Moore, "Black Militancy in Readjuster Virginia, 1879–1883," *JSH*, 51 (May 1975), 167–86; Carl N. Degler, *The Other South: Southern Dissenters in the Nineteenth Century* (New York, 1974), 270–304; Escott, *Many Excellent People*, 247–53.

a time flexible and somewhat indeterminate. From church services to sporting events, blacks and whites went their separate ways; "there is generally an entire dissociation between the two races," commented a white North Carolina lawyer in the early 1880s. Yet blacks could still gain admission to theaters, bars, and a few hotels, and obtain equal seating on some streetcars and railroads. A new etiquette of race relations slowly took shape under the Redeemers, as more reserved kinds of everyday behavior replaced the black assertiveness so evident during Reconstruction. But not until the 1890s did the system of racial segregation become embedded in Southern law.[52]

In other realms, however, change was immediate. Once Southern whites were allowed "free play" by the North, a South Carolina writer predicted early in 1877, they would "go as far as they dare in restricting colored liberty . . . without actually reestablishing personal servitude." And throughout the South, Democrats upon assuming office rewrote the statute books so as to reinforce planters' control over their labor force. Broad new vagrancy laws allowed the arrest of virtually any person without a job and "antienticement" laws made it a criminal offense to offer employment to an individual already under contract, or to leave a job before a contract had expired. Laws also prohibited the nighttime sale of seed (that is, unginned) cotton and other farm products. Justified as a means of combating theft, these "sunset" measures severely limited the former slaves' economic rights. "If a man commits a crime he ought to be punished," commented former black Congressman James T. Rapier, "but every man ought to have a right to dispose of his own property. . . . I may raise as much cotton as I please in the seed, but I am prohibited by law from selling it to anybody but the landlord." Although these laws harked back to the Black Codes of Presidential Reconstruction, because of the Fourteenth Amendment they ostensibly affected black and white equally. Many, however, only applied to counties with black majorities. At any rate, a Tennessee black convention noted, "a single instance of punishment of whites under these acts has never occurred, and is not expected."[53]

Meanwhile, Southern criminal laws increased sharply the penalty for petty theft. Since violence against blacks generally went unpunished

52. William P. Vaughan, Schools for All: The Blacks and Public Education in the South, 1865–1877 (Lexington, Ky., 1974), 70, 100; C. Vann Woodward, The Strange Career of Jim Crow, 3rd ed. (New York, 1974), 17–48; Howard N. Rabinowitz, "From Exclusion to Segregation: Southern Race Relations, 1865–1900," JAH, 63 (September 1976), 325–50; 46th Congress, 2d Session, Senate Report 693, pt. 1:404; Tindall, South Carolina Negroes, 292–94; Charles E. Wynes, Race Relations in Virginia 1870–1902 (Charlottesville, Va., 1961), 68–78.
53. [Townsend] "Political Condition of South Carolina," 192; Alrutheus A. Taylor, The Negro in Tennessee, 1865–1880 (Washington, D.C., 1941), 137–38; Perman, Road to Redemption, 243–44; 46th Congress, 2d Session, Senate Report 693, pt. 2:23, 466–67; Charles L. Flynn, Jr., White Land, Black Labor: Caste and Class in Late Nineteenth-Century Georgia (Baton Rouge, 1983), 85, 94–95; Nashville Union and American, May 20, 1875.

(partly because state laws designed to suppress the Ku Klux Klan had been repealed), the sole concern of law enforcement seemed to be to protect property owned by whites. South Carolina made arson a capital offense, mandated life imprisonment for burglary, and increased drastically the penalty for the theft of livestock. In North Carolina and Virginia, charged one black spokesman, "they send [a man] to the penitentiary if he steals a chicken." Mississippi's famous "pig law" defined the theft of any cattle or swine as grand larceny punishable by five years in jail. "It looks to me," commented a black resident of the state, "that the white people are putting in prison all that they can get their hand on." One result was to facilitate a vast expansion of the convict lease system. Within two months of Redemption, South Carolina's legislature authorized the hiring out of virtually every convict in the state, as did Florida after dismantling its penitentiary. Railroads, mining and lumber companies, and planters vied for access to this new form of involuntary labor, the vast majority of whom were blacks imprisoned for petty crimes. Republicans were not far wrong when they charged of this system, "the courts of law are employed to reenslave the colored race."[54]

New laws also redefined in the interest of the planter the terms of credit and the right to property—the essence of economic power in the rural South. Lien laws now gave a landlord's claim to his share of the crop precedence over a laborer's for wages or a merchant's for supplies, thus shifting much of the risk of farming from employer to employee. North Carolina's notorious Landlord and Tenant Act of 1877 placed the entire crop in the planter's hands until rent had been paid and allowed him full power to decide when a tenant's obligation had been fulfilled—thus making the landlord "the court, sheriff, and jury," complained one former slave. Beginning in 1872 with *Appling* v. *Odum* in Georgia, moreover, a series of court decisions defined the sharecropper not as a "partner" in agriculture or a renter with a property right in the growing crop, but as a wage laborer possessing "only a right to go on the land to plant, work, and gather the crop." At the same time, the process of enclosing the open range (thus preventing those without land from owning livestock) and restricting trespassing and hunting, suspended during Republican rule, now resumed. So as not to arouse hostility among the yeomanry, such laws initially applied only to black belt counties, but eventually they

54. *Alabama Acts 1875*, 84–85; Alrutheus A. Taylor, *The Negro in South Carolina During the Reconstruction* (Washington, D.C., 1924), 283–64; Perman, *Road to Redemption*, 242–43; 46th Congress, 2d Session, Senate Report 693, pt. 1:130, pt. 3:490–91; Wharton, *Negro in Mississippi*, 237; Gilbert Horton to Blanche K. Bruce, May 16, 1877, Bruce Papers; Tindall, *South Carolina Negroes*, 267; Mildred L. Fryman, "Career of a Carpetbagger: Malachai Martin in Florida," *FlHQ*, 56 (January 1978), 317, 333; Fletcher M. Green, "Some Aspects of the Convict Lease System in the Southern States," in Fletcher M. Green, ed., *Essays in Southern History* (Chapel Hill, 1949), 115–20; *Colored Men, Read! How Your Friends Are Treated!* broadside, July 1876, R. C. Martin Papers, LSU.

spread to the upcountry as well. Such uses of the state's powers, complained a memorial to Congress by Alabama's Republican legislators, were "utterly hostile to the spirit and letter of the laws of . . . free labor communities."[55]

Blacks, moreover, were all but excluded from the machinery of law enforcement. Few remained on local police forces or in state militias, whose budgets were exempted from the Redeemers' parsimony. Except in a few localities, blacks no longer served on Southern juries. (Blacks' names, commented a Georgia freedmen, seemed to be "nailed to the bottom" of the box from which lists of jurors were chosen.) And, if necessary, the Redeemer state stood ready to employ brutal force in the interest of labor control. Although urban workers were, on occasion, able to organize effectively after the end of Reconstruction, collective action by rural laborers became all but impossible. Time and again during the 1880s and 1890s, Southern sheriffs, backed by state militias, crushed efforts to organize agricultural workers. An 1887 strike for higher wages on Louisiana sugar plantations led to a massacre of over 100 blacks by the militia and groups of white vigilantes. Four years later, fifteen leaders of an Arkansas cotton pickers' strike were killed, including nine lynched after being arrested.[56]

None of this should suggest that Redeemer rule proved completely successful in controlling the black labor force. The law alone cannot compel men and women to work in a disciplined manner. In the post-Reconstruction South, complaints about a "labor shortage" continued, and blacks clung to whatever day-to-day autonomy they could wrest from the sharecropping system and to the ability to move from plantation to plantation (a right, needless to say, inconceivable before emancipation). But the balance of power between social classes in the South had been

55. Jonathan M. Wiener, *Social Origins of the New South: Alabama 1860–1885* (Baton Rouge, 1978), 98–103; Perman, *Road to Redemption*, 247–51; Joseph H. Taylor, "The Great Migration from North Carolina in 1879," *NCHR*, 31 (January 1954), 26–30; Harold D. Woodman, "Post-Civil War Southern Agriculture and the Law," *AgH*, 53 (January 1979), 319–37; Foner, *Nothing But Freedom*, 61, 66–67; Flynn, *White Land, Black Labor*, 117–21, 131–36; J. Crawford King, "The Closing of the Southern Range: An Exploratory Study," *JSH*, 48 (February 1982), 55–64; Steven Hahn, *The Roots of Southern Populism: Yeoman Farmers and the Transformation of the Georgia Upcountry, 1850–1890* (New York, 1983), 239–68; 43d Congress, 2d Session, Senate Miscellaneous Document 107, 1–4.

56. Philip J. Wood, *Southern Capitalism: The Political Economy of North Carolina, 1880–1980* (Durham, 1986), 25–26, 108–10; Harold D. Woodman, "Postbellum Social Change and Its Effect on Marketing the South's Cotton Crop," *AgH* (January, 1982), 215–30; Howard N. Rabinowitz, "The Conflict Between Blacks and the Police in the Urban South, 1865–1900," *Historian*, 39 (November 1976), 64–65; "List of Expenditures for Arms for Militia," October 9, 1873, Georgia Governor's Papers, UGa; *Report of Committee on Labor and Capital*, 4:616; Thomas W. Kremm and Diane Neal, "Challenges to Subordination: Organized Black Agricultural Protest in South Carolina, 1886–1895," *SAQ*, 77 (Winter 1978), 98–112; Philip S. Foner and Ronald L. Lewis, eds., *The Black Worker: A Documentary History from Colonial Times to the Present* (Philadelphia, 1978–84), 3:143–242, 367–404; William F. Holmes, "The Demise of the Colored Farmers' Alliance," *JSH*, 41 (May 1975), 187–200.

fundamentally transformed—a process already visible soon after Reconstruction ended. "This year," reported a New York business journal at the end of 1877, "labor is under control for the first season since the war."[57]

The policies of Redeemer governments not only helped to reshape Southern class relations, but affected the course of regional economic development in the last quarter of the nineteenth century. Partly because of Redeemer rule, the South emerged as a peculiar hybrid—an impoverished colonial economy integrated into the national capitalist marketplace yet with its own distinctive system of repressive labor relations. While the region's new upper class of planters, merchants, and industrialists prospered, the majority of Southerners of both races sank deeper and deeper into poverty.

For the South's yeomanry, the restoration of white supremacy brought few economic rewards. As cotton prices fell and world demand stagnated, cotton farming continued to drive upcountry yeomen into indebtedness and tenancy. By 1890, fewer than half the farms in Upper Piedmont Georgia were cultivated by their owners. The economic decline of some yeoman families and the outright dispossession of others created a growing labor force of men, women, and children for the rapidly expanding cotton textile industry, whose mills now dotted upcountry rivers and streams. But since the industry's main attractions for prospective investors were the South's low wage scale and the availability of convict labor in the event of strikes, its growth did little to halt the cycle of poverty that engulfed upcountry families. Indeed in the new mill villages the Redeemers "solved" the problem that would bedevil the rulers of other societies (like twentieth-century South Africa) with labor markets strictly segmented along racial lines: how to establish low wages for whites as well as blacks.[58]

At first glance, the textile industry's expansion appeared to herald the long-awaited advent of a modernized New South, especially when coupled with the influx of Northern investment attracted to the region by the end of the depression and the restoration of "home rule." Most of this development took place in the Upper South, where the Redeemers were most closely tied to the new entrepreneurial elite, and in the Piedmont counties of the cotton states, where a white industrial work force had come into being. Yet instead of ushering in an era of progress and prosperity, economic change under the Redeemers fastened the bonds of

57. Wharton, *Negro in Mississippi*, 95–96; Flynn, *White Land, Black Labor*, 9, 72–73, 110–11; Gavin Wright, *Old South, New South: Revolutions in the Southern Economy Since the Civil War* (New York, 1986), 64; New York *Commercial and Financial Chronicle*, in *AC*, 1877, 231.

58. Hahn, *Roots of Southern Populism*, 151–69; Grady McWhiney, "The Revolution in Nineteenth-Century Alabama Agriculture," *Alabama Review*, 31 (January 1978), 4; Wood, *Southern Capitalism*, 43; John W. Cell, *The Highest Stage of White Supremacy: The Origins of Segregation in South Africa and the American South* (New York, 1982), 127–30.

poverty and outside economic control ever more firmly upon the region. The enhancement of rural labor control eliminated any incentive for the mechanization of agriculture. With much Northern investment going into extractive enterprises (like mining and lumbering) that made no lasting contribution to regional development, and with the poverty of rural Southerners precluding the rise of a substantial home market, the economic revolution proved remarkably meager. Economic change took place in piecemeal fashion and within a colonial framework, and low taxes and the weakness of the Redeemer state allowed outside corporations to exploit Southern resources and carve up the environment without contributing to regional development. The Deep South stagnated entirely, its per capita income showing no increase at all between 1880 and 1900, and even the industrializing Piedmont remained overwhelmingly rural. As late as 1900, only 6 percent of the Southern labor force worked in manufacturing.[59]

One historian has described the New South as "a miserable landscape dotted only by a few rich enclaves that cast little or no light upon the poverty surrounding them." Poverty afflicted white Southerners as well as black, and tenancy, and the increasingly oppressive lien system, were regional rather than racial institutions. But blacks, confronting a unique combination of legal and extralegal coercions, were more vulnerable to fraud, and found it more difficult to obtain alternative employment. The Upper South did offer a modicum of opportunity: mines, iron furnaces, and tobacco factories employed black laborers, and increasing numbers of black farmers managed to acquire their own land, although mostly tiny plots of poor soil, incapable of sustaining a family. But in the Gulf States, planters were able to block industrial development altogether, while in the southeastern cotton states, the expanding textile work force remained entirely white, testimony to an implicit understanding between planters and industrialists that economic development must not threaten control of the rural black labor force. Blacks in the cotton South owned a smaller percentage of the land in 1900 than they had at the end of Reconstruction and possessed few options other than moving from plantation to plantation each January in search of improved conditions. Their fortunes varied from year to year, according to vicissitudes of the weather and cotton prices, but rare was the black farmer who could sustain upward mobility

59. Ronald D. Eller, *Miners, Millhands, and Mountaineers: Industrialization of the Appalachian South, 1880–1930* (Knoxville, 1982), 47–93; Gail W. O'Brien, *The Legal Fraternity and the Making of a New South Community, 1848–1882* (Athens, Ga., 1986), 144–46; Jack P. Maddex, Jr., *The Virginia Conservatives 1867–1879* (Chapel Hill, 1970); Rupert B. Vance, *Human Geography of the South* (Chapel Hill, 1932), 274–80; David R. Goldfield, "The Urban South: A Regional Framework," *AHR*, 86 (December 1981), 1029–33; Louis Ferleger, "Farm Mechanization in the Southern Sugar Sector After the Civil War," *LaH*, 23 (Winter 1982), 34; Wright, *Old South, New South*, 159–63; Roger L. Ransom and Richard Sutch, *One Kind of Freedom: The Economic Consequences of Emancipation* (New York, 1977), 8–9, 176; Paul M. Gaston, *The New South Creed* (New York, 1970), 203.

or escape dependence upon white landowners and merchants. Spurred by the growth of black colleges and the development of segregated neighborhoods in Southern cities, the black professional and entrepreneurial class slowly expanded. But most black urbanites remained trapped in personal service and unskilled labor.[60]

Thus, blacks in the Redeemers' New South found themselves enmeshed in a seamless web of oppression, whose interwoven economic, political, and social strands all reinforced one another. In illiteracy, malnutrition, inadequate housing, and a host of other burdens, blacks paid the highest price for the end of Reconstruction and the stagnation of the Southern economy. *"Poverty,"* commented one observer a few years after the Redemption of Arkansas, "is a word inadequate to describe the condition of these people. And it seems impossible for them to extricate themselves." With politics eliminated as an avenue to power, and displays of militancy likely to be met by overwhelming force, ambitious and talented men in the black community found other outlets—education, business, the church, and the professions. During Reconstruction, political involvement, economic self-help, and family and institution building had all formed parts of a coherent ideology of community advancement. With the end of Reconstruction, this program separated into its component parts. Severed from any larger political purpose, economic self-help, especially among the emerging black middle class, became an alternative to involvement in public life. "All my politics, gentlemen," the black owner of a Raleigh livery stable told a Congressional committee, "is that if a man has got 25 cents I will take him uptown on my omnibus." In general, black activity turned inward. Assuming a defensive cast, it concentrated on strengthening the black community rather than on directly challenging the new status quo.[61]

One index of the narrowed possibilities for change was the revival of interest, all but moribund during Reconstruction, in emigration to Africa or the West. The spate of black public meetings and letters to the American Colonization Society favoring emigration in the immediate aftermath

60. Robert L. Brandfon, *Cotton Kingdom of the New South* (Cambridge, Mass., 1967), viii; Allen W. Moger, *Virginia: Bourbonism to Byrd 1870–1925* (Charlottesville, Va., 1968), 79–83; Claude F. Oubre, *Forty Acres and a Mule: The Freedmen's Bureau and Black Landownership* (Baton Rouge, 1978), 178–80; Shifflett, *Patronage and Poverty*, 18–23; Stanley B. Greenberg, *Race and State in Capitalist Development: Comparative Perspectives* (New Haven, 1980), 215–27; Roger Ransom and Richard Sutch, "Sharecropping: Market Response or Mechanism of Race Control," in David E. Sansing, ed., *What Was Freedom's Price?* (Jackson, 1978), 67; W.E.B. Du Bois, *The Negro in Business* (Atlanta, 1899), 7–13, 19–24; Richard J. Hopkins, "Occupational and Geographic Mobility in Atlanta, 1870–1896," *JSH*, 34 (May 1968), 200–13.

61. A. L. Stanford to William Coppinger, November 1, 1877, American Colonization Society Papers, LC; Samuel D. Proctor, "Survival Techniques and the Black Middle Class," in Rhoda L. Goldstein, ed., *Black Life and Culture in the United States* (New York, 1971), 281–85; *Report of Committee on Labor and Capital*, 4:560, 612, 620, 789; 46th Congress, 2d Session, Senate Report 693, pt. 1:245; Elizabeth Bethel, *Promiseland: A Century of Life in a Negro Community* (Philadelphia, 1981), 144.

of Redemption reflected less an upsurge of nationalist consciousness than the collapse of hopes invested in Reconstruction and the arousal of deep fears for the future by the restoration of white supremacy. Henry Adams, the former soldier and Louisiana political organizer, claimed in 1877 to have enrolled the names of over 60,000 "hard laboring people" eager to leave the South. "This is a horrible part of the country," he wrote the Colonization Society from Shreveport, "and our race can not get money for our labor. . . . It is impossible for us to live with these slaveholders of the South and enjoy the right as they enjoy it."[62]

The most extensive organizing for Liberian emigration took place in South Carolina. Understandably, interest seemed most intense in what freedman Harrison N. Bouey, a young schoolteacher and former probate judge, called the "miserable county of Edgefield." Many of his neighbors seem to have shared Bouey's conviction that "the colored man has no home in America." Black lawyer John Mardenborough forwarded to the Colonization Society a list of 400 prospective emigrants ready to embark at a moment's notice. Ranging in age from early childhood to eighty and in occupation from laborer to artisan and teacher, Mardenborough's "colonists" exemplified both the enduring accomplishments of Reconstruction (since nearly all were identified as members of families, and many parents noted that their children had learned to read and write), and the bitter disillusionment that followed Redemption. In April 1878, the *Azor*, chartered by the black-run Liberian Exodus Joint Stock Steamship Company, left Charleston for Africa, carrying 200 South Carolina blacks, among them Bouey. Rev. Henry M. Turner, convinced by his experiences in Reconstruction Georgia that white America would never allow his people to "rise above a state of serfdom," gave the ship his benediction.[63]

Despite widespread interest in Liberian emigration, few blacks actually sailed for Africa, and some who did, like Harrison Bouey, soon returned. Financial exigencies partly explain this, since neither the Colonization Society nor the federal government heeded calls to advance funds to emigrants, and the Exodus Steamship Company quickly went bankrupt. But it also reflected the fact that most blacks were not prepared to surrender their claim to citizenship and equal rights. "We are not Africans now,

62. 46th Congress, 2d Session, Senate Report 693, pt. 2:393–95; Nashville *Union and American*, May 20, 1875; Mose J. Pringle to William Coppinger, May 1, 1876, Isaac Skinner to Coppinger, July 30, 1877, Nicholas Dill to Coppinger, August 11, 1877, Henry Adams to John H. B. LaTrobe, August 31, 1877, Henry Adams to William Coppinger, September 24, 1877, American Colonization Society Papers.

63. Nelson Davis to William Coppinger, August 23, 1877, Samuel J. Lee to Coppinger, September 19, 1877, Jasper Smith to Coppinger, October 5, 1877, J. J. Russell to Coppinger, October 15, 1877, Harrison N. Bouey to Henry M. Turner, May 23, 1877, John Mardenborough to William Coppinger, April 30, June 6, 1877, American Colonization Society Papers; Tindall, *South Carolina Negroes*, 154–60; Edwin S. Redkey, ed., *Respect Black: The Writings and Speeches of Henry McNeal Turner* (New York, 1971), 42–43.

but colored Americans, and are entitled to American citizenship," wrote a correspondent of Sen. Blanche K. Bruce in 1877. If blacks could not enjoy this status "among the white people," the government should set aside "one of the States or Territories" for their settlement. Although Congress evinced no interest in doing any such thing, blacks throughout the South took up the cry of internal "emigration" in the late 1870s. A group of Mississippi freedmen proposed to move en masse to New Mexico or Arizona. North Carolina blacks circulated pamphlets describing land available in Nebraska under the Homestead Act. But the greatest interest centered on Kansas, which became the destination of tens of thousands of refugees from "oppression and bondage."[64]

Nationally prominent black leaders, including Frederick Douglass, opposed the Kansas migration, fearing it amounted to a tacit abandonment of the struggle for citizenship rights in the South. But the movement generated immense excitement among ordinary blacks. The name they gave it—the Exodus—suggested that it tapped deep religious convictions. The freedmen, a Montgomery black convention had declared shortly after the Democrats' 1874 victory in Alabama, might yet be compelled "to repeat the history of the Israelites" and "seek new homes . . . beyond the reign and rule of Pharaoh." To countless blacks, Kansas offered the prospect of political equality, freedom from violence, access to education, economic opportunity, and liberation from the presence of the old slaveowning class—in sum the *"practical independence"* that Reconstruction had failed to secure. Those promoting the movement (including a real estate company organized by former slave Benjamin "Pap" Singleton) circulated lithographs depicting black farmers surrounded by livestock and abundant crops, and even lassoing buffalo on the open range. Few refugees found life in Kansas this idyllic. Lacking the funds or experience to take up plains farming, most settled for menial jobs in the state's towns. But few succumbed to disappointment sufficiently to return to the South. In the words of one minister active in the movement, "we had rather sufer and be free."[65]

Until Northern employers dismantled the color bar that restricted nearly all industrial jobs to whites, the vast majority of blacks were destined to remain in the South. Yet the Kansas Exodus not only revealed the disillusionment that followed the end of Reconstruction, but testified

64. William J. Simmons, *Men of Mark: Eminent, Progressive and Rising* (Cleveland, 1887), 953–54; Tindall, *South Carolina Negroes*, 160–61; L. W. Ballard to Blanche K. Bruce, November 2, 1877, C. A. Sullivan to Bruce, October 1, 1877, Bruce Papers; Taylor, "Great Migration," 23–24; Painter, *Exodusters*, 191.

65. Painter, *Exodusters*, 108–15, 142, 178–80, 247–50; 46th Congress, 2d Session, Senate Report 693, pt. 2:118, 243–44, 252, 400, 528, pt. 3:379; John M. Langston, *Freedom and Citizenship* (Washington, D.C., 1883), 238; Morgan D. Peoples, " 'Kansas Fever' in North Louisiana," *LaH*, 11 (Spring 1970), 127–28; Robert G. Athearn, *In Search of Canaan: Black Migration to Kansas, 1879–80* (Lawrence, Kans., 1978), 69–75, 81, 255–78; Rice, *Negro in Texas*, 202–204; 46th Congress, 2d Session, Senate Report 693, pt. 1:90.

to the fact that blacks' apparent quiescence did not imply consent to Redeemer hegemony. "We have no enemy in our front," commented Mississippi's Lucius Q. C. Lamar soon after Hayes's inauguration. "But the negroes are almost as well disciplined in their silence and inactivity as they were before in their aggressiveness." In black eyes, the entire system fashioned by the Redeemers bore the mark of illegitimacy. Conviction of crime in Southern courts carried little onus in the black community. "Colored men who have been in the Penitentiary," commented a Tennessee editor, "come back and live among their fellows without receiving contumely or social disgrace." In fact, remarked another white, many were "rather lionized." Nor did the New South's political system command black respect. "We are taxed without representation. . . ." wrote Charles Harris, a Union Army veteran and former member of Alabama's legislature:

> We obey laws; others make them. We support state educational institutions, whose doors are virtually closed against us. We support asylums and hospitals, and our sick, deaf, dumb, or blind are met at the doors by invidious distinctions and unjust discriminations. . . . From these and many other oppressions . . . our people long to be *free*.

In 1879, on the final day of his nine-year Congressional career, Joseph H. Rainey offered a fitting commentary on Redeemer rule: "Doubtless [Reconstruction government] was more extravagant. . . . [But] can the saving of a few thousand or hundreds of thousands of dollars compensate for the loss of the political heritage of American citizens?" As under slavery, blacks in the Redeemers' New South never acknowledged the justice of the social order under which they were forced to live.[66]

66. Lucius Q. C. Lamar to William H. Trescot, July 24, 1877, William H. Trescot Papers, USC; Edward L. Ayers, *Vengeance and Justice: Crime and Punishment in the Nineteenth-Century American South* (New York, 1984), 178–79; Steven V. Ash, *Middle Tennessee Society Transformed, 1860–1870: War and Peace in the Upper South* (Baton Rouge, 1987), Chapter 9; Charles Harris to William Coppinger, August 28, 1877, American Colonization Society Papers; *CR*, 45th Congress, 3d Session, Appendix, 267.

Epilogue

"The River Has Its Bend"

THUS, in the words of W.E.B. Du Bois, "the slave went free; stood a brief moment in the sun; then moved back again toward slavery." The magnitude of the Redeemer counterrevolution underscored both the scope of the transformation Reconstruction had assayed and the consequences of its failure. To be sure, the era of emancipation and Republican rule did not lack enduring accomplishments. The tide of change rose and then receded, but it left behind an altered landscape. The freedmen's political and civil equality proved transitory, but the autonomous black family and a network of religious and social institutions survived the end of Reconstruction. Nor could the seeds of educational progress planted then be entirely uprooted. While wholly inadequate for pupils of both races, schooling under the Redeemers represented a distinct advance over the days when blacks were excluded altogether from a share in public services.[1]

If blacks failed to achieve the economic independence envisioned in the aftermath of the Civil War, Reconstruction closed off even more oppressive alternatives than the Redeemers' New South. The post-Reconstruction labor system embodied neither a return to the closely supervised gang labor of antebellum days, nor the complete dispossession and immobilization of the black labor force and coercive apprenticeship systems envisioned by white Southerners in 1865 and 1866. Nor were blacks, as in twentieth-century South Africa, barred from citizenship, herded into labor reserves, or prohibited by law from moving from one part of the country to another. As illustrated by the small but growing number of black landowners, businessmen, and professionals, the doors of economic opportunity that had opened could never be completely closed. Without Reconstruction, moreover, it is difficult to imagine the establish-

1. W.E.B. Du Bois, *Black Reconstruction in America* (New York, 1935), 30; Howard N. Rabinowitz, *Race Relations in the Urban South 1865–1890* (New York, 1978), 152.

ment of a framework of legal rights enshrined in the Constitution that, while flagrantly violated after 1877, created a vehicle for future federal intervention in Southern affairs. As a result of this unprecedented redefinition of the American body politic, the South's racial system remained regional rather than national, an outcome of great importance when economic opportunities at last opened in the North.

Nonetheless, whether measured by the dreams inspired by emancipation or the more limited goals of securing blacks' rights as citizens and free laborers, and establishing an enduring Republican presence in the South, Reconstruction can only be judged a failure. Among the host of explanations for this outcome, a few seem especially significant. Events far beyond the control of Southern Republicans—the nature of the national credit and banking systems, the depression of the 1870s, the stagnation of world demand for cotton—severely limited the prospects for far-reaching economic change. The early rejection of federally sponsored land reform left in place a planter class far weaker and less affluent than before the war, but still able to bring its prestige and experience to bear against Reconstruction. Factionalism and corruption, although hardly confined to Southern Republicans, undermined their claim to legitimacy and made it difficult for them to respond effectively to attacks by resolute opponents. The failure to develop an effective long-term appeal to white voters made it increasingly difficult for Republicans to combat the racial politics of the Redeemers. None of these factors, however, would have proved decisive without the campaign of violence that turned the electoral tide in many parts of the South, and the weakening of Northern resolve, itself a consequence of social and political changes that undermined the free labor and egalitarian precepts at the heart of Reconstruction policy.

For historians, hindsight can be a treacherous ally. Enabling us to trace the hidden patterns of past events, it beguiles us with the mirage of inevitability, the assumption that different outcomes lay beyond the limits of the possible. Certainly, the history of other plantation societies offers little reason for optimism that emancipation could have given rise to a prosperous, egalitarian South, or even one that escaped a pattern of colonial underdevelopment. Nor do the prospects for the expansion of scalawag support—essential for Southern Republicanism's long-term survival—appear in retrospect to have been anything but bleak. Outside the mountains and other enclaves of wartime Unionism, the Civil War generation of white Southerners was always likely to view the Republican party as an alien embodiment of wartime defeat and black equality. And the nation lacked not simply the will but the modern bureaucratic machinery to oversee Southern affairs in any permanent way. Perhaps the remarkable thing about Reconstruction was not that it failed, but that it was attempted at all and survived as long as it did. Yet one can, I think, imagine alternative scenarios and modest successes: the Republican party establishing itself as a permanent fixture on the Southern landscape, the North summoning the resolve to insist that the Constitution must be respected. As the experiences of Readjuster Virginia and Populist-Republican North Carolina suggest, even Redemption did not entirely foreclose the possibility of biracial politics, thus raising the question of how Southern life might have been affected had Deep South blacks enjoyed genuine political freedoms when the Populist movement swept the white counties in the 1890s.

Here, however, we enter the realm of the purely speculative. What remains certain is that Reconstruction failed, and that for blacks its failure was a disaster whose magnitude cannot be obscured by the genuine accomplishments that did endure. For the nation as a whole, the collapse of Reconstruction was a tragedy that deeply affected the course of its future development. If racism contributed to the undoing of Reconstruction, by the same token Reconstruction's demise and the emergence of blacks as a disenfranchised class of dependent laborers greatly facilitated racism's further spread, until by the early twentieth century it had become more deeply embedded in the nation's culture and politics than at any time since the beginning of the antislavery crusade and perhaps in our entire history. The removal of a significant portion of the nation's laboring population from public life shifted the center of gravity of American politics to the right, complicating the tasks of reformers for generations to come. Long into the twentieth century, the South remained a one-party region under the control of a reactionary ruling elite who used the same violence and fraud that had helped defeat Reconstruction to stifle internal dissent. An enduring consequence of Reconstruction's failure, the Solid South helped define the contours of American politics and weaken the prospects not simply of change in racial matters but of progressive legislation in many other realms.

The men and women who had spearheaded the effort to remake Southern society scattered down innumerable byways after the end of Reconstruction. Some relied on federal patronage to earn a livelihood. The unfortunate Marshall Twitchell, armless after his near-murder in 1876, was appointed U.S. consul at Kingston, Ontario, where he died in 1905. Some fifty relatives and friends of the Louisiana Returning Board that had helped make Hayes President received positions at the New Orleans Custom House, and Stephen Packard was awarded the consulship at Liverpool—compensation for surrendering his claim to the governorship. John Eaton, who coordinated freedmen's affairs for General Grant during the war and subsequently took an active role in Tennessee Reconstruction, served as federal commissioner of education from 1870 to 1886, and organized a public school system in Puerto Rico after the island's conquest in the Spanish-American War. Most carpetbaggers returned to the North, often finding there the financial success that had eluded them in the South. Davis Tillson, head of Georgia's Freedman's Bureau immediately after the war, earned a fortune in the Maine granite business. Former South Carolina Gov. Robert K. Scott returned to Napoleon, Ohio, where he became a successful real estate agent—"a most fitting occupation" in view of his involvement in land commission speculations. Less happy was the fate of his scalawag successor, Franklin J. Moses, Jr., who drifted north, served prison terms for petty crimes, and died in a Massachusetts rooming house in 1906.[2]

Republican governors who had won reputations as moderates by courting white Democratic support and seeking to limit blacks' political influence found the Redeemer South remarkably forgiving. Henry C. Warmoth became a successful sugar planter and remained in Louisiana until his death in 1931. James L. Alcorn retired to his Mississippi plantation, "presiding over a Delta domain in a style befitting a prince" and holding various local offices. He remained a Republican, but told one Northern visitor that Democratic rule had produced "good fellowship" between the races. Even Rufus Bullock, who fled Georgia accused of every kind of venality, soon reentered Atlanta society, serving, among other things, as president of the city's chamber of commerce. Daniel H. Chamberlain left South Carolina in 1877 to launch a successful New York City law practice, but was well received on his numerous visits to the state. In retrospect, Chamberlain altered his opinion of Reconstruction: a "frightful experiment" that sought to "lift a backward or inferior race" to political equality, it had inevitably produced "shocking and unbearable misgovernment." "Governor Chamberlain," commented a Charleston newspaper, "has lived and learned."[3]

Not all white Republicans, however, abandoned Reconstruction ideals. In 1890, a group of reformers, philanthropists, and religious leaders gathered at the Lake Mohonk Conference on the Negro Question, chaired by former President Hayes. Amid a chorus of advice that blacks

2. Ted Tunnell, *Crucible of Reconstruction: War, Radicalism, and Race in Louisiana 1862–1877* (Baton Rouge, 1984), 209; William I. Hair, *Bourbonism and Agrarian Protest: Louisiana Politics 1877–1900* (Baton Rouge, 1969), 20; Frank B. Williams, "John Eaton, Jr., Editor, Politician, and School Administrator," *THQ*, 10 (December 1951), 292; William C. Harris, *Day of the Carpetbagger: Republican Reconstruction in Mississippi* (Baton Rouge, 1979), 717–20; Paul A. Cimbala, "The 'Talisman Power': Davis Tillson, the Freedmen's Bureau, and Free Labor in Reconstruction Georgia, 1865–1866," *CWH*, 28 (June 1982), 171; Carol Bleser, *The Promised Land: The History of the South Carolina Land Commission, 1869–1890* (Columbia, S.C., 1969), 57; Francis B. Simkins and Robert H. Woody, *South Carolina During Reconstruction* (Chapel Hill, 1932), 545.

3. Richard N. Current, *Three Carpetbag Governors* (Baton Rouge, 1967), 41; interview with James L. Alcorn, Southern Notebook B, Frederic Bancroft Papers, Columbia University; Alan Conway, *The Reconstruction of Georgia* (Minneapolis, 1966), 211–14; William McKinley to Rufus Bullock, November 13, 1893, Rufus Bullock Collection, HL; Simkins and Woody, *South Carolina*, 544; Daniel H. Chamberlain, "Reconstruction in South Carolina," *Atlantic Monthly*, 87 (April 1901), 476–78; Charleston *News and Courier*, August 1, 1904.

eschew political involvement and concentrate on educational and economic progress and remedying their own character deficiencies, former North Carolina Judge Albion W. Tourgée, again living in the North, voiced the one discordant note. There was no "Negro problem," Tourgee observed, but rather a "white" one, since "the hate, the oppression, the injustice, are all on our side." The following year, Tourgée established the National Citizens' Rights Association, a short-lived forerunner of the National Association for the Advancement of Colored People, devoted to challenging the numerous injustices afflicting Southern blacks. Adelbert Ames, who left Mississippi in 1875 to join his father's Minnesota flour-milling business and who later settled in Massachusetts, continued to defend his Reconstruction record. In 1894 he chided Brown University President E. Benjamin Andrews for writing that Mississippi during his governorship had incurred a debt of $20 million. The actual figure, Ames pointed out, was less than 3 percent of that amount, and he found it difficult to understand how Andrews had made "a $19,500,000 error in a $20,000,000 statement." Ames lived to his ninety-eighth year, never abandoning the conviction that "caste is the curse of the world." Another Mississippi carpetbagger, Massachusetts-born teacher and legislator Henry Warren, published his autobiography in 1914, still hoping that one day, "possibly in the present century," America would live up to the ideal of "equal political rights for all without regard to race."[4]

For some, the Reconstruction experience became a springboard to lifetimes of social reform. The white voters of Winn Parish in Louisiana's hill country expressed their enduring radicalism by supporting the Populists in the 1890s, Socialism in 1912, and later their native son Huey Long. Among the female veterans of freedmen's education, Cornelia Hancock founded Philadelphia's Children's Aid Society, Abby May became prominent in the Massachusetts women's suffrage movement, Ellen Collins turned her attention to New York City housing reform, and Josephine Shaw Lowell became a supporter of the labor movement and principal founder of New York's Consumer League. Louis F. Post, a New Jersey-

4. Isabel C. Barrows, ed., *First Mohonk Conference on the Negro Question* (Boston, 1891), 10–28, 108–10; Otto H. Olsen, *Carpetbagger's Crusade: The Life of Albion Winegar Tourgée* (Baltimore, 1965), 312; Harris, *Day of the Carpetbagger*, 721; Adelbert Ames to John R. Lynch n.d. [1914] (draft), Ames Family Papers, SC; Ames to E. Benjamin Andrews, May 24, 1895 (copy), Ames to R. H. DeKay, October 22, 1923 (copy), Adelbert Ames Papers, MDAH; Henry Warren, *Reminiscences of a Mississippi Carpetbagger* (Holden, Mass., 1914), 87–90. In 1901, a committee of the New York Union League Club met to consider ways of dealing with the "annoyance" to members caused by the fact that "the white and colored help have not harmonized." Four decades earlier, the Club had pressed relentlessly for emancipation, the enlistment of black soldiers, and black suffrage. Now, the committee recommended that all the black servants be dismissed. A group of "old timers" led by Wager Swayne, the Reconstruction head of Alabama's Freedmen's Bureau and subsequently a successful New York corporation lawyer, persuaded the full membership to overturn the decision, arguing "that such action was contrary to all the traditions of the organization." *New York Times*, April 25, 1901, December 19, 1902.

born carpetbagger who took stenographic notes for South Carolina's legislature in the early 1870s, became a follower of Henry George, attended the founding meeting of the NAACP, and as Woodrow Wilson's Assistant Secretary of Labor, sought to mitigate the 1919 Red Scare and prevent the deportation of foreign-born radicals. And Texas scalawag editor Albert Parsons became a nationally known Chicago labor reformer and anarchist, whose speeches drew comparisons between the plight of Southern blacks and Northern industrial workers, and between the aristocracy resting on slavery the Civil War had destroyed and the new oligarchy based on the exploitation of industrial labor it had helped to create. Having survived the perils of Texas Reconstruction, Parsons met his death on the Illinois gallows after being wrongfully convicted of complicity in the Haymarket bombing of 1886.[5]

Like their white counterparts, many black veterans of Reconstruction survived on federal patronage after the coming of "home rule." P. B. S. Pinchback and Blanche K. Bruce held a series of such posts and later moved to Washington, D.C., where they entered the city's privileged black society. Richard T. Greener, during Reconstruction a professor at the University of South Carolina, combined a career in law, journalism, and education with various government appointments, including a stint as American commercial agent at Vladivostok. Long after the destruction of his low country political machine by disenfranchisement, Robert Smalls served as customs collector for the port of Beaufort, dying there in 1915. Mifflin Gibbs held positions ranging from register of Little Rock's land office to American consul at Madagascar. Other black leaders left the political arena entirely to devote themselves to religious and educational work, emigration projects, or personal advancement. Robert G. Fitzgerald continued to teach in North Carolina until his death in 1919; Edward Shaw of Memphis concentrated on activities among black Masons and the AME Church; Richard H. Cain served as president of a black college in Waco, Texas; and Francis L. Cardozo went on to become principal of a Washington, D.C., high school. Aaron A. Bradley, the militant spokesman for Georgia's lowcountry freedmen, helped publicize the Kansas Exodus and died in St. Louis in 1881, while Henry M. Turner, ordained an AME bishop in 1880, emerged as the late nineteenth century's most prominent advocate of black emigration to Africa. Former Atlanta councilman William Finch prospered as a tailor. Alabama Congressman Jeremiah Haralson engaged in coal

5. John M. Price, "Slavery in Winn Parish," *LaH*, 8 (Spring 1967), 146; Allis Wolfe, "Women Who Dared: Northern Teachers of the Southern Freedmen, 1862–1872" (unpub. diss., City University of New York, 1982), 195–98; Dominic Cadeloro, "Louis Post as a Carpetbagger in South Carolina: Reconstruction as a Forerunner of the Progressive Movement," *American Journal of Economics and Sociology*, 34 (October 1975), 423–32; Paul Avrich, *The Haymarket Tragedy* (Princeton, 1984), 13–25; Philip S. Foner, ed., *The Autobiographies of the Haymarket Martyrs* (New York, 1969), 27–36.

mining in Colorado, where he was reported "killed by wild beasts."[6] Other Reconstruction leaders found, in the words of a black lawyer, that "the tallest tree . . . suffers most in a storm." Former South Carolina Congressman and Lieut. Gov. Alonzo J. Ransier died in poverty in 1882, having been employed during his last years as a night watchman at the Charleston Custom House and as a city street sweeper. Robert B. Elliott, the state's most brilliant political organizer, found himself "utterly unable to earn a living owing to the severe ostracism and mean prejudice of my political opponents." He died in 1884 after moving to New Orleans and struggling to survive as a lawyer. James T. Rapier died penniless in 1883, having dispersed his considerable wealth among black schools, churches, and emigration organizations. Most local leaders sank into obscurity, disappearing entirely from the historical record. Although some of their children achieved distinction, none of Reconstruction's black officials created a family political dynasty—one indication of how Redemption aborted the development of the South's black political leadership. If their descendants moved ahead, it was through business, the arts, or the professions. T. Thomas Fortune, editor of the New York *Age*, was the son of Florida officeholder Emanuel Fortune; Harlem Renaissance writer Jean Toomer, the grandson of Pinchback; renowned jazz pianist Fletcher Henderson, the grandson of an official who had served in South Carolina's constitutional convention and legislature.[7]

By the turn of the century, as soldiers from North and South joined to take up the "white man's burden" in the Spanish-American War, Recon-

6. Joel Williamson, *New People: Miscegenation and Mulattoes in the United States* (New York, 1980), 148; Vernon L. Wharton, *The Negro in Mississippi 1865–1890* (Chapel Hill, 1947), 161; Allison Blakeley, "Richard T. Greener and the 'Talented Tenth's' Dilemma," *JNH*, 59 (October 1974), 307–309; Okon E. Uya, *From Slavery to Public Service: Robert Smalls 1839–1915* (New York, 1971), 152–55; Tom W. Dillard, " 'Golden Prospects and Fraternal Amenities': Mifflin W. Gibbs' Arkansas Years," *ArkHQ*, 35 (Winter 1976), 313–17; Pauli Murray, *Proud Shoes* (New York, 1956), 275; Lester C. Lamon, *Blacks in Tennessee 1791–1970* (Knoxville 1981), 27; William J. Simmons, *Men of Mark: Eminent, Progressive and Rising* (Cleveland, 1887), 430, 867; Joseph P. Reidy, "Aaron A. Bradley: Voice of Black Labor in the Georgia Lowcountry," in Howard N. Rabinowitz, ed., *Southern Black Leaders of the Reconstruction Era* (Urbana, Ill., 1982), 300; Edwin S. Redkey, *Black Exodus: Black Nationalist and Back-to-Africa Movements, 1890–1910*, 24; Clarence A. Bacote, "William Finch, Negro Councilman and Political Activities in Atlanta During Early Reconstruction," *JNH*, 40 (October 1955), 363–64; Virginia Hamilton, *Alabama* (New York, 1977), 78.

7. "Ridout and Thompson" to William Coppinger, May 29, 1877, American Colonization Society Papers, LC; Terry L. Seip, *The South Returns to Congress: Men, Economic Measures, and Intersectional Relationships, 1868–1879* (Baton Rouge, 1983), 28; Peggy Lamson, *The Glorious Failure: Black Congressman Robert Brown Elliott and the Reconstruction in South Carolina* (New York, 1973), 270–87; Robert B. Elliott to John Sherman, June 23, 1879, John Sherman Papers, LC; Loren Schweninger, "James T. Rapier of Alabama and the Noble Cause of Reconstruction," in Rabinowitz, ed., *Southern Black Leaders*, 94; Darwin T. Turner, ed., *The Wayward and the Seeking: A Collection of Writings by Jean Toomer* (Washington, D.C., 1980), 22–25; Henry L. Suggs, ed., *The Black Press in the South, 1865–1979* (Westport, Conn., 1983), 104; "Papers of the Fletcher Hamilton Henderson Family," *Amistad Log*, 2 (February 1985), 8–9.

struction was widely viewed as little more than a regrettable detour on the road to reunion. To the bulk of the white South, it had become axiomatic that Reconstruction had been a time of "savage tyranny" that "accomplished not one useful result, and left behind it, not one pleasant recollection." Black suffrage, wrote Joseph Le Conte, who had fled South Carolina for a professorship at the University of California to avoid teaching black students, was now seen by "all thoughtful men" as "the greatest political crime ever perpetrated by any people." In more sober language, many Northerners, including surviving architects of Congressional policy, concurred in these judgments. "Years of thinking and observation" had convinced O. O. Howard "that the restoration of their lands to the planters provided for [a] future better for the negroes." John Sherman's recollections recorded a similar change of heart: "After this long lapse of time I am convinced that Mr. Johnson's scheme of reorganization was wise and judicious. . . . It is unfortunate that it had not the sanction of Congress."[8]

This rewriting of Reconstruction's history was accorded scholarly legitimacy—to its everlasting shame—by the nation's fraternity of professional historians. Early in the twentieth century a group of young Southern scholars gathered at Columbia University to study the Reconstruction era under the guidance of Professors John W. Burgess and William A. Dunning. Blacks, their mentors taught, were "children" utterly incapable of appreciating the freedom that had been thrust upon them. The North did "a monstrous thing" in granting them suffrage, for "a black skin means membership in a race of men which has never of itself succeeded in subjecting passion to reason, has never, therefore, created any civilization of any kind." No political order could survive in the South unless founded on the principle of racial inequality. The students' works on individual Southern states echoed these sentiments. Reconstruction, concluded the study of North Carolina, was an attempt by "selfish politicians, backed by the federal government . . . to Africanize the State and deprive the people through misrule and oppression of most that life held dear." The views of the Dunning School shaped historical writing for generations, and achieved wide popularity through D. W. Griffith's film *Birth of a Nation* (which glorified the Ku Klux Klan and had its premiere at the White House during Woodrow Wilson's Presidency), James Ford Rhodes's popular multivolume chronicle of the Civil War era, and the national best-seller *The Tragic Era* by Claude G. Bowers. Southern whites, wrote Bowers, "literally were put to the torture" by "emissaries of hate"

8. I. W. Avery, *The History of the State of Georgia From 1850 to 1881* (New York, 1881), 335; Hilary A. Herbert, ed., *Why the Solid South?* (Baltimore, 1890), 139; William D. Armes, ed., *The Autobiography of Joseph Le Conte* (New York, 1903), 238–39; [Oliver O. Howard] *The Autobiography of Oliver Otis Howard* (New York, 1907), 2:244; John Sherman, *Recollections of Forty Years in the House, Senate and Cabinet* (Chicago, 1895), 1:361.

who inflamed "the negroes' egotism" and even inspired "lustful assaults" by blacks upon white womanhood.[9]

Few interpretations of history have had such far-reaching consequences as this image of Reconstruction. As Francis B. Simkins, a South Carolina-born historian, noted during the 1930s, "the alleged horrors of Reconstruction" did much to freeze the mind of the white South in unalterable opposition to outside pressures for social change and to any thought of breaching Democratic ascendancy, eliminating segregation, or restoring suffrage to disenfranchised blacks. They also justified Northern indifference to the nullification of the Fourteenth and Fifteenth Amendments. Apart from a few white dissenters like Simkins, it was left to black writers to challenge the prevailing orthodoxy. In the early years of this century, none did so more tirelessly than former Mississippi Congressman John R. Lynch, then living in Chicago, who published a series of devastating critiques of the racial biases and historical errors of Rhodes and Bowers. "I do not hesitate to assert," he wrote, "that the Southern Reconstruction Governments were the best governments those States ever had." In 1917, Lynch voiced the hope that "a fair, just, and impartial historian will, some day, write a history covering the Reconstruction period, [giving] the actual facts of what took place."[10]

Only in the family traditions and collective folk memories of the black community did a different version of Reconstruction survive. Growing up in the 1920s, Pauli Murray was "never allowed to forget" that she walked in "proud shoes" because her grandfather, Robert G. Fitzgerald, had "fought for freedom" in the Union Army and then enlisted as a teacher in the "second war" against the powerlessness and ignorance inherited from slavery. When the Works Progress Administration sent agents into the black belt during the Great Depression to interview former slaves, they found Reconstruction remembered for its disappointments and betrayals, but also as a time of hope, possibility, and accomplishment. Bitterness still lingered over the federal government's failure to distribute land or protect blacks' civil and political rights. "The Yankees helped free us, so they say," declared eighty-one-year old former slave Thomas Hall, "but they let us be put back in slavery again." Yet coupled with this disillusionment were proud, vivid recollections of a time when "the col-

9. John W. Burgess, *Reconstruction and the Constitution 1866–1876* (New York, 1902), 44–45, 133, 244–46; William A. Dunning, *Essays on the Civil War and Reconstruction* (New York, 1904), 384–85; J. G. de Roulhac Hamilton, *Reconstruction in North Carolina* (New York, 1914), 667; Vernon L. Wharton, "Reconstruction," in Arthur S. Link and Rembert W. Patrick, eds., *Writing Southern History* (Baton Rouge, 1965), 297–306; Claude G. Bowers, *The Tragic Era* (Cambridge, Mass., 1929), vi, 198–200, 307–308.

10. Francis B. Simkins, "New Viewpoints of Southern Reconstruction," *JSH*, 5 (February 1939), 50–51; John R. Lynch, *The Facts of Reconstruction* (New York, 1913); Lynch, "Some Historical Errors of James Ford Rhodes," *JNH*, 2 (October 1917), 345–68; Lynch, "More About the Historical Errors of James Ford Rhodes," *JNH*, 3 (April 1918), 144–45; Lynch, "The Tragic Era," *JNH*, 16 (January 1931), 103–20.

ored used to hold office." Some pulled from their shelves dusty scrapbooks of clippings from Reconstruction newspapers; others could still recount the names of local black leaders. "They made pretty fair officers," remarked one elderly freedman; "I thought them was good times in the country," said another. Younger blacks spoke of being taught by their parents "about the old times, mostly about the Reconstruction, and the Ku Klux." "I know folks think the books tell the truth, but they shore don't," one eighty-eight-year old former slave told the WPA.[11]

For some blacks, such memories helped to keep alive the aspirations of the Reconstruction era. "This here used to be a good county," said Arkansas freedman Boston Blackwell, "but I tell you it sure is tough now. I think it's wrong—exactly wrong that we can't vote now." "I does believe that the negro ought to be given more privileges in voting," echoed Taby Jones, born a slave in South Carolina in 1850, "because they went through the reconstruction period with banners flying." For others, Reconstruction inspired optimism that better times lay ahead. "The Bible says, 'What has been will be again'," said Alabama sharecropper Ned Cobb. Born in 1885, Cobb never cast a vote in his entire life, yet he never forgot that outsiders had once taken up the black cause—an indispensable source of hope for one conscious of his own weakness in the face of overwhelming and hostile local power. When radical Northerners ventured South in the 1930s to help organize black agricultural workers, Cobb seemed almost to have been waiting for them: "The whites came down to bring emancipation, and left before it was over. . . . Now they've come to finish the job." The legacy of Reconstruction affected the 1930s revival of black militancy in other ways as well. Two leaders of the Alabama Share Croppers Union, Ralph and Thomas Gray, claimed to be descended from a Reconstruction legislator. (Like many nineteenth-century predecessors, Ralph Gray paid with his life for challenging the South's social order—he was killed in a shootout with a posse while guarding a union meeting.)[12]

Twenty more years elapsed before another generation of black Southerners launched the final challenge to the racial system of the New South. A few participants in the civil rights movement thought of themselves as following a path blazed after the Civil War. Discussing the reasons for his involvement, one black Mississippian spoke of the time when "a few Negroes was admitted into the government of the State of Mississippi and

---

11. Murray, *Proud Shoes*, 9–10, 24; George P. Rawick, ed., *The American Slave: A Composite Autobiography* (Westport, Conn., 1972–79), 9, pt. 3:30, 81, pt. 4:42, pt. 5:220, 14:145, 361, Supplement 1, 5:1650, 6:134, 7:567, Supplement 2, 1650, 4015; Paul D. Escott, *Slavery Remembered* (Chapel Hill, 1979), 153.

12. Rawick, ed., *American Slave*, 8, pt. 1:172, Supplement 2, 6:2152; Theodore Rosengarten, *All God's Dangers: The Life of Nate Shaw* (New York, 1974), 7; Mark D. Naison, "Black Agrarian Radicalism in the Great Depression: The Threads of a Lost Tradition," *Journal of Ethnic Studies*, 1 (Fall 1973), 53–55.

to the United States." Reconstruction's legacy was also evident in the actions of federal judge Frank Johnson, who fought a twelve-year battle for racial justice with Alabama Gov. George Wallace. Johnson hailed from Winston County, a center of Civil War Unionism, and his great-grandfather had served as a Republican sheriff during Reconstruction.[13] By this time, however, the Reconstruction generation had passed from the scene and even within the black community, memories of the period had all but disappeared. Yet the institutions created or consolidated after the Civil War—the black family, school, and church—provided the base from which the modern civil rights revolution sprang. And for its legal strategy, the movement returned to the laws and amendments of Reconstruction.

"The river has its bend, and the longest road must terminate."[14] Rev. Peter Randolph, a former slave, wrote these words as the dark night of injustice settled over the South. Nearly a century elapsed before the nation again attempted to come to terms with the implications of emancipation and the political and social agenda of Reconstruction. In many ways, it has yet to do so.

13. Sally Belfrage, *Freedom Summer* (New York, 1965), 210; Jack Bass, *Unlikely Heroes* (New York, 1981), 66.
14. Peter Randolph, *From Slave Cabin to the Pulpit* (Boston, 1893), 131.

Acknowledgments

IN nine years spent researching and writing this book, I incurred innumerable obligations to librarians, scholars, research assistants, students, friends, and editors. Rather than thank them individually (as I have done personally) and thereby enlarge an already lengthy volume, I wish here to express my deep appreciation to all those who so generously assisted my research, directed me to sources, shared ideas and information about the Reconstruction era, allowed me to read unpublished writings, offered critical comments on drafts of this work, and in many other ways offered help and encouragement.

Portions of this book were delivered as the Paley Lectures at Hebrew University, Jerusalem, in 1986, and excerpts have appeared in the *Radical History Review*, and *Journal of American History*. Research was facilitated by a fellowship from the National Endowment for the Humanities as well as grants from the Faculty Research Award Program of the City University of New York and the Columbia University History Department's Dunning Fund.

The book's dedication acknowledges a special bond that I treasure very deeply.

Selected Bibliography

Manuscript Collections

D. Wyatt Aiken Papers, South Caroliniana Library, University of South Carolina
Amos T. Akerman Papers, University of Virginia
Alabama Governor's Papers, Alabama State Department of Archives and History
American Colonization Society Papers, Library of Congress
American Missionary Association Archives, Amistad Research Center, Tulane University
Ames Family Papers, Sophia Smith Collection, Smith College
Edward Atkinson Papers, Massachusetts Historical Society
Frederic Bancroft Papers, Columbia University
Samuel L. M. Barlow Papers, Huntington Library
Albert A. Batchelor Papers, Louisiana State University
Blair Family Papers, Library of Congress
Benjamin H. Bristow Papers, Library of Congress
Blanche K. Bruce Papers, Howard University
John E. Bryant Papers, Duke University
Bryant-Godwin Papers, New York Public Library
Rufus Bullock Collection, Huntington Library
Benjamin F. Butler Papers, Library of Congress
William E. Chandler Papers, Library of Congress
Salmon P. Chase Papers, Library of Congress
William W. Clapp Papers, Library of Congress
Cobb-Erwin-Lamar Papers, University of Georgia
Cornelius Cole Papers, University of California, Los Angeles
Lemuel P. Conner Family Papers, Louisiana State University
Francis P. Corbin Papers, New York Public Library
Cave Johnson Couts Papers, Huntington Library
John Covode Papers, Library of Congress
Henry L. Dawes Papers, Library of Congress
Francis W. Dawson Papers, Duke University
James R. Doolittle Papers, State Historical Society of Wisconsin
Stephen Duncan Papers, Natchez Trace Collection, Eugene C. Barker Texas History Center, University of Texas

John Eaton Papers, Tennessee State Library and Archives
Lucius Fairchild Papers, State Historical Society of Wisconsin
Henry P. Farrow Papers, University of Georgia
Robert G. Fitzgerald Papers, Schomburg Center for Research in Black Culture
James A. Garfield Papers, Library of Congress
James W. Garner Papers, Mississippi Department of Archives and History
Georgia Governor's Papers, University of Georgia
George A. Gillespie Papers, Huntington Library
Horace Greeley Papers, New York Public Library
Hargrett Collection, University of Georgia.
Rutherford B. Hayes Papers, Hayes Memorial Library
Benjamin S. Hedrick Papers, Duke University
Heyward Family Papers, South Caroliniana Library, University of South Carolina
Houston H. Holloway Autobiography, Miscellaneous Manuscript Collections, Library of Congress
Mark Howard Papers, Connecticut Historical Society
O. O. Howard Papers, Bowdoin College
Timothy O. Howe Papers, State Historical Society of Wisconsin
Andrew Johnson Papers, Library of Congress
George W. Julian Papers, Indiana State Library
William P. Kellogg Papers, Louisiana State University
Elisha W. Keyes Papers, State Historical Society of Wisconsin
Basil G. Kiger Papers, Natchez Trace Collection, Eugene C. Barker Texas History Center, University of Texas
Kincaid-Anderson Papers, South Caroliniana Library, University of South Carolina
Le Conte Family Papers, Bancroft Library, University of California, Berkeley
Alexander Long Papers, Cincinnati Historical Society
Manton Marble Papers, Library of Congress
Maryville Union League Minute Book, McClung Collection, Lawson McGee Library
Robert McKee Papers, Alabama State Department of Archives and History
William N. Mercer Papers, Louisiana State University
Middleton Papers, Langdon Cheves Collection, South Carolina Historical Society
Mississippi Governor's Papers, Mississippi Department of Archives and History
Loring Moody Papers, Boston Public Library
Joshua B. Moore Diary, Alabama State Department of Archives and History
Thomas A. R. Nelson Papers, McClung Collection, Lawson McGee Library
James P. Newcomb Papers, Eugene C. Barker Texas History Center, University of Texas
North Carolina Governor's Papers, North Carolina Division of Archives and History
E. O. C. Ord Papers, Bancroft Library, University of California, Berkeley
J. M. Perry Family Papers, Atlanta Historical Society
Edwards Pierrepont Papers, Yale University
Charles H. Ray Papers, Huntington Library
Records of the Bureau of Refugees, Freedmen, and Abandoned Lands, Record Group 105, National Archives

Records of the United States Army Continental Commands, Record Group 393, National Archives
John W. A. Sanford Papers, Alabama State Department of Archives and History
Rufus and S. Willard Saxton Papers, Yale University
Robert B. Schenck Papers, Hayes Memorial Library
Jacob Schirmer Diary, South Carolina Historical Society
Carl Schurz Papers, Library of Congress
Robert K. Scott Papers, Ohio Historical Society
Thomas Settle Papers, Southern Historical Collection, University of North Carolina
William H. Seward Papers, University of Rochester
Isaac Sherman Papers, Private Collection
John Sherman Papers, Library of Congress
Augustine T. Smythe Letters, South Caroliniana Library, University of South Carolina
South Carolina Governor's Papers, South Carolina Department of Archives
William Sprague Papers, Columbia University
State Superintendent of Education Papers, South Carolina Department of Archives
Alexander H. Stephens Papers, Manhattanville College
Thaddeus Stevens Papers, Library of Congress
Charles Sumner Papers, Houghton Library, Harvard University
Wager Swayne Papers, Alabama State Department of Archives and History
Oliver P. Temple Papers, University of Tennessee
Tennessee Governor's Papers, Tennessee State Library and Archives
Samuel J. Tilden Papers, New York Public Library
William H. Trescot Papers, South Caroliniana Library, University of South Carolina
Lyman Trumbull Papers, Library of Congress
U.S. Senate, Select Committee to Investigate Elections in Mississippi Papers, New York Public Library
Henry C. Warmoth Papers, Southern Historical Collection, University of North Carolina
Elihu B. Washburne Papers, Library of Congress
Henry Watson, Jr., Papers, Duke University
Gideon Welles Papers, Huntington Library
Richard Yates Papers, Illinois State Historical Library

Government Documents and Publications

Congressional Globe
Congressional Record
39th Congress, 1st Session, House Executive Document 70: Orders Issued by the Freedmen's Bureau, 1865–66
39th Congress, 1st Session, House Report 30: Report of the Joint Committee on Reconstruction
39th Congress, 1st Session, Senate Executive Document 27: Reports of Assistant Commissioners of the Freedmen's Bureau, 1865–66

39th Congress, 2d Session, Senate Executive Document 6: Reports of Freedmen's Bureau Assistant Commissioners and Laws in Relation to the Freedmen

40th Congress, 3d Session, House Miscellaneous Document 52: Condition of Affairs in Georgia

42d Congress, 2d Session, House Report 22: Testimony Taken by the Joint Committee to Enquire into the Condition of Affairs in the Late Insurrectionary States (Ku Klux Klan Hearings)

43d Congress, 2d Session, House Report 261: Condition of Affairs in the South (Louisiana)

43d Congress, 2d Session, House Report 262: Affairs in Alabama

43d Congress, 2d Session, House Report 265: Vicksburg Troubles

44th Congress, 1st Session, Senate Report 527: Mississippi in 1875

44th Congress, 2d Session, House Miscellaneous Document 31: Recent Election in South Carolina

44th Congress, 2d Session, Senate Miscellaneous Document 45: Mississippi

44th Congress, 2d Session, Senate Miscellaneous Document 48: South Carolina in 1876

46th Congress, 2d Session, Senate Report 693: Report and Testimony of the Select Committee of the United States Senate to Investigate the Causes of the Removal of the Negroes from the Southern States to the Northern States

Report of the Commissioner of Agriculture for the Year 1876. Washington, D.C., 1877.

Report of the Committee of the Senate upon the Relations Between Labor and Capital, and Testimony Taken by the Committee, 4 vols. Washington, D.C., 1885.

Richardson, James D., ed., A Compilation of the Messages and Papers of the Presidents 1789–1897, 10 vols. Washington, D.C. 1896–99.

Thorpe, Francis L., ed., The Federal and State Constitutions, 7 vols. Washington, D.C., 1909.

U. S. Bureau of the Census, Historical Statistics of the United States, Colonial Times to 1870. Washington, D.C., 1975.

Proceedings of the constitutional conventions of the Southern states

Sessional laws of the Southern states

Newspapers, Periodicals, and Annuals

American Freedman
Annual Cyclopedia
Atlanta Constitution
Atlantic Monthly
Augusta Loyal Georgian
Charleston Daily Republican
Charleston News and Courier
Charleston South Carolina Leader
Christian Recorder
Cincinnati Commercial
Columbia Daily Phoenix

Harper's Weekly
Huntsville *Advocate*
Jackson *Mississippi Pilot*
Journal of Social Science
Knoxville *Whig*
Macon *American Union*
Mobile *Nationalist*
Mobile *Register*
Montgomery *Alabama State Journal*
Nashville *Colored Tennessean*
Nashville *Daily Press and Times*
National Anti-Slavery Standard
National Freedman
New National Era
New Orleans *Louisianian*
New Orleans *Tribune*
New York *Herald*
New York *Journal of Commerce*
New York Times
New York *Tribune*
New York *World*
North American Review
Raleigh *Daily Sentinel*
Raleigh *Standard*
Richmond *Dispatch*
Richmond *New Nation*
Rural Carolinian
Rutherford *Star*
St. Landry *Progress*
Savannah *Daily News and Herald*
Savannah *Freemen's Standard*
Savannah *Weekly Republican*
Selma *Southern Argus*
Southern Cultivator
Southern Field and Factory
Springfield *Republican*
The Great Republic (Washington, D.C.)
The Nation
Tribune Almanac
Washington *Chronicle*

Contemporary Publications and Published Documents

Abbott, Martin, ed., "A Southerner Views the South, 1865: Letters of Harvey M. Watterson," *Virginia Magazine of History and Biography*, 68 (October 1960), 478–89.

Allen, Walter, *Governor Chamberlain's Administration in South Carolina.* New York, 1888.

Ames, Blanche B., ed., *Chronicles from the Nineteenth Century: Family Letters of Blanche Butler and Adelbert Ames,* 2 vols. Clinton, Mass., 1957.

Andrews, Eliza F., *The War-Time Journal of a Georgia Girl.* New York, 1908.

Andrews, Sidney, *The South Since the War.* Boston, 1866.

Atkinson, Edward, *On Cotton.* Boston, 1865.

Avery, I. W., *The History of the State of Georgia from 1850 to 1881.* New York, 1881.

Bancroft, Frederic, ed., *Speeches, Correspondence and Political Papers of Carl Schurz,* 6 vols. New York, 1913.

Banks, Nathaniel P., *Emancipated Labor in Louisiana.* N.p., 1864.

Barrow, David C., Jr., "A Georgia Plantation," *Scribner's Monthly,* 21 (March 1881), 830–36.

Beale, Howard K., ed., *Diary of Gideon Welles,* 3 vols. New York, 1960.

Bellows, Henry W., *Historical Sketch of the Union League Club of New York,* New York, 1879.

[Benham, George C.] *A Year of Wreck.* New York, 1880.

Berlin, Ira et al., eds., *Freedom: A Documentary History of Emancipation.* New York, 1982– .

———, eds., "The Terrain of Freedom: The Struggle over the Meaning of Free Labor in the U.S. South," *History Workshop,* 22 (Autumn 1986), 108–30.

Brooks, Aubrey L., and Hugh T. Lefler, eds., *The Papers of Walter Clark,* 2 vols. Chapel Hill, 1948–50.

Brown, Henry J., and Frederick D. Williams, eds., *The Diary of James A. Garfield,* 3 vols. East Lansing, Mich., 1967–73.

Burnham, W. Dean, *Presidential Ballots 1836–1892.* Baltimore, 1955.

Childs, Arney R., ed., *The Private Journal of Henry William Ravenel, 1859–1887.* Columbia, S.C., 1947.

Clemenceau, Georges, *American Reconstruction.* Edited by Fernand Baldensperger. Translated by Margaret MacVeagh. New York, 1928.

"Colloquy with Colored Ministers," *Journal of Negro History,* 16 (January 1931), 88–94.

Cox, LaWanda and John H., eds., *Reconstruction, the Negro, and the New South.* Columbia, S.C., 1973.

DeForest, John W., *A Union Officer in the Reconstruction.* Edited by James H. Croushore and David M. Potter. New Haven, 1948.

Dennett, John R., *The South As It Is: 1865–1866.* Edited by Henry M. Christman. New York, 1965.

DuBois, Ellen C., ed., *Elizabeth Cady Stanton, Susan B. Anthony: Correspondence, Writings, Speeches.* New York, 1981.

Easterby, J. H., ed., *The South Carolina Rice Plantation as Revealed in the Papers of Robert F. W. Allston.* Chicago, 1945.

Foner, Philip S., ed., *The Life and Writings of Frederick Douglass,* 4 vols. New York, 1950–55.

———, and Ronald L. Lewis, eds., *The Black Worker: A Documentary History from Colonial Times to the Present,* 8 vols. Philadelphia, 1978–84.

———, and George E. Walker, eds., *Proceedings of the Black National and State Conventions, 1865–1900.* Philadelphia, 1986– .

————, and George E. Walker, eds., *Proceedings of the Black State Conventions, 1840–1865*, 2 vols. Philadelphia, 1979.

"George W. Julian's Journal—The Assassination of Lincoln" *Indiana Magazine of History*, 11 (December, 1915), 324–37.

Graf, Leroy P., and Ralph W. Haskins, eds., *The Papers of Andrew Johnson*. Knoxville, 1967– .

Hamilton, J. G. de Roulhac, ed., *The Correspondence of Jonathan Worth*, 2 vols. Raleigh, 1909.

————, ed., *The Papers of Randolph Abbott Shotwell*, 3 vols. Raleigh, 1929–36.

————, ed., *The Papers of Thomas Ruffin*, 4 vols. Raleigh, 1918–20.

————, and Max R. Williams, eds., *The Papers of William Alexander Graham*. Raleigh, 1957– .

Hinsdale, Mary L., ed., *Garfield-Hinsdale Letters*. Ann Arbor, 1949.

Holland, Rupert S., ed., *Letters and Diary of Laura M. Towne*. Cambridge, Mass., 1912.

Howe, M. A. De Wolfe, ed., *Home Letters of General Sherman*. New York, 1909.

Hyman, Harold M., ed., *The Radical Republicans and Reconstruction 1861–1870*. Indianapolis, 1967.

Kendrick, Benjamin B., *The Journal of the Joint Committee of Fifteen on Reconstruction*. New York, 1914.

King, Edward, *The Southern States of North America*. London, 1875.

Langston, John M., *Freedom and Citizenship*. Washington, 1883.

Loring, F. W., and C. F. Atkinson, *Cotton Culture and the South, Considered with Reference to Emigration*. Boston, 1869.

Macrae, David, *The Americans at Home*. New York, 1952 (orig. pub. 1870).

Mahaffey, Joseph H., ed., "Carl Schurz's Letters From the South," *Georgia Historical Quarterly*, 35 (September 1951), 222–56.

McPherson, Edward, *The Political History of the United States of America During the Period of Reconstruction*. Washington, D.C., 1875.

McPherson, Elizabeth G., ed., "Letters from North Carolina to Andrew Johnson," *North Carolina Historical Review*, 27 (July 1950), 336–63, (October 1950), 462–90, 28 (January 1951), 63–87, (April 1951), 219–37, (July 1951), 362–75, (October 1951), 486–516, 29 (January 1952), 104–19, (April 1952), 259–69, (July 1952), 400–31, (October 1952), 569–78.

Merrill, Walter M., and Louis Ruchames, eds., *The Letters of William Lloyd Garrison*, 6 vols. Cambridge, Mass., 1971–81.

Moore, John H., ed., *The Juhl Letters to the "Charleston Courier."* Athens, Ga., 1974.

Mordell, Albert, ed., *Civil War and Reconstruction: Selected Essays by Gideon Welles*. New York, 1959.

Nordhoff, Charles, *The Cotton States in the Spring and Summer of 1875*. New York, 1876.

Oliphant, Mary C. et al, eds., *The Letters of William Gilmore Simms*, 5 vols. Columbia, S.C., 1952–56.

Olsen, Otto H. and Ellen Z. McGrew, "Prelude to Reconstruction: The Correspondence of State Senator Leander Sams Gash, 1866–67," *North Carolina Historical Review*, 60 (January 1983), 37–88, (July 1983), 333–66.

Pearson, Elizabeth W., ed., *Letters From Port Royal Written at the Time of the Civil War*. Boston, 1906.

Pike, James S., *The Prostrate State.* New York, 1874.

Rawick, George P., ed., *The American Slave: A Composite Autobiography,* 39 vols. Westport, Conn., 1972–79.

Reconstruction. Speech of the Hon. Thaddeus Stevens Delivered in the City of Lancaster, September 7, 1865. Lancaster, Pa., 1865.

Reid, Whitelaw, *After the War: A Southern Tour.* Cincinnati, 1866.

Schafer, Joseph, ed., *Intimate Letters of Carl Schurz 1841–1869.* Madison, Wis., 1928.

Sefton, James E., ed., "Aristotle in Blue and Braid: General John M. Schofield's Essays on Reconstruction," *Civil War History,* 17 (March 1971), 45–57.

Smith, Daniel E. Huger et al., eds., *Mason Smith Family Letters 1860–1868.* Columbia, S.C., 1950.

Somers, Robert, *The Southern States Since the War 1870–71.* London, 1871.

Stealey, John E., III, ed., "Reports of Freedmen's Bureau Operations in West Virginia: Agents in the Eastern Panhandle," *West Virginia History,* 43 (Fall 1980–Winter 1981), 94–129.

Sterling, Dorothy, ed., *The Trouble They Seen.* Garden City, N.Y., 1976.

———, ed., *We Are Your Sisters: Black Women in the Nineteenth Century.* New York, 1984.

Stewart, Edgar A., ed., "The Journal of James Mallory, 1834–1877," *Alabama Review,* 13 (July 1961), 219–32.

Sufferings of the Rev. T. G. Campbell and His Family, in Georgia. Washington, D.C., 1877.

Swint, Henry L., ed., *Dear Ones at Home.* Nashville, 1966.

———, ed., "Reports from Educational Agents of the Freedmen's Bureau in Tennessee, 1865–1870," *Tennessee Historical Quarterly,* 1 (March 1942), 51–80, (June 1942), 152–70.

The Southern Loyalists Convention. New York, 1866.

The Works of Charles Sumner, 15 vols. Boston, 1870–83.

Trowbridge, J. T., *The South: A Tour of Its Battle-Fields and Ruined Cities.* Hartford, Conn., 1866.

Tyler, Ronnie C. and Lawrence R. Murphy, eds., *The Slave Narratives of Texas.* Austin, 1974.

Wagandt, Charles L., ed., "The Civil War Journal of Dr. Samuel A. Harrison," *Civil War History,* 13 (July 1967), 131–46.

Wainwright, Nicholas B., ed., *A Philadelphia Perspective: The Diary of Sidney George Fisher Covering the Years 1834–1871.* Philadelphia, 1967.

Williams, Charles R., ed., *Diary and Letters of Rutherford Birchard Hayes,* 5 vols. Columbus, 1922–26.

Memoirs, Reminiscences, and Autobiographies

Blaine, James G., *Twenty Years of Congress,* 2 vols. Norwich, Conn., 1884–86.

Brinckerhoff, Isaac W., "Missionary Work Among the Freed Negroes. Beaufort, South Carolina, St. Augustine, Florida, Savannah, Georgia," manuscript, American Baptist Historical Society.

Brinkerhoff, Roeliff, *Recollections of a Lifetime.* Cincinnati, 1904.

Burkhead, L. S., "History of the Difficulties of the Pastorate of the Front Street Methodist Church, Wilmington, N.C., for the Year 1865," *Trinity College Historical Society Historical Papers,* 8 (1908–09), 35–118.

Chamberlain, Hope S., *Old Days in Chapel Hill.* Chapel Hill, 1926.

Clayton, Powell, *Aftermath of the Civil War in Arkansas.* New York, 1915.

Cole, Cornelius, *Memoirs of Cornelius Cole.* New York, 1908.

Cox, Samuel S., *Three Decades of Federal Legislation.* Providence, R.I., 1885.

Cullom, Shelby M., *Fifty Years of Public Service.* Chicago, 1911.

[Douglass, Frederick], *Life and Times of Frederick Douglass.* New York, 1962 ed.

Eaton, John, *Grant, Lincoln and the Freedmen.* New York, 1907.

Eppes, Susan B., *Through Some Eventful Years.* Macon, Ga., 1926.

Gibbs, Mifflin W., *Shadow and Light: An Autobiography.* Washington, D.C., 1902.

Green, John P., *Fact Stranger Than Fiction.* Cleveland, 1920.

Hardy, W. H., "Recollections of Reconstruction in East and Southeast Mississippi," *Publications of the Mississippi Historical Society,* 4 (1901), 105–32, 7 (1903), 199–215, 8 (1904), 137–51.

Houzeau, Jean-Charles, *My Passage at the New Orleans "Tribune": A Memoir of the Civil War Era.* Edited by David C. Rankin. Translated by Gerard F. Denault. Baton Rouge, 1984.

[Howard, Oliver O.], *Autobiography of Oliver Otis Howard,* 2 vols. New York, 1907.

Julian, George W., *Political Recollections, 1840 to 1872.* Chicago, 1884.

Lynch, John R., *Reminiscences of an Active Life: The Autobiography of John Roy Lynch.* Edited by John Hope Franklin. Chicago, 1970.

Morgan, A. T., *Yazoo: or, On the Picket Line of Freedom in the South.* Washington, D.C., 1884.

Randolph, Peter, *From Slave Cabin to the Pulpit.* Boston, 1893.

Schurz, Carl, *The Reminiscences of Carl Schurz,* 3 vols. New York, 1907–8.

Sherman, John, *Recollections of Forty Years in the House, Senate and Cabinet,* 2 vols. Chicago, 1895.

[Sherman, William T.], *Memoirs of General W. T. Sherman, Written by Himself,* 2 vols. New York, 1891.

Warmoth, Henry C., *War, Politics, and Reconstruction.* New York, 1930.

Warren, Henry W., *Reminiscences of a Mississippi Carpetbagger.* Holden, Mass., 1914.

Withers, Robert E., *Autobiography of an Octogenarian.* Roanoke, Va., 1907.

Books

Abbott, Martin, *The Freedmen's Bureau in South Carolina, 1865–1872.* Chapel Hill, 1967.

Abbott, Richard H., *The Republican Party and the South, 1855–1877: The First Southern Strategy.* Chapel Hill, 1986.

Alexander, Roberta S., *North Carolina Faces the Freedmen: Race Relations During Presidential Reconstruction, 1865–67.* Durham, 1985.

Alexander, Thomas B., *Political Reconstruction in Tennessee.* Nashville, 1950.

Allen, James S., *Reconstruction: The Battle for Democracy*. New York, 1937.

Ash, Stephen V., *Middle Tennessee Society Transformed, 1860–1870: War and Peace in the Upper South*. Baton Rouge, 1987.

Ayers, Edward L., *Vengeance and Justice: Crime and Punishment in the Nineteenth-Century American South*. New York, 1984.

Baum, Dale, *The Civil War Party System: The Case of Massachusetts, 1848–1876*. Chapel Hill, 1984.

Beale, Howard K., *The Critical Year: A Study of Andrew Johnson and Reconstruction*. New York, 1930.

Beard, Charles A. and Mary R., *The Rise of American Civilization*. New York, 1933 ed.

Belz, Herman, *A New Birth of Freedom: The Republican Party and Freedmen's Rights 1861–1866*. Westport, Conn., 1976.

———. *Reconstructing the Union: Theory and Practice During the Civil War*. Ithaca, N.Y., 1969.

Benedict, Michael L., *A Compromise of Principle: Congressional Republicans and Reconstruction 1863–1869*. New York, 1974.

———. *The Impeachment and Trial of Andrew Johnson*. New York, 1973.

Bentley, George R., *A History of the Freedmen's Bureau*. Philadelphia, 1955.

Berwanger, Eugene H., *The West and Reconstruction*. Urbana, Ill., 1981.

Bethel, Elizabeth, *Promiseland: A Century of Life in a Negro Community*. Philadelphia, 1981.

Billings, Dwight B., Jr., *Planters and the Making of a "New South": Class, Politics, and Development in North Carolina, 1865–1900*. Chapel Hill, 1979.

Blassingame, John W., *Black New Orleans 1860–1880*. Chicago, 1973.

Bleser, Carol R., *The Promised Land: The History of the South Carolina Land Commission, 1869–1890*. Columbia, S.C., 1969.

Bogue, Allan G., *The Earnest Men: Republicans of the Civil War Senate*. Ithaca, N.Y., 1981.

Bond, Horace Mann, *Negro Education in Alabama: A Study in Cotton and Steel*. Washington, D.C., 1939.

Bowers, Claude G., *The Tragic Era*. Cambridge, Mass., 1929.

Bradley, Erwin S., *The Triumph of Militant Republicanism*. Philadelphia, 1964.

Bremner, Robert H., *The Public Good: Philanthropy and Welfare in the Civil War Era*. New York, 1980.

Brock, W. R., *An American Crisis*. London, 1963.

Brooks, Robert P., *The Agrarian Revolution in Georgia, 1865–1912*. Madison, Wis., 1914.

Brown, Dee, *Bury My Heart at Wounded Knee*. New York, 1970.

Bruce, Robert V., *1877: Year of Violence*. Indianapolis, 1959.

Buechler, Steven M., *The Transformation of the Woman Suffrage Movement: The Case of Illinois, 1850–1920*. New Brunswick, N.J., 1986.

Burgess, John W., *Reconstruction and the Constitution 1866–1876*. New York, 1902.

Burton, Orville V., *In My Father's House Are Many Mansions: Family and Community in Edgefield, South Carolina*. Chapel Hill, 1985.

———, and Robert C. McMath, eds., *Toward a New South? Studies in Post-Civil War Southern Communities*. Westport, Conn., 1982.

Butchart, Ronald E., *Northern Schools, Southern Blacks, and Reconstruction: Freedmen's Education, 1862–1875*. Westport, Conn., 1980.

Campbell, Randolph B., *A Southern Community in Crisis: Harrison County, Texas, 1850–1880*. Austin, 1983.

Carlton, David L., *Mill and Town in South Carolina, 1880–1920*. Baton Rouge, 1982.

Carter, Dan T., *When the War Was Over: The Failure of Self-Reconstruction in the South, 1865–1867*. Baton Rouge, 1985.

Cell, John W., *The Highest Stage of White Supremacy: The Origins of Segregation in South Africa and the American South*. New York, 1982.

Chandler, Alfred D., *The Visible Hand: The Managerial Revolution in American Business*. Cambridge, Mass., 1977.

Chesson, Michael B., *Richmond After the War 1865–1890*. Richmond, 1981.

Cimprich, John, *Slavery's End in Tennessee*. University, Ala., 1986.

Cochran, Thomas C., and William Miller, *The Age of Enterprise: A Social History of Industrial America*. New York, 1942.

Conway, Alan, *The Reconstruction of Georgia*. Minneapolis, 1966.

Cooper, Jerry M., *The Army and Civil Disorder*. Westport, Conn., 1980.

Coulter, E. Merton, *The Civil War and Readjustment in Kentucky*. Chapel Hill, 1926.

———. *The South During Reconstruction 1865–1877*. Baton Rouge, 1947.

———. *William G. Brownlow, Fighting Parson of the Southern Highlands*. Chapel Hill, 1937.

Cox, LaWanda, *Lincoln and Black Freedom: A Study in Presidential Leadership*. Columbia, S.C., 1981.

———, and John H., *Politics, Principle, and Prejudice 1865–1866*. New York, 1963.

Currie, James T., *Enclave: Vicksburg and Her Plantations, 1863–1870*. Jackson, Miss., 1980.

Curry, Richard O., ed., *Radicalism, Racism, and Party Realignment: The Border States During Reconstruction*. Baltimore, 1969.

Davis, Ronald F., *Good and Faithful Labor: From Slavery to Sharecropping in the Natchez District, 1860–1890*. Westport, Conn., 1982.

Davis, William W., *The Civil War and Reconstruction in Florida*. New York, 1913.

Dawson Joseph G., III, *Army Generals and Reconstruction: Louisiana, 1862–1877*. Baton Rouge, 1982.

DeSantis, Vincent P., *Republicans Face the Southern Question*. Baltimore, 1959.

Donald, David, *Charles Sumner and the Rights of Man*. New York, 1970.

———. *The Politics of Reconstruction 1863–1867*. Baton Rouge, 1965.

Drago, Edmund L., *Black Politicians and Reconstruction in Georgia*. Baton Rouge, 1982.

DuBois, Ellen C., *Feminism and Suffrage: The Emergence of an Independent Women's Movement in America, 1848–1869*. Ithaca, N.Y., 1978.

Du Bois, W.E.B., *Black Reconstruction in America*. New York, 1935.

Duncan, Russell, *Freedom's Shore: Tunis Campbell and the Georgia Freedmen*. Athens, Ga., 1986.

Dunning, William A., *Essays on the Civil War and Reconstruction*. New York, 1904.

———. *Reconstruction, Political and Economic 1865–1877*. New York, 1907.

Engs, Robert F., *Freedom's First Generation: Black Hampton, Virginia, 1861–1890*. Philadelphia, 1979.

Escott, Paul D., *Many Excellent People: Power and Privilege in North Carolina, 1850–1900*. Chapel Hill, 1985.

——. *Slavery Remembered*. Chapel Hill, 1979.

Evans, Frank B., *Pennsylvania Politics, 1872–1877: A Study in Political Leadership*. Harrisburg, Penn., 1966.

Evans, W. McKee, *Ballots and Fence Rails: Reconstruction on the Lower Cape Fear*. Chapel Hill, 1967.

Fairman, Charles, *Reconstruction and Reunion 1864–1888: Part One*. New York, 1971.

Field, Phyllis F., *The Politics of Race in New York: The Struggle for Black Suffrage in the Civil War Era*. Ithaca, N.Y., 1982.

Fields, Barbara J., *Slavery and Freedom on the Middle Ground: Maryland During the Nineteenth Century*. New Haven, 1985.

Fischer, Roger A., *The Segregation Struggle in Louisiana 1862–77*. Urbana, Ill., 1974.

Fleming, Walter L., *Civil War and Reconstruction in Alabama*. New York, 1905.

Flick, Alexander C., *Samuel Jones Tilden: A Study in Political Sagacity*. New York, 1939.

Flynn, Charles L., Jr., *White Land, Black Labor: Caste and Class in Late Nineteenth-Century Georgia*. Baton Rouge, 1983.

Foner, Eric, *Free Soil, Free Labor, Free Men: The Ideology of the Republican Party Before the Civil War*. New York, 1970.

——. *Nothing but Freedom: Emancipation and Its Legacy*. Baton Rouge, 1983.

——. *Politics and Ideology in the Age of the Civil War*. New York, 1980.

Fraser, Walter J., Jr., and Winfred B. Moore, Jr., eds., *From Old South to New: Essays on the Transitional South*. Westport, Conn., 1981.

——, eds., *The Southern Enigma: Essays on Race, Class, and Folk Culture*. Westport, Conn., 1983.

Fredrickson, George M., *The Inner Civil War: Northern Intellectuals and the Crisis of the Union*. New York, 1965.

——, ed., *A Nation Divided*. Minneapolis, 1975.

Gambill, Edward L., *Conservative Ordeal: Northern Democrats and Reconstruction, 1865–1868*. Ames, Iowa, 1981.

Garner, James W., *Reconstruction in Mississippi*. New York, 1901.

Gaston, Paul M., *The New South Creed*. New York, 1970.

Gerber, David A., *Black Ohio and the Color Line 1860–1915*. Urbana, Ill., 1976.

Gerteis, Louis S., *From Contraband to Freedman: Federal Policy Toward Southern Blacks, 1861–1865*. Westport, Conn., 1973.

Gilchrist, David T., and W. David Lewis, eds., *Economic Change in the Civil War Era*. Greenville, Del., 1965.

Gillette, William, *Retreat from Reconstruction 1869–1879*. Baton Rouge, 1979.

——. *The Right to Vote*. Baltimore, 1969 ed.

Glymph, Thavolia, and John J. Kushma, eds., *Essays on the Postbellum Southern Economy*. College Station, Tex., 1985.

Goodrich, Carter, *Government Promotion of American Canals and Railroads 1800–1890*. New York, 1960.

Grossman, Lawrence, *The Democratic Party and the Negro: Northern and National Politics 1868–92.* Urbana, Ill., 1976.

Gutman, Herbert G., *The Black Family in Slavery and Freedom 1750–1925.* New York, 1976.

Hahn, Steven, *The Roots of Southern Populism: Yeoman Farmers and the Transformation of the Georgia Upcountry, 1850–1890.* New York, 1983.

Hair, William I., *Bourbonism and Agrarian Protest: Louisiana Politics 1877–1900.* Baton Rouge, 1969.

Hamilton, J. G. de Roulhac, *Reconstruction in North Carolina.* New York, 1914.

Harding, Vincent, *There Is a River: The Black Struggle for Freedom in America.* New York, 1981.

Harris, William C., *Presidential Reconstruction in Mississippi.* Baton Rouge, 1967.

———. *The Day of the Carpetbagger: Republican Reconstruction in Mississippi.* Baton Rouge, 1979.

Haskins, James, *Pinckney Benton Stewart Pinchback.* New York, 1973.

Henderson, William D., *The Unredeemed City: Reconstruction in Petersburg, Virginia: 1865–1874.* Washington, 1977.

Hermann, Janet S., *The Pursuit of a Dream.* New York, 1981.

Hesseltine, William B., *Ulysses S. Grant, Politician.* New York, 1935.

Hobsbawm, Eric, *The Age of Capital 1848–1875.* London, 1975.

Holt, Thomas, *Black over White: Negro Political Leadership in South Carolina During Reconstruction.* Urbana, Ill., 1977.

Hoogenboom, Ari, *Outlawing the Spoils: A History of the Civil Service Reform Movement 1865–1883.* Urbana, Ill., 1961.

Howard, Victor B., *Black Liberation in Kentucky: Emancipation and Freedom, 1862–1884.* Lexington, Ky., 1983.

Hutchinson, William T., *Cyrus Hall McCormick,* 2 vols. New York, 1930–35.

Hyman, Harold M., *A More Perfect Union: The Impact of the Civil War and Reconstruction on the Constitution.* New York, 1973.

———, ed., *New Frontiers of the American Reconstruction.* Urbana, Ill., 1966.

———, and William M. Wiecek, *Equal Justice Under Law: Constitutional Development 1835–1875.* New York, 1982.

James, Joseph B., *The Framing of the Fourteenth Amendment.* Urbana, Ill., 1956.

Jaynes, Gerald D., *Branches Without Roots: Genesis of the Black Working Class in the American South, 1862–1882.* New York, 1986.

Jellison, Charles A., *Fessenden of Maine.* Syracuse, 1962.

Jentz, John B., and Richard Schneirov, *The Origins of Chicago's Industrial Working Class* (forthcoming).

Jones, Jacqueline, *Labor of Love, Labor of Sorrow: Black Women, Work and the Family, from Slavery to the Present.* New York, 1985.

———. *Soldiers of Light and Love: Northern Teachers and Georgia Blacks, 1865–1873.* Chapel Hill, 1980.

Kaczorowski, Robert, *The Politics of Judicial Interpretation: The Federal Courts, Department of Justice and Civil Rights, 1866–1876.* New York, 1985.

Katz, Irving, *August Belmont: A Political Biography.* New York, 1968.

Katzman, David M., *Before the Ghetto: Black Detroit in the Nineteenth Century.* Urbana, Ill., 1973.

Keiser, John H., *Building for the Centuries: Illinois, 1865 to 1898.* Urbana, Ill., 1977.

Keller, Morton, *Affairs of State: Public Life in Late Nineteenth Century America.* Cambridge, Mass., 1977.

Kibler, Lillian A., *Benjamin F. Perry: South Carolina Unionist.* Durham, 1946.

King, Willard L., *Lincoln's Manager: David Davis.* Cambridge, Mass., 1960.

Kirkland, Edward C., *Industry Comes of Age: Business, Labor and Public Policy, 1860–1897.* New York, 1961.

Klingman, Peter D., *Josiah Walls.* Gainesville, Fla., 1976.

Kolchin, Peter, *First Freedom: The Responses of Alabama's Blacks to Emancipation and Reconstruction.* Westport, Conn., 1972.

Kousser, J. Morgan, and McPherson, James M., eds., *Region, Race, and Reconstruction: Essays in Honor of C. Vann Woodward.* New York, 1982.

Krug, Mark M., *Lyman Trumbull: Conservative Radical.* New York, 1965.

Lamon, Lester C., *Blacks in Tennessee 1791–1970.* Knoxville, 1981.

Lamson, Peggy, *The Glorious Failure: Black Congressman Robert Brown Elliott and the Reconstruction in South Carolina.* New York, 1973.

Litwack, Leon F., *Been in the Storm So Long: The Aftermath of Slavery.* New York, 1979.

Logsdon, Joseph, *Horace White: Nineteenth Century Liberal.* Westport, Conn., 1971.

Lonn, Ella, *Reconstruction in Louisiana After 1868.* New York, 1918.

Lynch, John R., *The Facts of Reconstruction.* New York, 1913.

Maddex, Jack P., Jr., *The Virginia Conservatives, 1867–1879.* Chapel Hill, 1970.

Magdol, Edward, *A Right to the Land: Essays on the Freedmen's Community.* Westport, Conn., 1977.

Maslowski, Peter, *Treason Must Be Made Odious: Military Occupation and Wartime Reconstruction in Nashville, Tennessee, 1862–1865.* Millwood, N.Y., 1978.

McCrary, Peyton, *Abraham Lincoln and Reconstruction: The Louisiana Experiment.* Princeton, 1978.

McFeely, William S., *Grant: A Biography.* New York, 1981.

———. *Yankee Stepfather: General O. O. Howard and the Freedmen.* New Haven, 1968.

McJimsey, George T., *Genteel Partisan: Manton Marble, 1834–1917.* Ames, Iowa, 1971.

McKinney, Gordon B., *Southern Mountain Republicans 1865–1900.* Chapel Hill, 1978.

McKitrick, Eric L., *Andrew Johnson and Reconstruction.* Chicago, 1960.

McMillan, Malcolm C., *Constitutional Development in Alabama 1798–1901: A Study in Politics, the Negro, and Sectionalism.* Chapel Hill, 1955.

McPherson, James M., *The Struggle for Equality: Abolitionists and the Negro in the Civil War and Reconstruction.* Princeton, 1964.

Messner, William F., *Freedmen and the Ideology of Free Labor: Louisiana 1862–1865.* Lafayette, La., 1978.

Milton, George F., *The Age of Hate: Andrew Johnson and the Radicals.* New York, 1930.

Mohr, James C., *The Radical Republicans and Reform in New York During Reconstruction.* Ithaca, N.Y., 1973.

———, ed., *Radical Republicans in the North: State Politics During Reconstruction.* Baltimore, 1976.

Moneyhon, Carl H., *Republicanism in Reconstruction Texas.* Austin, 1980.

Montgomery, David, *Beyond Equality: Labor and the Radical Republicans 1862–1872.* New York, 1967.

Morris, Robert C., *Reading, 'Riting, and Reconstruction: The Education of Freedmen in the South, 1861–1870.* Chicago, 1981.

Mushkat, Jerome, *The Reconstruction of the New York Democracy, 1861–1874.* Rutherford, N.J., 1981.

Nathans, Elizabeth S., *Losing the Peace: Georgia Republicans and Reconstruction, 1865–1871.* Baton Rouge, 1968.

Nevins, Allan, *Abram S. Hewitt, with Some Account of Peter Cooper.* New York, 1935.

———. *The Emergence of Modern America 1865–1878.* New York, 1927.

Nieman, Donald G., *To Set the Law in Motion: The Freedmen's Bureau and the Legal Rights of Blacks, 1865–1868.* Millwood, N.Y., 1979.

Novack, Daniel A., *The Wheel of Servitude: Black Forced Labor After Slavery.* Lexington, Ky., 1978.

Nugent, Walter T. K., *Money and American Society 1865–1880.* New York, 1968.

O'Brien, Gail W., *The Legal Fraternity and the Making of a New South Community, 1848–1882.* Athens, Ga., 1986.

Olsen, Otto H., *Carpetbagger's Crusade: The Life of Albion Winegar Tourgée.* Baltimore, 1965.

———, ed., *Reconstruction and Redemption in the South.* Baton Rouge, 1980.

Osthaus, Carl R., *Freedmen, Philanthropy, and Fraud: A History of the Freedman's Savings Bank.* Urbana, Ill., 1976.

Oubre, Claude F., *Forty Acres and a Mule: The Freedmen's Bureau and Black Landownership.* Baton Rouge, 1978.

Overy, David H., Jr., *Wisconsin Carpetbaggers in Dixie.* Madison, Wis., 1961.

Paisley, Clifton, *From Cotton to Quail: An Agricultural Chronicle of Leon County, Florida 1860–1967.* Gainesville, Fla., 1968.

Parks, Joseph H., *Joseph E. Brown of Georgia.* Baton Rouge, 1976.

Parrish, William E., *Missouri Under Radical Rule 1865–1870* Columbia, Mo., 1965.

Pereyra, Lillian A., *James Lusk Alcorn: Persistent Whig.* Baton Rouge, 1966.

Perman, Michael, *Reunion Without Compromise: The South and Reconstruction 1865–1868.* New York, 1973.

———. *The Road to Redemption: Southern Politics, 1869–1879.* Chapel Hill, 1984.

Peterson, Norma L., *Freedom and Franchise: The Political Career of B. Gratz Brown.* Columbia, Mo., 1965.

Polakoff, Keith I., *The Politics of Inertia: The Election of 1876 and the End of Reconstruction.* Baton Rouge, 1973.

Powell, Lawrence N., *New Masters: Northern Planters During the Civil War and Reconstruction.* New Haven, 1980.

Rabinowitz, Howard N., *Race Relations in the Urban South 1865–1890.* New York, 1978.

———, ed., *Southern Black Leaders of the Reconstruction Era.* Urbana, Ill., 1982.

Rable, George C., *But There Was No Peace: The Role of Violence in the Politics of Reconstruction.* Athens, Ga., 1984.

Rachleff, Peter J., *Black Labor in the South: Richmond, Virginia 1865–1890.* Philadelphia, 1984.

Ramsdell, Charles W., *Reconstruction in Texas.* New York, 1910.

Ransom, Roger L., and Richard Sutch, *One Kind of Freedom: The Economic Consequences of Emancipation.* New York, 1977.

Rice, Lawrence D., *The Negro in Texas 1874–1900.* Baton Rouge, 1971.

Richardson, Joe M., *Christian Reconstruction: The American Missionary Association and Southern Blacks, 1861–1890*. Athens, Ga., 1986.

———. *The Negro in the Reconstruction of Florida, 1865–1877*. Tallahassee, 1965.

Richardson, Leon B., *William E. Chandler, Republican*. New York, 1940.

Riddleberger, Patrick W., *George Washington Julian: Radical Republican*. Indianapolis, 1966.

Ripley, C. Peter, *Slaves and Freedmen in Civil War Louisiana*. Baton Rouge, 1976.

Roark, James L., *Masters Without Slaves: Southern Planters in the Civil War and Reconstruction*. New York, 1977.

Robinson, Armstead, *Bitter Fruits of Bondage: The Demise of Slavery and the Collapse of the Confederacy* (forthcoming).

Rogers, William W., *The One-Gallused Rebellion: Agrarianism in Alabama 1865–1896*. Baton Rouge, 1970.

Rose, Willie Lee, *Rehearsal for Reconstruction: The Port Royal Experiment*. Indianapolis, 1964.

Ross, Steven J., *Workers on the Edge: Work, Leisure, and Politics in Industrializing Cincinnati, 1788–1890*. New York, 1985.

Satcher, Buford, *Blacks in Mississippi Politics 1865–1900*. Washington, 1978.

Schweninger, Loren, *James T. Rapier and Reconstruction*. Chicago, 1978.

Sefton, James E., *Andrew Johnson and the Uses of Constitutional Power*. Boston, 1980.

———. *The United States Army and Reconstruction 1865–1877*. Baton Rouge, 1967.

Seip, Terry L., *The South Returns to Congress: Men, Economic Measures, and Intersectional Relationships, 1868–1879*. Baton Rouge, 1983.

Sellin, J. Thorsten, *Slavery and the Penal System*. New York, 1976.

Sharkey, Robert P., *Money, Class, and Party: An Economic Study of Civil War and Reconstruction*. Baltimore, 1967 ed.

Shifflett, Crandall A., *Patronage and Poverty in the Tobacco South: Louisa County, Virginia, 1860–1900*. Knoxville, 1982.

Shofner, Jerrell H., *Nor Is It Over Yet: Florida in the Era of Reconstruction, 1863–1877*. Gainesville, Fla., 1974.

Silbey, Joel, *A Respectable Minority: The Democratic Party in the Civil War Era*. New York, 1977.

Simkins, Francis B., and Robert H. Woody, *South Carolina During Reconstruction*. Chapel Hill, 1932.

Sitterson, J. Carlyle, *Sugar Country*. Lexington, Ky., 1953.

Skowronek, Stephen, *Building a New American State*. New York, 1982.

Smallwood, James W., *Time of Hope, Time of Despair: Black Texans During Reconstruction*. Port Washington, N.Y., 1981.

Sproat, John G., *"The Best Men": Liberal Reformers in the Gilded Age*. New York, 1968.

Stampp, Kenneth M., *The Era of Reconstruction 1865–1877*. New York, 1965.

Stewart, James B., *Wendell Phillips: Liberty's Hero*. Baton Rouge, 1986.

Stover, John F., *The Railroads of the South 1865–1900*. Chapel Hill, 1955.

Suggs, Henry L., ed., *The Black Press in the South, 1865–1979*. Westport, Conn., 1983.

Summers, Mark W., *Railroads, Reconstruction, and the Gospel of Prosperity: Aid Under the Radical Republicans, 1865–1877*. Princeton, 1984.

Taylor, Alrutheus A., *The Negro in South Carolina During the Reconstruction*. Washington, D.C., 1924.

————. *The Negro in Tennessee, 1865–1880.* Washington, D.C., 1941.

————. *The Negro in the Reconstruction of Virginia.* Washington, D.C., 1926.

Taylor, Arnold H., *Travail and Triumph: Black Life and Culture in the South Since the Civil War.* Westport, Conn., 1976.

Taylor, Joe G., *Louisiana Reconstructed, 1863–1877.* Baton Rouge, 1974.

Thomas, Benjamin P., and Harold M. Hyman, *Stanton: The Life and Times of Lincoln's Secretary of War.* New York, 1962.

Thompson, C. Mildred, *Reconstruction in Georgia,* New York, 1915.

Thompson, George H., *Arkansas and Reconstruction.* Port Washington, N.Y., 1976.

Thompson, Margaret S., *The "Spider Web": Congress and Lobbying in the Age of Grant.* Ithaca, N.Y., 1985.

Tilley, Nannie M., *The Bright-Tobacco Industry 1860–1929.* Chapel Hill, 1948.

Tindall, George B., *South Carolina Negroes 1877–1900.* Columbia, S.C., 1952.

Trefousse, Hans L., *Benjamin Franklin Wade: Radical Republican from Ohio.* New York, 1963.

————. *The Radical Republicans: Lincoln's Vanguard for Racial Justice.* New York, 1969.

Trelease, Allen W., *White Terror: The Ku Klux Klan Conspiracy and Southern Reconstruction.* New York, 1971.

Tunnell, Ted, *Crucible of Reconstruction: War, Radicalism, and Race in Louisiana 1862–1877.* Baton Rouge, 1984.

Ullmann, Victor, *Martin Delany: The Beginnings of Black Nationalism.* Boston, 1971.

Unger, Irwin, *The Greenback Era: A Social and Political History of American Finance 1865–1879.* Princeton, 1964.

Uya, Okon E., *From Slavery to Public Service: Robert Smalls 1839–1915.* New York, 1971.

Vaughan, William P., *Schools for All: The Blacks and Public Education in the South, 1865–1877.* Lexington, Ky., 1974.

Vincent, Charles, *Black Legislators in Louisiana During Reconstruction.* Baton Rouge, 1976.

Wagandt, Charles L., *The Mighty Revolution: Negro Emancipation in Maryland, 1862–1864.* Baltimore, 1964.

Walker, Clarence G., *A Rock in a Weary Land: The African Methodist Episcopal Church During the Civil War and Reconstruction.* Baton Rouge, 1982.

Wallenstein, Peter, *From Slave South to New South: Public Policy in Nineteenth-Century Georgia.* Chapel Hill, 1987.

Washington, James M., *Frustrated Fellowship: The Black Baptist Quest for Social Power.* Macon, Ga., 1986.

Wayne, Michael, *The Reshaping of Plantation Society: The Natchez District, 1860–1880.* Baton Rouge, 1983.

Webb, Ross A., *Benjamin Helm Bristow: Border State Politician.* Lexington, Ky., 1969.

Wharton, Vernon L., *The Negro in Mississippi 1865–1890.* Chapel Hill, 1947.

White, Howard A., *The Freedmen's Bureau in Louisiana.* Baton Rouge, 1970.

Wiener, Jonathan M., *Social Origins of the New South: Alabama 1860–1885.* Baton Rouge, 1978.

Wiggins, Sarah W., *The Scalawag in Alabama Politics, 1865–1881.* University, Ala., 1977.

Wiley, Bell I., *Southern Negroes 1861–1865.* New Haven, 1938.

Williamson, Joel, *After Slavery: The Negro in South Carolina During Reconstruction, 1861–1877.* Chapel Hill, 1965.

Wood, Forrest G., *Black Scare: The Racist Response to Emancipation and Reconstruction.* Berkeley, 1968.

Wood, Philip J., *Southern Capitalism: The Political Economy of North Carolina, 1880–1980.* Durham, 1986.

Woodward, C. Vann, *Origins of the New South, 1877–1913.* Baton Rouge, 1951.

———. *Reunion and Reaction: The Compromise of 1877 and the End of Reconstruction,* rev. ed. Garden City, N.Y., 1956.

Wright, Gavin, *The Political Economy of the Cotton South.* New York, 1978.

Yearley, C. K., *The Money Machines: The Breakdown and Reform of Governmental and Party Finance in the North, 1860–1920.* Albany, 1970.

Zuber, Richard L., *Jonathan Worth.* Chapel Hill, 1965.

Articles

Alexander, Thomas B., "Persistent Whiggery in the Confederate South, 1860–1877," *Journal of Southern History,* 27 (August 1961), 305–29.

Anderson, George L., "The South and Problems of Post-Civil War Finance," *Journal of Southern History,* 9 (May 1943), 205–16.

Aptheker, Herbert, "Mississippi Reconstruction and the Negro Leader, Charles Caldwell," *Science and Society,* 11 (Fall 1947), 340–71.

Arcanti, Stephen J., "To Secure the Party: Henry L. Dawes and the Politics of Reconstruction," *Historical Journal of Western Massachusetts,* 5 (Spring 1977), 33–45.

Auman, William T., and David D. Scarboro, "The Heroes of America in Civil War North Carolina," *North Carolina Historical Review,* 58 (Autumn 1981), 327–63.

Baggett, James A., "Origins of Early Texas Republican Party Leadership," *Journal of Southern History,* 40 (August 1974), 441–54.

———. "Origins of Upper South Scalawag Leadership," *Civil War History,* 29 (March 1983), 53–73.

Balanoff, Elizabeth, "Negro Leaders in the North Carolina General Assembly, July, 1868–February, 1872," *North Carolina Historical Review,* 49 (Winter 1972), 22–55.

Barr, Alwyn, "Black Legislators of Reconstruction Texas," *Civil War History,* 32 (December 1986), 340–52.

Baum, Dale, "Woman Suffrage and the 'Chinese Question': The Limits of Radical Republicanism in Massachusetts, 1865–1876," *New England Quarterly,* 56 (March 1983), 60–77.

Beale, Howard K., "On Rewriting Reconstruction History," *American Historical Review,* 45 (July 1940), 807–27.

Belz, Herman, "Origins of Negro Suffrage During the Civil War," *Southern Studies,* 17 (Summer 1978), 115–30.

Benedict, Michael L., "Preserving Federalism: Reconstruction and the Waite Court," *Supreme Court Review,* 1978, 39–79.

———. "Preserving the Constitution: The Conservative Basis of Radical Reconstruction," *Journal of American History,* 61 (June 1974), 65–90.

————. "Southern Democrats in the Crisis of 1876–1877: A Reconsideration of *Reunion and Reaction,*" *Journal of Southern History,* 46 (November 1980), 489–524.

Bernstein, Samuel, "American Labor in the Long Depression, 1873–1878," *Science and Society,* 20 (Winter 1956), 59–83.

Binning, F. Wayne, "The Tennessee Republicans in Decline, 1869–1876," *Tennessee Historical Quarterly,* 39 (Winter 1980), 471–84, 40 (Spring 1981), 68–84.

Blain, William T., "Banner Unionism in Mississippi, Choctaw County 1861–1869," *Mississippi Quarterly,* 29 (September 1976), 207–20.

Blassingame, John W., "Before the Ghetto: The Making of the Black Community in Savannah, Georgia, 1865–1880," *Journal of Social History,* 6 (Summer 1973), 463–88.

Bridges, Roger D., "Equality Deferred: Civil Rights for Illinois Blacks, 1865–1885," *Journal of the Illinois State Historical Society,* 74 (Spring 1981), 82–108.

Brock, Euline W., "Thomas W. Cardozo: Fallible Black Reconstruction Leader," *Journal of Southern History,* 47 (May 1981), 183–206.

Brown, Ira V., "Pennsylvania and the Rights of the Negro, 1865–1887," *Pennsylvania History,* 28 (January 1961), 45–57.

————. "William D. Kelley and Radical Reconstruction," *Pennsylvania Magazine of History and Biography,* 85 (July 1961), 316–29.

Burton, Vernon, "Race and Reconstruction: Edgefield County, South Carolina," *Journal of Social History,* 12 (Fall 1978), 31–56.

Chandler, Robert J., "Friends in Time of Need: Republicans and Black Civil Rights in California During the Civil War Era," *Arizona and the West,* 22 (Winter 1982), 319–40.

Cimbala, Paul A., "The 'Talisman Power': Davis Tillson, the Freedmen's Bureau, and Free Labor in Reconstruction Georgia, 1865–1866," *Civil War History,* 28 (June 1982), 153–72.

Cimprich, John, "Military Governor Johnson and Tennessee Blacks, 1862–65," *Tennessee Historical Quarterly,* 39 (Winter 1980), 459–70.

Coben, Stanley, "Northeastern Business and Radical Reconstruction: A Re-Examination," *Mississippi Valley Historical Review,* 46 (June 1959), 67–90.

Coelho, Philip R. P. and Shepherd, James F., "Regional Differences in Real Wages: The United States, 1851–1880," *Explorations in Economic History,* 13 (April 1976), 203–30.

Cohen, William, "Negro Involuntary Servitude in the South, 1865–1940: A Preliminary Analysis," *Journal of Southern History,* 42 (February 1976), 31–60.

Cox, John H. and LaWanda, "General O. O. Howard and the 'Misrepresented Bureau'," *Journal of Southern History,* 19 (November 1953), 427–56.

Cox, LaWanda, "The Promise of Land for the Freedmen," *Mississippi Valley Historical Review,* 45 (December 1958), 413–40.

Crouch, Barry A., "A Spirit of Lawlessness: White Violence, Texas Blacks, 1865–1868," *Journal of Social History,* 18 (Winter 1984), 217–32.

————, and Larry Madaras, "Reconstructing Black Families: Perspectives from the Texas Freedmen's Bureau Records," *Prologue,* 18 (Summer 1986), 109–22.

Current, Richard N., "Carpetbaggers Reconsidered," in David H. Pinkney and Theodore Ropp, eds., *A Festschrift for Frederick B. Artz.* Durham, 1964.

Curtis, Michael K., "The Fourteenth Amendment and the Bill of Rights," *Connecticut Law Review*, 14 (Winter 1982), 237–306.

Davis, J. R., "Reconstruction in Cleveland County," *Trinity College Historical Society Historical Papers*, 10 (1914), 5–31.

Donald, David H., "The Scalawag in Mississippi Reconstruction," *Journal of Southern History*, 10 (November 1944), 447–60.

Driggs, Orval T., Jr., "The Issues of the Powell Clayton Regime, 1868–1871," *Arkansas Historical Quarterly*, 8 (Spring 1949), 1–75.

Du Bois, W.E.B., "Reconstruction and Its Benefits," *American Historical Review*, 15 (July 1910), 781–99.

Durrill, Wayne K., "Producing Poverty: Local Government and Economic Development in a New South County, 1874–1884," *Journal of American History*, 71 (March 1985), 764–81.

Easterlin, Richard A., "Regional Income Trends, 1840–1950," in Seymour E. Harris, ed., *American Economic History*. New York, 1961.

Ellem, Warren A., "Who Were the Mississippi Scalawags?" *Journal of Southern History*, 38 (May 1972), 217–40.

Farber Daniel A., and John E. Muench, "The Ideological Origins of the Fourteenth Amendment," *Constitutional Commentary*, 1 (Summer 1984), 235–79.

Fields, Barbara J., "The Nineteenth-Century American South: History and Theory," *Plantation Society in the Americas*, 2 (April 1983), 7–28.

Fishel, Leslie H., Jr., "Repercussions of Reconstruction: The Northern Negro, 1870–1883," *Civil War History*, 14 (December 1968), 325–45.

Foner, Eric, "Reconstruction and the Black Political Tradition," in Richard L. McCormick, ed., *Political Parties and the Modern State*. New Brunswick, N.J., 1983.

———. "Reconstruction Revisited," *Reviews in American History*, 10 (December 1982), 82–100.

Foner, Laura, "The Free People of Color in Louisiana and St. Domingue," *Journal of Social History*, 3 (Summer 1970), 406–30.

Foner, Philip S., "The Battle to End Discrimination Against Negroes on Philadelphia Streetcars," *Pennsylvania History*, 40 (July 1973), 261–90, (October 1973), 355–79.

Forbath, William E., "The Ambiguities of Free Labor: Labor and the Law in the Gilded Age," *Wisconsin Law Review*, 1985, 767–817.

Ford, Lacy K., "Rednecks and Merchants: Economic Development and Social Tensions in the South Carolina Upcountry, 1865–1900," *Journal of American History*, 71 (September 1984), 294–318.

Foster, Gaines M., "The Limitations of Federal Health Care for Freedmen, 1862–1868," *Journal of Southern History*, 48 (August 1982), 349–72.

Fraser, Walter J., Jr., "Black Reconstructionists in Tennessee," *Tennessee Historical Quarterly*, 34 (Winter 1975), 362–82.

Fuke, Richard P., "A Reform Mentality: Federal Policy Toward Black Marylanders, 1864–1868," *Civil War History*, 22 (September 1976), 214–35.

Gates, Paul W., "Federal Land Policy in the South, 1866–1888," *Journal of Southern History*, 6 (August 1940), 303–30.

Gerofsky, Milton, "Reconstruction in West Virginia," *West Virginia History*, 6 (July 1945), 295–360, 7 (October 1945), 5–39.

Goldfield, David R., "The Urban South: A Regional Framework," *American Historical Review*, 86 (December 1981), 1009–34.

Goodrich, Carter, "Public Aid to Railroads in the Reconstruction South," *Political Science Quarterly*, 71 (September 1956), 407–42.

Gottlieb, Manuel, "The Land Question in Georgia During Reconstruction," *Science and Society*, 3 (Summer 1939), 356–88.

Gutman, Herbert G., "The Failure of the Movement by the Unemployed for Public Works in 1873," *Political Science Quarterly*, 80 (June 1965), 254–76.

———. "The Tompkins Square 'Riot' in New York City on January 13, 1874: A Re-examination of Its Causes and Its Aftermath," *Labor History*, 6 (Winter 1965), 44–70.

Hall, Robert L., "Tallahassee's Black Churches, 1865–1885," *Florida Historical Quarterly*, 58 (October 1979), 185–96.

Harlan, Louis R., "Desegregation in New Orleans Public Schools During Reconstruction," *American Historical Review*, 67 (April 1962), 663–75.

Harris, William C., "James Lynch: Black Leader in Southern Reconstruction," *Historian*, 34 (November 1971), 40–61.

———. "The Creed of the Carpetbaggers: The Case of Mississippi," *Journal of Southern History*, 40 (May 1974), 199–224.

Hart, John F., "The Role of the Plantation in Southern Agriculture," *Proceedings, Tall Timbers Ecology and Management Conference*, 16 (1979), 1–20.

Hennessey, Melinda M., "Racial Violence During Reconstruction: The 1876 Riots in Charleston and Cainhoy," *South Carolina Historical Magazine*, 86 (April 1985), 100–12.

———. "Reconstruction Politics and the Military: The Eufaula Riot of 1874," *Alabama Historical Quarterly*, 38 (Summer 1976), 112–25.

Hesseltine, William B., "Economic Factors in the Abandonment of Reconstruction," *Mississippi Valley Historical Review*, 22 (September 1935), 191–210.

Highsmith, William E., "Louisiana Landholding During War and Reconstruction," *Louisiana Historical Quarterly*, 37 (January 1955), 39–54.

———. "Some Aspects of Reconstruction in the Heart of Louisiana," *Journal of Southern History*, 13 (November 1947), 460–91.

Hine, William C., "Black Organized Labor in Reconstruction Charleston," *Labor History*, 25 (Fall 1984), 504–17.

———. "Black Politicians in Reconstruction Charleston, South Carolina: A Collective Study," *Journal of Southern History*, 49 (November 1983), 555–84.

Hopkins, Richard J., "Occupational and Geographic Mobility in Atlanta, 1870–1896," *Journal of Southern History*, 34 (May 1968), 200–213.

Horowitz, Robert F., "Seward and Reconstruction: A Reconsideration," *Historian*, 47 (May 1985), 382–401.

Huffman, Frank J., "Town and Country in the South, 1850–1880: A Comparison of Urban and Rural Social Structures," *South Atlantic Quarterly*, 76 (Summer 1977), 366–81.

Hume, Richard L., "Carpetbaggers in the Reconstruction South: A Group Portrait of Outside Whites in the 'Black and Tan' Constitutional Conventions," *Journal of American History*, 64 (September 1977), 313–30.

Hyman, Harold M., "Johnson, Stanton, and Grant: A Reconsideration of the Army's Role in the Events Leading to Impeachment," *American Historical Review*, 66 (October 1960), 85–100.

Joshi, Manoj K., and Joseph P. Reidy, " 'To Come Forward and Aid in Putting Down This Unholy Rebellion': The Officers of Louisiana's Free Black Native Guard During the Civil War Era," *Southern Studies*, 21 (Fall 1982), 326–42.

Kaczorowski, Robert J., "Searching for the Intent of the Framers of the Fourteenth Amendment," *Connecticut Law Review*, 5 (Winter 1972–73), 368–98.

Kellogg, John, "The Evolution of Black Residential Areas in Lexington, Kentucky, 1865–1887," *Journal of Southern History*, 48 (February 1982), 21–52.

Kelly, Alfred H., "The Congressional Controversy Over School Segregation, 1867–1875," *American Historical Review*, 64 (April 1959), 537–63.

King, J. Crawford, "The Closing of the Southern Range: An Exploratory Study," *Journal of Southern History*, 48 (February 1982), 53–70.

Kolchin, Peter, "The Business Press and Reconstruction," *Journal of Southern History*, 33 (May 1967), 183–96.

Kornell, Gary L., "Reconstruction in Nashville, 1867–1869," *Tennessee Historical Quarterly*, 30 (Fall 1971), 277–87.

Lebsock, Suzanne, "Radical Reconstruction and the Property Rights of Southern Women," *Journal of Southern History*, 43 (May 1977), 195–216.

Lerner, Eugene, "Southern Output and Agricultural Income, 1860–1880," *Agricultural History*, 33 (July 1959), 117–25.

Levine, Richard R., "Indian Fighters and Indian Reformers: Grant's Indian Peace Policy and the Conservative Consensus," *Civil War History*, 31 (December 1985), 329–50.

Lovett, Bobby L., "Memphis Riots: White Reactions to Blacks in Memphis, May 1865–July 1866," *Tennessee Historical Quarterly*, 38 (Spring 1979), 9–33.

———. "Some 1871 Accounts for the Little Rock, Arkansas, Freedman's Savings and Trust Company," *Journal of Negro History*, 66 (Winter 1981–82), 322–28.

Lowe, Richard, "Another Look at Reconstruction in Virginia," *Civil War History*, 32 (March 1986), 56–76.

Lowrey, Walter McG., "The Political Career of James Madison Wells," *Louisiana Historical Quarterly*, 31 (October 1948), 995–1123.

May, J. Thomas, "The Freedmen's Bureau at the Local Level: A Study of a Louisiana Agent," *Louisiana History*, 9 (Winter 1968), 5–20.

McDonald, Forrest, and Grady McWhiney, "The South from Self-Sufficiency to Peonage: An Interpretation," *American Historical Review*, 85 (December 1980), 1095–1118.

McGerr, Michael E., "The Meaning of Liberal Republicanism: The Case of Ohio," *Civil War History*, 28 (December 1982), 307–23.

McPherson, James M., "Grant or Greeley? The Abolitionist Dilemma in the Election of 1872," *American Historical Review*, 71 (October 1965), 43–61.

McWhiney, Grady, "The Revolution in Nineteenth-Century Alabama Agriculture," *Alabama Review*, 31 (January 1978), 3–32.

Meier, August, "Negroes in the First and Second Reconstructions of the South," *Civil War History*, 13 (June 1967), 114–30.

Mering, John V., "Persistent Whiggery in the Confederate South: A Reconsideration," *South Atlantic Quarterly*, 69 (Winter 1970), 124–43.

Moneyhon, Carl H., "Black Politics in Arkansas During the Gilded Age, 1876–1900," *Arkansas Historical Quarterly*, 44 (Autumn 1985), 222–45.

Montgomery, David, "Radical Republicanism in Pennsylvania, 1866–1873," *Pennsylvania Magazine of History and Biography*, 85 (October 1961), 439–57.

Moore, James T., "Black Militancy in Readjuster Virginia, 1879–1883," *Journal of Southern History*, 41 (May 1975), 167–86.

Morris-Crowther, Jayne, "An Economic Study of the Substantial Slaveholders of Orangeburg County, 1860–1880," *South Carolina Historical Magazine*, 86 (October 1985), 296–314.

Nieman, Donald G., "Andrew Johnson, the Freedmen's Bureau, and the Problem of Equal Rights, 1865–1866," *Journal of Southern History*, 44 (August 1978), 399–420.

O'Brien, John T., "Factory, Church, and Community: Blacks in Antebellum Richmond," *Journal of Southern History*, 44 (November 1978), 509–36.

———. "Reconstruction in Richmond: White Restoration and Black Protest, April–June 1865," *Virginia Magazine of History and Biography*, 89 (July 1981), 259–81.

Odom, E. Dale, "The Vicksburg, Shreveport and Texas: The Fortunes of a Scalawag Railroad," *Southwestern Social Science Quarterly*, 44 (December 1963), 277–85.

Olsen, Otto H., "Reconsidering the Scalawags," *Civil War History*, 12 (December 1966), 304–20.

Otto, John S., "Southern 'Plain Folk' Agriculture," *Plantation Society in the Americas*, 2 (April 1983), 29–36.

Peek, Ralph L., "Lawlessness in Florida, 1868–1871," *Florida Historical Quarterly*, 40 (October 1961), 164–85.

Peskin, Allan, "Was There a Compromise of 1877?" *Journal of American History*, 60 (June 1973), 63–75.

Pope, Christie F., "Southern Homesteads for Negroes," *Agricultural History*, 44 (April 1970), 201–12.

Powell, Lawrence N., "The American Land Company and Agency: John A. Andrew and the Northernization of the South," *Civil War History*, 21 (December 1975), 293–308.

Price, Charles L., "The Railroad Schemes of George W. Swepson," *East Carolina Publications in History*, 1 (1964), 32–50.

Rabinowitz, Howard N., "From Exclusion to Segregation: Southern Race Relations, 1865–1890," *Journal of American History*, 63 (September 1976), 325–50.

———. "Half a Loaf: The Shift from White to Black Teachers in the Negro Schools of the Urban South, 1865–1890," *Journal of Southern History*, 40 (November 1974).

———. "The Conflict Between Blacks and the Police in the Urban South, 1865–1900," *Historian*, 39 (November 1976), 62–76.

Rable, George, "Republican Albatross: The Louisiana Question, National Politics, and the Failure of Reconstruction," *Louisiana History*, 23 (Spring 1982), 109–30.

————. "Southern Interests and the Election of 1876: A Reappraisal," *Civil War History*, 26 (December 1980), 347–61.

Rankin, David C., "The Impact of the Civil War on the Free Colored Community of New Orleans," *Perspectives in American History*, 11 (1977–78), 379–416.

————. "The Origins of Black Leadership in New Orleans During Reconstruction," *Journal of Southern History*, 40 (August 1974), 417–40.

Richardson, Joe M., "Jonathan C. Gibbs: Florida's Only Black Cabinet Member," *Florida Historical Quarterly*, 42 (April 1964), 363–68.

Robinson, Armstead L., "Beyond the Realm of Social Consensus: New Meanings of Reconstruction for American History," *Journal of American History*, 68 (September 1981), 276–97.

Roediger, David, "Ira Steward and the Anti-Slavery Origins of American Eight-Hour Theory," *Labor History*, 27 (Summer 1986), 410–26.

Rousey, Dennis C., "Black Policemen in New Orleans During Reconstruction," *Historian*, 49 (February 1987), 223–43.

St. Clair, Kenneth E., "Debtor Relief in North Carolina During Reconstruction," *North Carolina Historical Review*, 18 (July 1941), 215–35.

St. Hilaire, Joseph M., "The Negro Delegates in the Arkansas Constitutional Convention," *Arkansas Historical Quarterly*, 33 (Spring 1974), 38–69.

Savitt, Todd L., "Politics in Medicine: The Georgia Freedmen's Bureau and the Organization of Health Care," *Civil War History*, 28 (March 1982), 45–64.

Schneirov, Richard, "Class Conflict, Municipal Politics, and Governmental Reform in Gilded Age Chicago," in Hartmut Keil and John B. Jentz, eds., *German Workers in Industrial Chicago, 1850–1910: A Comparative Perspective.* DeKalb, Ill., 1983.

Schweninger, Loren, "Black Citizenship and the Republican Party in Reconstruction Alabama," *Alabama Review*, 29 (April 1976), 83–101.

Scott, John A., "Justice Bradley's Evolving Concept of the Fourteenth Amendment from the Slaughterhouse Case to the Civil Rights Cases," *Rutgers Law Review*, 25 (Summer 1971), 552–69.

Scott, Rebecca, "The Battle over the Child: Child Apprenticeship and the Freedmen's Bureau in North Carolina," *Prologue*, 10 (Summer 1978), 101–13.

Scroggs, Jack B., "Carpetbagger Constitutional Reform in the South Atlantic States, 1867–1868," *Journal of Southern History*, 27 (November 1961), 475–93.

Shlomowitz, Ralph, "The Origins of Southern Sharecropping," *Agricultural History*, 53 (July 1979), 557–75.

Shortreed, Margaret, "The Anti-Slavery Radicals: From Crusade to Revolution 1840–1868," *Past and Present*, 16 (November 1959), 65–87.

Simkins, Francis B., "New Viewpoints of Southern Reconstruction," *Journal of Southern History*, 5 (February 1939), 49–61.

Sitterson, J. Carlyle, "The Transition from Slave to Free Economy on the William J. Minor Plantations," *Agricultural History*, 17 (January 1943), 216–24.

Small, Sandra E., "The Yankee Schoolmarm in Freedmen's Schools: An Analysis of Attitudes," *Journal of Southern History*, 45 (August 1979), 381–402.

Smallwood, James, "Perpetuation of Caste: Black Agricultural Workers in Reconstruction Texas," *Mid-America*, 61 (January 1979), 2–24.

Smith, George W., "Some Northern Wartime Attitudes Toward the Post-Bellum South," *Journal of Southern History*, 10 (August 1944), 353–74.

Somers, Dale A., "Black and White in New Orleans: A Study in Urban Race Relations, 1865–1900," *Journal of Southern History*, 40 (February 1974), 19–42.

———. "James P. Newcomb: The Making of a Radical," *Southwestern Historical Quarterly*, 72 (April 1969), 449–69.

Spackman, S.G.F., "American Federalism and the Civil Rights Act of 1875," *Journal of American Studies*, 10 (December 1976), 313–28.

Sproat, John G., "Blueprint for Radical Reconstruction," *Journal of Southern History*, 23 (February 1957), 25–44.

Swinney, Everette, "Enforcing the Fifteenth Amendment, 1870–1877," *Journal of Southern History*, 28 (May 1962), 202–18.

Stagg, J. C. A., "The Problem of Klan Violence: The South Carolina Up-Country, 1868–1871," *Journal of American Studies*, 8 (December 1974), 303–18.

Taylor, A. Elizabeth, "The Origins and Development of the Convict Lease System in Georgia," *Georgia Historical Quarterly*, 26 (June 1942), 113–28.

Thomas, Herbert A., Jr., "Victims of Circumstance: Negroes in a Southern Town, 1865–1880," *Register of the Kentucky Historical Society*, 71 (July 1973), 253–71.

Thornbery, Jerry, "Northerners and the Atlanta Freedmen, 1865–69," *Prologue*, 6 (Winter 1974), 236–51.

Trelease, Allen W., "Who Were the Scalawags?" *Journal of Southern History*, 29 (November 1963), 445–68.

Vincent, Charles, "Aspects of the Family and Public Life of Antoine Dubuclet: Louisiana's Black State Treasurer, 1868–1878," *Journal of Negro History*, 66 (Spring 1981), 29–36.

Wagstaff, Thomas, " 'Call Your Old Master—"Master" ': Southern Political Leaders and Negro Labor During Presidential Reconstruction," *Labor History*, 10 (Summer 1969), 323–45.

Waller, Altina L., "Community, Class and Race in the Memphis Riot of 1866," *Journal of Social History*, 18 (Winter 1984), 233–46.

Weisberger, Bernard A., "The Dark and Bloody Ground of Reconstruction Historiography," *Journal of Southern History*, 25 (November 1959), 427–47.

West, Earle H., "The Harris Brothers: Black Northern Teachers in the Reconstruction South," *Journal of Negro Education*, 48 (Spring 1979), 126–38.

White, Kenneth B., "The Alabama Freedmen's Bureau and Black Education: The Myth of Opportunity," *Alabama Review*, 34 (April 1981), 107–24.

Wiecek, William M., "The Reconstruction of Federal Judicial Power, 1863–1875," *American Journal of Legal History*, 13 (October 1969), 333–59.

Wiley, B. I., "Vicissitudes of Early Reconstruction Farming in the Lower Mississippi Valley," *Journal of Southern History*, 3 (November 1937), 441–52.

Williams, Frank B., "John Eaton, Jr., Editor, Politician, and School Administrator, 1865–1870," *Tennessee Historical Quarterly*, 10 (December 1951), 291–319.

Williams, John A., "The New Dominion and the Old: Ante-Bellum and Statehood Politics as the Background of West Virginia's 'Bourbon Democracy'," *West Virginia History*, 33 (July 1972), 317–407.

Williams, T. Harry, "An Analysis of Some Reconstruction Attitudes," *Journal of Southern History*, 12 (November 1946), 469–86.

———. "The Louisiana Unification Movement of 1873," *Journal of Southern History*, 11 (August 1945), 349–69.

Woodman, Harold D., "Postbellum Social Change and Its Effect on Marketing the South's Cotton Crop," *Agricultural History*, 56 (January 1982), 215–30.
——. "Post-Civil War Southern Agriculture and the Law," *Agricultural History*, 53 (January 1979), 319–37.

Unpublished Dissertations and Papers

Bailey, Richard, "Black Legislators During the Reconstruction of Alabama, 1867–1878," Kansas State University, 1984.

Benson, Harry K., "The Public Career of Adelbert Ames, 1861–1876," University of Virginia, 1975.

Blank, Charles, "The Waning of Radicalism: Massachusetts Republicans and Reconstruction Issues in the Early 1870's," Brandeis University, 1972.

Carrier, John P., "A Political History of Texas During the Reconstruction, 1865–1874," Vanderbilt University, 1971.

Cohen, Roger A., "The Lost Jubilee: New York Republicans and the Politics of Reconstruction and Reform, 1867–1878," Columbia University, 1975.

Fitzgerald, Michael W., "The Union League Movement in Alabama and Mississippi: Politics and Agricultural Change in the Deep South during Reconstruction," University of California, Los Angeles, 1986.

Frankel, Noralee, "Workers, Wives, and Mothers: Black Women in Mississippi, 1860–1870," George Washington University, 1983.

Gilmour, Robert A., "The Other Emancipation: Studies in the Society and Economy of Alabama Whites During Reconstruction," Johns Hopkins University, 1972.

Graves, John W., "Town and Country: Race Relations and Urban Development in Arkansas 1865–1905," University of Virginia, 1978.

Hansen, Stephen L., "Principles, Politics, and Personalities: Voter and Party Identification in Illinois, 1850–1876," University of Illinois, Chicago, 1978.

Harrison, Robert, "The Structure of Pennsylvania Politics, 1876–1880," Cambridge University, 1970.

Hume, Richard L., "The 'Black and Tan' Constitutional Conventions of 1867–1869 in Ten Former Confederate States: A Study of Their Membership," University of Washington, 1969.

Lancaster, James L., "The Scalawags of North Carolina, 1850–1868," Princeton University, 1974.

McTigue, Geraldine, "Forms of Racial Interaction in Louisiana, 1860–1880," Yale University, 1975.

Miller, Wilbur E., "Reconstruction as a Police Problem," paper, annual meeting of Organization of American Historians, 1978.

Mittelman, Amy, "The Politics of Alcohol Production: The Liquor Industry and the Federal Government 1862–1900," Columbia University, 1986.

Moore, Ross H., "Social and Economic Conditions in Mississippi During Reconstruction," Duke University, 1937.

Phillips, Paul D., "A History of the Freedmen's Bureau in Tennessee," Vanderbilt University, 1964.

Powell, Lawrence N., "Southern Republicanism During Reconstruction: The Contradictions of State and Party Formation," paper, annual meeting of Organization of American Historians, 1984.

Reidy, Joseph P., "Masters and Slaves, Planters and Freedmen: The Transition from Slavery to Freedom in Central Georgia, 1820–1880," Northern Illinois University, 1982.

———. "Sugar and Freedom: Emancipation in Louisiana's Sugar Parishes," paper, annual meeting of American Historical Association, 1980.

Richardson, Barbara A., "A History of Blacks in Jacksonville, Florida, 1860–1895: A Socio-Economic and Political Study," Carnegie-Mellon University, 1975.

Spackman, S. G. F., "National Authority in the United States: A Study of Concepts and Controversy in Congress 1870–1875," Cambridge University, 1970.

Stanley, Amy Dru, "Status or Free Contract: Marriage in the Age of Reconstruction," paper, annual meeting of American Historical Association, 1985.

Index

Georgia, racism/segregation in: and civil rights, 423–24; and the Democratic party, 441; depth of, 70; and the legal system, 423–24, 430; and politics, 331, 347, 423, 441; and public accommodations, 321, 369–70; and Redemption, 423–24; and social/sports activities, 371; and suffrage, 423

Georgia, state constitution in: and apportionment, 195, 323; and Atlanta as the state capitol, 325; and black politics, 195, 314, 317, 318, 318n70, 324, 325–26, 331; and compensated emancipation, 194; and constitutional conventions, 314, 317, 318, 318n70; and industry, 325–26; and office holders, 324, 331; and ratification, 332; and suffrage, 323, 324

Gerrymandering, 590

Gibbs, Jonathan C., 27, 102, 286, 352, 353, 359

Gibbs, Mifflin, 359, 607

Gibson, Hamilton, 358

Gideon's Band, 52, 97

Gleaves, Richard H., 353

Gleed, Robert, 546

Godkin, E. L.: and class conflict, 518–19; and economic policies, 233; and the election of 1872, 503; and the labor movement, 518; and the Liberal Republicans, 488, 489, 492, 496–97, 500, 503; and political democracy, 518–19; and racism, 496–97, 526, 526n27; and the tariff issue, 489, 503; and violence, 499

Gompers, Samuel, 514

Goodloe, Daniel R., 209

Gordon, John B., 281, 332, 400, 432–33

Gospel of prosperity, 379–92, 394–95, 541

Gould, Jay, 382, 458, 462, 468, 491, 568

Government. See Federal intervention; Federalism/states rights; name of specific topic, e.g. Business community

Governors: and civil rights, 370; Johnson's appointments as,

Governors (cont.) 187–89, 192–93; as leaders of industrial development, 210–12; Redemption and the Republican, 605. See also State government; name of specific governor

Graham, David, 572

Graham, North Carolina, 344

Graham, William A., 202, 293–94, 433

Grange [Patrons of Husbandry], 474, 490, 516, 548–49

Grant, Ulysses S.: as an army officer, 333; and Arkansas politics, 528; and black/civil rights, 531, 532; business community support for, 337, 341; Cabinet of, 444–45; congressional relations of, 333, 445; and corruption, 565–66; and currency issues, 522; and the Dominican Republic treaty, 494–97; during the Civil War, 59; and the election of 1866, 264; and the electoral crisis, 579; and federal intervention, 551, 554–55, 558–63, 586; and the Fifteenth Amendment, 577; and the Great Strike of 1877, 586; inaugural address of, 444; and patronage, 528, 565–66; popularity of, 499–500; professional background of, 337; and Reconstruction issues, 444–46, 528, 577; and the Republican party, 337, 341, 494–97. See also Elections: of 1868; Elections: of 1872

Grant Parish, Louisiana, 437

Gray, Ralph, 611

Gray, Thomas, 611

Great Strike of 1877, 583–85

Greeley, Horace, 234, 310, 501–10, 518

Green, John T., 543

Green, Virginia C., 100

Greenbackism. See Currency issues

Greene County, Alabama, 363, 427, 442

Greene County, Georgia, 426

Greene County, North Carolina, 118

Greener, Richard T., 607

Greenville, South Carolina, 97

Gregory, Ovide, 321

Grey, William H., 319

Griffing, Josephine, 503

Griffith, D. W., 609